Interdisciplinary Biotechnological Advances

Series Editors

Jayanta Kumar Patra, . Research Institute of Integrative Life Sciences, Dongguk University, Ilsandong, Goyang, Korea (Republic of)

Gitishree Das, Research Institute of Integrative Life Sciences, Dongguk University, Goyang, Korea (Republic of)

This series is an authoritative resource that explains recent advances and emerging directions in biotechnology, reflecting the forefront of research clearly and reliably, without excessive hype. Each volume is written by authors with excellent reputations and acknowledged expertise in the topic under discussion. The volumes span the entire field from an interdisciplinary perspective, covering everything from biotechnology principles and methods to applications in areas including genetic engineering, transgenic plants and animals, environmental problems, genomics, proteomics, diagnosis of disease, gene therapy, and biomedicine. The significance of these applications for the achievement of UN Sustainable Development Goals is highlighted. The series will be highly relevant for Master's and PhD students in Biotechnology, Nanochemistry, Biochemical Engineering, and Microbiology, medical students, academic and industrial researchers, agricultural scientists, farmers, clinicians, industry personnel, and entrepreneurs.

Aman Dekebo
Editor

Bioactive Secondary Metabolites from Medicinal Plants of Africa

Springer

Editor
Aman Dekebo
Department of Applied Chemistry
Adama Science and Technology
University
Adama, Ethiopia

ISSN 2730-7069 ISSN 2730-7077 (electronic)
Interdisciplinary Biotechnological Advances
ISBN 978-981-96-8589-9 ISBN 978-981-96-8590-5 (eBook)
https://doi.org/10.1007/978-981-96-8590-5

This Springer imprint is published by the registered company Springer Nature Singapore Pte Ltd.
The registered company address is: 152 Beach Road, #21-01/04 Gateway East, Singapore 189721, Singapore

Preface

The African continent is a treasure trove of medicinal plants, offering a diverse array of bioactive compounds with immense therapeutic potential. Africa is endowed with up to 45,000 plant species—about 25% of the world's plant genetic resources. More than 5000 plant species from this enormous African resource are used in traditional medicines. The use of medicinal plants and herbal treatments is widespread globally, particularly in developing regions such as Africa, Asia, and South America. In Africa, traditional medicine represents a fundamental healthcare system, partly organized into herbalism and the utilization of medicinal plants and their derivatives. Traditional healers provide healthcare services grounded in their beliefs, attitudes, knowledge, religious practices, and the cultural norms of their communities.

The plants used in herbal medicines contain bioactive compounds known as phytochemicals, or secondary metabolites, which form the basis of their therapeutic properties. Secondary metabolites are organic compounds produced by organisms to protect themselves against various biotic and abiotic factors, such as diseases, pests, pathogens, herbivores, and environmental stresses. While these compounds do not play a role in the normal development of the organism or influence primary metabolic activities like growth and reproduction, they include significant classes such as phenolics, alkaloids, tannins, saponins, lignins, glycosides, terpenoids, etc.

Historically, herbal medicine was often dismissed as ineffective and primitive by practitioners of Western medicine. However, scientific studies have significantly advanced our understanding of its therapeutic properties and efficacy. Today, numerous pharmaceuticals, medicines, and nutraceuticals are modeled on phytochemicals derived from medicinal plants. Despite this progress, challenges persist in using medicinal plants in Africa and other developing regions, including concerns about safety, poor quality control, and unverified claims by sellers. Additionally, traditional medicine practices remain partly undocumented, with limited reports on adverse reactions and side effects.

Nevertheless, the future of African traditional medicine is promising. Its contributions to employment creation, economic development, healthcare coverage, and drug development underscore its significance. The formal integration and recognition of herbal medicines within modern medical systems hold great potential. Across Africa, traditional healers and medicinal products derived from native plants play critical roles in healthcare, with over 80% of the population relying on them as their primary healthcare source.

Traditional methods for screening and isolating natural products were once ineffective, costly, and cumbersome. However, the advent of modern analytical and high-throughput techniques has revolutionized the isolation and characterization of natural products, opening up exciting opportunities for the pharmaceutical industry in drug discovery and development.

This book entitled *Bioactive Secondary Metabolites from Medicinal Plants of Africa* aims to present cutting-edge research in the field of drug discovery and development from bioactive secondary metabolites derived from African medicinal plants. It provides readers with insights into how African natural product research continues to make significant contributions to global drug discovery efforts.

The book comprises sixteen chapters, primarily review articles authored by experts in their respective research areas. It covers ethnobotanical knowledge, bioactivities, pharmacology, chemical profiling, and structure–activity relationships of various bioactive natural products reported from African medicinal plants. This book serves as a valuable resource for researchers working to identify and characterize new bioactive agents from diverse flora, enabling the discovery of novel therapeutic leads against various diseases. It will also benefit graduate and undergraduate students, educators, industry professionals, and healthcare practitioners involved in natural products and therapeutic research.

Adama, Ethiopia Aman Dekebo
March 2025

Acknowledgments

The editor wishes to thank all the contributors to this book who took the time to write their valuable book chapters. Without their contributions, this book would not have been possible. The editor extends heartfelt thanks to Prof. Veeranoot Nissapatorn from the School of Allied Health Sciences, Research Excellence Center for Innovation and Health Products (RECIHP), Walailak University, Thailand, for her invaluable support during the book project development stage. He also expresses his gratitude to his family for their patience, encouragement, and unwavering support.

The editor sincerely thank Springer for publishing this book and all the Springer staff members who were ever ready to answer our queries about the book. He would especially like to place on record his appreciation and thanks to Dr. Jayanta Kumar Patra, Associate Professor, Dongguk University, Republic of Korea, as well as Nandhini Natchatra and Michelle Harding, Production Editors, who handled the correspondence very efficiently and provided all the needed assistances to make this book a reality.

Peer reviewing is an important component of technical writing. The editor would like to express his sincere thanks to all the peer reviewers who took time to provide critical comments on several manuscripts. Their feedback helped us maintain a high standard for the chapters. He would especially like to thank Prof. Ananda Murthy, School of Applied Sciences, Papua New Guinea University of Technology, Papua New Guinea; Dr. Tekalign Begna and Dr. Daniel Bisrat, Department of Plant Medicals, Andong National University, Republic of Korea; Dr. Hadush Gebrehiwot, Department of Chemistry, Wachamo University, Ethiopia; Dr. Tsegu Kiros, Department of Chemistry, Haramaya University, Ethiopia; Dr. Getachew Tegegn, Department of Chemistry, Salale University, Ethiopia; Dr. Teshome Degfe, Institute of Pharmaceutical Sciences, Adama Science and Technology University, Ethiopia; Gashaw Nigussie and Eyob Debebe, Armauer Hansen Research Institute,

Department of Molecular Biology, Ethiopia; Dr. Tariku Nefo, Department of Materials Science and Engineering, National Taiwan University of Science and Technology, Taiwan; Dr. Seifu Juneidi, Department of Applied Biology, Adama Science and Technology University, Ethiopia; and Dr. Kesatebrhan Haile, UNC Eshelman School of Pharmacy, USA.

Contents

Editor and Contributors

About the Editor

Aman Dekebo holds a PhD in Bioorganic Chemistry from Ehime University, Japan (2007), and a PhD in Natural Products Chemistry from Addis Ababa University, Ethiopia (2002). His research areas include Natural Products Chemistry, Medicinal Chemistry, Organic Synthesis, and Chemical Ecology. He is a member of various professional organizations, including the Chemical Society of Ethiopia, the Japanese Society of Applied Entomology and Zoology, the Korea Society of Applied Entomology, the Korean Apiculture Society, the Japan Alumni Association, and the Global Association of Traditional Medicine.

He has extensive teaching experience in Organic Chemistry, Natural Product Chemistry, Synthesis, and other courses at both undergraduate and postgraduate levels. He began his academic career in 1990 as a graduate assistant at Gondar College of Medical Sciences, Addis Ababa University. He was later promoted to Lecturer and served the college until 2002. From 2002 to 2007, he worked as a Research Fellow at Ehime University, Japan, based in Kochi University, Japan. He then served as Assistant Professor of Chemistry at Ambo University (2008–2012). In February 2012, he joined Adama Science and Technology University as an Assistant Professor and was subsequently promoted to Associate Professor and Professor. During his sabbatical leave (February 2017–February 2019), he worked as a researcher at Andong National University, South Korea.

He has supervised 36 MSc and 5 PhD students as a major supervisor in the Departments of Applied Chemistry and Applied Biology. Currently, he supervises 6 PhD candidates in the Departments of Applied Chemistry, Applied Biology, and Chemical Engineering at Adama Science and Technology University.

He has held several administrative roles, including Head of the Chemistry Department (2012–2014), Associate Dean for Consultancy Services in the Research Park (2015–2016), and Associate Dean of the Science, Technology, Engineering, and Mathematics (STEM) Center (2019–2021). Since 2019, he has served as the Associate Editor of the *Ethiopian Journal of Science and Sustainable Development*.

He also a peer reviewer for numerous reputable international journals and a board member of the Natural Products Research Network for Eastern and Central Africa (NAPRECA). He is also the chairperson of the NAPRECA Ethiopia chapter.

He has edited two books, *Plant Extracts* (2019) and *Capsicum* (2020), published by IntechOpen, UK. He also co-authored the book *Phytochemistry of Turmeric Rhizomes (Curcuma longa): Isolation and Characterization of Compounds from Roots of Turmeric Rhizomes (Curcuma longa)*, published by Lambert Academic Publishing (2012).

He has published over 95 peer-reviewed journal articles and has presented his work at 33 national and international conferences, workshops, and symposiums. He was awarded best researcher awards in Adama Science and Technology University and during international conferences.

He has served as the principal investigator (PI) for three research projects funded by Adama Science and Technology University, the Korea Engineering Research Center, and The World Academy of Sciences (TWAS), all of which were successfully completed.

Contributors

Negera Abdissa Traditional and Modern Medicine Research and Development Directorate, Armauer Hansen Research Institute, Addis Ababa, Ethiopia

Siri Abihudi Department of Medical Botany, Agronomy and Plant Breeding, Institute of Traditional Medicine, Muhimbili University of Health and Allied Sciences, Dar es Salaam, Tanzania

Sameh AbouZid Department of Pharmacognosy, Faculty of Pharmacy, Beni-Suef University, Beni-Suef, Egypt

Madina Mohamed Adia Department of Chemistry, College of Natural Sciences, Makerere University, Kampala, Uganda

Yonatan Alebachew Department of Pharmaceutical Chemistry and Pharmacognosy, School of Pharmacy, College of Health Sciences, Addis Ababa University, Addis Ababa, Ethiopia

Miora Henintsoa Andrianarivelo Institut of Higher Education Antsirabe Vakinankaratra, Vakinankaratra, Madagascar

Godwin Anywar Department of Plant Sciences, Microbiology & Biotechnology, Makerere University, Kampala, Uganda

Mathewos Anza Department of Chemistry, College of Natural and Computational Science, Wolaita Sodo University, Soddo, Ethiopia

Savina Asiimwe Department of Plant Sciences, Microbiology & Biotechnology, Makerere University, Kampala, Uganda

Abba Babandi Department of Biochemistry, Bayero University, PMB 3011-Kano, Nigeria

Medical Biochemistry Unit, Faculty of Basic Medical Sciences, Federal University Dutse, PMB 7156-Dutse, Jigawa State, Nigeria

Dina M. Bahgat Department of Pharmacognosy, Faculty of Pharmacy, Ain Shams University, Cairo, Egypt

Gaetan Bayiha Ba Njock Department of Chemistry, Faculty of Science, University of Maroua, Maroua, Cameroon

Imalingat Betty Department of Chemistry, College of Natural Sciences, Makerere University, Kampala, Uganda

Daniel Buyinza Department of Chemistry, Faculty of Science, Kabale University, Kyambogo-Kampala, Uganda

Robert Byamukama Department of Chemistry, College of Natural Sciences, Makerere University, Kampala, Uganda

Teshome Degfie Institute of Pharmaceutical Sciences, Adama Science and Technology University, Adama, Ethiopia

Aman Dekebo Department of Applied Chemistry, Adama Science and Technology University, Adama, Ethiopia

Institute of Pharmaceutical Sciences, Adama Science and Technology University, Adama, Ethiopia

Moïse Diguir Department of Chemistry, Faculty of Science, University of Maroua, Maroua, Cameroon

Dalia El Amir Department of Pharmacognosy, Faculty of Pharmacy, Beni-Suef University, Beni-Suef, Egypt

Omayma A. Eldahshan Department of Pharmacognosy, Faculty of Pharmacy, Ain Shams University, Cairo, Egypt

Esraa A. Elhawary Department of Pharmacognosy, Faculty of Pharmacy, Ain Shams University, Cairo, Egypt

Milkyas Endale Traditional and Modern Medicine Research and Development Directorate, Armauer Hansen Research Institute, Addis Ababa, Ethiopia

Albert Fouda Atangana Department of Chemistry, Faculty of Science, University of Maroua, Maroua, Cameroon

Isaac Silvere Gade Department of Chemistry, Faculty of Science, University of Maroua, Maroua, Cameroon

Ivan Gumula Department of Chemistry, Faculty of Science, Kyambogo University, Kyambogo-Kampala, Uganda

Mazuru Gundidza Department of Biomedical Sciences, Herbal Medicine Unit, School of Medical and Health Sciences, Great Zimbabwe University, Masvingo, Zimbabwe

Ester Innocent Department of Biological and Pre-clinical Studies, Institute of Traditional Medicine, Muhimbili University of Health and Allied Sciences, Dar es Salaam, Tanzania

Namukobe Jane Department of Chemistry, College of Natural Sciences, Makerere University, Kampala, Uganda

Emmanuel Kademeteme Department of Biomedical Sciences, Physiology Unit, School of Medical and Health Sciences, Great Zimbabwe University, Masvingo, Zimbabwe

Esezah K. Kakudidi Department of Plant Sciences, Microbiology & Biotechnology, Makerere University, Kampala, Uganda

Patrick Kareru Chemistry Department, Jomo Kenyatta University of Agriculture and Technology, Nairobi, Kenya

Sammy Kiprotich Department of Chemistry, College of Natural Sciences, Makerere University, Kampala, Uganda

Tsegu Kiros Chemistry Department, College of Natural and Computational Sciences, Haramaya University, Dire Dawa, Ethiopia

Gideon Kirui Chemistry Department, Jomo Kenyatta University of Agriculture and Technology, Nairobi, Kenya

Christine Kyarimpa Department of Chemistry, Faculty of Science, Kyambogo University, Kyambogo-Kampala, Uganda

Edwin Madivoli Chemistry Department, Jomo Kenyatta University of Agriculture and Technology, Nairobi, Kenya

David Makhanu Department of Biological and Physical Sciences, Karatina University, Karatina, Kenya

Alphonce Ignace Marealle Department of Clinical Pharmacy and Pharmacology, School of Pharmacy, Muhimbili University of Health and Allied Sciences, Dar es Salaam, Tanzania

Yadessa Melaku Department of Applied Chemistry, Adama Science and Technology University, Adama, Ethiopia

Asfaw Meresa Traditional and Modern Medicine Research and Development Directorate, Armauer Hansen Research Institute, Addis Ababa, Ethiopia

Enas Mohamed Department of Pharmacognosy, Faculty of Pharmacy, Beni-Suef University, Beni-Suef, Egypt

Heba Mohammed Department of Pharmacognosy, Faculty of Pharmacy, Beni-Suef University, Beni-Suef, Egypt

Nada M. Mostafa Department of Pharmacognosy, Faculty of Pharmacy, Ain Shams University, Cairo, Egypt

Ian Mutasa Department of Biomedical Sciences, Physiology Unit, School of Medical and Health Sciences, Great Zimbabwe University, Masvingo, Zimbabwe

Vincent Mwangi Chemistry Department, Jomo Kenyatta University of Agriculture and Technology, Nairobi, Kenya

Jane Namukobe Department of Chemistry, College of Natural Sciences, Makerere University, Kampala, Uganda

Temesgen Negassa Traditional and Modern Medicine Research and Development Directorate, Armauer Hansen Research Institute, Addis Ababa, Ethiopia

Gashaw Nigussie Traditional and Modern Medicine Research and Development Directorate, Armauer Hansen Research Institute, Addis Ababa, Ethiopia

Elliot Nyagumbo National Pathology Research and Diagnostic Centre, Midlands State University, Gweru, Zimbabwe

Dennis Nzilu Chemistry Department, Jomo Kenyatta University of Agriculture and Technology, Nairobi, Kenya

Patrick Onen Department of Chemistry, Faculty of Science, Kyambogo University, Kyambogo-Kampala, Uganda

Department of Chemistry, University of Kerala, Karyavattom Campus, Thiruvananthapuram, Kerala, India

Sekandi Peter Department of Chemistry, College of Natural Sciences, Makerere University, Kampala, Uganda

William Pote Department of Biomedical Sciences, Physiology Unit, School of Medical and Health Sciences, Great Zimbabwe University, Masvingo, Zimbabwe

Vonjison Rakotoarimanana Faculty of Sciences, University of Antananarivo, Antananarivo, Madagascar

Falitiana Marrino Rakotoarisoa CNARP Centre National d'Application de Recherches Pharmaceutiques, Antananarivo, Madagascar

Stéphan Richard Rakotonandrasana CNARP Centre National d'Application de Recherches Pharmaceutiques, Antananarivo, Madagascar

Andriamalala Rakotondrafara CNARP Centre National d'Application de Recherches Pharmaceutiques, Antananarivo, Madagascar

Christian Marius Rakotonirina CNARP Centre National d'Application de Recherches Pharmaceutiques, Antananarivo, Madagascar

Rianasoambolanoro Rakotosaona CNARP Centre National d'Application de Recherches Pharmaceutiques, Antananarivo, Madagascar

Laboratoire de Recherche en Sciences Alimentaires et Santé, Ecole Supérieure Polytechnique University of Antananarivo, Antananarivo, Madagascar

Vincent Emile Rasamison Institut of Higher Education Antsirabe Vakinankaratra, Vakinankaratra, Madagascar

Byamukama Robert Department of Chemistry, College of Natural Sciences, Makerere University, Kampala, Uganda

Kiprotich Sammy Department of Chemistry, Ben-Gurion University of the Negev, Be'er Sheva, Israel

Peter Sekandi Department of Chemistry, College of Natural Sciences, Makerere University, Kampala, Uganda

Muhammad Ajmal Shah Department of Pharmacy, Hazara University, Mansehra, Pakistan

Albert Shayo Department of Medical Botany, Agronomy and Plant Breeding, Institute of Traditional Medicine, Muhimbili University of Health and Allied Sciences, Dar es Salaam, Tanzania

Abdel Nasser Singab Department of Pharmacognosy, Faculty of Pharmacy, Ain Shams University, Cairo, Egypt

Patience Tugume Department of Plant Sciences, Microbiology & Biotechnology, Makerere University, Kampala, Uganda

Herbet Zirima Department of Psychiatry and Behavioural Sciences, School of Medical and Health Sciences, Great Zimbabwe University, Masvingo, Zimbabwe

Chapter 1
Research Progress on the Ethnobotanical, Biological, and Phytochemical Studies of Medicinal Plants of Ethiopian Flora—Part 1: A Six-Decade (1960–2020) Review

Tsegu Kiros and Aman Dekebo

Abstract The beginning of scientific research in Ethiopia is expected following the establishment of Addis Ababa University in 1950 as the University College of Addis Ababa. Despite this milestone, the progress of biochemical studies on Ethiopian plants remains unclear. This comprehensive review aims to collect original scientific research on ethnobotanical, biological, and phytochemical studies of Ethiopian medicinal plants published in 1960–2020. The review was conducted using PubMed and ScienceDirect databases and certain appropriate descriptors. The ethnobotanical, biological, and phytochemical studies of Ethiopian plants were started in 2003, 1990, and 1975, respectively. Of the 3092 papers retrieved for ethnobotanical studies, 75 focused on Ethiopian medicinal plants, mainly *Allium species*, *Hagenia abyssinica*, *Vernonia amygdalina*, *Ocimum* species, *Rhamnus prinoides*, *Solanum incanum*, *Azadirachta indica*, and *Phytolacea dodecandra*. Out of 2346 papers, 115 were specific to biological activities of Ethiopian plants, with 35 discussing antimicrobial and 22 highlighting antimalarial activities. Ninety-five out of 2082 papers reported the isolation of compounds from Ethiopian plants, with 40% exploring their biological activities. *Aloe*, *Artemisia*, and *Vernonia* species were frequently mentioned in the phytochemical studies. In conclusion, this review can be used as preliminary data to dereplicate already-reported works and support the ongoing establishment of an Ethiopian pharmacopoeia.

T. Kiros (✉)
Chemistry Department, College of Natural and Computational Sciences, Haramaya University, Dire Dawa, Ethiopia

A. Dekebo
Department of Applied Chemistry, Adama Science and Technology University, Adama, Ethiopia

Institute of Pharmaceutical Sciences, Adama Science and Technology University, Adama, Ethiopia

A. Dekebo (ed.), *Bioactive Secondary Metabolites from Medicinal Plants of Africa*, Interdisciplinary Biotechnological Advances,
https://doi.org/10.1007/978-981-96-8590-5_1

Keywords Ethiopian flora · Ethiopian plants · Ethiopian education · Research progress · Ethiopian universities

1.1 Introduction

1.1.1 The Historical Background of Traditional Medicine in Ethiopia

The use of medicinal plants for traditional remedies against human and animal ailments in Ethiopia dates back to ancient times and is not evidently known. This is because, most of the time, the indigenous knowledge of plant-based traditional medicine is preserved orally within each ethnic group and transferred from one generation to the next without any written pharmacopeias. However, there are few written documents found in Ethiopian religious places, such as those of the Islamic, Christian, and other indigenous religions, that describe the traditional medicinal uses of various plant species. For instance, in the Orthodox Church, there are many medico-religious books written in the "*Geez*" language during the fifteenth century, according to (Abate, 1989; Abebe & Ayehu, 1993) reports. There is also another ancient book locally known as *Metsehafe Fews*, meaning a book of remedy, written in the seventeenth century, which contains many medicinal plants along with their wide range of traditional medicinal descriptions (Fullas, 2001). Traditional medicinal practices involving medicinal plants are deeply intertwined with the socio-cultural characteristics of specific ethnic groups, forming an integral part of their daily efforts to combat various diseases.

The traditional medicinal practices in Ethiopia are complex with great diversity and differ between ethnic groups. Most of the practices have been carried out based on the back-track of each disease and depend on both the "mystical" and "natural" roots of the illness, and an all-inclusive method is employed to treat it (Bishaw, 1991). The traditional healers, locally called "traditional medical doctors," are those who implement therapeutic functions using herbal formulations, spiritual curing, bone-setting, and slight surgical-like procedures (Gall & Shenkute, 2009). Even though modern medicine has been incorporated into the Ethiopian healthcare system since 1865–1913, the preference for traditional medicine in Ethiopia is still continuous even in this modern era. It is expected that nearly 70% and 90% of the human and livestock populations, respectively, depend on traditional medicines like those of other developing countries, mainly those of sub-Saharan African countries (Gall & Shenkute, 2009). Besides, 95% of the traditional medical preparations in Ethiopia are of plant origin. This is believed due to the deep-rooted socio-cultural practices, inaccessibility and unaffordability of modern medicine, strong belief in local healers and the efficacy of medicinal plants, the medical comfortability of traditional remedies, and their comparatively low cost (Gall & Shenkute, 2009). It is believed that nearly 6500–7000 higher plant species are expected to be found in the Ethiopian flora, among which about 12–19% are recognized as endemic and 1000 are identified as medicinally valued species, making the country one of the top six botanically rich African countries (Ayalew et al., 2022).

1.1.2 Beginning of Modern Education and Scientific Research in Ethiopia: A Short Highlight

The history of indigenous or "traditional" education in Ethiopia is as old as the history of the country and directly related to various religious educations provided by Islamic (Quranic education) and Christian (Orthodox monastic education) religions. These religious educations played a pivotal role in the introduction of modern education to Ethiopia, though there is still a blurred understanding about the extent of their contribution due to the lack of reliable documents. The history of secular and western types of modern education in Ethiopia was believed to be traced back to the reign of Emperor Menelik II, in which a school known as Menelik II School was opened in the capital Addis Ababa in 1908 for the first time. Before that, there were no government or private modern schools, despite some missionaries, such as Catholic and Jesuits, attempting to provide modern education in the sixteenth century. And then, the development of modern education in Ethiopia was further strengthened during the reign of Emperor Haile Selassie I in 1930, though it was disrupted during the period 1936–1941 following the Italian invasion and occupation (Bekele, 1966; Wondemetegegn, 2016).

In the history of higher learning institutions, Addis Ababa University is the oldest and first institution, which was initially founded as the University College of Addis Ababa in 1950 with only two faculties, including the faculty of science. Then, in 1961, the University College of Addis Ababa was renamed as Haile Selassie I University, and the faculty of science was restructured and upgraded into many departments. Finally, Haile Selassie I University was changed to its current name, Addis Ababa University, in 1975. Presently, there are many colleges, including the College of Natural and Computational Sciences and the College of Health Sciences, with many departments and schools such as the Departments of Chemistry, Biology, Physics, Mathematics, and Statistics; and the Schools of Medicine, Pharmacy, and Public Health (Addis Ababa University website, http://www.aau.edu.et/cns/, http://www.aau.edu.et/chs/). The establishment of Addis Ababa University and many other higher learning and research Universities and institutions has laid a cornerstone for the scientific research studies of many medicinal plants in Ethiopia. Initially, the research studies conducted on various plant species in universities and research institutions were restricted to the ethnobotanical aspects of the plants, which focused mainly on the surveying and listing of plenty of plant species along with their traditional medicinal usages. Later, some researchers have made an attempt to do research aimed at investigating the biological activities and chemical constituents of a few plants (Bekele, 2007).

But the questions now are: (1) By now, which aspect (the ethnobotanical, the biological, and/or the phytochemical) of the Ethiopian medicinal plants has extensively been studied? (2) Which plant species have mostly been investigated? (3) Which part of the country has mostly been explored for its medicinal plant sources? (4) Which institutions have taken the Lion's share in studying the ethnobotanical, biological, and/or phytochemical features of the Ethiopian medicinal plants? (5) And finally, in which period was the scientific study done widely on medicinal plants? Based on our knowledge and best search, there is no comprehensive and compiled document that tries to answer the above questions, except a very few online databases, including the Ethiopian traditional medicine database (ETM-DB), which was developed to provide online information about the "prescription, compound, phenotype, and gene/protein" of Ethiopian medicinal herbs (Bultum et al., 2019). Hence, the intention of this comprehensive review is to collect the published scientific research works done on the ethnobotanical uses, biological activities, and chemical constituents of Ethiopian medicinal plants for the last six decades (1960–2020). The output of this six-decade review may be expected to address the above questions and provide important information about the entire scientific research progress on Ethiopian medicinal plants. And at the end, it may pave the way for researchers and other stakeholders in the country to think about the way forward regarding the overall scientific studies of Ethiopian medicinal plants to establish an Ethiopian pharmacopoeia that is currently initiated.

1.2 Methodologies

In this comprehensive review, original research works done on Ethiopian medicinal plants and published over the last six decades (1960–2020) were appraised. The topics covered in this review were the ethnobotanical uses, biological activities, and phytochemical constituents of the Ethiopian medicinal plants. The PubMed and ScienceDirect websites (https://pubmed.ncbi.nlm.nih.gov/31126296/ and https://www.sciencedirect.com/search) were used as search engines or electronic databases to collect the articles compiled herein. The descriptors used in this review were directly related to the ethnobotanical uses, biological activities, and phytochemical constituents of Ethiopian medicinal plants. These were: "ethnobotanical studies in Ethiopia," "medicinal plants of Ethiopia," "flora of Ethiopia," "Ethnomedicine of Ethiopia," and "traditional medicinal plants of Ethiopia" for the ethnobotanical uses; "biological activity of medicinal plants of Ethiopia," "bioactivity of Ethiopian medicinal plants," "studies on biological activity of Ethiopian medicinal plants," "activity of medicinal plants from Ethiopia," and "biological activity evaluation of plants in Ethiopia" for the biological activity; and "isolation of compounds from plants of Ethiopia," "compounds from Ethiopian plants," and "natural compounds from plants of Ethiopia" for the phytochemical constituents of the Ethiopian medicinal plants. Those published articles that did not align with the aforementioned list of keywords, were literature reviews, or were written in a language other than English were excluded from this review paper. The detailed information on the obtained results of corresponding topics was presented in tables, and the numbers of reported papers versus the year of publication were expressed in the form of bar graphs generated using Microsoft Excel 2010.

1.3 Results and Discussion

In this comprehensive and vast review, the original research works reported on the ethnobotanical, biological activity, and phytochemical studies of Ethiopian medicinal plants over the last six decades (1960–2020) were collected using the search tools and keywords mentioned in the methodology part (see Sect. 1.2). The detailed information about the obtained reports on the ethnobotanical study, biological activity study, and phytochemical study of Ethiopian plants is presented in Tables 1.1, 1.2, and 1.3, respectively. And the corresponding numbers of reported papers versus publication years are depicted in Figs. 1.1, 1.2, and 1.3. In each case, the title and abstract of extracted papers were taken into consideration and those duplicates were excluded. The discussion of each obtained result was presented as follows:

1.3.1 A Six-Decade (1960–2020) Ethnobotanical Study of Ethiopian Medicinal Plants

According to the search results, a total of 3092 articles were extracted, out of which 75 were found to be directly related to the ethnobotanical studies of Ethiopian medicinal plants conducted over the last six decades (1960–2020). The overall information (number of plants, study area, institution responsible for the study, publication year, and references) about the searched ethnobotanical studies is presented in Table 1.1. The number of reported papers versus the publication year is also depicted in Fig. 1.1. Even though the start of scientific research in Ethiopia is expected since the opening of the higher education institution, Addis Ababa University, in 1950 as the University College of Addis Ababa, the scientific reports dealing with the ethnobotanical studies of Ethiopian plants were not found until 2002. As shown in Fig. 1.1, the ethnobotanical study report was found since 2003, in which only two articles were obtained. Unfortunately, again, no reports were found in 2004 and 2005. The highest number of publications (9) was recorded in 2009, followed by 2018 and 2020, which had eight publications each. A total of 5641 plants, regardless of their frequency and type, were reported in the 75 papers collected. The highest number of plant species was surveyed in the provinces of Odo-Bulu and Demaro of the Oromia region (294), Sheka zone of Southern Ethiopia (266), and Mana Angetu district of Southeastern Ethiopia (230). The next more than 100 plant species were reported from the areas Ganta Afeshum district of Northern Ethiopia (173), Libo Kemkem district of Northwest Ethiopia (163), Wonago of Southern Ethiopia (155), Ankober district of Amhara region (135), Gubalafto district of Northern Ethiopia (135), Delanta of Northern Ethiopia (133), Ada'a district of Oromia region (131), Maale and Debub Ari districts of Southern Ethiopia (128), Wayu Tuka district of Oromia region (126), Konta of Southern Ethiopia (120), Kilte Awulaelo district of Northern Ethiopia (114), Sebeta-Awas district of Oromia

Table 1.1 A six-decade (1960–2020) ethnobotanical study of Ethiopian medicinal plants

S/N	Study area/location	# of plants reported	Institution type	References
1	Shinile district/Somali region	27	University	Mesfin et al. (2012)
2	Dirre Sheikh Hussein heritage/Southeastern Ethiopia	87	University and research institute	Demie et al. (2018)
3	Northern Ethiopia	34	University	Giday et al. (2016)
4	Dek Island/Lake Tana	60	University	Teklehaymanot (2009)
5	Bench-Maji Zone/Southwest Ethiopia	51	University	Giday et al. (2009a)
6	Dheeraa town/Arsi Zone	83	University	Wondimu et al. (2007)
7	Lake Ziway	33	University	Giday et al. (2003)
8	Kofe kebele/Jimma zone	9	University	Karunamoorthi et al. (2009a)
9	Badeye Kebele/Western Hararghe zone	13	University	Karunamoorthi and Husen (2012)
10	Gindeberet district/Western Ethiopia	26	University	Zerabruk and Yirga (2012)
11	Sheko District/Southwest Ethiopia	71	University	Giday et al. (2010)
12	Harar region	31	University	Bussa and Belayneh (2019)
13	Bale Mountains National Park	74	"	Yineger et al. (2007)
14	Koka Negewo village/Oromia region	4	University/ institute	Dugassa et al. (2009)
15	lower Omo River Valley/Southern Ethiopia	57	University	Teklehaymanot and Giday (2010b)
16	Dibatie and Guangua/Northwestern Ethioipa	76	"	Giday et al. (2007)
17	Debre Libanos monastery	80	"	Teklehaymanot et al. (2007)
18	Awetu mendera Kebele/Jimma town	16	"	Karunamoorthi and Tsehaye (2012)
19	Maale and Debub Ari/Southern Ethiopia	46	University/ institute	Kidane et al. (2014b)
20	Abergelle, Sekota and Lalibela districts/ Northern Ethiopia	53	University/ institute	Assefa and Bahiru (2018)

(continued)

Table 1.1 (continued)

S/N	Study area/location	# of plants reported	Institution type	References
21	Butajira district/Southern Ethiopia	25	University	Gedif and Hahn (2003)
22	Addis Zemen/Amhara region	15	University	Karunamoorthi et al. (2009b)
23	Jimma zone/Southwestern Ethiopia	28	"	Suleman et al. (2018)
24	Hawassa district/Southern Ethiopia.	105	"	Tefera and Kim (2019)
25	Gubalafto district/Northern Ethiopia	135	"	Chekole (2017)
26	Bale Robe, Goba, Bahir Dar, Filiklik, Gewane, Wondo Genet, Sodo, Doyo Gena, North Bench, Mizan Aman, and Shako	53	"	Tesfaye et al. (2020)
27	Ganta Afeshum district/Northern Ethiopia	173	"	Kidane et al. (2018)
28	Raya Kobo district/Northeastern Ethiopia	91	"	Osman et al. (2020)
29	Sheka zone/Southern Ethiopia	266	"	Kassa et al. (2020)
30	Shinasha and Agew-awi/Northwest Ethiopia	76	"	Giday et al. (2007)
31	Ambo district/Oromia region	55	"	Berhanu et al. (2020)
32	Wonago/Southern Ethiopia	155	"	Mesfin et al. (2009)
33	Yalo/Afar region	106	"	Teklehaymanot (2017)
34	Jimma zone/Southwest Ethiopia	72	"	Siraj et al. (2020)
35	Dega Damot district/Northwestern Ethiopia	60	"	Wubetu et al. (2017)
36	Ada'a district/Oromia region	131	"	Kefalew et al. (2015)
37	Delanta/Northern Ethiopia	133	"	Meragiaw et al. (2016)
38	Zegie Peninsula/Northwestern Ethiopia	67	"	Teklehaymanot and Giday (2007)
39	Fiche/Oromia region	73	"	D'avigdor et al. (2014)
40	Maale and Debub Ari districts/Southern Ethiopia	128	Institute	Kidane et al. (2014a)
41	Kilte Awulaelo district/Northern Ethiopia	114	University	Teklay et al. (2013)
42	Hamer/ southwestern Ethiopia	60	"	Paulos et al. (2016)

(continued)

Table 1.1 (continued)

S/N	Study area/location	# of plants reported	Institution type	References
43	Ankober district/Amhara region	51	"	Lulekal et al. (2014a)
44	Boricha district/Southern Ethiopia	42	"	Asnake et al. (2016)
45	Yabelo, Liben, Wondo Genet, and Kochere districts	49	"	Eshetu et al. (2015)
46	Berbere district/Oromia region	70	"	Jima and Megersa (2018)
47	South Omo/Southern Ethiopia	91	"	Tolossa et al. (2013)
48	Ghimbi district/Southwest Ethiopia	49	"	Abera (2014)
49	Mana Angetu district/Southeastern Ethiopia	230	"	Lulekal et al. (2008)
50	Gozamin/Amhara region	93	"	Amsalu et al. (2018)
51	Babile Wereda/Eastern Ethiopia	51	"	Belayneh et al. (2012)
52	Bench district/Southwest Ethiopia	35	"	Giday et al. (2009b)
53	Eastern Shewa and Arsi zones/Oromia region	21	"	Scantlebury et al. (2013)
54	Libo Kemkem district/Northwest Ethiopia	163	"	Chekole et al. (2015)
55	Wayu Tuka district/Oromia region	126	"	Megersa et al. (2013)
56	Konta/Southern Ethiopia	120	"	Bekalo et al. (2009)
57	Seharti Samre district/Tigray region	90	"	Araya et al. (2015)
58	Ankober district/Amhara region	135	"	Lulekal et al. (2013)
59	Somali region	13	"	Gebeyew et al. (2020)
60	Sebeta-Awas district/Oromia Region	113	"	Mekonen et al. (2015)
61	Dalle district/Southern Ethiopia	71	"	Tuasha et al. (2018)
62	Ada'ar district/Afar region	49	"	Giday and Teklehaymanot (2013)
63	Omo River Valley/Southern Ethiopia	38	"	Teklehaymanot and Giday (2010a)
64	Harla town and Dengego valleys/Eastern Ethiopia	83	"	Belayneh and Bussa (2014)

(continued)

Table 1.1 (continued)

S/N	Study area/location	# of plants reported	Institution type	References
65	Fafan Zone/Eastern Ethiopia	49	"	Feyera et al. (2017)
66	Burka Jato Kebele/West Ethiopia	52	"	Gari et al. (2015)
67	Gurage and Silti zones/South central Ethiopia	80	"	Teka et al. (2020a)
68	Derashe and Kucha districts/Southern Ethiopia	66	"	Balemie and Kebebew (2006)
69	Jimma zone/Oromia region	74	"	Yigezu et al. (2014)
70	Shinle/Somali region	31	"	Kebede et al. (2018)
71	Sekoru district/Oromia region	27	"	Yineger and Yewhalaw (2007)
72	Assosa Zone/Western Ethiopia	40	"	Flatie et al. (2009)
73	Omo Nada, Kersa, and Tiro Afeta districts/ Oromia region	67	"	Yineger et al. (2008)
74	Odo-Bulu and Demaro/Oromia region	294	Institute	Bussmann et al. (2011)
75	Genta Meyche/Southern Ethiopia	26	University/ institute	Guadie et al. (2020)

Region (113), Yalo of Afar region (106), and Hawassa district of Southern Ethiopia (105). The bar graph (Fig. 1) indicated that no consistent progress, either in increasing or decreasing numbers of published papers, was shown across the year of publications. The higher education institutions (universities) did the Lion's share (95%) of the ethnobotanical studies.

1.3.2 *A Six-Decade (1960–2020) Biological Activity Study of Ethiopian Medicinal Plants*

A total of 115 papers (Table 1.2 and Fig. 1.2), out of the 2346 total numbers of searched papers, were found to be directly related to the biological activity studies of Ethiopian plants. These biological activity reports were found starting in 1990, and only two papers were obtained. Based on the searched data (Fig. 1.2), a huge difference (fourfold), in terms of reported papers, was observed between the

Table 1.2 A six-decade (1960–2020) biological activity study of Ethiopian medicinal plants

S/N	Plants studied	Samples tested	Activities	Institution type	References
1	*Solanum incanum* L., *Harrisonia abyssinica*, *Hydnora Abyssinia*, and *Leucas aspera*	Roots and leaves organic solvent extracts	In vitro antibacterial	University/ research institute	Guadie et al. (2020)
2	*Lavandula angustifolia*, *Cymbopogon citratus*, *Mentha piperita*, *Chenopodium ambrosioides*, *Aloe ankoberensis*, and *Aloe pulcherima*	Essential oils and extracts	In vitro antimicrobial	University	Gishen et al. (2020)
3	*Carissa spinarum*, *Dodonaea angustifolia*, *Jasminum abyssinicum*, *Rumex nepalensis*, *Rubus steudneri*, and *Verbascum sinaiticum*	Ethanol extracts	In vitro antioxidant and antiproliferative	University/ research institute	Tauchen et al. (2015)
4	*Achyranthes aspera*, *Ficus caria*, *Malvi parviflora*, *Vernonia* species, *Solanum hastifolium*, *Calpurinia aurea*, *Nicotiana tabacum*, *Ziziphus spina-christi*, and *Croton macrostachys*	Hydroalcoholic extracts	In vitro antimicrobial	University/ research institute	Kalayou et al. (2012)
5	*Acokanthera schimperi*, *Calpurnia aurea*, *Kalanchoe petitiana*, *Lippia adoensis*, *Malva parviflora*, *Olinia rochetiana*, *Phytolacca dodecandra*, and *Verbascum sinaiticum*	Hydroalcoholic extracts	In vitro antimicrobial	University/ research institute	Tadeg et al. (2005)
6	*Albizzia gummifera*, *Trichila emetic*, *Dovyalis abyssinica*, and 64 other plant species	Petroleum ether, methanol and aqueous extracts	In vitro antimicrobial	Research institute	Geyid et al. (2005)
7	*Moringa oleifera*, *Azadirachta indica*, and *Lepidium sativum*	Methanol and aqueous leaves extracts	In vitro antibacterial	University	Ibrahim and Kebede (2020)
8	*Ocimum lamiifolium*, *Rosmarinus officinalis* L, *Catharanthus roseus* Linn., *Azadirachta indica*, and *Moringa stenopetala*	Organic solvents, leaf extracts	In vitro antibacterial	University	Manilal et al. (2020)
9	*Aloe* sp., *Azadirachta indica*, and *Tamarindus indica*	Ethanol and aqueous extracts	In vivo antimalarial	University	Mesfin et al. (2012)

10	Four *Artemisia* species	Different leaves and aerial part extracts	In vitro antimalarial and cytotoxicity	University	Nibret and Wink (2010a, b)
11	*Lippia adoensis*, *Nigella sativa*, *Piper capense*, *Thymus schimperi*, and *Trachyspermum ammi*	Methanol and aqueous extracts	In vitro antioxidant	University	Sasikumar et al. (2020)
12	*Juniperus procera*	Essential oils	Larvicidal	University	Karunamoorthi et al. (2014)
13	*Rumex abyssinicus*	Ethanol and aqueous extracts	*In vitro* antioxidant and antibacterial	University/research institute	Mohammed et al. (2017)
14	*Glinus lotoides*, *Plumbago zeylanica*, *Rumex steudelii*, and *Thymus schimperi*	Hydroalcoholic extracts	Genotoxicity	University	Demma et al. (2009)
15	*Plumbago zeylanica*	Hydromethanolic root extract	In vivo dermatotoxicity	University	Teshome et al. (2008)
16	*Xylopia aethiopica* and *Piper guineense*	Acidic-methanol extracts	In vitro and in vivo antioxidant	University	Adefegha and Oboh (2012)
17	*Phytolacca dodecandra*	*n*-Butanol and aqueous berry extracts	In vitro and in vivo antifungal	University	Mekonnen et al. (2012)
18	*Glinus lotoides*	Methanol seed extract	In vivo cytotoxicity	University	Demma et al. (2007)
19	*Plectranthus carinus* Roth	Essential oil from aerial parts	In vitro antimicrobial and anti-oxidant, and in vivo anti-inflammatory	University	Tadesse et al. (2011)
20	*Cheilanthes farinosa*	Different extracts	In vivo anti-inflammatory and antinociceptive	University/research institute	Yonathan et al. (2006)
21	*Rumex steudelii*	Methanol root extracts	In vivo antifertility	University/research institute	Gebrie et al. (2005)
22	*Achyranthes aspera L.*	Methanol leaves extract	In vivo wound healing	University/research institute	Fikru et al. (2012)

(continued)

Table 1.2 (continued)

S/N	Plants studied	Samples tested	Activities	Institution type	References
23	*Avicennia marina*	Different extracts	In vitro antimicrobial	University	Manilal et al. (2016)
24	*Rhizophora mucronata*	Different extracts	In vitro antimicrobial	University	Manilal et al. (2015)
25	*Taverniera abyssinica*	Aqueous root extracts	In vivo anti-acetylcholine and histamine	University	Noamesi et al. (1990)
26	*Allium sativum*	Extract along with honey	In vitro antibacterial	University	Andualem (2013)
27	*Holigarna arnottiana*	Organic extracts	In vitro antimicrobial	University	Manilal and Idhayadhulla (2014)
28	*Embelia schimperi*	Hydroalcoholic extract	In vitro and in vivo anthelmintic	Research institute	Debebe et al. (2015)
29	*Verbascum sinaiticum* Benth	Aqueous and methanol leaves extracts	In vivo antitrypanosomal	University	Mergia et al. (2016)
30	*Artemisia absinthium* and *Echinops kebericho*	Essential oils	In vitro antileishmanial and toxicity	University	Tariku et al. (2011)
31	*Ajuga remota* Benth	Aqueous and hydromethanolic leaf extracts	In vivo diuretic	University	Hailu and Engidawork (2014)
32	*Osyris quadripartita*	Aqueous, chloroform, and methanol leaf extracts	In vivo antimalarial	University	Girma et al. (2015b)
33	*Echinops kebericho*	Methanol root extract	In vitro and in vivo antitrypanosomal	University	Abdeta et al. (2020)
34	*Ritchiea albersii Gilg* and *Cynoglossum amplifolium*	Hydromethanolic, chloroform, and acetone leaf extracts	In vitro antibacterial	University	Engiso et al. (2020)

35	*Artemisia abyssinica*	Dichloromethane and hydromethanol aerial part extracts	In vivo antitrypanosomal	University	Feyera et al. (2014)
36	*Nicotiana tabacum* L.	Different organic solvent extracts	In vitro antibacterial	University	Ameya et al. (2017)
37	*Hibiscus micranthus* Linn	Hydromethanolic leaf extracts	In vitro antibacterial and in vivo wound healing	University	Begashaw et al. (2017)
38	*Combretum molle*	Hydromethanolic stem bark extract	In vivo antimalarial	University/ research institute	Mulaw et al. (2019)
39	*Clematis longicauda* and *Clematis burgensis*	Methanol and petroleum ether leaf extracts	In vitro antimicrobial	University	Hawaze et al. (2012)
40	*Aloe megalacantha*	Leaves exudate	In vitro antimicrobial	University	Asmerom et al. (2020b)
41	*Leonotis ocymifolia*	Hydroalcoholic leaf extract	In vivo antimalarial	University	Teklu et al. (2020)
42	*Echinops kebericho*	Ethanol root extract	In vivo antiplasmodial	University	Toma et al. (2015)
43	*Aloe schelpei*	Leaf latex	In vitro antioxidant	University	Teka and Kassahun (2020)
44	*Acokanthera schimperi*, *Euclea schimperi*, *Inula confertiflora*, *Melilotus elegans*, and *Plumbago zeylanica*	Hydromethanolic extracts	In vitro antiviral	University	Gebre-Mariam et al. (2006)
45	*Aloe adigratana* Reynolds	Leaf latex	In vitro antimicrobial	University	Asmerom et al. (2020a)
46	*Olea europaea* Linn subsp. *cuspidata*	Hydromethanolic stem bark extract	In vivo antimalarial	University	Hailesilase et al. (2020)

(continued)

Table 1.2 (continued)

S/N	Plants studied	Samples tested	Activities	Institution type	References
47	*Vernonia amygdalina, Calpurnia aurea, Schinus molle, Ricinus communis, Croton macrostachyus*, and *Nicotiana tabacum*	Methanolic extract of leaves	In vitro acaricidal	University	Kemal et al. (2020)
48	*Embelia schimperi, Rubus steudneri, Ocimum lamiifolium*, and 16 other plants	Ethanol extracts	In vitro antimicrobial	University	Lulekal et al. (2014b)
49	*Cissus quadrangularis, Commelina benghalensis, Euphorbia heterophylla, Euphorbia prostrate, Momordica schimperiana, Trianthema* spp., and *Solanum incanum*	Methanol extracts	In vitro antibacterial	University	Mummed et al. (2018)
50	*Calpurnia aurea, Croton macrostachyu*, and *Erythrina brucei*	Organic solvent extracts	In vitro antibacterial	University	Romha et al. (2018)
51	*Ajuga integrifolia, Clerodendrum myricoides, Melia azedarach, Peponium vogelii*, and *Premna schimperi*	Hydromethanolic extracts	Antimalarial	University/ research institute	Asnake et al. (2015)
52	*Salvadora persica* and *Balanites rotundifolia*	Aqueous and organic extracts	In vivo antimalarial	University	Gebrehiwot et al. (2019)
53	*Cymbopogon citratus, Foeniculum vulgare, Eucalyptus globulus*, and *Calpurnia aurea*	Essential oils and aqueous extract	Insecticidal	Research institute	Gemeda et al. (2014)
54	*Calpurnia aurea, Ocimum basilicum, Artemisia abyssinica, Croton macrostachyus*, and *Eucalyptus camaldulensis*	Methanol extracts	In vitro antimycobacterial	University	Gemechu et al. (2013)
55	*Achyranthes aspera, Brucea antidysenterica, Datura stramonium, Croton macrostachyus, Acokanthera schimperi, Phytolacca dodecandra, Millettia ferruginea*, and *Solanum incanum*	Methanol and water extracts	In vitro antimicrobial	Research institute	Taye et al. (2011)
56	*Verbena officinalis, Myrtus communis*, and *Melilotus elegans*	Hydromethanolic extract	In vitro antibacterial	University	Sisay et al. (2019)

57	*Allium sativum, Cinnamomum zeylanicum, Syzygium aromaticum, Persea americana, Rosmarinus officinalis*, and *Argemone Mexicana*	Various extracts	In vitro antimoraxella	University	Rasheed and Thajuddin (2011)
58	*Justicia schimperiana* and *Verbascum sinaiticum*	Hydroalcoholic extracts	In vivo hepatoprotective	University	Umer et al. (2010)
59	*Combretum molle* and six other plants	Fifteen extracts	In vitro antimycobacterial	University	Asres et al. (2001a)
60	*Cymbopogon citratus* and *Croton macrostachyus*	Methanol extracts	Larvicidal	University	Karunamoorthi and Ilango (2010)
61	*Moringa stenopetala*	Organic extracts	In vitro antitrypanosomal	University	Mekonnen et al. (1999)
62	*Albizia gummifera* and *Ferula communis*	Methanol extracts	In vitro antibacterial	College	Unasho et al. (2009)
63	*Bersama abyssinica*	Crude extract and solvent fractions	In vivo antimalarial	University/ hospital	Alehegn et al. (2020)
64	*Rhamnus prinoides*	Aqueous and organic extracts	In vitro antibacterial	University	Molla et al. (2016)
65	*Capsicum frutescens*	Hydromethanolic extract	In vivo antimalarial	University	Habte and Assefa (2020)
66	*Brucea antidysenterica* and *Ocimum lamiifolium*	Aqueous and organic extracts	In vivo antimalarial	University	Kefe et al. (2016)
67	*Croton macrostachyus*	Aqueous and organic extracts	In vivo antidiarrheal	University	Degu et al. (2016)
68	*Maesa lanceolata* and *Plectranthus punctatus*	Aqueous and hydro-alcoholic extracts	In vitro ovicidal and larvicidal activity	University	Tadesse et al. (2009)

(continued)

Table 1.2 (continued)

S/N	Plants studied	Samples tested	Activities	Institution type	References
69	*Solanum incanum* L.	Ethanol and aqueous extracts of different parts	In vitro antimicrobial	University	Sbhatu and Abraha (2020)
70	*Schinus molle Linnaeus*	Aqueous and hydromethanol crude extracts and fractions	In vivo antimalarial	University	Habte et al. (2020)
71	*Grewia ferruginea*	Hydromethanolic extract	In vivo laxative	University/ research institute	Tessema et al. (2020)
72	*Strychnos mitis*	Aqueous and hydro-methanolic extracts	In vivo antimalarial	University	Fentahun et al. (2017)
73	*Evolvulus alsinoides*	Ethanolic extract	In vitro antioxidant	University	Gomathi et al. (2014)
74	*Terminalia brownii Fresen.*	Methanol and aqueous extracts	In vivo antimalarial	University	Biruk et al. (2020)
75	*Cordia africana*	Organic extracts and fractions	In vivo antimalarial	University	Wondafrash et al. (2019)
76	*Senna singueana*	Methanol extract	In vivo antinociceptive	University	Hishe et al. (2018)
77	*Gardenia ternifolia*	Hydromethanolic extract and solvent fractions	In vivo antimalarial	University	Nureye et al. (2018)
78	*Pterolobium stellatum*, *Otostegia integrifolia*, and *Vernonia amygdalina*	Organic extracts	Genotoxicity	University	Kahaliw et al. (2018)
79	*Calpurnia aurea (Ait.) Benth.* subsp. *aurea*	Hydromethanol extract	In vivo antihypertensive	University	Getiye et al. (2016)
80	*Calpurnia aurea*	Hydromethanolic	In vivo hypoglycemic, antihyperglycemic, and antihyperlipidemic	University	Belayneh et al. (2019)

81	*Echinops kebericho* M.	Hydromethanolic extract	In vivo analgesic and anti-inflammatory	University	Yimer et al. (2020)
82	*Maerua subcordata*	Hydroethanol extract	In vivo antidiarrheal	University	Palla et al. (2020)
83	*Brucea antidysenterica*	Hydromethanolic root extract	In vivo antidiarrheal and in vitro antibacterial	University	Zewdie et al. (2020)
84	*Coriandrum sativum*	Aqueous and hydroalcoholic seeds extracts	In vitro anthelmintic	University/ research institute	Eguale et al. (2007)
85	*Calpurnia aurea*	Hydromethanolic leaves extract	In vivo antidiabetic	University	Belayneh and Birru (2018)
86	*Brassica carinata*	Hydromethanol extract and fractions	In vitro antioxidant and in vivo wound healing	University	Alemu et al. (2020)
87	*Moringa stenopetala*	Organic solvent extracts	In vivo anti-inflammatory and analgesic	University	Tamrat et al. (2017)
88	*Cassia didymobotrya*	Aqueous and ethanol extract	In vivo antimalarial	University	Nagappan (2012)
89	*Rumex abyssinicus*	Hydromethanolic rhizome extracts	In vivo hepatoprotective and in vitro antioxidant	University	Adamu et al. (2020)
90	*Croton macrostachyus*	Hydroethanolic extract and solvent fractions	In vitro antioxidant and in vivo hepatoprotective	University	Tafere et al. (2020)
91	*Agave sisalana, Birbira*, and *Hypericum revolutum*	Methanol extracts	In vitro antibacterial and lethality to brine shrimp	University/ College	Kassu et al. (1999)
92	*Zingiber officinale Roscoe* and *Echinops Kebericho* Mesfin	Hydromethanolic extract	In vivo antiplasmodial	University	Biruksew et al. (2018)
93	*Acanthus polystachyus*	Hydromethanolic extract	Hypoglycemic and antihyperglycemic	University/ College	Derebe et al. (2020)
94	*Moringa stenopetala*	Ethanol extract	Effect against guinea-pig and mouse smooth muscle	University	Mekonnen (1999)

(continued)

Table 1.2 (continued)

S/N	Plants studied	Samples tested	Activities	Institution type	References
95	*Phytolacca dodecandra*	Aqueous extract	Cercariacidal and miracidiacidal	University	Birrie et al. (n.d.)
96	*Taverniera abyssinica*	Crude extracts	Antipyretic and analgesic	University	Dagne et al. (1990b)
97	*Acanthus polystachyus*	Hydromethanol extract	In vivo wound healing	University	Demilew et al. (2018)
98	*Artemisia herbaalba* and *Punica granatum*	Methanolic extracts	In vitro anthelmintic	University	Ahmed et al. (2020)
99	*Osyris quadripartita*	Hydromethanolic extract	In vivo antidiarrheal	University/ research institute	Teferi et al. (2019)
100	*Euphorbia abyssinica*	Hydromethanolic extract	In vivo antimalarial	University	Muluye et al. (2019)
101	*Lantana camara*	Aqueous extract	In vivo antidiarrheal	University	Tadesse et al. (2017)
102	*Dodonaea viscosa, Rumex nervosus*, and *Rumex abyssinicus*	Hydromethanolic extracts	Antimicrobial and anti-inflammatory	University	Getie et al. (2003)
103	*Aloe pirottae*	Aqueous extract of the latex	In vivo antimalarial	University	Dibessa et al. (2020)
104	*Ocimum suave, Ocimum lamiifolium, Lippia adoensis,* and *Ajuga remota*	Aqueous and ethanol extracts	In vivo antianalgesic	University	Makonnen et al. (2003)
105	*Aloe weloensis*	Leaves latex	In vivo antiplasmodium berghei	University	Teka et al. (2020b)
106	*Ocimum sauve, Ocimum lamiifolium, Lippia adoensis*, and *Ajuga remota*	Aqueous and ethanol extracts	In vivo antianalegesic	University	Debella et al. (2003)
107	*Ruta chalepensis* and *Vernonia amygdalina*	Hydromethanolic extracts	In vivo antidiarrheal	University	Degu et al. (2020)

108	*Bersama abyssinica, Combretum paniculatum, Dodonaea angustifolia,* and *Ximenia americana*	Polar and nonpolar organic extracts	In vitro antiviral activity	University	Asres et al. (2001b)
109	*Idigofera spicata*	Methanolic extract	In vivo antidiarrheal	University	Birru et al. (2016)
110	*Terminalia brownii*	Crude extracts and solvent fractions	In vivo antidiabetic	University	Alema et al. (2020)
111	*Blepharis cuspidata, Boswellia ogadensis*, and *Thymus schimper*	Essential oils	In vitro antibacterial	University	Gadisa et al. (2019)
112	*Senna occidentalis, Leonotis ocymifolia, Leucas martinicensis, Rumex abyssinicus*, and *Albizia schimperiana*	Aqueous and hydro-alcoholic extracts	In vitro anthelmintic	University	Tadesse et al. (2011)
113	*Dovyalis abyssinica, Albizia schimperiana, Ocimum urticifolium, Acokanthera schimperi, Chenopodium ambrosioides*, and 25 other plants	Chloroform and methanol extracts	In vitro trypanocidal and cytotoxic	University	Nibret and Wink (2011)
114	*Vernonia leopoldi, Sideroxylon oxyacanthum*, and *Clematis sinensis*	Organic solvent fractions	In vitro antibreast cancer	University	Tuasha et al. (2020)
115	*Asparagus africanus, Guizotia schimperi, Lippia adoensis*, and *Premna schimperi*	Ethanolic extracts	In vitro antimicrobial	University	Teka et al. (2015)

Table 1.3 A six-decade (1960–2020) phytochemical study of Ethiopian medicinal plants

S/N	Plant name	Part(s) studied	Compound (s) reported	Biological activity	Institution type	References
1	*Brassica carinata*, *Lathyrus sativus*, and *Phaseolus vulgaris*	Seeds	Glucosylceramides	*	University	Tessema et al. (2017b)
2	*Warburgia ugandensis*	Stem bark	Two sesquiterpenes (nerolidol and muzigadial)	*	University	Gonfa et al. (2020)
3	*Aframomum corrorima*	Leaves and rhizomes	Essential oils with 38 and 52 components, respectively (mainly β-caryophyllene, γ-terpinene, and β-pinene)	*	University/ research institute	Eyob et al. (2008)
4	*Laggera tomentosa*	Leaves	Essential oil with components, (1*S*,5*R*)-(−)-2,4,4 trimethylbicyclo [3.1.1]hept-2-en-6-one, chrysanthenone, thymoquinol dimethyl ether, and dehydrocitral	*	University	Asfaw et al. (1999)
5	*Avena abyssinica*	Grain	Glucosylceramides	*	University	Tessema et al. (2017a)
6	*Aloe yavellana*	Latex	Aloin A/B and microdontin A/B	Larvicidal effect against *Amblyomma variegatum*	University	Hailesillassie et al. (2018)
7	*Cineraria abyssinica*	Leaves	Rutin	In vivo hepatoprotective and in vitro antioxidant	University	Sintayehu et al. (2012)
8	Five *Vernonia* species *(V. leopoldi, V. myriantha, V. syringifolia, V. filigera*, and *V. cinerascens)*	Aerial parts	Sesquiterpene lactones	*	University	Abegaz et al. (1994b)
9	*Thymus schimperi*	Flowers	Flavone polyphenols, mainly flavones and luteolin	*	University	Desta et al. (2017)

10	*Calpurnia aurea*	Leaves	Alkaloids, including 3*β*,4*α*,13*α*-trihydroxylupanine, 3*β*,4*α*-dihydroxy 13*α*-*O*-(2'-pyrrolylcarbonyl)-lupanine, Lupinine, epilupinine, Calpurmenine, and calpurmenine pyrrolecarboxylic acid ester	*	University	Kaleab Asres et al. (1986)
11	Khat (*Catha edulis*)	Leaves	Phenylpropylamino alkaloids (cathinone, cathine, and norephedrine)	In vitro permeation across oral and intestinal mucosal tissues	University	Atlabachew et al. (2016)
12	*Erythrina brucei*	Stem and root bark	Two isoflavanones (erybrucein A and B) and 19 other compounds	*	University	Gurmessa et al. (2018)
13	*Laggera tomentosa*	Leaves	Chrysanthenone, filifolone, and (*Z*)-isogeranic acid as artifacts of hydrodistilled oil	*	University	Asfaw et al. (2001)
14	*Hagenia abyssinica*, *Leonotis ocymifolia*, and *Moringa stenopetala*	Flowers, aerial parts, and seeds, respectively	Many components of essential oils	Trypanocidal against *Trypanosoma b. brucei* and antileukemic against HL-60)	University	Nibret and Wink (2010a, b)
15	*Melilotus elegans*	Leaves	Flavonoids, mainly kaempferol	*	University	Tsige Gebre-Mariam et al. (2005)
16	*Discopodium penninervium*	Stem bark	Withanolide, sesquiterpene, and 5 other compounds	In vitro leukotriene and cycloxygenase-1 and -2 inhibition	University	Wube et al. (2008)
17	*Tephrosia pumila*	Seed pods	Pumilaisoflavones A and B	*	University	Dagne et al. (1988)
18	*Aloe marlothii* and *Aloe rupestris*	Leaves exudate	Chromone (7-*O* methylaloeresin A), and anthrone (5-hydroxyaloin A 6′-*O*-acetate)	*	University	Bisrat et al. (2000)
19	*Millettia ferruginea*	Root bark and seeds	Chalcone and seven isoflavones	*	University	Dagne et al. (1990a)

(continued)

Table 1.3 (continued)

S/N	Plant name	Part(s) studied	Compound (s) reported	Biological activity	Institution type	References
20	*Aloe berhana*	Roots	Aloesaponol I, laccaic acid D methyl ester, aloesaponol III, aloesaponarin I, aloechrysone	*	University	Dagne et al. (1992a)
21	*Senna didymobotrya*	Pods part	Two bianthraquinones and others	*	University	Alemayehu et al. (1996)
22	*Lathyrus sativus*	Seeds	Quantification of β-*N*-oxalyl-L-α,β-diaminopropionic acid	*	University	Yigzaw et al. (2001)
23	*Millettia griffoniana*	Root bark	*O*-geranylated isoflavone and 7 other compounds	*	University	Yankep et al. (1997)
24	*Senna multiglandulosa*	Seeds	Torosachrysone, physcion, floribundone-1, anhydrophlegmacin, and bianthraquinone (isosengulone)	*	University	Alemayehu and Abegaz (1996)
25	*Aloe littoralis*	Leaf exudate	Littoraloside, littoraloin, Deacetyllittoraloin, and 10-hydroxyaloin B	*	University	Dagne et al. (1998)
26	*Kniphofia foliosa*	Stems	Isoknipholone, isoknipholone anthrone, foliosone, isofoliosone, and three other compounds	*	University	Yenesew et al. (1994)
27	*Warburgia ugandensis*	Stem bark	12 Sesquiterpenes	Antimicrobial activity	University	Wube et al. (2005)
28	*Teclea nobilis*	Leaves and fruits	Nobiline and 7 other quinoline alkaloids	*	University	Yenesew and Dagne (1988)
29	*Senna multiglandulosa*	Seeds	Emodin and five bianthraquinones	*	University	Abegaz et al. (1994a)

30	*Albizia gummifera*	Stem bark	Many triterpenoid saponins and sapogenin lactones	*	University/ research institute	Debella et al. (2000)
31	*Senna longiracemosa*	leaves and root bark	Eleven anthracene and naphthalene derivatives	*	University	Alemayehu et al. (1993)
32	*Tephrosia polyphylla*	Roots	4′-Demethyltoxicarol isoflavone, toxicarol isoflavone, and 7-methylglabranin	*	University	Dagne (1992b)
33	*Teclea simplicifolia*	Leaves	Seven alkaloids, including isohaplopine and its 3,3-dimethylallylether	*	University	Wondimu et al. (1988)
34	*Asparagus africanus*	Roots	Two monodesmosidic spirostanosides and bisdesmosidic furostanol glycoside	*	University/ research institute	Debella et al. (1999)
35	*Senna sophera*	Seeds	Presengulone physcion, physcion bianthrone, xanthorin, floribundone-1, isosengulone, sengulone, and anhydrophlegmacin-9,10-quinones A_2 and B_2	*	University	Alemayehu et al. (1998)
36	*Salsola somalensis*	Roots	Three isoflavones	*	University	Woldu and Abegaz (1990)
37	*Kniphofia foliosa*	Roots	Chrysophanol and knipholone	*	University	Dagne and Steglich (1984b)
38	*Kniphofia foliosa*	Stems	Knipholone anthrone	*	University	Dagne and Yenesew (1993)

(continued)

Table 1.3 (continued)

S/N	Plant name	Part(s) studied	Compound (s) reported	Biological activity	Institution type	References
39	*Lupinus angustifolius*	Whole seeds	Four triterpene glycosides	Antifungal and hemolytic activity	University	Woldemichael and Wink (2002)
40	*Erythrina brucei*	Flowers	An erythrina alkaloid (8-oxoerythrinine.) and other compounds	*	University	Dagne and Steglich (1984a)
41	*Echinops pappii*	Roots	Six polyacetylenic thiophenes and terpenoids	*	University	Abegaz (1991)
42	*Artemisia schimperi*	Aerial parts	*nor*-Monoterpene (1,5-octadien-3-hydroxy-3-methyl-7-one), artemisyl and lyratyl acetates, and quercetin 3,3′,4′-trimethyl ether	*	University	Abegaz and Herz (1991)
43	*Discopodium penninervium*	Leaves	Three 16-oxygenated withanolides and jaborosala	*	University	Habtemariam et al. (1993)
44	*Aloe microstigma*	Leaves exudate	5-Hydroxyaloin A and microstigmin A	*	University	Dagne et al. (1997)
45	*Millettia ferruginea*	Immature seeds	Three rotenoids, four pyrano-isoflavones, and three novel *C*-prenylated isoflavones	*	University	Dagne and Bekele (1990)
46	*Gasteria bicolor*	Stem and leaves	Three dihydroanthracenones	*	University	Dagne et al. (1996a)
47	*Kalanchoe marmorata* and *Kalanchoe petitiana*	Aerial parts	24-Ethylsterols, 24-ethyl-25-dehydrocholesterol, and clionastanol	*	University	Akihisa et al. (1992)
48	*Buxus hildebrandtii*	Leaves	Three steroidal alkaloids	*	University	Akihisa et al. (1992)

49	*Aloe littoralis*	Leaves exudate	Two C-glucoside oxanthrones and 10-hydroxyaloin B	*	University	Dagne et al. (1996b)
50	*Araliorhamnus vaginata*	Bark	17 Natural quinone pigments	*	University	Mammo et al. (1992)
51	*Cassia floribunda*	Leaves	N^1,N^8-dibenzoylspermidine, physcion, 5,7′-biphyscion, and 5,7′--physcion-physcionanthrone	*	University	Alemayehu et al. (1988)
52	*Tephrosia pentaphylla* and *Tephrosia pentaphylla*	Roots, leaves, and pods	6-Oxygenated rotenoids and rotenoids	*	University	Dagne et al. (1989)
53	*Dorstenia barniniana*	Tubers	Three styrenes	*	University	Woldu et al. (1988)
54	*Araliorhamnus vaginata*	Bark	Three anthraquinone lactones (araliorhamnone A, B, and C)	*	University	Mammo et al. (1990)
55	*Artemisia rehan*	Essential oil of leaves	Twenty-two components with davanone and camphor as major compounds	*	University	Abegaz and Yohannes (1982)
56	*Salsola somalensis*	Roots	Nine isoflavones	*	University	Abegaz and Woldu (1991)
57	*Dodonaea angustifolia*	Leaves	Pinocembrin, santin, and clerodane diterpene 2-hydroxy-15,16-epoxyceloda-3,13(16),14-trien-18-oic acid	In vivo antiplasmodial	University	Melaku et al. (2017)
58	*Vernonia amygdalina*	Flowers	Tricosane, vernolide, isorhamnetin and luteolin	In vitro antibacterial and antioxidant	University	Habtamu and Melaku (2018)
59	*Embelia schimperi*	Stem bark	Epicatechin and its close flavan derivative	In vitro antibacterial	University	Guyasa et al. (2018)

(continued)

Table 1.3 (continued)

S/N	Plant name	Part(s) studied	Compound (s) reported	Biological activity	Institution type	References
60	*Aloe megalacantha*	Roots	1,8-dimethoxynepodinol, aloesaponarin III, 10-*O*-methylchrysalodin, methyl-26-*O*-feruloyl-oxyhexacosanate, and ten other compounds	Cytotoxic activity	University	Abdissa et al. (2017c)
61	*Echinops hoehnelii*	Roots	Five acetylenic thiophenes	In vivo antimalarial	University	Bitew et al. (2017)
62	*Plectranthus punctatus*	Roots	Four abietane-type diterpenoids and 13 diterpenoids	Antibacterial and cytotoxic activity	University	Abdissa et al. (2017b)
63	*Aloe pulcherrima*	Roots	Chrysophanol, aloesaponarin I, and aloesaponarin II	In vitro antibacterial and antiplasmodial	University	Abdissa et al. (2017a)
64	*Gnidia involucrata*	Root bark	Tetratriacontanyl caffeate, 12-*O*-dodeca-2,4-dienoylphorbol-13-acetate, and naringenin	In vitro antibacterial and antioxidant	University	Kalbessa et al. (2019)
65	*Artemisia abyssinica* and *Satureja punctata* ssp. *punctata*	Leaves	Sixty-seven components of essential oils, respectively.	Antileishmanial and cytotoxicity	University	Tariku et al. (2010)
66	*Artemisia absinthium* and *Echinops kebericho*	Leaves	Sixty-five and 43 components of essential oils, respectively	Antileishmanial and cytotoxicity	University	Tariku et al. (2011)
67	*Combretum molle*	Stem bark	Two tannins and two oleanane-type pentacyclic triterpene glycosides	In vitro antiprotozoal	University	Asres et al. (2001c)
68	*Premna schimperi* and *Premna oligotricha*	Leaves	Three diterpenes	In vitro antileishmanial	University	Habtemariam (2003)
69	*Aloe calidophila*	Leaves latex	Three anthrones (aloinoside, aloin, and microdontin)	In vitro antileishmanial	University	Abeje et al. (2014)
70	*Pycnostachys abyssinica* and *Pycnostachys eminii*	Stems, roots, and leaves	Many components of essential oils	*	University	Hussien et al. (2010)

71	*Hagenia abyssinica*, *Leonotis ocymifolia*, and *Moringa stenopetala*	Flowers, aerial part and seeds, respectively	Many components of essential oils	Trypanocidal and antileukemic	University	Nibret and Wink (2010a, b)
72	*Aloe trigonantha*	Leaves latex	Aloesin, 8-*O*-methy-7-hydroxyaloin A/B, aloin A/B, and aloin-6'-*O*-acetate A/B	In vitro antimicrobial	University	Megeressa et al. (2015)
73	*Taverniera abyssinica*	Roots	Medicarpin and 4-hydroxymedicarpin	Antinematicidal, cytotoxic, and antimicrobial activities	University	Stadler et al. (1994)
74	*Warburgia ugandensis*	Stem bark	Twelve drimane and coloratane sesquiterpenes	In vitro 12(S)-HETE and leukotriene inhibitory	University	Wube et al. (2006)
75	*Podocarpus falcatus*	Roots	16-hydroxynagilactone F, 2β,16-dihydroxynagilactone F, 7*β*-hydroxymacrophyllic acid, and 7 other compounds	Antiproliferative against human colorectal adenocarcinoma (HT-29) cell line	University	Addo et al. (2015)
76	*Artemisia absinthium*, *A. abyssinica*, *A. afra* and *A. annua*	Leaves and aerial parts	Volatile components	In vitro antitrypanosomal and cytotoxic activities	University	Nibret and Wink (2010a, b)
77	*Aloe pulcherrima*	Leaves latex	Nataloin and 7-hydroxyaloin	In vivo antimalarial	University	Teka et al. (2016)
78	*Melilotus elegans*	Leaves	Azukisaponin V	In vivo anti-inflammatory	University	Asres et al. (2005)
79	*Cadia purpurea*	Leaves and twigs	13-Hydroxylupanin, amino acid pinitol, alpha-amyrin, 1-spartein, lupanin, cadiamin, and dihydroxyluparzin	*	University	Eijk and Radema (1975)
80	*Trichilia emetica*	Stem bark	Nymania 1, drageana 4, trichilin A, rohituka 3, Tr-B, and seco-A protolimonoid	Selective inhibitory activity toward DNA repair-deficient yeast mutants	University	Gunatilaka et al. (1998)

(continued)

Table 1.3 (continued)

S/ N	Plant name	Part(s) studied	Compound (s) reported	Biological activity	Institution type	References
81	*Gum olibanum*	Gummy exudate	Twenty-two components of essential oil	*	Research institute	Shi et al. (2002)
82	*Commiphora kua*	Resin	15 alpha-hydroxymansumbinone, 28-acetoxy-15 alpha-hydroxymansumbinone, mansumbinone, mansumbinol, dihydroxydammar-24-en-3-one, and T-cadinol	*	University	Dekebo et al. (2002a)
83	*Erythrina burana*	Bark	Phaseollidin and cristacarpin	Selective activity toward DNA repair-deficient yeast mutants	University	Dagne et al. (1993)
84	*Commiphora sphaerocarpa*	Resin	Furanosesquiterpene and five other compounds	*	University	Dekebo et al. (2002b)
85	*Vernonia amygdalina*	Leaves	14-*O*-methylvernolide, 3'-deoxyvernodalol, and vernomygdalin	Thiol trapping potential	University	Sinisi et al. (2015)
86	*Trachyspermum ammi*	Fruits	Twenty-two components of essential oil	Insecticidal toxicities	University	Bisrat and Jung (2020)
87	*Aloe citrina*	Leaves latex	Homonataloin A/B	In vivo antimalarial	University	Girma et al. (2015a)
88	*Commiphora erlangeriana*	Resin	Erlangerin A, erlangerin B, erlangerin C, and erlangerin D	*	University	Dekebo et al. (2002c)
89	*Combretum molle*	Stem bark	Punicalagin, arjunglucoside, and sericoside	Antimycobacterial	University	Asres et al. (2001a)
90	*Embelia schimperi*	Fruit seeds	Embelin and nine fatty acids	*	University	Atlabachew et al. (2017)

91	*Pentas parvifolia and Pentas bussei*	Roots	Many naphthalene derivatives, including parvinaphthols A, B, and C	Cytotoxicity against human triple-negative breast cancer (MDA-MB-231)	University	Abdissa et al. (2016)
92	*Aloe harlana*	Latex	Anthrone (aloin) and chromone (7-*O*-methylaloeresin A)	In vitro antimicrobial and antioxidant	University	Asamenew et al. (2011)
93	*Echinops ellenbeckii*	Roots	Seven acetylenic thiophenes, seven fatty acid esters, and three long-chain hydrocarbons	*	University	Hymete et al. (2005)
94	*Crinum abyscinicum*	Bulbs	6-Hydroxycrinamine and lycorine	Antiproliferative	University	Abebe et al. (2020)
95	*Clausena anisata*	Roots	Heptazoline and three coumarins, including imperatorin and chalepin	Antibacterial	University	Tamene and Endale (2019)

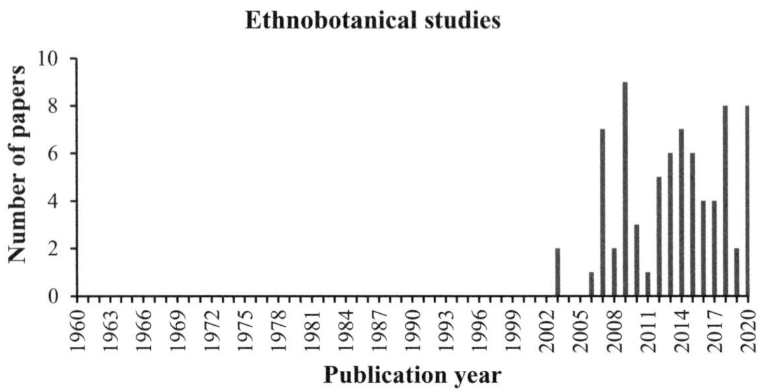

Fig. 1.1 Number of papers versus publication year of ethnobotanical study of Ethiopian plants reported over the last six decades (1960–2020)

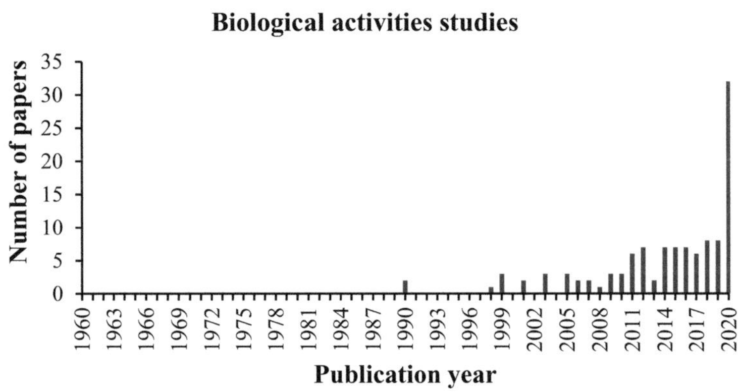

Fig. 1.2 Number of papers versus publication year of biological activity studies of Ethiopian plants reported over the last six decades (1960–2020)

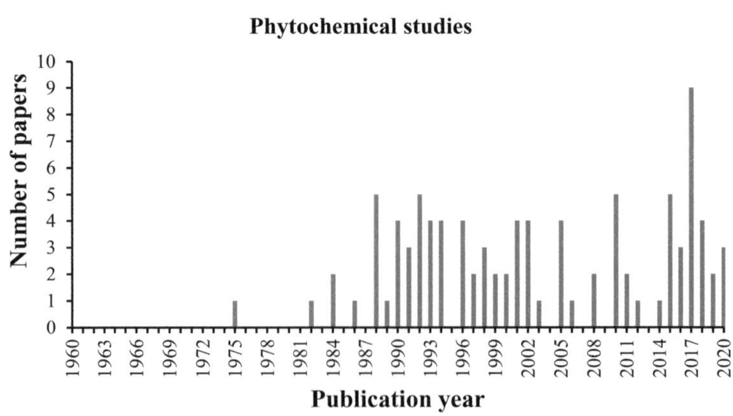

Fig. 1.3 Number of papers versus publication year of phytochemical studies of Ethiopian plants reported over the last six decades (1960–2020)

publication year 2020 and the preceding years 2018 and 2019. That is, in 2020, highest number of papers (32) was extracted, whereas the next highest number of papers (8) was obtained in 2018 and 2019. A consecutively increasing number of papers were observed since 2014 up to 2020, except in 2017, in which a slight decline occurred. Among the 115 papers, 35 of them reported the antimicrobial activities, and 22 of them reported the antimalarial activities of the Ethiopian plant species. The majority of the reported works (82%) were done by universities.

1.3.3 A Six-Decade (1960–2020) Phytochemical Study of Ethiopian Medicinal Plants

A total of 95 papers, out of the total of 2082 papers, were found to be directly related to the isolation of phytochemicals from Ethiopian plants (Table 1.3 and Fig. 1.3). According to the obtained reports, unlike the ethnobotanical and biological activity studies, the phytochemical investigation of Ethiopian plants was started early in 1975, and one paper was found. The largest numbers of papers (9) were reported in 2017, followed by 1988, 1992, 2010, and 2015 with five papers each. The obtained result indicated no consistent decrease or increment in the number of reported works across the specified years. Among the total number of reported phytochemical studies (95), 40% of them also reported the various in vitro and in vivo biological activities of isolated compounds. Among the plant species studied for their phytochemical constituents, *Aloe* species were reported predominantly in 14 papers, followed by *Artemisia* species and *Vernonia* species, which were reported in 8 and 7 papers, respectively. Almost all of the reported works (96%) were found to be done by universities, of which Addis Ababa University took the Lion's share.

1.4 Conclusion and Future Prospects

Scientific reports on ethnobotanical, biological, and phytochemical studies of Ethiopian plants were found starting in 2003, 1990, and 1975, respectively. Regarding the ethnobotanical study, of the 75 reported papers, the highest number of plant species was reported from Odo-Bulu and Demaro of the Oromia region, the Sheka zone of Southern Ethiopia, and Mana Angetu district of Southeastern Ethiopia. *Allium* species, *Hagenia abyssinica*, *Vernonia amygdalina*, *Ocimum* species, *Rhamnus prinoides*, *Solanum incanum*, *Azadirachta indica*, and *Phytolacea dodecandra* were mentioned in various papers for their traditional medicinal uses. With respect to the biological activity study, among the collected 115 papers, the majority reported the antimicrobial and antimalarial activities of Ethiopian plant species. Among the 95 total phytochemical studies reported, 40% also reported various in vitro and in vivo biological activities of isolated compounds. *Aloe* species were reported predominantly in 14 papers, followed by *Artemisia* and *Vernonia* species reported in 8 and 7 papers, respectively, for phytochemical studies. In all cases, no consistent research progress, either in decreasing or increasing numbers of reported works, was observed across the years 1960–2020. Universities did the Lion's share of the studies. In conclusion, this review can be used as preliminary data for researchers to explore unexplored areas and plant species. It is also important to repurpose the biological and phytochemical studies of some of the reported plants, based on their earlier remarkable and exclusive biochemical features, to bring their medical uses from the shelf to the market. To do so, the present compiled data may serve as a quick reference to pick up the target plants and compounds in a simple and easy way. Besides, this review may encourage academicians, researchers, and other stakeholders to establish an Ethiopian pharmacopoeia that is currently initiated. The limitation of this review was the use of limited search words and search engines, which might result in the exclusion of some published articles. So, it is recommended to compile further original research works using many more words and search tools to establish big data with full information.

Competing Interests The authors have no conflicts of interest to declare that are relevant to the content of this chapter.

References

Abate, G. (1989). *Etse debdabe: Ethiopian traditional medicine* (pp. 34–40). Addis Ababa University.

Abdeta, D., Kebede, N., Giday, M., Terefe, G., & Abay, S. M. (2020). *In vitro* and *in vivo* antitrypanosomal activities of methanol extract of *Echinops kebericho* roots. *Evidence-based Complementary and Alternative Medicine: Ecam, 2020*, 8146756.

Abdissa, N., Pan, F., Gruhonjic, A., Gräfenstein, J. R., Fitzpatrick, P. A., Landberg, G., & Erdelyi, M. (2016). Naphthalene derivatives from the roots of *Pentas parvifolia* and *Pentas bussei*. *Journal of Natural Products, 79*(9), 2181–2187.

Abdissa, D., Geleta, G., Bacha, K., & Abdissa, N. (2017a). Phytochemical investigation of *Aloe pulcherrima* roots and evaluation for its antibacterial and antiplasmodial activities. *PLoS One, 12*(3), e0173882.

Abdissa, N., Frese, M., & Sewald, N. (2017b). Antimicrobial abietane-type diterpenoids from *Plectranthus punctatus*. *Molecules, 22*(11), 1919.

Abdissa, N., Gohlke, S., Frese, M., & Sewald, N. (2017c). Cytotoxic compounds from *Aloe megalacantha*. *Molecules, 22*(7), 1136.

Abebe, D., & Ayehu, A. (1993). Medicinal plants and enigmatic health practices of Northern Ethiopia.

Abebe, B., Tadesse, S., Hymete, A., & Bisrat, D. (2020). Antiproliferative effects of alkaloids from the bulbs of *Crinum abyscinicum* Hochst. ExA. Rich. *Evidence-based Complementary and Alternative Medicine, 2020*, 2529730.

Abegaz, B. M. (1991). Polyacetylenic thiophenes and terpenoids from the roots of *Echinops pappii*. *Phytochemistry, 30*(3), 879–881.

Abegaz, B. M., & Herz, W. (1991). A nor-monoterpene from *Artemisia schimperi*. *Phytochemistry, 30*(3), 1011–1012.

Abegaz, B. M., & Woldu, Y. (1991). Isoflavonoids from the roots of *Salsola somalensis*. *Phytochemistry, 30*(4), 1281–1284.

Abegaz, B., & Yohannes, P. G. (1982). Constituents of the essential oil of *Artemisia rehan*. *Phytochemistry, 21*(7), 1791–1793.

Abegaz, B. M., Bezabeh, M., Alemayehu, G., & Duddeck, H. (1994a). Anthraquinones from *Senna multiglandulosa*. *Phytochemistry, 35*(2), 465–468.

Abegaz, B. M., Keige, A. W., Diaz, J. D., & Herz, W. (1994b). Sesquiterpene lactones and other constituents of *Vernonia* species from Ethiopia. *Phytochemistry, 37*(1), 191–196.

Abeje, F., Bisrat, D., Hailu, A., & Asres, K. (2014). Phytochemistry and antileishmanial activity of the leaf latex of *Aloe calidophila* Reynolds. *Phytotherapy Research, 28*(12), 1801–1805.

Abera, B. (2014). Medicinal plants used in traditional medicine by Oromo people, Ghimbi District, Southwest Ethiopia. *Journal of Ethnobiology and Ethnomedicine, 10*, 1–15.

Adamu, B. A., Emiru, Y. K., Sintayehu, B., Araya, E. M., Periasamy, G., & Gebrelibanos Hiben, M. (2020). *In vivo* hepatoprotective and *in vitro* radical scavenging activities of extracts of *Rumex abyssinicus* Jacq. rhizome. *Journal of Experimental Pharmacology, 12*, 221–231.

Addo, E. M., Chai, H.-B., Hymete, A., Yeshak, M. Y., Slebodnick, C., Kingston, D. G., & Rakotondraibe, L. H. (2015). Antiproliferative constituents of the roots of Ethiopian *Podocarpus falcatus* and structure revision of 2α-hydroxynagilactone F and nagilactone I. *Journal of Natural Products, 78*(4), 827–835.

Adefegha, S., & Oboh, G. (2012). Effect of diets supplemented with Ethiopian pepper [*Xylopia aethiopica* (Dun.) A. Rich (Annonaceae)] and Ashanti pepper [*Piper guineense* Schumach. et Thonn (Piperaceae)] on some biochemical parameters in normal rats. *Asian Pacific Journal of Tropical Biomedicine, 2*(2), S558–S566.

Ahmed, A. H., Ejo, M., Feyera, T., Regassa, D., Mummed, B., & Huluka, S. A. (2020). *In vitro* anthelmintic activity of crude extracts of *Artemisia herba-alba* and *Punica granatum* against Haemonchus contortus. *Journal of Parasitology Research, 2020*, 4950196.

Akihisa, T., Gebreyesus, T., Hayashi, H., & Tamura, T. (1992). Co-occurrence of C-24 epimeric 24-ethylsterols possessing and lacking a Δ25-bond in*Kalanchoe petitiana*. *Phytochemistry, 31*(1), 163–166.

Alehegn, A. A., Yesuf, J. S., & Birru, E. M. (2020). Antimalarial activity of crude extract and solvent fractions of the leaves of *Bersama abyssinica* fresen.(Melianthaceae) against *Plasmodium berghei* infection in Swiss albino mice. *Evidence-based Complementary and Alternative Medicine, 2020*, 9467359.

Alema, N. M., Periasamy, G., Sibhat, G. G., Tekulu, G. H., & Hiben, M. G. (2020). Antidiabetic activity of extracts of *Terminalia brownii* Fresen. Stem bark in mice. *Journal of Experimental Pharmacology, 12*, 61–71.

Alemayehu, G., & Abegaz, B. M. (1996). Bianthraquinones from the seeds of *Senna multiglandulosa*. *Phytochemistry, 41*(3), 919–921.

Alemayehu, G., Abegaz, B., Snatzke, G., & Duddeck, H. (1988). Bianthraquinones and a spermidine alkaloid from *Cassia floribunda*. *Phytochemistry, 27*(10), 3255–3258.

Alemayehu, G., Abegaz, B., Snatzke, G., & Duddeck, H. (1993). Bianthrones from *Senna longiracemosa*. *Phytochemistry, 32*(5), 1273–1277.

Alemayehu, G., Hailu, A., & Abegaz, B. M. (1996). Bianthraquinones from *Senna didymobotrya*. *Phytochemistry, 42*(5), 1423–1425.

Alemayehu, G., Abegaz, B., & Kraus, W. (1998). A 1, 4-anthraquinone-dihydroanthracenone dimer from *Senna sophera*. *Phytochemistry, 48*(4), 699–702.

Alemu, B. K., Ayalew Getahun, K., & Kahaliw, W. (2020). *In vitro* antioxidant and i*n viv*o wound healing activities of the 80% methanol extract and solvent fractions of seeds of *Brassica carinata* a. Braun (brassicaceae) in mice. *Journal of Experimental Pharmacology, 12*, 463–474.

Ameya, G., Manilal, A., & Merdekios, B. (2017). *In vitro* antibacterial activity and phytochemical analysis of *Nicotiana tabacum* L. extracted in different organic solvents. *The Open Microbiology Journal, 11*, 352.

Amsalu, N., Bezie, Y., Fentahun, M., Alemayehu, A., & Amsalu, G. (2018). Use and conservation of medicinal plants by indigenous people of Gozamin Wereda, East Gojjam Zone of Amhara region, Ethiopia: An ethnobotanical approach. *Evidence-based Complementary and Alternative Medicine, 2018*, 2973513.

Andualem, B. (2013). Combined antibacterial activity of stingless bee (*Apis mellipodae*) honey and garlic (*Allium sativum*) extracts against standard and clinical pathogenic bacteria. *Asian Pacific Journal of Tropical Biomedicine, 3*(9), 725–731.

Araya, S., Abera, B., & Giday, M. (2015). Study of plants traditionally used in public and animal health management in Seharti Samre District, Southern Tigray, Ethiopia. *Journal of Ethnobiology and Ethnomedicine, 11*, 1–25.

Asamenew, G., Bisrat, D., Mazumder, A., & Asres, K. (2011). *In vitro* antimicrobial and antioxidant activities of anthrone and chromone from the latex of *Aloe harlana* Reynolds. *Phytotherapy Research, 25*(12), 1756–1760.

Asfaw, N., Storesund, H. J., Skattebøl, L., & Aasen, A. J. (1999). (1S, 5R)-(−)-2, 4, 4-Trimethylbicyclo [3.1. 1] hept-2-en-6-one, from the essential oil of the Ethiopian plant *Laggera tomentosa*. *Phytochemistry, 52*(8), 1491–1494.

Asfaw, N., Storesund, H. J., Skattebøl, L., & Aasen, A. J. (2001). Coexistence of chrysanthenone, filifolone and (Z)-isogeranic acid in hydrodistillates. Artefacts! *Phytochemistry, 58*(3), 489–492.

Asmerom, D., Hailu, G. S., Yimer, E. M., Bitew, H., & Kahsay, G. (2020a). Antimicrobial evaluation of latex and TLC fractions from the leaves of *Aloe adigratana* reynolds. *Evidence-based Complementary and Alternative Medicine, 2020*, 8312471.

Asmerom, D., Kalay, T. H., & Tafere, G. G. (2020b). Antibacterial and antifungal activities of the leaf exudate of *Aloe megalacantha* Baker. *International Journal of Microbiology, 2020*, 8840857.

Asnake, S., Teklehaymanot, T., Hymete, A., Erko, B., & Giday, M. (2015). Evaluation of the antiplasmodial properties of selected plants in Southern Ethiopia. *BMC Complementary and Alternative Medicine, 15*(1), 1–12.

Asnake, S., Teklehaymanot, T., Hymete, A., Erko, B., & Giday, M. (2016). Survey of medicinal plants used to treat malaria by Sidama People of Boricha District, Sidama Zone, South Region of Ethiopia. *Evidence-based Complementary and Alternative Medicine, 2016*, 9690164.

Asres, K., Gibbons, W. A., Phillipson, J. D., & Mascagni, P. (1986). Alkaloids of Ethiopian *Calpurnia aurea* subsp. aurea. *Phytochemistry, 25*(6), 1443–1447.

Asres, K., Bucar, F., Edelsbrunner, S., Kartnig, T., Höger, G., & Thiel, W. (2001a). Investigations on antimycobacterial activity of some Ethiopian medicinal plants. *Phytotherapy Research, 15*(4), 323–326.

Asres, K., Bucar, F., Kartnig, T., Witvrouw, M., Pannecouque, C., & De Clercq, E. (2001b). Antiviral activity against human immunodeficiency virus type 1 (HIV-1) and type 2 (HIV-2) of ethnobotanically selected Ethiopian medicinal plants. *Phytotherapy Research: An International Journal Devoted to Pharmacological and Toxicological Evaluation of Natural Product Derivatives, 15*(1), 62–69.

Asres, K., Bucar, F., Knauder, E., Yardley, V., Kendrick, H., & Croft, S. (2001c). *In vitro* antiprotozoal activity of extract and compounds from the stem bark of *Combretum molle*. *Phytotherapy Research, 15*(7), 613–617.

Asres, K., Gibbons, S., & Nachname, V. (2005). Anti-inflammatory activity of extracts and a saponin isolated from *Melilotus elegans*. *Die Pharmazie-An International Journal of Pharmaceutical Sciences, 60*(4), 310–312.

Assefa, A., & Bahiru, A. (2018). Ethnoveterinary botanical survey of medicinal plants in Abergelle, Sekota and Lalibela districts of Amhara region, Northern Ethiopia. *Journal of Ethnopharmacology, 213*, 340–349.

Atlabachew, M., Combrinck, S., Viljoen, A. M., Hamman, J. H., & Gouws, C. (2016). Isolation and *in vitro* permeation of phenylpropylamino alkaloids from Khat (*Catha edulis*) across oral and intestinal mucosal tissues. *Journal of Ethnopharmacology, 194*, 307–315.

Atlabachew, M., Mehari, B., Combrinck, S., & McCrindle, R. (2017). Single-step isolation of embelin using high-performance countercurrent chromatography and determination of the fatty acid composition of seeds of *Embelia schimperi*. *Biomedical Chromatography, 31*(12), e4018.

Ayalew, H., Tewelde, E., Abebe, B., Alebachew, Y., & Tadesse, S. (2022). Endemic medicinal plants of Ethiopia: Ethnomedicinal uses, biological activities and chemical constituents. *Journal of Ethnopharmacology, 293*, 115307.

Balemie, K., & Kebebew, F. (2006). Ethnobotanical study of wild edible plants in Derashe and Kucha Districts, South Ethiopia. *Journal of Ethnobiology and Ethnomedicine, 2*, 1–9.

Begashaw, B., Mishra, B., Tsegaw, A., & Shewamene, Z. (2017). Methanol leaves extract of *Hibiscus micranthus* Linn exhibited antibacterial and wound healing activities. *BMC Complementary and Alternative Medicine, 17*(1), 1–11.

Bekalo, T. H., Woodmatas, S. D., & Woldemariam, Z. A. (2009). An ethnobotanical study of medicinal plants used by local people in the lowlands of Konta Special Woreda, southern nations, nationalities and peoples regional state, Ethiopia. *Journal of Ethnobiology and Ethnomedicine, 5*, 1–15.

Bekele, M. (1966). *A study of modern education in Ethiopia: Its foundations, its development, its future, with emphasis on primary education*. Columbia University.

Bekele, E. (2007). Study on actual situation of medicinal plants in Ethiopia. 2007. *Document prepared for JAICAF, Addis Ababa, Ethiopia*.

Belayneh, Y. M., & Birru, E. M. (2018). Antidiabetic activities of hydromethanolic leaf extract of *Calpurnia aurea* (Ait.) Benth. Subspecies aurea (Fabaceae) in mice. *Evidence-based Complementary and Alternative Medicine, 2018*, 3509073.

Belayneh, A., & Bussa, N. F. (2014). Ethnomedicinal plants used to treat human ailments in the prehistoric place of Harla and Dengego valleys, eastern Ethiopia. *Journal of Ethnobiology and Ethnomedicine, 10*(1), 1–17.

Belayneh, A., Asfaw, Z., Demissew, S., & Bussa, N. F. (2012). Medicinal plants potential and use by pastoral and agro-pastoral communities in Erer Valley of Babile Wereda, Eastern Ethiopia. *Journal of Ethnobiology and Ethnomedicine, 8*(1), 1–11.

Belayneh, Y. M., Birru, E. M., & Ambikar, D. (2019). Evaluation of hypoglycemic, antihyperglycemic and antihyperlipidemic activities of 80% methanolic seed extract of *Calpurnia aurea* (Ait.) Benth. (Fabaceae) in mice. *Journal of Experimental Pharmacology, 11*, 73–83.

Berhanu, M., Tintagu, T., Fentahun, S., & Giday, M. (2020). Ethnoveterinary survey of medicinal plants used for treatment of animal diseases in Ambo District of Oromia Regional State of Ethiopia. *Evidence-based Complementary and Alternative Medicine, 2020*, 8816227.

Birrie, H., Balcha, F., Erko, B., Bezuneh, A., & Gemeda, N. (n.d.). Investigation into the cercariacidal and miracidiacidal properties of endod. *Phytolacca dodecandra. East African Medical Journal, 75*, 311–314.

Birru, E. M., Asrie, A. B., Adinew, G. M., & Tsegaw, A. (2016). Antidiarrheal activity of crude methanolic root extract of *Idigofera spicata* Forssk.(Fabaceae). *BMC Complementary and Alternative Medicine, 16*, 1–7.

Biruk, H., Sentayehu, B., Alebachew, Y., Tamiru, W., Ejigu, A., & Assefa, S. (2020). *In vivo* antimalarial activity of 80% methanol and aqueous bark extracts of *Terminalia brownii* fresen. (Combretaceae) against *Plasmodium berghe*i in mice. *Biochemistry Research International, 2020*, 9749410.

Biruksew, A., Zeynudin, A., Alemu, Y., Golassa, L., Yohannes, M., Debella, A., & Suleman, S. (2018). *Zingiber Officinale* Roscoe and *Echinops Kebericho* Mesfin showed antiplasmodial activities against *Plasmodium berghei* in a dosedependent manner in Ethiopia. *Ethiopian Journal of Health Sciences, 28*(5), 655.

Bishaw, M. (1991). Promoting traditional medicine in Ethiopia: A brief historical review of government policy. *Social Science & Medicine, 33*(2), 193–200.

Bisrat, D., & Jung, C. (2020). Insecticidal toxicities of three main constituents derived from *Trachyspermum ammi* (L.) Sprague ex Turrill fruits against the small hive beetles, *Aethina tumida* Murray. *Molecules, 25*(5), 1100.

Bisrat, D., Dagne, E., van Wyk, B.-E., & Viljoen, A. (2000). Chromones and anthrones from *Aloe marlothii* and *Aloe rupestris*. *Phytochemistry, 55*(8), 949–952.

Bitew, H., Mammo, W., Hymete, A., & Yeshak, M. Y. (2017). Antimalarial activity of acetylenic thiophenes from *Echinops hoehnelii* Schweinf. *Molecules, 22*(11), 1965.

Bultum, L. E., Woyessa, A. M., & Lee, D. (2019). ETM-DB: Integrated Ethiopian traditional herbal medicine and phytochemicals database. *BMC Complementary and Alternative Medicine, 19*, 1–11.

Bussa, N., & Belayneh, A. (2019). Traditional medicinal plants used to treat cancer, tumors and inflammatory ailments in Harari Region, Eastern Ethiopia. *South African Journal of Botany, 122*, 360–368.

Bussmann, R. W., Swartzinsky, P., Worede, A., & Evangelista, P. (2011). Plant use in odo-bulu and demaro, Bale region, Ethiopia. *Journal of Ethnobiology and Ethnomedicine, 7*, 1–21.

Chekole, G. (2017). Ethnobotanical study of medicinal plants used against human ailments in Gubalafto District, Northern Ethiopia. *Journal of Ethnobiology and Ethnomedicine, 13*(1), 1–29.

Chekole, G., Asfaw, Z., & Kelbessa, E. (2015). Ethnobotanical study of medicinal plants in the environs of Tara-gedam and Amba remnant forests of Libo Kemkem District, northwest Ethiopia. *Journal of Ethnobiology and Ethnomedicine, 11*, 1–38.

D'avigdor, E., Wohlmuth, H., Asfaw, Z., & Awas, T. (2014). The current status of knowledge of herbal medicine and medicinal plants in Fiche, Ethiopia. *Journal of Ethnobiology and Ethnomedicine, 10*(1), 1–33.

Dagne, E., & Bekele, A. (1990). C-prenylated isoflavones from *Millettia ferruginea*. *Phytochemistry, 29*(8), 2679–2682.

Dagne, E., & Steglich, W. (1984a). 8-Oxoerythrinine: An alkaloid from *Erythrina brucei*. *Phytochemistry, 23*(2), 449–451.

Dagne, E., & Steglich, W. (1984b). Knipholone: A unique anthraquinone derivative from *Kniphofia foliosa*. *Phytochemistry, 23*(8), 1729–1731.

Dagne, E., & Yenesew, A. (1993). Knipholone anthrone from *Kniphofia foliosa*. *Phytochemistry, 34*(5), 1440–1441.

Dagne, E., Dinku, B., Gray, A. I., & Waterman, P. G. (1988). Pumilaisoflavones A and B from the seed pods of *Tephrosia pumila*. *Phytochemistry, 27*(5), 1503–1505.

Dagne, E., Yenesew, A., & Waterman, P. G. (1989). Flavonoids and isoflavonoids from *Tephrosia fulvinervis* and *Tephrosia pentaphylla*. *Phytochemistry, 28*(11), 3207–3210.

Dagne, E., Bekele, A., Noguchi, H., Shibuya, M., & Sankawa, U. (1990a). *O*-Geranylated and *O*-prenylated flavonoids from *Millettia ferruginea*. *Phytochemistry, 29*(8), 2671–2673.

Dagne, E., Yenesew, A., Capasso, F., Mascolo, N., & Pinto, A. (1990b). Preliminary studies on antipyretic and analgesic properties of *Taverniera abyssinica*. *Ethiopian Medical Journal, 28*(4), 155–161.

Dagne, E., Casser, I., & Steglich, W. (1992a). Aloechrysone, a dihydroanthracenone from *Aloe berhana*. *Phytochemistry, 31*(5), 1791–1793.

Dagne, E., Mammo, W., & Sterner, O. (1992b). Flavonoids of *Tephrosia polyphylla*. *Phytochemistry, 31*(10), 3662–3663.

Dagne, E., Gunatilaka, A. L., Kingston, D. G., Alemu, M., Hofmann, G., & Johnson, R. K. (1993). Two bioactive pterocarpans from *Erythrina burana*. *Journal of Natural Products, 56*(10), 1831–1834.

Dagne, E., Van Wyk, B.-E., Mueller, M., & Steglich, W. (1996a). Three dihydroanthracenones from *Gasteria bicolor*. *Phytochemistry, 41*(3), 795–799.

Dagne, E., Van Wyk, B.-E., Stephenson, D., & Steglich, W. (1996b). Three oxanthrones from *Aloe littoralis*. *Phytochemistry, 42*(6), 1683–1687.

Dagne, E., Bisrat, D., Van Wyk, B.-E., Viljoen, A., Hellwig, V., & Steglich, W. (1997). Anthrones from *Aloe microstigma*. *Phytochemistry, 44*(7), 1271–1274.

Dagne, E., Bisrat, D., Codina, C., & Bastida, J. (1998). AC, *O*-diglucosylated oxanthrone from *Aloe littoralis*. *Phytochemistry, 48*(5), 903–905.

Debebe, Y., Tefera, M., Mekonnen, W., Abebe, D., Woldekidan, S., Abebe, A., & Tesfaye, B. (2015). Evaluation of anthelmintic potential of the Ethiopian medicinal plant *Embelia schimperi* Vatke *in vivo* and in vitro against some intestinal parasites. *BMC Complementary and Alternative Medicine, 15*(1), 1–6.

Debella, A., Haslinger, E., Kunert, O., Michl, G., & Abebe, D. (1999). Steroidal saponins from *Asparagus africanus*. *Phytochemistry, 51*(8), 1069–1075.

Debella, A., Haslinger, E., Schmid, M. G., Bucar, F., Michl, G., Abebe, D., & Kunert, O. (2000). Triterpenoid saponins and sapogenin lactones from *Albizia gummifera*. *Phytochemistry, 53*(8), 885–892.

Debella, A., Makonnen, E., Abebe, D., Teka, F., & Kidanemariam, A. (2003). Pain management in mice using the aqueous and ethanol extracts of four medicinal plants. *East African Medical Journal, 80*(8), 435–439.

Degu, A., Engidawork, E., & Shibeshi, W. (2016). Evaluation of the anti-diarrheal activity of the leaf extract of *Croton macrostachyus* Hocsht. ex Del.(Euphorbiaceae) in mice model. *BMC Complementary and Alternative Medicine, 16*, 1–11.

Degu, A., Kefale, B., Alemayehu, D., & Tegegne, G. T. (2020). Evaluation of the antidiarrheal activity of hydromethanol crude extracts of *Ruta chalepensis* and *Vernonia amygdalina* in mice. *Evidence-based Complementary and Alternative Medicine, 2020*, 8318713.

Dekebo, A., Dagne, E., Hansen, L. K., Gautun, O. R., & Aasen, A. J. (2002a). Two octanordammarane triterpenes from *Commiphora kua*. *Phytochemistry, 59*(4), 399–403.

Dekebo, A., Dagne, E., & Sterner, O. (2002b). Furanosesquiterpenes from *Commiphora sphaerocarpa* and related adulterants of true myrrh. *Fitoterapia, 73*(1), 48–55.

Dekebo, A., Lang, M., Polborn, K., Dagne, E., & Steglich, W. (2002c). Four lignans from *Commiphora erlangeriana*. *Journal of Natural Products, 65*(9), 1252–1257.

Demie, G., Negash, M., & Awas, T. (2018). Ethnobotanical study of medicinal plants used by indigenous people in and around Dirre Sheikh Hussein heritage site of South-eastern Ethiopia. *Journal of Ethnopharmacology, 220*, 87–93.

Demilew, W., Adinew, G. M., & Asrade, S. (2018). Evaluation of the wound healing activity of the crude extract of leaves of *Acanthus polystachyus* Delile (Acanthaceae). *Evidence-based Complementary and Alternative Medicine, 2018*, 2047896.

Demma, J., Gebre-Mariam, T., Asres, K., Ergetie, W., & Engidawork, E. (2007). Toxicological study on *Glinus lotoides*: A traditionally used taenicidal herb in Ethiopia. *Journal of Ethnopharmacology, 111*(3), 451–457.

Demma, J., Engidawork, E., & Hellman, B. (2009). Potential genotoxicity of plant extracts used in Ethiopian traditional medicine. *Journal of Ethnopharmacology, 122*(1), 136–142.

Derebe, D., Wubetu, M., & Alamirew, A. (2020). Hypoglycemic and antihyperglycemic activities of 80% methanol root extract of *Acanthus polystachyus* delile (Acanthaceae) in type 2 diabetic rats. In *Clinical pharmacology: Advances and applications* (pp. 149–157).

Desta, K. T., Kim, G. S., Abd El-Aty, A., Raha, S., Kim, M.-B., Jeong, J. H., & Shim, J.-H. (2017). Flavone polyphenols dominate in *Thymus schimperi* Ronniger: LC–ESI–MS/MS characterization and study of anti-proliferative effects of plant extract on AGS and HepG2 cancer cells. *Journal of Chromatography B, 1053*, 1 8.

Dibessa, T. T., Engidawork, E., Nedi, T., & Teklehaymanot, T. (2020). Antimalarial activity of the aqueous extract of the latex of *Aloe pirottae* Berger.(Aloaceae) against *Plasmodium berghei* in mice. *Journal of Ethnopharmacology, 255*, 112763.

Dugassa, S., Medhin, G., Balkew, M., Seyoum, A., & Gebre-Michael, T. (2009). Field investigation on the repellent activity of some aromatic plants by traditional means against *Anopheles arabiensis* and *An. pharoensis* (Diptera: Culicidae) around Koka, central Ethiopia. *Acta Tropica, 112*(1), 38–42.

Eguale, T., Tilahun, G., Debella, A., Feleke, A., & Makonnen, E. (2007). *In vitro* and *in vivo* anthelmintic activity of crude extracts of *Coriandrum sativum* against *Haemonchus contortus*. *Journal of Ethnopharmacology, 110*(3), 428–433.

Eguale, T., Tadesse, D., & Giday, M. (2011). *In vitro* anthelmintic activity of crude extracts of five medicinal plants against egg-hatching and larval development of *Haemonchus contortus*. *Journal of Ethnopharmacology, 137*(1), 108–113.

Eijk, J. L. V., & Radema, M. H. (1975). Lupanin alkaloids and other compounds from *Cadia purpurea*. *Planta Medica, 28*(06), 139–142. https://doi.org/10.1055/s-0028-1097842

Engiso, H., Worku, T., Nureye, D., Salahaddin, M., Woldeselassie, W., Hambisa, S., & Sharief, N. (2020). Antibacterial activity of *Ritchiea albersii* Gilg and *Cynoglossum amplifolium* leaves extracts against selected bacteria. *Saudi Journal of Medicine & Medical Sciences, 8*(3), 201.

Eshetu, G. R., Dejene, T. A., Telila, L. B., & Bekele, D. F. (2015). Ethnoveterinary medicinal plants: Preparation and application methods by traditional healers in selected districts of Southern Ethiopia. *Veterinary World, 8*(5), 674.

Eyob, S., Appelgren, M., Rohloff, J., Tsegaye, A., & Messele, G. (2008). Traditional medicinal uses and essential oil composition of leaves and rhizomes of korarima (*Aframomum corrorima* (Braun) PCM Jansen) from Southern Ethiopia. *South African Journal of Botany, 74*(2), 181–185.

Fentahun, S., Makonnen, E., Awas, T., & Giday, M. (2017). *In vivo* antimalarial activity of crude extracts and solvent fractions of leaves of *Strychnos mitis* in *Plasmodium berghe*i infected mice. *BMC Complementary and Alternative Medicine, 17*, 1–12.

Feyera, T., Terefe, G., & Shibeshi, W. (2014). Evaluation of *in vivo* antitrypanosomal activity of crude extracts of *Artemisia abyssinica* against a*Trypanosoma congolense* isolate. *BMC Complementary and Alternative Medicine, 14*(1), 1–9.

Feyera, T., Mekonnen, E., Wakayo, B. U., & Assefa, S. (2017). Botanical ethnoveterinary therapies used by agro-pastoralists of Fafan zone, Eastern Ethiopia. *BMC Veterinary Research, 13*, 1–11.

Fikru, A., Makonnen, E., Eguale, T., Debella, A., & Mekonnen, G. A. (2012). Evaluation of *in vivo* wound healing activity of methanol extract of *Achyranthes aspera* L. *Journal of Ethnopharmacology, 143*(2), 469–474.

Flatie, T., Gedif, T., Asres, K., & Gebre-Mariam, T. (2009). Ethnomedical survey of Berta ethnic group Assosa Zone, Benishangul-Gumuz regional state, mid-west Ethiopia. *Journal of Ethnobiology and Ethnomedicine, 5*(1), 1–11.

Fullas, F. (2001). Ethiopian traditional medicine: Common medicinal plants in perspective: F. Fullas.

Gadisa, E., Weldearegay, G., Desta, K., Tsegaye, G., Hailu, S., Jote, K., & Takele, A. (2019). Combined antibacterial effect of essential oils from three most commonly used Ethiopian traditional medicinal plants on multidrug resistant bacteria. *BMC Complementary and Alternative Medicine, 19*, 1–9.

Gall, A., & Shenkute, Z. (2009). Ethiopian traditional medications and their interactions with conventional drugs.

Gari, A., Yarlagadda, R., & Wolde-Mariam, M. (2015). Knowledge, attitude, practice, and management of traditional medicine among people of Burka Jato Kebele, West Ethiopia. *Journal of Pharmacy & Bioallied Sciences, 7*(2), 136.

Gebeyew, K., Abera, B., Bajigo, A., Gebresilassie, G., Martínez, Y., & Adebowale, T. (2020). Indigenous medicinal uses, toxicity, and chemical composition of browsing plant used by camel in Ethiopia Somali Regional State: A survey. *Tropical Animal Health and Production, 52*, 1459–1466.

Gebrehiwot, S., Shumbahri, M., Eyado, A., & Yohannes, T. (2019). Phytochemical screening and *in vivo* antimalarial activity of two traditionally used medicinal plants of Afar region, Ethiopia, against *Plasmodium berghe*i in Swiss Albino mice. *Journal of Parasitology Research, 2019*, 1–8.

Gebre-Mariam, T., Asres, K., Getie, M., Endale, A., Neubert, R., & Schmidt, P. C. (2005). *In vitro* availability of kaempferol glycosides from cream formulations of methanolic extract of the leaves of *Melilotus elegans*. *European Journal of Pharmaceutics and Biopharmaceutics, 60*(1), 31–38.

Gebre-Mariam, T., Neubert, R., Schmidt, P., Wutzler, P., & Schmidtke, M. (2006). Antiviral activities of some Ethiopian medicinal plants used for the treatment of dermatological disorders. *Journal of Ethnopharmacology, 104*(1–2), 182–187.

Gebrie, E., Makonnen, E., Debella, A., & Zerihun, L. (2005). Phytochemical screening and pharmacological evaluations for the antifertility effect of the methanolic root extract of *Rumex steudelii*. *Journal of Ethnopharmacology, 96*(1–2), 139–143.

Gedif, T., & Hahn, H.-J. (2003). The use of medicinal plants in self-care in rural central Ethiopia. *Journal of Ethnopharmacology, 87*(2–3), 155–161.

Gemechu, A., Giday, M., Worku, A., & Ameni, G. (2013). *In vitro* anti-mycobacterial activity of selected medicinal plants against *Mycobacterium tuberculosis* and *Mycobacterium bovis* strains. *BMC Complementary and Alternative Medicine, 13*, 1–6.

Gemeda, N., Mokonnen, W., Lemma, H., Tadele, A., Urga, K., Addis, G., & Yirsaw, K. (2014). Insecticidal activity of some traditionally used Ethiopian medicinal plants against sheep ked *Melophagus ovinus*. *Journal of Parasitology Research, 2014*, 978537.

Getie, M., Gebre-Mariam, T., Rietz, R., Höhne, C., Huschka, C., Schmidtke, M., & Neubert, R. (2003). Evaluation of the anti-microbial and anti-inflammatory activities of the medicinal plants *Dodonaea viscosa, Rumex nervosus* and *Rumex abyssinicus*. *Fitoterapia, 74*(1–2), 139–143.

Getiye, Y., Tolessa, T., & Engidawork, E. (2016). Antihypertensive activity of 80% methanol seed extract of *Calpurnia aurea* (Ait.) Benth. subsp. *aurea* (Fabaceae) is mediated through calcium antagonism induced vasodilation. *Journal of Ethnopharmacology, 189*, 99–106.

Geyid, A., Abebe, D., Debella, A., Makonnen, Z., Aberra, F., Teka, F., & Biza, T. (2005). Screening of some medicinal plants of Ethiopia for their anti-microbial properties and chemical profiles. *Journal of Ethnopharmacology, 97*(3), 421–427.

Giday, M., & Teklehaymanot, T. (2013). Ethnobotanical study of plants used in management of livestock health problems by Afar people of Ada'ar District, Afar Regional State, Ethiopia. *Journal of Ethnobiology and Ethnomedicine, 9*, 1–10.

Giday, M., Asfaw, Z., Elmqvist, T., & Woldu, Z. (2003). An ethnobotanical study of medicinal plants used by the Zay people in Ethiopia. *Journal of Ethnopharmacology, 85*(1), 43–52.

Giday, M., Teklehaymanot, T., Animut, A., & Mekonnen, Y. (2007). Medicinal plants of the Shinasha, Agew-awi and Amhara peoples in northwest Ethiopia. *Journal of Ethnopharmacology, 110*(3), 516–525.

Giday, M., Asfaw, Z., & Woldu, Z. (2009a). Medicinal plants of the Meinit ethnic group of Ethiopia: An ethnobotanical study. *Journal of Ethnopharmacology, 124*(3), 513–521.

Giday, M., Asfaw, Z., Woldu, Z., & Teklehaymanot, T. (2009b). Medicinal plant knowledge of the Bench ethnic group of Ethiopia: An ethnobotanical investigation. *Journal of Ethnobiology and Ethnomedicine, 5*(1), 1–10.

Giday, M., Asfaw, Z., & Woldu, Z. (2010). Ethnomedicinal study of plants used by Sheko ethnic group of Ethiopia. *Journal of Ethnopharmacology, 132*(1), 75–85.

Giday, K., Lenaerts, L., Gebrehiwot, K., Yirga, G., Verbist, B., & Muys, B. (2016). Ethnobotanical study of medicinal plants from degraded dry afromontane forest in northern Ethiopia: Species, uses and conservation challenges. *Journal of Herbal Medicine, 6*(2), 96–104.

Girma, B., Bisrat, D., & Asres, K. (2015a). Antimalarial evaluation of the leaf latex of *Aloe citrina* and its major constituent. *Ancient Science of Life, 34*(3), 142.

Girma, S., Giday, M., Erko, B., & Mamo, H. (2015b). Effect of crude leaf extract of *Osyris quadripartita* on *Plasmodium berghei* in Swiss albino mice. *BMC Complementary and Alternative Medicine, 15*(1), 1–9.

Gishen, N. Z., Taddese, S., Zenebe, T., Dires, K., Tedla, A., Mengiste, B., & Lulekal, E. (2020). *In vitro* antimicrobial activity of six Ethiopian medicinal plants against *Staphylococcus aureus, Escherichia coli* and *Candida albicans*. *European Journal of Integrative Medicine, 36*, 101121.

Gomathi, D., Ravikumar, G., Kalaiselvi, M., Devaki, K., & Uma, C. (2014). Antioxidant activity and functional group analysis of *Evolvulus alsinoides*. *Chinese Journal of Natural Medicines, 12*(11), 827–832.

Gonfa, T., Fisseha, A., & Thangamani, A. (2020). Isolation, characterization and drug-likeness analysis of bioactive compounds from stem bark of *Warburgia ugandensis* Sprague. *Chemical Data Collections, 29*, 100535.

Guadie, A., Dakone, D., Unbushe, D., Wang, A., & Xia, S. (2020). Antibacterial activity of selected medicinal plants used by traditional healers in Genta Meyche (Southern Ethiopia) for the treatment of gastrointestinal disorders. *Journal of Herbal Medicine, 22*, 100338.

Gunatilaka, A. L., da Bolzani, S. V., Dagne, E., Hofmann, G. A., Johnson, R. K., McCabe, F. L., & Kingston, D. G. (1998). Limonoids showing selective toxicity to DNA repair-deficient yeast and other constituents of *Trichilia emetica*. *Journal of Natural Products, 61*(2), 179–184.

Gurmessa, G. T., Kusari, S., Laatsch, H., Bojase, G., Tatolo, G., Masesane, I. B., & Majinda, R. R. (2018). Chemical constituents of root and stem bark of *Erythrina brucei*. *Phytochemistry Letters, 25*, 37–42.

Guyasa, B., Melaku, Y., & Endale, M. (2018). Antibacterial activity of two flavans from the stem bark of *Embelia schimperi*. *Advances in Pharmacological Sciences, 2018*, 5870161.

Habtamu, A., & Melaku, Y. (2018). Antibacterial and antioxidant compounds from the flower extracts of *Vernonia amygdalina*. *Advances in Pharmacological and Pharmaceutical Sciences, 2018*, 4083736.

Habte, G., & Assefa, S. (2020). *In vivo* antimalarial activity of crude fruit extract of *Capsicum frutescens* Var. Minima (Solanaceae) against *Plasmodium berghei*-Infected mice. *BioMed Research International, 2020*, 1320952.

Habte, G., Nedi, T., & Assefa, S. (2020). Antimalarial activity of aqueous and 80% methanol crude seed extracts and solvent fractions of *schinus molle* linnaeus (anacardiaceae) in *plasmodium berghei*-infected mice. *Journal of Tropical Medicine, 2020*, 9473250.

Habtemariam, S. (2003). *In vitro* antileishmanial effects of antibacterial diterpenes from two Ethiopian *Premna* species: P. schimperi and P. oligotricha. *BMC Pharmacology, 3*, 1–6.

Habtemariam, S., Gray, A. I., & Waterman, P. G. (1993). 16-Oxygenated withanolides from the leaves of *Discopodium penninervium*. *Phytochemistry, 34*(3), 807.

Hailesilase, G. G., Rajeshwar, Y., Hailu, G. S., Sibhat, G. G., & Bitew, H. (2020). *In vivo* antimalarial evaluation of crude extract, solvent fractions, and TLC-isolated compounds from *Olea europaea* Linn subsp. cuspidata (Oleaceae). *Evidence-based Complementary and Alternative Medicine, 2020*, 6731485.

Hailesillassie, T., Bisrat, D., & Asres, K. (2018). Larvicidal effect of the leaf latex of *Aloe yavellana* Reynolds and its major compounds against *Amblyomma variegatum* (Ixodidae). *Veterinary Parasitology, 263*, 23–26.

Hailu, W., & Engidawork, E. (2014). Evaluation of the diuretic activity of the aqueous and 80% methanol extracts of *Ajuga remota* Benth (Lamiaceae) leaves in mice. *BMC Complementary and Alternative Medicine, 14*(1), 1–8.

Hawaze, S., Deti, H., & Suleman, S. (2012). *In vitro* antimicrobial activity and phytochemical screening of *Clematis species* indigenous to Ethiopia. *Indian Journal of Pharmaceutical Sciences, 74*(1), 29.

Hishe, H. Z., Ambech, T. A., Hiben, M. G., & Fanta, B. S. (2018). Anti-nociceptive effect of methanol extract of leaves of *Senna singueana* in mice. *Journal of Ethnopharmacology, 217*, 49–53.

Hussien, J., Hymete, A., & Rohloff, J. (2010). Volatile constituents and biological activities of *Pycnostachys abyssinica* and *Pycnostachys eminii* extracts. *Pharmaceutical Biology, 48*(12), 1384–1391.

Hymete, A., Rohloff, J., Kjøsen, H., & Iversen, T.-H. (2005). Acetylenic thiophenes from the roots of *Echinops ellenbeckii* from Ethiopia. *Natural Product Research, 19*(8), 755–761.

Ibrahim, N., & Kebede, A. (2020). *In vitro* antibacterial activities of methanol and aqueous leave extracts of selected medicinal plants against human pathogenic bacteria. *Saudi Journal of Biological Sciences, 27*(9), 2261–2268.

Jima, T. T., & Megersa, M. (2018). Ethnobotanical study of medicinal plants used to treat human diseases in Berbere District, Bale Zone of Oromia Regional State, South East Ethiopia. *Evidence-based Complementary and Alternative Medicine, 2018*, 8602945.

Kahaliw, W., Hellman, B., & Engidawork, E. (2018). Genotoxicity study of Ethiopian medicinal plant extracts on HepG2 cells. *BMC Complementary and Alternative Medicine, 18*(1), 1–8.

Kalayou, S., Haileselassie, M., Gebre-Egziabher, G., Tiku'e, T., Sahle, S., Taddele, H., & Ghezu, M. (2012). *In–vitro* antimicrobial activity screening of some ethnoveterinary medicinal plants traditionally used against mastitis, wound and gastrointestinal tract complication in Tigray Region, Ethiopia. *Asian Pacific Journal of Tropical Biomedicine, 2*(7), 516–522.

Kalbessa, A., Dekebo, A., Tesso, H., Abdo, T., Abdissa, N., & Melaku, Y. (2019). Chemical constituents of root barks of *Gnidia involucrata* and evaluation for antibacterial and antioxidant activities. *Journal of Tropical Medicine, 2019*, 8486214.

Karunamoorthi, K., & Husen, E. (2012). Knowledge and self-reported practice of the local inhabitants on traditional insect repellent plants in Western Hararghe zone, Ethiopia. *Journal of Ethnopharmacology, 141*(1), 212–219.

Karunamoorthi, K., & Ilango, K. (2010). Larvicidal activity of *Cymbopogon citratus*. *European Review for Medical and Pharmacological Sciences, 14*, 57–62.

Karunamoorthi, K., & Tsehaye, E. (2012). Ethnomedicinal knowledge, belief and self-reported practice of local inhabitants on traditional antimalarial plants and phytotherapy. *Journal of Ethnopharmacology, 141*(1), 143–150.

Karunamoorthi, K., Ilango, K., & Endale, A. (2009a). Ethnobotanical survey of knowledge and usage custom of traditional insect/mosquito repellent plants among the Ethiopian Oromo ethnic group. *Journal of Ethnopharmacology, 125*(2), 224–229.

Karunamoorthi, K., Mulelam, A., & Wassie, F. (2009b). Assessment of knowledge and usage custom of traditional insect/mosquito repellent plants in Addis Zemen Town, South Gonder, North Western Ethiopia. *Journal of Ethnopharmacology, 121*(1), 49–53.

Karunamoorthi, K., Girmay, A., & Fekadu, S. (2014). Larvicidal efficacy of Ethiopian ethnomedicinal plant *Juniperus procera* essential oil against Afrotropical malaria vector *Anopheles arabiensis* (Diptera: Culicidae). *Asian Pacific Journal of Tropical Biomedicine, 4*, S99–S106.

Kassa, Z., Asfaw, Z., & Demissew, S. (2020). An ethnobotanical study of medicinal plants in sheka zone of southern nations nationalities and peoples regional state, Ethiopia. *Journal of Ethnobiology and Ethnomedicine, 16*, 1–15.

Kassu, A., Dagne, E., Abate, D., Castro, A., & Van Wyk, B. (1999). Ethnomedical aspects of the commonly used toothbrush sticks in Ethiopia. *East African Medical Journal, 76*(11), 651–653.

Kebede, E., Mengistu, M., & Serda, B. (2018). Ethnobotanical knowledge of pastoral community for treating livestock diseases in Somali regional state, eastern Ethiopia. *Tropical Animal Health and Production, 50*, 1379–1386.

Kefalew, A., Asfaw, Z., & Kelbessa, E. (2015). Ethnobotany of medicinal plants in Ada'a District, East Shewa Zone of Oromia regional state, Ethiopia. *Journal of Ethnobiology and Ethnomedicine, 11*(1), 1–28.

Kefe, A., Giday, M., Mamo, H., & Erko, B. (2016). Antimalarial properties of crude extracts of seeds of *Brucea antidysenterica* and leaves of *Ocimum lamiifolium*. *BMC Complementary and Alternative Medicine, 16*(1), 1–8.

Kemal, J., Zerihun, T., Alemu, S., Sali, K., Nasir, M., Abraha, A., & Feyera, T. (2020). *In vitro* acaricidal activity of selected medicinal plants traditionally used against ticks in eastern Ethiopia. *Journal of Parasitology Research, 2020*, 7834026.

Kidane, B., van Andel, T., van der Maesen, L. J. G., & Asfaw, Z. (2014a). Use and management of traditional medicinal plants by Maale and Ari ethnic communities in Southern Ethiopia. *Journal of Ethnobiology and Ethnomedicine, 10*, 1–15.

Kidane, B., Van Der Maesen, L., van Andel, T., & Asfaw, Z. (2014b). Ethnoveterinary medicinal plants used by the Maale and Ari ethnic communities in Southern Ethiopia. *Journal of Ethnopharmacology, 153*(1), 274–282.

Kidane, L., Gebremedhin, G., & Beyene, T. (2018). Ethnobotanical study of medicinal plants in ganta afeshum district, eastern zone of tigray, northern Ethiopia. *Journal of Ethnobiology and Ethnomedicine, 14*, 1–19.

Lulekal, E., Kelbessa, E., Bekele, T., & Yineger, H. (2008). An ethnobotanical study of medicinal plants in Mana Angetu District, southeastern Ethiopia. *Journal of Ethnobiology and Ethnomedicine, 4*, 1–10.

Lulekal, E., Asfaw, Z., Kelbessa, E., & Van Damme, P. (2013). Ethnomedicinal study of plants used for human ailments in Ankober District, North Shewa Zone, Amhara region, Ethiopia. *Journal of Ethnobiology and Ethnomedicine, 9*, 1–13.

Lulekal, E., Asfaw, Z., Kelbessa, E., & Van Damme, P. (2014a). Ethnoveterinary plants of Ankober district, north Shewa zone, Amhara region, Ethiopia. *Journal of Ethnobiology and Ethnomedicine, 10*(1), 1–19.

Lulekal, E., Rondevaldova, J., Bernaskova, E., Cepkova, J., Asfaw, Z., Kelbessa, E., & Van Damme, P. (2014b). Antimicrobial activity of traditional medicinal plants from Ankober district, north Shewa Zone, Amhara region, Ethiopia. *Pharmaceutical Biology, 52*(5), 614–620.

Makonnen, E., Debella, A., Abebe, D., & Teka, F. (2003). Analgesic properties of some Ethiopian medicinal plants in different models of nociception in mice. *Phytotherapy Research: An International Journal Devoted to Pharmacological and Toxicological Evaluation of Natural Product Derivatives, 17*(9), 1108–1112.

Mammo, W., Dagne, E., Casser, I., & Steglich, W. (1990). Unusual anthraquinone lactones from *Araliorhamnus vaginata*. *Phytochemistry, 29*(8), 2637–2640.

Mammo, W., Dagne, E., & Steglich, W. (1992). Quinone pigments from *Araliorhamnus vaginata*. *Phytochemistry, 31*(10), 3577–3581.

Manilal, A., & Idhayadhulla, A. (2014). Potential *in vitro* antimicrobial efficacy of *Holigarna arnottiana* (Hook F). *Asian Pacific Journal of Tropical Biomedicine, 4*(1), 25–29.

Manilal, A., Merdekios, B., Idhayadhulla, A., Muthukumar, C., & Melkie, M. (2015). An *in vitro* antagonistic efficacy validation of *Rhizophora mucronata*. *Asian Pacific Journal of Tropical Disease, 5*(1), 28–32.

Manilal, A., Tsalla, T., Zerdo, Z., Ameya, G., Merdekios, B., & John, S. E. (2016). Evaluating the antibacterial and anticandidal potency of mangrove, *Avicennia marina*. *Asian Pacific Journal of Tropical Disease, 6*(2), 136–140.

Manilal, A., Sabu, K. R., Shewangizaw, M., Aklilu, A., Seid, M., Merdekios, B., & Tsegaye, B. (2020). *In vitro* antibacterial activity of medicinal plants against biofilm-forming methicillin-resistant *Staphylococcus aureus*: Efficacy of *Moringa stenopetala* and *Rosmarinus officinalis* extracts. *Heliyon, 6*(1), e03303.

Megeressa, M., Bisrat, D., Mazumder, A., & Asres, K. (2015). Structural elucidation of some antimicrobial constituents from the leaf latex of *Aloe trigonantha* LC Leach. *BMC Complementary and Alternative Medicine, 15*, 1–7.

Megersa, M., Asfaw, Z., Kelbessa, E., Beyene, A., & Woldeab, B. (2013). An ethnobotanical study of medicinal plants in Wayu Tuka district, east Welega zone of oromia regional state, West Ethiopia. *Journal of Ethnobiology and Ethnomedicine, 9*(1), 1–18.

Mekonen, T., Giday, M., & Kelbessa, E. (2015). Ethnobotanical study of homegarden plants in Sebeta-Awas District of the Oromia Region of Ethiopia to assess use, species diversity and management practices. *Journal of Ethnobiology and Ethnomedicine, 11*, 1–13.

Mekonnen, Y. (1999). Effects of ethanol extract of *Moringa stenopetala* leaves on Guinea-pig and mouse smooth muscle. *Phytotherapy Research, 13*(5), 442–444.

Mekonnen, Y., Yardley, V., Rock, P., & Croft, S. (1999). *In vitro* antitrypanosomal activity of *Moringa stenopetala* leaves and roots. *Phytotherapy Research: An International Journal Devoted to Pharmacological and Toxicological Evaluation of Natural Product Derivatives, 13*(6), 538–539.

Mekonnen, N., Makonnen, E., Aklilu, N., & Ameni, G. (2012). Evaluation of berries of *Phytolacca dodecandra* for growth inhibition of *Histoplasma capsulatum* var. *farciminosum* and treatment of cases of epizootic lymphangitis in Ethiopia. *Asian Pacific Journal of Tropical Biomedicine, 2*(7), 505–510.

Melaku, Y., Worku, T., Tadesse, Y., Mekonnen, Y., Schmidt, J., Arnold, N., & Dagne, E. (2017). Antiplasmodial compounds from leaves of *Dodonaea angustifolia*. *Current Bioactive Compounds, 13*(3), 268–273.

Meragiaw, M., Asfaw, Z., & Argaw, M. (2016). The status of ethnobotanical knowledge of medicinal plants and the impacts of resettlement in Delanta, northwestern Wello, northern Ethiopia. *Evidence-based Complementary and Alternative Medicine, 2016*, 5060247.

Mergia, E., Shibeshi, W., Terefe, G., & Teklehaymanot, T. (2016). Antitrypanosomal activity of *Verbascum sinaiticum* Benth.(Scrophulariaceae) against *Trypanosoma congolense* isolates. *BMC Complementary and Alternative Medicine, 16*(1), 1–9.

Mesfin, F., Demissew, S., & Teklehaymanot, T. (2009). An ethnobotanical study of medicinal plants in Wonago Woreda, SNNPR, Ethiopia. *Journal of Ethnobiology and Ethnomedicine, 5*(1), 1–18.

Mesfin, A., Giday, M., Animut, A., & Teklehaymanot, T. (2012). Ethnobotanical study of antimalarial plants in Shinile District, Somali Region, Ethiopia, and *in vivo* evaluation of selected ones against *Plasmodium berghei. Journal of Ethnopharmacology, 139*(1), 221–227.

Mohammed, S. A., Panda, R. C., Madhan, B., & Demessie, B. A. (2017). Extraction of bio-active compounds from Ethiopian plant material *Rumex abyssinicus* (mekmeko) root—A study on kinetics, optimization, antioxidant and antibacterial activity. *Journal of the Taiwan Institute of Chemical Engineers, 75*, 228–239.

Molla, Y., Nedi, T., Tadesse, G., Alemayehu, H., & Shibeshi, W. (2016). Evaluation of the *in vitro* antibacterial activity of the solvent fractions of the leaves of *Rhamnus prinoides* L'Herit (Rhamnaceae) against pathogenic bacteria. *BMC Complementary and Alternative Medicine, 16*(1), 1–9.

Mulaw, T., Wubetu, M., Dessie, B., Demeke, G., & Molla, Y. (2019). Evaluation of antimalarial activity of the 80% methanolic stem bark extract of *Combretum molle* against *plasmodium berghe*i in mice. *Journal of Evidence-Based Integrative Medicine, 24*, 2515690X19890866.

Muluye, A. B., Desta, A. G., Abate, S. K., & Dano, G. T. (2019). Anti-malarial activity of the root extract of *Euphorbia abyssinica* (Euphorbiaceae) against *Plasmodium berghei* infection in mice. *Malaria Journal, 18*(1), 1–8.

Mummed, B., Abraha, A., Feyera, T., Nigusse, A., & Assefa, S. (2018). *In vitro* antibacterial activity of selected medicinal plants in the traditional treatment of skin and wound infections in eastern Ethiopia. *BioMed Research International, 2018*, 1862401.

Nagappan, R. (2012). Evaluation of aqueous and ethanol extract of bioactive medicinal plant, *Cassia didymobotrya* (Fresenius) Irwin & Barneby against immature stages of filarial vector, *Culex quinquefasciatus* Say (Diptera: Culicidae). *Asian Pacific Journal of Tropical Biomedicine, 2*(9), 707–711.

Nibret, E., & Wink, M. (2010a). Trypanocidal and antileukaemic effects of the essential oils of *Hagenia abyssinica, Leonotis ocymifolia, Moringa stenopetala*, and their main individual constituents. *Phytomedicine, 17*(12), 911–920.

Nibret, E., & Wink, M. (2010b). Volatile components of four Ethiopian *Artemisia* species extracts and their *in vitro* antitrypanosomal and cytotoxic activities. *Phytomedicine, 17*(5), 369–374.

Nibret, E., & Wink, M. (2011). Trypanocidal and cytotoxic effects of 30 Ethiopian medicinal plants. *Zeitschrift für Naturforschung C, 66*(11–12), 541–546.

Noamesi, B., Bogale, M., & Dagne, E. (1990). Intestinal smooth muscle spasmolytic actions of the aqueous extract of the roots of *Taverniera abyssinica. Journal of Ethnopharmacology, 30*(1), 107–113.

Nureye, D., Assefa, S., Nedi, T., & Engidawork, E. (2018). *In vivo* antimalarial activity of the 80 Methanolic root bark extract and solvent fractions of *Gardenia ternifolia* Schumach. & Thonn.(Rubiaceae) against *Plasmodium berghei. Evidence based Complementary and Alternative Medicine, 2018*, 9217835.

Osman, A., Sbhatu, D. B., & Giday, M. (2020). Medicinal plants used to manage human and livestock ailments in Raya Kobo District of Amhara Regional State, Ethiopia. *Evidence-based Complementary and Alternative Medicine, 2020*, 1329170.

Palla, A. H., Sibhat, G. G., Karim, A., Rehman, N. U., & Hiben, M. G. (2020). Multiple pathway–mediated gut-modulatory effects of *Maerua subcordata* (Gilg) DeWolf. *Journal of Experimental Pharmacology, 12*, 203–211.

Paulos, B., Fenta, T. G., Bisrat, D., & Asres, K. (2016). Health seeking behavior and use of medicinal plants among the Hamer ethnic group, South Omo zone, southwestern Ethiopia. *Journal of Ethnobiology and Ethnomedicine, 12*(1), 1–13.

Rasheed, M. U., & Thajuddin, N. (2011). Effect of medicinal plants on *Moraxella cattarhalis*. *Asian Pacific Journal of Tropical Medicine, 4*(2), 133–136.

Romha, G., Admasu, B., Hiwot Gebrekidan, T., Aleme, H., & Gebru, G. (2018). Antibacterial activities of five medicinal plants in Ethiopia against some human and animal pathogens. *Evidence-based Complementary and Alternative Medicine, 2018*, 2950758.

Sasikumar, J., Erba, O., & Egigu, M. C. (2020). *In vitro* antioxidant activity and polyphenolic content of commonly used spices from Ethiopia. *Heliyon, 6*(9), e05027.

Sbhatu, D. B., & Abraha, H. B. (2020). Preliminary antimicrobial profile of *Solanum incanum* L.: A common medicinal plant. *Evidence-based Complementary and Alternative Medicine, 2020*, 3647065.

Scantlebury, C. E., Peachey, L., Hodgkinson, J., Matthews, J. B., Trawford, A., Mulugeta, G., & Pinchbeck, G. L. (2013). Participatory study of medicinal plants used in the control of gastrointestinal parasites in donkeys in Eastern Shewa and Arsi zones of Oromia region, Ethiopia. *BMC Veterinary Research, 9*, 1–12.

Shi, S.-M., Tian, J.-G., & Wang, B.-Q. (2002). Study on the detecting methods of the imported materia medica--Olibanum. *China Journal of Chinese Materia Medica, 27*(3), 170–173.

Sinisi, A., Millán, E., Abay, S. M., Habluetzel, A., Appendino, G., Muñoz, E., & Taglialatela-Scafati, O. (2015). Poly-electrophilic sesquiterpene lactones from *Vernonia amygdalina*: New members and differences in their mechanism of thiol trapping and in bioactivity. *Journal of Natural Products, 78*(7), 1618–1623.

Sintayehu, B., Bucar, F., Veeresham, C., & Asres, K. (2012). Hepatoprotective and free radical scavenging activities of extracts and a major compound isolated from the leaves of *Cineraria abyssinica* Sch. Bip. exA. Rich. *Pharmacognosy Journal, 4*(29), 40–46.

Siraj, J., Belew, S., & Suleman, S. (2020). Ethnobotanical assessment and physicochemical properties of commonly used medicinal plants in Jimma zone, Southwest Ethiopia: Traditional healers based cross-sectional study. *Journal of Experimental Pharmacology, 2020*, 665–681.

Sisay, M., Bussa, N., Gashaw, T., & Mengistu, G. (2019). Investigating i*n vitro* antibacterial activities of medicinal plants having folkloric repute in Ethiopian traditional medicine. *Journal of Evidence-Based Integrative Medicine, 24*, 2515690X19886276.

Stadler, M., Dagne, E., & Anke, H. (1994). Nematicidal activities of two phytoalexins from *Taverniera abyssinica*. *Planta Medica, 60*(06), 550–552.

Suleman, S., Tufa, T. B., Kebebe, D., Belew, S., Mekonnen, Y., Gashe, F., & De Spiegeleer, B. (2018). Treatment of malaria and related symptoms using traditional herbal medicine in Ethiopia. *Journal of Ethnopharmacology, 213*, 262–279.

Tadeg, H., Mohammed, E., Asres, K., & Gebre-Mariam, T. (2005). Antimicrobial activities of some selected traditional Ethiopian medicinal plants used in the treatment of skin disorders. *Journal of Ethnopharmacology, 100*(1–2), 168–175.

Tadesse, D., Eguale, T., Giday, M., & Mussa, A. (2009). Ovicidal and larvicidal activity of crude extracts of *Maesa lanceolata* and *Plectranthus punctatus* against *Haemonchus contortus*. *Journal of Ethnopharmacology, 122*(2), 240–244.

Tadesse, S., Mazumder, A., Bucar, F., Veeresham, C., & Asres, K. (2011). Chemical composition and biological activities of the essential oil of *Plectranthus caninus* Roth. *Pharmacognosy Journal, 3*(26), 1–7.

Tadesse, E., Engidawork, E., Nedi, T., & Mengistu, G. (2017). Evaluation of the anti-diarrheal activity of the aqueous stem extract of *Lantana camara* Linn (Verbenaceae) in mice. *BMC Complementary and Alternative Medicine, 17*(1), 1–8.

Tafere, G. G., Tuem, K. B., Gebre, A. K., & Balasubramaniam, R. (2020). *In vitro* antioxidant and *in vivo* hepatoprotective activities of root bark extract and solvent fractions of *Croton macrostachyus* hochst. Ex del.(Euphorbiaceae) on paracetamol-induced liver damage in mice. *Journal of Experimental Pharmacology, 12*, 301–311.

Tamene, D., & Endale, M. (2019). Antibacterial activity of coumarins and carbazole alkaloid from roots of *Clausena anisata*. *Advances in Pharmacological and Pharmaceutical Sciences, 2019*, 5419854.

Tamrat, Y., Nedi, T., Assefa, S., Teklehaymanot, T., & Shibeshi, W. (2017). Anti-inflammatory and analgesic activities of solvent fractions of the leaves of *Moringa stenopetala* Bak.(Moringaceae) in mice models. *BMC Complementary and Alternative Medicine, 17*(1), 1–10.

Tariku, Y., Hymete, A., Hailu, A., & Rohloff, J. (2010). Essential-oil composition, antileishmanial, and toxicity study of *Artemisia abyssinica* and *Satureja punctata* ssp. *punctata* from Ethiopia. *Chemistry & Biodiversity, 7*(4), 1009–1018.

Tariku, Y., Hymete, A., Hailu, A., & Rohloff, J. (2011). *In vitro* evaluation of antileishmanial activity and toxicity of essential oils of *Artemisia absinthium* and *Echinops kebericho*. *Chemistry & Biodiversity, 8*(4), 614–623.

Tauchen, J., Doskocil, I., Caffi, C., Lulekal, E., Marsik, P., Havlik, J., & Kokoska, L. (2015). *In vitro* antioxidant and anti-proliferative activity of Ethiopian medicinal plant extracts. *Industrial Crops and Products, 74*, 671–679.

Taye, B., Giday, M., Animut, A., & Seid, J. (2011). Antibacterial activities of selected medicinal plants in traditional treatment of human wounds in Ethiopia. *Asian Pacific Journal of Tropical Biomedicine, 1*(5), 370–375.

Tefera, B. N., & Kim, Y.-D. (2019). Ethnobotanical study of medicinal plants in the Hawassa Zuria District, Sidama zone, Southern Ethiopia. *Journal of Ethnobiology and Ethnomedicine, 15*, 1–21.

Teferi, M. Y., Abdulwuhab, M., & Yesuf, J. S. (2019). Evaluation of *in vivo* antidiarrheal activity of 80% methanolic leaf extract of *Osyris quadripartita* Decne (Santalaceae) in Swiss Albino Mice. *Journal of Evidence-Based Integrative Medicine, 24*, 2515690X19833340.

Teka, T., & Kassahun, H. (2020). Characterization and evaluation of antioxidant activity of *Aloe schelpei* Reynolds. *Drug Design, Development and Therapy, 14*, 1003–1008.

Teka, A., Rondevaldova, J., Asfaw, Z., Demissew, S., Van Damme, P., Kokoska, L., & Vanhove, W. (2015). *In vitro* antimicrobial activity of plants used in traditional medicine in Gurage and Silti Zones, south central Ethiopia. *BMC Complementary and Alternative Medicine, 15*, 1–7.

Teka, T., Bisrat, D., Yeshak, M. Y., & Asres, K. (2016). Antimalarial activity of the chemical constituents of the leaf latex of *Aloe pulcherrima* Gilbert and Sebsebe. *Molecules, 21*(11), 1415.

Teka, A., Asfaw, Z., Demissew, S., & Van Damme, P. (2020a). Medicinal plant use practice in four ethnic communities (Gurage, Mareqo, Qebena, and Silti), south central Ethiopia. *Journal of Ethnobiology and Ethnomedicine, 16*, 1–12.

Teka, T., Awgichew, T., & Kassahun, H. (2020b). Antimalarial activity of the leaf latex of *Aloe weloensis* (aloaceae) against *plasmodium berghe*i in mice. *Journal of Tropical Medicine, 2020*, 1397043.

Teklay, A., Abera, B., & Giday, M. (2013). An ethnobotanical study of medicinal plants used in Kilte Awulaelo District, Tigray Region of Ethiopia. *Journal of Ethnobiology and Ethnomedicine, 9*, 1–23.

Teklehaymanot, T. (2009). Ethnobotanical study of knowledge and medicinal plants use by the people in Dek Island in Ethiopia. *Journal of Ethnopharmacology, 124*(1), 69–78.

Teklehaymanot, T. (2017). An ethnobotanical survey of medicinal and edible plants of Yalo Woreda in Afar regional state, Ethiopia. *Journal of Ethnobiology and Ethnomedicine, 13*, 1–26.

Teklehaymanot, T., & Giday, M. (2007). Ethnobotanical study of medicinal plants used by people in Zegie Peninsula, Northwestern Ethiopia. *Journal of Ethnobiology and Ethnomedicine, 3*(1), 1–11.

Teklehaymanot, T., & Giday, M. (2010a). Ethnobotanical study of wild edible plants of Kara and Kwego semi-pastoralist people in Lower Omo River Valley, Debub Omo Zone, SNNPR, Ethiopia. *Journal of Ethnobiology and Ethnomedicine, 6*(1), 1–8.

Teklehaymanot, T., & Giday, M. (2010b). Quantitative ethnobotany of medicinal plants used by Kara and Kwego semi-pastoralist people in lower Omo River Valley, Debub Omo zone, southern nations, nationalities and peoples regional state, Ethiopia. *Journal of Ethnopharmacology, 130*(1), 76–84.

Teklehaymanot, T., Giday, M., Medhin, G., & Mekonnen, Y. (2007). Knowledge and use of medicinal plants by people around Debre Libanos monastery in Ethiopia. *Journal of Ethnopharmacology, 111*(2), 271–283.

Teklu, T., Engidawork, E., Nedi, T., Teklehaymanot, T., & Gebremeskel, L. (2020). Evaluation of the antimalarial activity of the hydroalcoholic extract of leaf of *Leonotis ocymifolia* (Burm. f.) Iwarsson (Lamiaceae) against p*lasmodium berghei* in mice. *Evidence-based Complementary and Alternative Medicine: Ecam, 2020*, 5384804.

Tesfaye, S., Belete, A., Engidawork, E., Gedif, T., & Asres, K. (2020). Ethnobotanical study of medicinal plants used by traditional healers to treat cancer-like symptoms in eleven districts, Ethiopia. *Evidence-based Complementary and Alternative Medicine, 2020*, 7683450.

Teshome, K., Gebre-Mariam, T., Asres, K., Perry, F., & Engidawork, E. (2008). Toxicity studies on dermal application of plant extract of *Plumbago zeylanica* used in Ethiopian traditional medicine. *Journal of Ethnopharmacology, 117*(2), 236–248.

Tessema, E. N., Gebre-Mariam, T., Lange, S., Dobner, B., & Neubert, R. H. (2017a). Potential application of oat-derived ceramides in improving skin barrier function: Part 1. Isolation and structural characterization. *Journal of Chromatography B, 1065*, 87–95.

Tessema, E. N., Gebre-Mariam, T., Schmelzer, C. E., & Neubert, R. H. (2017b). Isolation and structural characterization of glucosylceramides from Ethiopian plants by LC/APCI-MS/MS. *Journal of Pharmaceutical and Biomedical Analysis, 141*, 241–249.

Tessema, M. Y., Wubneh, Z. B., & Asrie, A. B. (2020). Laxative activities of 80% methanolic extract of the leaves of *Grewia ferruginea* hochst ex a rich in mice. *Journal of Evidence-Based Integrative Medicine, 25*, 2515690X20926922.

Tolossa, K., Debela, E., Athanasiadou, S., Tolera, A., Ganga, G., & Houdijk, J. G. (2013). Ethnomedicinal study of plants used for treatment of human and livestock ailments by traditional healers in South Omo, Southern Ethiopia. *Journal of Ethnobiology and Ethnomedicine, 9*, 1–15.

Toma, A., Deyno, S., Fikru, A., Eyado, A., & Beale, A. (2015). *In vivo* antiplasmodial and toxicological effect of crude ethanol extract of *Echinops kebericho* traditionally used in treatment of malaria in Ethiopia. *Malaria Journal, 14*(1), 1–5.

Tuasha, N., Petros, B., & Asfaw, Z. (2018). Medicinal plants used by traditional healers to treat malignancies and other human ailments in Dalle District, Sidama Zone, Ethiopia. *Journal of Ethnobiology and Ethnomedicine, 14*(1), 1–21.

Tuasha, N., Seifu, D., Gadisa, E., Petros, B., & Oredsson, S. (2020). Solvent fractions of selected Ethiopian medicinal plants used in traditional breast cancer treatment inhibit cancer stem cells in a breast cancer cell line. *BMC Complementary Medicine and Therapies, 20*(1), 1–9.

Umer, S., Asres, K., & Veeresham, C. (2010). Hepatoprotective activities of two Ethiopian medicinal plants. *Pharmaceutical Biology, 48*(4), 461–468.

Unasho, A., Geyid, A., Melaku, A., Debela, A., Mekasha, A., Girma, S., & Mamo, K. (2009). Investigation of antibacterial activities of *Albizia gummifera* and *Ferula communis* on *Streptococcus pneumoniae* and *Streptoccus pyogenes*. *Ethiopian Medical Journal, 47*(1), 25–32.

Woldemichael, G. M., & Wink, M. (2002). Triterpene glycosides of *Lupinus angustifolius*. *Phytochemistry, 60*(4), 323–327.

Woldu, Y., & Abegaz, B. (1990). Isoflavonoids from *Salsola somalensis*. *Phytochemistry, 29*(6), 2013–2015.

Woldu, Y., Abegaz, B., Botta, B., Delle Monache, G., & Delle Monache, F. (1988). Styrenes from *Dorstenia barnimiana*. *Phytochemistry, 27*(4), 1227–1228.

Wondafrash, D. Z., Bhoumik, D., Altaye, B. M., Tareke, H. B., & Assefa, B. T. (2019). Antimalarial activity of *Cordia africana* (Lam.)(Boraginaceae) leaf extracts and solvent fractions in *Plasmodium berghei*-infected mice. *Evidence-based Complementary and Alternative Medicine, 2019*, 8324596.

Wondemetegegn, S. A. (2016). The historic move, contemporary challenges and opportunities in Ethiopian education. *International Journal of African and Asian Studies, 26*, 56–66.

Wondimu, A., Dagne, E., & Waterman, P. G. (1988). Quinoline alkaloids from the leaves of *Teclea simplicifolia*. *Phytochemistry, 27*(3), 959–960.

Wondimu, T., Asfaw, Z., & Kelbessa, E. (2007). Ethnobotanical study of medicinal plants around 'Dheeraa'town, Arsi Zone, Ethiopia. *Journal of Ethnopharmacology, 112*(1), 152–161.

Wube, A. A., Bucar, F., Gibbons, S., & Asres, K. (2005). Sesquiterpenes from *Warburgia ugandensis* and their antimycobacterial activity. *Phytochemistry, 66*(19), 2309–2315.

Wube, A. A., Gibbons, S., Asres, K., Streit, B., Adams, M., Bauer, R., & Bucar, F. (2006). *In vitro* 12 (S)-HETE and leukotriene metabolism inhibitory activity of sesquiterpenes of *Warburgia ugandensis*. *Planta Medica, 72*(08), 754–756.

Wube, A. A., Wenzig, E.-M., Gibbons, S., Asres, K., Bauer, R., & Bucar, F. (2008). Constituents of the stem bark of *Discopodium penninervium* and their LTB4 and COX-1 and-2 inhibitory activities. *Phytochemistry, 69*(4), 982–987.

Wubetu, M., Abula, T., & Dejenu, G. (2017). Ethnopharmacologic survey of medicinal plants used to treat human diseases by traditional medical practitioners in Dega Damot district, Amhara, Northwestern Ethiopia. *BMC Research Notes, 10*(1), 1–13.

Yankep, E., Fomum, Z., & Dagne, E. (1997). An *O*-geranylated isoflavone from *Millettia griffoniana*. *Phytochemistry, 46*(3), 591–593.

Yenesew, A., & Dagne, E. (1988). Alkaloids of *Teclea nobilis*. *Phytochemistry, 27*(2), 651–653.

Yenesew, A., Dagne, E., Müller, M., & Steglich, W. (1994). An anthrone, an anthraquinone and two oxanthrones from *Kniphofia foliosa*. *Phytochemistry, 37*(2), 525–528.

Yigezu, Y., Haile, D. B., & Ayen, W. Y. (2014). Ethnoveterinary medicines in four districts of Jimma zone, Ethiopia: Cross sectional survey for plant species and mode of use. *BMC Veterinary Research, 10*(1), 1–12.

Yigzaw, Y., Larsson, N., Gorton, L., Ruzgas, T., & Solomon, T. (2001). Liquid chromatographic determination of total and β-*N*-oxalyl-l-α, β-diaminopropionic acid in *Lathyrus sativus* seeds using both refractive index and bioelectrochemical detection. *Journal of Chromatography A, 929*(1–2), 13–21.

Yimer, T., Birru, E. M., Adugna, M., Geta, M., & Emiru, Y. K. (2020). Evaluation of analgesic and anti-inflammatory activities of 80% methanol root extract of *Echinops kebericho* M. (Asteraceae). *Journal of Inflammation Research, 13*, 647–658.

Yineger, H., & Yewhalaw, D. (2007). Traditional medicinal plant knowledge and use by local healers in Sekoru District, Jimma Zone, Southwestern Ethiopia. *Journal of Ethnobiology and Ethnomedicine, 3*(1), 1–7.

Yineger, H., Kelbessa, E., Bekele, T., & Lulekal, E. (2007). Ethnoveterinary medicinal plants at bale mountains national park, Ethiopia. *Journal of Ethnopharmacology, 112*(1), 55–70.

Yineger, H., Yewhalaw, D., & Teketay, D. (2008). Ethnomedicinal plant knowledge and practice of the Oromo ethnic group in southwestern Ethiopia. *Journal of Ethnobiology and Ethnomedicine, 4*(1), 1–10.

Yonathan, M., Asres, K., Assefa, A., & Bucar, F. (2006). *In vivo* anti-inflammatory and anti-nociceptive activities of *Cheilanthes farinosa*. *Journal of Ethnopharmacology, 108*(3), 462–470.

Zerabruk, S., & Yirga, G. (2012). Traditional knowledge of medicinal plants in Gindeberet district, Western Ethiopia. *South African Journal of Botany, 78*, 165–169.

Zewdie, K. A., Bhoumik, D., Wondafrash, D. Z., & Tuem, K. B. (2020). Evaluation of *in-vivo* antidiarrhoeal and *in-vitro* antibacterial activities of the root extract of *Brucea antidysenterica* JF Mill (Simaroubaceae). *BMC Complementary Medicine and Therapies, 20*(1), 1–11.

Chapter 2
Rose-Scented Geranium, an Important African Plant for the Perfume and Cosmetic Industries

Heba Mohammed, Enas Mohamed, Dalia El Amir, and Sameh AbouZid

Abstract Rose-scented geranium, *Pelargonium graveolens* L'Hérit. ex Aiton: syn. *P. roseum* Willd (Geraniaceae), is a perennial aromatic and medicinal plant native to the southern part of Africa. It is commercially cultivated in several African countries for the production of its essential oil which is used extensively in the industries of cosmetics and perfumes as an alternative to rose oil. In the mid-1800s, as *Rosa damascena* (a species from the rose family) became scarce and its essential oil became expensive, attention shifted toward *P. graveolens* as an alternative. The essential oil, commercially obtained by steam distillation from the leaves and stems of the plant, is used in high-quality perfumes having a floral rosy scent. Solvent extraction and supercritical fluid extraction with carbon dioxide have been investigated as alternative methods for extraction of essential oil. Analysis of the essential oil by gas chromatography coupled with mass spectrometry is performed for qualitative and quantitative determination of its constituents. The major compounds are citronellol, geraniol, linalool, iso-menthone, citronellyl formate, and geranyl formate. High geraniol content imparts a rosy smell to the essential oil. The remarkable bioactive profile of geranium essential oil presents a multifaceted opportunity for industrial innovation across various critical sectors.

Keywords Rose-scented geranium · *Pelargonium graveolens* · Essential oil · Biological activity

Abbreviations

Bax	Bcl-2 associated X-protein
CUPRAC	Cupric ion reducing capability assay
COX	Cyclooxygenase

H. Mohammed · E. Mohamed · D. El Amir · S. AbouZid (✉)
Department of Pharmacognosy, Faculty of Pharmacy, Beni-Suef University, Beni-Suef, Egypt
e-mail: dalia.elamir@pharm.bsu.edu.eg; sameh.zaid@pharm.bsu.edu.eg

A. Dekebo (ed.), *Bioactive Secondary Metabolites from Medicinal Plants of Africa*, Interdisciplinary Biotechnological Advances,
https://doi.org/10.1007/978-981-96-8590-5_2

COVID-19	Coronavirus Disease 2019
DPPH	2,2′-Diphenyl-1-picrylhydrazyl
EC_{50}	Half maximal effective concentration
ERK	Extracellular signal-regulated kinase
FRAP	Ferric reducing ability of plasma assay
GC/MS	Gas chromatography/mass spectrometry
GRAS	Generally recognized as safe
ICU	Intensive care unit
iNOS	Induced nitric oxide synthase
JAK2	Janus kinase 2
LOX	Lipoxygenase
MIC	Minimum inhibitory concentration
NO	Nitric oxide
PPHG	Postprandial hyperglycemia
RIP	Receptor interacting protein
SARS-CoV-2	Severe acute respiratory syndrome coronavirus 2
SCF	Supercritical fluid extraction
STAT3	Signal transducer and activator of transcription 3
syn.	Synonym
TE	Trolox equivalent
TNF-α	Tumor necrosis factor alpha

2.1 Introduction

Essential oils have been extremely popular since ancient Egyptian times. Tutankhamun used more than 35 vessels with over 350 liters of oils in his afterlife (https://nilescribes.org). The most popular aromatic plants used by ancient Egyptians were white and blue water lily, henna, frankincense, myrrh, cinnamon, thyme, and coriander. The ancient Egyptians used to prepare the essential oil from aromatic plants by pressing the flowers and plants together directly. They also used the fats saturated with the flower perfume or dipped flowers and plants into the fats and heated and then cooled it.

Rose-scented geranium, *Pelargonium graveolens* L'Herit. ex Aiton: syn. *P. roseum* Willd (Geraniaceae), is a perennial aromatic and medicinal herb native to South Africa. The genus *Pelargonium* contains about 750 species growing in temperate and subtropical climates. Most of these species are indigenous to South Africa. *P. graveolens* is characterized by deeply incised rose-scented leaves and hairy herbaceous stems that become woody with age (Fig. 2.1). The plant is widely grown as an ornamental plant for its fragrant leaves. Its economic importance is derived mainly from its essential oil. The International Organization for Standardization defines geranium oil as "The oil obtained by steam distillation of the fresh or slightly withered herbaceous parts of *P. graveolens*, *P. roseum* and other undefined hybrids which have given rise to differing ecotypes in the various geographical areas" (Lis-Balchin, 2002).

Fig. 2.1 Rose-scented geranium cultivated at the experimental botanical garden of the Department of Pharmacognosy, Faculty of Pharmacy, Beni-Suef University, Egypt

2.2 Rose-Scented Geranium Essential Oil

The essential oil prepared from *P. graveolens* is widely used in the perfumery and cosmetic industries for imparting rose-like fragrance. It is used as a substitution for the expensive rose (*Rosa damascena* Mill., Rosaceae) oil, known as "poor man's rose oil" (Narnoliya et al., 2019). The oil is also used for the commercial extraction of rhodinol which is a high-value aroma fraction used widely in fine-grade perfumes (Babu & Kaul, 2005). Rhodinol is isolated by fractional distillation, and it is composed mainly of citronellol, geraniol, and other free alcohols (Rao, 2009). In some literature, rhodinol refers to linalool, citronellol, and geraniol (Fig. 2.2). These monoterpenoid alcohols together with their esters are referred to as total rhodinol (Ravindra & Kulkarni, 2015).

P. graveolens essential oil, prized for its distinctive floral rose scent with green, herbal, fresh, minty, and earthy undertones, plays a significant role in perfumery. In men's fragrances, it serves as a heart note, particularly enhancing Green and Fougère compositions, as exemplified in iconic fragrances like Kouros by Yves Saint Laurent (1981) and Polo Blue by Ralph Lauren (2003). Its use in women's perfumes is more limited, but it still features as a heart note in classical chypre fragrances such as Clinique's Aromatic Elixir (1975) and Dior Essence (1979), as well as in floral scents like Paris by Yves Saint Laurent (1983) and Paul Smith Women (2000) (Gomes et al., 2007).

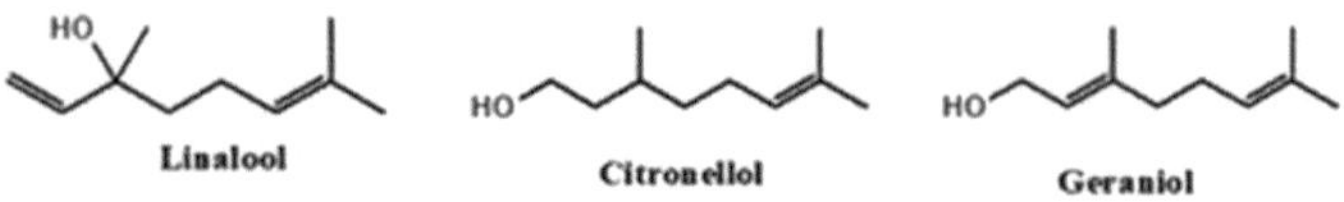

Fig. 2.2 Chemical structures of linalool, citronellol, and geraniol

The composition of the essential oil of *P. graveolens* differs according to many factors, including cultivar, growth location, date of transplantation, fertilizers, age of the plant, photo period, intensity of light, temperature, seasonal variation, plant part used, harvesting time, method of preparation, wilting period, and storage conditions (Babu & Kaul, 2005). The Spring/Summer harvesting season was reported to be superior to the Autumn/Winter harvesting season in *P. graveolens* oil production in South Africa (Motsa et al., 2006). Moreover, concentration of the essential oil is greatly improved via water stress with limited contribution to total yield of the essential oil (Eiasu et al., 2009). In addition, plant population density was reported to affect the yield and content of the oil cultivated in Ethiopia (Kassahun et al., 2012).

Hundreds of compounds have been identified in the essential oil of *P. graveolens*. Gas chromatography coupled to mass spectrometry (GC-MS) is the method of choice for qualitative and quantitative determination of these compounds. Terpenoid compounds constitute a major part of the essential oil of *P. graveolens*. The main terpenoid compounds in the essential oil which are important for the perfumery industry are citronellol, geraniol, linalool, citronellyl formate, geranyl formate, guaia-6,9-diene, and rose-oxide (Sedibe & Allemann, 2012). Citronellol, geraniol, linalool, and their esters constitute more than 60% of the total essential oil and determine its odor (Narnoliya et al., 2019). Citronellol content is important for the determination of the quality of *P. graveolens* essential oil (Babu & Kaul, 2005).

In a study aimed at investigating the geranium essential oil in Algeria, it has been revealed that 45 compounds constituting about 94% of the essential oil were annotated where the main constituents were citronellol (30.2%), citronellyl formate (9.3%), and geraniol (7.6%) (Boukhatem et al., 2013), while GC/MS analysis of Tunisian oil revealed the main components as citronellol (21.9%), citronellyl formate (13.2%), geraniol (11.1%), 10-epi-γ-eudesmol (7.9%), geranyl formate (6.2%), and linalool (5.6%) (Boukhris et al., 2013) and in Morocco, majority of compounds detected were 10-epi-γ-eudesmol (16.6%), geraniol (12.5%), citronellol (12.3%), citronellyl formate (7.7%), and geranyl tiglate (5.2%) (M'hamdi et al., 2024). The main constituents in the essential oil of *P. graveolens* growing in Egypt have been reported to be citronellol (27.6%), cis-menthone (10.2%), linalool (10.0%), 10-epi-γ-eudesmol (9.4%), geranyl formate 6.8%, and rose oxide (5.7%) (SM Abd El-Kareem et al., 2020).

Geranium is cultivated in various parts of the world for the production of essential oil. The main cultivars are coming from China, North Africa (Egypt and Morocco), Reunion Island, and India. The chemical composition of essential oils varies depending on their country of origin, allowing for unique characterization based on geographic source. The annual global production of *P. graveolens* oil ranges from 500 to 750 tons (Mazeed et al., 2023). China is the biggest producer of geranium oil with a share of 61% of the supply (Gomes et al., 2007). The Chinese oil is characterized by high content of citronellyl formate and large citronellol to geraniol ratio. The North Africa oil is characterized by high content of 10-epi-γ-eudesmol and citronellol to geraniol ratio that is close to 1. Bourbon-type oil obtained from Reunion Island contains no 10-epi-γ-eudesmol. This oil contains high content of guaia-6,9-diene and citronellol to geraniol ratio that is close to 1. In India, the total annual production of geranium oil is about 20 tons. The production is inadequate to meet the growing demand of the Indian perfumery industry, so additional 20 tons of oil is imported annually (Lothe & Verma, 2023).

2.3 Preparation of Geranium Essential Oil

2.3.1 Distillation

Distillation is the traditional and the most widely used method for the preparation of geranium essential oil. According to one reported method, the conditions are: extraction of 20 g plant (green and fresh) using de-ionized water (600 mL) for 2 hours (Gomes et al., 2004). The yield is calculated using Eq. (2.1).

$$\text{Extraction yield} = \frac{\text{Mass of essential oil obtained}}{\text{Mass of fresh plant material used}} \quad (2.1)$$

Babu and Kaul (2005) reported that water distillation (the herb is totally immersed in water) of the herb gave a higher oil yield (0.16–0.22%) than the water–steam distillation (the herb is loaded on a perforated grid placed above the distillation tank) (0.09–0.12%) and steam distillation methods (direct steam is injected in the distillation tank) (0.06–0.18%) (Babu & Kaul, 2005). The quality of the essential oil produced from *P. graveolens* varies according to the plant part used for distillation (Ravindra & Kulkarni, 2015). The yield of essential oil prepared from the plant using microwave-assisted hydro-distillation was 1.75% (Aydınlık & Yüceer, 2021). Citronellyl formate, iso-menthone, linalool, citronellol, and geraniol were the main constituents.

2.3.2 Solvent Extraction

Fekri et al. (2021) compared the composition of "absolute" prepared by solvent extraction from *P. graveolens* to the essential oil prepared by distillation. It is worth noting that absolute is a highly concentrated form of plant aroma, and it is most frequently prepared via extraction using organic solvent such as *n*-hexane followed

Table 2.1 Comparison between different methods used for preparation of rose-scented geranium essential oil

Method	Advantages	Disadvantages
Hydro-distillation	Low cost	Low yield as it delivers only volatile material High content of linalool as an artifact of geraniol
Solvent extraction	High yield	The extract includes pigment and waxes (concrete)
Supercritical fluid extraction	Low temperature Versatility	High cost Safety precautions Extraction of nonvolatile compounds

by absolute ethanol. The major identified components were citronellol (29.30%), geraniol (7.54%), geranial (12.90%), iso-menthone (9.87%), and epi-α-cadinol (8.79%). Linalool was absent in the absolute (Fekri et al., 2021). Linalool is claimed to be formed from geraniol under high temperature in the presence of water vapor during the hydro-distillation process (Gomes et al., 2007).

2.3.3 Supercritical Fluid Extraction (SCF)

This method produces extracts that are generally recognized as safe (GRAS). There is an increasing tendency to use SCF due to regulatory restrictions on organic solvent residue trace amounts. It is used to obtain ultra-pure products. In addition, SCF uses low temperature, so the essential oil produced is of superior quality. The operating conditions in SCF can be selected to control the composition of the oil. The following operating parameters affect SCF: pressure, temperature, processing of plant material, time of extraction, dimensions of extractor, flow rate of solvent, and solute physical properties (vapor pressure, polarity, and molecular weight) (Gomes et al., 2007). Table 2.1 compares the three main methods used for the preparation of *P. graveolens* essential oil.

2.4 Biotechnological Manipulations

Many biotechnological approaches have been followed to enhance the productivity of *P. graveolens* utilization. These approaches are mainly aimed at enhancing the production of essential oil. It is essential to thoroughly study biosynthetic pathways and their regulation leading to the major constituents of the essential oil. For example, Narnoliya et al. (2017) reported sequencing of *P. graveolens* leaf transcriptome, transcript assembly, annotation, expression profiling, and their validation (Narnoliya et al., 2017). Moreover, a transcriptome dataset of *P. graveolens* was established and used to functionally characterize four terpene synthases (Blerot et al., 2018). Researchers found a significant relationship between the expression of four genes responsible for terpene synthases and the levels of their associated chemical products across various *Pelargonium* cultivars and species. In addition, two monoterpene biosynthetic pathways were reported to be involved in the formation of monoterpenes in *P. graveolens* (Bergman et al., 2019).

Plant tissue culture offers an effective tool for rapid propagation of desirable clones of geranium plant containing high and desirable essential oil content. Saxena et al. (2007) established hairy root culture from *P. graveolens*. Hairy root culture is a type of transgenic root culture system mediated by transformation with *Agrobacterium rhizogenes* (Saxena et al., 2007). Regenerated transgenic plants were produced from the established hairy roots. The regenerated plants exhibited distinctive characteristics, including more extensive branching, a greater number of leaves with pronounced serrated edges, a compact and rounded form, a highly ramified root system, and smooth leaf surfaces without any wrinkling. Two transgenic plants demonstrated elevated levels of geraniol and geranyl esters, indicating an enhanced essential oil quality in relation to citronellol:geraniol ratio.

El-Sonbaty et al. (2022) reported that supplementing the plant with iron nanoparticles and humic acid increased plant growth and essential oil production (El-Sonbaty et al., 2022). The effects of abscisic acid and methyl jasmonate on essential oil production in simulated hail-damaged geranium were studied (Khetsha et al., 2024), and it was concluded that the application of methyl jasmonate may improve the essential oil biosynthesis in hail-damaged geranium plants that could be a possible technique for enhancing the biosynthesis of geranium oil after hail induced damage. Moreover, it was reported that wounding and moisture stress closer to the harvest with cytokinin application may improve the essential oil yield, quality, and most of the major essential oil compounds for rose-scented geranium from Reunion and Madagascar (Khetsha, 2024).

2.5 Biological Activities of *P. graveolens* and Its Essential Oil

P. graveolens has a wide range of biological activities that extend beyond its traditional uses. Numerous scientific studies have confirmed and expanded upon its antimicrobial, antioxidant, anti-inflammatory, anticancer, and psychological effects, underscoring its potential as a natural therapeutic agent.

2.5.1 Antimicrobial Activity

One of the thoroughly studied biological activities of *P. graveolens* extract and essential oil is their antimicrobial action. Recent studies have revealed their efficacy against a broad spectrum of microorganisms (Table 2.2).

The antimicrobial activity of alcoholic and aqueous extracts of leaves and flowers of the plant is most likely attributed to their phenolic content which can alter the permeability of cell membrane and interrupt membrane functions, including nutrient uptake, enzyme activity, nucleic acid synthesis, and electron transport (Boukhris et al., 2013).

The antimicrobial efficacy of essential oil from *P. graveolens* is closely related to its chemical composition. This essential oil is rich in diverse compounds, including monoterpenes, sesquiterpenes, and phenolic components, which contribute to its broad-spectrum antimicrobial properties. Key constituents like citronellol, geraniol, linalool, and iso-menthone play crucial roles in determining the oil's activity against many microorganisms.

Citronellol and geraniol These two compounds are among the most abundant in essential oil and are primarily responsible for its antimicrobial properties (Inouye et al., 2001; Lorenzi et al., 2009). Citronellol and geraniol have been stated to interrupt the integrity of microbial cell membrane, causing augmented permeability and cell lysis. These monoterpenes exhibited strong antibacterial effect against bacteria such as *Staphylococcus aureus* and *Escherichia coli* due to their ability to disrupt lipid membranes, which compromises cellular functions (Dorman & Deans, 2000). Geraniol, geranyl acetate, and citronellol have antifungal action on *Candida albicans* that has been attributed to damaging membrane integrity. At low concentrations, they have the ability to prevent germ tube induction and stop the fungus cell cycle (Zore et al., 2011).

Linalool It is another significant component that has revealed powerful antifungal and antibacterial activity. It has been found that linalool effectively inhibits the growth of *C. albicans* and other fungi, likely due to its ability to interfere with the synthesis of ergosterol, a key component of fungal cell membranes. Its synergistic effects with other terpenes in geranium oil can enhance the overall antimicrobial efficacy (Prabuseenivasan et al., 2006).

The antimicrobial effectiveness of geranium oil is not solely dependent on the presence of individual components but also on their concentration and synergistic interactions. The synergistic interaction between compounds like citronellol, geraniol, and linalool enhances the antimicrobial activity of geranium oil, and these interactions can result in a more powerful disruption of microbial cells compared to individual compounds (Burt, 2004). The efficacy of geranium oil can vary depending on its chemical composition, which is influenced by factors such as the geographical location of cultivation, method of distillation, and different phenological stages, in addition to seasonal variations (Szutt et al., 2020). Comparative studies found significant differences in the chemical composition with differences

Table 2.2 List of microorganisms susceptible to *P. graveolens* essential oil of different geographical sources*

Microorganism			Geographical source						
			Iran[a]	Morocco[b]	Egypt[c]	Tunisia[d]	Rwanda[e]	India[f]	Turkey[g]
Bacteria	**Gram-positive**	*Staphylococcus aureus*	+	+	+	+			
		S. epidermis		+		+			
		S. constellatus			+				
		MRSA		+					
		Bacillus cereus				+			
		B. subtilis	+	+		+			
		Enterococcus faecalis				+			
		Micrococcus luteus				+			
		Listeria monocytogenes		+		+			
		Priestia megaterium			+				
	Gram-negative	*Escherichia coli*	+	+		+	+		
		Pseudomonas aeruginosa	+	+	+	+			
		Salmonella enteritidis	+			+			
		S. typhimurium		+			+		
		Klebsiella pneumoniae		+	+	+			
		Proteus mirabilis							
		Shigella sonnei			+		+		
		S. flexnerie					+		
		Helicobacter pylori			+				
		Acinetobacter baumannii							+
Fungi		*Candida albicans*	+	+		+	+		
		Aspergillus niger				+			
		A. flavus				+		+	
		A. fumigatus						+	

(continued)

Table 2.2 (continued)

Microorganism		Geographical source						
		Iran[a]	Morocco[b]	Egypt[c]	Tunisia[d]	Rwanda[e]	India[f]	Turkey[g]
	A. terreus						+	
	Fusarium graminearum				+			
	F. oxysporum				+		+	
	Alternaria alternata				+		+	
	Cladosporium herbarum						+	
	Curvularia lunata							
	Helminthosporium oryzae						+	
	Trichoderma viride						+	
	Malassezia spp.		+					
	Coniophora puteana		+					
	Coriolus versicolor		+					
	Poria placenta		+					
	Gloeophyllum trabeum		+					

*All essential oils reported here were prepared using a distillation method
[a]Ghannadi et al. (2012)
[b]Moutaouafiq et al. (2019), Al-Mijalli et al. (2022), EL-Otmani et al. (2024)
[c]Kačániová et al. (2023), Wafa et al. (2023)
[d]Hsouna and Hamdi (2012), M'hamdi et al. (2024)
[e]Justin et al. (2013)
[f]Kujur et al. (2020)
[g]Boztaş et al. (2024)

in the relative concentrations of key constituents like citronellol and geraniol (Sharma, 2013; Verma et al., 2013; Abd El-Wahab et al., 2016; Fekri et al., 2021; Al-Mijalli et al., 2022; Kačániová et al., 2023). This underscores the need for standardized extraction processes to ensure consistent antimicrobial efficacy.

2.5.1.1 Antibacterial Effect

The essential oil of *P. graveolens* has shown potent antibacterial activity against a panel of microorganisms: Gram-positive (G^+) and Gram-negative (G^-) bacteria and fungi (Ghannadi et al., 2012; Dimitrova et al., 2015; M'hamdi et al., 2024). Geranium oil extracted by hydro-distillation suppressed the growth of MRSA (methicillin-resistant *Staphylococcus aureus*) more than ciprofloxacin and ampicillin (MIC: 1.56 μg/mL) (Jaradat et al., 2022), and it exhibited antimycobacterial action against *Mycobacterium tuberculosis*, multidrug-resistant *M. tuberculosis*, *M. fortuitum*, and *M. kansasii* (MIC: 78–156 μg/mL) (Kardan-Yamchi et al., 2020). Furthermore, it possesses antimicrobial action on colistin-resistant and colistin-susceptible *Acinetobacter baumannii* clinical isolates (MIC: 5–20 μl/mL) (Kafa et al., 2022). Studies reported that the oil is more selective on G^+ than G^- bacteria and attributed that to the resistance of Gram-negative bacteria owing to their unique restricting outer membrane acting as a barrier halting entry into the cell (Gilles et al., 2010; Saxena et al., 2023).

In the vapor phase, the oil, which has been extracted by steam distillation, has been proven to inhibit the growth of a number of G^+, G^-, and biofilm-forming bacteria on infected fruits. Furthermore, *P. graveolens* essential oil is capable of disintegrating the biofilm of *Salmonella enterica* (MIC: 169 μg/mL) and suppressing its growth on surfaces such as plastic and stainless steel (Kačániová et al., 2023).

The plant extracts possess antibacterial activity as well. For instance, aqueous and alcoholic extracts of the plant revealed antibacterial effect against *Escherichia coli*, *Staphylococcus epidermis*, *S. aureus* (inhibition zones: 13–16 mm) (Ennaifer et al., 2020), *Pseudomonas aeruginosa* (MIC: 470 μg/mL) (El Aanachi et al., 2020), *Bacillus subtilis*, *Listeria monocytogenes*, and *L. innocua* (Dimitrova et al., 2015). Antibacterial potential of ethanolic extract of geranium against clinical isolates of *S. aureus* has been proved (MIC: 200–800 μg/mL) (Mahboubi et al., 2011).

2.5.1.2 Antifungal Effect

The essential oil has also demonstrated broad antifungal activity, particularly against *Aspergillus* and *Fusarium* species (steam distilled oil, MIC: $300–2 \times 10^3$ μg/mL) (Hsouna & Hamdi, 2012). It revealed superior antifungal action over synthetic fungicides (Singh et al., 2008). Moreover, the oil was reported to have anti-aflatoxigenic effect. These effects make the oil a potential natural alternative for managing fungal infections, especially those resistant to conventional antifungal treatments (Singh et al., 2008).

P. graveolens oil exhibits anti-dermatophyte action against dermatophytes as *Microsporum gypseum*, *M. canis*, *Trichophyton mentagrophytes*, *T. rubrum*, and *T. schoenleinii* and anticandidal activity against *C. albicans* clinical isolates (Mahdizadeh & Tabar, 2018). Citronellol and geraniol were confirmed to be responsible for the activity (Mahboubi & Valian, 2019).

Alcoholic and aqueous extracts of *P. graveolens* exhibited activity against *C. albicans*, *C. neoformans*, and *A. niger* (MIC: 470–1.8×10^3 μg/mL) (El Aanachi et al., 2020).

2.5.1.3 Combination of Geranium Oil and Other Antimicrobials

Since combination of more than one drug involves many targets at the same time, it is better than the single target approach. The synergism between geranium oil and ciprofloxacin in the management of urinary tract infections or amphotericin B in the management of *Candida* infections has been reported (Rosato et al., 2008; Malik et al., 2011). On the other hand, the oil potentiates the anti-Helicobacter effect of clarithromycin (Ibrahim et al., 2021). It exerts a strong synergistic effect upon combination with colistin against various strains of *Acinetobacter baumannii* (Kafa et al., 2022). The synergism allowed to minimize the active dose of the antimicrobial drug and reduce its adverse effects and resistance of the microorganism. Geranium oil can be used in combination with lemongrass oil to yield a vapor spray effective against airborne bacteria (Ibrahim et al., 2021).

A list of antimicrobial investigations on essential oil from *P. graveolens* in different geographical locations are presented in Table 2 where in Tunisia and Morocco, geranium oil was tested and exhibited antimicrobial effect against a wide range of G+ and G− bacteria and fungi. In Iran and Egypt, most antimicrobial studies were concerned with bacteria more than fungi. On the other hand, antimicrobial studies in India were concentrated on antifungal effects of the oil.

2.5.2 *Antiviral Activity*

Although less studied, some research has suggested that *P. graveolens* may possess antiviral properties. The essential oil in vapor phase has antiviral action against influenza virus where it completely inactivated the virus following 30 minutes exposure without any adverse effects on epithelial cells (Vimalanathan & Hudson, 2014); it also showed antiviral action on SARS-CoV-2, and this action was attributed to its main component, citronellol (EC_{50}: 0.2 μg/mL) (Torres Neto et al., 2022).

2.5.3 Antioxidant Activity

P. graveolens has garnered attention for its robust antioxidant properties. Antioxidants play a crucial role in neutralizing free radicals, thereby protecting cells from oxidative damage linked to aging and various chronic diseases (Ahmadinejad et al., 2017).

Antioxidant capability of *P. graveolens* solvent extract and essential oil has been assessed by many authors via DPPH radical scavenging assay, and they revealed antioxidant potential comparable to that of ascorbic acid (Mnif et al., 2011; Cavar & Maksimovic, 2012; Dzamic et al., 2014; Asgarpanah & Ramezanloo, 2015). Other techniques have been used to evaluate the antioxidant capacity of *P. graveolens* including 2,2′-azino-bis(3-ethylbenzothiazoline-6-sulfonic acid) assay, ferric reducing antioxidant power assay, and cupric reducing antioxidant capacity, where the essential oil extracted by hydro-distillation demonstrated significant antioxidant activity, comparable to that of synthetic antioxidants like butylated hydroxytoluene (Boukhris et al., 2013; Dimitrova et al., 2015) (Table 2.3).

Beyond in vitro assays, the antioxidant properties of *P. graveolens* have also been observed in cellular models. The plant extract reduced oxidative stress markers in human cells, indicating potential protective effects against oxidative damage in biological systems (Al-Sayed et al., 2015). Hydro-distillation prepared oil prevented deltamethrin-induced oxidative damage of the male reproductive system in mice comparable to alpha tocopherol (Ben Slima et al., 2013).

Antioxidant activity of the plant is attributed to its major constituents, including phenolic compounds, its essential oil, and other phytochemicals.

Phenolic Compounds Phenolics, including flavonoids and tannins, are well known for their antioxidant activities. These compounds can act as reducing agents, hydrogen donors, and singlet oxygen quenchers, effectively scavenging free radicals. Studies have identified a variety of phenolic compounds, including flavonoids, in *P. graveolens*, contributing significantly to its overall antioxidant potential (Latté & Kolodziej, 2004; Li et al., 2024).

Table 2.3 Antioxidant activity of *Pelargonium graveolens* essential oil and plant solvent extract via different antioxidant assays (Ali et al., 2020; Gevrenova et al., 2024)

Antioxidant assay	Antioxidant activity of *P. graveolens*
DPPH	**IC_{50}** Essential oil: 0.7–1.2 mg/mL Plant solvent extract: 12–45 μg/mL
ABTS	Essential oil: 15–18 μg TE/mg Plant solvent extract: 131–242 μg TE/mg
FRAP	Plant solvent extract: 292 μg TE/mg
CUPRAC	Plant solvent extract: 431 μg TE/mg

DPPH 2,2′-Diphenyl-1-picrylhydrazil radical assay, *FRAP* Ferric reducing ability of plasma assay, *CUPRAC* Cupric ion reducing capability assay, *TE* Trolox equivalent

Essential Oil The essential oil extracted from *P. graveolens* contains major components such as citronellol, geraniol, and linalool, all of which exhibit strong antioxidant properties (Jagdale et al., 2015; Prasad & Muralidhara, 2017; Jabir et al., 2018; El Azab et al., 2022).

Other Phytochemicals In addition to phenolics and essential oil, *P. graveolens* contains alkaloids, and other phytochemicals that further enhance its antioxidant capacity (Cavar & Maksimovic, 2012).

These findings imply that *P. graveolens* could be a valuable source of natural antioxidants in various fields such as skincare and cosmetics where the antioxidant capacity alongside the antibacterial potential of *P. graveolens* make it of a great impact in skincare formulations aimed at combating oxidative stress and aging (Dinu et al., 2024).

Due to its natural antioxidant properties, extracts from *P. graveolens* are being explored as natural preservatives to prevent oxidative degradation in food products, extending their shelf life and maintaining nutritional quality (Kačániová et al., 2023).

2.5.4 Anti-Inflammatory Activity

The anti-inflammatory properties of rose-scented geranium are another area of significant research interest. Inflammation is a critical component of many diseases, including arthritis, cardiovascular diseases, and cancer. Numerous studies have explored the anti-inflammatory effects of *P. graveolens* in both in vitro and in vivo settings. In vivo, animal models of inflammation, such as carrageenan-induced paw edema, acetic acid-induced colitis (Ghanizadeh et al., 2015; Bastani et al., 2019), croton oil-induced ear edema, and histological examination of cutaneous inflammation (Boukhatem et al., 2013), have shown significant reductions in inflammation when treated with *P. graveolens* extract and its steam-distilled essential oil.

The essential oil components have been studied for their anti-inflammatory effects and have shown significant results in various in vitro and in vivo models. Citronellol and geraniol, the predominant components, are particularly noted for their ability to inhibit pro-inflammatory cytokines and enzymes such as COX and LOX, which play critical roles in the inflammatory pathway (Karadağ et al., 2023).

Although some studies reported the selectivity of geranium oil to COX-1 than COX-2, its main constituent, geraniol, has been stated to be a selective inhibitor of COX-2 (Karadağ et al., 2023). Computational studies revealed that geraniol interacts with the main residues of the active sites of both COX-1 and COX-2 enzymes similarly to non-steroidal anti-inflammatory drugs (Karadağ et al., 2023).

Studies have demonstrated that *P. graveolens* essential oil prepared via steam distillation can significantly inhibit the production of NO (a critical inflammation mediator) (70% reduction of induced NO by 20 μg/mL geranium oil) and reduce the expression of proinflammatory enzymes including iNOS and COX-2 in activated microglial cells (immune cells in brain, activation of which is involved in neurodegenerative diseases) (Elmann et al., 2010).

The enzymatic inhibition of COX and LOX pathways is another key mechanism through which *P. graveolens* exerts its anti-inflammatory effects. These enzymes are responsible for the synthesis of prostaglandins and leukotrienes, which are potent mediators of inflammation. By inhibiting these enzymes, *P. graveolens* reduces the production of these inflammatory mediators, thereby alleviating the inflammatory response (Karadağ et al., 2023).

These findings support the potential of *P. graveolens* as a natural anti-inflammatory agent that can be utilized in the development of therapeutic interventions. The translation of these findings into clinical applications requires further investigation to determine the efficacy, safety, and appropriate dosages of *P. graveolens* essential oil in human subjects. Nevertheless, its use in traditional medicine and as a complementary therapy for inflammatory conditions suggests a favorable safety profile and potential benefits.

2.5.4.1 Clinical Applications

While preclinical studies provide promising evidence of the anti-inflammatory effects of *P. graveolens*, clinical studies are still limited. A clinical study reported the effectiveness of geranium oil in the form of topical gel formulation in the management of denture stomatitis and in reducing the local inflammation (Sabzghabaee et al., 2011).

2.5.5 *Anticancer Activity*

The plant essential oil and extracts have been investigated for their ability to inhibit cancer cell growth and induce apoptosis (Table 2.4). Several in vitro studies have demonstrated that *P. graveolens* essential oil (by hydro-distillation) exhibits cytotoxic effects against various cancer cell lines, including breast, lung, prostate, gastric, liver, colon, and melanoma cell lines (Paltiyan et al., 2017; Boukhatem et al., 2021; Fahmy et al., 2023).

The anticancer effects are largely attributed to the presence of compounds like geraniol and citronellol, which have been shown to interfere with cancer cell proliferation and promote cell cycle arrest. Geraniol induces apoptosis via downregulation of mRNA expressions of apoptotic mediators, and it also upregulates mRNA levels of caspase-3 and Bax and reduces JAK2, STAT3, and ERK (Yang et al., 2022), while citronellol induces apoptosis through upregulation of RIP1/RIP3 and TNF-α and downregulation of caspase3 and caspase 8 (Yu et al., 2019).

Table 2.4 Anticancer activity of *Pelargonium graveolens* essential oil against different cancer cell lines (Boukhatem et al., 2021; Fahmy et al., 2023)

Cancer type	Anticancer activity of *P. graveolens* (% inhibition (at dose of 4 μL/mL) or IC_{50} (μg/mL)
Gastric adenocarcinoma AGS	92.87%
Melanoma cell line MV3	88.76%
Hepatocellular carcinoma HEPg2	6.2 μg/mL
Prostatic adenocarcinoma PC3	16.8 μg/mL
Pulmonary adenocarcinoma A549	13.3 μg/mL
Colon cancer cell line HCT116	25.2 μg/mL
Breast cancer cell line MCF7	61.9 μg/mL

2.5.6 Antidiabetic Activity

Several animal studies have demonstrated the antidiabetic effect of *P. graveolens*. Studies reported that both the solvent extract and essential oil of the plant prepared by hydro-distillation exert antihyperglycemic action on induced diabetic rats and protective effect from diabetic complications due to oxidative stress (Boukhris et al., 2012; Hajri et al., 2016; Al-Hamdani & Ali, 2021; Fadwa et al., 2022).

The antidiabetic activity of *P. graveolens* can be attributed to several mechanisms:

1. *Enhancement of insulin sensitivity.* In a study investigating the effect of geranium oil on high fat-high sucrose diet-induced diabetes and associated complications, it was found that the oil reduced blood glucose levels and ameliorated insulin resistance (Kshirsagar et al., 2015).
2. *Modulation of carbohydrate metabolism.* α-Glucosidase and α-amylase enzymes are located in gastrointestinal tract for breakdown of disaccharides and polysaccharides. Their inhibition reduces the sudden rise of blood glucose levels after meals (PPHG) controlling which is one of the strategies in the management of type 2 diabetes. *P. graveolens* extract and essential oil inhibit α-glucosidase (IC_{50}: 93.7 μg/mL) and α-amylase (IC_{50}: 29.5 μg/mL) enzymes with efficacy comparable to that of acarbose (Afifi et al., 2013; Ahamad & Uthirapathy, 2021; Jaradat et al., 2022; Rahhal et al., 2024).

2.5.7 Wound Healing and Skin Health

P. graveolens has long been used in traditional medicine for its beneficial effects on skin health and wound healing. Recent studies have validated these traditional uses, highlighting the plant's potential in promoting tissue repair and enhancing skin regeneration. *P. graveolens* essential oil has been shown to accelerate the healing of wounds suggesting its potential for use in dermatological products aimed at treating cuts, burns, and other skin injuries. The hydroalcoholic extract of *P. graveolens* in the form of lipogel formulation revealed wound healing properties in animal models (Asgarirad et al., 2023), while the essential oil exhibited wound healing potential in the management of diabetic foot ulcer in rats in combination with *Oliveria decombens* essential oil (Mahboubi et al., 2018).

The combined antimicrobial and anti-inflammatory properties of *P. graveolens* make it an excellent natural antiseptic, useful in preventing wound infections and reducing inflammation. Its application in skin care may extend to treating acne, eczema, and other inflammatory skin conditions.

P. graveolens essential oil has a sebostatic activity (inhibition of sebaceous gland activity) for long periods of time that may reach two hours after application to the skin (Kozlowska et al., 2017).

2.5.8 Hepatoprotective Activity

Researchers reported the hepatoprotective effect of *P. graveolens* extract against CCl_4-induced liver damage in mice where it inhibited the induced increase in oxidative stress markers; meanwhile, it increased the activity of antioxidant enzymes. Moreover, it ameliorated the infiltration of inflamed cells and the hydropic degeneration. It revealed reduction in fatty changes and necrosis induced by CCl_4 in a comparable manner to silymarin (Al-Sayed et al., 2015).

2.5.9 Aromatherapy and Psychological Effects

Beyond its physical health benefits, *P. graveolens* is widely used in aromatherapy for its calming and mood-enhancing effects. The pleasant aroma of the essential oil is believed to have anxiolytic and antidepressant properties.

- In a clinical study investigated the effect of *P. graveolens* essential oil on the fatigue and sleep quality of nurses working during the COVID-19 pandemic in the ICU, it was found that aromatherapy with the oil decreased the nurses' fatigue (Karimi et al., 2024).
- Clinical studies reported the effectiveness of aromatherapy with the oil in reducing postoperative pain in hospitalized children (Sirousfard et al., 2014), reducing anxiety during labor (Fakari et al., 2015), and ameliorating pain, anxiety, and stress in lumbar and spinal stenosis patients (Seo et al., 2023).
- Furthermore, aromatherapy massage with the oil (2% in almond oil) decreased physical and mental symptoms of premenstrual syndrome (Lotfipur-Rafsanjani et al., 2018).

Continued research into the plant's phytochemical properties and mechanisms of action will further elucidate its potential applications in medicine, cosmetics, and aromatherapy. As interest in natural and alternative therapies grows, rose-scented geranium is poised to play an increasingly important role in integrative health practices.

2.6 Future Prospects for Industrial Applications of *P. graveolens* Essential Oil

The remarkable bioactive profile of geranium essential oil presents a multifaceted opportunity for industrial innovation across several critical sectors, including natural antimicrobial agents, anti-inflammatory products with reduced side effects, and complementary oncology support formulations. Advanced cosmeceutical technologies could be utilized for introducing geranium oil in wound healing, tissue regeneration, and skin care products. Moreover, it can be utilized in providing agricultural and crop protection biological solutions for agricultural challenges and in food and nutritional development applications such as natural food preservatives, antioxidant-rich formulations, or natural flavor enhancement techniques.

References

Abd El-Kareem, S. M., et al. (2020). Mass spectral fragmentation of *Pelargonium graveolens* essential oil using GC–MS semi-empirical calculations and biological potential. *PRO, 8*(2), 128.

Abd El-Wahab, M., et al. (2016). Effect of different planting locations in Egypt on volatile oil of geranium (*Pelargonium graveolens* L.) plant. *Journal of Basic and Applied Research in Biomedicine, 2*(4), 522–533.

Afifi, F., et al. (2013). Chemical composition and in vitro studies of the essential oil and aqueous extract of *Pelargonium graveolens* growing in Jordan for hypoglycaemic and hypolipidemic properties. *European Journal of Medicinal Plants, 4*(2), 220–233.

Ahamad, J., & Uthirapathy, S. (2021). Chemical characterization and antidiabetic activity of essential oils from *Pelargonium graveolens* leaves. *Aro - The Scientific Journal of Koya University, 9*(1), 109–113.

Ahmadinejad, F., et al. (2017). Molecular mechanisms behind free radical scavengers function against oxidative stress. *Antioxidants, 6*(3), 51.

Al-Hamdani, M. M., & Ali, S. A.-J. (2021). Antidiabetic property of ethanol extract of *Pelargonium graveolens* with constituents chemical screening. *Journal of Clinical Biochemistry and Nutrition, 69*(2), 197–207.

Ali, I. B. E., et al. (2020). Bioactive compounds from Tunisian *Pelargonium graveolens* (L'Hér.) essential oils and extracts: α-amylase and acethylcholinesterase inhibitory and antioxidant, antibacterial and phytotoxic activities. *Industrial Crops and Products, 158*, 112951.

Al-Mijalli, S. H., et al. (2022). Chemical profiling and biological activities of *Pelargonium graveolens* essential oils at three different phenological stages. *Plants, 11*(17), 2226.

Al-Sayed, E., et al. (2015). Protective effect of *Pelargonium graveolens* against carbon tetrachloride-induced hepatotoxicity in mice and characterization of its bioactive constituents by HPLC–PDA–ESI–MS/MS analysis. *Medicinal Chemistry Research, 24*, 1438–1448.

Asgarirad, H., et al. (2023). Wound healing properties of *Pelargonium graveolens* L'Hér extract lipogel: *In-vivo* evaluation in an animal burn model. *Current Drug Delivery, 20*(5), 601–607.

Asgarpanah, J., & Ramezanloo, F. (2015). An overview on phytopharmacology of *Pelargonium graveolens* L. *Indian Journal of Traditional Knowledge, 14*(4), 558–563.

Aydınlık, P., & Yüceer, Y. K. (2021). Characteristic some properties of *Pelargonium graveolens* (rose-scented geranium) essential oil obtained by microwave-assisted hydrodistilation method. *Gida the Journal of Food, 46*(5), 1117–1131.

Babu, K. G., & Kaul, V. (2005). Variation in essential oil composition of rose-scented geranium (*Pelargonium* sp.) distilled by different distillation techniques. *Flavour and Fragrance Journal, 20*(2), 222–231.

Bastani, M., et al. (2019). Biochemical and histopathological evidence for beneficial effects of *Pelargonium graveolens* essential oil on the rat model of inflammatory bowel disease. *Research Journal of Pharmacognosy, 6*(2), 77–84.

Ben Slima, A., et al. (2013). Antioxidant properties of *Pelargonium graveolens* L'Her essential oil on the reproductive damage induced by deltamethrin in mice as compared to alpha-tocopherol. *Lipids in Health and Disease, 12*, 1–9.

Bergman, M. E., et al. (2019). Distinct metabolic pathways drive monoterpenoid biosynthesis in a natural population of *Pelargonium graveolens* (rose scented geranium). *Journal of Experimental Botany, 71*(1), 258–271.

Blerot, B., et al. (2018). Functional analysis of four terpene synthases in rose-scented *Pelargonium* cultivars (*Pelargonium*× *hybridum*) and evolution of scent in the *Pelargonium* genus. *Frontiers in Plant Science, 9*, 1435.

Boukhatem, M. N., et al. (2013). Rose geranium essential oil as a source of new and safe anti-inflammatory drugs. *Libyan Journal of Medicine, 8*(1), 1–7.

Boukhatem, M., et al. (2021). Rose-scented geranium essential oil from Algeria (*Pelargonium graveolens* L'Hérit.): Assessment of antioxidant, anti-inflammatory and anticancer properties against different metastatic cancer cell lines. *Annales Pharmaceutiques Françaises, 80*, 383.

Boukhris, M., et al. (2012). Hypoglycemic and antioxidant effects of leaf essential oil of *Pelargonium graveolens* L'Hér. in alloxan induced diabetic rats. *Lipids in Health and Disease, 11*, 1–10.

Boukhris, M., et al. (2013). Chemical composition and biological activities of polar extracts and essential oil of rose-scented geranium, *Pelargonium graveolens*. *Phytotherapy Research, 27*(8), 1206–1213.

Boztaş, K., et al. (2024). Microwave hydrodistillation of *Pelargonium graveolens* L'Her leaves: Essential oil profile, phytochemical composition of wastewater, histo-anatomical structure. *Journal of Essential Oil-Bearing Plants, 27*(3), 659–677.

Burt, S. (2004). Essential oils: Their antibacterial properties and potential applications in foods—A review. *International Journal of Food Microbiology, 94*(3), 223–253.

Ćavar, S., & Maksimović, M. (2012). Antioxidant activity of essential oil and aqueous extract of *Pelargonium graveolens* L'Her. *Food Control, 23*(1), 263–267.

Dimitrova, M., et al. (2015). Phenolic profile, antibacterial and antioxidant activity of *Pelargonium graveolens* leaves' extracts. *Scientific Bulletin, 21*, 130–135.

Dinu, M., et al. (2024). Natural sources of therapeutic agents used in skin conditions. *Life, 14*(4), 492.

Dorman, H. D., & Deans, S. G. (2000). Antimicrobial agents from plants: Antibacterial activity of plant volatile oils. *Journal of Applied Microbiology, 88*(2), 308–316.

Dzamic, A. M., et al. (2014). Chemical composition, antifungal and antioxidant activity of *Pelargonium graveolens* essential oil. *Journal of Applied Pharmaceutical Science, 4*(3), 1–5.

Eiasu, B. K., et al. (2009). Rose-scented geranium (*Pelargonium capitatum*× *P. radens*) growth and essential oil yield response to different soil water depletion regimes. *Agricultural Water Management, 96*(6), 991–1000.

El Aanachi, S., et al. (2020). Phenolic contents and in vitro investigation of the antioxidant, enzyme inhibitory, photoprotective, and antimicrobial effects of the organic extracts of *Pelargonium graveolens* growing in Morocco. *Biocatalysis and Agricultural Biotechnology, 29*, 101819.

El Azab, E. F., et al. (2022). New insights into geraniol's antihemolytic, anti-inflammatory, antioxidant, and anticoagulant potentials using a combined biological and in silico screening strategy. *Inflammopharmacology, 30*(5), 1811–1833.

Elmann, A., et al. (2010). Anti-neuroinflammatory effects of geranium oil in microglial cells. *Journal of Functional Foods, 2*(1), 17–22.

EL-Otmani, N., et al. (2024). Assessment of *Pelargonium graveolens* flower essential oil: Antimicrobial, antioxidant, enzyme inhibition and in vivo topical analgesic and anti-inflammatory efficacy as treatment for atopic dermatitis. *F1000Research, 13*, 1366.

El-Sonbaty, A. E., et al. (2022). Enhancement of rose scented geranium plant growth, secondary metabolites, and essential oil components through foliar applications of iron (nano, sulfur and chelate) in alkaline soils. *Agronomy, 12*(9), 2164.

Ennaifer, M., et al. (2020). Phytochemicals, antioxidant, anti-acetyl-cholinesterase, and antimicrobial activities of decoction and infusion of *Pelargonium graveolens*. *Natural Product Research, 34*(18), 2634–2638.

Fadwa, E.-O., et al. (2022). Effect of *Pelargonium graveolens* on glucose metabolism in streptozotocin-induced diabetic rats. *Cardiovascular & Hematological Disorders, 22*(1), 18–26.

Fahmy, M. A., et al. (2023). Essential oil from *Pelargonium graveolens* shows anticancer activity and ameliorates the DNA damage and genotoxicity of doxorubicin in mice. *Egyptian Journal of Chemistry, 66*(2), 543–552.

Fakari, F. R., et al. (2015). Effect of inhalation of aroma of geranium essence on anxiety and physiological parameters during first stage of labor in nulliparous women: A randomized clinical trial. *Journal of Caring Sciences, 4*(2), 135.

Fekri, N., et al. (2021). Studies on essential oil from rose-scented geranium, *Pelargonium graveolens* L'Hérit.(Geraniaceae). *Natural Product Research, 35*(15), 2593–2597.

Gevrenova, R., et al. (2024). *Pelargonium graveolens*: Towards in-depth metabolite profiling, antioxidant and enzyme-inhibitory potential. *Plants, 13*(18), 2612.

Ghanizadeh, B., et al. (2015). The anti-inflammatory effect of *Pelargonium graveolens* methanolic extract in acetic acid-induced ulcerative colitis in rat model. *Clinical Therapeutics, 37*(8), e51.

Ghannadi, A., et al. (2012). Antibacterial activity and composition of essential oils from *Pelargonium graveolens* L'Her and *Vitex agnus-castus* L. *Iranian Journal of Microbiology, 4*(4), 171.

Gilles, M., et al. (2010). Chemical composition and antimicrobial properties of essential oils of three Australian *Eucalyptus* species. *Food Chemistry, 119*(2), 731–737.

Gomes, P. B., et al. (2004). Characterization of Portuguese-grown geranium oil (*Pelargonium* sp.). *Journal of Essential Oil Research, 16*(5), 490–495.

Gomes, P. B., et al. (2007). Production of rose geranium oil using supercritical fluid extraction. *Journal of Supercritical Fluids, 41*(1), 50–60.

Hajri, A., et al. (2016). Protective effect of *Pelargonium graveolens* essential oil against alloxan-induced diabetes and oxidative stress in rats. *Journal of Biologically Active Products from Nature, 6*(4), 299–314.

Hsouna, A. B., & Hamdi, N. (2012). Phytochemical composition and antimicrobial activities of the essential oils and organic extracts from *Pelargonium graveolens* growing in Tunisia. *Lipids in Health and Disease, 11*, 1–7.

Ibrahim, M. A., et al. (2021). Potentiation of anti-*Helicobacter pylori* activity of clarithromycin by *Pelargonium graveolens* oil. *Arab Journal of Gastroenterology, 22*(3), 224–228.

Inouye, S., et al. (2001). Antibacterial activity of essential oils and their major constituents against respiratory tract pathogens by gaseous contact. *The Journal of Antimicrobial Chemotherapy, 47*(5), 565–573.

Jabir, M. S., et al. (2018). Antioxidant activity of Linalool. *Engineering And Technology Journal, 36*(1), 64–67.

Jagdale, A. D., et al. (2015). Citronellol: A potential antioxidant and aldose reductase inhibitor from *Cymbopogon citratus*. *International Journal of Pharmacy and Pharmaceutical Sciences, 7*, 203–209.

Jaradat, N., et al. (2022). Chemical markers and pharmacological characters of *Pelargonium graveolens* essential oil from Palestine. *Molecules, 27*(17), 5721.

Justin, K., et al. (2013). Chemical composition and antimicrobial effect of the essential oil of *Pelargonium graveolens* (Geranium Rosat) grown in Butare (Rwanda) towards formulation of plant-based antibiotics. *Journal of Microbiology, 3*(2), 87–91.

Kačániová, M., et al. (2023). An in-depth study on the chemical composition and biological effects of *Pelargonium graveolens* essential oil. *Food, 13*(1), 33.

Kafa, A. H. T., et al. (2022). Antimicrobial synergism and antibiofilm activities of *Pelargonium graveolens, Rosemary officinalis*, and *Mentha piperita* essential oils against extreme drug-resistant *Acinetobacter baumannii* clinical isolates. *Zeitschrift für Naturforschung. Section C, 77*(3–4), 95–104.

Karadağ, A. E., et al. (2023). Comparative in vitro and in silico enzyme inhibitory screening of *Rosa x damascena* and *Pelargonium graveolens* essential oils and geraniol. *Plants, 12*(18), 3296.

Kardan-Yamchi, J., et al. (2020). The chemical composition and anti-mycobacterial activities of *Trachyspermum copticum* and *Pelargonium graveolens* essential oils. *Recent Patents on Anti-Infective Drug Discovery, 15*(1), 68–74.

Karimi, N., et al. (2024). The effect of aromatherapy with *Pelargonium graveolens* (*P. graveolens*) on the fatigue and sleep quality of critical care nurses during the COVID-19 pandemic: A randomized controlled trial. *Explore, 20*(1), 82–88.

Kassahun, B. M., et al. (2012). Yield and yield components of rose scented Geranium (*Pelargonium graveolens*) as influenced by plant population density in Ethiopia. *International Journal of Medicinal and Aromatic Plants, 2*(1), 60–68.

Khetsha, Z. (2024). Using cytokinin to enhance essential oil biosynthesis of two rose geranium (*Pelargonium graveolens* L.) varieties: Reunion and Madagascar-type. *Scientific Papers. Series B. Horticulture, 68*(1), 786–805.

Khetsha, Z., et al. (2024). Comparative effects of two elicitors on the essential oil biosynthesis of simulated hail-damaged rose geranium (*Pelargonium graveolens* L'Hér. Cv.'Bourbon'). *African Journal of Food, Agriculture, Nutrition and Development, 24*(5), 26441–26463.

Kozlowska, J., et al. (2017). Evaluation of sebostatic activity of *Juniperus communis* fruit oil and *Pelargonium graveolens* oil compared to niacinamide. *Cosmetics, 4*(3), 36.

Kshirsagar, R. P., et al. (2015). Geranium oil ameliorates endothelial dysfunction in high fat high sucrose diet induced metabolic complications in rats. *Journal of Functional Foods, 15*, 284–293.

Kujur, A., et al. (2020). Antifungal and aflatoxin B1 inhibitory efficacy of nanoencapsulated *Pelargonium graveolens* L. essential oil and its mode of action. *Lwt, 130*, 109619.

Latté, K. P., & Kolodziej, H. (2004). Antioxidant properties of phenolic compounds from *Pelargonium reniforme*. *Journal of Agricultural and Food Chemistry, 52*(15), 4899–4902.

Li, W., et al. (2024). Flavonoids from the aerial part of *Pelargonium graveolens* and their antoxidant activity. *Chemistry of Natural Compounds, 60*(4), 729–731.

Lis-Balchin, M. (2002). Geranium essential oil: Standardisation, ISO; adulteration and its detection using GC, enantiomeric columns, toxicity and bioactivity. In *Geranium and Pelargonium* (pp. 196–204). CRC Press.

Lorenzi, V., et al. (2009). Geraniol restores antibiotic activities against multidrug-resistant isolates from Gram-negative species. *Antimicrobial Agents and Chemotherapy, 53*(5), 2209–2211.

Lotfipur-Rafsanjani, S. M., et al. (2018). Effects of geranium aromatherapy massage on premenstrual syndrome: A clinical trial. *International Journal of Preventive Medicine, 9*(1), 98.

Lothe, N. B., & Verma, R. K. (2023). A study on geranium (*Pelargonium graveolens* L′ Herit ex Aiton) cultivars' productivity and economics as intervening by diverse climatic conditions of the western peninsular region of India. *Industrial Crops and Products, 200*, 116882.

Mahboubi, M., & Valian, M. (2019). Anti-dermatophyte activity of *Pelargonium graveolens* essential oils against dermatophytes. *Clinical Phytoscience, 5*, 1–5.

Mahboubi, M., et al. (2011). Antimicrobial activity of *Pelargonium graveolens* and *Oliveria decumbens* extracts against clinical isolates of *Staphylococcus aureus*. *Journal of Biologically Active Products from Nature, 1*(2), 105–111.

Mahboubi, M., et al. (2018). The wound healing effects of herbal cream containing *Oliveria decumbens* and *Pelargonium graveolens* essential oils in diabetic foot ulcer model. *World Journal of Plastic Surgery, 7*(1), 45.

Mahdizadeh, E., & Tabar, R. H. (2018). The anti-candidal activity of *Pelargonium graveolens* essential oils against clinical isolates of *Candida albicans*. *Infection, 22*(1), 9–12.
Malik, T., et al. (2011). Potentiation of antimicrobial activity of ciprofloxacin by *Pelargonium graveolens* essential oil against selected uropathogens. *Phytotherapy Research, 25*(8), 1225–1228.
Mazeed, A., et al. (2023). Enhancing productivity, quality, and economics of rose scented geranium (*Pelargonium graveolens* L.) through a novel integrated approach to phosphorus application. *Industrial Crops and Products, 204*, 117293.
M'hamdi, Z., et al. (2024). Chemical composition and antibacterial activity of essential oil of *Pelargonium graveolens* and its fractions. *Arabian Journal of Chemistry, 17*(1), 105375.
Mnif, W., et al. (2011). Characterization of leaves essential oil of *Pelargonium graveolens* originating from Tunisia: Chemical composition, antioxidant and biological activities. *Journal of Essential Oil-Bearing Plants, 14*(6), 761–769.
Motsa, N. M., et al. (2006). Plant shoot age and temperature effects on essential oil yield and oil composition of rose-scented geranium (*Pelargonium* sp.) grown in South Africa. *Journal of Essential Oil Research, 18*(sup1), 106–110.
Moutaouafiq, S., et al. (2019). Antifungal activity of *Pelargonium graveolens* essential oil and its fractions against wood decay fungi. *Journal of Essential Oil-Bearing Plants, 22*(4), 1104–1114.
Narnoliya, L. K., et al. (2017). *De novo* transcriptome analysis of rose-scented geranium provides insights into the metabolic specificity of terpene and tartaric acid biosynthesis. *BMC Genomics, 18*, 1–14.
Narnoliya, L. K., et al. (2019). The phytochemical composition, biological effects and biotechnological approaches to the production of high-value essential oil from geranium. *Essential Oil Research, 9*, 327.
Paltiyan, J. C., et al. (2017). Phytochemical analysis, larvicidal activity and cytotoxic properties of malvarosa *(Pelargonium graveolens*) leaf extract. *International Journal of Agriculture and Technology, 13*, 2281–2295.
Prabuseenivasan, S., et al. (2006). In vitro antibacterial activity of some plant essential oils. *BMC Complementary and Alternative Medicine, 6*, 1–8.
Prasad, S. N., & Muralidhara, M. (2017). Analysis of the antioxidant activity of geraniol employing various in-vitro models: Relevance to neurodegeneration in diabetic neuropathy. *Asian Journal of Pharmaceutical and Clinical Research, 10*(7), 101–105.
Rahhal, B., et al. (2024). Antioxidative and antidiabetic iInvestigations of the aqueous extract of *Pelargonium graveolens* grown in Palestine. *Palestinian Medical and Pharmaceutical Journal, 9*(2), 137–143.
Rao, B. R. (2009). Chemical composition and uses of Indian rose-scented Geranium (*Pelargonium* species) essential oil-a review. *Journal of Essential Oil-Bearing Plants, 12*(4), 381–394.
Ravindra, N., & Kulkarni, R. (2015). Essential oil yield and quality in rose-scented geranium: Variation among clones and plant parts. *Scientia Horticulturae, 184*, 31–35.
Rosato, A., et al. (2008). The inhibition of *Candida* species by selected essential oils and their synergism with amphotericin B. *Phytomedicine, 15*(8), 635–638.
Sabzghabaee, A. M., et al. (2011). Clinical evaluation of the essential oil of *Pelargonium graveolens* for the treatment of denture stomatitis. *Dental Research Journal, 8*(Suppl1), S105–S108.
Saxena, G., et al. (2007). Rose-scented geranium (*Pelargonium* sp.) generated by *Agrobacterium rhizogenes* mediated Ri-insertion for improved essential oil quality. *Plant Cell, Tissue and Organ Culture, 90*, 215–223.
Saxena, D., et al. (2023). Tackling the outer membrane: Facilitating compound entry into Gram-negative bacterial pathogens. *npj Antimicrobials and Resistance, 1*(1), 17.
Sedibe, M., & Allemann, J. (2012). Yield and quality response of rose geranium (*Pelargonium graveolens* L.) to sulphur and phosphorus application. *South African Journal of Plant and Soil, 29*(3–4), 151–156.

Seo, E., et al. (2023). Inhalation of *Pelargonium graveolens* essential oil alleviates pain and related anxiety and stress in patients with lumbar spinal stenosis and moderate to severe pain. *Pharmaceuticals, 17*(1), 1–12.

Sharma, V. (2013). Comparative studies on essential oil compositions of rose scented geranium "*Pelargonium graveolens*" from different higher altitude ranges of northIndian Himalayas. *International Journal of Recent Scientific Research, 4*(6), 742–746.

Singh, P., et al. (2008). Assessment of *Pelargonium graveolens* oil as plant-based antimicrobial and aflatoxin suppressor in food preservation. *Journal of the Science of Food and Agriculture, 88*(14), 2421–2425.

Sirousfard, M., et al. (2014). The effect of aromatherapy with Pelargonium graveolens on postoperative pain intensity in hospitalized children: A clinical randomized trial. *Journal of Clinical Nursing and Midwifery, 3*(2), 67–76.

Szutt, A., et al. (2020). Evaluation of chemical composition of essential oils derived from different species leaves. *Ecological Chemistry and Engineering S, 26*(4), 807–816.

Torres Neto, L., et al. (2022). Essential oils block cellular entry of SARS-CoV-2 delta variant. *Scientific Reports, 12*(1), 20639.

Verma, R. S., et al. (2013). Essential oil composition of *Pelargonium graveolens* L'Her ex Ait. cultivars harvested in different seasons. *Journal of Essential Oil Research, 25*(5), 372–379.

Vimalanathan, S., & Hudson, J. (2014). Anti-influenza virus activity of essential oils and vapors. *American Journal of Essential Oils and Natural Products, 2*(1), 47–53.

Wafa, S. S. A. E., et al. (2023). Optimization of oil yield of *Pelargonium graveolens* L'Hér using Box-Behnken design in relation to its antimicrobial activity and *in silico* study. *Scientific Reports, 13*(1), 19887.

Yang, H., et al. (2022). Apoptosis-Mediated anticancer activity of geraniol inhibits NF-κB, MAPK, and JAK-STAT-3 signaling pathways in human thyroid cancer cells. *Pharmacognosy Magazine, 18*(80), 1183–1189.

Yu, W.-N., et al. (2019). Citronellol induces necroptosis of human lung cancer cells via TNF-α pathway and reactive oxygen species accumulation. *In Vivo, 33*(4), 1193–1201.

Zore, G. B., et al. (2011). Evaluation of anti-Candida potential of geranium oil constituents against clinical isolates of *Candida albicans* differentially sensitive to fluconazole: Inhibition of growth, dimorphism and sensitization. *Mycoses, 54*(4), e99–e109.

Chapter 3
Fruits and Vegetables Used in Treating Respiratory and Other Troubles in Africa: Ethnopharmacology and Phytochemistry

Ivan Gumula, Daniel Buyinza, Patrick Onen, and Christine Kyarimpa

Abstract Respiratory troubles, which encompass conditions such as cough, asthma, tuberculosis, bronchitis, common cold, and chronic obstructive pulmonary disease, pose significant health challenges globally. This chapter explores the diverse array of biologically active agents (BAA) found in vegetables and fruit plants used not only as sources of food but also for treatment of respiratory troubles in Africa. A total of 38 vegetables and fruit plants from 22 families were encountered. The plants are sources of antimicrobials that help the lungs to resist infection as well as boost the immunity of the host; expectorants which stimulate the coughing up of mucus; anti-inflammatory agents and demulcents which soothe irritated membranes, while others are spasmolytic which relax the bronchial muscles. The BAA include allicin from garlic, quercetin from onions, theobromine from cacao, and anethole from fennel. BAA, such as bromelain in pineapple, beta-carotene in carrots, vitamin C in lemon, and glucosinolates in radish, play pivotal roles in mucolytic activities and enhancement of immune function as well as repair of the bronchial membrane. Some of the vegetables/fruits are a source of mineral salts that are vital to the integrity and functioning of lungs. Thus, *Cleome gynandra* contains selenium, a key component that inhibits HIV-associated tuberculosis. The available research evidence

I. Gumula (✉) · C. Kyarimpa
Department of Chemistry, Faculty of Science, Kyambogo University, Kyambogo-Kampala, Uganda
e-mail: igumula@kyu.ac.ug; ckyarimpa@kyu.ac.ug

D. Buyinza
Department of Chemistry, Faculty of Science, Kabale University, Kyambogo-Kampala, Uganda
e-mail: dbuyinza@kab.ac.ug

P. Onen
Department of Chemistry, Faculty of Science, Kyambogo University, Kyambogo-Kampala, Uganda

Department of Chemistry, University of Kerala, Karyavattom Campus, Thiruvananthapuram, Kerala, India
e-mail: onenpatrick@keralauniversity.ac.in

A. Dekebo (ed.), *Bioactive Secondary Metabolites from Medicinal Plants of Africa*, Interdisciplinary Biotechnological Advances,
https://doi.org/10.1007/978-981-96-8590-5_3

underscores the importance of integrating the BAA into human daily diets as a prophylactic approach and therapeutic strategy against respiratory troubles.

Keywords Cough · Asthma · Bronchitis · Flavonoids · Acetogenins · Limonoids · Anti-inflammatory · Antimicrobial · Antineoplastic

3.1 Introduction

Respiratory disorders such as bronchitis, pneumonia, asthma, and obstructive pulmonary disease represent a significant global health burden, particularly in low-resource settings such as Africa and Asia (Labaki & Han, 2020). These conditions are exacerbated by environmental factors, especially polluted air or exposure to biomass smoke. Despite these, most indigenous communities still have limited access to health services in developing countries. Thus, such communities rely on traditional medicine as the basic means of managing respiratory ailments (Mutombo et al., 2023). In African ethnopharmacology, vegetables and fruit plants occupy a central role, offering a rich source of remedies with both nutritional and therapeutic benefits.

Africa's vast biodiversity provides a wealth of plant species traditionally used to alleviate respiratory troubles. Many of these plants are not only consumed as part of the diet but are also integral to indigenous healing systems. Ethnobotanical studies have documented a variety of vegetables and fruit plants with antitussive, anti-inflammatory, expectorant, and antimicrobial properties (Parham et al., 2020). These plants are often administered as decoctions, infusions, or poultices and are valued for their efficacy, accessibility, and cultural significance.

The therapeutic potential of the plants can be largely ascribed to their phytochemical constituents. Flavonoids, alkaloids, terpenoids, phenolic derivatives, and saponins have been identified in many species, exhibiting activities that align with the management of respiratory conditions. Moreover, ongoing scientific research has begun to make contributions toward validating the traditional uses of the plants, uncovering relevant phytochemicals, their mechanisms of action, and potential applications for modern drug development (Nasim et al., 2022).

This chapter explored the ethnopharmacological significance and phytochemical profiles of vegetables and fruit plants used in managing respiratory disorders across Africa. It highlights the interplay between traditional knowledge and modern pharmacology, offering insights into how these natural resources can contribute to addressing respiratory health challenges. By bridging traditional practices with scientific evidence, this chapter underscores the potential of African flora in global health contexts while advocating for sustainability in the utilization of medicinal plants and conservation of biodiversity.

3.2 Literature Search Strategy

We retrieved literature on edible vegetables and fruit plants used for treating respiratory troubles from published articles in multidisciplinary databases (Scopus, Science Direct, Web of Science, and Google Scholar) dated until March 13, 2025. Information of interest was limited to use in the treatment of respiratory disorders plus any other ethnopharmacological use in Africa. Data on the plant part (not necessarily the edible part) used in the management of respiratory diseases was also captured. In addition, phytochemical and pharmacological data on the featured plant materials included information sourced from Africa and beyond. Other traditional uses in addition to management of respiratory disorders were also included in the review. Key words/phrases used in the searches included: vegetables and fruits used against cough, asthma, tuberculosis, bronchitis, pneumonia, chest pain, and/or flu; antitussive vegetables and fruits; vegetables and fruits used against influenza; plants used for treating respiratory diseases.

3.3 Fruit Plants and Vegetables Used in Treating Respiratory Problems

The plants were categorized into five sub-sections as those that are normally consumed in forms of (1) fruits, (2) seeds, berries or pods, (3) leafy vegetables, (4) bulbs, and (5) roots.

3.3.1 Fruits

Photos of some fruits commonly consumed are shown in Fig. 3.1.

3.3.1.1 Pineapple (*Ananas comosus* (L.) Merr)

Ananas comosus (Bromeliaceae) (Fig. 3.1A) is commonly called pineapple and is a herbaceous, biennial crop growing up to 1.0–1.5 m high and is widely distributed in Africa and Asia (Spengler, 2019). It has fleshy fruits, edible especially when ripe, with the pulp that is white to yellow in color. The unripe fruit is macerated and prepared as a single-component juice for the treatment of cough associated with respiratory diseases in Nigeria (Lawal et al., 2020). It is effective in clearing bronchial passages in cases of bronchitis, colds, pneumonia, and sinusitis (Abubakar et al., 2022). The fruit is also indicated in cases of indigestion. In addition to respiratory diseases, the fruit/juice is also used to treat hypertension, type II diabetes, overweight, and constipation, in India (Kandasamy & Shanmugapriya, 2015;

Fig. 3.1 Fruits of (**A**) *Ananas comosus*, (**B**) *Annona muricata* Linn, (**C**) *Annona senegalensis* Pers, (**D**) *Annona squamosa* L, (**E**) *Annona stenophylla* Engl. & Diels, (**F**) *Carica papaya* L, (**G**) *Citrus limon* (L.) Burm, (**H**) *Citrus reticulata*, (**I**) *Citrus sinensis*, L. Osbeck**,** (**J**) *Cocos nucifera*, (**K**) *Dennettia tripetala* Barker (F.), (**L**) *Ficus carica* L., (**M**) *Hexalobus monopetalus* (A. Rich.) Engl. & Diels, (**N**) *Malus domestica* Borkh, (**P**) *Musa paradiasiaca* L, (**Q**) *Phoenix dactylifera* L. (**R**) *Uvaria chamae* P. Beauv

Shawarb et al., 2023; Aiyeloja & Bello, 2006; Lawin et al., 2015; Magassouba et al., 2007; Emmanuel & Didier, 2012). Tannins, cardenolides, cardiac glycosides, dienolides, and flavonoids are present in the juice. Among the key constituents of *A. comosus* is bromelain (fruit bromelain, **1**), a proteolytic enzyme responsible for softening the mucus, hence facilitating its expulsion, and it also functions as an immune modulator (Wang, 2016). In a murine model experimentation of acute asthma, it was observed that bromelain caused a decrease in the lung inflammation markers (including infiltration by leucocytes and eosinophils) as well as moderation in the local airway immunity aspects (Secor-Jr et al., 2005, 2008). Bromelain has beneficial effects when used in the treatment of sinusitis (Seltzer, 1967; Braun et al., 2005). Patients with sinusitis experiencing breathing difficulties and with inflammation of the nasal mucosa demonstrated complete resolution in their breathing upon administration of bromelain (Zatuchni & Colombi, 1967; Ahle & Hamlet, 1987). Furthermore, it has been documented that bromelain has inhibitory activities against cancer metastasis (Kleef et al., 1996; Gerard, 1972; Rathnavelu et al., 2016).

3.3.1.2 Soursop (*Annona muricata* Linn)

Annona muricata (Annonaceae) (Fig. 3.1B) is distributed throughout the tropics around the world. Its leaf extracts and decoctions from its stem bark and roots are traditionally used in managing colds, flu, asthma, hypertension, epilepsy, fever, headache, cystitis, diabetes, headaches, insomnia, cancer, relaxing nerves, heart pain, hypertension, and mumps (Gavamukulya et al., 2017). The flowers have pectoral and febrifuge properties, and it is used in hot herbal preparations to treat influenza and bronchial catarrh (Emmanuel & Didier, 2012; Suroowan et al., 2019; Konkon et al., 2017; Omara et al., 2020; Moghadamtousi et al., 2015; Kasole et al., 2019; Anywar et al., 2020). The main secondary metabolites in *A. muricata* include acetogenins, terpenoids, alkaloids, saponins, flavonoids, lactones, coumarins, anthraquinones, cardiac glycosides, tannins, phytosterols, and phenols. Some of the isolated compounds include the cytotoxic annonaceous acetogenins muricins J (**2**) and K (**3**), annopentocins A (**4**), muricapentocin (**5**), annomuricin A (**6**), and compounds **7–9**. Argentinine (**9**) displayed significant cytotoxicity against human lung carcinoma A549 cells, human colon cancer HT-29 cells, the murine leukemia P388 cells, and the human KB cells, with ED_{50} values of 33.5, 9.9, 15, and 15.5 μM, respectively (Wu et al., 1993). The annonaceous acetogenin **5** isolated from the leaf extract of the plant exhibited potent cytotoxicity against human lung carcinoma A549 cells with an ED_{50} value of 0.193 μg/ml (Kim et al., 1998).

3.3.1.3 Wild Soursop (*Annona senegalensis* Pers.)

Annona senegalensis (Annonaceae) (Fig. 3.1C) is commonly referred to as wild soursop or African wild custard apple. It is a tree that thrives within the sub-humid and semi-arid parts of Africa (Orwa et al., 2009). Different parts of the tree are, traditionally, used against tuberculosis, inflammations, pain, malaria, convulsions, and trypanosomiasis, and the fruit is used as food/food additive (Okhale et al., 2016; Tabuti et al., 2010; Bunalema et al., 2014). In Ivory Coast pharmacopoeia, the plant is used against asthma and cough (Yeo et al., 2011). The fruit is also used in the treatment of kwashiorkor and marasmus (Okhale et al., 2016). The plant is largely fragrant and a source of essential oils. The ethanol extract from the leaves of the plant collected from Ivory Coast significantly induced reduction in the number of inflammatory cells, which supports the use of the plant in combating respiratory diseases (Yeo et al., 2011). Similarly, the methanol extract from the stem bark of *A. senegalensis* exhibited significant dose and time-dependent analgesic and anti-inflammatory activities in mice at oral doses of 100, 400, and 1000 mg/kg (Suleiman et al., 2014). Some of the compounds identified and isolated from the seeds of *A. senegalensis* include anonaine (**10**), isoboldine (**11**), and epicatechin (**12**). Anonaine and epicatechin are reported to have exhibited strong antioxidant properties, whereas isoboldine is reported to have suppressed the transcription of the pro-inflammatory cytokines IL-1β and IL-6 (Costa et al., 2012; Yilmaz & Toledo, 2004; Chou et al., 2014).

3.3.1.4 Custard Apple (*Annona squamosa* L.)

Annona squamosa (Annonaceae) (Fig. 3.1D) is also native to the tropical regions. It has deciduous leaves and dark brown, quilted fruits. The pulp is reddish yellow, sweet, and very soft, making it a desired ingredient in juices (Tabuti et al., 2010; Syamsuhidayat & Hutapea, 1991; Karou et al., 2011). It is used in treating cough, catarrhal inflammation, dysentery, diabetes, diarrhea, eczema, bladder problems, inflammation rheumatism, and skin eruptions, and is used as an astringent, emetic, tonic, febrifuge, aphrodisiac agent, anti-parasitic, anti-rheumatic, and analgesic and antispasmodic (Anywar et al., 2020; Karou et al., 2011; Nugraha et al., 2019; Dos Santos & Sant'Ana, 2000). In addition to the cytotoxic annonaceous acetogenins as those obtained from other *Annona* species, *A. squamosa* has yielded a wide range of biologically active compounds. Thus, caryophyllene oxide (**13**) and acetoxy-ent-kaur-16-ene (**14**) isolated from the stem bark of the plant showed significant analgesic and anti-inflammatory activities at doses of 12.5 and 25 mg/kg body weight in mice model tests (Chavan et al., 2010, 2011). Furthermore, a cyclic peptide cyclosquamosin D (**15**) isolated from the seeds of *A. squamosa* inhibits the production of pro-inflammatory cytokines (Yang et al., 2008). A number of annonaceous acetogenins have been reported from *A. squamosa* including annosquacin-I (**16**) and uvarigrandin-A (**17**) which exhibited cytotoxicity against human lung carcinoma A-549 cells with IC_{50} values of 0.012 and 0.08 μg/ml, respectively (Chen et al., 2011).

3.3.1.5 Dwarf Custard Apple (*Annona stenophylla* Engl. & Diels)

Annona stenophylla (Fig. 3.1E) is one of the Annonaceae species whose distribution spans from Central Arica, through Eastern Africa, to Southern Africa. The fruits of *A. stenophylla* have pineapple-like flavor and a pleasant taste. The local people eat them in raw form, cooked, or, in some cases, when preserved (Maroyi, 2019). The roots and leaves of the plant and its aerial parts are used as a herbal remedy for cold, bronchitis, asthma, fever, infertility, malaria, skin infections and stomach problems anemia, blood and stomach cleansing, scoliosis abdominal, muscle pains, anemia, malaria, stomach ache and sexually transmitted diseases and to induce (Maroyi, 2019; Dzoyem et al., 2016; Lautenschläger et al., 2018; Maroyi, 2024; Nyagumbo et al., 2022). In Zimbabwe, *A. stenophylla is* used in treating chest pain and pneumonia (Nyagumbo et al., 2022; Maroyi, 2011; Shopo et al., 2022). In Namibia, the roots of *A. stenophylla* are mixed with the roots of *Diospyros chamaethamnus* Mildbr., *Strychnos pungens* Soler., *Diplorhynchus condylocarpon* (Müll. Arg.) Pichon, and *Bobgunnia madagascariensis* (Desv.), and the concoction is used against malaria and psychological disorders, though the plant parts and proportions are not stated (Maroyi, 2019; Shopo et al., 2022). The plant is generally considered to be non-toxic with $LD_{50} > 2000$ mg/kg in rat models (Nyagumbo et al., 2022). The plant has exhibited the following bioactivities: anti-inflammatory, antifungal, antibacterial hypoglycemic, and antioxidant activities (Maroyi, 2019; Nyagumbo et al., 2022). Phytochemical studies aimed at isolating compounds remain scant.

3.3.1.6 Pawpaw (*Carica papaya* L.)

Pawpaw (*Carica papaya* L.) (Fig. 3.1F) is a herbaceous succulent plant and is commonly found in India and South-East Asia, as well as in the tropical and sub-tropical parts of the world (Tan et al., 2022). The latex from the leaves and the unripe fruit/stem of *C. papaya* is employed in the treatment of whooping cough. In Nigeria, fruit juice from *Citrus aurantifolia* is mixed with the extracts from *Allium sativum* and the powdered seeds of *C. papaya,* and the resultant concoction is used to manage tuberculosis, whereas the dried and cured pawpaw leaves are smoked to treat asthma in Uganda (Adesina et al., 2017). *C. papaya* is used to treat asthma in children and other respiratory diseases in the Democratic Republic of Congo and to manage cough in South Africa (Mhlongo & Van Wyk, 2019). The latex and the leaves are used in individuals lacking digestive juice, pancreatitis, and against intestinal parasites, especially tapeworms, in the Mediterranean region. Consumption of pawpaw is also recommended for colitis, irritable bowel syndrome, and constipation (Muss et al., 2013). The latex contains a protease enzyme called papain, which has properties similar to those of bromelain (**1**), found in pineapples. Thus, papain may be responsible for emulsifying the mucus and facilitating its expulsion from the tracheal system. The fruit is rich in vitamins A (**18**) and C (**19**). The leaves of *C. papaya* contain an alkaloid, carpaine (**20**), with cardiovascular, antimalarial, anti-inflammatory, antioxidant, and antimicrobial properties. The compound is reported to have inhibited *Mycobacterium tuberculosis* with an in vitro minimum inhibitory concentration of 10^{-4} M (Ramasawamy & Sirsi, 1960). Other isolated phytochemicals present include loliolide (**21**), nicotiflorin (**22**), and isorhamnetin-3-*O*-*β*-D-glucopyranoside (**23**) (Hamed et al., 2022).

3.3.1.7 Lemon (*Citrus limon* (L.) Burm.)

Lemon *Citrus limon* (L.) Burm.) (Fig. 3.1G) is one of the tree species in the family Rutaceae, subfamily *Aurantioideae.* It is used in the management of many diseases including common colds, cough, obesity, urinary tract infections, gall bladder, kidney stones (renal lithiasis), acne scars, and pimples (Lawal et al., 2020). The fruits and the peels were part of several recipes that featured in the fight against COVID-19 (Mahmood et al., 2019). It is also used in treating circulatory disorders like anemia, arterial hypertension, coronary disease, angina pectoris, arrhythmia, atherosclerosis, thrombosis, edema, and chronic venous insufficiency (Khan et al., 2010). Lemon is rich in vitamin C (**19**), citric acid (**24**), flavonoids like neohesperidin (**25**), and diosmin (**26**), which are excellent immune-boosters and antioxidants. It also contains minerals like iron, potassium, and calcium, making the fruit a good candidate in cases requiring remineralization such as anemia and osteoporosis. The essence from the fruit peels and leaves contains D-limonene (**27**) which is an excellent detoxifier and with antioxidant, antimicrobial, antiviral, antiseptic, and anti-inflammatory properties (Mahmood et al., 2019; Rafique et al., 2020; Yi et al., 2017; Lv et al., 2015). Other compounds identified from lemon include limonoids such as limonin (**28**), which do have a wide range of pharmacological properties including anti-inflammatory, antioxidant, antiviral, anticancer, antibacterial, liver-protective, and analgesic (Fan et al., 2019; Roy & Saraf, 2006).

3.3.1.8 Tangerine (*Citrus reticulata*)

Tangerine (*Citrus reticulata*) (Fig. 3.1H) belongs to the citrus (Rutaceae) family (Shi et al., 2024). Tangerine is a common choice in the food processing and the pharmaceutical manufacturing industry because of the unique flavor it has. Ethnopharmacologically, *C. reticulata* is recommended in conditions such as sore throat, bronchitis, allergic reactions, sinusitis, gastritis, among others (Shi et al., 2024). Like other citrus fruits, it is reputed for high concentrations of biologically active ingredients and having health-promoting properties (John, 2018), and its major active components can be very effective against respiratory ailments and cardiovascular diseases (Shi et al., 2024). Some of the compounds identified from *C. reticulata* fruit juice include vitamin C (**19**) and quercetin (**30**) as well as limonoids (John, 2018). Quercetin is one of the most valuable constituents in fruits and vegetables because of its potent activities such as antioxidant (particularly radical scavenging), immune stimulation, anti-inflammatory, anti-histamine, antiviral, and cytotoxicity. The compound suppresses the release of pro-inflammatory cytokines from mast cells and it inhibits the release of histamine from basophil cells (Yang et al., 2020; Mlcek et al., 2016). Quercetin, further, exhibited antiviral properties against influenza virus H1N1, dengue virus, and Zika virus with IC_{50} values of 7.75, 28.9, and 2.3 μM, respectively (Johari et al., 2012; Zou et al., 2020).

Furthermore, quercetin was among the phytochemicals that exhibited promising activities against COVID-19 with an IC_{50} value of 23.8 μM (Solnier & Fladerer, 2021).

3.3.1.9 Oranges (*Citrus sinensis,* L. Osbeck)

Oranges or sweet orange (*Citrus sinensis,* L. Osbeck) (Fig. 3.1I) belongs to the family *Rutaceae*, subfamily *Aurantioideae.* Its origin is South-East Asia, but it is now grown in all the tropics, subtropic, and temperate regions of the world (Esan et al., 2024). Sweet oranges are important for both fresh markets and processing industries. In addition to their role in food and nutrition, consumption of oranges is associated with health benefits. Having oranges as part of a daily menu is recommended as a preventive measure against allergic reactions like rhinitis, bronchial asthma, flu, colds, and various forms of infections (Turner & Burri, 2013). It is also used, traditionally, in the prevention of risks associated with thrombosis, arteriosclerosis, heart diseases, kidney stones, excess uric acid, osteoporosis, rickets, and cancer (Esan et al., 2024; Dongre et al., 2023). Some of the vital constituents present in oranges include vitamin C (**19**), responsible for the fruit's wide biological activities such as immune-boosting; flavonoids, like rutin (**31**) and sinensetin (**32**), with excellent anti-inflammatory and antioxidant properties; as well as limonoids and glycosides with antitumor effects. Carotenoids are also present in significant amounts as well as minerals like potassium, calcium, magnesium, and phosphorus, hence making the fruit to be suitable in cases of demineralization. The inner peel (white mesocarp) of the fruit is rich in soluble fibers, particularly pectin, that reduce cholesterol levels (Dongre et al., 2023; Miles & Calder, 2021).

3.3.1.10 Coconut (*Cocos nucifera* Linn)

Coconut (*Cocos nucifera* Linn) (Fig. 3.1J), commonly called the "coconut tree," belongs to Arecaceae family. It is adapted to a wide range of habitats. The plant originated from India, dispersed to the East African coast and the interior, then later distributed to Western Africa and to the rest of the tropical regions of the world (Lima et al., 2015). Coconut fruit is consumed for energy from kernel and refreshment from the "coconut water" which provides minerals to compensate for any losses brought about by strenuous exercises (Moreno et al., 2020). Coconut is traditionally used in managing cough, tuberculosis, bronchitis, asthma, sore throat, flu, urinary infections, colitis, bacterial and viral infections, among other diseases (Lima et al., 2015; Nagata et al., 2011). The fruit of *C. nucifera* has mucolytic, demulcent, antiviral, antibiotic, immuno-stimulant, antioxidant, antihelminthic, antifungal, anti-inflammatory, antitumor, and analgesic activities. The edible portion of *C. nucifera* contains a range of nutrients including magnesium (32 mg/100 g), calcium (14 mg/100 g), potassium (356 mg/100 g), iron (2.43 mg/100 g), phosphorus (113 mg/100 g) as well as medium chain triglycerides, tannins, phenolics, anthocyanidins, terpenes, flavonoids, alkaloids, and sterols (Lima et al., 2015; Nagata et al., 2011). The coconut water is also rich in minerals, and with the advantage that it does not contain saturated fats. In addition, the coconut water is rich in *L*-arginine (**33**) (30 mg/dL) and ascorbic acid (**19**) (15 mg/100 mL), both of which possess strong antioxidant properties (Das et al., 2001; Salil & Rajamohan, 2001). Other isolated from the fruit pulp include lauric fatty acid (**34**) which, upon ingestion, is bio-transformed into an effective antimicrobial derivative, 1-lauroyl-glycerol (**35**).

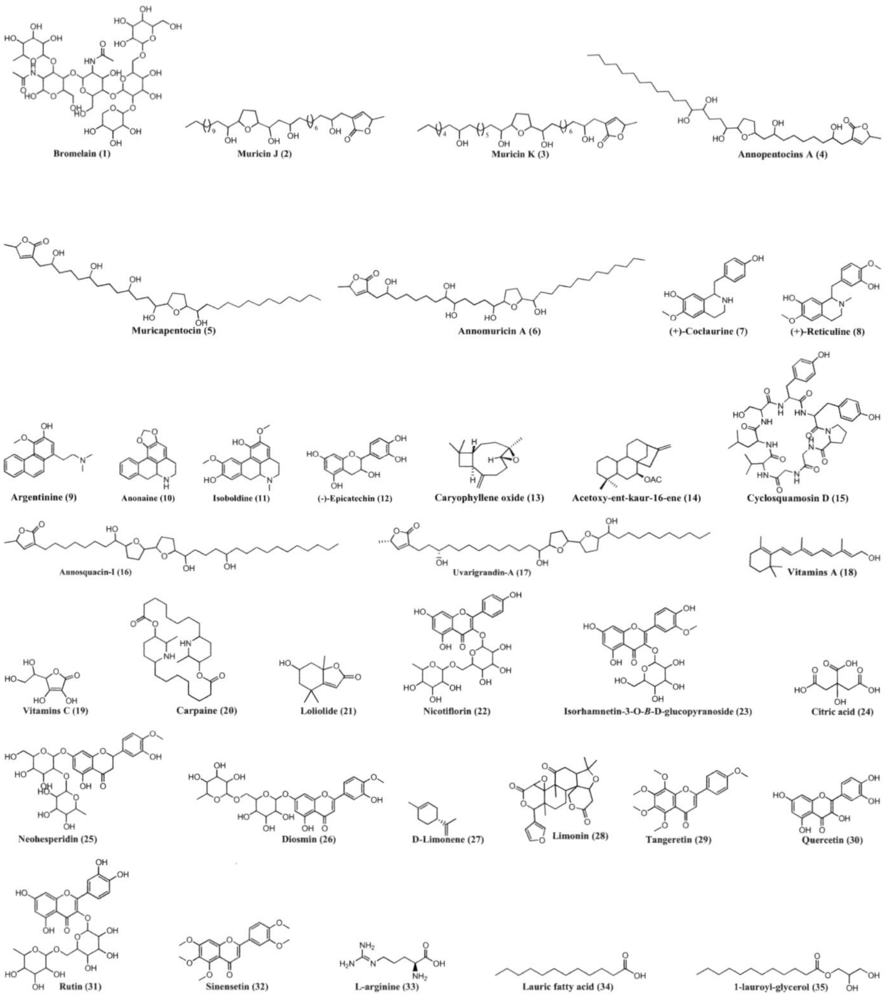

3.3.1.11 Pepper Fruit (*Dennettia tripetala*) Barker (F.)

The leaves and fruits of *Dennettia tripetala* (Fig. 3.1K) (Annonaceae) are edible and are used as appetizers, stimulants, or as a spice in West Africa (Kandasamy & Shanmugapriya, 2015). The leaves and seeds are used by traditional health practitioners in West Africa as remedies for cough, asthma, catarrh, fevers, toothache, nausea, and rheumatism, among other diseases. The phenolic acid and essential oil extracts from the plant have antimicrobial activities against *Staphylococcus aureus*,

Escherichia coli, and *Salmonella* sp. among other bacterial strains, and the leaves are effective against *Sclerotium rolfsii* fungus (Ejechi & Akpomedaye, 2005; Nwachukwu & Osuji, 2008). The root of *D. tripetala* furnishes vanillin (**36**), dennetine (2, 6-dimethoxychromone) (**37**), and alkaloids such as uvariopsine (**38**). Compound **38** improves the functioning of the liver and facilitates the secretion of bile (López-Martín et al., 2002a). The presence of vanillin (**36**) explains the use of the plant in boosting appetite (Ogawa et al., 2018). The fruit fragrance containing 2-phenylnitroethane (**39**) and linalool (**40**) (72.4 and 18.0%, respectively) as the major components has antimicrobial, anti-inflammatory, muscle relaxant, and antifungal properties (Adesina et al., 2017; López-Martín et al., 2002b).

3.3.1.12 Figs (*Ficus carica* L.)

Ficus carica (Fig. 3.1L), (Moraceae) is native to the Middle East; Mediterranean regions, Northern India, Morocco, Algeria, and Egypt (Mazzeo et al., 2024). It is used for making jellies, jams, and for garnishing baked food products. Dried fig fruits that have been rehydrated or boiled with milk are used to treat cough in North Africa and in Spain. It is, further, utilized in treating fever, chronic bronchitis, respiratory tract infections, inflammation, diabetes, intestinal worms, anemia, constipation, smallpox, paralysis, ringworm, leprosy, herpes, and dysenteric gonorrhea. The fruits possess diaphoretic, aphrodisiac, astringent, laxative, demulcent, and emollient properties. They have a pectoral action that fights infections. They soothe the respiratory tract, relieve cough, and facilitate expectoration (Lazreg-Aref et al., 2011). The figs contain monosaccharides and disaccharides, alkaloids, flavonoids, sterols, coumarins, triterpenes, and anthocyanins; quinine (**41**), rutin (**31**), 4′,5′--dihydropsoralen (**42**), umbelliferone (**43**), and psoralen (**44**) (Badgujar et al., 2014). Umbellifone (**43**) exhibited anti-inflammatory activities by inhibiting the production of nitric oxide, NO (46.10% inhibition at 0.5 mg/ml dosage), and antioxidant activity with radical scavenging properties (IC_{50} = 36.28 μg/ml) against $ONOO^-$ radicals (Ali et al., 2019).

3.3.1.13 Baboon's Breakfast (*Hexalobus monopetalus* (A. Rich.) Engl. & Diels)

Hexalobus (Fig. 3.1M) (Annonaceae) is restricted to the tropical savannah of Africa. The genus comprises only six species, which are either small trees or shrubs (Orwa et al., 2009). *Hexalobus monopetalus* is commonly used traditionally by the rural populations of Mwanza, Iringa, Mbeya, Kigoma, Tabora, and Ruvuma sub-regions in Tanzania for the treatment of fevers and stomach troubles (Kasole et al., 2019; Malebo et al., 2014). It is also used against colds, gastric ulcer, sexually transmitted diseases, diarrhea, stomach pains, snakebites, and headaches (Dzoyem et al., 2016; Nyagumbo et al., 2022; Kayode & Kayode, 2008). The stems, roots, bark, leaves, and fruits are utilized in Zimbabwe for the treatment of colds, pulmonary disorders, and bronchitis (Dzoyem et al., 2016). Some of the bioactive compounds isolated from the fruits of *H. monopetalus* include the indoline alkaloids 3,5-hexalobine C (**45**) and 3,5-hexalobine D (**46**), which exhibited mild antifungal properties (zones of inhibition = 5 and 14 mm, respectively) against *Candida albicans* (Malebo et al., 2014).

3.3.1.14 Apples (*Malus domestica* Borkh.)

Apple (*Malus domestica* Borkh.) (Fig. 3.1N) is a domesticated tree from Rosaceae family. It is also used commercially for vinegar, juice, jelly, and apple sauce (Palmatier, 2000). This plant is widely grown globally, and it is prone to drought, which affects its growth, quality, and in some places, yield (Spengler, 2019). Apple fruit/juice is used to treat tuberculosis, dry cough, chronic bronchitis, type II diabetes, neuropsychiatric disorders, obesity, several microbial infections, digestive problems, inflammations and ulcers, coronary heart diseases, cancer, palpitation, and dysentery (Kandasamy & Shanmugapriya, 2015; Hammad & Malik, 2021; Opyd et al., 2017). It is a nutritious fruit having a high dietary fiber content, total soluble solids, vitamins, polysaccharides (pectin), polyphenols, sterols, and triterpenes (Patocka et al., 2020). Some of the compounds present in apples and apple peels include chlorogenic acid (**47**), epicatechin (**12**), *p*-coumaric acid (**48**), and quercetin-3-rhamnoside (**49**) (Opyd et al., 2017; Lu & Foo, 1998; Tsao et al., 2005). Compounds **47** and **48** obtained from other plant materials have exhibited antioxidant, antibacterial, antiviral, and anti-inflammatory properties (Palmatier, 2000). Quercetin-3-rhamnoside (**49**) has exhibited antiviral activities against influenza A (causing a titer reduction of 6 logs ($p < 0.01$) at a concentration of 150 μg/ml) (Mehrbod et al., 2018).

3.3.1.15 Bananas (*Musa paradiasiaca* L.)

Bananas (*Musa paradiasiaca* L.) (Fig. 3.1P) are cultivated in Asia, South America, and Africa (Ahmed et al., 2021; Galani, 2019). The fruits are majorly consumed fresh, but some varieties are eaten after cooking. Bananas are industrially processed jellies, syrup, and banana vinegar (Galani, 2019; Tugume et al., 2016; Begashaw et al., 2023). In Nigeria, the leaves and fruits from the crop are utilized in treating cough (Lawal et al., 2020). In Pakistan, it is used to treat whooping cough (Lawal et al., 2020). The people from Eastern Uganda use the ash from the petiole of *M. paradiasiaca* var *paradiasiaca* and a decoction of the inflorescence from *M. paradiasiaca* var. *sapientum* against cough (Tabuti et al., 2003). It has been revealed that the *M. paradiasiaca* has numerous pharmacological activities such as anti-diarrheal, diuretic, antiulcer, anti-dysentery, antimalarial, vermifuge, and antimicrobial activities (Begashaw et al., 2023). Phytochemicals obtained from the leaves of *M. paradisiaca* include phenolics, cardiac glycosides, flavonoids, tannins, terpenoids, saponins, anthraquinones, and alkaloids, viz, rutin (**31**) and naringenin

(**50**) (Behiry et al., 2019). Rutin has anti-inflammatory and neuroprotective properties (Aruna et al., 2014; Park et al., 2014). Naringenin has shown activity against Chikungunya virus and Zika virus (IC_{50} = 6.18 and 58.79 μM, respectively) (Zou et al., 2020; Ahmadi et al., 2016).

3.3.1.16 Dates (*Phoenix dactylifera* L.)

Dates (*Phoenix dactylifera* L.) (Fig. 3.1Q), scientifically known as *Phoenix dactylifera* (Arecaceae, photo 16), were first domesticated in the Gulf of Persia and then later introduced to Northern Africa (Najafiniya & Rahkhodaei, 2024).

Dates do contain high concentrations of total polyphenols and significant amounts of antioxidants, and this has been attributed to the intense heat from the sun that the dates are exposed to (Al-Karmadi & Okoh, 2024). Traditionally, it is used in the treatment of the respiratory system, hypertension, and skin and skin-related diseases (Cámara-Leret et al., 2014). The phytochemical compounds isolated from the leaves of *P. dactylifera* include rutin (**31**), quercetin-3-*O*-glucoside (**51**), vanillyl alcohol (**52**), and oleanolic acid (**53**) (Suleiman et al., 2021).

3.3.1.17 Bush Banana or Finger Root (*Uvaria chamae* P. Beauv)

Uvaria chamae (Fig. 3.1R), (Annonaceae) is a small tree and sometimes a climbing shrub distributed across West Africa, through Central Africa up to the Eastern Africa region (Guisan et al., 2013). The fruit of *U. chamae* is edible and widely eaten. The plant is used in treating asthma, fevers, wounds, sores, and yellow fever in Nigeria (Abu et al., 2018), and it has antibiotic properties (Adesina et al., 2017). The methanol extract from the stem bark of the plant exhibited antibacterial activity against *Escherichia coli*, methicillin-resistant *Staphylococcus aureus*, *Proteus* spp., *Klebsiella* spp., *Staphylococcus aureus* ATCC25923, *Escherichia coli* ATCC25922, and *Pseudomonas aeruginosa* ATCC27853 with MIC of 5.62, 200, 12.59, 35.48, 5.62, 28.18, and 19.95 mg/ml, respectively (Oluremi et al., 2010). It is also reported to have exhibited anticancer activity against human nasopharyngeal carcinoma (in vitro) and mouse P-388 lymphocytic leukemia (in vivo). The extract further yielded the cytotoxic C-benzylated flavonoids including chamanetin (**54**), dichamanetin (**55**), and uvaretin (**56**), among others (Lasswell & Hufford, 1977). The crude root extract of *U. chamae* exhibited significant antiplasmodial activity in mice when orally administered at 300–900 mg/kg/day (Okokon et al., 2006).

3.3.2 Fruits Consumed as Pods, Berries, and/or Seeds

Photos of vegetables consumed as pods, berries, and/or seeds were shown in Fig. 3.2.

3.3.2.1 Okra (*Abelmoschus esculentus* L. Moench)

Okra (*Abelmoschus esculentus* L. Moench) (Fig. 3.2A), belonging to Malvaceae family, is sometimes called lady's finger. It is grown in many parts of Africa, Asia, America, and Southern Europe (Islam, 2019). Okra pods, which are approximately 15–20 cm in length, are valued for their edibility.

The pods, fruits, together with the seeds as well as the leaves of the plant have various applications due to their composition and properties especially in treating bronchitis, tuberculosis, and heart diseases in Zimbabwe (Nyagumbo et al., 2022; Chimponda & Mukanganyama, 2010), as well as pneumonia, diabetes, syphilis, anemia, catarrhal infections, dysuria, gonorrhea, ulcer, antispasmodic in the rest of Africa (Omara et al., 2020; Uwimbabazi et al., 2023; Obakiro et al., 2020). The extract from the seed and peel (100 mg/Kg) caused significant ($P < 0.05$) reduction in blood glucose level (96.84 ± 9.09 mg/dL) compared to that of the standard drug (metformin, 182.70 ± 34.81 mg/dL) 15 days after alloxan-induction (80 mg/Kg) of diabetes in Winstar rats (Abi et al., 2017). Since one of the symptoms of diabetes is an irritant cough (Wang et al., 2008), it is most likely that *A. esculentus* alleviates respiratory tract disorders through the regulation of blood sugar levels. The plant is generally considered safe with $LD_{50} > 2000$ mg/kg, and phytochemically, *A. esculentus* has been found to contain high amounts of flavonoids and moderate concentrations of sterols, terpenes, and phenolic compounds (Taiwo et al., 2021; Roy et al., 2014). These phytochemicals possess pharmacological properties; antioxidant, antidiabetic, hypolipidemic, antiulcer, neuroprotective activities; and are very effective in the healing of wounds. They also modulate the immune system, inhibit cancer and microbial growth, and protect the liver, the cardiovascular system,

Fig. 3.2 Aerial parts of (**A**) *Abelmoschus esculentus* L. Moench, (**B**) *Coffea canephora* Pierra ex A. Froehner, (**C**) *Helianthus annuus* L, (**D**) *Nigella sativa*, (**E**) *Phaseolus vulgaris* L, (**F**) *Physalis alkenkengi* L, (**G**) *Sesamum indicum* L, (**H**) *Theobroma cacoa* L.

and the gastro-intestinal tract (Prabhune et al., 2017; Chowdhury et al., 2019). Some of the isolated compounds from the plant include isoquercitrin (**57**) and quercetrin diglucoside (**58**) (Taiwo et al., 2021; Lin et al., 2014). The two compounds **57** and **58** showed strong antioxidant properties against intracellular reactive oxygen species (ROS), in a 2′,7′-dichlorofluorescein diacetate assay, with IC_{50} values of 2.7 ± 0.5 and 1.9 ± 0.2 μg/ml, respectively (3 hours after treatment) and 2.0 ± 0.8 and 1.5 ± 0,4 μg/ml, respectively (24 hours after treatment) (Taiwo et al., 2021).

3.3.2.2 Coffee (*Coffea canephora* Pierra ex A. Froehner)

Coffee (*Coffea canephora* Pierra ex A. Froehner) (Fig. 3.2B) is a shrubby plant, in the Rubiaceae family, distributed from tropical West Africa to Ethiopia and northern Angola, as well as in countries along the equator. In various countries like Ethiopia, Sudan, Indonesia, and Jamaica, coffee is used as a source of ethnomedicines, especially as a leaf-based beverage (Mewaba et al., 2022). In different parts of Africa, the seeds of *C. canephora* are used against asthmatic attacks (Moyo et al., 2015). In countries like Cuba, Peru, Nicaragua, and Mexico, coffee leaves are used as an emollient when managing flu, fever, and cough (Chen et al., 2018), as well as intestinal troubles and diarrhea in Africa (Mewaba et al., 2022). In addition to being a stimulant, the beverage obtained from coffee beans is believed to alleviate several disorders, like fever, anemia, and intestinal pain (Campa & Petivallet, 2018). The extract from the beans is reported to have caused a reduction in fasting glucose, insulin resistance, serum leptin, inflammatory tendencies in adipose tissues, among other effects, in atherogenic-diet fed ApoE−/− mice at 220 mg/kg intra-gastric dosage, thrice a week for 14 weeks (de Almeida et al., 2019; Caro-Gómez et al., 2019). A diversity of classes of secondary metabolites have been identified from *C. canephora*, namely, alkaloids, flavonoids, and phenolic compounds. Most of these compounds have anti-inflammatory, antioxidant, lipid-lowering, and immunomodulatory activities. Thus, caffeic acid (**59**), chlorogenic acid (**47**), mangiferin (**60**), and caffeine (**61**) are among the secondary metabolites identified from different parts the crop. Consumption of foods (and beverages) containing compounds **47** and **59** is associated with health-promoting attributes such as antioxidant and anti-inflammatory properties of the compounds (Liang & Kitts, 2015). Compound **47** is reported to have significantly reduced pulmonary eosinophilia and expression of interleukin 4 (IL-4), interleukin 5 (IL-5), and the tumor necrosis factor (TNF-α) in the lungs of mice with ovalbumin-induced allergic asthma upon intraperitoneal administration of the compound at a concentration range of 0.5–25 mg/Kg body weight (Kim et al., 2010). However, the roasting process during the production of brewed coffee beans tends to reduce the concentration of compound **47** (Mondolot et al., 2006). Relatedly, compound **60**, which was first isolated from the leaves, stem bark, and the peel of *Mangifera indica*, is reputed for a wide range of pharmacological properties including anti-inflammatory, antihyperlipidemic, antidiabetic, neuroprotective, antioxidant, and antimicrobial activities (Du et al., 2018). The xanthine alkaloids, including caffeine, present in various species of coffee have bronchodilatory properties and, thus, help in relieving pain in the respiratory tract (Zhang et al., 2024; Shi et al., 2016); however, high concentrations of caffeine are associated with insomnia, palpitations (tachycardia), tremors, headache and anxiety in some sensitive individuals. So it is advisable that coffee be consumed but in moderation (O'Keefe et al., 2018).

3.3.2.3 Sunflower (*Helianthus annuus* L.)

Helianthus annuus (Fig. 3.2C) (Asteraceae) is considered to be the most important oilseed in the world, producing dietary fiber in addition to premium oil (Adeleke & Babalola, 2020). Traditionally, in some communities, different parts of sunflower are used to treat asthma, chronic obstructive disease, bronchitis and laryngitis, migraine, and seasonal allergic fever, and also used as an expectorant (Sharif et al., 2022). The seeds contain flavonoids, steroids, polyphenols, monounsaturated and polyunsaturated fatty acids, amino acids, especially L-arginine (**33**) with antidiabetic and antihypertensive properties, as well as minerals like potassium which contribute to the well-being of the consumers. The phenolic compounds in *H. annuus* have exhibited very good antioxidant properties (Gai et al., 2020). Some of the compounds identified from the aerial parts and the non-oil fractions of the seeds of the plant include 1,5-Di-*O*-caffeoylquinic acid (**61**) and 4,5-Di-*O*-caffeoylquinic acid (**62**). The caffeoylquinic acid derivatives isolated from other plant materials exhibited significant anti-inflammatory, expectorant, and antitussive properties in a three-day oral administration experiment, with doses of 10 and 20 mg/kg per day, in a mouse model (Ye et al., 2015). The bioactivity of the caffeoylquinic acid derivatives, in this regard, may be associated with the catechol moiety similar to what is found in β_2-agonists, with smooth muscle relaxant and bronchodilation properties (Marques & Vale, 2022).

3.3.2.4 Black Seeds (*Nigella sativa*)

Black seed (*Nigella sativa*) (Fig. 3.2D) is from Ranunculaceae plant family, and considered to be a miracle herb with an enriched ritualistic and historical background (Yimer et al., 2019; Javed et al., 2010). The seed oil is used in Unani and Ayurveda medicine practices. According to Prophet Muhammad, "black seed" was considered to cure almost all ailments (Hosseinzadeh et al., 2007). The seeds are commonly prescribed in cases of asthma, cough, digestive diseases, bacterial infections, diabetes, and viral infections, among others (Javed et al., 2010; Ferizi et al., 2023). *N. sativa* has a wide range of biological activities including bronchodilation, diuretic, antihypertensive, anti-inflammatory, analgesic, antidiabetic, antioxidant, antimicrobial, contraceptive, and antifertility, among others. The ethanol extract of *N. sativa* exhibited was active against the methicillin-resistant *Staphylococcus aureus* isolates with MIC values of 0.2–05 mg/ml (Forouzanfar et al., 2014). Some of the isolated compounds from black seeds include nigellicine (**63**), nigellidine (**64**), thymohydroquinone (**65**), and thymoquinone (**66**) (Ferizi et al., 2023; Mehta et al., 2009). Thymoquinone induced a significant reduction in the levels of pro-inflammatory mediators (TNF-γ, IL-6, IL-1γ, PGE-2, IFN-γ) and raised the level of IL-10 in collagen-induced arthritic Wistar rats (Umar et al., 2012).

3.3.2.5 Kidney Beans (*Phaseolus vulgaris* L.)

Phaseolus vulgaris (Fig. 3.2E) (Fabaceae) is an herbaceous annual plant. In Africa and parts of Asia, the green leaves and the young pods are consumed as vegetables and sometimes as salads, and the dry beans, mainly in cooked form, are a good source of protein (Ganesan & Xu, 2017). The beans are recognized for their special health attributes such as their high concentration of proteins and dietary fiber, low levels of saturated fats, presence of minerals and vitamins, as well as phytochemicals, making them to be suitable options in diet replacement formulations. In Uganda, bean pods/husks are also used to treat respiratory disorders, in particular tuberculosis (Bunalema et al., 2014), and in Zimbabwe, they are used to manage bilharzia and postpartum conditions (Maroyi, 2023). Beans contain nonheme iron, potassium, flavonoids such as 3,5,7,4′-tetrahydroxyflavone (**67**), 5,7,4′-trihydroxyflavone (**68**), europinidin (**69**), and aurantinidin (**70**), as well as amino acids including *L*-arginine (**32**), which is associated with the antidiabetic properties of the bean pods (Olowookere & Malomo, 2024).

3.3.2.6 Winter Cherry (*Physalis alkenkengi* L)

The red ripened berries of Winter cherry (*Physalis alkenkengi* L) (Fig. 3.2F) are edible. In Uganda, where it is known as "Entuntunu" (Luganda), the berries are used for the management of colds and flu as well as for improvement of female fertility (Orwa et al., 2009; Namukobe et al., 2011). The berries also have diuretic, depurative, and uricosuric properties. They are, further, used in cases of urinary lithiasis, gout, and uric arthritis. Winter cherry berries are rich in vitamin C (**19**) (its concentration is higher than in lemon), malic acid (**71**), citric acid (**24**), and carotene/provitamin A (**72**) and traces of alkaloids (Pamploma-Roger, 2006).

3.3.2.7 Sesame (*Sesamum indicum* L.)

Sesame (*Sesamum indicum* L.) (Fig. 3.2G) is an herbaceous plant from Pedaliaceae family, whose seeds are edible and a source of vegetable oil. It originated in Pakistan from where it was distributed to the Indo-China region, Malaysia, and Myanmar, and it is currently cultivated in several countries worldwide (Hou & Meng, 2022). Ethnomedicinally, sesame has been used as food, medicine, and in cultural rituals in Egypt, the Middle East, and India. Sesame oil was used to prepare perfumes and medicines in the Babylonian empire, and for the purification ceremony in ancient Egypt (Mili et al., 2021). The fruits are used against cough and constipation (Dzoyem et al., 2014). The oil is traditionally used in cases of tuberculosis and cough, and it has demulcent and antitussive properties. Other medicinal uses of the oil include treatment of eye diseases, ulcers, migraines, hair loss, constipation, and induction of breast-milk production as well as promotion of menstruation (Peter,

2012; Ross, 2007). The leaves are used in treating catarrh, inflammations in the mouth cavity, dysenteric diarrhea, cystitis, type II diabetes, rabies, skin dandruff, and urinary tract diseases (Peter, 2012; Anis & Iqbal, 1994). The roots are also used to treat asthma and cough, promote hair growth, and prevent premature growth of gray hair (Peter, 2012). It was reported that application of the seed oil into the nasal cavity could inhibit the proliferation of SARS coronavirus in humans (Allam et al., 2021). Sesame seeds are rich in terpenoids, phenols, anthraquinones, lignans, steroids, naphthoquinones, and glycosides, having many nutritive and health benefits. The crop has important pharmacological properties like anti-inflammatory, antihypertensive, antioxidant, anticancer, antipyretic, and hepatoprotective activities (Hou & Meng, 2022; Mili et al., 2021). The following are some of the isolated compounds from sesame leaves: sesamin (**73**), pinoresinol (**74**), sesamolin (**75**), and spicatolignan (**76**) (Allam et al., 2021). Sesamin has exhibited immunomodulatory and anti-inflammatory properties. Thus, 30-minute pre-treatment of murine BV-2 microglial cells with sesamin (10–100 μg/ml) followed by co-treatment with lipopolysaccharide and sesamin (10–100 μg/ml) for 24 hours caused inhibition in the production of the pro-inflammatory cytokines, IL-6, and TNF-γ in a concentration-dependent manner (Jeng et al., 2005). Structure activity studies show that the catechol moiety is responsible for the anti-inflammatory activities of sesamin (Abe-Kanoh et al., 2019).

3.3.2.8 Cocoa (*Theobroma cacoa* L.)

Cocoa/Cacao (*Theobroma cacoa* L.) (Fig. 3.2H) (family Malvaceae) is a tree that originated from Central America, particularly in the Guyana Plateau and the Amazon basin (Zhang et al., 2011). It is one of the major economic crops in Western Africa with the highest producer of the beans being Ivory Coast. A concoction from the leaves of *T. cacao* is used to treat cough associated with respiratory diseases in Nigeria (Lawal et al., 2020). A decoction from the pod rinds is used against cough and as a stimulant in Mexico (Zhang et al., 2024). The cocoa beans are used in the preparation of "teas" and help to fight high blood pressure. It is also believed to protect nerves from injury and inflammation, as well as the skin against oxidative damage caused by UV radiation from the sun. Furthermore, it is believed to improve on mood and cognitive abilities. The beverage and extract from the beans protect against heart disease, reduce the viscosity of blood, and exhibit anticancer properties (Zhang et al., 2011; Rojo-Poveda et al., 2020). It has diuretic, sedative, emollient, and wound-healing properties. The health benefits gained from the consumption of chocolate and cocoa may be due to the presence of polyphenolic compounds, flavonoids, and procyanidins. *T. cacao* contains catechin (**77**), 3′-*O*-methyl epicatechin (**78**), and procyanidin B_2 (**79**) (Erlejman et al., 2008), as well as xanthine alkaloids such as theobromine (**80**) and theophylline (**81**). Compound **79** is reported to have suppressed lipopolysaccharide-induced inflammation and apoptosis in human alveolar epithelial cells and lung fibroblasts by inhibiting nuclear factor kappa B (NF-kB) and NOD-like receptor family, pyrin domain containing

3 (NLRP3) inflammasomes (Jiang et al., 2020). Catechin exhibited strong antioxidant properties with a strong ferric reducing antioxidant power (FRAP) of 0.793 ± 0.004 mol trolox equivalents per mol (Grzesik et al., 2018). The xanthine alkaloids have bronchodilation properties and hence help in relieving cough, but they are also associated with the risk of insomnia, anxiety, and tachycardia in some individuals. *T. cacao* is commonly applied in the following cases: lip sores, nipple sores, burns, skin rashes, hemorrhoids and vaginitis, and for hair growth.

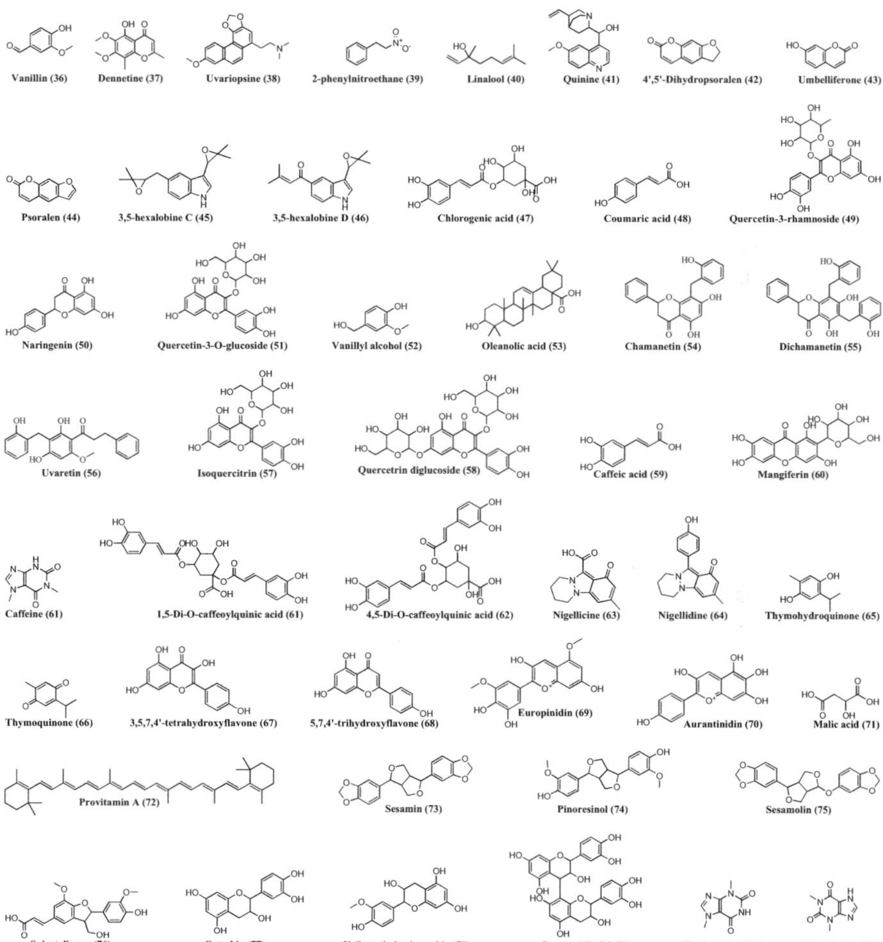

3.3.3 *Leafy Vegetables*

Photos of leafy vegetables were shown in Fig. 3.3.

3.3.3.1 Spiny Amaranth or Spiny Pigweed (*Amaranthus spinosus* L.)

Amaranthus spinosus (Fig. 3.3A) (Amaranthaceae) is an annual herb found throughout Africa. A decoction of the aqueous extract from the leaves is used against respiratory diseases in Nigeria (Lawal et al., 2020; Dele & Ojone, 2023). It is also used against bronchitis in Pakistan (Mahmood et al., 2011). In addition, the plant is used traditionally against anemia, hepatitis B, hemorrhoids, venereal diseases, eczemas, abdominal troubles, intestinal pain and bladder inflammation, vomiting and as a laxative, an emollient poultice, and it might have antimalarial, antioxidant, antimicrobial, anti-inflammatory, antidiuretic and abortifacient properties (Suroowan et al., 2019; Peter & Gandhi, 2017; Soladoye et al., 2010; Asiimwe et al., 2014). Isolated compounds from the different parts of *A. spinosus* include amaranthoside (**82**), trigonelline (**83**) (Bulbul et al., 2011; Chaudhary et al., 2012), and spinoside (**84**) (Sarker & Oba, 2019; Tanmoy et al., 2014).

Fig. 3.3 Aerial parts of leafy vegetables, (**A**) *Amaranthus spinosus* L, (**B**) *Chenopodium ambrosioides* L, (**C**) *Chenopodium opulifolium*, (**D**) *Cleome gynandra* L, (**E**) *Gongronema latifolium* Benth et Hook, (**F**) *Solanum nigrum* L, (**G**) *Vernonia amygdalina* (Delile)

3.3.3.2 Mexican Tea or Wormseed (*Chenopodium ambrosioides* L.)

Chenopodium ambrosioides (Fig. 3.3B) belongs to the family Chenopodiaceae and is spread worldwide. It is used for managing several diseases in both humans and livestock (Kasali et al., 2021). In Nigeria, where it is used as a condiment, the herb is traditionally prescribed against bronchitis, pneumonia, catarrh, asthma, and tuberculosis, among other ailments (Kasali et al., 2021). In Ghana, water extracts from the leaves of *C. ambrosioides* are used against tuberculosis (Nguta et al., 2015). The whole plant is said to possess anti-asthmatic, analgesic, carminative, vermifuge, and stomachic properties (Kokanova-Nedialkova et al., 2009). The essential oil is effective against a number of parasites, such as dysenteric amoeba, but is not effective against tapeworms. Similarly, across the Pacific Ocean, in Southern America, the essential oil treats insect bites and athlete's foot (plants, P.f.a.F.D.o.e.a.u, 2024). Small quantities of the leaves are added to cooked bean dishes as a flavoring because of their carminative activity and, thus, the ability to reduce flatulence (plants, P.f.a.F. D.o.e.a.u, 2024). In Brazil, this plant is commonly used as an antiprotozoal agent against *Trypanosoma, Leishmania*, and *Trichomonas* species (Monzote-Fidalgo et al., 2004; Okuyama et al., 1993). The plant is also used for the treatment of enteroparasites, including *Trichuris trichiura, Ascaris lumbricoides*, and *Ancylostoma duodenale* (Nakazawa, 1996). The wide range of traditional uses not only in Africa but also in other continents is an indication that medicinal products can be developed from the herb. Some of the chemical constituents identified from *C. ambrosioides* include 1-piperoylpiperidine (**85**), 1,2-benzopyrone (**86**), patuletin (**87**), and afzelin (**88**) (Kasali et al., 2021; Ghareeb et al., 2016; Hammoda et al., 2015). The essence from the plant is rich in monoterpenes such as ascaridoles (Kokanova-Nedialkova et al., 2009). Ascaridole **88a** is reported to show an hypothermic effect and an analgesic effect after oral administration at a dose of 100 mg/Kg to acetic acid-induced writhing mice (Okuyama et al., 1993). The methanol extract of the dried leaves of *C. ambrosioides* exhibited anti-inflammatory activity by producing a dose-dependent inhibition (300–700 mg/Kg) of cotton pellet-induced granuloma and carrageenan-induced paw edema in rats (Ibironke & Ajiboye, 2007). Herbal teas made from the water extract of the plant showed moderate antioxidant activity in the trolox equivalent antioxidant capacity (TEAC) test (Speisky et al., 2006), whereas the essential oil from the plant exhibited potent antioxidant activity when tested using the 2,2′-azino-bis(3-ethylbenzthiazoline-6-sulphonic acid) ABTS method (Kumar et al., 2007). The essential oil from the leaves of *C. ambrosioides* exhibited an in vitro antifungal activities by inhibiting the mycelial growth of *Aspergillus flavus* at a concentration of 100 μg/ml and broad fungitoxicity against *Aspergillus niger, A. fumigatus, Botryodiplodia theobromae, Cladosporium cladosporioides, Helminthosporium oryzae*, and *Pythium debaryanum* at the same concentration (Kumar et al., 2007). Ascaridole **88a** is also reported to have exhibited in vitro antineoplastic activity against a range of tumor cell lines including human acute lymphoblastic leukemia CCRF-CEM cells, human promyelocytic leukemia HL60 cells, and human breast cancer MDA-MB-231 cells (Efferth et al., 2002). The compound has also shown strong antiplasmodial activity against *Plasmodium falciparum*. It exhibited, further, potent inhibition of *P. falciparum* by arresting the growth of the microorganism at a concentration of 0.05 μM and caused total elimination of the parasite from the culture at 0.1 μM (Pollack et al., 1990). The antimicrobial, antioxidant, and anti-inflammatory properties, as well as antineoplastic activities of the extracts and the constituent compounds, may be major contributory factors to the efficacy of the plant against respiratory diseases.

3.3.3.3 Grey Goosefoot (*Chenopodium opulifolium*)

Chenopodium opulifolium (Fig. 3.3C), (Chenopodiaceae) is commonly used as a vegetable in African local cuisines and for treating cough (Esezah et al., 2015) and sore throat in Uganda (Tugume et al., 2016). A decoction from the plant relieves pain during menstruation and quickens childbirth (Ajayi et al., 2016). Some of the constituents identified from *C. opulifolium* include flavone derivatives, phenolics, saponins, and triterpenoids with a wide range of properties: antiviral, antimicrobial, pain-relieving, anti-inflammatory and immunomodulatory activities (Ajayi et al., 2016). Some of the isolated bioactive compounds include allantoin (**89**) and decan-2-one (**90**). Allantoin exhibited promising antiplasmodial activity against *Plasmodium falciparum* clinical field isolates KOM_234 and KDH_204, with IC_{50} values of 2.29 ± 0.98 and 10.94 ± 1.51 μg/ml, respectively (Juma, 2022). Allantoin is also reported to boost wound-healing properties through a mechanism of regulating inflammatory response and stimulus to fibroblastic proliferation as well as extracellular matrix synthesis (Araújo et al., 2010). This could be one of the major contributory factors to the effectiveness of *C. opulifolium* in the healing of the pulmonary system. The compound has antioxidant properties and biological properties similar to those of ascorbic acid, which has been attributed to the closeness in the thermodynamic properties of the two compounds (Gus Kov et al., 2004; Shestopalov et al., 2006). Compound **89**, furthermore, plays the role of being a barrier in the placenta for primates, thus protecting the embryo from sources of free radicals, and it is considered to be a dominant embryonic antioxidant. It is also assumed to be a non-specific immune stimulant (Gus Kov et al., 2004).

3.3.3.4 Spiderwisp (*Cleome gynandra* L.)

Cleome gynandra (Fig. 3.3D) (Capparaceae) is an upright herbaceous vegetable distributed in sub-Saharan and in some South-East Asian countries. It is also known as African spider flower, traditionally used as a leafy vegetable throughout Africa

(Chataika et al., 2020; Mashamaite et al., 2022). *C. gynandra* is used in the management of HIV-associated tuberculosis (Otieno et al., 2016), due to the presence of selenium which boosts the defense mechanism of cells against infections and oxidative stress (Hussain et al., 2019; Gupta et al., 2017). The leaves of *C. gynandra* are used as a source of livelihood for vendors in Southern Africa (Thovhogi et al., 2021). The whole plant has been used by the local communities in Africa and Asia in treating a range of medical conditions including chest pains, anemia, scurvy, arthritis, hepatic disorders, stomach ache, irritable bowel syndrome, piles, diabetes, malaria, cardiovascular diseases, constipation, epilepsy, tumors, rheumatism, earaches, and cancer. The root decoction is drunk to facilitate childbirth and is against stomach ache, diarrhea, severe infestation of threadworms, and constipation (Omara et al., 2020; Obakiro et al., 2020; Chand et al., 2022). The plant is rich in vitamins, minerals, and proteins. A number of studies have led to the identification of several health-promoting compounds from *C. gynandra* including alkaloids, flavonoids, proanthocyanidins, glucosinolates, carbonyls, phenolics, saponins, tannins, terpenes, and sterols (Chand et al., 2022). Fractionation of an extract from the aerial part of the plant led to the isolation and identification of cleogynones A (**91**), B (**92**) and (hexadecahydro-17-(tetrahydro-5-(2-hydroxypropan-2-yl)-2-methylfuran-2-yl)-4,4,8,14-tetramethyl-3-oxo-1H-cyclopenta[a] phenanthren-10-yl)methyl acetate (**93**). Compound **92** was moderately active against human lung cancer A549 cells, and all the three compounds were moderately cytotoxic against human breast cancer MDA-MB-468 cells, whereas compounds **92** and **93** were moderately active against colorectal cancer (HCT-116 and HCT-15) cells (Mzondo et al., 2021). This information further supports and validates the use of *C. gynandra* in treating respiratory diseases and other ailments.

3.3.3.5 Amaranth Globe (*Gongronema latifolium* Benth et Hook)

Gongronema latifolium (Fig. 3.3E) (Asclepiadaceae), with a common name of "arokeke" or "utazi" in Nigeria, grows in tropical rainforests. It is used as a vegetable and a spice (Ugochukwu et al., 2003; Ugochukwu & Babady, 2002). It is, further, used as an anti-asthmatic and antitussive in herbal medicine and for the treatment of viral hepatitis, cough, diarrhea, among other infections. The vegetable also finds use in the management of cough in poultry (Essien et al., 2007). Extracts from *G. latifolium* have exhibited antibacterial activities (Edim et al., 2012). From the ethanolic leaf extract, pregnane derivatives such as iloneoside(**94**), marsectohexol (**95**), 3-*O*-[6-deoxy-3-*O*-methyl-*β*-D-allopyranosyl-(1→4)-*β*-D-oleandropyranosyl-17*β*-marsdenin (**96**), 3-O-[6-deoxy-3-O-methyl-*β*-D-allopyranosyl-(1→4)-*β*-D-canaropyranosyl]-17*β*-marsdenin(**97**), as well as ajugoside (**98**) were isolated.

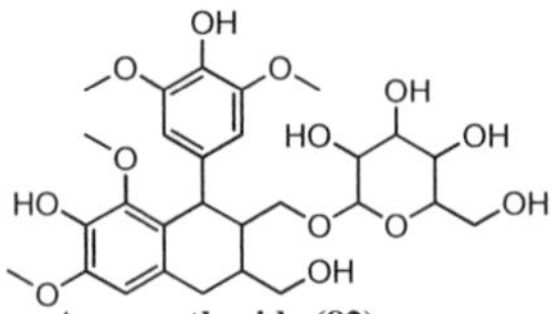
Amaranthoside (82)

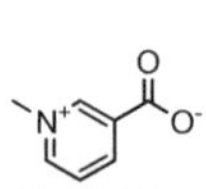
Trigonelline (83)

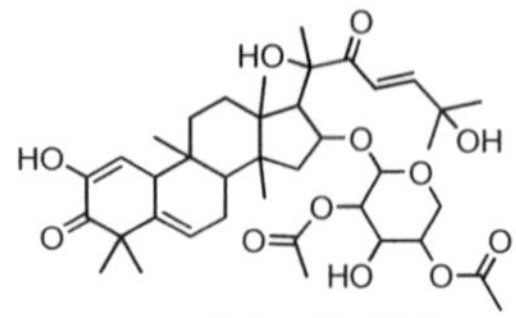
Spinoside (84)

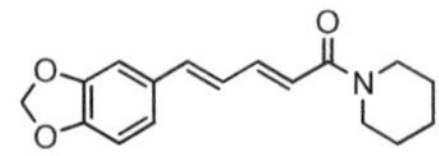
1-piperoylpiperidine (85)

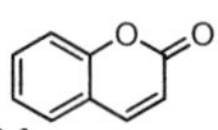
1,2-benzopyrone (86)

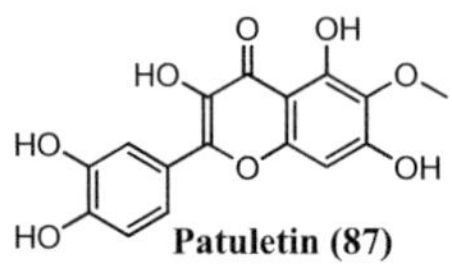
Patuletin (87)

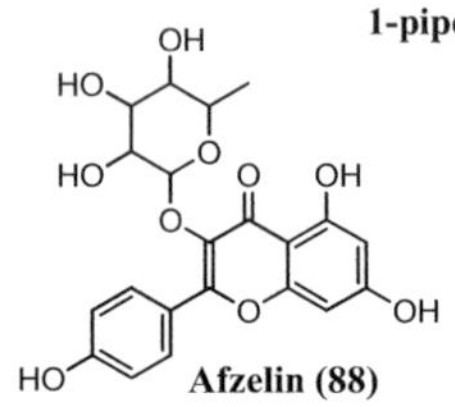
Afzelin (88)

Ascaridole 88a

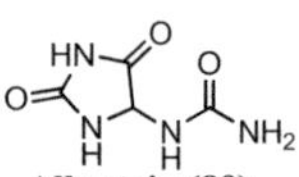
Allantoin (89)

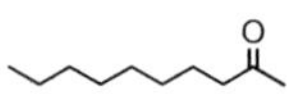
Decan-2-one (90)

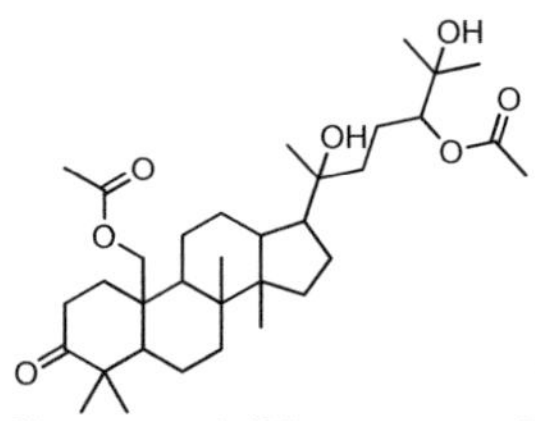
Cleogynone A (91)

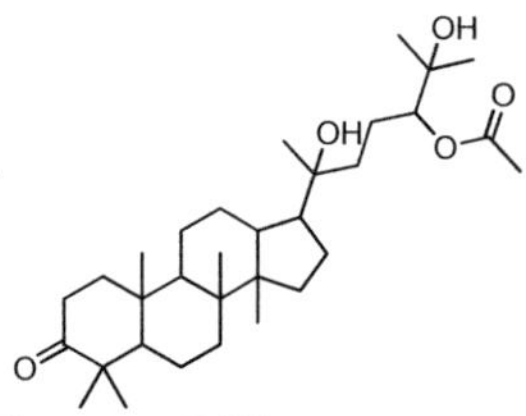
Cleogynone B (92)

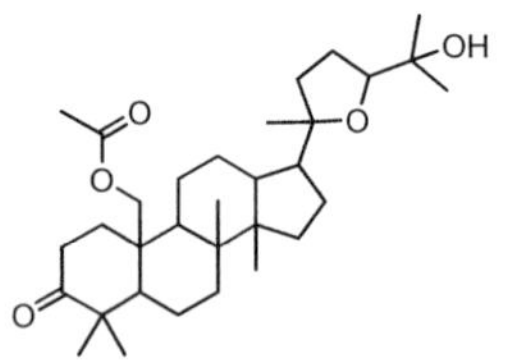
(hexadecahydro-7-(tetrahydro-5-(2-hydroxypan-2-yl)-2-methyfuran-2-yl)-4,4,8,14-tetramethyl-3-oxo-1H-cyclopenta[a]phenanthren-10-yl) methyl acetate (93)

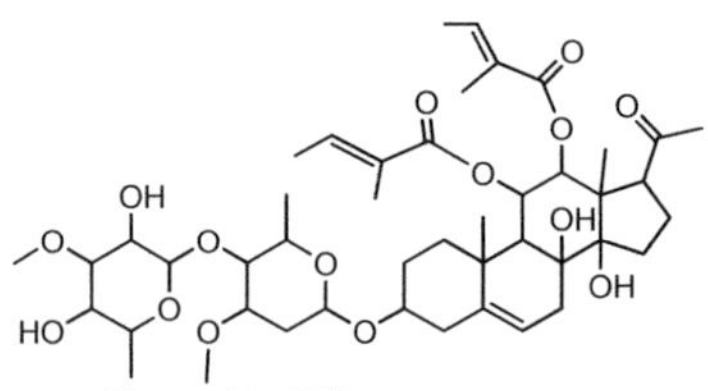
Iloneoside (94)

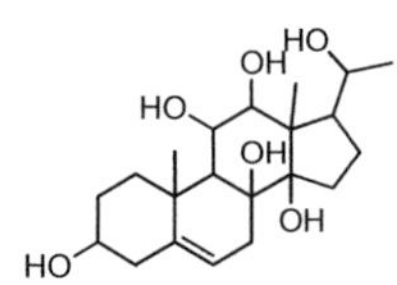
Marsectohexol (95)

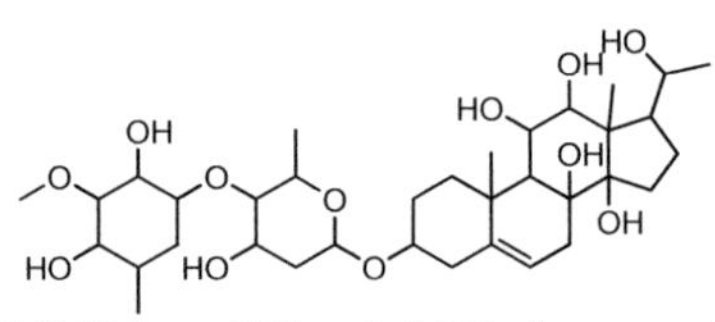
3-O-[6-deoxy-3-O-methyl-β-D-allopyranosyl-(1›4)-β-D-canaropyranosyl]-17β-marsdenin (97)

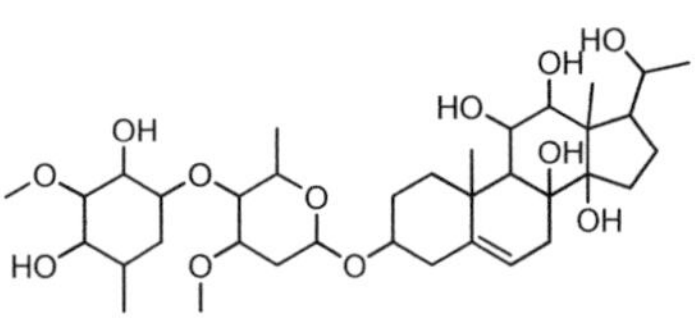
3-O-[6-deoxy-3-O-methyl-β-Dallopyranosyl-(1›4)-β-D-oleandropyranosyl]-17β-marsdenin (96)

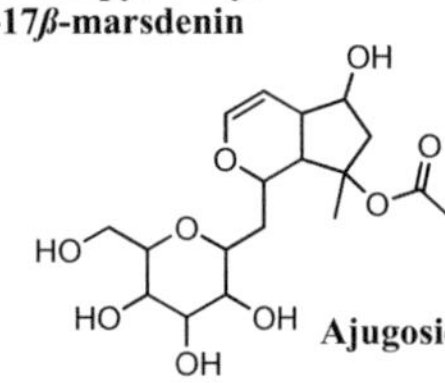
Ajugoside (98)

The compounds exhibited antiproliferative activities (at a concentration of 50 μM) against human leukemia KL-60 cells with inhibitions of 4.7 ± 0.3, 22.3 ± 1.2, 23.6 ± 3.7, 27.6 ± 1.0, and 18.4 ± 0.4%, respectively (Gyebi et al., 2018).

3.3.3.6 Black Nightshade (*Solanum nigrum* Linn)

Solanum nigrum L. (Fig. 3.3F), (Solanaceae) is an herb native to tropical Africa as well as India, the Mediterranean region, and Southern America. Its leaves are edible as vegetables (after cooking) in addition to being used in alleviating chronic bronchitis and sore throat (Maroyi, 2024; Tugume et al., 2016; Wang et al., 2017; Younis et al., 2018). In South Africa, it is used against asthma, fever, and tuberculosis (Thinyane & Maroyi, 2019). Juice extracted from the roots is used against asthma and whooping cough in India (Sikdar & Dutta, 2008).

The fresh leaves are applied (externally) as lotions, compresses or poultices to alleviate pruritus (itching) on the vagina or the anus and relieve skin itching caused by scabies, herpes, ringworm, warts, or any other form of rashes and burns (https://safelizwellness.com/en/product-view/enciclopedia-de-las-plantas-medicinales/; Moshi et al., 2009; Boulos, 1983). A diluted infusion of the berries is used against eye diseases: blindness, conjunctivitis, glaucoma, trachoma, and cataract (Boulos, 1983). The leaf paste is applied to dog-bite wounds to cause healing and to treat rabies in India (Hills, 2010; Muthu et al., 2006). The roots are boiled with sugar to increase fertility in women (Upadhyay et al., 2007). Different studies have revealed that the plant is antimicrobial, anticancer, anti-inflammatory, radical scavenging, anti-epileptic (anticonvulsant), and anti-neurodegenerative, both in vivo and in vitro (Chen et al., 2022). Both the crude extracts and individual components from *S. nigrum* have exhibited antiproliferative properties against several cancer cell lines. A phytoglycoprotein (with molecular mass of 150 kDa) isolated from the plant is reported to possess antiproliferative activities against human cervical cancer HeLa cells, human breast cancer MCF-7 cells, human colon cancer HCT-116 cells, human female colorectal adenocarcinoma HT29 cells, and liver cancer Hep3B cells (Son et al., 2003; Lee & Lim, 2008; Lim, 2005). Compounds identified from the plant include flavonoids, steroidal compounds, alkaloids, propanoids, and glycosides (Chen et al., 2022). One of the major constituents of the plant is a glycoalkaloid solanine (**99**), which is so powerful on the nervous system, causing muscular relaxation, analgesia, and narcosis. It has antispasmodic, antiproliferative, analgesic, and sudorific properties but is toxic (https://safelizwellness.com/en/product-view/enciclopedia-de-las-plantas-medicinales/).

3.3.3.7 Bitter Leaf (*Vernonia amygdalina* (Delile))

Vernonia amygdalina (Fig. 3.3G), Asteraceae) is a perennial herbaceous shrub indigenous to sub-Saharan Africa (Kamatenesi et al., 2013). The leaves of the shrub are eaten in Nigeria (Nwosu et al., 2013). Different plant parts of *V. amygdalina* are used against tuberculosis, asthma, and cough in Nigeria (Mann et al., 2007; Ofukwu et al., 2008), diabetes mellitus, intestinal worms, tonsillitis, hepatitis, snake bites, urine retention, ulcers, and malaria among many other ailments in much of Africa (Aiyeloja & Bello, 2006; Tugume et al., 2016; Uwimbabazi et al., 2023; Namukobe et al., 2011; Tabuti, 2008; Betti, 2004; Cos et al., 2002). In Uganda, it is used against respiratory infections, cough, and tuberculosis in addition to malaria (Anywar et al., 2020; Bunalema et al., 2014; Mugisha et al., 2014). It has antioxidant, antimicrobial, and antiviral properties (Cos et al., 2002). The ethanol and methanol crude extracts demonstrated a 4-day parasitemia suppression of 67% against *Plasmodium berghia* at doses of 500 mg/Kg and 1000 mg/Kg, respectively, whereas the aqueous extract reduced the parasitemia by 63% at 125 mg/Kg in a mice model test (Abosi & Raseroka, 2003; Iwalokun, 2008; Challand & Willcox, 2009). Toxicological studies have revealed that the plant is safe (Tijjani et al., 2017). Chemical analyses have indicated the presence of alkaloidal compounds, saponins, flavones, tannin derivatives, terpenes, lactones, steroids, as well as phenols. Isolated phytochemicals from different parts of *V. amygdalina* include vernolide (**100**), vernodalol (**101**), vernodalin (**102**), and vernomygdin (**103**) (Erasto et al., 2007), Vernolide and vernodalin exhibited in vitro antiplasmodial activities with IC_{50} values of 1.87 and 0.52 µg/ml, respectively, against blood stage *P. falciparum* (Chukwujekwu et al., 2009). Vernodalol (**101**) has exhibited antitumor properties and studies demonstrated that the compound suppresses acute promyelocytic leukemia APL cell proliferation and induces cell cycle arrest in the G2/M phase through the upregulation of p21 and cell division cycle 25. The compound also induced cellular apoptosis via the mitochondrial pathway (Wu et al., 2018). Furthermore, vernomygdin (**103**), vernodalin (**102**), and vernolide (**100**) isolated from the chloroform extract of *V. amygdalina* exhibited significant in vitro cytotoxicity against human nasopharyngeal (KB) carcinoma cells (ED_{50} = 1.5, 1.8, and 2.0 µg/ml, respectively) (Kupchan et al., 1969), thus validating the usage of the plant against respiratory ailments.

3.3.4 Vegetables in Form of Bulbs

Figure 3.4 shows vegetables in form of bulbs.

Fig. 3.4 Vegetables consumed as bulbs: (**A**) *Allium cepa* L, (**B**) *Allium sativum* L, (**C**) *Allium tuberosum* Rottl. ex Spreng, and (**D**) *Foeniculum vulgare* Mill

3.3.4.1 Onions (*Allium cepa* L.)

Onion (*Allium cepa* L.) (Fig. 3.4A), scientifically known as *Allium cepa* (Amaryllidaceae), is envisaged to have originated from Afghanistan, Iran, and Russia, then spread to more than 175 countries around the world.

It is rich in dietary nutrients, and it is reputed for its numerous health benefits. Some of the indications are in ailments including cough, respiratory catarrh, sinusitis, pharyngitis, bronchial asthma, lung emphysema, tuberculosis, sore throat, cholera, hypertension, blood thinning, hepatic disorders (hepatitis, fatty liver degeneration, and cirrhosis), among others. It has mucolytic, antibiotic, diuretic, depurative, vermifuge, hypoglycemic, hypotensive, anti-inflammatory, and invigorating properties (Shawarb et al., 2023; Mahmood et al., 2011; Teshika et al., 2019; Chakraborty et al., 2022; Marwat & Rehman, 2011; Ganesan & Kumaresan, 2017). Phytochemicals isolated from *A. cepa* include the antioxidant, anti-inflammatory, and antiviral quercetin (**30**) and its derivatives: quercetin 4′-glucoside (**104**), 3-Hydroxyquinine (**105**) (Fossen et al., 1998).

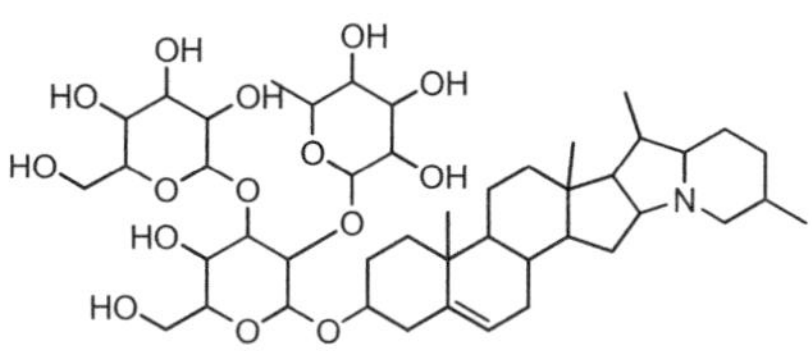

Solanine (99)

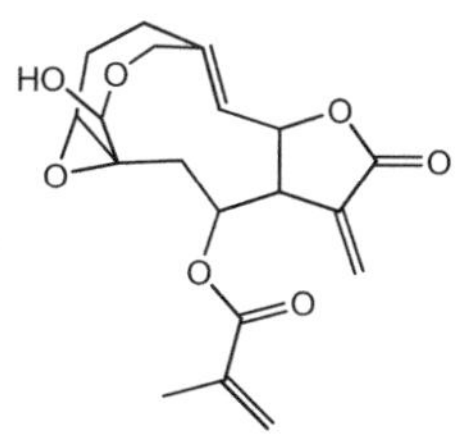

Vernolide (100)

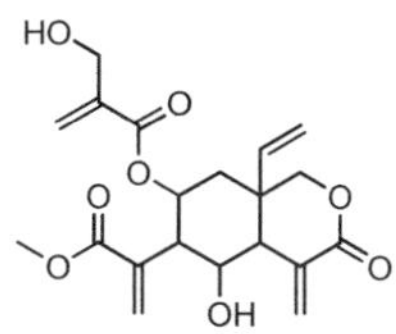

Vernodalol (101)

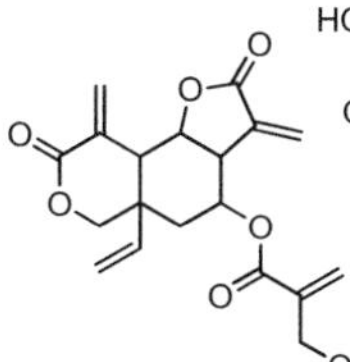

Vernodalin (102)

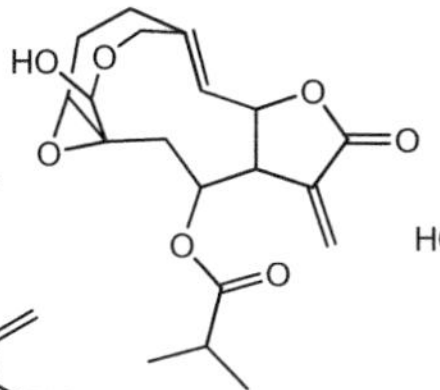

Vernomygdin (103)

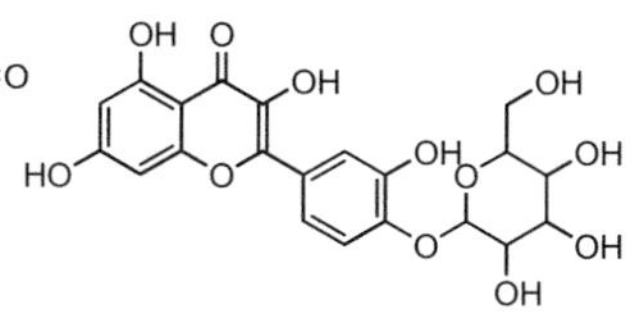

Quercetin 4'-glucoside (104)

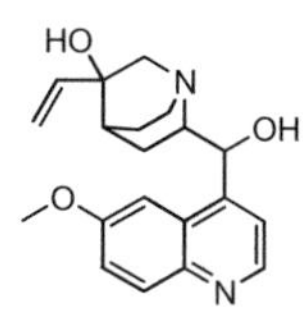

3-Hydroxyquinine (105)

3.3.4.2 Garlic (*Allium sativum* L.)

Garlic (*Allium sativum* L.) (Fig. 3.4B) (Amaryllidaceae) originated from Asia. All the parts of *A. sativum* are utilized in the preparation of decoction mixtures to treat several ailments such as respiratory disorders (Tudu et al., 2022). The garlic cloves are a common spice and a food additive in various dishes around the world. Ethnopharmacologically, garlic is used against acute and chronic bronchitis, whooping cough, tuberculosis, asthma, lung cancer, blood pressure, high cholesterol, COVID-19. It is an expectorant, antibiotic and antiseptic, antiviral, hypoglycemic, anti-carcinogenic, immune-stimulant, hypotensive, hypolipidemic, antidiabetic, blood thinner, and vermifuge (Shawarb et al., 2023; Omara et al., 2020; Anywar et al., 2020; Obakiro et al., 2020; Mailu et al., 2020; Cock & Van-Vuuren, 2020; Villena-Tejada et al., 2021). Garlic contains bioactive compounds including allicin (**106**), *trans*-1-propenyl-thiosulfinate (**107**), E/Z-ajoene (**108**), alliin (**109**), quercetin (**30**) (Shang et al., 2019), diosgenin (**110**), and gitogenin (**111**) (Diretto et al., 2017).

3.3.4.3 Leek (*Allium tuberosum* Rottl. ex Spreng)

Allium tuberosum (Fig. 3.4C), (Amaryllidaceae) is a perennial herb quite popular within the genus *Allium* originating from Southeast and East Asia (Liu et al., 2015). It also grows in several parts of Africa. The crop is reputed for good medicinal and nutritional values. The main edible part of leek are the leaves and the bulb is edible only after processing (Wang et al., 2023). The roots are effective in reducing on blood stasis and improving blood circulation. It is recommended as a remedy for bronchitis, sinusitis, inflammation, constipation, uratic arthritis, gout, kidney disorders, ulcers, and cerebrovascular and cardiovascular diseases (Liu et al., 2015). It has mucolytic properties and, thus, it enhances fluidization of mucus. Phytochemical studies of leek have yielded flavonoids, lignans, phenols, saponins, and sulfur-containing compounds. Studies by Ma et al. (2012) as well as Liu et al. (2015) resulted in the isolation of pharmacologically active compounds which exhibited expectorant, antibacterial, vascular, antioxidant, and endothelial protection activities (Liu et al., 2015; Ma, 2012). These compounds include buddlenol (**112**), daucosterol (**113**), 3,4-dihydroxyacetophenone (**114**), alliin (**109**), and (*E*)-1-allyl-2-(3-(allylsulfonyl)prop-1-en-1-yl) disulfane (**115**) (Wang et al., 2023).

3.3.4.4 Fennel (*Foeniculum vulgare* Mill.)

Fennel, scientifically called *Foeniculum vulgare* (Fig. 3.4D) (Apiaceae or Umbelliferae), is ethnobotanically used to treat respiratory tract diseases, hormonal disorders, gastrointestinal tract pain, and reproductive and cardiovascular-related diseases (Jadid et al., 2023). Fennel has anti-inflammatory, gastro-protective, antiviral, antimicrobial, antioxidant, anti-anxiety, estrogenic, antidiabetic, anti-mutagenic, anticancer activity, hepato-protective, and neuro-protective properties (Jadid et al., 2023).

Isolated compounds include furano-coumarins like bergapten (**116**) and xanthotoxin (**117**) as well as antiviral flavonols like quercetin (**30**) (Nassar et al., 2010; Riska, 2002). Compounds **116** and **117** exhibited antibacterial activity against the gram-positive bacterial strain of *Staphylococcus aureus* (MIC = 1.87 and 3.75 mg/ml, respectively) (Golfakhrabadi et al., 2016). Xanthotoxin has exhibited antifungal activities against the plant pathogens *Rhizoctonia solani*, *Fusarium graminearum*, and *Magnaporthe oryzae* (88, 68, and 83% inhibition, respectively) at a 30 μg/ml dose and an EC_{50} of 14.6 μg/ml against *Coletotrichum capsica* (Song et al., 2017). It is, therefore, possible that such compounds could inhibit pathogenic microbes within the human pulmonary tract. Bergapten also exhibited anti-carcinogenic activity (IC_{50} of 40.05 μM) against osteosarcoma (Saos 2) cell lines, whereas xanthotoxin inhibited neuroblastoma (SK-N-AS) and metastatic colon cancer (SW640) cells (IC_{50} = 56.3 and 88.5 μM, respectively) (Bartnik et al., 2017; Bartnik et al., 2023).

3.3.5 Underground Vegetables (Roots)

Photos of underground vegetables (roots) are shown in Fig. 3.5.

3.3.5.1 Carrot (*Daucus carota* subsp. *sativus* L.)

Daucus carota subsp. *sativus* (Fig. 3.5A), (Apiaceae) is used in South Africa to treat tuberculosis (Que et al., 2019). Juice made from a mixture of carrot, fennel bulbs, onions, and honey (or molasses) is used to treat cough within the Mediterranean region (Pamploma-Roger, 2006). Carrot is also used in treating arteriosclerosis, liver inflammation, heartburn, liquid retention, diseases of the retina, and in attempts to quit smoking (Pamploma-Roger, 2006). The characteristic orange color is attributed to the presence of carotenoids, and thus carrots are considered to be rich in lycopene (**118**), crocin (**119**), falcarinol (**120**), falcarindiol (**121**), daucol (**122**), and carotol (**123**) (Horbowicz et al., 2008; Kumarasamy et al., 2005). These compounds are beneficial in fighting against lung cancer, emphysema, asthma, fibrosis, acute lung

A B

Fig. 3.5 Vegetables consumed in the form of roots, (**A**) *Daucus carota* subsp. *sativus* L and (**B**) *Raphanus sativus* L injury, and COPD. The carotenoids do activate the NRF2/HO-1 pathway and inhibit the MAPK, NFK-JAK/STAT-3, and PI3K/AKT pathways. Epidemiological investigations have revealed that when carotenoids are included in the diet, the risk of lung complications is lowered (Manochkumar et al., 2022).

3.3.5.2 Radish (*Raphanus sativus* L.)

The root part of *Raphanus sativus* (Fig. 3.5B), (Brassicaceae) is the main edible organ. The root tuber of *R. sativus* may appear in many color shades like white, purple, or red, due to anthocyanin accumulation (Gamba et al., 2021). Radishes, traditionally, find use in the treatment of bronchial catarrh, bronchitis, laryngitis, sinusitis, tobacco addiction, and hepatic and biliary diseases. It has mucolytic, expectorant, antibiotic, anticancer, cell-protecting, choleretic, cholagogue, liver regeneration, and mood-stabilization properties (Manivannan et al., 2019). Radish is reputed for the presence of sulfur-containing glucosinolates, particularly glucoraphanin (**124**), which is responsible for the sharp pungent taste. Upon enzyme action, glucoraphanin is hydrolyzed to raphanin (**125**), which has mucolytic properties, hence softening the mucus/phlegm and facilitating its expulsion from the pulmonary airways. Compound **125** is also active against a wide range of microorganisms, including *Corynebacterium diphtheria, Mycobacterium* sp., *Staphylococcus aureus*, *Streptococcus pyogenes*, *Bordetella pertussis*, and some pathogenic fungi (Ivànovics & Horvath, 1947). Other compounds present include pelargonidin (**126**), sinapine (**127**), and scopoletin (**128**) (Noreen et al., 2023; Gutiérrez & Perez, 2004).

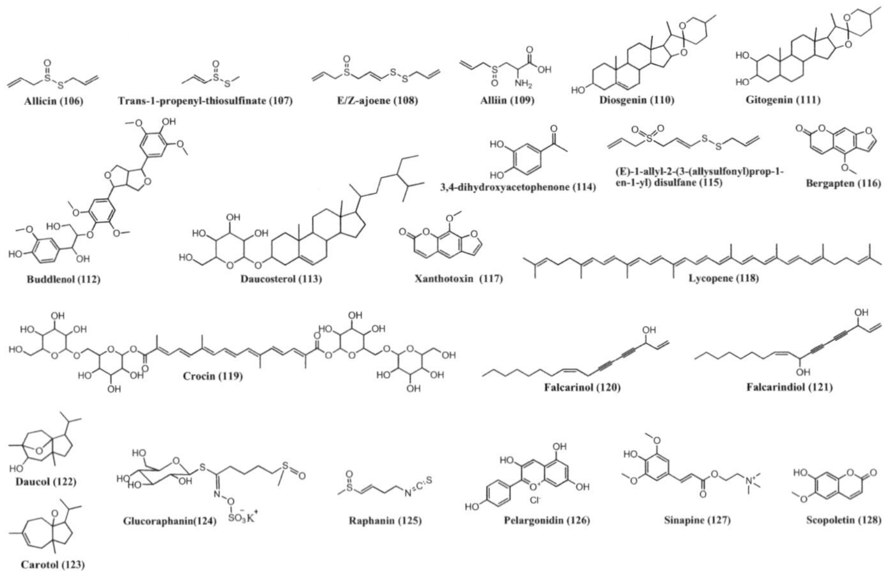

3.4 Future Prospects

In the future, with the general public moving toward maximizing utilization of resources plus tendencies toward circular economy, it can be forecasted that farmers will be growing vegetables and fruits not only for food, but also as sources of nutraceuticals such as those against respiratory troubles. Several vegetables and fruits will be used as supplements to the available conventional drugs in treating health conditions. Various parts of the food crops, other than the edible parts, will be considered valuable, not to be discarded but to be processed for utilization in medicine and other applications. Given the vast scientific knowledge, individuals will be making informed choices of what to consume depending on their health status and motivation, which will greatly contribute to longevity and increased lifespan. With the sensitization on the importance of particular plants, there will be an increase in the demand for vegetables and fruits which will translate into domestication of what are currently considered to be wild plants and commercialization of the vegetables and fruits. Furthermore, it is expected that, with time, the fruits and vegetables will be subjected to targeted processing in order to extract the active ingredients, standardize the biomass and where possible the ingredients will be synthesized chemically or by use of biotechnological tools.

3.5 Conclusions

This chapter concludes by highlighting the vital role that a variety of naturally occurring dietary fruits, vegetables, and plants play in promoting respiratory health. Based on both traditional and modern ethnopharmacological research, this chapter highlights the unique bioactive compounds present in these medicinal plants that have shown promise in easing symptoms and providing therapeutic aid for respiratory illnesses. Compounds like flavonoids, alkaloids, terpenoids, phenolics, saponins, vitamins, and amino acids have the potential to exhibit antimicrobial, anti-inflammatory, and immune-boosting properties. As a result, they are helpful allies in the treatment of respiratory conditions, from bronchitis and asthma to flu, colds, and coughs. The findings demonstrate the importance of including naturally occurring foods with therapeutic properties into regular meals as a preventative strategy and as an adjunct to existing treatments for respiratory conditions. It is crucial to recognize that more clinical studies are required to validate the safety and efficacy of these plants in a variety of settings and individuals.

References

Abe-Kanoh, N., et al. (2019). Sesamin catechol glucuronides exert anti-inflammatory effects by suppressing interferon β and inducible nitric oxide synthase expression through deconjugation in macrophage-like J774. 1 cells. *Journal of Agricultural and Food Chemistry, 67*(27), 7640–7649.

Abi, E., Abi, I., & Ladan, M. J. (2017). Hypoglycaemic effect of Abelmoschus Esculentus extracts in alloxan-induced diabetic wistar rats. *Endocrinology and Diabetes Research, 3*, 1.

Abosi, A. O., & Raseroka, B. H. (2003). In vivo antimalarial activity of Vernonia amygdalina. *British Journal of Biomedical Science, 60*(2), 89–91.

Abu, T., Rex-Ogbuku, E., & Idibiye, K. (2018). A review: Secondary metabolites of Uvaria chamae p. Beauv.(Annonaceae) and their biological activities. *International Journal of Agriculture Environment and Food Sciences, 2*(4), 177–185.

Abubakar, I. B., et al. (2022). Traditional medicinal plants used for treating emerging and re-emerging viral diseases in northern Nigeria. *European Journal of Integrative Medicine, 49*, 102094.

Adeleke, B. S., & Babalola, O. O. (2020). Oilseed crop sunflower (Helianthus annuus) as a source of food: Nutritional and health benefits. *Food Science & Nutrition, 8*(9), 4666–4684.

Adesina, S. K., Johnny, I. I., & Olayiwola, G. (2017). Plants in respiratory disorders II antitussives. *British Journal of Pharmaceutical Research, 16*(3), 1–21.

Ahle, N. W., & Hamlet, M. P. (1987). Enzymatic frostbite eschar debridement by bromelain. *Annals of Emergency Medicine, 16*(9), 1063–1065.

Ahmadi, A., et al. (2016). Inhibition of chikungunya virus replication by hesperetin and naringenin. *RSC Advances, 6*(73), 69421–69430.

Ahmed, O. M., et al. (2021). Musa paradisiaca L. leaf and fruit peel hydroethanolic extracts improved the lipid profile, glycemic index and oxidative stress in nicotinamide/streptozotocin-induced diabetic rats. *Veterinary Medicine and Science, 7*(2), 500–511.

Aiyeloja, A. A., & Bello, O. A. (2006). Ethnobotanical potentials of common herbs in Nigeria: A case study of Enugu state. *Educational Research and Reviews, 1*(1), 16.

Ajayi, A. M., et al. (2016). Antinociceptive and anti-inflammatory effects of aqueous extract of Chenopodium opulifolium schrad leaves. *Journal of Intercultural Ethnopharmacology, 6*(1), 14–21.

Ali, M. Y., et al. (2019). Ethnobotany, phytochemistry, and pharmacology of Angelica decursiva Fr. et Sav. *Natural Product Sciences, 25*(3), 181–199.

Al-Karmadi, A., & Okoh, A. (2024). An overview of date (Phoenix dactylifera) fruits as an important global food resource. *Food, 13*(7), 1024.

Allam, A. E., et al. (2021). In silico study of natural compounds from sesame against COVID-19 by targeting M pro, PL pro and RdRp. *RSC Advances, 11*(36), 22398–22408.

Anis, M., & Iqbal, M. (1994). Medicinal plantlore of Aligarh, India. *International Journal of Pharmacognosy, 32*(1), 59–64.

Anywar, G., et al. (2020). Indigenous traditional knowledge of medicinal plants used by herbalists in treating opportunistic infections among people living with HIV/AIDS in Uganda. *Journal of Ethnopharmacology, 246*, 112205.

Araújo, L. U., et al. (2010). Profile of wound healing process induced by allantoin. *Acta Cirurgica Brasileira, 25*, 460–461.

Aruna, R., Geetha, A., & Suguna, P. (2014). Expression of caspase activation recruitment and pyrin domain levels of apoptosis-associated speck-like protein complex in the pancreas of rats subjected to experimental pancreatitis: Influence of rutin administration. *Human & Experimental Toxicology, 33*(9), 940–948.

Asiimwe, S., et al. (2014). Documentation and consensus of indigenous knowledge on medicinal plants used by the local communities of western Uganda. *Journal of Natural Product and Plant Resource, 4*(1), 34–42.

Badgujar, S. B., et al. (2014). Traditional uses, phytochemistry and pharmacology of Ficus carica: A review. *Pharmaceutical Biology, 52*(11), 1487–1503.

Bartnik, M., et al. (2017). 8-methoxypsoralen reduces AKT phosphorylation, induces intrinsic and extrinsic apoptotic pathways, and suppresses cell growth of SK-N-AS neuroblastoma and SW620 metastatic colon cancer cells. *Journal of Ethnopharmacology, 207*, 19–29.

Bartnik, M., et al. (2023). Evaluation of the biological effect of non-UV-activated Bergapten on selected human tumor cells and the insight into the molecular mechanism of its action. *International Journal of Molecular Sciences, 24*(21), 15555.

Begashaw, T., Dagne, A., & Yibeltal, D. (2023). Review on phytochemistry, medicinal properties, and toxicities of the genus Musa. *Traditional Medicine, 4*(2), 1–24.

Behiry, S. I., et al. (2019). Antifungal and antibacterial activities of Musa paradisiaca L. Peel extract: HPLC analysis of phenolic and flavonoid contents. *PRO, 7*, 215.

Betti, J. L. (2004). An ethnobotanical study of medicinal plants among the Baka pygmies in the Dja biosphere reserve, Cameroon. *African Study Monographs, 25*(1), 1–27.

Boulos, L. (1983). *Medicinal plants of North Africa*. Michigan: Reference Publications, Inc.

Braun, J. M., Schneider, B., & Beuth, H. J. (2005). Therapeutic use, efficiency and safety of the proteolytic pineapple enzyme Bromelain-POS® in children with acute sinusitis in Germany. *In Vivo, 19*(2), 417–421.

Bulbul, I. J., et al. (2011). Antibacterial, cytotoxic and antioxidant activity of chloroform, n-hexane and ethyl acetate extract of plant Amaranthus spinosus. *International Journal of PharmTech Research, 3*(3), 1675–1680.

Bunalema, L., et al. (2014). Knowledge on plants used traditionally in the treatment of tuberculosis in Uganda. *Journal of Ethnopharmacology, 151*(2), 999–1004.

Cámara-Leret, R., et al. (2014). Ethnobotanical knowledge is vastly under-documented in northwestern South America. *PLoS One, 9*(1), e85794.

Campa, C., & Petivallet, A. (2018). *Achieving sustainable cultivation of coffee. Beneficial compounds from coffee leaves*. Burleigh Science Publishing.

Caro-Gómez, E., et al. (2019). Green coffee extract improves cardiometabolic parameters and modulates gut microbiota in high-fat-diet-fed ApoE−/-mice. *Nutrients, 11*(3), 497.

Chakraborty, A. J., et al. (2022). Allium cepa: A treasure of bioactive phytochemicals with prospective health benefits. *Evidence-based Complementary and Alternative Medicine, 2022*(18), 4586318.

Challand, S., & Willcox, M. (2009). A clinical trial of the traditional medicine Vernonia amygdalina in the treatment of uncomplicated malaria. *The Journal of Alternative and Complementary Medicine, 15*(11), 1231–1237.

Chand, J., et al. (2022). Phytochemistry and polypharmacology of cleome species: A comprehensive ethnopharmacological review of the medicinal plants. *Journal of Ethnopharmacology, 282*, 114600.

Chataika, B. Y., et al. (2020). Utilisation of spider plants (Gynandropsis gynandra, l. briq) amongst farming households and consumers of northern Namibia. *Sustainability, 12*, 6604.

Chaudhary, M. A., et al. (2012). Evaluation of gut modulatory and bronchodilator activities of Amaranthus spinosus Linn. *BMC Complementary and Alternative Medicine, 12*, 1–11.

Chavan, M. J., Wakte, P. S., & Shinde, D. B. (2010). Analgesic and anti-inflammatory activity of Caryophyllene oxide from Annona squamosa L. bark. *Phytomedicine, 17*(2), 149–151.

Chavan, M. J., Wakte, P. S., & Shinde, D. B. (2011). Analgesic and anti-inflammatory activities of 18-acetoxy ent kaur 16 ene from Annona squamosa L. bark. *Inflammopharmacology, 19*, 111–115.

Chen, Y., Chen, J. W., & Li, X. (2011). Cytotoxic bistetrahydrofuran annonaceous acetogenins from the seeds of Annona squamosa. *Journal of Natural Products, 74*(11), 2477–2481.

Chen, X. M., Ma, Z., & Kitts, D. D. (2018). Effects of processing method and age of leaves on phytochemical profiles and bioactivity of coffee leaves. *Food Chemistry, 249*, 143–153.

Chen, X., et al. (2022). Solanum nigrum Linn.: An insight into current research on traditional uses, phytochemistry, and pharmacology. *Frontiers in Pharmacology, 16*(13), 918071.

Chimponda, T., & Mukanganyama, S. (2010). Antimycobacterial activities of selected medicinal plants from Zimbabwe against Mycobacterium aurum and Corynebacterium glutamicum. *Tropical Biomedicine, 27*(3), 595–610.

Chou, D. J., et al. (2014). Alkaloids of Linderae Radix suppressed the lipopolysaccharide-induced expression of cytokines in cultured macrophage RAW 264.7 cells. *Cell Medicine, 4*(4), 28–21.

Chowdhury, N. S., et al. (2019). A review on ethnomedicinal, pharmacological, phytochemical and pharmaceutical profile of lady's finger (Abelmoschus esculentus L.) plant. *Pharmacology & Pharmacy, 10*(02), 94–108.

Chukwujekwu, J. C., et al. (2009). Antiplasmodial and cytotoxic activity of isolated sesquiterpene lactones from the acetone leaf extract of Vernonia colorata. *South African Journal of Botany, 75*(1), 176–179.

Cock, I. E., & Van-Vuuren, S. F. (2020). The traditional use of southern African medicinal plants for the treatment of bacterial respiratory diseases: A review of the ethnobotany and scientific evaluations. *Journal of Ethnopharmacology, 263*, 113204.

Cos, P., et al. (2002). Further evaluation of Rwandan medicinal plant extracts for their antimicrobial and antiviral activities. *Journal of Ethnopharmacology, 79*(2), 155–163.

Costa, E. V., et al. (2012). Antioxidant and antimicrobial activities of aporphinoids and other alkaloids from the bark of Annona salzmannii A. DC. (Annonaceae). *Natural Product Research, 27*(11), 1002–1006.

Das, K. K., Das, S. N., & Gupta, S. D. (2001). Influence of ascorbic acid against nickel-induced hepatic lipid peroxidation in rats. *Journal of Basic and Clinical Physiology and Pharmacology, 12*(3), 187–196.

de Almeida, R. F., et al. (2019). Nutraceutical compounds: Echinoids, flavonoids, xanthones and caffeine identified and quantitated in the leaves of Coffea arabica trees from three regions of Brazil. *Food Research International, 115*, 493–503.

Dele, O. F., & Ojone, A. S. (2023). Ethno-medicinal plants used for respiratory disorders among the Igala and Yoruba (Okun) people in Kogi State, Nigeria. *Journal of Biology and Genetic Research, 9*(1), 18–36.

Diretto, G., et al. (2017). Tissue-specific accumulation of sulfur compounds and saponins in different parts of garlic cloves from purple and white ecotypes. *Molecules, 22*(8), 1359.

Dongre, P., et al. (2023). Botanical description, chemical composition, traditional uses and pharmacology of Citrus sinensis: An updated review. *Pharmacological Research - Modern Chinese Medicine, 8*, 100272.

Dos Santos, A. F., & Sant'Ana, A. E. G. (2000). The molluscicidal activity of plants used in Brazilian folk medicine. *Phytomedicine, 6*(6), 431–438.

Du, S., et al. (2018). Mangiferin: An effective therapeutic agent against several disorders. *Molecular Medicine Reports, 18*(6), 4775–4786.

Dzoyem, J. P., et al. (2014). The 15-lipoxygenase inhibitory, antioxidant, antimycobacterial activity and cytotoxicity of fourteen ethnomedicinally used African spices and culinary herbs. *Journal of Ethnopharmacology, 156*, 1–8.

Dzoyem, J. P., et al. (2016). Antimycobacterial activity against different pathogens and selectivity index of fourteen medicinal plants used in southern Africa to treat tuberculosis and respiratory ailments. *South African Journal of Botany, 102*, 70–74.

Edim, E. H., et al. (2012). A review on Gongronema latifolium (Utasi): A novel antibiotic against Staphylococcus aureus related infections. *International Journal of Biochemistry and Biotechnology, 1*(8), 204–208.

Efferth, T., et al. (2002). Activity of ascaridol from the anthelmintic herb Chenopodium anthelminticum L. against sensitive and multidrug-resistant tumor cells. *Anticancer Research, 22*(6C), 4221–4224.

Ejechi, B. O., & Akpomedaye, D. E. (2005). Activity of essential oil and phenolic acid extracts of pepperfruit (Dennetia tripetala G. Barker; Anonaceae) against some food-borne microorganisms. *African Journal of Biotechnology, 4*, 258–261.

Emmanuel, M. M., & Didier, D. S. (2012). Traditional knowledge on medicinal plants use by ethnic communities in Douala, Cameroon. *European Journal of Medicinal Plants, 2*, 159.

Erasto, P., Grierson, D. S., & Afolayan, A. J. (2007). Antioxidant constituents in Vernonia amygdalina. Leaves. *Pharmaceutical Biology, 45*(3), 195–199.

Erlejman, A. G., et al. (2008). TNF alpha-induced NF-kappaB activation and cell oxidant production are modulated by hexameric procyanidins in Caco-2 cells. *Archives of Biochemistry and Biophysics, 476*(2), 186–195.

Esan, V. I., et al. (2024). Ethnobotanical study of sweet orange (Citrus sinensis) and mango (Mangifera indica L.) varieties in Nigeria. *Ethnobotany Research and Applications, 29*, 1–14.

Esezah, K., et al. (2015). Antifungal medicinal plants used by communities adjacent to Bwindi impenetrable National Park, South Western Uganda. *European Journal of Medicinal Plants, 7*, 184–192.

Essien, J. P., Ebong, G. A., & Akpan, E. J. (2007). Antioxidant and antitussive properties of Gongronema latifolium leaves used locally for the treatment of fowl cough in Nigeria. *Journal of Applied Sciences & Environmental Management, 11*(4), 47–50.

Fan, S., et al. (2019). Limonin: A review of its pharmacology, toxicity, and pharmacokinetics. *Molecules, 24*(20), 3679.

Ferizi, R., Ramadan, M. F., & Maxhuni, Q. (2023). Black seeds (Nigella sativa) medical application and pharmaceutical perspectives. *Journal of Pharmacy & Bioallied Sciences, 15*(2), 63–67.

Forouzanfar, F., Bazzaz, B. S. F., & Hosseinzadeh, H. (2014). Black cumin (Nigella sativa) and its constituent (thymoquinone): A review on antimicrobial effects. *Iranian Journal of Basic Medical Sciences, 17*(12), 929.

Fossen, T., Pedersen, A. T., & Andersen, Ø. M. (1998). Flavonoids from red onion (Allium cepa). *Phytochemistry, 47*(2), 281–285.

Gai, F., et al. (2020). Sunflower (Helianthus annuus L.) plants at various growth stages subjected to extraction—Comparison of the antioxidant activity and phenolic profile. *Antioxidants, 9*(6), 535.

Galani, V. J. (2019). Musa paradisiaca Linn. - A comprehensive review. *Scholars International Journal of Traditional and Complementary Medicine, 2*(4), 45–56.

Gamba, M., et al. (2021). Nutritional and phytochemical characterization of radish (Raphanus Sativus): A systematic review. *Trends in Food Science & Technology, 113*, 205–218.

Ganesan, C. M., & Kumaresan, G. (2017). Ethnomedicinal approaches for treating various disease by Irulas tribals, Konbanur village, Anaikatti hills, the western Ghats, Coimbatore district. *Kongunadu Research Journal, 4*(2), 1–8.

Ganesan, K., & Xu, B. (2017). Polyphenol-rich dry common beans (Phaseolus vulgaris L.) and their health benefits. *International Journal of Molecular Sciences, 11*(18), 2331.

Gavamukulya, Y., Wamunyokoli, F., & El-Shemy, H. A. (2017). Annona muricata: Is the natural therapy to most disease conditions including cancer growing in our backyard? A systematic review of its research history and future prospects. *Asian Pacific Journal of Tropical Medicine, 10*, 835–848.

Gerard, G. (1972). Anti-cancer therapy with bromelain. *Agressologie, 3*, 261–274.

Ghareeb, M. A., et al. (2016). A new kaempferol glycoside with antioxidant activity from Chenopodium ambrosioides growing in Egypt. *Oriental Journal of Chemistry, 32*, 3053–3061.

Golfakhrabadi, F., et al. (2016). In vitro antimicrobial and acetylcholinesterase inhibitory activities of coumarins from Ferulago carduchorum. *Medicinal Chemistry Research, 25*, 1623–1629.

Grzesik, M., et al. (2018). Antioxidant properties of catechins: Comparison with other antioxidants. *Food Chemistry, 241*, 480–492.

Guisan, A., et al. (2013). Predicting species distributions for conservation decisions. *Ecology Letters, 16*(12), 1424–1435.

Gupta, A., et al. (2017). Dietary intake of minerals, vitamins, and trace elements among geriatric population in India. *Biological Trace Element Research, 180*(1), 28–38.

Gus Kov, E. P., et al. (2004). Allantoin as a vitamin. In *Doklady biochemistry & biophysics* (Vol. 398).

Gutiérrez, R. M. P., & Perez, R. L. (2004). Raphanus sativus (Radish): Their chemistry and biology. *The Scientific World Journal, 4*(1), 811–837.

Gyebi, G. A., et al. (2018). Iloneoside: A cytotoxic ditigloylated pregnane glycoside from the leaves of Gongronema latifolium Benth. *Natural Product Research, 32*(24), 2882–2886.

Hamed, A. N. E., et al. (2022). Chemical constituents from Carica papaya Linn. leaves as potential cytotoxic, EGFRwt and aromatase (CYP19A) inhibitors; a study supported by molecular docking. *RSC Advances, 12*(15), 9154–9162.

Hammad, A. R., & Malik, R. (2021). A review of apple and its medicinal uses in Unani Medicine. *International Journal of Unani and Integrative Medicin, 5*(2), 165–169.

Hammoda, H. M., et al. (2015). Two new flavone glycosides from Chenopodium ambrosioides growing wildly in Egypt. *Records of Natural Products, 9*, 609–613.

Hills, D. (2010). Ethnopharmacological studies on the medicinal plants used by tribal inhabitants of Kottur. *Environment & We: An International Journal of Science and Technology, 2010*(5), 57–64.

Horbowicz, M., et al. (2008). Anthocyanins of fruits and vegetables—Their occurrence analysis and role in human nutrition. *Vegetable Crops Research Bulletin, 68*, 5–22.

Hosseinzadeh, H., et al. (2007). Effect of Nigella sativa seed extracts on ischemia-reperfusion in rat skeletal muscle. *Pharmacology, 2*, 326–335.

Hou, G., & Meng, Q. (2022). Sesame (Sesamum indicum L.): A comprehensive review of nutritional value, phytochemical composition, health benefits, development of food, and industrial applications. *Nutrients, 14*(19), 4079.

https://safelizwellness.com/en/product-view/enciclopedia-de-las-plantas-medicinales/

Hussain, M. I., et al. (2019). Immune modulatory and anti-oxidative effect of selenium against pulmonary tuberculosis. *Pakistan Journal of Pharmaceutical Sciences, 32*(2), 779–784.

Ibironke, G. F., & Ajiboye, K. I. (2007). Studies on the anti-inflammatory and analgesic properties of Chenopodium ambrosioides leaf extract in rats. *International Journal of Pharmacology, 3*(1), 111–115.

Islam, M. T. (2019). Phytochemical information and pharmacological activities of Okra (Abelmoschus esculentus): A literature-based review. *Phytotherapy Research, 33*(1), 72–80.

Ivànovics, G., & Horvath, S. (1947). Raphanin, an antibacterial principle of the radish (Raphanus sativus). *Nature, 160*, 297–298.

Iwalokun, B. A. (2008). Enhanced antimalarial effects of chloroquine by aqueous Vernonia amygdalina leaf extract in mice infected with chloroquine resistant and sensitive Plasmodium berghei strains. *African Health Sciences, 8*(1), 25–35.

Jadid, N., et al. (2023). The medicinal Umbelliferae plant Fennel (Foeniculum vulgare Mill.): Cultivation, traditional uses, phytopharmacological properties, and application in animal husbandry. *Arabian Journal of Chemistry, 16*(3), 104541.

Javed, S., et al. (2010). Nutritional, phytochemical potential and pharmacological evaluation of Nigella Sativa (Kalonji) and Trachyspermum Ammi (Ajwain). *Journal of Medicinal Plants Research, 6*(5), 768–775.

Jeng, K. C. G., et al. (2005). Sesamin inhibits lipopolysaccharide-induced cytokine production by suppression of p38 mitogen-activated protein kinase and nuclear factor-κB. *Immunology Letters, 97*(1), 101–106.

Jiang, Y., et al. (2020). Procyanidin B2 suppresses lipopolysaccharides-induced inflammation and apoptosis in human type II alveolar epithelial cells and lung fibroblasts. *Journal of Interferon & Cytokine Research, 40*(1), 54–63.

Johari, J., et al. (2012). Antiviral activity of baicalein and quercetin against the Japanese encephalitis virus. *International Journal of Molecular Sciences, 13*(12), 16785–16795.

John, O. O. (2018). Comparative studies of phytochemistry, proximate, mineral and vitamin compositions of Citrus tangerina and Citrus sinensis crude fruit peel extracts. *Acta Scientific Pharmaceutical Sciences, 2*, 22–26.

Juma, S. (2022). Anti-plasmodial activity and phytochemical analysis of leaf extract of Chenopodium Opulifolium Schrad. *EX W.D.J. Koch and Ziz used in Uganda.*

Kamatenesi, M. M., et al. (2013). Ethno-pharmacological screening of Vernonia amygdalina and Cleome gynandra traditionally used in childbirth in Western Uganda. In *NAPRECA symposium book of proceedings*.
Kandasamy, P., & Shanmugapriya, C. (2015). Medicinal and nutritional characteristics of fruits in human health. *Journal of Medicinal Plants Studies, 4*(4), 124–131.
Karou, S. D., et al. (2011). Ethnobotanical study of medicinal plants used in the management of diabetes mellitus and hypertension in the Central Region of Togo. *Pharmaceutical Biology, 49*(12), 1286–1297.
Kasali, F. M., et al. (2021). Ethnomedical uses, chemical constituents, and evidence-based pharmacological properties of Chenopodium ambrosioides L.: Extensive overview. *Future Journal of Pharmaceutical Sciences, 7*, 153.
Kasole, R., Martin, H. D., & Kimiywe, J. (2019). Traditional medicine and its role in the management of diabetes Mellitus: "Patients" and herbalists' perspectives. *Evidence Based Complementary Alternative Medicines, 2019*, 2835691.
Kayode, J., & Kayode, G. (2008). Ethnomedicinal survey of botanicals used in treating sexually transmitted diseases in Ekiti State, Nigeria. *Ethnobotanical Leaflets, 12*, 44–55.
Khan, Y., et al. (2010). Evaluation of hypolipidemic effect of citrus lemon. *Journal of Basic and Applied Sciences, 6*(1), 39–43.
Kim, G. S., et al. (1998). Two new mono-tetrahydrofuran ring acetogenins, annomuricin E and muricapentocin, from the leaves of Annona muricata. *Journal of Natural Products, 61*(4), 432–436.
Kim, H. R., et al. (2010). Chlorogenic acid suppresses pulmonary eosinophilia, IgE production, and Th2-type cytokine production in an ovalbumin-induced allergic asthma: Activation of STAT-6 and JNK is inhibited by chlorogenic acid. *International Immunopharmacology, 10*(10), 1242–1248.
Kleef, R., Delohery, T. M., & Bovbjerg, D. H. (1996). Selective modulation of cell adhesion molecules on lymphocytes by bromelain protease. *Pathobiology, 64*, 339–346.
Kokanova-Nedialkova, Z., Nedialkov, P., & Nikolov, S. (2009). The genus Chenopodium: Phytochemistry, ethnopharmacology and pharmacology. *Pharmacognosy Reviews, 3*(6), 280.
Konkon, N. G., et al. (2017). Medicinal plants used for treatment of diabetes by traditional practitioners in the markets of Abidjan district in Côte d'Ivoire. *Journal of Medicinal Plants Studies, 5*(2), 39–48.
Kumar, R., et al. (2007). Evaluation of Chenopodium ambrosioides oil as a potential source of antifungal, antiaflatoxigenic and antioxidant activity. *International Journal of Food Microbiology, 115*(2), 159–164.
Kumarasamy, Y., et al. (2005). The assessment of biological activities associated with the major constituents of the methanol extract of "Wild Carrot" (Daucus carota L.) seeds. *Journal of Herbal Pharmacotherapy, 5*, 61–72.
Kupchan, S. M., et al. (1969). Tumor inhibitors. XLVII. Vernodalin and vernomygdin, two new cytotoxic sesquiterpene lactones from Vernonia amygdalina Del. *The Journal of Organic Chemistry, 34*(12), 3908–3911.
Labaki, W. W., & Han, M. K. (2020). Chronic respiratory diseases: A global view. *The Lancet. Respiratory Medicine, 8*, 531–533.
Lasswell, J. W. L., & Hufford, C. D. (1977). Cytotoxic C-benzylated flavonoids from Uvaria chamae. *The Journal of Organic Chemistry, 42*(8), 1295–1302.
Lautenschläger, T., et al. (2018). First large-scale ethnobotanical survey in the province of Uíge, northern Angola. *Journal of Ethnobiology and Ethnomedicine, 14*, 1–73.
Lawal, I. O., et al. (2020). Ethnobotanical survey of plants used for treating cough associated with respiratory conditions in Ede South local government area of Osun State, Nigeria. *Plants, 9*(5), 647.
Lawin, I. F., et al. (2015). Ethnobotanical assessment of the plant species used in the treatment of diabetes in the Sudano-Guinean zone of Benin. *Journal of Animal and Plant Sciences, 26*(1), 4108–4123.

Lazreg-Aref, H., et al. (2011). In vitro cytotoxic and antiviral activities of Ficus carica latex extracts. *Natural Product Research, 25*(3), 310–319.

Lee, S. J., & Lim, K. T. (2008). Cell death signal by glycine-and proline-rich plant glycoprotein is transferred from cytochrome c and nuclear factor kappa B to caspase 3 in Hep3B cells. *The Journal of Nutritional Biochemistry, 19*(3), 166–174.

Liang, N., & Kitts, D. D. (2015). Role of chlorogenic acids in controlling oxidative and inflammatory stress conditions. *Nutrients, 8*(1), 16.

Lim, K. T. (2005). Glycoprotein isolated from Solanum nigrum L. kills HT-29 cells through apoptosis. *Journal of Medicinal Food, 8*(2), 215–226.

Lima, E. B. C., et al. (2015). Cocos nucifera (L.)(Arecaceae): A phytochemical and pharmacological review. *Brazilian Journal of Medical and Biological Research, 48*, 953–964.

Lin, Y., et al. (2014). Content determination of the flavonoids in the different parts and different species of Abelmoschus esculentus L. by reversed phase-high performance liquid chromatograph and colorimetric method. *Pharmacognosy Magazine, 10*(39), 278–284.

Liu, C. X., et al. (2015). Laboratory evaluation of Larvicidal activity of the essential oil of Allium tuberosum roots and its selected major constituent compounds against Aedes albopictus (Diptera: Culicidae). *Journal of Medical Entomology, 52*(3), 437–441.

López-Martín, J., et al. (2002a). Chromone and phenanthrene alkaloids from Dennettia tripetala. *Chemical & Pharmaceutical Bulletin, 50*, 1613–1615.

López-Martín, J., et al. (2002b). Chromone and phenanthrene alkaloids from Dennettia tripetala. *Chemical and Pharmaceutical Bulletin, 50*(12), 1613–1615.

Lu, Y., & Foo, L. Y. (1998). Constitution of some chemical components of apple seed. *Food Chemistry, 61*(1–2), 29–33.

Lv, X., et al. (2015). Citrus fruits as a treasure trove of active natural metabolites that potentially provide benefits for human health. *Chemistry Central Journal, 9*, 1–14.

Ma, Y. (2012). Study on the chemical composition of Allium tuberosum roots.

Magassouba, F. B., et al. (2007). Ethnobotanical survey and antibacterial activity of some plants used in Guinean traditional medicine. *Journal of Ethnopharmacology, 114*(1), 44–53.

Mahmood, A., et al. (2011). Indigenous medicinal knowledge of common plants used by local people of Hattian Bala District, Azad Jammu and Kashmir (AJK), Pakistan. *Journal of Medicinal Plants Research, 5*(23), 5517–5521.

Mahmood, S., et al. (2019). A critical review on nutritional and medicinal importance of lemon. *Acta Scientific Agriculture, 3*(6), 95–97.

Mailu, J. K., et al. (2020). Medicinal plants used in managing diseases of the respiratory system among the Luo community: An appraisal of Kisumu East Sub-County, Kenya. *Chinese Medicine, 15*, 95.

Malebo, H. M., et al. (2014). Diprenylated indole alkaloids from fruits of Hexalobus monopetalus. *Natural Products and Bioprospecting, 4*, 101–105.

Manivannan, A., et al. (2019). Deciphering the nutraceutical potential of Raphanus sativus—A comprehensive overview. *Nutrients, 11*(2), 402.

Mann, A., et al. (2007). An ethnobotanical survey of indigenous flora for treating tuberculosis and other respiratory diseases in Niger State, Nigeria. *Journal of Phytomedicine and Therapeutics, 12*, 1–21.

Manochkumar, J., et al. (2022). Untapping the protective role of carotenoids against respiratory diseases. *Phytomedicine, 104*, 154286.

Maroyi, A. (2011). An ethnobotanical survey of medicinal plants used by the people in Nhema communal area. Zimbabwe. *Journal of ethnopharmacology, 136*(2), 347–354.

Maroyi, A. (2019). Annona stenophylla Engl. & Diels: A review of its botany, medicinal uses and biological activities. *Journal of Pharmaceutical Sciences and Research, 11*, 3385–3390.

Maroyi, A. (2023). Medicinal uses of the Fabaceae Family in Zimbabwe: A review. *Plants, 12*(6), 1255.

Maroyi, A. (2024). Plants used for the management of paediatric ailments in Zimbabwe: A narrative review. *Journal of Medicinal Plants for Economic Development, 8*(1), 9.

Marques, L., & Vale, N. (2022). Salbutamol in the management of asthma: A review. *International Journal of Molecular Sciences, 23*(22), 14207.

Marwat, S. K., & Rehman, F. U. (2011). Medicinal folk recipes used as traditional phytotherapies in district Dera Ismail Khan, KPK, Pakistan. *Pakistan Journal of Botany, 43*(3), 1453–1462.

Mashamaite, C. V., Manyevere, A., & Chakauya, E. (2022). Cleome gynandra: A wonder climate-smart plant for nutritional security for millions in semi-arid areas. *Frontiers in Plant Science, 13*, 1003080.

Mazzeo, A., Magarelli, A., & Ferrara, G. (2024). The fig (Ficus carica L.): Varietal evolution from Asia to Puglia region, southeastern Italy. *CABI Agriculture and Bioscience, 5*, 57.

Mehrbod, P., et al. (2018). Immunomodulatory properties of quercetin-3-O-α-L-rhamnopyranoside from Rapanea melanophloeos against influenza a virus. *BMC Complementary and Alternative Medicine, 18*, 184.

Mehta, B. K., Pandit, V., & Gupta, M. (2009). New principles from seeds of Nigella sativa. *Natural Product Research, 23*(2), 138–148.

Mewaba, S. G., et al. (2022). Antibacterial compounds of the cultivated Robusta coffee: *Coffea canephora* Pierre ex A. Froehner. *Scientific African, 16*, e01274.

Mhlongo, L. S., & Van Wyk, B. E. (2019). Zulu medicinal ethnobotany: New records from the Amandawe area of KwaZulu-Natal, South Africa. *South African Journal of Botany, 122*, 266–290.

Miles, E. A., & Calder, P. C. (2021). Effects of citrus fruit juices and their bioactive components on inflammation and immunity: A narrative review. *Frontiers in Immunology, 12*, 712608.

Mili, A., et al. (2021). A comprehensive review on Sesamum indicum L.: Botanical, ethnopharmacological, phytochemical, and pharmacological aspects. *Journal of Ethnopharmacology, 281*, 114503.

Mlcek, J., et al. (2016). Quercetin and its anti-allergic immune response. *Molecules, 21*(5), 623.

Moghadamtousi, S. Z., et al. (2015). Annona muricata (Annonaceae): A review of its traditional uses, isolated Acetogenins and biological activities. *International Journal of Molecular Sciences, 16*, 15625–15658.

Mondolot, L., et al. (2006). Evolution in caffeoylquinic acid content and histolocalization during Coffea canephora leaf development. *Annals of Botany, 98*(1), 33–40.

Monzote-Fidalgo, L., et al. (2004). Propiedades antiprotozoarias de aceites esenciales extraídos de plantas cubanas. *Revista Cubana de Medicina Tropical, 56*(3), 230–233.

Moreno, M., Kuwornu, J., & Szabo, S. (2020). Overview and constraints of the coconut supply chain in The Philippines. *International Journal of Fruit Science, 20*, S524–S541.

Moshi, M. J., et al. (2009). The Ethnomedicine of the Haya people of Bugabo ward, Kagera Region, north western Tanzania. *Journal of Ethnobiology and Ethnomedicine, 5*, 1–5.

Moyo, M., Aremu, A. O., & van Staden, J. (2015). Ethnopharmacology in sub-Sahara Africa: Current trends and future perspectives. In *Ethnopharmacology* (pp. 263–278).

Mugisha, M. K., et al. (2014). Ethnobotanical study of indigenous knowledge on medicinal and nutritious plants used to manage opportunistic infections associated with HIV/AIDS in western Uganda. *Journal of Ethnopharmacology, 155*(1), 194–202.

Muss, C., Mosgoeller, W., & Endler, T. (2013). Papaya preparation (Caricol®) in digestive disorders. *Neuroendocrinology Letters, 34*(1), 38–46.

Muthu, C., et al. (2006). Medicinal plants used by traditional healers in Kancheepuram District of Tamil Nadu, India. *Journal of Ethnobiology and Ethnomedicine, 2*, 1–10.

Mutombo, P. N., et al. (2023). Experiences and challenges of African traditional medicine: Lessons from COVID-19 pandemic. *BMJ Global Health, 8*(8), e010813.

Mzondo, B., et al. (2021). Dammarane-type triterpenoids with anti-cancer activity from the leaves of Cleome gynandra. *Phytochemistry Letters, 43*, 16–22.

Nagata, J. M., et al. (2011). Medical pluralism on Mfangano Island: Use of medicinal plants among persons living with HIV/AIDS in Suba District, Kenya. *Journal of Ethnopharmacology, 135*, 501–509.

Najafiniya, M., & Rahkhodaei, E. (2024). A review on date palm (Phoenix dactylifera) fruit bunch wilting disorder in Iran. *Indian Phytopathology, 77*, 1–11.
Nakazawa, R. A. G. (1996). Traditional medicine in the treatment of enteroparasitosis. *Revista de Gastroenterologia del Peru, 16*(3), 197–202.
Namukobe, J., et al. (2011). Traditional plants used for medicinal purposes by local communities around the Northern sector of Kibale National Park, Uganda. *Journal of Ethnopharmacology, 136*(1), 236–245.
Nasim, N., Sandeep, I. S., & Mohanty, S. (2022). Plant-derived natural products for drug discovery: Current approaches and prospects. *The Nucleus : an International Journal of Cytology and Allied Topics, 65*(3), 399–411.
Nassar, M. I., et al. (2010). Secondary metabolites and pharmacology of Foeniculum vulgare Mill. Subsp. Piperitum. *Revista latinoamericana de química, 38*(2), 103–112.
Nguta, J. M., et al. (2015). Medicinal plants used to treat TB in Ghana. *International Journal of Mycobacteriology, 4*(2), 116–123.
Noreen, S., et al. (2023). Application of radish pods in households and effect of their active components against different diseases: A review. *International Journal of Food Properties, 26*(1), 2039–2054.
Nugraha, A. S., et al. (2019). Anti-infective and anti-cancer properties of the Annona species: Their ethnomedicinal uses, alkaloid diversity, and pharmacological activities. *Molecules, 24*, 4419.
Nwachukwu, E., & Osuji, J. (2008). Evaluation of plant extracts for antifungal activity against Sclerotium rolfsii causing Cocoyam Cormel Rot in storage. *Research Journal of Agriculture and Biological Sciences, 4*, 784–787.
Nwosu, S. I., Stanley, H. O., & Okerentugba, P. O. (2013). Occurrence, types and location of calcium oxalate crystals in Vernonia amydalina Del (Asteraceae). *International Journal of Natural Sciences, 4*(3), 533–537.
Nyagumbo, E., et al. (2022). Medicinal plants used for the management of respiratory diseases in Zimbabwe: Review and perspectives potential management of COVID-19. *Physics and Chemistry of the Earth, Parts A/B/C, 128*, 103232.
Obakiro, S. B., et al. (2020). Ethnobotany, ethnopharmacology, and phytochemistry of traditional medicinal plants used in the management of symptoms of tuberculosis in East Africa: A systematic review. *Tropical Medicine and Health, 48*, 1–21.
Ofukwu, R. A., Ayoola, A., & Akwuobu, C. A. (2008). Medicinal plants used in the treatment of tuberculosis in humans and animals by Idom tribe of North Central Nigeria. *Nigerian Veterinary Journal, 29*(2), 25–30.
Ogawa, K., et al. (2018). Appetite-enhancing effects of vanilla flavours such as vanillin. *Journal of Natural Medicines, 72*, 798–802.
O'Keefe, J. H., DiNicolantonio, J. J., & Lavie, C. J. (2018). Coffee for cardioprotection and longevity. *Progress in Cardiovascular Diseases, 61*(1), 38–42.
Okhale, S. E., et al. (2016). Annona senegalensis Persoon (Annonaceae): A review of its ethnomedicinal uses, biological activities and phytocompounds. *Journal of Pharmacognosy and Phytochemistry, 5*(2), 211–219.
Okokon, J. E., Ita, B. N., & Udokpoh, A. E. (2006). The in-vivo antimalarial activities of Uvaria chamae and Hippocratea africana. *Annals of Tropical Medicine & Parasitology, 100*(7), 585–590.
Okuyama, E., et al. (1993). Ascaridole as a pharmacologically active principle of" Paico", a medicinal Peruvian plant. *Chemical and Pharmaceutical Bulletin, 41*(7), 1309–1311.
Olowookere, D. A., & Malomo, S. A. (2024). Amino acid compositions, in vitro antioxidant and anti-diabetic properties of cookies from wheat and kidney beans (Phaseolus vulgaris L.) Flour blends. *European Journal of Nutrition & Food Safety, 16*(5), 32–47.
Oluremi, B. B., Osungunna, M. O., & Omafuma, O. O. (2010). Comparative assessment of antibacterial activity of Uvaria chamae parts. *African Journal of Microbiology Research, 4*(13), 1391–1394.

Omara, T., et al. (2020). Medicinal plants used in traditional management of cancer in Uganda: A review of ethnobotanical surveys, phytochemistry, and anticancer studies. *Evidence-based Complementary and Alternative Medicine, 2020*(1), 3529081.

Opyd, P. M., et al. (2017). Nutritional and health-related effects of a diet containing apple seed meal in rats: The case of amygdalin. *Nutrients, 9*(10), 1091.

Orwa, C., et al. (2009). Agroforestree Database: à tree reference and selection guide. Version 4.0. Available from: http://www.worldagroforestry.org/sites/treedbs/treedatabases.asp

Otieno, S. B., et al. (2016). Determinants of selenium intake in a high HIV prevalence fishing Community in Bondo District, Kenya. *International Journal of Medical and Health Sciences, 10*(9), 463–467.

Palmatier, R. (2000). *Food: A dictionary of literal and nonliteral terms*. Bloomsbury Publishing.

Pamploma-Roger, G. D. (2006). *Encycopedia of medicinal plants* (p. 585). Educational Safeliz.

Parham, S., et al. (2020). Antioxidant, antimicrobial and antiviral properties of herbal materials. *Antioxidants (Basel, Switzerland), 9*(12), 1309.

Park, S. E., et al. (2014). Rutin from Dendropanax morbifera Leveille protects human dopaminergic cells against rotenone induced cell injury through inhibiting JNK and p38 MAPK signaling. *Neurochemical Research, 39*, 707–718.

Patocka, J., et al. (2020). Malus domestica: A review on nutritional features, chemical composition, traditional and medicinal value. *Plants, 9*(11), 1408.

Peter, K. V. (2012). *Handbook of herbs and spices*. Woodhead Pub. Limited.

Peter, K., & Gandhi, P. (2017). Rediscovering the therapeutic potential of Amaranthus species: A review. *Egyptian Journal of Basic and Applied Sciences, 4*(3), 196–205.

plants, P.f.a.F.D.o.e.a.u. (2024). October 1, 2024.

Pollack, Y., Segal, R., & Golenser, J. (1990). The effect of ascaridole on the in vitro development of Plasmodium falciparum. *Parasitology Research, 76*, 570–572.

Prabhune, A., Sharma, M., & Ojha, B. (2017). *Abelmoschus esculentus (Okra) potential natural compound for prevention and management of Diabetes and diabetic induced hyperglycemia. Tamil Nadu.* International. *Journal of Herbal Medicine, 5*(2), 66–68.

Que, F., et al. (2019). Advances in research on the carrot, an important root vegetable in the Apiaceae family. *Horticulture Research, 6*, 69.

Rafique, S., et al. (2020). Biological attributes of lemon: A review. *Journal of Addiction Medicine and Therapeutic Science, 6*(1), 030–034.

Ramasawamy, A. M., & Sirsi, M. (1960). Antituberculosis activity of some chemical constituents from higher plants. *The Indian Journal of Pharmacy, 22*, 34–35.

Rathnavelu, V., et al. (2016). Potential role of bromelain in clinical and therapeutic applications. *Biomedical Reports, 5*(3), 283–288.

Riska, S. R. N. (2002). Overview of traditional use, phytochemical and pharmacological activities of Fennel (Foeniculum vulgare). *International Journal of Modern Pharmaceutical Research, 5*(1), 1–9.

Rojo-Poveda, O., et al. (2020). Cocoa bean shell-A by-product with nutritional properties and biofunctional potential. *Nutrients, 12*(4), 1123.

Ross, I. A. (2007). *Medicinal plants of the world, volume 3: Chemical constituents, traditional and modern medicinal uses* (Vol. 3). Springer Science & Business Media.

Roy, A., & Saraf, S. (2006). Limonoids: Overview of significant bioactive triterpenes distributed in plants kingdom. *Biological and Pharmaceutical Bulletin, 29*(2), 191–201.

Roy, A., Shrivastava, S. L., & Mandal, S. M. (2014). Functional properties of Okra Abelmoschus esculentus L.(Moench): Traditional claims and scientific evidences. *Plant science Today, 1*(3), 121–130.

Salil, G., & Rajamohan, T. (2001). Hypolipidemic and antiperoxidative effect of coconut protein in hypercholesterolemic rats. *Indian Journal of Experimental Biology, 39*, 1028–1034.

Sarker, U., & Oba, S. (2019). Nutraceuticals, antioxidant pigments, and phytochemicals in the leaves of Amaranthus spinosus and Amaranthus viridis weedy species. *Scientific Reports, 9*(1), 1–10.

Secor-Jr, E. R., et al. (2005). Bromelain exerts anti-inflammatory effects in an ovalbumin-induced murine model of allergic airway disease. *Cellular Immunology, 237*, 68–75.

Secor-Jr, E. R., et al. (2008). Oral Bromelain attenuates inflammation in an Ovalbumin-Induced Murine Model of asthma. *Evidence-based Complementary and Alternative Medicine, 5*(1), 61–69.

Seltzer, A. P. (1967). Adjunctive use of bromelains in sinusitis: A controlled study. *Eye, Ear, Nose & Throat Monthly, 46*(10), 1281–1288.

Shang, A., et al. (2019). Bioactive compounds and biological functions of garlic (Allium sativum L). *Food, 8*(7), 246.

Sharif, A., et al. (2022). Ethnomedicinal uses of plants for various diseases in the remote areas of Changa Manga Forest, Pakistan. *Brazilian Journal of Biology, 84*, e255916.

Shawarb, N., et al. (2023). An ethno-botanical study of medicinal plants used for the management of respiratory tract disorders in northern parts of Palestine. *BMC Complementary Medicine and Therapies, 23*(1), 387.

Shestopalov, A. V., et al. (2006). Biological functions of allantoin. *Biology Bulletin, 33*, 437–440.

Shi, W., et al. (2016). Molecular mechanisms underlying mangiferin-induced apoptosis and cell cycle arrest in A549 human lung carcinoma cells. *Molecular Medicine Reports, 13*(4), 3423–3432.

Shi, M., et al. (2024). Recent advances in the health benefits and application of tangerine peel (Citri Reticulatae Pericarpium): A review. *Food, 13*(13), 1978.

Shopo, B., Mapaya, R. J., & Maroyi, A. (2022). Ethnobotanical study of medicinal plants traditionally used in Gokwe South District, Zimbabwe. *South African Journal of Botany, 149*, 29–48.

Sikdar, M., & Dutta, U. (2008). Traditional phytotherapy among the Nath people of Assam. *Studies on Ethno-Medicine, 2*(1), 39–45.

Soladoye, M. O., et al. (2010). Ethnobotanical survey of plants used in the treatment of haemorrhoids in South-Western Nigeria. *Annals of Biological Research, 1*(4), 1–15.

Solnier, J., & Fladerer, J. P. (2021). Flavonoids: A complementary approach to conventional therapy of COVID-19? *Phytochemistry Reviews, 20*(4), 773–795.

Son, Y. O., et al. (2003). Ripe fruits of Solanum nigrum L. inhibits cell growth and induces apoptosis in MCF-7 cells. *Food and Chemical Toxicology, 41*(10), 1421–1428.

Song, P. P., et al. (2017). Evaluation of antifungal activities and structure–activity relationships of coumarin derivatives. *Pest Management Science, 73*(1), 94–101.

Speisky, H., et al. (2006). Antioxidant screening of medicinal herbal teas. *Phytotherapy Research, 20*(6), 462–467.

Spengler, R. N. (2019). Origins of the apple: The role of megafaunal mutualism in the domestication of Malus and rosaceous trees. *Frontiers in Plant Science, 10*, 617.

Suleiman, M. M., et al. (2014). Evaluation of analgesic and anti-inflammatory effects of the crude methanol extract of the stem-bark of Annona senegalensis Pers. *International Journal of Medicinal and Aromatic Plants, 4*(2), 88–96.

Suleiman, R. K., et al. (2021). New constituents from the leaves of date palm (Phoenix dactylifera L.) of Saudi origin. *Molecules, 26*(14), 4192.

Suroowan, S., Pynee, K. B., & Mahomoodally, M. F. (2019). A comprehensive review of ethnopharmacologically important medicinal plant species from Mauritius. *South African Journal of Botany, 122*, 189–213.

Syamsuhidayat, S., & Hutapea, J. R. (1991). *Inventaris Tanaman Obat Indonesia*. Departemen Kesehatan RI, Badan Penelitian dan Pengembangan Kesehatan.

Tabuti, J. R. (2008). Herbal medicines used in the treatment of malaria in Budiope county, Uganda. *Journal of Ethnopharmacology, 116*(1), 33–42.

Tabuti, J. R. S., Dhillion, S. S., & Lye, K. A. (2003). Traditional medicine in Bulamogi county, Uganda: Its practitioners, users and viability. *Journal of Ethnopharmacology, 85*(1), 119–129.

Tabuti, J. R., Kukunda, C. B., & Waako, P. J. (2010). Medicinal plants used by traditional medicine practitioners in the treatment of tuberculosis and related ailments in Uganda. *Journal of Ethnopharmacology, 127*(1), 130–136.

Taiwo, B. J., et al. (2021). Isolation and characterisation of two quercetin glucosides with potent anti-reactive oxygen species (ROS) activity and an olean-12-en triterpene glucoside from the fruit of Abelmoschus esculentus (L.) moench. *Chemistry & Biodiversity, 18*(1), e2000670.

Tan, G. H., Ali, A., & Siddiqui, Y. (2022). Current strategies, perspectives and challenges in management and control of postharvest diseases of papaya. *Scientia Horticulturae, 301*, 111139.

Tanmoy, G., et al. (2014). Pharmacological actions and phytoconstituents of Amaranthus spinosus Linn: A review. *International Journal of Pharmacognosy and Phytochemical Research, 6*, 405–413.

Teshika, J. D., et al. (2019). Traditional and modern uses of onion bulb (Allium cepaL.): A systematic review. *Critical Reviews in Food Science and Nutrition, 59*, S39–S70.

Thinyane, Z., & Maroyi, A. (2019). Medicinal plants used by the inhabitants of Alfred Nzo District Municipality in the Eastern Cape province, South Africa. *Journal of Pharmacy and Nutrition Sciences, 9*(3), 157–166.

Thovhogi, F., et al. (2021). Perceptions of end-users in Limpopo province (South Africa) about the spider plant (Cleome gynandra l.). *Genetic Resources and Crop Evolution, 68*, 605–614.

Tijjani, M. A., et al. (2017). Phytochemical analysis, analgesic and antipyretic properties of ethanolic leaf extract of Vernonia amygdalina Del. *Journal of Herbmed Pharmacology, 6*(3), 95–99.

Tsao, R., et al. (2005). Which polyphenolic compounds contribute to the total antioxidant activities of apple? *Journal of Agricultural and Food Chemistry, 53*(12), 4989–4995.

Tudu, C. K., et al. (2022). Traditional uses, phytochemistry, pharmacology and toxicology of garlic (Allium sativum), a storehouse of diverse phytochemicals: A review of research from the last decade focusing on health and nutritional implications. *Frontiers in Nutrition, 9*, 949554.

Tugume, P., et al. (2016). Ethnobotanical survey of medicinal plant species used by communities around Mabira Central Forest Reserve, Uganda. *Journal of Ethnobiology and Ethnomedicine, 12*, 1–28.

Turner, T., & Burri, B. J. (2013). Potential nutritional benefits of current citrus consumption. *Agriculture, 3*(1), 170–187.

Ugochukwu, N. H., & Babady, N. E. (2002). Antioxidant effects of Gongronema latifolium in hepatocytes of rat models of non-insulin dependent diabetes mellitus. *Fitoterapia, 73*(7–8), 612–618.

Ugochukwu, N. H., et al. (2003). The effect of Gongronema latifolium extracts on serum lipid profile and oxidative stress in hepatocytes of diabetic rats. *Journal of Biosciences, 28*, 1–5.

Umar, S., et al. (2012). Modulation of the oxidative stress and inflammatory cytokine response by thymoquinone in the collagen induced arthritis in Wistar rats. *Chemico-Biological Interactions, 197*(1), 40–46.

Upadhyay, B., Roy, S., & Kumar, A. (2007). Traditional uses of medicinal plants among the rural communities of Churu district in the Thar Desert, India. *Journal of Ethnopharmacology, 113*(3), 387–399.

Uwimbabazi, M., et al. (2023). Ethnobotanical survey of medicinal plants used for the treatment of diabetes in Uganda. *Ethnobotany Research and Applications, 26*, 1–14.

Villena Tejada, M., et al. (2021). Use of medicinal plants for COVID-19 prevention and respiratory symptom treatment during the pandemic in Cusco, Peru: A cross-sectional survey. *PLoS One, 16*(9), e0257165.

Wang, C. K. (2016). Passion fruit juice. In *Handbook of functional beverages and human health*. CRC Press.

Wang, X., et al. (2008). Increased prevalence of symptoms of gastroesophageal reflux diseases in type 2 diabetics with neuropathy. *World Journal of Gastroenterology, 14*(5), 709.

Wang, Y., et al. (2017). Potential anti-inflammatory steroidal saponins from the berries of solanum nigrum L. (European black nightshade). *Journal of Agricultural and Food Chemistry, 65*(21), 4262–4272.

Wang, H., et al. (2023). Research and application of leek roots in medicinal field. *Chinese Herbal Medicines, 15*(3), 391–397.

Wu, Y. C., et al. (1993). Cytotoxic alkaloids of Annona montana. *Phytochemistry, 33*(2), 497–500.

Wu, W., et al. (2018). Vernodalol mediates antitumor effects in acute promyelocytic leukemia cells. *Oncology Letters, 15*(2), 2227–2235.

Yang, Y. L., et al. (2008). New cyclic peptides from the seeds of Annona squamosa L. and their anti-inflammatory activities. *Journal of Agricultural and Food Chemistry, 56*(2), 386–392.

Yang, D., et al. (2020). Quercetin: Its main pharmacological activity and potential application in clinical medicine. *Oxidative Medicine and Cellular Longevity, 2020*(1), 8825387.

Ye, F., et al. (2015). Solvent effects on phenolic content, composition, and antioxidant activity of extracts from florets of sunflower (Helianthus annuus L.). *Industrial Crops and Products, 76*, 574–581.

Yeo, D., et al. (2011). Evaluation of the anti-inflammatory activity and phytochemical screening of Annona senegalensis leaves. *Thérapie, 66*(1), 73–80.

Yi, L., Ma, S., & Ren, D. (2017). Phytochemistry and bioactivity of Citrus flavonoids: A focus on antioxidant, anti-inflammatory, anticancer and cardiovascular protection activities. *Phytochemistry Reviews, 16*, 479–511.

Yilmaz, Y., & Toledo, R. T. (2004). Major flavonoids in grape seeds and skins: Antioxidant capacity of catechin, epicatechin, and gallic acid. *Journal of Agricultural and Food Chemistry, 52*(2), 255–260.

Yimer, E. M., et al. (2019). Nigella sativa L. (Black Cumin): A promising natural remedy for wide range of illnesses. *Evidence-based Complementary and Alternative Medicine, 2019*(12), 1528635.

Younis, W. A., et al. (2018). Traditional medicinal plants used for respiratory disorders in Pakistan: A review of the ethno-medicinal and pharmacological evidence. *Chinese Medicine, 13*, 1–29.

Zatuchni, G. I., & Colombi, D. J. (1967). Bromelains therapy for the prevention of episiotomy pain. *Obstetrics & Gynecology, 29*(2), 275–278.

Zhang, D., et al. (2011). Theobroma. In C. Kole (Ed.), *Wild crop relatives: Genomic and breeding resources* (pp. 277–296). Springer.

Zhang, M., et al. (2024). Health benefits and mechanisms of theobromine. *Journal of Functional Foods, 115*, 106126.

Zou, M., et al. (2020). Structure-activity relationship of flavonoid bifunctional inhibitors against Zika virus infection. *Biochemical Pharmacology, 177*, 113962.

Chapter 4
Secondary Metabolites Used in Pest Control

Patience Tugume, Godwin Anywar, Savina Asiimwe, and Esezah K. Kakudidi

Abstract Pests are a threat to agricultural productivity. The current pest management approaches with pesticides are riddled with a plethora of challenges, such as non-selective toxicity, pest resistance to pesticides and bioaccumulation in tissues. The pest challenge requires the search for eco-friendly and safer alternatives from natural sources. This chapter considers plant secondary metabolites and other natural sources as actual or potential pesticides. We discuss the categories of pesticides and pests of economic importance in Africa, with an emphasis on medicinal plants with promising bioactivities. We use a holistic approach to pest management while promoting the traditional use of plant diversity for the benefit of agriculture and the environment. This chapter further highlights the benefits that can be derived from bridging farmer pest management knowledge and scientific techniques. The most promising secondary metabolites from plants were saponins, terpenoids, anthocyanins, tannins, leucoanthocyanidins and emodins. Tannins through oxidation inhibit protein digestion, hinder insect growth and cause insect death. Alkaloids are neurotoxic and have anti-repellent and antifeedant effects, while terpenoids interfere with metamorphosis and other physiological processes in insects. The most promising secondary metabolites can be scientifically exploited and their use enhanced through identification, extraction, isolation, characterization, formulation and application in comprehensive pest management. The biopesticides with the most promising activity work through various mechanisms that deter certain pests' activities for their growth and survival. However, most of the pesticides extend their activity by inhibiting octopamine and tyramine receptors, interfering with the receptors for the juvenile hormone and inhibiting the action of gamma-aminobutyric acid. Some of the opportunities for using secondary metabolites in pest control include restrictions by most governments on production of chemical pesticides and promotion of green agricultural practices. Despite these prospects, some challenges in using secondary metabolites in pest control are eminent. These include high costs of production, inconsistent performance, limited shelf life, lack of large-scale production facilities,

P. Tugume (✉) · G. Anywar · S. Asiimwe · E. K. Kakudidi
Department of Plant Sciences, Microbiology & Biotechnology, Makerere University, Kampala, Uganda

A. Dekebo (ed.), *Bioactive Secondary Metabolites from Medicinal Plants of Africa*, Interdisciplinary Biotechnological Advances,
https://doi.org/10.1007/978-981-96-8590-5_4

lengthy regulation process and limited research to establish active ingredients of the plants used in pest control. This chapter is a valuable resource in exploring approaches to pest management while promoting the utilization of indigenous knowledge and biodiversity. In conclusion, an increase in agricultural productivity can be achieved if the challenge of pests is addressed using eco-friendly approaches, one of which is the use of biopesticides. This calls for further research to incorporate indigenous traditional knowledge in the development of safe and effective biopesticides through identification and separation of active ingredients.

Keywords Biopesticides · Secondary metabolites · Safety · Toxicity · Integrated pest management

Abbreviations

AcMNPV	Autographa Californica Multiple Nucleopolyhedrovirus
AChE	Acetyl cholinesterase enzyme
DAFF	Department of Agriculture, Forestry and Fisheries
DNA	Deoxyribonucleic acid
GABA	Gamma-aminobutyric acid
HPLC	High Pressure Liquid Chromatography
IPM	Integrated pest management
ITK	Indigenous traditional knowledge
NAFDAC	National Agency for Food and Drug Administration and Control
NMR	Nuclear magnetic resonance
MeOH	Methyl alcohol
OAR	Octopamine receptor
OECD	Organisation for Economic Cooperation and Development
PCPB	Pest Control Products Board
PEOs	Plant essential oils
SAR	Systemic acquired resistance
SDGs	Sustainable Development Goals
TCL	Thin-layer chromatography
TPRI	Tropical Pesticides Research Institute
UNACB	Uganda National Agricultural Chemicals Board
^{13}C-NMR	Carbon 13 nuclear magnetic resonance
^{15}N-NMR	Nitrogen 15 nuclear magnetic resonance
^{1}H-NMR	Proton nuclear magnetic resonance
^{31}P-NMR	Phosphorus-31 nuclear magnetic resonance

4.1 Introduction

4.1.1 Pest Management in Agriculture

The world has experienced exponential human population growth since the second half of the nineteenth century. This population explosion has led to an increased food requirement, which is estimated to double by the middle of this century (FAO, 2023; FAOSTAT, 2013). Population growth has brought many challenges, particularly low crop production, which has affected food security.

Food production is under threat from various fronts, including environmental degradation, climate change and different pests (Mafongoya et al., 2019; Skendžić et al., 2021). Pests, mainly invertebrates, are a huge problem and threat to agricultural productivity and food security in Africa (Pretty & Pervez Bharucha, 2015). Pests attack at various stages of crops in cultivation and storage, causing major losses (Sawicka & Egbuna, 2020). The crop protection methods that have been used since the middle of the twentieth century have mainly depended on synthetic agrochemicals in an attempt to boost food security (Grzywacz et al., 2014; Essiedu et al., 2020). Generally, plants provide their own protection against pests and diseases (War et al., 2015). However, some plants are not able to adequately defend themselves due to climatic changes. This has caused the geographical expansion of many hitherto confined pests, but also contributed to resistance of synthetic agrochemicals used in agriculture (Skendžić et al., 2021). The use of synthetic pesticides to manage pests is associated with pest resistance, accumulation of chemicals in various tissues of crops and the environment. This results in destruction of plants, animals, environment and consumers (Ayilara et al., 2023).

Various research projects have therefore advocated for the search for eco-friendly and safer alternatives such as biopesticides obtained from biodiversity and some minerals to tackle the pest problems (Chorol et al., 2021; Khursheed et al., 2022). Natural products are the preferred biopesticide candidates as they contain naturally occurring substances that do not present similar challenges as synthetic pesticides. Natural biopesticides, such as neem oil, garlic and pyrethrin, would reduce the pest-related challenges that have so far been encountered in agriculture (Chorol et al., 2021; Khursheed et al., 2022).

4.1.2 The Financial and Environmental Effects of Pests in Africa

Economic impacts of pests and pathogens worldwide are not fully understood despite the fact that they cause annual crop losses of about 20–40% (Mafongoya et al., 2019), although Oerke (2006) reported higher losses depending on the crops in question. The impact is worse especially in low-income and developing countries in Africa, where resources for intervention, extension services and research are limited

(Mafongoya et al., 2019; De Groote, 2022). Considering that farming contributes to a large percentage of Africa's population, crop infestation by pests impacts the domestic and export economies. Various studies have reported crop losses caused by pests, many of which are native while others were not originally known in Africa. In Kenya, Kibira et al. (2015) reported losses of fruits up to 80% due to infestation by fruit flies. In Egypt it was estimated that fruit flies caused revenue losses of approximately 177 million USD (Mahmoud et al., 2017). Masika et al. (2017) reported the virus infestation (15–80%) in fields of the cucumber family (Cucurbitaceae), thus limiting their production and economic potential by causing losses to farmers in Uganda. Bananas, which are consumed by over 40 million people globally, are increasingly vulnerable to diseases that are spreading extensively. While bananas are not native to Africa, they have become a major local staple food and for generating income especially in East Africa (Akankwasa et al., 2021; Drenth & Kema, 2021). Whereas the genetically improved bananas are less vulnerable to pests, the local varieties are preferred by many consumers for various reasons. However, the local varieties are highly susceptible to diseases and pests, and thus their productivity remains economically low (Akankwasa et al., 2021). Weeds were considered to cause higher crop loss, but recent reports show that pathogens and insect pests cause more damage and loss to crops in Africa. The incidence has been attributed to the increase in temperatures that has favoured their spread and severity (Mafongoya et al., 2019; Skendžić et al., 2021). For example, intensified farming has provided an ecosystem that has supported the escalation of plant pests in cultivated areas and thus contributed to the modification of plant communities and functioning (Schowalter et al., 2018). Generally, researchers have not fully comprehended the economics of agricultural pests although comprehensive management of pests is considered a more feasible solution to especially the smallholder farmer through the use of biopesticides as they provide positive ecosystem stability.

4.1.3 The Need for Sustainable and Bioactive Alternatives to Synthetic Pesticides

The utilization of synthetic pesticides negatively impacts on the environment (Kumar, 2012). For instance, persistent utilization of inorganic pesticides causes harm to targeted and non-targeted organisms, including flora and fauna, by inflicting toxicity, physiological deformities, diseases, population changes, genetic disorders, gene erosion and death (Fountain & Wratten, 2013; Kumar, 2012). In humans, toxic effects of pesticides result from bioaccumulation of contaminants in food chains (Lushchak et al., 2018). Dichlorodiphenyltrichloroethane and toxaphene persist in the environment over prolonged periods and end up accumulating in groundwater and aquatic biota (Carvalho, 2017). Some human diseases of the immune, nervous and endocrine systems are associated with ingestion or exposure to pesticides, and in some cases, they have been reported to cause metabolic diseases, cancer and infertility (Damalas & Koutroubas, 2016; Kumari et al., 2014). Given these negative side effects, sourcing for alternatives which are eco-friendly is pertinent, thus the need to explore the production and use of biopesticides.

Biopesticides do not accumulate in foods, and this has contributed to the renewed interest in their usage. This has also increased their demand in the management of pests in fruits and vegetables (Kumar, 2012). This demand is driven by the fact that their efficacy is comparable to inorganic pesticides when used in comprehensive pest management. Thus, the combined performance and environmental safety of biopesticides coupled with flexibility in application render them the best alternative to synthetic pesticides (Ayilara et al., 2023).

4.2 Methods

We reviewed literature on plants traditionally utilized to manage agricultural pests, their associated indigenous traditional knowledge and secondary metabolites responsible for their bioactivities. We further assessed published literature on bioprospecting in Africa for biopesticides, isolation and characterization of plant active compounds, their integration in IPM, challenges and future prospects of biopesticide use. The following were the source of information: books and reports, peer-reviewed articles in various databases such as Web of Science, Google Scholar, PubMed and Scopus from 1954 to 2024.

4.3 Indigenous Traditional Knowledge (ITK) of Pest Management

This is the accumulated wisdom, application and beliefs developed by Indigenous peoples through centuries of interaction with their environment in pest management. This knowledge is holistic, encompassing cultural, ecological and social dimensions. In the context of pest management, ITK offers sustainable and ecologically sound practices that can complement modern agricultural techniques (Melash et al., 2023; Tripura et al., 2024).

The management of pests using indigenous knowledge comes with several benefits such as environmental sustainability, cost-effectiveness, cultural preservation, as well as the promotion of biodiversity (Altieri, 1993; Tolo et al., 2014; He et al., 2019; Mushtaq et al., 2020). However, while indigenous pest management practices offer numerous benefits, integrating them with modern agricultural systems can be challenging. There is a need for greater recognition of ITK in scientific and policy-making communities. Collaborative efforts between indigenous communities, scientists and policymakers can foster an exchange of knowledge that enhances sustainable pest management (Thakur et al., 2020; Anandajayasekeram, 2022).

Several communities worldwide possess indigenous traditional knowledge on the management of pests. This knowledge is transmitted orally throughout generations. It is therefore worthwhile to document this knowledge with respect to pest management before it is lost given the fact that the main holders of this knowledge are the elderly. For instance, subsistence farmers in the Lake Victoria basin in Uganda (Kamatenesi et al., 2008), small holder farmers in Mpumalanga province South Africa (Shai et al., 2024) and the Manipuri community in India (Bhattacharjee & Ray, 2010) possess unique traditional knowledge of agricultural practices which are eco-friendly and compatible with other IPM.

Several practices have been used for millennia to manage the menace of pests, and they include the following:

(i) *Biodiversity and Crop Rotation:* Indigenous farmers often cultivate a variety of crops together, promoting biodiversity which naturally suppresses pest populations. Crop rotation prevents pests from establishing themselves in a single location, reducing the need for chemical interventions (Altieri, 1993; He et al., 2019).

(ii) *Botanical Pesticides:* Utilization of natural products from plants is common in ITK. For instance, neem leaves, tobacco and chili peppers can be used to create natural sprays that deter pests without harming beneficial insects (Mushtaq et al., 2020).

(iii) *Cultural Practices:* Techniques such as intercropping and polyculture create an environment that disrupts pest life cycles. Indigenous knowledge also includes planting certain crops at specific times to avoid peak pest populations (Altieri, 1993).

(iv) *Biological Control:* Indigenous communities often encourage the presence of natural predators. For example, certain birds, bats and insects are protected and promoted to naturally control pest populations (Altieri, 2019; Mafongoya et al., 2010).

(v) *Physical Barriers and Traps:* Traditional methods include using physical barriers such as nets or setting traps to capture pests. This reduces the dependence on inorganic pesticides and maintains ecosystem stability (Boiteau & Vernon, 2001).

(vi) *Traditional Wisdom and Observation:* Indigenous knowledge is deeply rooted in careful observation of natural cycles and patterns. This wisdom allows for timely interventions and adaptive management of pest issues (Tolo et al., 2014). Drying maize cobs with heat to decrease the water content of the seeds and minimize their invasion by storage pests (Asemu et al., 2020). Other practices in sub-Saharan Africa include mulching, weeding using a hoe, crop rotation and using pest-resistant seed varieties (Ajani et al., 2013). In Ghana, pests were controlled traditionally by dusting of ash to control aphids and ants in a potato field (Owusu, 2000). Ash is an insect repellent and discourages surface-feeding insects. Also burying of infected plants, applying a combination of salt and shea butter (*Vitellaria paradoxa*) and planting elephant grass (*Cenchrus purpureus*) around the garden (Maayiem et al., 2012) are important in pest control. Field burning is traditionally used to destroy pests and rodents in crop fields. This practice prevents the escalation of pests and increases crop output (Makundi et al., 1999).

(vii) Spraying cow urine over the plant to aphids (Tesfaye & Gautam, 2003).

(viii) Application of water onto plants exterminates pests amidst their outbreak (Grzywacz et al., 2014).

In many countries, farmers have applied traditional knowledge to manage army fall worm, such as handpicking and killing of larvae in Kenya (Kumela et al., 2019), adding soil to plant whorls, drenching tobacco extracts in Zambia and destroying ratoon host crops (Kansiime et al., 2019), early planting, deep ploughing to kill pupae (Chimweta et al., 2020) and burning stubbles after harvesting infected plants (Assefa, 2018).

In sub-Saharan Africa, post-harvest pest management practices included mixing herbs and ash with stored produce (Kitch et al., 1997). In the field, infected plants were hand-picked and, in some instances, deep ploughing was used to control different pests (Kumar et al., 2020). *Sida rhombifolia* was used to control nematodes and soil arthropods in rice fields (Lopes et al., 2020) while chopped pieces of citrus were applied in rice fields after transplanting to repel insects (Chou et al., 2022; Deka et al., 2006). The grounded bark of *Moringa oleifera* was applied in the field to control rice stem borer (Deka et al., 2006).

Indigenous Traditional Knowledge on Pesticides

Several plants are used in pest management and are not harmful to the environment. Deshmukh (2015) reported the use of several plant species to naturally control pests. Fresh leaves of *Artemisia* sp. are incorporated around the paddy field to protect rice from attack by field pests. Dried branches and leaves of *Artemisia* sp. are mixed with potato to protect them against tomato moth (Ansari et al., 2021). The same authors reported the use of dried and pounded leaves of *Goniothalamus sesquipedalis* and *Plectranthus ternifolius* against pests in stored grain. Several plant species are traditionally used to manage pests in different countries (Table 4.1).

4.4 Secondary Metabolites with Pesticidal Properties

Plant species are the main source of several natural products with organic properties including pesticidal (Tekman et al., 2024). Natural products are intermediate products of metabolism involved in structural formation. They also signal, stimulate and inhibit enzyme and catalytic activities in addition to defending and promoting interaction with other microorganisms (Jain et al., 2019). Some of the secondary metabolites include flavonoids, steroids and alkaloids (Jain et al., 2019). Medicinal plants contain chemical compounds which promote their biopesticidal effects (Hilal et al., 2024; Savithramma et al., 2011). These compounds protect plants against biotic and abiotic stresses (Hilal et al., 2024). These defence mechanisms are thus essential in formulation of biopesticides besides being important in discovery of drugs for management of human ailments.

Table 4.1 Plant species traditionally used to control pests

Plant species	Pests controlled	References	Part used	Mode of use	Country
1. *Melia azedarach* L.	Army worm, mustard saw-fly, tobacco caterpillar	Makaza and Mabhegedhe (2016) and Naharki and Jaishi (2020)		Not mentioned	Zimbabwe, Nepal
2. *Acorus calamus* L.	Ants, maize weevil	Naharki and Jaishi (2020)		Not mentioned	Nepal
3. *Lantana camara* L.	Maize weevil	Mugisha-Kamatenesi et al. (2008), Naharki and Jaishi (2020) and Ogendo et al. (2003)	Leaves and whole plant	Powder added to stored grain	Uganda, Nepal, Kenya
4. *Ageratum conyzoides* L.	Diamond black moth, stored grain pest	Naharki and Jaishi (2020)		Not mentioned	Nepal
5. *Azadirachta indica* A. Juss.	Rice pest, diamond black moth, stored grain pests	Mugisha-Kamatenesi et al. (2008) and Naharki and Jaishi (2020)	Leaves	Not mentioned	Uganda
6. *Tagetes erecta* L.	Maize weevil aphids	Mugisha-Kamatenesi et al. (2008) and Naharki & Jaishi, 2020)	Leaves and whole plant	Water extract	Uganda, Nepal
7. *Tagetes minuta* L.	Most insects, nematodes	Mwine et al. (2011)		Water extract and as a repellent crop	Uganda
8. *Ocimum basilicum* L.	Rice pest, beetles, fruit flies	Naharki and Jaishi (2020) and Maize weevil Makaza and Mabhegedhe (2016)	Leaves and branches	Mixed with grain in bag	Nepal, Zimbabwe
9. *Nicotiana tabacum* L.	Storage and soil pests	Mwine et al. (2011)	Leaves	Water extract, smoke and physical admixture	Uganda
10. *Spirostachys Africana* Sond.	Maize and cowpea weevil	Mvumi et al. (1995)	Wood chips	Placed at different positions in a bag containing grain	Zimbabwe

11. *Melia azedarach* L.	Maize weevil and grain borer	Makaza and Mabhegedhe (2016) and Mvumi et al. (1995)	Leaves and branches	Added to grain in bags	Zimbabwe
12. *Helianthus annuus* L.			Flower	Ash from sunflower stem mixed with grain in bag	
13. *Eleucine coracana* L.				Chuff mixed with grain in bag	
14. *Lippia javanica* (Burm.f.) Spreng.			Leaves and branches	Mixed with grain in sealed bags	
15. *Eucalyptus* spp.	Storage pests	Mugisha-Kamatenesi et al. (2008) and Mwine et al. (2011)	Leaves	Mixed with grain in the bag Oil extract applied Physical admixture	Uganda
16. *Allium cepa* L.	Field and storage pests	Kaluwa and Kruger (2012)	Bulbs and leaves Whole plant	Not mentioned	South Africa
17. *Allium sativum* L.	*Bemisia tabaci*, *Hellula undalis* and *Brevicoryne brassicae*	Kaluwa and Kruger (2012)	Whole plant	Intercropped with susceptible crops to repel pests	South Africa
18. *Aloe vera* L.	Plant bugs	Kaluwa and Kruger (2012)	Leaves	The gel obtained from *Aloe vera* leaf is mixed with water to make a spray for plants. Harmful insects are repelled or killed by the bitter taste	South Africa
19. *Tithonia diversifolia* Del.	Storage and field pests	Mugisha-Kamatenesi et al. (2008)	Leaves	Not mentioned	Uganda
20. *Canna indica* L.	Storage and field pests	Mugisha-Kamatenesi et al. (2008)	Leaves	Not mentioned	Uganda
21. *Tephrosia vogelli* Hook. F.	Maize army worm	Lusaka, (2018)	Leaves	Aqueous leaf extract sprayed on the plants	Zambia

(continued)

Table 4.1 (continued)

Plant species	Pests controlled	References	Part used	Mode of use	Country
	Bean weevil	Kamugire and Nshutiyayesu (2020)	Leaves	Seeds treated with leaf powder	Rwanda
	Maize weevil	Ogendo et al. (2003)	Whole plant	Powder mixed with stored grain	Kenya
22. *Tetradenia riparia* (Hochst).	Bean weevil	Kamugire and Nshutiyayesu (2020)	Leaves	Seeds treated with leaf powder	Rwanda
23. *Carica papaya* L.	Maize weevil	Mugisha-Kamatenesi et al. (2008) and Muzemu et al. (2013)	Leaves	Leaf powder added to stored grain	Uganda, Zimbabwe
24. *Capsicum frutescens* L.	Storage pests	Mwine et al. (2011)	Fruits	Water extract and crushed fruits added to stored produce	Uganda
25. *Cassia sophera* L.	*Rhyzopertha dominica*, *Callosobruchus maculatus*, *Sitophilus zeamais*, *Prostephanus truncatus*	Belmaina et al. (2001)	Leaves	Leaves/roots shade dried, milled to a fine powder and mixed with seeds/grain	Ghana
26. *Mitragyna inermis* (Willd) Kuntze.			Leaves		
27. *Securidaca longepedunculata* Fresen.			Roots		
28. *Synedrella nodiflora* L. (Gaertn).			Leaves		
29. *Ocimum americanum* L.			Leaves		

Plant secondary metabolites have proven pharmacological and biochemical actions against insects. The bioactivity depends on the chemical compounds that affect insects in different ways. These properties have increased the demand for these plants and their extracts in pest management practices. Some phytochemicals with toxic effects to insects have been reported by Ogbonna et al. (2016). Tannins cause oxidation by binding to certain semiquinone radicals, quinones and other reactive oxygen species, leading to insect death (Barbehenn & Constabel, 2011). They also inhibit protein digestion and hinder insect growth by causing inactivation of adhesion enzymes and cell development proteins resulting in antifeedant and astringent properties and finally death (Cowan, 1999). Plant tannins obstruct insect feeding by reducing food absorption in insect guts leading to stomach poisoning (Barbehenn & Constabel, 2011). Tannins from sorghum repelled pests in stored grain (Wongo, 1998). Tannins of *Khaya senegalensis* caused death of *Dinoderus porcellus* (Loko et al., 2017), *Schizaphis graminum*, *Myzus persicae* and *Heliothis virescens* in tobacco (Ukoroije & Otayor, 2020).

Alkaloids work synergistically with saponins and tannins to cause repellent and antifeedant effects in pests (Adeyemi, 2010; Eich, 2008). Alkaloids are neurotoxic and attack receptor sites of acetylcholinesterase by modifying their absorptive capacity, triggering nerve contractions, slow larval development and death of nymphs and adult insects (Celis et al., 2008; Eich, 2008). Several authors reported alkaloids cause death of larva and adult insects and limit insect feeding (Emam et al., 2009; Lee, 2000). *Ruta chalepensis* alkaloids resulted in death of larvae and deterred feeding in *Spodoptera littoralis* larvae (Emam et al., 2009; Pavela, 2005).

Terpenoids regulate moulting in insects by obstructing the synthesis of hydroxyecdysone, thus interfering with metamorphosis (Dubey et al., 2003). Terpenoids from *Gliricidium sepium* deterred feeding in larvae of *Helicoverpa armigera* (Jose & Sujatha, 2017; Truman & Riddiford, 2002). They also act as insect repellents and monoterpenes from flowers of *Philax paniculata* on ants (Junker et al., 2011) and sesquiterpenes from *Solanum habrochaites* on silverleaf whitefly (Bleeker et al., 2011; Kortbeek et al., 2019). Flavonoids influence insect behaviour, growth, feeding, inhibit growth and cause deaths (Gould & Lister, 2006; Mallikarjuna et al., 2004).

Glycosides are poisonous derivatives from some plant species. Though inactive in storage, they are hydrolysed and become available for use during insect feeding (Malka et al., 2020) to negatively impact on the feeding insect. Glycosides exhibited larvicidal activity and inhibited insect growth (Al-Rajhy et al., 2003). They also exhibited antifeedant activity (Bowers & Puttick, 1989; Makenzi et al., 2019). Digitoxin, azadirachtin and neem oil killed larvae and adults of the camel tick (Al-Rajhy et al., 2003). Also, flavan glycosides such as viscutin inhibited the cotton growth of *Pectinophora gossypiella* (Kubo et al., 1987; Makenzi et al., 2019). Cyanogenic glycosides in cassava, bamboo and flax have insecticidal properties and were used as fumigants against stored-product insects (Dave & Ledwani, 2012).

Saponins have a bitter and acrid taste which cause insect feeding deterrence (Qasim et al., 2020). These compounds exhibit several activities against insects such as antifeedants, disturbance of moulting and growth regulation and in some instances may result in mortality (Ellen et al., 2007; Harmatha, 2000). Saponins interfered with behaviour, regulated growth and induced high toxicity on *Tetranychus urticae*, *Spodoptera littoralis*, *Culex pipiens* (Adel et al., 2000) and Plutella *xylostella* (Tian et al., 2021). The toxicity of saponins results from their ability to break down erythrocytes (Francis et al., 2002) and, therefore, are essential in protection of plants against diseases and pests (Zaynab et al., 2021).

The biological hazards caused by synthetic pesticides (Karunamoorthi et al., 2008) have led to evolution of new pest control techniques that are eco-friendly (Chorol et al., 2021; Khursheed et al., 2022), for instance, the use of plant products (Pretty & Pervez Bharucha, 2015; Khursheed et al., 2022; Salvadores et al., 2007). Repeatedly, plant products have been traditionally used for pest control in China (Liu et al., 2007), Egypt, Greece and India (Isman, 2006) and therefore have for centuries sustained crop production (Godlewska et al., 2021), as well as in other parts of the world. The ability of some plants in controlling pests is attributed to possession of secondary metabolites (García-Mateos et al., 2004). These metabolites confer on plants several activities such as insectistatic or insecticidal. Insectistatic effects inhibit insect's growth and activity (Celis et al., 2008) which comprise of repellence (Viglianco et al., 2006), anti-feeding activity (Eriksson et al., 2008), growth regulation (Wheeler & Isman, 2001), feed and oviposition obstacles (Banchio et al., 2003). Based on these activities, plants can be exploited to produce biopesticides (Ukoroije & Otayor, 2020).

4.4.1 Mechanisms of Action Underlying Their Pesticidal Effects

Biopesticides exhibit different mechanisms of action that promote their effectiveness in pest control. Essential oils from *Juniperus communis*, *Juniperus oxycedrus*, *Juniperus pygmaea* and *Juniperus sibirica* significantly repelled and killed two aphid species at concentrations of 1%, 2.5% and 5% (Semerdjieva et al., 2021). Such properties make biopesticides an important component in pest management and control. For instance, the pesticidal activity of *Cassia occidentalis* occurs through restraining the activity of acetyl cholinesterase in target pests which disrupts ionic composition (Okonkwo & Ohaeri, 2018). Acetyl cholinesterase enzyme (AChE) inhibition results in its pile up at nerve junctions (Sakthivel et al., 2022), causing increased nervous excitement and eventually insect death. *Piper guineense* seed oil offers protection against insect pests by deterring oviposition (Nzelu et al., 2020). Other essential oils that inhibit AChE activity include Oregano oil (*Origanum* spp.) and *Artemisia nakaii* on larvae of *Aedes aegypti* and *Spodoptera litura*, respectively (Almadiy & Nenaah, 2022; Liu et al., 2021). A remarkable reduction in AChE activity was caused by lemon grass (*Cymbopogon citratus*) and citrus oils in *Callosobruchus maculatus* (Campos et al., 2019; Oboh et al., 2017).

Plant essential oils (PEOs) are toxic to insects. For instance, *Laurus nobilis* oil was toxic to insect larvae of *Trogoderma granarium* (Tayoub et al., 2012). Low concentrations of *Melissa officinalis* oil disrupted feeding in *Tribolium castaneum* in a dose-dependent manner (Ebadollahi et al., 2016). *Laurelia sempervirens* oil showed a strong repellent activity against *Tribolium castaneum* (Zapata & Smagghe, 2010). Karabörklü and Ayvaz (2023) reported the repellent activity of essential oil compounds which helps to keep pests out of the produce during storage. Mint essential oil exhibited a strong repellent effect on *Lasioderma serricorne* (Zhang et al., 2015). The use of *Tephrosia vegelli* as a pesticide plant is attributed to presence of deguelin, tephrosin, α-toxicarol, and sarcolobine, though their toxicity and effectiveness vary (Belmain et al., 2012).

Essential oils from aerial parts of *Artemisia judaica* repelled the cowpea weevil, *Callosobruchus maculatus* (Ebadollahi et al., 2016). Essential oil of *Pogostemon cablin* exhibited strong insectistatic activities against *Spodoptera litura* and *Spodoptera exigua* (Abd-Elhady, 2012). Essential oils from several plants inhibited feeding in the fourth instar larvae and adults of *Leptinotarsa decemlineata.* These included *Satureja khuzistanica*, *Ocimum basilicum*, *Myrtus communis*, *Thymus daenensis*, *Mentha spicata* and *Eugenia caryophyllus* (Saroukolai et al., 2014). *Cymbopogon citratus*, *Myristica fragrans* and *Mentha citrata* oils significantly limited the growth of the potato tuber moth (Aziza et al., 2014).

Biopesticides also inhibit octopamine and tyramine receptors which are important for different physiological processes in insects. For instance, octopamine is important in memory and conditioning in honeybees (Evans & Maqueira, 2005), and modifies muscle performance, synthesis and degradation of lipids (Roeder, 2005). Clove oil blocks octopamine receptor system (OAR) in corn leaf aphid and the non-target lady beetle (Toledo-Hernández et al., 2022). Cinnamon (*Cinnamomum* spp.) oil damages antennal receptors in red worker ants resulting in a reduction octopamine. This disrupts the social behaviour and capacity to identify mates (Orchard, 1982).

Some essential oils interfere with stages of insect life cycle through interacting with receptors for juvenile growth hormone (Corrêa et al., 2023). Nerolidol from the essential oils of *Stevia rebaudiana* (Benelli et al., 2020), jasmine, lavender, ginger and cannabis (Ghoneim et al., 2021) deters larval development of some insects. This interruption in the development stage leads to malformed individuals and may cause death if the essential oils are at high enough concentrations (Corrêa et al., 2023).

Biopesticides also inhibit gamma-aminobutyric acid (GABA) receptors which can cause insect death (Tong & Coats, 2010). This is achieved by opening chloride channels and enabling chloride ions to enter neurons, thus inhibiting the nervous system (Devrnja et al., 2022), for instance, in *Caenorhabditis elegans* and *Drosophila.*

4.5 Isolation and Characterization of Bioactive Compounds

Extraction is a critical starting point in the analysis of medicinal plants before separation and characterization of their compounds (Sasidharan et al., 2011). Isolation of natural products involves several steps (Sticher, 2008). Accepted natural products extraction procedures include Soxhlet extraction, maceration, percolation, turbo-extraction and sonication (Sticher, 2008). Advanced techniques include solid-phase, supercritical-fluid, pressurized-liquid, microwave-assisted and surfactant-mediated extraction (Sasidharan et al., 2011).

Hydrophilic compounds are extracted using polar solvents (Sasidharan et al., 2011), whereas lipophilic compounds are extracted using dichloromethane or mixtures of different compounds. At times, extraction is done using hexane to remove chlorophyll (Cos et al., 2006). Extraction of bioactive compounds can be achieved through thin layer chromatography (TLC), flash column chromatography, Sephadex and high-pressure liquid chromatography (HPLC), in order to obtain pure compounds (Sasidharan et al., 2011).

Isolation techniques: Separations in liquid chromatography are performed using different techniques used including thin-layer chromatography, preparative thin-layer chromatography and high-pressure liquid chromatography (HPLC) (Sticher, 2008).

Characterization of bioactive compounds: Nuclear magnetic resonance (NMR) is mainly used to identify and characterize the structure of organic compounds. The technology provides information about chemical structure, reaction state and dynamics of the molecules (Sharma et al., 2021). Various types of NMR methods used include ^{1}H-NMR, ^{13}C-NMR, ^{15}N-NMR and ^{31}P-NMR (Claridge, 2016).

Biopesticides can be broadly categorized as: microbial, biochemical and plant-incorporated protectants. They constitute only 5% of the global pesticide market and are predominantly microbial biopesticides (Pathma et al., 2021). Some of the biopesticides that have been developed include bacterial biopesticides. These have been widely used because of their low cost (Sarwar, 2015). *Bacillus thuringiensis* is the commonest microbial pesticide (Singh et al., 2019) with about 100 well-known subspecies (Glare et al., 2012). They are active against Lepidoptera, Diptera (Nematocera) and Coleoptera (Chrysomelidae and Scarabaeidae) (Van Frankenhuyzen, 2009).

Other biopesticides are derived from *Beauveria* sp., *Metarhizium* sp., *Isaria fumosorosea* and *Lecanicillium* spp. (Singh et al., 2019). Baculovirus (BV) has since received a lot of attention in development of biopesticides at commercial level (Moscardi et al., 2011). Genetically enhanced isolates of *Autographa californica* nuclear polyhedrosis virus (AcMNPV) and *Tegenaria agrestis* (vAcTaITX-1 and vAcDTX9.2) were commercially evaluated against lepidopteran insects (Hughes et al., 1997) and are currently produced on small scale (Lacey, 2017). The pesticide from *Oryctes nudivirus* lowered the population of coconut palm rhinoceros beetle and their larvae (Singh et al., 2019).

Botanical pesticides: Over 6000 plant species have been identified to possess insecticidal properties (Nawaz et al., 2016). From these, botanical pesticides have penetrated the market.

(i) Limonene and linalool are bioactive compounds from different plant species which target fleas, aphids, mites and ants (Mazid et al., 2011). Azadirachtin is among the best known limonoids (Koul, 2012). The main bioactive compounds in neem oil are azadirachtin, dihydroazadirachtin and triterpenoids such as nimbin, and salannin. Neem acts as moulting inhibitor or ecdysone antagonist, an antifeedant/repellent, and causes physical smothering and desiccation. Neem is also marketed under several trade names globally (Copping & Duke, 2007; Isman, 2006; Isman & Paluch, 2011). Neem seed kernel extract effectively decreased *Helicoverpa armigera* larvae in chickpea (Bhushan et al., 2011). Neem extract also kills eggs of peach fruit fly (Saunders) (Kumar, 2015).
(ii) *Ryania* sp. (*Ryania speciosa* Vahl) contains alkaloids such as ryanodine, ryania, and 9,21-didehydroryanodine which are the main bioactive compounds. These compounds work as neuromuscular poisons (calcium channel agonist) and are marketed under several trade names (Copping & Duke, 2007; Isman, 2006; Isman & Paluch, 2011). These compounds target several pests, especially caterpillars and thrips (Mazid et al., 2011).
(iii) Several plant species such as *Derris, Lonchocarpus* and *Tephrosia* are sources of rotenone and isoflavonoids such as deguelin. Rotenone is a mitochondrial cytotoxin and is also commercially produced under various trade names (Copping & Duke, 2007; Isman, 2006; Isman & Paluch, 2011).
(iv) Nicotine is the main bioactive compound from the tobacco plant *Nicotiana* spp. Nicotine exists as three isomers of nicotine sulphate. Nicotine damages the nervous system and is a widely used botanical pesticide (Copping & Duke, 2007; Isman, 2006; Isman & Paluch, 2011; Mathew, 2016).
(v) The main bioactive compounds in *Schoenocaulon officinale* are a mixture of alkaloids such as cevadine, and veratridine from the seeds. The mixture works as an axonic poison (Copping & Duke, 2007; Isman, 2006; Isman & Paluch, 2011).

4.5.1 Case Studies Illustrating Successful Isolation and Application of Biopesticides

Calceolaria talcana extracts possess significant antifeedant, antifungal, insecticidal and antibacterial activities on insect pest models (Munoz et al., 2013). The n-hexane extract of the aerial parts of *C. talcana* contained diterpenes, triterpenes and naphthoquinones. Fractions on n-hexane resulted in either death or reduced growth of *Spodoptera frugiperda* and *Drosophila melanogaster* at dosages between 35 and 50 ppm. This indicates their selective effects on insect metabolism at low concentrations (Céspedes et al., 2014). The results are comparable to natural insect growth inhibitors which makes them potential insecticidal agents (Munoz et al., 2013).

4.5.2 *Efficacy Studies of Biopesticides*

Even though synthetic pesticides are known to be very effective, biopesticides may perform better when applied in the right concentrations and appropriate frequencies (Shah et al., 2013). Utilization of biopesticides which are eco-friendly contributes towards sustainable agriculture (Bharti & Ibrahim, 2020; Nawaz et al., 2016) and could be integrated with crop rotation (Ruiu, 2015).

Biopesticides are highly specific to targeted pests without harming humans or other beneficial organisms (Kumar et al., 2021; Oguh et al., 2019). On the contrary, synthetic pesticides have a broad spectrum action and may affect non-target organisms (Kumar, 2015). Additionally, the use of biopesticides greatly decreases dependence on chemical pesticides, while achieving almost the same level of crop yield (Kumar, 2015). A formulation of *Allium cepa* and *Zingiber officinale* controlled 70%–80% of tomato fruit worm subsequently increasing yield in the treated plants compared to untreated ones (Arora et al., 2012). Muzemu et al.'s (2011) study reported a reduction of *Brevicoryne brassicae* and *Tetranychus evans* by more than 50% when powder extracts of *Lippia javanica* and *Solanum delaguense* were used.

A combination of tea extracts and poultry litter reduced severity of bacterial wilt on *Solanum melongena* with a higher efficacy when the compost tea extract was applied as a soil drench resulting in healthy plants and improved yield (Islam et al., 2014). The same formulation reduced the incidence and severity of *Phytophtora infestans* on potato when the compost tea was applied as a foliar spray (Islam et al., 2013). These findings thus justify that biopesticides are as effective as chemical pesticides.

4.6 Applications in Integrated Pest Management (IPM)

The use of organic pesticides as an element of IPM scheme is paramount in minimizing the drawbacks of chemical insecticides. Biopesticides are biodegradable and self-perpetuating, less harmful on beneficial pests, mostly host specific and have a shorter lifespan (Matyjaszczyk, 2015). Integrated pest management entails the use of chemical and organic pesticides and organic prevention methods for crop protection (Céspedes et al., 2014).

Utilization of microorganisms is a more practical approach compared to chemical pesticides given that it is eco-friendly (Koul & Cuperus, 2008). Biopesticides have been used on specific pests in contrast to chemical pesticides which may affect a wide variety of organisms in addition to the pest (Rizvi et al., 2012). When used in IPM programs, biopesticides are able to minimize the use of chemical pesticides and at the same time increase crop yield (Kumar, 2015; Rizvi et al., 2012). For instance, entomopoxviruses are important in pest control due to their presence in a number of insect pest orders (Rizvi et al., 2012). The release of pheromones is instrumental in appropriate timing and application of pesticides (O'Brien et al., 2009). These chemicals are released by insects to attack mates. They are mimicked for insect control in IPM programmes. They effectively alter insect behaviour and thus reduce their fecundity (Gonzalez-Coloma et al., 2004).

4.6.1 Integration of Natural Products into IPM Strategies

The use of plants' secondary metabolites such as terpenes, phenolics and alkaloids in IPM is on the rise. This trend has necessitated the need for bioprospecting in a bid to discover natural products that have new modes of action (Meinwald, 2011) and features that enhance their value as practical pesticides (Akhtar et al., 2008). Secondary metabolites are found in different plant parts which are often targeted as sources of unique biopesticides (Munoz et al., 2013). Plant defences can either be constitutive or induced. Plant defences as part of IPM can be enhanced by combining them with biological control and cultivation methods (Tortorici et al., 2022).

The inducers of plant resistance hinder pest development through stimulation of defence responses of plants (Tortorici et al., 2022). Thus, their use in IPM is commended (Siah et al., 2018). Natural products are instrumental in plant protection against pathogens through induction of systemic acquired resistance (SAR). These SAR-inducing compounds and preparations are termed elicitors (Dayan et al., 2009). Through this indirect activity, pathogen resistance cannot occur which makes these products useful in integrated disease management (Garcia-del-Pino et al., 2013). Plant inducers act on a wide variety of plant species, fungi and viruses. Their effectiveness is affected by environmental factors, genes, physiology of the plants and crop nutrition (Regnault-Roger, 2011).

4.6.2 Synergies with Conventional Pest Control Methods

Cultural practices such as strip harvesting of alfalfa can favour biological pest control, and this was promoted during the birth of IPM (Stern et al., 1964). Of recent, this method has been evaluated during the making of alfalfa hay in Australia (Hossain et al., 2002). Predators painted with fluorescent dye were released into alfalfa plants before mowing. A majority of predators survived mowing by relocating within a short distance to uncut strips (Gurr & Kvedaras, 2010). Biological control is one of several pest management methods (Gurr & Kvedaras, 2010). The innate resistance of the host plant may contribute towards making the susceptibility of pests to insecticides as a result of reduction in their growth rate (Gurr & Kvedaras, 2010). Increased pest susceptibility may permit utilization of low insecticide concentrations that may favour survival of natural enemies while killing the pest (Emden, 1990). Such an interaction would facilitate pest control while keeping the within-crop community of natural enemies for protection from future pest establishment (Gurr & Kvedaras, 2010).

4.6.3 Successful Implementation in Agricultural Settings

The initial traded Bt biopesticide Sporeine® was formulated about a century ago (Angus, 1954; Lacey, 2017). Nonetheless, by the early 1990s, only three Bt biopesticide formulations were traded on the Brazilian market. These were Bt *kurstaki* (Dipel®, Thuricide® and Bactospeine®) (Habib et al., 1991). In the 2000s, the number of biopesticides marketed and recommended for pest control was only six (Polanczyk et al., 2008). By 2020, the number had increased to 29 Bt biopesticides registered in Brazil (Agrofit, 2019). The surge in numbers was attributed to *Helicoverpa armigera* outbreaks during the 2012/2013 crop season, resulting in fundamental losses in soybean and cotton (Pomari-Fernandes et al., 2015). Dipel® effectively controlled *H. armigera* compared to synthetic insecticides available at the time. This was a milestone in the importance of microbial biopesticides in insect pest control (do Nascimento et al., 2022). Consequently, Brazil is the largest user of Bt biopesticides globally, with a sprayed area of approximately nine million hectares (Polanczyk et al., 2017).

4.7 Ecological Implications and Environmental Sustainability

Assessment of the Ecological Impact of Using Plant-Derived Biopesticides. Generally, plant-derived biopesticides are safe to humans, the planet and other organisms. However, some studies have indicated the contrary. For instance, rotenone derived from roots of *Lonchocarpus* species which is commonly used as a pesticide was reported to be harmful to fish and toxic to humans following ingestion or inhalation (NIOSH, 1994). Pyrethrum, which is extracted from *Chrysanthemum* sp., is highly toxic. It paralyses and kills the insects by changing their mode of transmission of electrical impulses to the nervous system (Ware & Whitacre, 2004). Thus, the release of pyrethrum and rotenone into the ecosystem in high doses can be harmful to aquatic and terrestrial organisms including humans.

Nicotine, sabadilla, pyrethrins and fluoroacetate impede functioning of the nervous system of insects (Oguh et al., 2019) but are plant-based pesticides (Bateman et al., 2018; Chengala & Singh, 2017). Essential oils-based insecticides negatively affect the pests' ability to respire and reduces their parasitization rates (Giunti et al., 2022). Also, the essential oil-based pesticides reduce the speed and distance travelled by pollinators. So, there is need to investigate the effect of aromatic oils on toxicity of non-pest insects so as to understand their mode of action.

On a positive note, the use of biopesticides increases crop productivity (Kumari et al., 2014) and minimise harmful effects on the environment compared to conventional pesticides (Bailey et al., 2010) whose repeated application leads to loss of biodiversity. They also accumulate in the soil ultimately leaching to ground and surface water. Tinklein (2015) reported ecological advantages of biopesticides including being less toxic to humans and animals, having a narrow toxicity range, effectiveness at relatively low concentration and usually degrading rapidly.

Consideration of Environmental Sustainability and Biodiversity Conservation. Biopesticides are eco-friendly and thus do not cause pollution of water bodies and land (Essiedu et al., 2020). Biopesticides serve as one of the renewable stocks and products used in agriculture that enable continuous use of land for food security of the populace. The sustainable use of biopesticides result in increased crop productivity through attack of crop pests but sparing non-pest organisms useful in pollination (Ahirwar et al., 2020). The application of biopesticides is harmless to the environment and improves productivity which contributes to attainment of sustainable development goals (SDGs): 3, 6, 2, 14 and 15. Such farming techniques ensure food security by enhancing food productivity using eco-friendly and cost-effective products to control pests (Sethi et al., 2022).

Biopesticides are environmentally friendly a characteristic attributed to their ability to decompose quickly (Kawalekar, 2013). In addition to being eco-friendly, biopesticides may be more effective against crop pests compared to synthetic pesticides in some cases (Leng et al., 2011). Furthermore, biopesticides have short preharvest intervals and target specific pests, thus are safe for use on non-target insects (Tadele Shiberu, 2016).

Though biopesticides are generally considered to be safe, their continued use requires to be assessed to identify potential risks as a result of prolonged use (Ashaolu et al., 2022). Thus, further investigations need to be done on the different plants and their metabolites to determine safe levels for use in controlling pests and which would also not have negative effects on human health.

4.8 Challenges and Future Prospects

Though biopesticides are considered safe, their use has been associated with challenges minimizing their uptake by farmers in different countries. For instance, in India, the biopesticides market has not developed to its full potential due to unwillingness of farmers to use them instead of chemical pesticides. Reasons advanced for this were high costs, uneven performance results in fields, minimum life span, lack of infrastructure for large-scale production and a costly bureaucratic registration process (Yadav, 2022).

Formulation of plant-based biopesticides has encountered challenges regarding their registration, accepted formulation, adoption and commercialization (Upadhyay et al., 2020). As documented by several studies, the use of neem biopesticide is widespread (George et al., 2014; Huang et al., 2020; Pavela & Benelli, 2016). However, despite its varied biological activities, the effectiveness of neem biopesticide is short-lived because it is easily destabilized by sunlight. Neem oil was reported to be phytotoxic to tomato in high concentrations (Nawaz et al., 2016).

The challenge of using pesticidal plants in developing countries is hindered by lack of enough research to establish their active ingredients. Though efforts to characterize some phytochemicals have been done, ascertaining accurate doses to use is still challenging.

Another hindrance is the lack of specific regulations for their risk assessment of biopesticides leading to overreliance by Organisation for Economic Co-operation and Development (OECD) countries on only existing protocols for chemical pesticides (OECD, 1998). Another neglected issue in biopesticides' safety assessment is the fact that tests have been done on a few commercially known pollinators such as *Apis mellifera* and *Bombus terrestris* (Franklin & Raine, 2019; Klinger et al., 2019) without testing toxicity on non-marketable pollinators (Borges et al., 2021).

The safety of pesticides from *Aconitum* sp. and *Ricinus communis* (castor bean) is not yet confirmed except for a report on high toxicity to humans. *Tephrosia vogelii* was also reported to adversely affect fish (Stevenson et al., 2012).

The use of biopesticides encounters various challenges regarding their commercialization. For instance, trade in these products confers little or no profit to the sellers due to their high selectivity levels (Gelernter, 2005). Though selectivity is beneficial, the niche market nature of biopesticides makes them less profitable. Low or lack of profitability for biopesticides is perpetuated by lack of economies of scale during their production which discourages early adopters. Additionally, most farmers are risk-averse and would rather use chemical pesticides which they have used for a long period and are conversant with their results than shift to a new technology of using biopesticides, whose results on crops are not properly known (Yaron et al., 1992).

The utilization of biopesticides in future is expected to increase due to the fact that many governments are implementing strict regulations regarding production and use of chemical pesticides. Thus, companies involved in biopesticide production will help fill the gap regardless of the challenges they are likely to face, such as limited resources for research and development, product registration and promotion.

Utilizing biopesticides promotes green agricultural practices, markedly protecting both human health and environment. Globally, the use of biopesticides has gained attention owing to their beneficial effects. Active substances derived from plants have been discovered promoting their potential for use in production of new pest control drugs. Consequently, over 175 biopesticides are registered globally (Moosavi & Zare, 2016; Samada & Tambunan, 2020; Yunus et al., 2019).

There is potential to improve the precision of characterization of active phytochemicals from pesticide plants and their concentration in final products through the use of the new field of nanotechnology which could enhance their shelf life, release rate and potency (Chaudhary et al., 2017; Perlatti et al., 2013). Fusion protein and recombinant DNA technologies have been recently employed to increase the efficacy of biopesticides. In future, omics technologies could be used to elucidate novel toxin discovery (Nakasu et al., 2014).

There is minimal demand and availability of biopesticides which discourages their production and application (Ayilara et al., 2023). Thus, future production and use of biopesticides calls for provision of grants to boost their manufacture and to understand how their stability can be enhanced to promote long shelf-life.

In order for biopesticide production to be successful, formulation processes used have to be of high quality, leading to safe, effective, easy-to-use products with a long shelf life. However, most biocontrol products rarely possess these properties failure of which restricts their entry into the market. In Africa, Kenya, Uganda, Tanzania, Ghana and South Africa have developed guidelines on biopesticides production and usage (Arora et al., 2016. The bodies and regulations responsible for this function are: the Pest Control Products Board (PCPB) and the Agriculture Act Chapter Cap 38 in Kenya (Kimani, 2014); the National Agency for Food and Drug Administration and Control (NAFDAC) in Nigeria, the Department of Agriculture, Forestry and Fisheries (DAFF) in South Africa (DAFF, 2010), the Tropical Pesticides Research Institute (TPRI) in Tanzania (Stadlinger et al., 2013) and the Uganda National Agricultural Chemicals Board (UNACB) and the Agriculture Chemical Control Committee (ACCTC) in Uganda. Despite the presence of such bodies, registration of biopesticides is still a challenge. For instance, in Nigeria, applicants interested in registering biopesticides do not finalize the process due to inability to comply with the requirements of the National Biosafety Management Agency (Ashaolu et al., 2022).

4.8.1 Opportunities for Collaboration and Knowledge Exchange

Worldwide acceptance in the use of biopesticides requires combined efforts of scientists from all over the world to collaborate and share knowledge to address any challenges biopesticides may cause. Several factors that hamper research and development of biopesticides in different countries have been recorded, and these could serve as starting points for research collaboration and knowledge sharing. For instance, Nigeria has developed risk assessment grounds for biopesticides (Ashaolu et al., 2022). This calls for coordination within the region so as to promote scientific knowledge for regulation, production and marketing of biopesticides. Promotion of the use of biopesticides requires farmer awareness about integrated pest management (IPM) tools. Though the public is highly aware of the residual effects of chemical pesticides, there is a paucity of information regarding the use of biopesticides. Thus, the need for knowledge exchange between biopesticide producers and farmers is paramount.

4.9 Conclusions

There are numerous threats to the food security of many poor people in the developing world. Pests particularly limit agricultural productivity. Their challenge is compounded by several factors including climatic changes that have expanded geographical ranges of some pests, as well as application of chemical pesticides that are highly toxic and indiscriminate in action. Plant products provide a potential solution to this challenge due to their various advantages such as their specificity, biodegradability and being safe. Therefore, the exploitation of hitherto unknown plant species and their compounds as well as the use of modern techniques to enhance the delivery parameters of some older, or previously known compounds, has the potential to boost agricultural activity, improve food security by providing better control measures for pests. Indigenous traditional knowledge provides valuable insights into sustainable pest management. By acknowledging and integrating these practices, modern agriculture can become more resilient, environmentally friendly and culturally inclusive. Embracing ITK not only supports biodiversity and ecosystem health but also honours the wisdom and heritage of Indigenous peoples. Therefore, bioprospecting for biopesticides based on African indigenous traditional knowledge offers a promising pathway for developing sustainable and environmentally friendly pest management solutions. By integrating traditional wisdom with scientific research, it is possible to create effective biopesticides that support both agricultural productivity and ecological health. Collaboration between scientists, policymakers and indigenous communities is essential to unlock the full potential of this valuable knowledge.

References

Abd-Elhady, H. (2012). Insecticidal activity and chemical composition of essential oil from *Artemisia judaica* L. against *Callosobruchus maculatus* (F.) (Coleoptera: Bruchidae). *Journal of Plant Protection Research, 52*(3).

Adel, M. M., Sehnal, F., & Jurzysta, M. (2000). Effects of alfalfa saponins on the moth *Spodoptera littoralis*. *Journal of Chemical Ecology, 26*, 1065–1078.

Adeyemi, M. H. (2010). The potential of secondary metabolites in plant material as deterents against insect pests: A review. *African Journal of Pure and Applied Chemistry, 4*(11), 243–246.

Agrofit – Sistema de Agrotóxicos Fitossanitários – Ministério da Agricultura, Pecuária e Abastecimento – Coordenação Geral de Agrotóxicos e Afins/DFIA/DAS (2019). Available from http://agrofit.agricultura.gov.br/agrofit_cons/principal_agrofit_cons. Accessed 6 July 2024.

Ahirwar, S., Malik, M. S., Ahirwar, R., & Shukla, J. (2020). Identification of suitable sites and structures for artificial groundwater recharge for sustainable groundwater resource development and management. *Groundwater for Sustainable Development, 11*, 100388.

Ajani, E., Mgbenka, R., & Okeke, M. (2013). Use of indigenous knowledge as a strategy for climate change adaptation among farmers in sub-Saharan Africa: Implications for policy. *Asian Journal of Agricultural Extension, Economics & Sociology, 2*, 23.

Akhtar, Y., Yeoung, Y. R., Isman, M. B. (2008). Comparative bioactivity of selected extracts from Meliaceae and some commercial botanical insecticides against two noctuid caterpillars, Trichoplusia ni and Pseudaletia unipuncta. Phytochemistry reviews, 7, 77–88.

Akankwasa, K., Marimo, P., Tumuhimbise, R., Asasira, M., Khakasa, E., Mpirirwe, I., et al. (2021). The East African highland cooking bananas 'Matooke'preferences of farmers and traders: Implications for variety development. *International Journal of Food Science & Technology, 56*(3), 1124–1134.

Almadiy, A. A., & Nenaah, G. E. (2022). Essential oil of *Origanum vulgare*, its nanoemulsion and bioactive monoterpenes as eco-friendly novel green pesticides for controlling *Aedes aegypti*, the common vector of Dengue virus. *Journal of Essential Oil Research, 34*(5), 424–438.

Al-Rajhy, D. H., Alahmed, A. M., Hussein, H. I., & Kheir, S. M. (2003). Acaricidal effects of cardiac glycosides, azadirachtin and neem oil against the camel tick, *Hyalomma dromedarii* (Acari: Ixodidae). *Pest Management Science: Formerly Pesticide Science, 59*(11), 1250–1254.

Altieri, M. A. (1993). Ethnoscience and biodiversity: Key elements in the design of sustainable pest management systems for small farmers in developing countries. *Agriculture, Ecosystems & Environment, 46*(1–4), 257–272.

Altieri, M. A. (2019). Designing and improving pest management systems for subsistence farmers. In *Crop protection strategies for subsistence farmers* (pp. 1–20). CRC Press.

Anandajayasekeram, P. (2022). The role of agricultural R&D within the agricultural innovation systems framework. In *Innovation in small-farm agriculture* (pp. 75–87). CRC Press.

Angus, T. (1954). A bacterial toxin paralysing silkworm larvae. *Nature, 173*(4403), 545–546.

Ansari, M. A., Sharmaa, S., Roy, S., Ramakrishna, Y., Datt, S., Ningombama, A., et al. (2021). Documenting the agriculture based indigenous traditional knowledge in Manipur State of North Eastern India. *Indian Journal of Traditional Knowledge (IJTK), 20*(4), 1065–1074.

Arora, S., Kanojia, A. K., Kumar, A., Mogha, N., Sahu, V. (2012). Biopesticide formulation to control tomato lepidopteran pest menace. Current science, 1051–1057.

Arora, N. K., Verma, M., Prakash, J., & Mishra, J. (2016). Regulation of biopesticides: Global concerns and policies. In N. Arora, S. Mehnaz, & R. Balestrini (Eds.), *Bioformulations: For sustainable agriculture* (pp. 283–299). Springer.

Asemu, A. M., Habtu, N. G., Subramanyam, B., Delele, M. A., Kalsa, K. K., & Alavi, S. (2020). Effects of grain drying methods on postharvest insect infestation and physicochemical characteristics of maize grain. *Journal of Food Process Engineering, 43*(7), e13423.

Ashaolu, C. A., Okonkwo, C. O., Njuguna, E., & Ndolo, D. (2022). Recommendations for effective and sustainable regulation of biopesticides in Nigeria. *Sustainability, 14*(5), 2846.

Assefa, F. (2018). Status of fall armyworm (*Spodoptera frugiperda*), biology and control measures on maize crop in Ethiopia: A review. *International Journal of Entomological Research, 6*(2), 75–85.

Ayilara, M. S., Adeleke, B. S., Akinola, S. A., Fayose, C. A., Adeyemi, U. T., Gbadegesin, L. A., et al. (2023). Biopesticides as a promising alternative to synthetic pesticides: A case for microbial pesticides, phytopesticides, and nanobiopesticides. *Frontiers in Microbiology, 14*, 1040901.

Aziza, S., Abdel Rahman, H., Abdel-Aziz, S., & Moawad, S. (2014). Natural plant oils and terpenes as protectors for the potato tubers against *Phthorimaea operculella* (Zeller) infestation by different application methods. *Egyptian Journal Of Biological Pest Control, 24*(1).

Bailey, K., Boyetchko, S., & Längle, T. (2010). Social and economic drivers shaping the future of biological control: A Canadian perspective on the factors affecting the development and use of microbial biopesticides. *Biological Control, 52*(3), 221–229.

Banchio, E., Valladares, G., Defago, M., Palacios, S., & Carpinella, C. (2003). Effects of *Melia azedarach*, (Meliaceae) fruit extracts on the leafminer *Liriomyza huidobrensis*, (Diptera, Agromyzidae): Assessment in laboratory and field experiments. *Annals of Applied Biology, 143*(2), 187–193.

Barbehenn, R. V., & Constabel, C. P. (2011). Tannins in plant–herbivore interactions. *Phytochemistry, 72*(13), 1551–1565.

Bateman, M. L., Day, R. K., Luke, B., Edgington, S., Kuhlmann, U., & Cock, M. J. (2018). Assessment of potential biopesticide options for managing fall armyworm (*Spodoptera frugiperda*) in Africa. *Journal of Applied Entomology, 142*(9), 805–819.

Belmain, S. R., Amoah, B. A., Nyirenda, S. P., Kamanula, J. F., & Stevenson, P. C. (2012). Highly variable insect control efficacy of *Tephrosia vogelii* chemotypes. *Journal of Agricultural and Food Chemistry, 60*(40), 10055–10063.

Belmaina, S., Nealb, G. E., Rayb, D., & Goloba, P. (2001). Insecticidal and vertebrate toxicity associated with ethnobotanicals used as post-harvest protectants in Ghana. *Food and Chemical Toxicology, 39*, 287ą291.

Benelli, G., Pavela, R., Drenaggi, E., Desneux, N., & Maggi, F. (2020). Phytol, (E)-nerolidol and spathulenol from *Stevia rebaudiana* leaf essential oil as effective and eco-friendly botanical insecticides against *Metopolophium dirhodum. Industrial Crops and Products, 155*, 112844.

Bharti, V., & Ibrahim, S. (2020). Biopesticides: Production, formulation and application systems. *International Journal of Current Microbiology and Applied Science, 9*(10), 3931–3946.

Bhattacharjee, P. P., & Ray, D. C. (2010). Pest management beliefs and practices of Manipuri rice farmers in Barak Valley, Assam.

Bhushan, S., Singh, R. P., Shanker, R. (2011). Bioefficacy of neem and Bt against pod borer, Helicoverpa armigera in chickpea. *Journal of Biopesticides, 4*(1), 87.

Bleeker, P. M., Diergaarde, P. J., Ament, K., Schütz, S., Johne, B., Dijkink, J., et al. (2011). Tomato-produced 7-epizingiberene and R-curcumene act as repellents to whiteflies. *Phytochemistry, 72*(1), 68–73.

Boiteau, G., & Vernon, R. S. (2001). Physical barriers for the control of insect pests. In *Physical control methods in plant protection* (pp. 224–247). Springer.

Borges, S., Alkassab, A. T., Collison, E., Hinarejos, S., Jones, B., McVey, E., et al. (2021). Overview of the testing and assessment of effects of microbial pesticides on bees: Strengths, challenges and perspectives. *Apidologie, 52*, 1–22.

Bowers, M. D., & Puttick, G. M. (1989). Iridoid glycosides and insect feeding preferences: Gypsy moths (*Lymantria dispar*, Lymantriidae) and buckeyes (Junonia coenia, Nymphalidae). *Ecological Entomology, 14*(3), 247–256.

Campos, I., DdMCd, B., Cardoso, C., & Pontes, E. (2019). Souza MAAd Efficacy of lemongrass essential oil and citral in controlling *Callosobruchus maculatus* (Coleoptera: Chrysomelidae), a post-harvest cowpea insect pest. *Crop Protection, 119*, 191–196.

Carvalho, F. P. (2017). Pesticides, environment, and food safety. *Food and Energy Security, 6*(2), 48–60.

Celis, Á., Mendoza, C., Pachón, M., Cardona, J., Delgado, W., & Cuca, L. E. (2008). Extractos vegetales utilizados como biocontroladores con énfasis en la familia Piperaceae. *Una revisión. Agronomía colombiana, 26*(1), 97–106.

Céspedes, C. L., Salazar, J. R., Ariza-Castolo, A., Yamaguchi, L., Ávila, J. G., Aqueveque, P., Kubo, I., Alarcón, J. (2014). Biopesticides from plants: Calceolaria integrifolia sl. *Environmental Research, 132*, 391–406.

Chaudhary, S., Kanwar, R. K., Sehgal, A., Cahill, D. M., Barrow, C. J., Sehgal, R., & Kanwar, J. R. (2017). Progress on *Azadirachta indica* based biopesticides in replacing synthetic toxic pesticides. *Frontiers in Plant Science, 8*, 226969.

Chengala, L., & Singh, N. (2017). Botanical pesticides—A major alternative to chemical pesticides: A review. *International Journal of Life Sciences, 5*(4), 722–729.

Chimweta, M., Nyakudya, I. W., Jimu, L., & Bray Mashingaidze, A. (2020). Fall armyworm [*Spodoptera frugiperda* (JE Smith)] damage in maize: Management options for flood-recession cropping smallholder farmers. *International Journal of Pest Management, 66*(2), 142–154.
Chorol, U., Lhamo, N., & Gupta, V. (2021). Biopesticides: Eco-friendly, biodegradable and a key to sustainable agriculture. *Good Agriculture Practices In Cold Arid Region*, 79–89.
Chou, C., Hadi, B. A., Chiba, S., Sato, I., Choi, I. R., & Tanaka, T. (2022). An entomopathogenic fungus and a natural extract benefit rice (*Oryza sativa*) by suppressing populations of insect pests while keeping high populations of their natural enemies. *Biological Control, 165*, 104793.
Claridge, T.D.W. (2016). High-resolution NMR techniques in Organic Chemistry. 3rd Edition, Elsevier.
Copping, L. G., & Duke, S. O. (2007). Natural products that have been used commercially as crop protection agents. *Pest Management Science: Formerly Pesticide Science, 63*(6), 524–554.
Corrêa, E. J. A., Carvalho, F. C., de Castro Oliveira, J. A., Bertolucci, S. K. V., Scotti, M. T., Silveira, C. H., et al. (2023). Elucidating the molecular mechanisms of essential oils' insecticidal action using a novel cheminformatics protocol. *Scientific Reports, 13*(1), 4598.
Cos, P., Vlietinck, A.J., Vanden Berghe, D., Maes, L. (2006). Anti-infective potential of natural products: How to develop a stronger in vitro 'proof-of-concept'. *Journal of Ethnopharmacology, 106*(3), 290–302.
Cowan, M. M. (1999). Plant products as antimicrobial agents. *Clinical Microbiology Reviews, 12*(4), 564–582.
DAFF (2010). Abstract of Agricultural Statistics. Directorate: Agricultural Statistics, Department of Agriculture, Forestry and Fisheries, Pretoria.
Damalas, C. A., & Koutroubas, S. D. (2016). *Farmers' exposure to pesticides: Toxicity types and ways of prevention* (Vol. 4, p. 1). MDPI.
Dave, H., & Ledwani, L. (2012). A review on anthraquinones isolated from *Cassia* species and their applications.
Dayan, F. E., Cantrell, C. L., & Duke, S. O. (2009). Natural products in crop protection. *Bioorganic & Medicinal Chemistry, 17*(12), 4022–4034.
De Groote, H. (2022). Economic analysis of pest problems in agriculture and food chains in Africa. *Current Opinion in Insect Science, 54*, 100969.
Deka, M. K., Bhuyan, M., & Hazarika, L. K. (2006). Traditional pest management practices of Assam. *Indian Journal of Traditional Knowledge, 5*(1), 75–78.
Deshmukh, P. S. (2015). A case study: Traditional methods of pest control in some villages of Kolhapur District. *Online International Interdisciplinary Research Journal, 5*(3), 87–92.
Devrnja, N., Milutinović, M., & Savić, J. (2022). When scent becomes a weapon—plant essential oils as potent bioinsecticides. *Sustainability, 14*(11), 6847.
do Nascimento, J., Goncalves, K. C., Dias, N. P., de Oliveira, J. L., Bravo, A., & Polanczyk, R. A. (2022). Adoption of Bacillus thuringiensis-based biopesticides in agricultural systems and new approaches to improve their use in Brazil. *Biological Control, 165*, 104792.
Drenth, A., & Kema, G. (2021). The vulnerability of bananas to globally emerging disease threats. *Phytopathology, 111*(12), 2146–2161.
Dubey, V. S., Bhalla, R., & Luthra, R. (2003). An overview of the non-mevalonate pathway for terpenoid biosynthesis in plants. *Journal of Biosciences, 28*, 637–646.
Ebadollahi, A., Ashrafi Parchin, R., & Farjaminezhad, M. (2016). Phytochemistry, toxicity and feeding inhibitory activity of *Melissa officinalis* L. essential oil against a cosmopolitan insect pest; *Tribolium castaneum* Herbst. *Toxin Reviews, 35*(3–4), 77–82.
Eich, E. (2008). *Solanaceae and convolvulaceae: Secondary metabolites: Biosynthesis, chemotaxonomy, biological and economic significance (a handbook).* Springer Science & Business Media.
Ellen, D., Ellen, L., Danny, G., & Guy, S. (2007). Novel advances with plant saponins as natural insecticides to control pest insects. *Pest Technology, 1*(2), 96–105.

Emam, A. M., Swelam, E. S., & Megally, N. Y. (2009). Furocoumarin and quinolone alkaloid with larvicidal and antifeedant activities isolated from *Ruta chalepensis* leaves. *Journal of Natural Products, 2*, 10–22.

Emden, H. V. (1990). The interaction of host plant resistance to insects with other control measures.

Eriksson, C., Månsson, P. E., Sjödin, K., & Schlyter, F. (2008). Antifeedants and feeding stimulants in bark extracts of ten woody non-host species of the pine weevil, *Hylobius abietis*. *Journal of Chemical Ecology, 34*, 1290–1297.

Essiedu, J. A., Adepoju, F. O., & Ivantsova, M. N. (2020). Benefits and limitations in using biopesticides: A review. Paper presented at the AIP conference proceedings.

Evans, P. D., & Maqueira, B. (2005). Insect octopamine receptors: A new classification scheme based on studies of cloned Drosophila G-protein coupled receptors. *Invertebrate Neuroscience, 5*, 111–118.

FAO. (2023). *The state of food and agriculture 2023. Revealing the true cost of food to transform agrifood systems*. FAO Documents.

FAOSTAT. (2013). *Food and agriculture organization of the United Nations*. Statistical Database.

Fountain, E., & Wratten, S. (2013). Conservation biological control and biopesticides in agricultural. *Encyclopedia of Ecology, 1*, 377–381.

Francis, G., Kerem, Z., Makkar, H. P., & Becker, K. (2002). The biological action of saponins in animal systems: A review. *British Journal of Nutrition, 88*(6), 587–605.

Franklin, E. L., & Raine, N. E. (2019). Moving beyond honeybee-centric pesticide risk assessments to protect all pollinators. *Nature Ecology & Evolution, 3*(10), 1373–1375.

Garcia-del-Pino, F., Alabern, X., & Morton, A. (2013). Efficacy of soil treatments of entomopathogenic nematodes against the larvae, pupae and adults of *Tuta absoluta* and their interaction with the insecticides used against this insect. *BioControl, 58*, 723–731.

García-Mateos, R., Pérez-Pacheco, R., Rodríguez-Hernández, C., & Soto-Hernández, M. (2004). Toxicidad de alcaloides de *Erythrina americana* en larvas de mosquito Culex quinquefasciatus. *Revista Fitotecnia Mexicana, 27*(4), 297–297.

Gelernter, W. (2005). *Biological control products in a changing landscape*. Paper presented at the proceedings of the BCPC international congress—crop science and technology. Glasgow.

George, D. R., Finn, R. D., Graham, K. M., & Sparagano, O. A. (2014). Present and future potential of plant-derived products to control arthropods of veterinary and medical significance. *Parasites & Vectors, 7*, 1–12.

Ghoneim, K., Hamadah, K., Selim, S., & Waheeb, H. (2021). Biopesticidal potential of Nerolidol, a sesquiterpene compound, and its drastic impact on growth and metamorphosis of the cotton leafworm *Spodoptera littoralis* (Lepidoptera: Noctuidae). *Scholars Academic Journal of Biosciences, 9*(2), 36–57.

Glare, T., Caradus, J., Gelernter, W., Jackson, T., Keyhani, N., Köhl, J., Marrone, P., Morin, L., Stewart, A. (2012). Have biopesticides come of age? *Trends in biotechnology, 30*(5), 250–258.

Giunti, G., Benelli, G., Palmeri, V., Laudani, F., Ricupero, M., Ricciardi, R., et al. (2022). Non-target effects of essential oil-based biopesticides for crop protection: Impact on natural enemies, pollinators, and soil invertebrates. *Biological Control, 176*, 105071.

Godlewska, K., Ronga, D., & Michalak, I. (2021). Plant extracts-importance in sustainable agriculture. *Italian Journal of Agronomy, 16*(2), 1851.

González-Coloma, A., Reina, M., Medinaveitia, A., Guadano, A., Santana, O., Martínez-Díaz, R., De la Fuente, G. (2004). Structural diversity and defensive properties of norditerpenoid alkaloids. *Journal of chemical Ecology, 30*, 1393–1408.

Gould, K. S., & Lister, C. (2006). Flavonoid functions in plants.

Grzywacz, D., Stevenson, P., Mushobozi, W., Belmain, S., & Wilson, K. (2014). The use of indigenous ecological resources for pest control in Africa. *Food Security, 6*, 71.

Gurr, G., & Kvedaras, O. (2010). Synergizing biological control: Scope for sterile insect technique, induced plant defences and cultural techniques to enhance natural enemy impact. *Biological Control, 52*(3), 198–20786.

Habib, M. E., de Carvalho, G. A., de Souza, C. E. P., & Hofling, J. C. (1991). Patogenicidade de dois formulados à base de Baclllus Thuringiensis' var Kurstak (H-3a: 3b) em larvas de Anagasta Kljhnlella (Zeller, 1879) (Lep.: Pyralidae). Bioikos–Título não-corrente, 5(2).

Harmatha, J. (2000). *Chemo-ecological role of spirostanol saponins in the interaction between plants and insects Saponins in food, feedstuffs and medicinal plants* (pp. 129–141). Springer.

He, H., Liu, L., Munir, S., Bashir, N. H., Yi, W., Jing, Y., & Li, C. (2019). Crop diversity and pest management in sustainable agriculture. *Journal of Integrative Agriculture, 18*(9), 1945–1952.

Hilal, B., Khan, M. M., & Fariduddin, Q. (2024). Recent advancements in deciphering the therapeutic properties of plant secondary metabolites: Phenolics, terpenes, and alkaloids. *Plant Physiology and Biochemistry, 211*, 108674.

Hossain, Z., Gurr, G. M., Wratten, S. D., & Raman, A. (2002). Habitat manipulation in lucerne Medicago sativa: Arthropod population dynamics in harvested and'refuge' crop strips. *Journal of Applied Ecology, 39*, 445–454.

Huang, J., Rücker, A., Schmidt, A., Gleixner, G., Gershenzon, J., Trumbore, S., & Hartmann, H. (2020). Production of constitutive and induced secondary metabolites is coordinated with growth and storage in Norway spruce saplings. *Tree Physiology, 40*(7), 928–942.

Hughes, P. R., Wood, H. A., Breen, J. P., Simpson, S. F., Duggan, A. J., & Dybas, J. A. (1997). Enhanced bioactivity of recombinant baculoviruses expressing insect-specific spider toxins in lepidopteran crop pests. *Journal of Invertebrate Pathology, 69*(2), 112–118.

Islam, M. R., Mondal, C., Hossain, I., Meah, M. B. (2014). Compost tea as soil drench: an alternative approach to control bacterial wilt in brinjal. Archives of Phytopathology and Plant Protection, 47(12), 1475–1488.

Islam, M. R., Mondal, C., Hossain, I., Meah, M. B. (2013). Organic management: an alternative to control late blight of potato and tomato caused by Phytophthora infestans. International Journal of Theoretical & Applied Sciences, 5(2), 32–42.

Isman, M. B. (2006). Botanical insecticides, deterrents, and repellents in modern agriculture and an increasingly regulated world. *Annual Review of Entomology, 51*, 45–66.

Isman, M. B., Paluch, G. (2011). Needles in the haystack: exploring chemical diversity of botanical insecticides, pp. 248–265. In: Green trends in insect control (Lopez O., Fernandez-Bolanos J. G., Eds).- Royal Society of Chemistry, Cambridge, UK.

Jain, C., Khatana, S., & Vijayvergia, R. (2019). Bioactivity of secondary metabolites of various plants: A review. *International Journal of Pharmaceutical Sciences and Research, 10*(2), 494–504.

Jose, S., & Sujatha, K. (2017). Antifeedant activity of different solvent extracts of *Gliricidia sepium* against third instar larvae of *Helicoverpa armigera* (Hubner) (Lepidoptera: Noctuidae). *International Journal of Advanced Research in Biological Sciences, 4*(4), 201–204.

Junker, R. R., Gershenzon, J., & Unsicker, S. B. (2011). Floral odor bouquet loses its ant repellent properties after inhibition of terpene biosynthesis. *Journal of Chemical Ecology, 37*, 1323–1331.

Kaluwa, K., & Kruger, E. (2012). Natural pest and disease control handbook. KwaZulu-Natal Department of Agriculture and Rural Development Private Bag X9059 Pietermaritzburg, 3200.

Kamugire, D. B., & Nshutiyayesu, S. (2020). Assessing the effectiveness of indigenous plants on postharvest pests control in Rwanda: Case of *Tetradenia riparia* and *Tephrosia Vogelii* on Bean Weevil (*Acanthoscelides obtectus*). *East African Journal of Science and Technology, 10*(10).

Kansiime, M. K., Mugambi, I., Rwomushana, I., Nunda, W., Lamontagne-Godwin, J., Rware, H., et al. (2019). Farmer perception of fall armyworm (*Spodoptera frugiderda* JE Smith) and farm-level management practices in Zambia. *Pest Management Science, 75*(10), 2840–2850.

Karabörklü, S., & Ayvaz, A. (2023). A comprehensive review of effective essential oil components in stored-product pest management. *Journal of Plant Diseases and Protection, 130*(3), 449–481.

Karunamoorthi, K., Ramanujam, S., & Rathinasamy, R. (2008). Evaluation of leaf extracts of *Vitex negundo* L.(Family: Verbenaceae) against larvae of *Culex tritaeniorhynchus* and repellent activity on adult vector mosquitoes. *Parasitology Research, 103*, 545–550.

Kawalekar, J. S. (2013). Role of biofertilizers and biopesticides for sustainable agriculture. *Journal of Biological Innovations, 2*(3), 73–78.

Khursheed, A., Rather, M. A., Jain, V., Rasool, S., Nazir, R., Malik, N. A., & Majid, S. A. (2022). Plant based natural products as potential eco-friendly and safer biopesticides: A comprehensive overview of their advantages over conventional pesticides, limitations and regulatory aspects. *Microbial Pathogenesis, 173*, 105854.

Kibira, M., Affognon, H., Njehia, B., Muriithi, B., Mohamed, S., & Ekesi, S. (2015). Economic evaluation of integrated management of fruit fly in mango production in Embu County, Kenya. *African Journal of Agricultural and Resource Economics, 10*(4), 343–353.

Kimani, V. (2014). biopesticides development, use and regulation in Kenya. In *Paper presented at the Regional Experts Workshop on Development, Regulation and Use of biopesticides in East Africa*.

Kitch, L., Bottenberg, H., & Wolfson, J. (1997). Indigenous knowledge and cowpea pest management in sub-Saharan Africa. *Advances in Cowpea Research, 292*.

Klinger, E. G., Camp, A. A., Strange, J. P., Cox-Foster, D., & Lehmann, D. M. (2019). Bombus (Hymenoptera: Apidae) microcolonies as a tool for biological understanding and pesticide risk assessment. *Environmental Entomology, 48*(6), 1249–1259.

Kortbeek, R. W., van der Gragt, M., & Bleeker, P. M. (2019). Endogenous plant metabolites against insects. *European Journal of Plant Pathology, 154*, 67–90.

Koul, O. P. E. N. D. E. R., Cuperus, G. W. (2008). Current approaches to areawide pest management: obstacles and future directions. In Areawide pest management: theory and implementation, pp. 560–572). Wallingford UK: CABI.

Koul, O. (2012). Plant biodiversity as a resource for natural products for insect pest management. In *Biodiversity and insect pests: Key issues for sustainable management* (pp. 85–105). Wiley.

Kubo, I., Kim, M., & Naoki, H. (1987). New insect growth inhibitory flavan glycosides from *Viscum tuberculatum. Tetrahedron Letters, 28*(9), 921–924.

Kumar, S. (2012). Biopesticides: A need for food and environmental safety. *Journal of Biofertiliser and Biopesticides, 3*(4), 1–3.

Kumar, V. (2015). A review on efficacy of biopesticides to control the agricultural insect's pest. *International Journal of Agricultural Science Research, 4*(9), 168–179.

Kumar, V., Kumar, V., & Yadav, P. (2020). Importance of summer ploughing for sustainable agriculture. *International Journal of Trend in Research and Development, 7*(3).336-337.

Kumar, J., Ramlal, A., Mallick, D., & Mishra, V. (2021). An overview of some biopesticides and their importance in plant protection for commercial acceptance. *Plants, 10*(6), 1185.

Kumari, K., Kumar, K. R., & Rao, C. N. (2014). Adverse effects of chemical fertilizers and pesticides on human health and environment. *Journal of Chemical and Pharmaceutical Sciences, 3*, 150–151.

Kumela, T., Simiyu, J., Sisay, B., Likhayo, P., Mendesil, E., Gohole, L., & Tefera, T. (2019). Farmers' knowledge, perceptions, and management practices of the new invasive pest, fall armyworm (*Spodoptera frugiperda*) in Ethiopia and Kenya. *International Journal of Pest Management, 65*(1), 1–9.

Lacey, L. (2017). Entomopathogens used as microbial control agents. In *Microbial control of insect and mite pests* (pp. 3–12). Elsevier.

Lee, S. E. (2000). Mosquito larvicidal activity of pipernonaline, a piperidine alkaloid derived from long pepper, Piper longum. *Journal of the American Mosquito Control Association-Mosquito News, 16*(3), 245–247.

Leng, P., Zhang, Z., Pan, G., & Zhao, M. (2011). Applications and development trends in biopesticides. *African Journal of Biotechnology, 10*(86), 19864–19873.

Liu, Z. L., Goh, S. H., & Ho, S. H. (2007). Screening of Chinese medicinal herbs for bioactivity against *Sitophilus zeamais* Motschulsky and *Tribolium castaneum* (Herbst). *Journal of Stored Products Research, 43*(3), 290–296.

Liu, J., Hua, J., Qu, B., Guo, X., Wang, Y., Shao, M., & Luo, S. (2021). Insecticidal terpenes from the essential oils of *Artemisia nakaii* and their inhibitory effects on acetylcholinesterase. *Frontiers in Plant Science, 12*, 720816.

Loko, L. Y., Alagbe, O., Dannon, E. A., Datinon, B., Orobiyi, A., Thomas-Odjo, A., et al. (2017). Repellent effect and insecticidal activities of *Bridelia ferruginea*, *Blighia sapida*, and *Khaya senegalensis* leaves powders and extracts against *Dinoderus porcellus* in infested dried yam chips. *Psyche: A Journal of Entomology, 2017*(1), 5468202.

Lopes, A. P. M., Soares, M. R. C., Chidichima, L. P. S., & Dias-Arieira, C. R. (2020). Weed hosts of *Meloidogyne* spp. and the effect of aqueous weed extracts on egg hatching. *Weed Research, 60*(2), 142–149.

Lusaka, Z. (2018). Effectiveness of Tephrosia vogelii on Armyworms control in Maize (Zea mays) crops: A case study of smallscale farmers in Mkushi District (Farmers Training Centre Area). The International Journal of Multi-Disciplinary Research. ISSN:3471-7102, ISBN: 978-9982-70-318-5; Paper-ID: CFP/979/2018.

Lushchak, V. I., Matviishyn, T. M., Husak, V. V., Storey, J. M., & Storey, K. B. (2018). Pesticide toxicity: A mechanistic approach. *EXCLI Journal, 17*, 1101.

Maayiem, D., Bernard, B. N., & Irunuoh, A. O. (2012). Indigenous knowledge of termite control: A case study of five farming communities in Gushegu District of Northern Ghana. *Journal of Entomology and Nematology, 4*(6), 58–64.

Mafongoya, P., Gubba, A., Moodley, V., Chapoto, D., Kisten, L., & Phophi, M. (2019). Climate change and rapidly evolving pests and diseases in Southern Africa. In *New frontiers in natural resources management in Africa* (pp. 41–57). Springer.

Mahmoud, M., Osman, M., El-Hussiny, M., Elsebae, A., Hassan, S., & Said, M. (2017). Low environmental impact method for controlling the peach fruit fly, *Bactrocera zonata* (Saunders) and the Mediterranean fruit fly, *Ceratitis capitata* (Wied.), in mango orchards in Egypt.

Makaza, K., & Mabhegedhe, M. (2016). Smallholder farmers' indigenous knowledge of maize storage pests and pesticidal plant use: The case of wards 9 and 10 in Bikita District, Masvingo Province, Zimbabwe. *African Journal of Agricultural Research, 11*(47), 4831–4839.

Makenzi, A. M., Manguro, L. O. A., Owour, P., & Opiyo, S. A. (2019). Flavonol glycosides with insecticidal activity from methanol extract of *Annona mucosa* Jacq. leaves.

Makundi, R. H., Oguge, N. O., & Mwanjabe, P. S. (1999). Rodent pest management in East Africa—An ecological approach: Rodent Pest Management in East Africa.

Malka, O., Easson, M. L., Paetz, C., Götz, M., Reichelt, M., Stein, B., et al. (2020). Glucosylation prevents plant defense activation in phloem-feeding insects. *Nature Chemical Biology, 16*(12), 1420–1426.

Mallikarjuna, N., Kranthi, K., Jadhav, D., Kranthi, S., & Chandra, S. (2004). Influence of foliar chemical compounds on the development of Spodoptera litura (Fab.) in interspecific derivatives of groundnut. *Journal of Applied Entomology, 128*(5), 321–328.

Masika, F. B., Kisekka, R., Alicai, T., & Tugume, A. K. (2017). Incidence of viruses and virus-like diseases of watermelons and pumpkins in Uganda, a hitherto none investigated pathosystem.

Mathew, L. K. (2016). Botanicals as biopesticides: A review. *International Journal of Advanced Research, 4*(3), 1734–1739.

Matyjaszczyk, E. (2015). Products containing microorganisms as a tool in integrated pest management and the rules of their market placement in the European Union. Pest management science, 71(9), 1201–1206.

Mazid, M., Khan, T., & Mohammad, F. (2011). Role of secondary metabolites in defense mechanisms of plants.

Melash, A. A., Bogale, A. A., Migbaru, A. T., Chakilu, G. G., Percze, A., Ábrahám, É. B., & Mengistu, D. K. (2023). Indigenous agricultural knowledge: A neglected human based resource for sustainable crop protection and production. *Heliyon, 9*(1), e12978. https://doi.org/10.1016/j.heliyon.2023.e12978

Moosavi, M. R., & Zare, R. (2016). Present status and the future prospects of microbial biopesticides in Iran. In *Agriculturally important microorganisms: Commercialization and regulatory requirements in Asia* (pp. 293–305). Springer.

Meinwald, J. (2011). Natural products as molecular messengers. *Journal of Natural Products, 74*(3), 305–309.

Moscardi, F., De Souza, M. L., De Castro, M. E. B., Lara Moscardi, M., & Szewczyk, B. (2011). Baculovirus pesticides: Present state and future perspectives Microbes and microbial technology. In *Agricultural and environmental applications* (pp. 415–445). Springer.

Mugisha-Kamatenesi, M., Deng, A., Ogendo, J., Omolo, E., Mihale, M., Otim, M., et al. (2008). Indigenous knowledge of field insect pests and their management around Lake Victoria basin in Uganda. *African Journal of Environmental Science and Technology, 2*(10), 342–348.

Munoz, E., Lamilla, C., Marin, J. C., Alarcon, J., & Cespedes, C. L. (2013). Antifeedant, insect growth regulatory and insecticidal effects of Calceolaria talcana (Calceolariaceae) on Drosophila melanogaster and Spodoptera frugiperda. *Industrial Crops and Products, 42*, 137–144.

Mushtaq, A., Pathania, S. S., Khan, Z. H., & Ahmad, M. O. (2020). Indigenous technical knowledge in pest management. *Journal of Entomology and Zoolofgy Study, 8*(5), 296–302.

Muzemu, S., Mvumi, B. M., Nyirenda, S. P. M., Sileshi, G. W., Sola, P., Chikukura, L., et al. (2011). Pesticidal effects of indigenous plants extracts against rape aphids and tomato red spider mites. In *African Crop Science Conference Proceedings* (Vol. 10, pp. 169–171).

Muzemu, S., Chitamba, J., & Mutetwa, B. (2013). Evaluation of Eucalyptus tereticornis, Tagetes minuta and Carica papaya as stored maize grain protectants against Sitophilus zeamais (Motsch.)(Coleoptera: Curculionidae). *Agriculture, Forestry and Fisheries, 2*(5), 196–201.

Mvumi, B., Giga, D., & Chiuswa, D. (1995). The maize (Zea mays L.) post-production practices of smallholder farmers in Zimbabwe: Findings from surveys.

Mwine, T. J., Van Damme, P., Gerard, K., & Charles, K. (2011). Ethnobotanical survey of pesticidal plants used in South Uganda: Case study of Masaka district. *Journal of Medicinal Plants Research, 5*(7), 1155–1163.

Naharki, K., & Jaishi, M. (2020). Documentation of indigenous technical knowledge and their application in pest management in western mid hill of Nepal. *SAARC Journal of Agriculture, 18*, 251.

Nakasu, E. Y., Williamson, S. M., Edwards, M. G., Fitches, E. C., Gatehouse, J. A., Wright, G. A., & Gatehouse, A. M. (2014). Novel biopesticide based on a spider venom peptide shows no adverse effects on honeybees. *Proceedings of the Royal Society B: Biological Sciences, 281*(1787), 20140619.

Nawaz, M., Mabubu, J. I., & Hua, H. (2016). Current status and advancement of biopesticides: Microbial and botanical pesticides. *Journal of Entomology and Zoology Studies, 4*(2), 241–246.

NIOSH. (1994). *Rotenone immediately dangerous to life and health*. National Institute for Occupational Safety and Health (NIOSH).

Nzelu, C. O., Emeasor, K. C., & Okonkwo, N. J. (2020). Insecticidal activity of *Piper guineense* (Schumach and Thonn) seed oil against *Callosobruchus maculatus* (F.) (Coleoptera: Chrysomelidae) in stored cowpea seeds. *International Journal of Research–granthaalayah, 8*(08), 262–270.

O'Brien, D. (2009). Combating the Brown Marmorated Stink Bug. *Agricultural Research, 57*(6), 14.

Oboh, G., Ademosun, A. O., Olumuyiwa, T. A., Olasehinde, T. A., Ademiluyi, A. O., & Adeyemo, A. C. (2017). Insecticidal activity of essential oil from orange peels (*Citrus sinensis*) against *Tribolium confusum*, *Callosobruchus maculatus* and *Sitophilus oryzae* and its inhibitory effects on acetylcholinesterase and Na+/K+-ATPase activities. *Phytoparasitica, 45*, 501–508.

OECD. (1998). Guidelines for the testing of chemicals, section 2-effects on biotic systems. *Honeybees, Acute Contact Toxicity Test, 214*, 7–7.

Oerke, E.-C. (2006). Crop losses to pests. *The Journal of Agricultural Science, 144*(1), 31–43.

Ogbonna, O., Ogbonna, P., & Dike, M. (2016). Phytochemical screening, Quantitative estimates of Bioactive compounds in *Spondias mombin* and *Azadirachta indica*. *Research Journal of Chemical Sciences. ISSN, 2231*, 606X.

Ogendo, J., Belmain, S., Deng, A., & Walker, D. (2003). Comparison of toxic and repellent effects of *Lantana camara* L. with *Tephrosia vogelii* Hook and A synthetic pesticide against *Sitophilus zeamais* Motschulsky (Coleoptera: Gurculionidae) in stored Maize Grain. *International Journal of Tropical Insect Science, 23*, 127–135.

Oguh, C., Okpaka, C., Ubani, C., Okekeaji, U., Joseph, P. S., & Amadi, E. (2019). Natural pesticides (biopesticides) and uses in pest management-a critical review. *Asian Journal of Biotechnology and Genetic Engineering, 2*(3), 1–18.

Okonkwo, C., & Ohaeri, O. (2018). Insecticidal efficacy and chemical composition of hexane oil extracts from the leaves of *Euphorbia milii* and *Cassia occidentalis*. *international Journal of Biochemistry Research & Review, 23*(3), 1–11.

Orchard, I. (1982). Octopamine in insects: Neurotransmitter, neurohormone, and neuromodulator. *Canadian Journal of Zoology, 60*(4), 659–669.

Owusu, E. O. (2000). Effect of some Ghanaian plant components on control of two stored-product insect pests of cereals. *Journal of Stored Products Research, 37*(1), 85–91.

Pathma, J., Kennedy, R. K., Bhushan, L. S., Shankar, B. K., & Thakur, K. (2021). Microbial biofertilizers and biopesticides: Nature's assets fostering sustainable agriculture. In *Recent developments in microbial technologies* (pp. 39–69).

Pavela, R. (2005). Insecticidal activity of some essential oils against larvae of *Spodoptera littoralis*. *Fitoterapia, 76*(7–8), 691–696.

Pavela, R., & Benelli, G. (2016). Essential oils as eco-friendly biopesticides? Challenges and constraints. *Trends in Plant Science, 21*(12), 1000–1007.

Perlatti, B., de Souza Bergo, P. L., Fernandes, J. B., & Forim, M. R. (2013). *Polymeric nanoparticle-based insecticides: A controlled release purpose for agrochemicals Insecticides-Development of safer and more effective technologies*. IntechOpen.

Polanczyk, R. A., Valicente, F. H., & Barreto, M. R. (2008). Utilização de Bacillus thuringiensis no controle de pragas agrícolas na América Latina. Controle Microbiano de Pragas na América Latina: avanços e desafios. Ed. FEALQ, Piracicaba, 111–136.

Polanczyk, R. A., van Frankenhuyzen, K., Pauli, G. (2017). The american Bacillus thuringiensis based biopesticides market. Bacillus thuringiensis and Lysinibacillus sphaericus: Characterization and use in the field of biocontrol, 173–184.

Pomari-Fernandes, A., de Freitas Bueno, A., & Sosa-Gómez, D. R. (2015). *Helicoverpa armigera:* Current status and future perspectives in Brazil. *Current Agricultural Science and Technology, 21*(1).

Pretty, J., & Pervez Bharucha, Z. (2015). Integrated pest management for sustainable intensification of agriculture in Asia and Africa. *Insects, 6*(1), 152–182.

Qasim, M., Islam, W., Ashraf, H. J., Ali, I., & Wang, L. (2020). Saponins in insect pest control. In J. M. Mérillon & K. Ramawat (Eds.), *Co-evolution of secondary metabolites* (pp. 897–924). Springer.

Regnault-Roger, C. (2011). *Trends for commercialization of biocontrol agent (biopesticide) products Plant defence: Biological control* (pp. 139–160). Springer.

Rizvi, R., Mahmood, I., Tiyagi, S. A., Khan, Z. (2012). Effect of some botanicals for the management of plantparasitic nematodes and soil-inhabiting fungi infesting chickpea. Turkish *Journal of Agriculture and Forestry, 36*(6), 710–719.

Roeder, T. (2005). Tyramine and octopamine: Ruling behavior and metabolism. *Annual Review of Entomology., 50*, 447–477.

Ruiu, L. (2015). Insect pathogenic bacteria in integrated pest management. Insects, 6(2), 352–367.

Sakthivel, S., Mohideen, H. S., Raman, C., & Mohamad, S. B. (2022). Potential acetylcholinesterase inhibitor acting on the pesticide resistant and susceptible cotton pests. *ACS Omega, 7*(24), 20515–20527.

Salvadores, Y., Silva, G., Tapia, M., & Hepp, R. (2007). Polvos de especias aromáticas para el control del gorgojo del maíz, Sitophilus zeamais Motschulsky, en trigo almacenado. *Agricultura Técnica, 67*(2), 147–154.
Samada, L. H., & Tambunan, U. S. F. (2020). Biopesticides as promising alternatives to chemical pesticides: A review of their current and future status. *OnLine Journal of Biological Sciences, 20*(2), 66–76.
Saroukolai, A. T., Nouri-Ganbalani, G., Rafiee-Dastjerdi, H., & Hadian, J. (2014). Antifeedant activity and toxicity of some plant essential oils to colorado potato beetle, *Leptinotarsa decemlineata* Say (Coleoptera: Chrysomelidae). *Plant Protection Science, 50*(4).
Sarwar, M. (2015). Biopesticides: An effective and environmental friendly insect-pests inhibitor line of action. *International Journal of Engineering and Advanced Research Technology, 1*(2), 10–15.
Sasidharan, S., Chen, Y., Saravanan, D., Sundram, K., & Latha, L. Y. (2011). Extraction, isolation and characterization of bioactive compounds from plants' extracts. *African Journal of Traditional, Complementary and Alternative Medicines, 8*(1), 1.
Savithramma, N., Rao, M. L., & Suhrulatha, D. (2011). Screening of medicinal plants for secondary metabolites. *Middle-East Journal of Scientific Research, 8*(3), 579–584.
Sawicka, B., & Egbuna, C. (2020). *Pests of agricultural crops and control measures Natural remedies for pest, disease and weed control* (pp. 1–16). Elsevier.
Schowalter, T. D., Noriega, J. A., & Tscharntke, T. (2018). *Insect effects on ecosystem services—Introduction* (Vol. 26, pp. 1–7). Elsevier.
Semerdjieva, I., Zheljazkov, V. D., Radoukova, T., Dincheva, I., Piperkova, N., Maneva, V., et al. (2021). Biological activity of essential oils of four juniper species and their potential as biopesticides. *Molecules, 26*(21), 6358.
Sethi, S., Dhakad, S., & Arora, S. (2022). The use of biopesticides for sustainable farming: Way forward toward Sustainable Development Goals (SDGs). In *Biotechnological innovations for environmental bioremediation* (pp. 571–596). Springer.
Shah, B. A., Shah, A. V., Jadav, P. Y. (2013). Extractive efficacy of microwave synthesized zeolitic material for acephate: Equilibrium and kinetics. *Journal of the Serbian Chemical Society, 78*(7), 1055–1077.
Sharma, D., Singh, V., Singh, R. K., Joshi, C., & Sharma, V. (2021). Isolation and characterization of bioactive compounds from natural resources. In *Metabolomics and molecular approaches evolutionary diversity as a source for anticancer molecules* (pp. 77–101). Elsevier.
Siah, A., Magnin-Robert, M., Randoux, B., Choma, C., Rivière, C., Halama, P., & Reignault, P. (2018). Natural agents inducing plant resistance against pests and diseases. In J. M. Mérillon & C. Riviere (Eds.), *Natural antimicrobial agents* (pp. 121–159). Springer.
Singh, T., Tiwari, Y., & Awasthi, G. (2019). Understanding the impact of Bacillus thuringiensis proteins on non-target organisms. *International Journal of Scientific Research in Biological Science, 6*, 2.
Skendžić, S., Zovko, M., Živković, I. P., Lešić, V., & Lemić, D. (2021). The impact of climate change on agricultural insect pests. *Insects, 12*(5), 440.
Stadlinger, N., Mmochi, A. J., & Kumblad, L. (2013). Weak governmental institutions impair the management of pesticide import and sales in Zanzibar. *Ambio, 42*, 72–82.
Stern, V., Bosch, R., & Leigh, T. (1964). Strip cutting alfalfa for lygus bug control. *California Agriculture, 18*(4), 4–6.
Stevenson, P. C., Nyirenda, S. P., Mvumi, B., Sola, P., Kamanula, J. F., Sileshi, G., & Belmain, S. R. (2012). *Pesticidal plants: A viable alternative insect pest management approach for resource-poor farming in Africa* (pp. 212–238). Biopesticides in Environment and Food Security, Scientific Publishers.
Sticher, O. (2008). Natural product isolation. *Natural Product Reports, 25*(3), 517–554.
Tadele Shiberu, E. G. (2016). Assessment of selected botanical extracts against Liriomyza species (Diptera: Agromyzidae) on tomato under glasshouse condition. *International Journal of Fauna and Biological Studies, 3*(1), 87–90.

Tayoub, G., Odeh, A., & Ghanem, I. (2012). Chemical composition and fumigation toxicity of Laurus nobilis L. and Salvia officinalis L. essential oils on larvae of khapra beetle (Trogoderma granarium Everts). *Herba Polonica, 58*(2).

Tekman, E., Asgarlı, T., Yuca, H., Atila, A., Çeçen, Ö., & Karakaya, S. (2024). Exploring quantitative biological major, trace, and ultratrace elements composition and qualitative primary-secondary metabolites in lamiaceae medicinal plants from Turkey. *Biological Trace Element Research*, 1–14.

Tesfaye, A., & Gautam, R. (2003). Traditional pest management practices and lesser exploited natural products in Ethiopia and India: Appraisal and revalidation.

Thakur, R., Rane, A. V., Harris, G., & Thakur, S. (2020). Chapter 18 – Future prospective and possible management of water resources in respect to indigenous technical knowledge in South Africa (P. Singh, Y. Milshina, K. Tian, D. Gusain, & J. P. B. T.-W. C. and W. T. in B. N. Bassin (eds.); pp. 353–363). Elsevier. https://doi.org/10.1016/B978-0-12-818339-7.00018-7

Tian, X., Li, Y., Hao, N., Su, X., Du, J., Hu, J., & Tian, X. (2021). The antifeedant, insecticidal and insect growth inhibitory activities of triterpenoid saponins from Clematis aethusifolia Turcz against Plutella xylostella (L.). *Pest Management Science, 77*(1), 455–463.

Tinklein, D. (2015). *Biopesticides: Eco-friendly pest control. Integrated pest management*. University of Missouri.

Toledo-Hernández, E., Peña-Chora, G., Hernandez-Velazquez, V. M., Lormendez, C. C., Toribio-Jiménez, J., Romero-Ramírez, Y., & León-Rodríguez, R. (2022). The stingless bees (Hymenoptera: Apidae: Meliponini): A review of the current threats to their survival. *Apidologie, 53*(1), 8.

Tolo, C. U., Majule, E. A., & Lejju, J. B. (2014). *Local and indigenous knowledge systems in subsistence agriculture, climate risk management, and mitigation of community vulnerability in changing climate, lake Victoria basin: A case study of Rakai and Isingiro districts, Uganda* (pp. 451–473). Nile River Basin: Ecohydrological Challenges, Climate Change and Hydropolitics.

Tong, F., & Coats, J. R. (2010). Effects of monoterpenoid insecticides on [3H]-TBOB binding in house fly GABA receptor and 36Cl− uptake in American cockroach ventral nerve cord. *Pesticide Biochemistry and Physiology, 98*(3), 317–324.

Tortorici, S., Biondi, A., Pérez-Hedo, M., Larbat, R., & Zappalà, L. (2022). Plant defences for enhanced integrated pest management in tomato. *Annals of Applied Biology, 180*(3), 328–337.

Tripura, P., Devi, C. A., Manzoor, U., & Kapri, A. (2024). Chapter 18- Indigenous technical knowledge integration in organic farming. In V. S. Meena, A. Rakshit, M. D. Meena, M. Baslam, I. M. R. Fattah, S. S. Lam, & J. S. B. T.-W. M. for S. and R. A. S. Kaba (Eds.), Agricultural soil degredation and restoration (pp. 371–386). Academic. https://doi.org/10.1016/B978-0-443-18486-4.00019-1

Truman, J. W., & Riddiford, L. M. (2002). Endocrine insights into the evolution of metamorphosis in insects. *Annual Review of Entomology, 47*(1), 467–500.

Ukoroije, R. B., & Otayor, R. A. (2020). Review on the bio-insecticidal properties of some plant secondary metabolites: Types, formulations, modes of action, advantages and limitations. *Asian Journal of Research in Zoology, 3*(4), 27–60.

Upadhyay, H., Mirza, A., & Singh, J. (2020). Impact of biopesticides in sustainable agriculture. Advances in plant microbiome and sustainable agriculture: Diversity and biotechnological applications, 281–296.

Van Frankenhuyzen, K. (2009). Insecticidal activity of *Bacillus thuringiensis* crystal proteins. *Journal of Invertebrate Pathology, 101*(1), 1–16.

Viglianco, A., Novo, R., Cragnolini, C., & Nassetta, M. (2006). Actividad biológica de extractos crudos de Larrea divaricata Cav. y Capparis atamisquea Kuntze sobre *Sitophilus oryzae* (L.). *Agriscientia, 23*(2), 83–89.

War, A.R., Paulraj, M.G., Ignacimuthu, S., Sharma, H.C. (2015). Inducedresistance to Helicoverpa armigera through exogenous application of jasmonic acid and salicylic acid in groundnut, Arachis hypogaea. *Pest Management Science, 71*,72–82.

Ware, G. W., & Whitacre, D. M. (2004). *An introduction to insecticides* (p. 6). The pesticide book.

Wheeler, D. A., & Isman, M. B. (2001). Antifeedant and toxic activity of *Trichilia americana* extract against the larvae of *Spodoptera litura. Entomologia Experimentalis et Applicata, 98*(1), 9–16.

Wongo, L. E. (1998). Biological activity of sorghum tannin extracts on the stored grain pests *Sitophilus oryzae* (L.), *Sitotroga cerealella* (Olivier) and *Tribolium castaneum* (Herbst). *International Journal of Tropical Insect Science, 18*(1), 17–23.

Yadav, R. (2022). Biopesticides: Current status and future prospects. *Proceedings of the International Academy of Ecology and Environmental Sciences, 12*(3), 211.

Yaron, D., Voet, H., & Dinar, A. (1992). Innovations on family farms: The Nazareth region in Israel. *American Journal of Agricultural Economics, 74*(2), 361–370.

Yunus, F.-U.-N., Raza, G., Makhdoom, R., & Zaheer, H. (2019). Genetic improvement of Bacillus thuringiensis against the cotton bollworm, *Earias vitella* (Fab.)(Lepidoptera: Noctuidae), to improve the cotton yield in Pakistan. *Egyptian Journal Of Biological Pest Control, 29*, 1–6.

Zapata, N., & Smagghe, G. (2010). Repellency and toxicity of essential oils from the leaves and bark of Laurelia sempervirens and Drimys winteri against *Tribolium castaneum. Industrial Crops and Products, 32*(3), 405–410.

Zaynab, M., Sharif, Y., Abbas, S., Afzal, M. Z., Qasim, M., Khalofah, A., et al. (2021). Saponin toxicity as key player in plant defense against pathogens. *Toxicon, 193*, 21–27.

Zhang, W. J., Yang, K., You, C. X., Wang, C. F., Geng, Z. F., Su, Y., et al. (2015). Contact toxicity and repellency of the essential oil from Mentha haplocalyx Briq. against *Lasioderma serricorne. Chemistry & Biodiversity, 12*(5), 832–839.

Chapter 5
African Endangered Plants: Their Phytochemistry, Ethnomedicine, and Conservation Strategies

Jane Namukobe, Sammy Kiprotich, Madina Mohamed Adia, Peter Sekandi, and Robert Byamukama

Abstract Africa is rich of plant biodiversity with at least 65,000 plant species equating to 18.5% of the global diversity. Natural products, including those derived from medicinal plants, are increasingly preferred in the treatment of diseases and general well-being. This has put pressure on plant biodiversity as a result of over exploitation. In addition, habitat destruction and climate change have severely threatened the survival of many plant species, making many of them endangered or threatened with extinction. Reports on bioactive phytochemicals identified from these medicinal plants have shown that natural products have promising pharmaceutical properties. In this chapter, rare and endangered medicinal plants including *Curtisia dentata, Milicia excelsa, Mondia whitei, Ocotea usambarensis, Prunus africana, Warburgia salutaris*, and *Warburgia ugandensis*, among others, are discussed with special emphasis on their distribution, phytochemical constituents, and medicinal significance. Conservation strategies that can be put in place to ensure their survival are also highlighted. The conservation of endangered plants in Africa is crucial not only for maintaining biodiversity but also for preserving the ethnomedicinal knowledge and phytochemical resources they provide.

Keywords Africa · Conservation · Endangered plants · Medicinal plants · Pharmacology · Phytochemistry

5.1 Introduction

Africa has a rich botanical heritage with at least 65,000 plant species equating to 18.5% of the global diversity (Qian et al., 2021; Antonelli et al., 2020). The rich plant diversity includes medicinal plants which have been found to be used in traditional medicine across the continent. Medicinal plants have now acquired an

J. Namukobe (✉) · S. Kiprotich (✉) · M. M. Adia · P. Sekandi · R. Byamukama
Department of Chemistry, College of Natural Sciences, Makerere University, Kampala, Uganda

A. Dekebo (ed.), *Bioactive Secondary Metabolites from Medicinal Plants of Africa*, Interdisciplinary Biotechnological Advances,
https://doi.org/10.1007/978-981-96-8590-5_5

important place as a source of remedies because they contain various phytochemicals and offer several health benefits and at times can be the sole form of primary healthcare in developing countries due to its accessibility, affordability, better tolerance profiles with fewer side effects, bioavailability, eco-friendly nature, safety margin from overdose and lower chances of abuse, or potential for development of resistance (Adebayo et al., 2021; Anywar et al., 2022; Chaachouay & Zidane, 2024). In recent days, herbal medicine has been highly favored for disease control and promoting well-being and natural health as evident from numerous herbal products in the market that are sold in traditional markets and health stores (Kloos, 2024). The ever-growing interest in medicinal plants has resulted in over-exploitation of the plant biodiversity leading to extinction of species that could be used in the future (Moyo et al., 2015). One study has estimated that a third of the tropical African flora could be potentially threatened with extinction (Catarino & Romeiras, 2020). Consequently, the continent has experienced the most extensive threat to its botanical heritage in terms of loss of medicinal plants resulting mainly from over-exploitation and partly from habitat destruction as well as climate change. Many of these plants are threatened as they are harvested unsustainably and also due to the fact that most of them come from wild populations (Muasya et al., 2024). These endangered plants hold the key to unlocking new bioactive compounds, with potential applications in medicine, agriculture, and beyond (Moyo et al., 2015; Abegaz & Mutanyatta, 2002). The loss of these plant species would not only destroy the cultural heritage of African communities but also deprive the world of potential life-saving medicines that could be derived from bioactive compounds of these species.

The objective of this review is to explore phytochemicals from some endangered medicinal plants of Africa, highlighting their potential to address pressing health, social, and environmental challenges. This chapter also examines the chemistry, pharmacology, and therapeutic applications of these compounds, as well as the conservation efforts aimed at protecting these valuable plant species.

5.2 Identification and Classification of Endangered Plant Species

The term endangered species refers to any species whose population suddenly decreases rapidly or whose habitat is lost resulting in extinction (Smith, 2016; Dublin, 2024). There are several signs that a species is endangered: its population is declining rapidly in just a few years, the decline lasts without a break, and its geographic range has also narrowed significantly (Smith, 2016). In order to classify endangered species of plants and animals, the International Union for Conservation of Nature and Natural Resources (IUCN) developed the following categories: (i) critically endangered—this is when the total plant population decreases by 90% in 10 years; (ii) endangered—this is when the total plant population reduces by half in 10 years; (iii) vulnerable—when there is 30% decrease in plant population in

10 years; (iv) near threatened—when species that are close to qualifying for vulnerable status but are not yet considered to be at risk of extinction; and (v) least concerned—species that are not currently at risk of extinction. The first three groups are broadly referred to as "threatened species" as they are at high risk of imminent extinction without adequate conservation measures (Dublin, 2024; Mace et al., 2008). Plant identification occurs most frequently through the use of leaves, which are stable in structure, easy to collect, and have a long shelf life (Barthélémy & Caraglio, 2007). Some research has been conducted on automatic leaf identification methods, which rely primarily on superficial characteristics such as color, shape, and texture to identify leaves (Saleem et al., 2019; Singh et al., 2018; Yang, 2021). Now that such species are endangered, it is crucial that information about them is collected, their vulnerability indices are calculated, laws and policies are enacted to protect their natural habitat, and plans are devised to conserve them through captive breeding and habitat restoration. Moreover, plants are sessile, and for the majority of species, the likelihood of discovery is determined by the total amount of time spent searching (during the proper season), not by the quantity of site visits or surveys conducted (Chen et al., 2012). Random quadrats, parallel transects, and random meanders are three frequently used techniques for surveying plant species (Little, 2013). Each of these strategies can have several variants that are frequently customized to a particular species or habitat (https://doi.org/10.1002/978144435583). Because the parallel field traverse method methodically covers the full area of acceptable habitat within a site and can be used for a varied range of species, habitats, and sites, it increases the detectability of vulnerable plants. When compared to parallel field transects, the usage of survey plots/quadrats or a random meander within an area of acceptable habitat is deemed insufficient to achieve the survey's aims (Merritt et al., 2017).

5.2.1 Phylogenetic Studies of Endangered Plant Species

In order to explain the structural, ecological, taxonomic, and functional features of biodiversity within an evolutionary framework, phylogenetic analysis is necessary (Hoveka et al., 2020; Kougioumoutzis et al., 2021). It is acknowledged that phylogenetic approaches offer a helpful foundation for projecting future losses and directing efforts for proactive conservation measures (Yessoufou et al., 2012). Researchers have characterized the distribution of extinction risk across taxonomic ranks and phylogeny using a reconstructed phylogenetic tree of angiosperm species in the Eastern Arc Mountains, a significant hotspot for biodiversity in Africa. They offered proof of phylogenetic and taxonomic selectivity in extinction risk. On the tree, vulnerable species showed a higher degree of relatedness than would be predicted by chance, but endangered and severely endangered species did not show a significant clustering. They proposed that vulnerable species, rather than always the most seriously threatened, are primarily responsible for the general observation of taxonomic and phylogenetic selectivity (i.e., phylogenetic signal, the tendency of closely related species to share similar features) in extinction risks (Yessoufou et al., 2012). The team also created a predictive model of at-risk species richness using data on altitudinal distribution and climate, and discovered that higher altitudes were associated with higher threatened species richness.

5.2.2 African Regions with a High Number of Endangered Plant Species

Africa has a myriad of plant species, many of which are widespread and face threats that place them in endangered categories (Wagner & Van Driesche, 2010). A number of African regions are notable for having a high concentration of threatened plant species: First and foremost, Madagascar is known for its distinctive flora and high levels of endemism. It is frequently referred to as a biodiversity hotspot (Fritz-Vietta et al., 2011). Secondly, South Africa is well known for its rich plant diversity, which includes a large number of endemic species. It is acknowledged as one of the six floral kingdoms of the globe. Known as ancient mountains, the Eastern Arc Mountains (Tanzania and Kenya) are a hotspot for biodiversity with a high concentration of unique plants, especially in alpine habitats and damp forests (Newmark, 2002). Ethiopia's highlands are home to a wide variety of unique plants that have adapted to the region's various altitudes and temperatures. The West African Rainforest stretches from Guinea to Nigeria, via Liberia, Ghana, and the Ivory Coast (Hall & Swaine, 2013). The succulent Karoo (South Africa and Namibia) is renowned for having distinct succulent plant species that are suited to dry environments (Whitman, 2017). Certain and frequently endangered plant species that are acclimated to semi-arid conditions can be found in the Sahara Desert Edge, the transition zone between the Sahara Desert and the Sahel (Brito et al., 2021). These regions' high endemism and the distinctive biological niches they sustain make them essential for efforts to conserve biodiversity worldwide. Many plant species in these regions are threatened by habitat loss due to agricultural expansion, deforestation, climate change, logging, overgrazing, mining activities, and desertification (Whitman, 2017). Conservation initiatives in these areas aim to mitigate threats and protect endangered plant species for future generations.

5.2.3 Ethnobotany of African Endangered Plants

Humans have co-existed with plants from the early days. Plants have been the main source of food, medicine, fiber, fuel, dye, and fodder, among others (Mushangwe, 2019). Knowledge has developed from hit or miss experiments, and where positive results were achieved, information was transferred from one generation to another and led to the advent of ethnobotany (Vallejo et al., 2024). Ethnobotany is the investigation of reciprocity between humans and plants with the purpose of inspecting the importance of plants that are used for performing rituals, medicines, as well as those with ecological roles (Cock & Van Vuuren, 2020). Approximately 80% of people still rely on herbal medicine for their primary health care (Asiimwe et al., 2021). This has even gained more recognition in Africa and Asia, and the number of plants recommended for use has increased. A number of surveys have documented several endangered plants such as *Hydrastis canadensis* L., used for treatment of digestive disorders (Evans & Arentz, 2022; Mandal et al., 2020; Rahaman et al., 2022); *Prunus africana* (Hook f.) Kalkman, used for treatment of prostate complications in Uganda, Kenya, and Ethiopia (Kudamba et al., 2024); *Taxus baccata* Thunb., reported in North-west Africa a source of paclitaxel a cancer drug (Yan-Hua et al., 2020); *Panax quinquefolius* L., used as energy and immunity booster (Kim et al., 2024); *Curtisia dentata* (Bum.f) C.A.Sm., used for treatment of dysentery (Adams et al., 2021); *Milicia ecelsa* (Welw.) C.C Berg., used to treat bacterial infection (Udegbunam et al., 2013); *Ephedra alata* Decne., used for treatment of asthma and colds (Oladeji et al., 2020); and *Warburgia ugandensis* Sprague, used for treating malaria (Okello et al., 2018), among others, as shown in Fig. 5.1.

5.2.4 Ecological Roles of Endangered Plants

Endangered plants play crucial roles in ecosystems such as biodiversity maintenance which enhances ecosystem stability and resilience, habitat, and shelter for various organisms (Rawat & Agarwal, 2015). Trees and shrubs, for example, offer nesting sites for birds and homes for insects and small mammals (Hails & Kavanagh, 2013). They often serve as food and medicinal sources for a range of animals and organisms, including insects, herbivores, pollinators, and humans (Gilgert & Vaughan, 2011). Endangered flowering plants are vital for pollinators such as bees, butterflies, and birds. Pollination is a critical ecological service that enables plant reproduction, which in turn supports food production for a wide array of species (Potts et al., 2016). Furthermore, plants aid in the development and health of soil because their roots reduce erosion and their decomposition provides organic matter to the soil, which enhances its fertility and structure (Twerski et al., 2021). They also play a significant in controlling water flow within the water cycle, maintaining the equilibrium between groundwater recharge and outflow, and purifying water by filtering contaminants (Yin, 2010). Climate regulation is another crucial role that endangered plants perform in the ecosystem. Through photosynthesis, they absorb carbon dioxide, thereby helping to mitigate climate change. Plant-rich areas also act as significant carbon sinks (Parmesan & Hanley, 2015). Therefore, the conservation of endangered plant species is essential, as they play a major role in maintaining a healthy and balanced ecosystem.

Fig. 5.1 Pictures of some African endangered plants

5.2.5 *Phytochemistry, Pharmacology, and Therapeutic Applications of Compounds Reported from Some Endangered Plants*

The ethnopharmacological approach, which is based on the traditional medical system, is the foundation for the development of phytopharmaceutical medications.

Plants have a myriad of compounds, some of which have been developed into drugs which are currently in use for treatment of various ailments. Drugs currently in use on the market that originate from endangered plants include: Artemisinin (**1**), used in the treatment of malaria and originally isolated from the sweet wormwood plant (*Artemisia annua*); reserpine (**2**), from the *Rauwolfia serpentina* plant and used to treat some mental illnesses and high blood pressure; ephedrine (**3**), derived from the plant *Ephedra sinica* and used as a decongestant and bronchodilator; codeine (**4**), from the opium poppy (*Papaver somniferum*) and used as a cough suppressant and pain reliever; atropine (**5**) produced from the lethal nightshade plant (*Atropa belladonna*) and used as an antidote for several poisonings and to treat bradycardia (slow heart rhythm); vincristine (**6**) and vinblastine (**7**), derived from the Madagascar periwinkle (*Catharanthus roseus*) are chemotherapy agents used in the treatment of various cancers; digoxin (**8**), derived from the foxglove plant (*Digitalis lanata*) and used for treatment of heart conditions, such as atrial fibrillation and heart failure; paclitaxel (taxol) (**9**), a chemotherapeutic drug that is taken from the bark of the Pacific yew tree (*Taxus brevifolia*) and used to treat cancer; morphine (**10**), produced from the opium poppy (*Papaver somniferum*), which is used to treat pain; aspirin (acetylsalicylic acid) (**11**), derived from salicin, a compound found in the bark of willow trees (genus *Salix*) is widely used for pain relief, as an anti-inflammatory, and antipyretic drug. The structures of these drugs are illustrated in Fig. 5.2. Plants are at the forefront of future medical advancements due to drug discovery, yet the loss of these plant species may result in the loss of possible illness cures and treatments (Table 5.1).

5.3 Threats of Endangered Plants

There has been a growing public concern about endangered plants which is still a concern up to date (Stohlgren & Kumar, 2013). Due to human activity, 90% of threatened and endangered species are in danger (Dublin, 2024). Human activity is putting tens of thousands of species in danger of going extinct (Tilman et al., 2017). Significant risks to biodiversity, particularly in regions with extraordinary plant diversity, mainly in the tropics, are frequently associated with large-scale industrial operations including mining, agriculture, large-scale plantations, and lumber harvesting (Tilman et al., 2017). Details of the threats are discussed below.

5.3.1 Deforestation

Deforestation and habitat loss have been attributed to a number of variables, some of which may have natural or anthropogenic causes (Allen & Barnes, 1985). Deforestation may occur naturally as a result of forest fires, droughts, floods, an overabundance of invasive species, and climate change. However, among the main drivers of worldwide deforestation and habitat loss are human activities; some

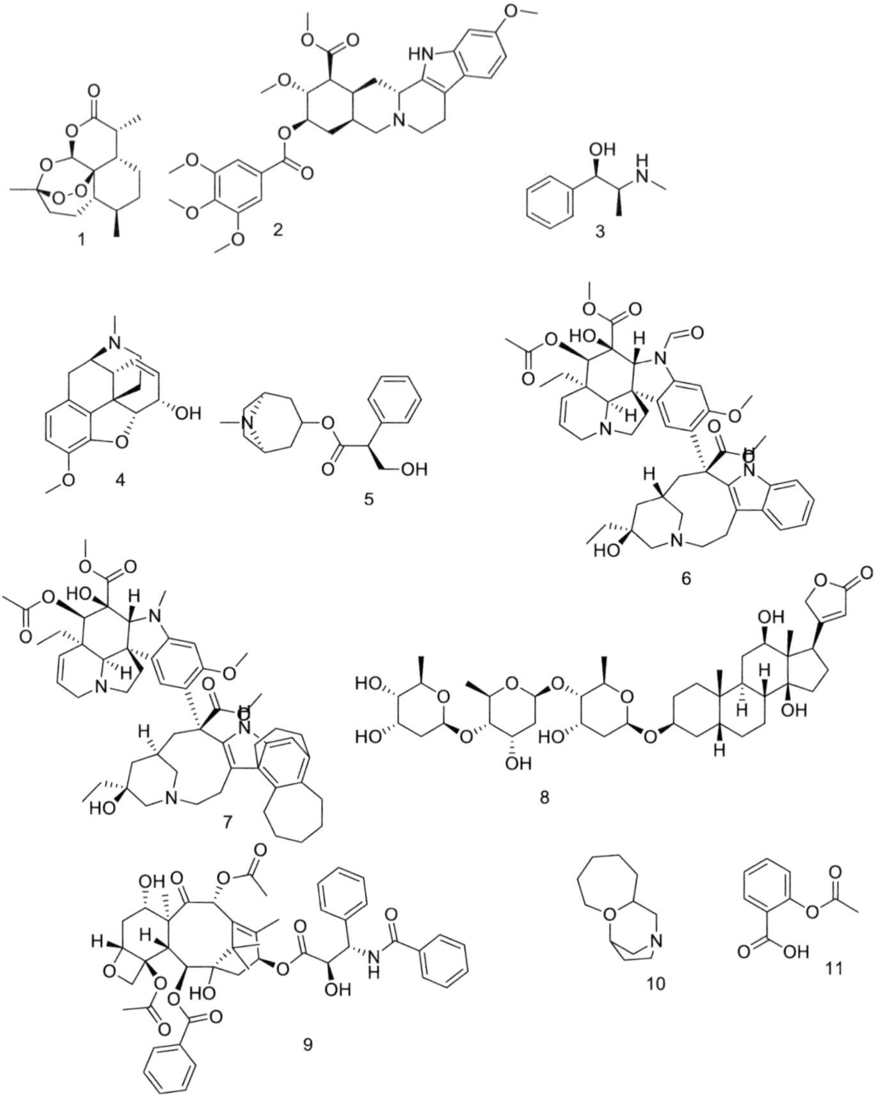

Fig. 5.2 Currently used drugs isolated from endangered plants

examples of these activities include the expansion of agriculture, the raising of cattle, the extraction of lumber, mining, the extraction of oil, the building of dams, and the development of infrastructure (Bodo et al., 2021). Large-scale human-caused deforestation and habitat loss have been attributed primarily to urbanization and agricultural practices. Natural environment is impacted by both urban and agricultural growth. Therefore, to provide food security while assuring little to no impact on natural habitat, farmland conservation rules must be improved (Tang et al., 2021).

Table 5.1 Compounds isolated from endangered plants of Africa and their medicinal uses

Plant name	Isolated compounds	Ethnomedicinal uses	Reference
Prunus africana (Hook.f) Kalkman (equatorial Africa)	β-sitosterol, β-sitosterone, ursolic acid, oleanolic acid, campesterol, stigmasterol, linoleic acid, crataegolic acid, epimanoyl oxide, atraric acid, ferulic acid, N-butylbenzenesulfonamide	Prostate issues, malaria, chest pain wounds	Kadu et al. (2012), Komakech & Kang (2019), Begeno et al. (2020), Deresa et al. (2022), and Bodeker et al. (2014)
Warburgia ugandensis Sprague (Kenya)	Warburganal, muzigadial, polygodial, ugandensiolide, muzigadiolide, quercetine, kaempferol, lupeol, cinnamic acid, betulinic acid	Malaria, toothache, respiratory infections	Opiyo (2023), Wube et al. (2005), Gonfa et al. (2020), Hordofa (2020), and Kairu et al. (2013)
Ocotea usambarensis Eng.	Usambrensine, β-carboline	Fever, cough	Nsolomo et al. (2000) and Kamboj et al. (2021)
Entandrophragma Meliaceae (Kenya, Uganda, Tanzania)	Gedunin, khivorin, swietenine, 3-oxo-olean-12-en-28-oic acid, 3β-hydroxy-olean-12-en-28-oic acid, quercetin, kaempferol, β-sitosterol, stigmasterol, tetrandrine, gallic acid, ellagic acid	Bacterial infections, malaria, cancer	Mouthé Happi et al. (2018), Happi et al. (2018), Kasongo Yakusu et al. (2021), and Mbatudde et al. (2013)
Milicia excelsa (Welw.) C.C. Berg. (Benin)	Cudraxanthone I, neocyclomorusin, acetoxylanosta lupenone, betulinic acid, β-sitosterol, stigmasterol, gallic acid, ellagic acid, emodin, chrysophanol	Inflammation, cancer malaria, skin diseases wounds	Oke-Altuntas et al. (2016) and Nantchouang Ouete et al. (2013)
Mondia whitei Skeels (South Africa)	β-sitosterol, kaempferol, quercetin, betulinic acid, oleanolic acid	Sexual dysfunction, stomach complications	Aremu et al. (2011), Onohuean et al. (2022), and Watcho et al. (2012)
Warburgia salutaris (G. Bertol.) Chiov. (South Africa)	Warburganal, muzigadial, polygodial, salutaridine, cinnamaldehyde, eugenol, methyleugenol, methyleugenol, spathulenol, caryophyllene, β-selinene	Malaria, stomach ulcers, chest pain	Dludlu et al. (2017)
Curtisia dentata (Burm.f.) C.A.Sm. (South Africa)	Betulinic acid, lupeol, betulin, oleanolic, catechins, epicatechin, quercetin, quercetin 3-rutinoside, kaempferol, kaempferol 3-rutinoside, proanthocyanidins	Dysentery, diarrhea	Doughari et al. (2012), Fadipe et al. (2017), and Maroyi (2019)
Bowiea volubilis Harv. ex Hook.f. Southern Africa	Bowiein, scillaren bowieaside, bufadienolides, kaempferol, quercetin	Heart disease, dropsy, wounds	Paul et al. (2012) and Thomsen et al. (2012)

(continued)

Table 5.1 (continued)

Plant name	Isolated compounds	Ethnomedicinal uses	Reference
Hypoxi hemerocallidea Fisch., C.A.Mey. & Avé-Lall. (South Africa)	Hypoxoside, rooperol, sitoindoside, diosgenin	Prostate issues	Nsibande et al. (2018)
Boswellia papyrifera (Caill.) Hochst. (Ethiopia)	Boswellic acid, incensole, incesole acetate, olibanic acid	Wounds, arthritis, digestive issues, asthma	Paul et al. (2012)
Hagenia abyssinica (Bruce) J.F.Gmel. (Ethiopia)	Kosins, 3,5-dihydroxy-4,6-dimethylphloroglucinol, quercetin, kaempferol	Tapeworm infections, bacterial infections	Thomsen et al. (2012)
Echinops kebericho Mesfin (Ethiopia)	Echinopiloid A, keberine, keberchol, keberic acid, limonene, stigmasterol, quercetin	Coughs, colds, malaria	Deyno et al. (2021) and Hymete et al. (2007)
Moringa stenopetala (Baker f.) Cufod. (Kenya, Ethiopia)	Moringinine, benzyl isothiocyanate, 1,4-(alpha-L-rhamnopyranosyloxy)-benzyl glucosinolate	Diabetes, malnutrition, inflammation	Adane et al. (2019)
Aloe pulcherrima M.G.Gilbert & Sebsebe (East Africa)	Emodin, kaempferol, quercetin, acemannan	Digestive disorders	Abdissa et al. (2017)
Osyris lanceolate Hochst. & Steud (East Africa)	Quercetin, kaempferol, gallic acid, ellagic acid	Skin healing, burns, wounds, inflammation	Yeboah et al. (2010)
Aloe killifiensis Christian	Aloin, aloe-emodin, aloin A, aloesin, emodin	Malaria, respiratory conditions, burns, cuts	Anywar et al. (2022)
Olea europaea L. (South Africa)	Oleuropein, hydroxytyrosol, oleanolic acid, maslinic acid, ligstroside	Oxidative stress, cardiovascular conditions	Pereira et al. (2007)
Strychnos innocua Delile. Benin	Brucine, curarine, strychnine	Muscle relaxant	Alagbe (2023)
Enantia chlorantha Oliv. (West Africa)	Cryptolepine, quercetin, kaempferol	Malaria, bacterial infections	Ejele et al. (2012) and Tcheghebe et al. (2016)
Garcinia kola Heckel	Garcinone, kolaviron, garcinol, kolaflavone	Cancer	Seanego & Ndip (2012)
Khaya Senegalensis (Desr.) A.Juss.	Khaysenegalensin, makonoside	Diarrhea, dysentery	Nganso et al. (2020), Allah et al. (2018), and Yuan et al. (2013)

(continued)

Table 5.1 (continued)

Plant name	Isolated compounds	Ethnomedicinal uses	Reference
Rauvolfia vomitoria Wennberg	Reserpine, serpentine, ajmaline, vohimbine	Insomnia, hypertension	Kumar & Singh (2022)
Ferula hermonis Boiss	Ferutinin, hermosine, umbelliprenin, sabinyl acetate, galbanic acid, 7,α-pinene, β-pinene	Respiratory problems, stomach crumps	Sattar & Iranshahi (2017)
Pergularia tomentosa L. (Egypt)	Calotropin, pergularoside, tomentoside, lupeol, pergularine, linoleic acid	Asthma, bronchitis	Al Hinai et al. (2018)
Ephedra alata Decne	Ephedrine, pseudopherine, quercetin, kaempferol, luteolin, catechin	Rheumatic pain, asthma, fever, colds	Ghanem & El-Magly (2008)
Salvadora persica L.	Salvadorine, myricetin, salvadoside, gallic acid, ellagic acid	Stomach ulcers, asthma, coughs	Ahmad & Rajagopal (2013)
Catharanthus roseus (L.) G.Don (Madagascar)	Vincristine, vindoline, lochnericine, tabersonine, vinblastine	Cancer	Koel et al. (2020)
Aloe polyphylla Schönland ex Pillans (Southern Africa)	Aloin, emodin, chrysophanol	Bacterial infections	Maliehe et al. (2022) and Abrie & Van Staden (2001)
Taxus baccata Thunb (North-Africa)	Paclitaxel	Cancer treatment	Yan-Hua et al. (2020)
Artemisia annua	Artemisin	Malaria	Willcox et al. (2004)
Rauwoflia serpen	Reserpine	Pain reliever	Ramarumo et al. (2019)
Ephedra sinica	Ephedrine	Decongestant	Ghanem & El-Magly (2008)
Opium poppy	Morphine	Pain	Benyhe (1994)
Atropa belladona	Atropine	Pain, bacterial infections	Tadayon et al. (2024)

Deforestation affects food security directly by reducing biodiversity, which is a source of food for humans, and indirectly by degrading soil and changing weather patterns, both of which lower agricultural production (Olagunju, 2015).

5.3.2 *Climate Change Patterns*

Research has demonstrated the detrimental impacts of climate change on species survival and population (Gómez et al., 2015). The outcomes of the simulation demonstrated that associated changes in the landscape and climate can either speed up or slow down (in the event of habitat loss or degradation) the localized extinction of species (Jeltsch et al., 2011). Heat waves, increasingly severe drought and flood cycles, and other environmental imbalances that impact Earth's complex climate dynamics are caused by rising greenhouse gas emissions. Such severe weather will cause significant alterations in a number of plant communities, leading to major dieback occurrences in forests, which may cause a significant decline in biodiversity across a number of global biomes (Menezes-Silva et al., 2019). Two of the primary factors that determine the emergence, dispersal, and survival of plant species worldwide are temperature and the availability of water (https://www.journals.uchicago.edu/doi/abs/10.1086/285186). Forest dieback, which is brought on by carbon starvation (CS) and the loss of nonstructural carbohydrates (NSCs), which creates a negative carbon balance, is caused by drought. This imbalance is caused by a complex web of interrelated variables, including decreased rates of photosynthesis owing to diffusive and biochemical constraints, which lowers carbon assimilation (https://www.journals.uchicago.edu/doi/abs/10.1086/285186), with the potential rise in respiration (R) and photorespiration rates, along with the increase in metabolic expenditures to manufacture defense chemicals and repair damaged components (Aerts et al., 2016). One of the primary causes of tree mortality during drought circumstances has been proposed to be CS (Bodo et al., 2021). The variables that contribute to forest dieback also seem to differ dramatically depending on how severe and long the drought is. Due to the negative carbon balance, plants exposed to moderate but protracted drought may not attain the crucial water potentials and may suffer a fatal drop in carbohydrate levels (Olagunju, 2015). Extremely high temperatures, which are frequently seen during heat waves, have an impact on plants as well. These effects seem to depend on a number of traits and modifications at the morpho-anatomical, physiological, and molecular levels (Onohuean et al., 2022). One of the primary consequences of human activity is the increasing rise in atmospheric CO_2 as a result of patterns of greenhouse gas emissions. Due to weather extremes that are directly linked to an increase in the concentration of CO_2 gas molecules, the rise in CO_2 may put many species at risk (Gurevitch et al., 2002). The most diverse biomes on Earth are home to a multitude of forest species, many of which are vulnerable to increasing degrees of extinction due to causes related to climate change (Gurevitch et al., 2002). The information gathered from these studies may be very helpful in optimizing models to track and forecast how climate change will affect plant species survival and distribution, as well as informing public policies about the management of forests and replanting (Aerts et al., 2016).

5.3.3 *Unsustainable Harvesting for Medicinal and Commercial Use*

The likelihood of over-exploitation of numerous valuable species has increased due to the current trend of large-scale commercial extraction of medicinal plants (MPs) to satisfy the expanding demand of the herbal industries (Ghimire, 2008). The world's supply of therapeutic plants has been continuously threatened, which worries resource consumers and conservationists alike (Mir et al., 2021). Because of the rising demand for medicinal plants in the pharmaceutical industry, for herbal treatments, and for other natural products, more plants are being harvested from the wild and used in an unsustainable way (Olagunju, 2015). In the case of medicinal plant resources, threats to genetic diversity and species survival have also increased due to overexploitation, habitat degradation, changes in land use, and other related issues (El-Banhawy et al., 2021). Additionally, there was an unsustainable trade in therapeutic herbs for financial gain (Butchart et al., 2010). In addition to endangering the availability of plant resources, habitat modification or destruction poses a serious risk to traditional community life, cultural diversity, and the understanding that certain endemic species have medicinal value. By 2005, human consumption of plants and the growing human population had depleted about 20% of the wild medicinal plant population. However, the kind, intensity, and impact of these variables, along with the level of threat, differ among nations. For this reason, it is critical to understand the biological traits and geographic distribution of medicinal plants in order to employ practical conservation techniques to protect them (Mir et al., 2021).

5.3.4 Impact of Exotic Species on Native Plant Populations

When introduced, exotic plants alter the habitats in which they are found. These changes, or parts of them, are commonly referred to as impacts. They can have a variety of effects and may be harmful to biodiversity and ecosystems. Numerous non-native species have become naturalized and accumulated within ecosystems as a result of global species migration. This is usually linked to habitat disruption and shifting environmental factors (Butchart et al., 2010). Non-native plants were always terrible neighbors, according to accounts of interactions between native and non-native species, and their detrimental effects on native plants were roughly twice as great as their effects on other non-native plants. In contrast, when a nearby native plant species was present, the performance of non-native plants was five times higher than when a nearby non-native plant species was present (Butchart et al., 2010). The impact of invasive species on native species is typically greater, and invading plants frequently seem to be more competitive than native ones (Vila & Weiner, 2004). Additionally, although they are not less productive than invasive species monocultures, mixes of native and invasive species are typically less productive than native species monocultures. Competition may play a significantly different role during the invasion process's founding and naturalization phases than it does during the "outbreak" stage (Vila & Weiner, 2004). There are reports of beneficial interactions between invading and native species in a variety of environments. These interactions can reorganize communities, have cascade effects across trophic levels, and even trigger evolutionary changes. Exotic species can help native species through trophic subsidy, pollination, competitive release, habitat change, and predatory release, among other mechanisms (Rodriguez, 2006). The majority of the time, invasive species have a facilitative impact on native species when they increase habitat complexity, supply a scarce resource, functionally replace a native species, or improve predation or competition (Rodriguez, 2006). The implications of invasive plant species, which include altered fire regimes, alterations to nutrient pools, and a decrease in native diversity, should be assessed and prioritized for control (Barney et al., 2013).

5.4 Conservation Strategies

Long-term procedures for the conservation and sustainable use of medicinal plants may include conservation strategies including cultivation, yield parameter optimization for productivity, and avoiding harmful harvesting. Hitherto, in situ and ex situ conservation have continued to be the most popular choices for plant conservation.

5.4.1 In Situ Conservation

In situ conservation of endangered plants in Africa involves preserving species within their natural habitats. A number of studies have reported several strategies and considerations for in situ conservation efforts in Africa, including (i) establishing and effectively managing protected areas and national parks for conserving biodiversity, consequently restricting human activities that could threaten their survival (Naughton-Treves et al., 2005); (ii) involving local communities in conservation efforts with programs such as sustainable harvesting practices, traditional ecological knowledge, and creating incentives for communities to protect their natural resources (Hanazaki et al., 2018); (iii) continuous monitoring of endangered plant populations in order to assess their status and trends. This helps in understanding the ecological requirements of these plants and informs conservation strategies (Hutton & Leader-Williams, 2003); (iv) rehabilitating degraded habitats and managing ecosystems by controlling invasive species, restoring water regimes, and combating habitat fragmentation to improve their resilience supports the survival of endangered plants (D'Antonio & Chambers, 2006); (v) implementing and enforcing laws that protect endangered species and their habitats with programs such as international agreements, national legislation, and local regulations that safeguard biodiversity (Gollin & Laird, 1996). Examples of successful in situ conservation initiatives in Africa include the protection of unique plant species found in diverse ecosystems such as tropical rainforests, savannas, and coastal regions (Akani, 2023). These efforts often require collaboration between governments, NGOs, local communities, and international organizations to be effective.

5.4.2 *Ex Situ Conservation*

Ex situ conservation refers to the preservation of threatened plant species away from their natural habitats, frequently in seed banks, botanical gardens, or other controlled settings. Ex situ conservation initiatives are essential to preserving Africa's wide variety of plant species from challenges like habitat loss, climate change, and overexploitation. Ex situ conservation employs different initiatives. For instance, in South Africa, botanical gardens and arboreta such as Kirstenbosch National Botanical Garden are well known for their extensive collection of South African flora, including many endangered species such as cycads and proteas (Duncan, 2011). In addition, Seed Banks and Germplasm Repositories (South African National Biodiversity Institute—SANBI) manages the South African National Seed Bank, which preserves seeds of threatened plant species across the country for conservation and research purposes (De Villiers et al., 2002). In Malawi, the National Botanic Gardens of Malawi houses a collection of Malawi's endemic and endangered plant species, promoting conservation awareness and education (Nyirenda, 2008). Moreover, in Kenya, the Kenya Forestry Research Institute (KEFRI) operates seed banks and germplasm repositories to conserve Kenya's diverse flora, including endangered species like tree species from the Eastern Arc Mountains (Mukhongo et al., 2011). Meanwhile, Conservation Organizations and Research Institutions such as the Ethiopian Biodiversity Institute (EBI) conducts research and implement conservation programs for Ethiopia's endemic and endangered plant species. These efforts include ex situ propagation and reintroduction initiatives aimed at preserving biodiversity (Aerts et al., 2016). In addition, the Gullele Botanical Garden in Ethiopia plays a significant role in enhancing biodiversity conservation. It provides a refuge and improved protection for threatened species, serves as a model ecosystem, preserves genetic and species diversity and contributes to global efforts in plant germplasm conservation. Aluka Project (Nigeria) collaborates with local communities and researchers to document and conserve Nigeria's biodiversity, including rare and endangered plant species. International collaborative efforts are also in place (Imarhiagbe et al., 2020). For instance, the Royal Botanic Gardens, Kew (UK) partners with African countries for conservation initiatives, including capacity-building in ex situ conservation techniques and managing global seed conservation projects like the Millennium Seed Bank Partnership (Ali & Trivedi, 2011). Botanic Gardens Conservation International (BGCI) collaborates with botanical gardens across Africa to develop conservation strategies, share best practices, and promote ex situ conservation of endangered plants. Community-Based Conservation Projects and Community nurseries and gardens initiatives aim to complement ex situ conservation efforts by providing a safety net for endangered plant species, ensuring their survival and potential reintroduction into the wild. By preserving genetic diversity and supporting research, ex situ conservation plays a vital role in mitigating the loss of biodiversity in Africa and globally.

5.4.3 Legislation and Policy

The need for strong phytosanitary laws and regulations is highlighted by the apparent rise in the introduction of species carried by this pathway as a result of the recent decades' significant expansion in global trade (Eschen et al., 2015). There are significant variations in how different nations manage the risk of introducing invasive alien species through international plant trade, but most countries base their phytosanitary regulations on the International Plant Protection Convention (IPPC) and the World Trade Organization (WTO) Agreement on Sanitary and Phytosanitary (SPS) Measures.

While Europe retains a universal license for plants for planting (P4P) imports, New Zealand and Australia have the tightest phytosanitary controls. The remaining nations have laws that fall between these two extremes. There is little information available about the effects of individual phytosanitary measures, their efficacy, and the quality of their implementation (Eschen et al., 2015). As nations seek stronger protections for their economic and environmental interests, particularly from the risks of weeds, pests, and diseases from imported non-native plants and plant parts, efforts have been made to strengthen their national laws regulating the import of live plants from other nations (Jenkins, 2005). It is necessary to balance the provisions of the international legal framework that permit trade restrictions with those that support a comparatively unconstrained plant commerce, in order to account for a nation's tolerable degree of risk from imported viruses, pests, and weeds (Rodriguez, 2006). Although invasive species frequently have terrible consequences on human health, agriculture, natural areas, man-made structures, and industry, certain environments are becoming increasingly vulnerable to the destructive effects of invasionism due to changes in land use and climate (Jenkins, 2005). The majority of the world's leading plant import and export nations are currently parties to trade-facilitating agreements that impose international norms on domestic regulatory measures, coinciding with the growing recognition of the issues surrounding freely imported plants (Jenkins, 2005). Regardless of whether it is a novel plant or one that has already been approved by that country, a country is not required by the World Trade Organization (WTO), the International Plant Protection Convention (IPPC), the application of Sanitary and Phytosanitary measures (SPS Agreement), or other international laws to permit the entry of a proposed plant import for which there is not sufficient proof of safety (Jenkins, 2005).

5.4.4 Conservation Challenges and Success Stories

It is evident that current strategies are insufficient to effectively stop the ongoing loss of biodiversity, despite the massive efforts that have been undertaken in recent years to conserve plant diversity worldwide (Heywood & Iriondo, 2003). Lists of endangered plant species keep expanding as global governments fall short of their targets for conserving biodiversity (Padulosi et al., 2002). Given the ongoing global decline in biodiversity, immediate action is needed. Conservation of endangered plants faces numerous challenges, which can be categorized into ecological, socio-economic, and logistical factors (Olah et al., 2016). Ecological challenges such as habitat loss and fragmentation involve urbanization, agricultural expansions, and climate change (Heywood & Iriondo, 2003). Plants, animals, and pathogens can outcompete, prey on, or bring diseases to native plant species, leading to declines in their populations (Gurevitch et al., 2002; Grubb, 1992). Air, water, and soil pollution can degrade habitats and harm plant health, leading to reduced growth and reproduction rates (Denny, 1993). Socio-economic challenges include lack of awareness and education, limited funding, and conflicting land use priorities (Adamo et al., 2022). Logistical challenges, such as difficulty in collecting and storing seeds from endagered plants, often arise because some seeds have short viability periods or require specific conditions and timing to germinate (Pessar, 2021). Because of their unique ecological needs and competition with other species, propagating endangered plants in nurseries and successfully reintroducing them into their natural habitats present additional challenges. However, preserving genetic diversity within small, isolated populations of endangered plants is essential to their long-term survival (Aerts et al., 2016). Small populations are more vulnerable to inbreeding, genetic drift, and other genetic issues (Lacy, 1997). Numerous tactics designed to preserve threatened plant species have shown remarkable results. Many plant species in Africa have been successfully conserved thanks to initiatives like recognizing and involving local communities in conservation efforts, creating national parks, protected areas, and legal frameworks for the preservation of biodiversity, conducting scientific research to understand the biology of the species, threats to it, and the needs of conservation, and working with international organizations and neighboring nations on transboundary conservation projects. Plants such as *Welwitschia mirabilis* found in the Namib desert of Namibia and Angola, African Baobab (*Adansonia* species) found across Africa, *Raffia* Palm (*Raffia* species) found in tropical Africa, Mountain Gorilla Habitat (*Virunga Massif*, Central Africa) found in Rwanda, Uganda, Democratic Republic of Congo, *Aloe vera* (*Aloe* species) found in various parts of Africa, Chimpanzee Nut (*Panda oleosa*) in west Africa, *Chilean Puya chilensis* and the South African *Encephalartos, brevifoliolatus* have all been conserved. The goal of the world's largest plant conservation effort, the Millennium Seed Bank Partnership, is to gather and preserve seeds from 25% of all plants (van Slageren, 2003).

5.5 Use of Technology to Monitor Plant Population

The use of technology for monitoring plant populations and habitats has revolutionized conservation efforts, providing researchers and conservationists with valuable tools to assess biodiversity, track changes over time, and implement targeted conservation strategies (Aravanopoulos, 2011). In contrast to other approaches, technology offers accurate data on plant populations and habitats at different temporal and spatial scales, lowers the time and cost of data collection when compared to traditional methods, makes it easier to identify threats early and take action to prevent biodiversity loss, and allows for ongoing monitoring over long periods of time to identify trends and evaluate the success of conservation efforts (Engelmann, 2011). Technologies for monitoring plant population include the following. (i) Satellite and aerial imagery drones, which help to monitor land cover changes and identify habitat loss or fragmentation, detect illegal logging, encroachment, and conduct rapid plant population surveys in inaccessible areas (Habibi et al., 2021). (ii) Geographic Information System (GIS) integrates data from field surveys and other sources to create detailed maps and spatial models of plant distributions, predicting suitable habitats for endangered plants based on environmental variables like climate, soil, and elevation and identifying priority areas for conservation based on habitat quality, species richness, and connectivity (Salem, 2003). (iii) Camera traps and acoustic monitoring monitor animal-plant interactions, such as seed dispersal by animals, which are crucial for plant population dynamics, capturing images or sounds of rare or elusive plant species for population assessment and conservation planning (Buxton et al., 2018). (iv) Environmental DNA (eDNA) Species Detection detects the presence of plant species through environmental samples such as soil and water without directly observing them and monitors invasive plant species by detecting their DNA in environmental samples, aiding early detection and rapid response efforts (Banerjee et al., 2022). (v) Mobile Apps and Citizen Science collect field data on plant species distributions, abundance, and phenology using standardized protocols, involving citizen scientists in monitoring efforts, increasing public awareness about plant conservation, and fostering stewardship of natural habitats (Kress et al., 2018). (vi) Sensor Networks and Internet of Things (IoT) monitor environmental variables like temperature, humidity, and light levels to understand plant habitat preferences and responses to climate change (Sanjeevi et al., 2020). By leveraging these technologies, conservationists can make informed decisions, implement adaptive management strategies, and effectively conserve plant diversity and habitats in the face of environmental challenges and human impacts.

5.6 Genetic Diversity and Resilience of Plants That Are Endangered

Understanding the genetic diversity and resilience of endangered plants is essential for creating effective conservation measures as well as for comprehending their flexibility and evolutionary potential (Van Rossum & Hardy, 2022). Numerous techniques have been employed in recent studies to try and comprehend this idea. Researchers evaluated genetic diversity both within and between populations of endangered plants using scientific methods including DNA sequencing (Brown & Weir, 1983). Higher genetic diversity generally correlates with greater adaptive potential, as it provides a broader range of traits for natural selection to act upon

(Ackerly et al., 2000). Populations with low genetic diversity may face challenges in adapting to environmental changes, such as climate change or habitat fragmentation (Pauls et al., 2013). According to other research, inbreeding depression can result in decreased reproductive success, decreased fitness, and heightened susceptibility to illnesses and environmental stressors (Keller & Waller, 2002). Understanding these genetic factors helps in developing breeding programs to maintain genetic diversity and improve population health. Fragmentation can lead to genetic isolation, reducing gene flow between populations and increasing the risk of inbreeding (Leimu et al., 2010). Genetic studies help identify corridors and habitat patches crucial for maintaining connectivity and genetic exchange, which are essential for long-term species survival (Brown & Weir, 1983). Genetic variation allows populations to adapt to new conditions, such as changing climates or emerging diseases. Species with greater genetic diversity are more likely to have individuals with traits that can survive and reproduce under changing environmental conditions (Schwartz et al., 2007).

5.7 Genetic Modification for Conservation of Endangered Plants

Genetic modification, also known as genetic engineering, holds potential for the conservation of endangered plants, although it is a topic that involves ethical, regulatory, and scientific considerations. Genetic engineering enhances disease resistance by introducing genes that confer resistance to diseases and pathogens that threaten endangered plant species. Consequently, this protects plants from devastating diseases that could otherwise wipe out entire populations (Montagu, 2019). This strategy could be specifically useful for species facing threats from emerging pathogens or invasive species (Lacy, 1997). Additionally, engineering plants enhances the resilience of endangered species to tolerate environmental stresses such as drought, extreme temperatures, or soil salinity and survive in changing climates or degraded habitats where natural populations struggle to thrive (Pauls et al., 2013). Furthermore, introducing genetic variation into small, isolated populations of endangered plants enhances the genetic diversity within populations, potentially improving their adaptability and evolutionary potential. This could reduce the risks associated with inbreeding depression and increase the long-term viability of populations (Lacy, 1997). Notwithstanding, facilitating reproduction and seed production is another advantage of genetic modification as it overcomes reproductive barriers or enhances seed production in endangered plants. Addressing reproductive limitations that contribute to population declines is another crucial concern (Brown & Weir, 1983). For example, modifying plants to improve pollination efficiency or seed germination rates could support efforts to increase population sizes and genetic diversity. Genetic modification rescues populations with extremely low genetic diversity or viability (Kramer & Havens, 2009).

5.8 Economic Importance of Endangered Plants

Endangered plants can hold significant economic importance, both directly and indirectly, despite their endangered status (Loomis & White, 1996). Endangered plants contain bioactive compounds used in traditional medicine or for developing pharmaceutical drugs. The pharmaceutical companies harness these compounds for treatments of various diseases, potentially leading to valuable patents and commercial products (Mir et al., 2021). Moreover, some endangered plants are sources of nutrition or traditional food items as local communities rely on these plants for food security (Olagunju, 2015) and cultural practices (Waylen et al., 2010), contributing to livelihoods and local economies through culinary heritage and tourism (Loomis & White, 1996). Loomis and White (2009) reported that endangered plants often possess unique genetic traits that can enhance crop resilience, productivity, and nutritional value improving traits like pest resistance and adaptation to climate change. Additionally, endangered plants frequently have a significant impact on ecosystem processes including seed dissemination by birds and animals and pollination by bees, butterflies, and other insects (Potts et al., 2016). Ecosystem services provided by endangered plants support agriculture, enhance soil fertility, and maintain biodiversity, which are essential for sustainable economic development (Twerski et al., 2021). Endangered plants often reside in unique ecosystems and habitats, attracting ecotourists and nature enthusiasts (Loomis & White, 1996). Conservation efforts to protect these plants and their habitats can generate income through ecotourism, guided tours, and sustainable recreational activities (Sarasan et al., 2011). Endangered plants may hold cultural significance for indigenous communities, contributing to cultural identity and traditional knowledge systems (Vallejo et al., 2024). Rare and beautiful plants attract botanical gardens, nurseries, and collectors, stimulating demand for horticultural and landscaping services (Mounce et al., 2017).

5.9 Raising Awareness about Endangered Plants

Raising awareness about endangered plants is crucial for their conservation (Erdogan et al., 2008). *Education campaigns* such as developing educational materials like brochures, posters, and videos that highlight the importance of endangered plants, their role in ecosystems, and the threats they face (Marlina et al., 2022). Organizing events like workshops, seminars, and field trips focused on endangered plants (Jacobson et al., 2015). Inviting experts to speak about their importance, conservation strategies, and how individuals can contribute to their protection. Sharing stories, facts, and updates about endangered plants through social media platforms (Bergman et al., 2022). Engaging content such as videos, infographics, and interactive posts to reach a wider audience. Partnering with journalists and media outlets to feature stories and articles about endangered plants and conservation efforts. This can include interviews with scientists, success stories, and profiles of local conservation heroes (Maher, 1997). Using botanical gardens and arboreta as educational hubs to showcase endangered plants. Create interpretive signage, guided tours, and interactive exhibits that inform visitors about these plants and their conservation status (Ballantyne et al., 2009). Engaging the public in monitoring and documenting endangered plants through citizen science initiatives (Dickinson et al., 2012). Developing apps or websites where volunteers can report sightings, monitor populations, and contribute data to research projects (Preece, 2017). Collaborating with local communities, conservation organizations, businesses, and government agencies to amplify awareness efforts (McAllister, 1999). Joint initiatives can include planting events, clean-up campaigns, and community forums (Mounce et al., 2017). Using art, music, and cultural events to raise awareness about endangered plants (Grant, 2014). Art exhibitions, performances, and competitions centered on plant conservation can attract diverse audiences and spark interest. Advocating for policies that protect endangered plants and their habitats. Engaging with policymakers, participate in public consultations, and support initiatives that promote sustainable land use and conservation practices (Aravanopoulos, 2011). Highlighting success stories and positive outcomes of conservation efforts. Showcasing successful recoveries or reintroductions of endangered plants can inspire hope and motivate continued support for conservation initiatives.

5.10 Conclusion

The study of endangered plants in Africa highlights a critical interplay between phytochemistry, ethnomedicine, and conservation strategies. The conservation of endangered plants in Africa is crucial not only for maintaining biodiversity but also for preserving the ethnomedicinal knowledge and phytochemical resources they provide. Collaborative efforts between governments, NGOs, and local communities are vital for the success of these initiatives.

5.11 Recommendation

Many of these endangered plants in Africa have a variety of bioactive compounds that have not been fully identified or characterized. These plants hold potential for medicinal uses and could benefit from further research to understand their pharmacological properties. This underscores the importance of further research into endangered African plants to identify and characterize bioactive compounds. Additionally, habitat restoration in terms of reforestation and afforestation, sustainable harvesting practices, cultivation and propagation initiatives such as agroforestry, legal protection and policy advocacy, community engagement and education, research and monitoring, and collaboration with indigenous knowledge. By combining scientific knowledge with traditional practices and sustainable economic incentives, we can effectively rehabilitate and preserve endangered plants for future generations.

Acknowledgments The authors wish to express their gratitude to Professor Aman Dekebo and reviewers for comments on earlier versions of the manuscript.

Conflicts of Interest The authors declare that they have no conflict of interest.

References

Abdissa, D., Geleta, G., Bacha, K., & Abdissa, N. (2017). Phytochemical investigation of Aloe pulcherrima roots and evaluation for its antibacterial and antiplasmodial activities. *PLoS One, 12*(3), e0173882.

Abegaz, B. M., & Mutanyatta, J. (2002). Bioactive compounds from some endangered plants of Africa. In *Biodiversity: Biomolecular aspects of biodiversity and innovative utilization* (pp. 71–78). Springer.

Abrie, A., & Van Staden, J. (2001). Micropropagation of the endangered Aloe polyphylla. *Plant Growth Regulation, 33*, 19–23.

Ackerly, D. D., Dudley, S. A., Sultan, S. E., Schmitt, J., Coleman, J. S., Linder, C. R., et al. (2000). The evolution of plant ecophysiological traits: Recent advances and future directions: New research addresses natural selection, genetic constraints, and the adaptive evolution of plant ecophysiological traits. *Bioscience, 50*(11), 979–995.

Adamo, M., Sousa, R., Wipf, S., Correia, R. A., Lumia, A., Mucciarelli, M., et al. (2022). Dimension and impact of biases in funding for species and habitat conservation. *Biological Conservation, 272*, 109636.

Adams, Z., Thekisoe, O., & Buwa-Komoreng, L. (2021). An ethnobotanical survey of traditional medicinal plants used against elephantiasis in the or Tambo District, eastern cape, South Africa. *Pharmacognosy Magazine, 17*(76), 915.

Adane, L., Teshome, M., & Tariku, Y. (2019). Isolation of compounds from root bark extracts of *Moringa stenopetala* and evaluation of their antibacterial activities. *Journal of Pharmacognosy and Phytochemistry, 8*(3), 4228–4244.

Adebayo, S. A., Amoo, S. O., Mokgehle, S. N., & Aremu, A. O. (2021). Ethnomedicinal uses, biological activities, phytochemistry and conservation of African ginger (Siphonochilus aethiopicus): A commercially important and endangered medicinal plant. *Journal of Ethnopharmacology, 266*, 113459.

Aerts, R., Overtveld, K. V., November, E., Wassie, A., Abiyu, A., Demissew, S., et al. (2016). Conservation of the Ethiopian church forests: Threats, opportunities and implications for their management. *Science of the Total Environment, 551–552*, 404–414.

Ahmad, H., & Rajagopal, K. (2013). Biological activities of Salvadora persica L. Meswak. *Medicinal & Aromatic Plants, 2*(4), 1–5.

Akani, G. C. (2023). Rationale behind conservation of Africa's biological resources. In *Sustainable utilization and conservation of Africa's biological resources and environment* (pp. 235–265). Springer.

Al Hinai, H. S., Al-Subhi, W. M., Al-Rubaiai, F. R. S., Hassan, S. I., Sherwani, N., & Fatope, M. O. (2018). Lupane and ursane-type triterpenoids from Pergularia tomentosa. *Chemistry of Natural Compounds, 54*, 790–792.

Alagbe, J. O. (2023). Bioactive compounds in ethanolic extract of Strychnos innocua root using gas chromatography and mass spectrometry (GC-MS). *Drug Discovery., 17*, e4dd1005.

Ali, N. S., & Trivedi, C. (2011). Botanic gardens and climate change: A review of scientific activities at the Royal Botanic Gardens, Kew. *Biodiversity and Conservation, 20*, 295–307.

Allah, M. O. W., Alrasheid, A. A., & Elamin, A. S. (2018). Phytochemical screening, chemical composition and antioxidant activity of leaves and bark extracts from Khaya senegalensis. *Advances in Biochemistry, 6*(4), 32–38.

Allen, J. C., & Barnes, D. F. (1985). The causes of deforestation in developing countries. *Annals of the Association of American Geographers, 75*(2), 163–184.

Antonelli, A., Smith, R., Fry, C., Simmonds, M. S., Kersey, P. J., & Pritchard, H. et al. (2020). State of the World's Plants and Fungi. (Doctoral dissertation, Royal Botanic Gardens (Kew); Sfumato Foundation).

Anywar, G., Tugume, P., & Kakudidi, E. K. (2022 Jul). A review of aloe species used in traditional medicine in East Africa. *South African Journal of Botany, 1*(147), 1027–1041.

Aravanopoulos, F. (2011). Genetic monitoring in natural perennial plant populations. *Botany, 89*(2), 75–81.

Aremu, A., Cheesman, L., Finnie, J., & Van Staden, J. (2011). Mondia whitei (Apocynaceae): A review of its biological activities, conservation strategies and economic potential. *The South African Journal of Botany, 77*(4), 960–971.

Asiimwe, J. B., Nagendrappa, P. B., Atukunda, E. C., Kamatenesi, M. M., Nambozi, G., Tolo, C. U., et al. (2021). Prevalence of the use of herbal medicines among patients with cancer: A systematic review and meta-analysis. *Evidence-Based Complementary and Alternative Medicine, 2021*(1), 9963038.

Ballantyne, R., Packer, J., & Hughes, K. (2009). Tourists' support for conservation messages and sustainable management practices in wildlife tourism experiences. *Tourism Management, 30*(5), 658–664.

Banerjee, P., Stewart, K. A., Dey, G., Antognazza, C. M., Sharma, R. K., Maity, J. P., et al. (2022). Environmental DNA analysis as an emerging non-destructive method for plant biodiversity monitoring: A review. *AoB Plants, 14*(4), plac031.

Barney, J. N., Tekiela, D. R., Dollete, E. S., & Tomasek, B. J. (2013). What is the "real" impact of invasive plant species? *Frontiers in Ecology and the Environment, 11*(6), 322–329.

Barthélémy, D., & Caraglio, Y. (2007). Plant architecture: A dynamic, multilevel and comprehensive approach to plant form, structure and ontogeny. *Annals of Botany, 99*(3), 375–407.

Begeno, T. A., Teka, A. E., Bafa, T. A., & Nassir, W. B. (2020). Phytochemical investigation and characterization on the leaf extract of prunus africana. *International Research Journal of Pure and Applied Chemistry, 21*, 47–57.

Benyhe, S. (1994). Morphine: New aspects in the study of an ancient compound. *Life Sciences, 55*(13), 969–979.

Bergman, J. N., Buxton, R. T., Lin, H. Y., Lenda, M., Attinello, K., Hajdasz, A. C., et al. (2022). Evaluating the benefits and risks of social media for wildlife conservation. *Facets, 7*(1), 360–397.

Bodeker, G., van't Klooster, C., & Weisbord, E. (2014). Prunus africana (Hook. f.) Kalkman: The overexploitation of a medicinal plant species and its legal context. *Journal of Alternative and Complementary Medicine, 20*(11), 810–822.

Bodo, T., Gimah, B. G., & Seomoni, K. J. (2021). Deforestation and habitat loss: Human causes, consequences and possible solutions. *Journal of Geographical Research, 4*(2), 22–30.

Brito, J., Del Barrio, G., Stellmes, M., Pleguezuelos, J., & Saarinen, J. (2021). Drivers of change and conservation needs for vertebrates in drylands: An assessment from global scale to Sahara-Sahel wetlands. *European Zoological Journal, 88*(1), 1103–1129.

Brown, A. H., & Weir, B. S. (1983). Measuring genetic variability in plant populations. *Isozymes in Plant Genetics and Breeding, Part A*, 219–239.

Butchart, S. H., Walpole, M., Collen, B., Van Strien, A., Scharlemann, J. P., Almond, R. E., et al. (2010). Global biodiversity: Indicators of recent declines. *Science, 328*(5982), 1164–1168.

Buxton, R. T., Lendrum, P. E., Crooks, K. R., & Wittemyer, G. (2018). Pairing camera traps and acoustic recorders to monitor the ecological impact of human disturbance. *Global Ecology and Conservation, 16*, e00493.
Catarino, L., & Romeiras, M. M. (2020). Biodiversity of vegetation and flora in tropical Africa. *Diversity, 12*(10), 369.
Chaachouay, N., & Zidane, L. (2024). Plant-derived natural products: A source for drug discovery and development. *Drugs Drug Candidates., 3*(1), 184–207.
Chen, G., Kéry, M., Plattner, M., Ma, K., & Gardner, B. (2012). Imperfect detection is the rule rather than the exception in plant distribution studies. *Journal of Ecology., 101*, 183–191. https://doi.org/10.1111/1365-2745.12021
Cock, I. E., & Van Vuuren, S. F. (2020). The traditional use of southern African medicinal plants for the treatment of bacterial respiratory diseases: A review of the ethnobotany and scientific evaluations. *Journal of Ethnopharmacology, 263*, 113204.
D'Antonio, C. M., & Chambers, J. C. (2006). Using ecological theory to manage or restore ecosystems affected by invasive plant species. *Foundations of Restoration Ecology*, 260–279.
De Villiers, A., Van Rooyen, M., & Theron, G. (2002). Seed bank classification of the strandveld succulent karoo, South Africa. *Seed Science Research, 12*(1), 57–67.
Denny, M. W. (1993). *Air and water: The biology and physics of life's media*. Princeton University Press.
Deresa, D. A., Abdissa, Z., Gurmessa, G. T., & Abdissa, N. (2022). Chemical constituents of the stem bark of Prunus africana and evaluation of their antibacterial activity. *Journal of the Turkish Chemical Society Section A: Chemistry, 9*(2), 395–414.
Deyno, S., Mtewa, A. G., Hope, D., Bazira, J., Makonnen, E., & Alele, P. E. (2021). Antibacterial activities of Echinops kebericho Mesfin tuber extracts and isolation of the most active compound, dehydrocostus lactone. *Frontiers in Pharmacology, 11*, 608672.
Dickinson, J. L., Shirk, J., Bonter, D., Bonney, R., Crain, R. L., Martin, J., et al. (2012). The current state of citizen science as a tool for ecological research and public engagement. *Frontiers in Ecology and the Environment, 10*(6), 291–297.
Dludlu, M. N., Dlamini, P. S., Sibandze, G. F., Vilane, V. S., & Dlamini, C. S. (2017). Distribution and conservation status of the endangered pepperbark tree Warburgia salutaris (Canellaceae) in Swaziland. *Oryx, 51*(3), 451–454.
Doughari, J. H., Ndakidemi, P. A., Human, I. S., & Benade, S. (2012). Antioxidant, antimicrobial and antiverotoxic potentials of extracts of Curtisia dentata. *Journal of Ethnopharmacology, 141*(3), 1041–1050.
Dublin, H. (2024). Endangered species. Encyclopedia Britannica. https://www.britannica.com/science/endangered-species. Accessed 27 June 2024.
Duncan, G. (2011). Kirstenbosch national botanical garden. *Veld and Flora, 97*(4), 159–163.
Ejele, A., Akpan, I., Ogukwe, C., Onyeocha, V., & Ukiwe, L. (2012). Bioassay-guided isolation and partial characterization of an antisickling compound from *Enantia chlorantha*. *International Journal of Bioscience, Biochemistry and Bioinformatics, 2*(7), 149–154.
El-Banhawy, A., Nour, I. H., Acedo, C., ElKordy, A., Faried, A., Al-Juhani, W., et al. (2021). Taxonomic revisiting and phylogenetic placement of two endangered plant species: Silene leucophylla Boiss. And Silene schimperiana Boiss. (Caryophyllaceae). *Plants, 10*(4), 740.
Engelmann, F. (2011). Use of biotechnologies for the conservation of plant biodiversity. *In Vitro Cellular & Developmental Biology – Plant, 47*, 5–16.
Erdogan, M., Erentay, N., Barss, M., & Nechita, A. (2008). Students' awareness of endangered species and threatened environments: A comparative case-study. *The International Journal on Hands-on Science, 1*(2), 46–53.
Eschen, R., Britton, K., Brockerhoff, E., Burgess, T., Dalley, V., Epanchin-Niell, R., et al. (2015). International variation in phytosanitary legislation and regulations governing importation of plants for planting. *Environmental Science & Policy, 51*, 228–237.
Evans, M., & Arentz, S. (2022). The influence of'hydrastis canadensis' on the gastrointestinal microbiota-beneficial or detrimental effects?: A critical review to update naturopathic and

western herbal clinical practice. *Australian Journal of Herbal and Naturopathic Medicine, 34*(4), 156–163.

Fadipe, V. O., Mongalo, N. I., Opoku, A. R., Dikhoba, P. M., & Makhafola, T. J. (2017). Isolation of anti-mycobacterial compounds from Curtisia dentata (Burm. f.) CA Sm (Curtisiaceae). *BMC Complementary and Alternative Medicine, 17*, 1–6.

Fritz-Vietta, N. V., Ferguson, H. B., Stoll-Kleemann, S., & Ganzhorn, J. U. (2011). Conservation in a biodiversity hotspot: Insights from cultural and community perspectives in Madagascar. In *Biodiversity hotspots: Distribution and protection of conservation priority areas* (pp. 209–233). Springer.

Ghanem, S., & El-Magly, U. I. (2008). Antimicrobial activity and tentative identification of active compounds from the medicinal Ephedra alata male plant. *The Journal of Taibah University Medical Sciences, 3*(1), 7–15.

Ghimire, S. K. (2008). Sustainable harvesting and management of medicinal plants in the Nepal Himalaya: Current issues, knowledge gaps and research priorities. *Medicinal Plants in Nepal: An Anthology of Contemporary Research*, 25–44.

Gilgert, W., & Vaughan, M. (2011). The value of pollinators and pollinator habitat to rangelands: Connections among pollinators, insects, plant communities, fish, and wildlife. *Rangelands, 33*(3), 14–19.

Gollin, M. A., & Laird, S. A. (1996). Global policies, local actions: The role of national legislation in sustainable biodiversity prospecting. *The Boston University Journal of Science & Technology Law, 2*, 9.

Gómez, J. M., González-Megías, A., Lorite, J., Abdelaziz, M., & Perfectti, F. (2015). The silent extinction: Climate change and the potential hybridization-mediated extinction of endemic high-mountain plants. *Biodiversity and Conservation, 24*, 1843–1857.

Gonfa, T., Fisseha, A., & Thangamani, A. (2020). Isolation, characterization and drug-likeness analysis of bioactive compounds from stem bark of Warburgia ugandensis Sprague. *Chemical Data Collections, 29*, 100535.

Grant, C. (2014). *Music endangerment: How language maintenance can help*. Oxford University Press.

Grubb, P. J. (1992). A positive distrust in simplicity–lessons from plant defences and from competition among plants and among animals. *Journal of Ecology, 80*(4), 585.

Gurevitch, J., Scheiner, S. M., & Fox, G. A. (2002). *The ecology of plants*. Sinauer Associates Sunderland.

Habibi, L. N., Watanabe, T., Matsui, T., & Tanaka, T. S. (2021). Machine learning techniques to predict soybean plant density using UAV and satellite-based remote sensing. *Remote Sensing, 13*(13), 2548.

Hails, C. J., & Kavanagh, M. (2013). Bring Back the birds. *The Raffles Bulletin of Zoology, 29*, 243–258.

Hall, J. B., & Swaine, M. D. (2013). *Distribution and ecology of vascular plants in a tropical rain forest: Forest vegetation in Ghana* (Vol. 1). Springer Science & Business Media.

Hanazaki, N., Zank, S., Fonseca-Kruel, V. S., & Schmidt, I. B. (2018). Indigenous and traditional knowledge, sustainable harvest, and the long road ahead to reach the 2020 global strategy for plant conservation objectives. *Rodriguesia, 69*(04), 1587–1601.

Happi, G. M., Talontsi, F. M., Laatsch, H., Zühlke, S., Ngadjui, B. T., Spiteller, M., et al. (2018). Seco-Tiaminic acids B and C: Identification of two novel 3, 4-seco-tirucallane triterpenoids isolated from the root of Entandrophragma congoënse (Meliaceae). *Fitoterapia, 124*, 17–22.

Heywood, V. H., & Iriondo, J. M. (2003). Plant conservation: Old problems, new perspectives. *Biological Conservation, 113*(3), 321–335.

Hordofa, T. G. (2020). Isolation and characterization of Sesquiterpenes from Stem Bark of Warburgia ugandensis Sprague. *Biology, Medicine, and Natural Product Chemistry, 9*(1), 39–46.

Hoveka, L. N., van Der Bank, M., Bezeng, B. S., & Davies, T. J. (2020). Identifying biodiversity knowledge gaps for conserving South Africa's endemic flora. *Biodiversity and Conservation, 29*(9), 2803–2819.

Hutton, J. M., & Leader-Williams, N. (2003). Sustainable use and incentive-driven conservation: Realigning human and conservation interests. *Oryx, 37*(2), 215–226.

Hymete, A., Rohloff, J., Iversen, T., & Kjøsen, H. (2007). Volatile constituents of the roots of *Echinops kebericho* Mesfin. *Flavour and Fragrance Journal, 22*(1), 35–38.

Imarhiagbe, O., Egboduku, W. O., & Nwankwo, B. J. (2020). A review of the biodiversity conservation status of Nigeria. *Journal Of Wildlife And Biodiversity, 4*(1), 73–83.

Jacobson, S. K., McDuff, M. D., & Monroe, M. C. (2015). *Conservation education and outreach techniques*. Oxford University Press.

Jeltsch, F., Moloney, K. A., Schwager, M., Körner, K., & Blaum, N. (2011). Consequences of correlations between habitat modifications and negative impact of climate change for regional species survival. *Agriculture, Ecosystems and Environment, 145*(1), 49–58.

Jenkins, P. T. (2005). International law related to precautionary approaches to national regulation of plant imports. *The Journal of World Trade, 39*, 895.

Kadu, C. A., Parich, A., Schueler, S., Konrad, H., Muluvi, G. M., Eyog-Matig, O., et al. (2012). Bioactive constituents in Prunus africana: Geographical variation throughout Africa and associations with environmental and genetic parameters. *Phytochemistry, 83*, 70–78.

Kairu, A., Gichuki, N., Kanya, J., & Kindt, R. (2013). Disappearing medicinal plants in mt. kenya forests, kenya: a case study of east african green heart (Warburgia ugandensis sprague). *Topclass Journal of Herbal Medicine, 2*(7), 159–165.

Kamboj, A., Sihag, B., Brar, D. S., Kaur, A., & Salunke, D. B. (2021). Structure activity relationship in β-carboline derived anti-malarial agents. *European Journal of Medicinal Chemistry, 221*, 113536.

Kasongo Yakusu, E., Louppe, D., Monthe, F. K., Hardy, O., Bola Mbele Lokanda, F., Hubau, W., et al. (2021). Improved management of species of the African Entandrophragma genus, now listed as vulnerable. *Bois Et Forets Des Tropiques, 349*(3), 65–84.

Keller, L. F., & Waller, D. M. (2002). Inbreeding effects in wild populations. *Trends in Ecology & Evolution, 17*(5), 230–241.

Kim, B., Jeon, H. J., Rhee, M. H., Kim, J. H., & Han, J. E. (2024). The effects of Panax ginseng on growth enhancement, innate immunity, and microbiome profiling in Penaeus vannamei. *Journal of Ginseng Research, 48*, 552.

Kloos, H. (2024). Traditional medicinal plants and plant products sold in open markets in Africa for COVID-19, their uses, medicinal properties, trade and sustainability: A review. *Journal of Medicinal Plants Studies, 12*, 56–62.

Koel, M., Kuhtinskaja, M., & Vaher, M. (2020). Extraction of bioactive compounds from Catharanthus roseus and Vinca minor. *Separation and Purification Technology, 252*, 117438.

Komakech, R., & Kang, Y. (2019). Ethnopharmacological potential of African cherry [Prunus africana]. *Journal of Herbal Medicine, 17*, 100283.

Kougioumoutzis, K., Kokkoris, I. P., Panitsa, M., Kallimanis, A., Strid, A., & Dimopoulos, P. (2021). Plant endemism centres and biodiversity hotspots in Greece. *Biology, 10*(2), 72.

Kramer, A. T., & Havens, K. (2009). Plant conservation genetics in a changing world. *Trends in Plant Science, 14*(11), 599–607.

Kress, W. J., Garcia-Robledo, C., Soares, J. V., Jacobs, D., Wilson, K., Lopez, I. C., et al. (2018). Citizen science and climate change: Mapping the range expansions of native and exotic plants with the mobile app Leafsnap. *Bioscience, 68*(5), 348–358.

Kudamba, A., Kasolo, J. N., Bbosa, G. S., Lugaajju, A., Wabinga, H., Kafeero, H. M., et al. (2024). Review of herbal medicinal plants used in the management of cancers in the East Africa region from 2019 to 2023. *Integrative Cancer Therapies, 23*, 15347354241235583.

Kumar, S., & Singh, B. (2022). An overview on bioactives and pharmacology of Rauvolfia vomitoria Afzel. *Bioactives and Pharmacology of Medicinal Plants*, 303–312.

Lacy, R. C. (1997 May). Importance of genetic variation to the viability of mammalian populations. *Journal of Mammalogy, 78*(2), 320–335.

Leimu, R., Vergeer, P., Angeloni, F., & Ouborg, N. J. (2010). Habitat fragmentation, climate change, and inbreeding in plants. *Annals of the New York Academy of Sciences, 1195*(1), 84–98.

Little, A. (2013). Sampling and analyzing wetland vegetation. In J. T. Anderson & C. A. Davis (Eds.), *Wetland techniques: Volume 1: Foundations* (pp. 273–324). Springer Netherlands. https://doi.org/10.1007/978-94-007-6860-4_5

Loomis, J. B., & White, D. S. (1996). Economic benefits of rare and endangered species: Summary and meta-analysis. *Ecological Economics, 18*(3), 197–206.

Mace, G. M., Collar, N. J., Gaston, K. J., Hilton-Taylor, C., Akçakaya, H. R., Leader-Williams, N., et al. (2008). Quantification of extinction risk: IUCN'S system for classifying threatened species. *Conservation Biology, 22*(6), 1424–1442.

Maher, T. M. (1997). How and why journalists avoid the population-environment connection. *Population and Environment, 18*(4), 339–372.

Maliehe, T. S., Ngidi, L. S., Shandu, J. S. E., Pooe, O. J., Selepe, T. N. (2022). Antibacterial activity and chemical profile of the bioactive compounds from *Aloe polyphylla* Schönland.

Mandal, S. K., Maji, A. K., Mishra, S. K., Ishfaq, P. M., Devkota, H. P., Silva, A. S., et al. (2020). Goldenseal (Hydrastis canadensis L.) and its active constituents: A critical review of their efficacy and toxicological issues. *Pharmacological Research, 160*, 105085.

Marlina, R., Oktavianty, E., Silitonga, H. T. M., Hamdani, H., & Afandi, A. (2022). Remote school students' perspectives on endemic and endangered animals: Knowledge, reasons, and solutions approach. *The Jurnal Pendidik Indonesia, 11*(3), 407–415.

Maroyi, A. (2019). Curtisia dentata: A review of its botany, medicinal uses, phytochemistry and biological activities. *Journal of Pharmaceutical Sciences and Research, 11*(10), 3404–3410.

Mbatudde, M., Majaliwa, G., Eilu, G., Kakudidi, E., & Dalitz, H. (2013). Potential distribution of vulnerable Entandrophragma angolense (Welw.) C. DC.(Meliaceae) in East Africa. *African Journal of Ecology, 51*(3), 471–481.

McAllister, S. T. (1999). Community-based conservation: Restructuring institutions to involve local communities in a meaningful way. *Colorado Journal of International Environmental Law and Policy, 10*, 195.

Menezes-Silva, P. E., Loram-Lourenço, L., Alves, R. D. F. B., Sousa, L. F., da Silva Almeida, S. E., & Farnese, F. S. (2019). Different ways to die in a changing world: Consequences of climate change for tree species performance and survival through an ecophysiological perspective. *Ecology and Evolution, 9*(20), 11979–11999.

Merritt, D. M., Manning, M. E., & Hough-Snee, N. (2017). *The national riparian core protocol: A riparian vegetation monitoring protocol for wadeable streams of the conterminous United States*. Gen Tech Rep RMRS-GTR-367 (p. 367). US Department of Agriculture, Forest Service, Rocky Mountain Research Station.

Mir, T. A., Jan, M., Khare, R. K., & Bhat, M. H. (2021). Medicinal plant resources: Threat to its biodiversity and conservation strategies. *Medicinal and Aromatic Plants: Healthcare and Industrial Applications*, 717–739.

Montagu, M. V. (2019). The future of plant biotechnology in a globalized and environmentally endangered world. *Genetics and Molecular Biology, 43*(1 suppl 2), e20190040.

Mounce, R., Smith, P., & Brockington, S. (2017). Ex situ conservation of plant diversity in the world's botanic gardens. *Nature Plants, 3*(10), 795–802.

Mouthé Happi, G., Tchaleu Ngadjui, B., Green, I. R., & Fogué, K. S. (2018). Phytochemistry and pharmacology of the genus Entandrophragma over the 50 years from 1967 to 2018: A 'golden'overview. *The Journal of Pharmacy and Pharmacology, 70*(11), 1431–1460.

Moyo, M., Aremu, A. O., & Van Staden, J. (2015). Medicinal plants: An invaluable, dwindling resource in sub-Saharan Africa. *Journal of Ethnopharmacology, 174*, 595–606.

Muasya, A. M., Slingsby, J. A., & Verboom, G. A. (2024). Preface to the special issue on African flora in a changing world: Integrating multiple dimensions of diversity. *Annals of Botany, 133*(5–6), 643–648.

Mukhongo, J. N., Kinyamario, J., Chira, R., & Musila, W. (2011). Assessment of soil seed bank from six different vegetation types in Kakamega forest, Western Kenya. *African Journal of Biotechnology, 10*(65), 14384–14391.

Mushangwe, C. T.. (2019). Human-plant interactions in semi-arid regions: An Archaeobotanical study of the iron age site of Mtanye, Southwestern Zimbabwe.

Nantchouang Ouete, J. L., Sandjo, L. P., Kapche, D. W., Liermann, J. C., Opatz, T., Simo, I. K., et al. (2013). A new flavone from the roots of Milicia excelsa (Moraceae). *Z Für Naturforschung C., 68*(7–8), 259–263.

Naughton-Treves, L., Holland, M. B., & Brandon, K. (2005). The role of protected areas in conserving biodiversity and sustaining local livelihoods. *Annual Review of Environment and Resources, 30*(1), 219–252.

Newmark, W. D. (2002). *Conserving biodiversity in east African forests: A study of the eastern Arc Mountains* (Vol. 155). Springer Science & Business Media.

Nganso, Y. O. D., Tchana, E. M. S., Kahouo, A. D., Abah, K., Fomena, H., & Mamoudou, H. (2020). Inhibitory effect and antimicrobial activity of secondary metabolites of Khaya senegalensis (Desr.) A. Juss.(Meliaceae). *The Journal of Chemical Sciences, 10*(1), 81–94.

Nsibande, B. E., Gustavsson, K. E., & Zhu, L. H. (2018). Analysis of health-associated phytochemical compounds in seven Hypoxis species. *American Journal of Plant Sciences, 9*(04), 571.

Nsolomo, V. R., Venn, K., & Solheim, H. (2000). The ability of some fungi to cause decay in the east African camphor tree, Ocotea usambarensis. *Mycological Research, 104*(12), 1473–1479.

Nyirenda, C. P. (2008). The conservation biology of Berberis holstii Engl. In *Nyika National Park*.

Oke-Altuntas, F., Kapche, G. D. F., Nantchouang Ouete, J. L., Demirtas, I., Koc, M. B., & Ngadjui, B. T. (2016). Bioactivity evaluation of cudraxanthone I, neocyclomorusin and (9βh)-3-β-acetoxylanosta-7, 24-diene isolated from Milicia excelsa Welw. CC Berg (Moraceae). *Medicinal Chemistry Research, 25*, 2250–2257.

Okello, D., Komakech, R., Matsabisa, M. G., & Kang, Y. (2018). A review on the botanical aspects, phytochemical contents and pharmacological activities of Warburgia ugandensis. *The Journal of Medicinal Plants Research, 12*(27), 448–455.

Oladeji, O. S., Adelowo, F. E., Oluyori, A. P., & Bankole, D. T. (2020). Ethnobotanical description and biological activities of Senna alata. *Evidence-Based Complementary and Alternative Medicine, 2020*(1), 2580259.

Olagunju, T. E. (2015). Impacts of human-induced deforestation, forest degradation and fragmentation on food security. *New York Science Journal, 8*(1), 10.

Olah, G., Butchart, S. H., Symes, A., Guzmán, I. M., Cunningham, R., Brightsmith, D. J., et al. (2016). Ecological and socio-economic factors affecting extinction risk in parrots. *Biodiversity and Conservation, 25*, 205–223.

Onohuean, H., Onohuean, F. E., Igbinoba, S. I., Ezeonwumelu, J. O. C., Agu, P. C., Ifie, J. E., et al. (2022). Elucidation of chemical profiles and molecular targets of Mondia whitei leave fractions bioactive as novel therapeutics: An in vitro and in silico assay. *Journal, Genetic Engineering & Biotechnology, 20*(1), 170.

Opiyo, S. A. (2023). Warburgia Ugandensis: A review of compounds and bioactivity. *International Journal of Pharmacognosy and Chemistry, 4*(2), 35.

Padulosi, S., Leaman, D., & Quek, P. (2002). Challenges and opportunities in enhancing the conservation and use of medicinal and aromatic plants. *Journal of Herbs Spices & Medicinal Plants, 9*(4), 243–267.

Parmesan, C., & Hanley, M. E. (2015). Plants and climate change: Complexities and surprises. *Annals of Botany, 116*(6), 849–864.

Paul, M., Brüning, G., Bergmann, J., & Jauch, J. (2012). A thin-layer chromatography method for the identification of three different olibanum resins (Boswellia serrata, Boswellia papyrifera and Boswellia carterii, respectively, *Boswellia sacra*). *Phytochemical Analysis, 23*(2), 184–189.

Pauls, S. U., Nowak, C., Bálint, M., & Pfenninger, M. (2013). The impact of global climate change on genetic diversity within populations and species. *Molecular Ecology, 22*(4), 925–946.

Pereira, A. P., Ferreira, I. C., Marcelino, F., Valentão, P., Andrade, P. B., Seabra, R., et al. (2007). Phenolic compounds and antimicrobial activity of olive (*Olea europaea* L. Cv. Cobrançosa) leaves. *Molecules., 12*(5), 1153–1162.
Pessar, D. (2021). Logistical and tax-related obstacles to coordinating conservation land assemblages. *Environmental Law, 51*(2), 437–484.
Potts, S. G., Imperatriz-Fonseca, V., Ngo, H. T., Biesmeijer, J. C., Breeze, T. D., Dicks, L. V., et al. (2016). The assessment report on pollinators, pollination and food production: Summary for policymakers. Secretariat of the Intergovernmental Science-Policy Platform on Biodiversity and Ecosystem Services
Preece, J. (2017). How two billion smartphone users can save species! *Interactions, 24*(2), 26–33.
Qian, H., Zhou, Y., Zhang, J., Jin, Y., Deng, T., & Cheng, S. (2021). A synthesis of botanical informatics for vascular plants in Africa. *Ecological Informatics, 64*, 101382.
Rahaman, M. M., Setzer, W. N., Sharifi-Rad, J., & Islam, M. T. (2022). Hydrastis canadensis (goldenseal) and Lawsonia inermis (henna). In *Herbs, shrubs, and trees of potential medicinal benefits* (pp. 151–162). CRC Press.
Ramarumo, L., Maroyi, A., & Tshisikhawe, M. (2019). Bowiea volubilis Harv. ex Hook. f. subsp. volubilis: A therapeutic plant species used by the traditional healers in the Soutpansberg region, Vhembe biosphere reserve, Limpopo Province, South Africa. *Journal of Pharmaceutical Sciences and Research, 11*(7), 2538–2542.
Rawat, U., & Agarwal, N. K. (2015). Biodiversity: Concept, threats and conservation. *Environment Conservation Journal, 16*(3), 19–28.
Rodriguez, L. F. (2006). Can invasive species facilitate native species? Evidence of how, when, and why these impacts occur. *Biological Invasions, 8*, 927–939.
Saleem, G., Akhtar, M., Ahmed, N., & Qureshi, W. S. (2019). Automated analysis of visual leaf shape features for plant classification. *Computers and Electronics in Agriculture, 157*, 270–280. https://doi.org/10.1016/j.compag.2018.12.038
Salem, B. (2003). Application of GIS to biodiversity monitoring. *Journal of Arid Environments, 54*(1), 91–114.
Sanjeevi, P., Prasanna, S., Siva Kumar, B., Gunasekaran, G., Alagiri, I., & Vijay, A. R. (2020). Precision agriculture and farming using internet of things based on wireless sensor network. *Transactions on Emerging Telecommunications Technologies, 31*(12), e3978.
Sarasan, V., Kite, G. C., Sileshi, G. W., & Stevenson, P. C. (2011). Applications of phytochemical and in vitro techniques for reducing over-harvesting of medicinal and pesticidal plants and generating income for the rural poor. *Plant Cell Reports, 30*, 1163–1172.
Sattar, Z., & Iranshahi, M. (2017). Phytochemistry and pharmacology of Ferula hermonis Boiss.–a review. *Drug Research, 67*(08), 437–446.
Schwartz, M. K., Luikart, G., & Waples, R. S. (2007). Genetic monitoring as a promising tool for conservation and management. *Trends in Ecology & Evolution, 22*(1), 25–33.
Seanego, C. T., & Ndip, R. N. (2012). Identification and antibacterial evaluation of bioactive compounds from Garcinia kola (Heckel) seeds. *Molecules, 17*(6), 6569–6584.
Singh, N. P., Nagahma, T., Yadav, P., & Yadav, D. (2018). Feature based leaf identification. Paper presented at the 2018 5th IEEE Uttar Pradesh section international conference on electrical, electronics and computer engineering (UPCON). https://doi.org/10.1109/UPCON.2018.8596921.
Smith, P. (2016). Building a global system for the conservation of all plant diversity: A vision for botanic gardens and botanic gardens Conservation International. *Sibbaldia: The International Journal of Botanic Garden Horticulture, 14*, 5–13.
Stohlgren, T. J., & Kumar, S. (2013). In S. A. L. Waltham (Ed.), *Endangered plants. Encyclopedia of biodiversity* (2nd ed., pp. 205–215). Academic.
Tadayon, N., Ramazani, A., Rezazadeh, S., Nohooji, M. G., & Angorani, H. R. (2024). Isolation and investigation of antibacterial activities and cytotoxicity of atropine and scopolamine. *Results in Chemistry, 10*, 101732.
Tang, L., Ke, X., Chen, Y., Wang, L., Zhou, Q., Zheng, W., & Xiao, B. (2021). Which impacts more seriously on natural habitat loss and degradation? Cropland expansion or urban expansion? *Land Degradation & Development, 32*(2), 946–964.

Tcheghebe, O. T., Tatong, F. N., & Seukep, A. J. (2016). Traditional uses, phytochemical and pharmacological profiles, and toxicity of *Enantia chlorantha* (Oliver): An overview. *Edorium Journal of Medicine, 3*, 12–18.

Thomsen, H., Reider, K., Franke, K., La, W., Keiser, J., Dagne, E., et al. (2012). Characterization of constituents and anthelmintic properties of *Hagenia abyssinica*. *Scientia Pharmaceutica, 80*(2), 433–446.

Tilman, D., Clark, M., Williams, D. R., Kimmel, K., Polasky, S., & Packer, C. (2017). Future threats to biodiversity and pathways to their prevention. *Nature, 546*(7656), 73–81.

Twerski, A., Fischer, C., & Albrecht, H. (2021). Effects of rare arable plants on plant diversity, productivity and soil fertility in agricultural fields. *Agriculture, Ecosystems and Environment, 307*, 107237.

Udegbunam, S., Nnaji, T., Udegbunam, R., Okafor, J., & Agbo, I. (2013). Evaluation of herbal ointment formulation of Milicia excelsa (Welw) CC berg for wound healing. *African Journal of Biotechnology, 12*(21), 3351–3359.

Vallejo, J. R., Baptista, G. C. S., Arco, H., González, J. A., Santos-Fita, D., & Postigo-Mota, S. (2024). Traditional knowledge and biocultural heritage about medicinal plants in a European transboundary area (La Raya: Extremadura, Spain—Alentejo, Portugal): Transdisciplinary research for curriculum Design in Health Sciences. *Heritage, 7*(1), 225–258.

Van Rossum, F., & Hardy, O. J. (2022). Guidelines for genetic monitoring of translocated plant populations. *Conservation Biology, 36*(1), e13670.

van Slageren, W. (2003). The millennium seed Bank: Building partnerships in arid regions for the conservation of wild species. *Journal of Arid Environments, 54*(1), 195–201.

Vila, M., & Weiner, J. (2004). Are invasive plant species better competitors than native plant species?–evidence from pair-wise experiments. *Oikos, 105*(2), 229–238.

Wagner, D. L., & Van Driesche, R. G. (2010). Threats posed to rare or endangered insects by invasions of nonnative species. *Annual Review of Entomology, 55*(1), 547–568.

Watcho, P., Zelefack, F., Ngouela, S., Nguelefack, T. B., Kamtchouing, P., Tsamo, E., et al. (2012). Enhancement of erectile function of sexually naïve rats by β–sitosterol and α–β–amyrin acetate isolated from the hexane extract of Mondia whitei. *Asian Pacific Journal of Tropical Biomedicine, 2*(3), S1266–S1269.

Waylen, K. A., Fischer, A., McGowan, P. J., Thirgood, S. J., & Milner-Gulland, E. J. (2010). Effect of local cultural context on the success of community-based conservation interventions. *Conservation Biology, 24*(4), 1119–1129.

Whitman, K. L. (2017). Improving conservation outcomes in a biodiversity hotspot: Alternative agriculture techniques in Maromizaha Forest.

Willcox, M., Bodeker, G., Bourdy, G., Dhingra, V., Falquet, J., Ferreira, J. F., et al. (2004). Artemisia annua as a traditional herbal antimalarial. *Traditional Medicinal Plants and Malaria, 4*, 43–59.

Wube, A. A., Bucar, F., Gibbons, S., & Asres, K. (2005). Sesquiterpenes from Warburgia ugandensis and their antimycobacterial activity. *Phytochemistry, 66*(19), 2309–2315.

Yang, C. (2021). Plant leaf recognition by integrating shape and texture features. *Pattern Recognition, 112*, 107809. https://doi.org/10.1016/j.patcog.2020.107809

Yan-Hua, Y., Jia-Wang, M., & Xiao-Li, T. (2020). Research progress on the source, production, and anti-cancer mechanisms of paclitaxel. *Chinese Journal of Natural Medicines, 18*(12), 890–897.

Yeboah, E. M., Majinda, R. R., Kadziola, A., & Muller, A. (2010). Dihydro-β-agarofuran sesquiterpenes and pentacyclic triterpenoids from the root bark of *Osyris lanceolata*. *Journal of Natural Products, 73*(6), 1151–1155.

Yessoufou, K., Daru, B. H., & Davies, T. J. (2012). Phylogenetic patterns of extinction risk in the eastern arc ecosystems, an African biodiversity hotspot. *Plos One, 7*, e47082.

Yin, C. Y. (2010). Emerging usage of plant-based coagulants for water and wastewater treatment. *Process Biochemistry, 45*(9), 1437–1444.

Yuan, C. M., Tang, G. H., Wang, X. Y., Zhang, Y., Guo, F., Liao, J. H., et al. (2013). Two new compounds from Khaya senegalensis. *Journal of Asian Natural Products Research, 15*(6), 638–643.

Chapter 6
In Vitro and In Silico Studies of Bioactive Compounds from Selected Ethiopian Plants

Milkyas Endale, Gashaw Nigussie, Mathewos Anza, Aman Dekebo, Teshome Degfie, Yadessa Melaku, Asfaw Meresa, Temesgen Negassa, and Negera Abdissa

Abstract Antibiotic resistance is a pressing issue in modern medicine, prompting the exploration of alternative treatments from natural sources. This study investigated the antimicrobial, antioxidant, and cytotoxicity properties of *Clausena anisata*, *Uvaria scheffleri*, *Zanthoxylum chalybeum*, *Vepris dainellii*, *Brucea antidysenterica*, *Embelia schimperi*, *Cyphostemma cyphopetalum*, and *Rhynchosia ferruginea*. Sequential extraction of the roots with *n*-hexane, dichloromethane/methanol (1:1), and methanol followed by silica gel column chromatography was employed to isolate compounds. From *C. anisata*, alkaloid (**1**) and coumarins (**2**–**4**) were identified. *V. dainellii* yielded alkaloids (**5**–**9**), while *E. schimperi* provided flavans (**10**–**11**). *B. antidysenterica* contained indole alkaloids (**12**–**14**), *C. cyphopetalum* gave compounds (**15**–**22**), *R. ferruginea* yielded isoflavans (**23**–

M. Endale (✉) · G. Nigussie · A. Meresa · T. Negassa · N. Abdissa
Traditional and Modern Medicine Research and Development Directorate, Armauer Hansen Research Institute, Addis Ababa, Ethiopia
e-mail: milkyas.endale@ahri.gov.et

M. Anza
Department of Chemistry, College of Natural and Computational Science, Wolaita Sodo University, Soddo, Ethiopia

A. Dekebo
Department of Applied Chemistry, Adama Science and Technology University, Adama, Ethiopia

Institute of Pharmaceutical Sciences, Adama Science and Technology University, Adama, Ethiopia

T. Degfie
Institute of Pharmaceutical Sciences, Adama Science and Technology University, Adama, Ethiopia

Y. Melaku
Department of Applied Chemistry, Adama Science and Technology University, Adama, Ethiopia

A. Dekebo (ed.), *Bioactive Secondary Metabolites from Medicinal Plants of Africa*, Interdisciplinary Biotechnological Advances,
https://doi.org/10.1007/978-981-96-8590-5_6

24), *Z. chalybeum* produced benzo[c]phenanthridine alkaloids (**25**–**28**), and *U. scheffleri* provided an indole alkaloid (**29**) and a chalcone (**30**).

Biological assays showed that gnetin H (**19**) had significant inhibition against *Pseudomonas aeruginosa* (12.55 mm at 50 μg/mL), outperforming chloramphenicol. Evodiamine (**8**) exhibited high cytotoxicity against the MDA-MB-231 cell line (IC_{50} = 1.73 μM). Chelerythrine (**25**) demonstrated strong activity against MDA-MB-231 cancer cells (IC_{50} = 3.62 μM). *Brucea antidysenterica* showed up to 87.5% antioxidant inhibition in the DPPH assay. Molecular docking analysis revealed promising binding affinities for several alkaloids, suggesting their potential as natural remedies for infectious diseases.

Keywords Antibacterial activity · Cytotoxicity · Indole alkaloids · Flavanoids · Radical scavenging · Molecular docking

Abbreviations

DCM	Dichloromethane
DMSO	Dimethyl sulfoxide
DPPH	2,2-Diphenyl-1-picrylhydrazyl
FT-IR	Fourier transform infrared spectrometer
IC_{50}	Half maximal inhibitory concentration
MCF-7	Michigan cancer foundation-7
MHA	Mueller hinton agar
MDA-MB-231	M. D. Anderson-Metastatic Breast 231
NMR	Nuclear magnetic resonance spectrometer
PDA	Potato dextrose agar
PTLC	Preparative thin layer chromatography
TLC	Thin layer chromatography
UV-Vis	Ultraviolet visible

6.1 Introduction

6.1.1 Background

Plants are foundational to traditional medicine in Africa, where they have been utilized for millennia due to their diverse biological and pharmacological properties (Ajayi et al., 2011). With the growing demand for innovative drugs with improved pharmacological attributes, there has been a corresponding emphasis on utilizing plants to extract bioactive compounds (Haruna & Yahaya, 2021). In Ethiopia, medicinal plants significantly contribute to human and livestock healthcare, especially in regions where modern public health services are limited or inaccessible.

Although medicinal plants play an essential function in primary healthcare, efforts to document and promote the associated indigenous knowledge in Ethiopia have been inadequate. Both the plants and the traditional knowledge are at risk due to various anthropogenic factors, underscoring the urgent need for effective conservation and sustainable management practices. Our group at Adama Science and Technology University (ASTU) is addressing this issue by studying the medicinal potential of various plants from Ethiopian flora. This review presents an overview of our ongoing research, focusing on the phytochemical analysis and biological activity of selected Ethiopian medicinal plants.

Brucea antidysenterica (Simaroubaceae) has been traditionally used to treat numerous diseases. Studies reveal that this plant contains a variety of bioactive substances, including quassinoids, alkaloids, triterpenes, steroids, coumarins, anthraquinones, and flavonoids (Barbosa et al., 2011). *Embelia schimperi*, a species from the Myrsinaceae family, is indigenous to Kenya (Beerntje, 1994) and Ethiopia, particularly in the Oromia, Gamo Gofa, and Amhara regions. The Maasai people of Kenya utilize its fruits, seeds, and roots to combat bacterial infections and intestinal parasites, notably tapeworms, as well as to treat diarrhea, fevers, and various skin and respiratory conditions (Cheikhyoussef et al., 2011; Kokwaro, 1993).

Vepris dainellii Pichi-Serm., which is native to Ethiopia (Dagne et al., 1988), has a traditional use for relieving abdominal cramps, treating intestinal worms, skin conditions, and toothaches (Tesfaye et al., 2009; Kebebew et al., 2013). The genus is known for its high content of alkaloids, limonoids, and flavonoids (Fouda et al., 2017). *Uvaria scheffleri* Diels (Annonaceae), known by the local name Boyyiniya (Wolaitigna), Ethiopia, and located widely across East Africa. The plant is traditionally employed in root decoctions for managing cough, tuberculosis, asthma, and sore throats (Wanzala et al., 2016). Additionally, its fruit is consumed in Ethiopia for treating skin infections (Kuma et al., 2015). The genus *Uvaria* is noted for producing uvarindole alkaloids and chalcones, which have diverse biological effects (Anza et al., 2021a, b, c). *Clausena anisata* (Willd) Hook. f. ex Benth, known in Ethiopia as "olmaa'ii" (Afan Oromo language), belongs to the Rutaceae family and is employed throughout Africa for treating bacterial and fungal skin infections (Hamza et al., 2006), as well as for malaria (Uwaifo, 1984). The genus *Zanthoxylum*, with around 549 species mainly found in tropical and temperate regions (Patiño et al., 2010), includes *Z. liebmanniaum* (Engelm.), used in traditional Mexican medicine for ailments such as stomach pains, amebiasis, intestinal parasites, and as a local anesthetic (Ross et al., 2004). The fruit of *Z. oxyphyllum* serves as an appetizer and is used for treating tumors, digestive issues, headaches, body pains, fever, and cough, while the stem bark is utilized for conditions like rheumatism, varicose ulcers, varicose veins, skin diseases, leg pain, inflammation, and low blood pressure (Munda & Bibhuti, 2017).

The genus *Embelia* includes approximately 130 species, mostly shrubby woody climbers native to tropical and subtropical regions such as Africa, Eastern Asia, the Pacific Islands, and Australia (Debebe et al., 2015). It is extensively utilized in traditional medicine for its anthelmintic and antimicrobial properties, proving particularly effective against bacteria and intestinal parasites (Chitra et al., 2003).

Cyphostemma cyphopetalum (Fresen.) is a medicinal plant found throughout tropical and subtropical Africa, including countries like Ethiopia, Cameroon, Sudan, Uganda, Kenya, Tanzania, Rwanda, Congo, and Zambia (Sousa et al., 2011). In Ethiopia, it is commonly referred to as "Milas Golgul" and "Gindosh" (Amharic) and is prevalent in regions such as Harerge, Afar, Ilubabor, Bale, Arsi, Welo, Shewa, Tigray, Gonder, Gojam, Welega, Sidamo, and Gamogofa (Teklehaymanot et al., 2007). Traditional Ethiopian uses of this plant include treating tumors, snake bites, abscesses, and eczema (atopic dermatitis) (Chekole, 2017; Teklehaymanot et al., 2007).

The genus *Rhynchosia* comprises approximately 300 species distributed across tropical and subtropical areas around the globe (Rungsung et al., 2015). Compounds derived from this genus and their extracts have exhibited strong antioxidant, anti-inflammatory, antifungal, antibacterial, and antiproliferative effects (Muhammad et al., 2019). *Rhynchosia ferruginea* A. Rich, known as Udusalim in Afan Oromo (Ethiopia), is traditionally utilized for treating a variety of ailments, including skin infections, wounds, stomach pains, and amoebiasis in the Barbare district of the Bale Zone in Oromia, Ethiopia (Jima and Megersa, 2018).

Over the past 4 years, our team has isolated and characterized 30 compounds from roots of plant extracts stated above, evaluating their antibacterial and antioxidant properties, and performing in silico molecular docking studies. This review presents a comprehensive summary of these studies, emphasizing the main results and impacts of our study.

6.2 Materials and Methods

6.2.1 Chemicals, Reagents, Apparatus, and Instruments

The following chemicals and reagents were utilized for the experiment: chloroform (Sigma-Aldrich, analytical grade), ethanol (Merck, analytical grade), n-hexane (Fisher Scientific, analytical grade), ethyl acetate (Acros Organics, analytical grade), dichloromethane (Sigma-Aldrich, analytical grade), acetic acid (Merck, analytical grade), DMSO (Dimethyl sulfoxide, Sigma-Aldrich, HPLC grade), methanol (Fisher Scientific, HPLC grade), anhydrous sodium sulfate (Sigma-Aldrich, analytical grade), sulfuric acid (Merck, analytical grade), silica gel (60–120 mesh, Sigma-Aldrich, analytical grade), hydrochloric acid (Fisher Scientific, analytical grade), lead acetate (Sigma-Aldrich, analytical grade), ferric chloride (Merck, analytical grade), acetic anhydride (Acros Organics, analytical grade), L-ascorbic acid (Sigma-Aldrich, analytical grade), DPPH (2,2-Diphenyl-1-picrylhydrazyl, Sigma-Aldrich, analytical grade), PDA (Potato Dextrose Agar, Merck, analytical grade), and MHA (Mueller Hinton Agar, Sigma-Aldrich, analytical grade).

The apparatus and instruments used in the experiment included a column (Pyrum, 50 × 2.5 cm), separatory funnel (Pyrex, 500 mL), oven (Thermo Fisher, model: Heratherm, 120 °C), TLC plates (Merck, Silica Gel 60 F254), Whatman filter paper

No. 1 (Whatman, 11 μm pore size), mortar and pestle (Fisher Scientific, ceramic), pipettes (Eppendorf, model: Research Plus, 1–1000 μL), water bath (Julabo, model: F12, 100 °C), UV lamp (UVP, model: UVGL-58), beakers (Pyrex, 250 mL, 500 mL), UV light cabinet (UVP, model: UVGL-55), electronic balance (Sartorius, model: CP224S, 220 g × 0.1 mg), conical flasks (Pyrex, 250 mL, 500 mL), measuring cylinder (Pyrex, 100 mL, 250 mL), rotary evaporator (Büchi, model: R-210), chromatographic chamber (Fisher Scientific, glass, 20 × 20 cm), PTLC apparatus (CAMAG, model: Linomat 5), Perkin-Elmer BX FT-IR spectrometer (400–4000 cm^{-1}, with KBr pellets), and a Bruker Avance NMR Spectrometer (400 MHz, model: Avance III).

6.2.2 Collection, Identification, and Preparation of Plant Materials

The roots of *E. schimperi* were gathered from Asela, Oromia Regional State, Ethiopia, in December 2018. The roots of *B. antidysenterica* were collected from Bahir Dar, Amhara Regional State, Ethiopia, in December 2019. *C. cyphopetalum* roots were obtained from Huruta town, Lode Hetosa Woreda, Arsi Zone, Oromia, Ethiopia, in September 2019. *R. ferruginea* roots were collected from the Berbere district in the Bale Zone, approximately 526 km southeast of Addis Ababa, located between 06°33′N and 06°75′N and 039°40′E and 040°29′E. The roots and fruits of *V. dainellii* were collected from Soula town in Gofa Zone, South Ethiopia Regional State, Ethiopia, in November 2018. *Z. chalybeum* roots were gathered from Yirgalem, Sidama Regional State, Ethiopia, in November 2018. *U. scheffleri* roots were obtained from Wolaita Zone, SNNPR, Ethiopia, in November 2018. The roots of *C. anisata* were harvested from the west Wollega Zone in the Oromia region, 541 km west of Addis Ababa, in March 2018. The botanical specimens of these plants were identified by botanist Dr. Shambel Alemu at the National Herbarium of Ethiopia, Addis Ababa University, where voucher specimens were deposited. Following collection, all samples were rinsed with distilled water, air-dried in the shade, and then pulverized using an electric grinder.

6.2.3 Extraction and Isolation

6.2.3.1 Extraction

The grounded root bark (300 g) of the respective medicinal plants was extensively extracted with *n*-hexane at room temperature for 72 h. The remaining residue was then extracted with 1.5 liters of a DCM/MeOH (1:1) and methanol, subsequently, for 72 h in each case at room temperature. The filtrates were concentrated under reduced pressure at 40 °C using a rotary evaporator.

For alkaloidal extraction, the ground roots (400 g) were macerated in 97% methanol (2 L) at room temperature for 24 h, with continuous shaking at 230 rpm using an electronic shaker. The solution obtained was filtered and concentrated with a vacuum rotary evaporator to obtain a crude extract. The methanol extract residue was dissolved in 100 mL of chloroform and then extracted with 5% aqueous H_2SO_4, adjusting the pH to 3–4 (3 × 50 mL) in a separatory funnel to remove lipids and other metabolites. The aqueous phases were combined, made alkaline with 5% NH_4OH to pH 11, and extracted with chloroform (3 × 50 mL). The chloroform layers were pooled, dried over anhydrous Na_2SO_4, filtered, and concentrated using a vacuum rotary evaporator to produce the alkaloidal extract.

6.2.3.2 Isolation

The crude DCM:MeOH (1:1) extract (8.5 g) of *C. anisata* was subjected to column chromatography with 150 g of silica gel as the stationary phase and eluted with increasing gradient of ethyl acetate in *n*-hexane, resulting in 65 fractions. Fractions with similar R_f values and characteristic colors on TLC were combined. Fractions 11–12, which showed a yellow spot under UV light with an R_f value of 0.56 in an *n*-hexane/ethyl acetate (8:2) solvent system, were concentrated and washed with *n*-hexane to yield compound **1** (Fig. 6.1a). Fractions 19–25, exhibiting a single spot with similar R_f values were combined, concentrated, and washed with *n*-hexane to obtain compound **2** (Fig. 6.1a). Fractions 29–34 (16 mg), based on their TLC profile showing a single red spot with *n*-hexane/ethyl acetate (6:4) as the mobile phase, were combined, concentrated, and washed repeatedly with *n*-hexane to afford imperatorin

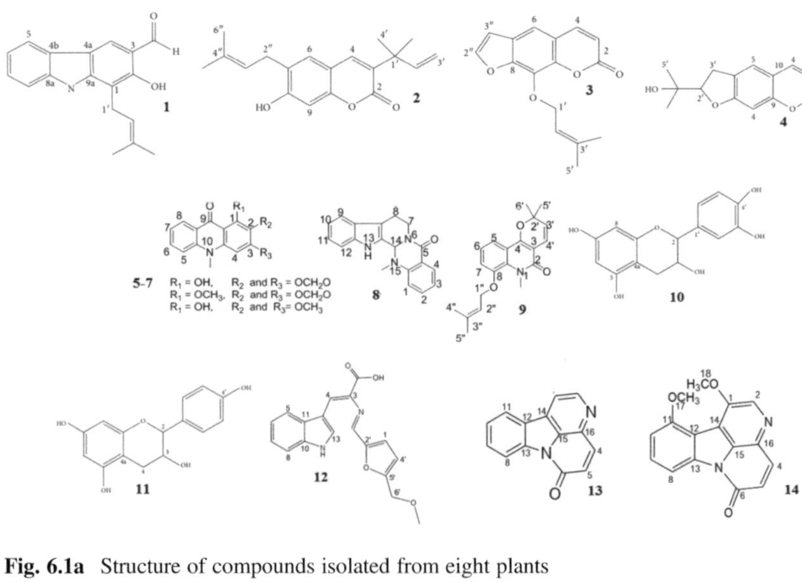

Fig. 6.1a Structure of compounds isolated from eight plants

(**3**, Fig. 6.1a). Fractions 46–50 (27 mg), identified by TLC as one spot using *n*-hexane/ethyl acetate (1:1) as the eluent, were combined and concentrated to obtain chalepin (**4**, Fig. 6.1a).

For the alkaloidal extract (5 g) of *V. dainellii* roots, silica gel column chromatography was performed using an increasing gradient of methanol in dichloromethane. Twenty-one fractions (50 mL each) were collected. Fraction 2, eluted with 100% dichloromethane, yielded compound **5** (6 mg, Fig. 6.1a) as colorless needles. Fraction 4, eluted with 2% methanol in dichloromethane, produced compound **6** (88 mg, Fig. 6.1a) as a yellow powder. Fractions 14–19, eluted with 5–8% methanol in dichloromethane, were combined (147 mg) and further purified by silica gel column chromatography, resulting in compound **7** as a yellow powder (46 mg, Fig. 6.1a).

The methanol extract of *V. dainellii* roots (11 g) underwent silica gel column chromatography, starting with an increasing gradient of ethyl acetate in *n*-hexane and followed by a gradient of methanol in dichloromethane. Fifty-three fractions (25 mL each) were collected. Fractions 20–24, eluted with 20% ethyl acetate in *n*-hexane, were pooled and subjected to further purification on a silica gel column with an increasing gradient of ethyl acetate in *n*-hexane. From this second column, 32 fractions (25 mL each) were collected. Fractions 10–16, eluted with 15% ethyl acetate in *n*-hexane, were combined and washed with *n*-hexane to isolate compound **9** (89 mg, Fig. 6.1a). Fractions 25–28, eluted with 30% ethyl acetate in *n*-hexane, were merged, and concentrated to obtain compound **8** (60 mg, Fig. 6.1a).

The crude methanol extract of *E. schimperi* roots (6.0 g) was adsorbed onto 6 g of silica gel and loaded onto a silica gel column (150 g as the stationary phase). The column was eluted with a gradient of ethyl acetate in *n*-hexane, followed by an increasing gradient of methanol in dichloromethane. Eighty-eight fractions (100 mL each) were collected. Fractions 54–58, eluted with 60% ethyl acetate in *n*-hexane, exhibited a single spot. After drying, these fractions were washed with *n*-hexane to isolate compound **10** (26 mg, Fig. 6.1a). Fractions 72–80, eluted with 15% methanol in dichloromethane, displayed one prominent spot. These fractions were combined and subjected to further purification on a silica gel column using a 60:40 ethyl acetate in *n*-hexane isocratic mode, resulting in compound **11** (20 mg, Fig. 6.1a).

The DCM/MeOH (1:1) root extract of *B. antidysenterica* (10 g) was subjected to column chromatography with 150 g of silica gel as the stationary phase and eluted with gradient of ethyl acetate in *n*-hexane, followed by a progressively increasing gradient of methanol in dichloromethane. Two hundred and twenty-one fractions were collected and concentrated to dryness. Fractions with similar R_f values and colors on TLC were combined. Fractions 29–31 yielded compound **12** (30 mg, Fig. 6.1a), which exhibited a single spot on TLC (ethyl acetate/n-hexane, 1:1, R_f 0.44). Fractions 51–54, showing a single spot on TLC (ethyl acetate/*n*-hexane 9:1, R_f 0.57), produced canthin-6-one (**13**, Fig. 6.1a, 25 mg). Fractions 84–87 provided a single spot on TLC (ethyl acetate/*n*-hexane 8:2, R_f 0.41), resulting in 1,11-dimethoxycanthin-6-one (**14**, Fig. 6.1a, 30 mg).

The DCM/MeOH (1:1) root extract (10 g) of *C. cyphopetalum* was subjected to silica gel column chromatography (180 g of silica gel as the stationary phase) and

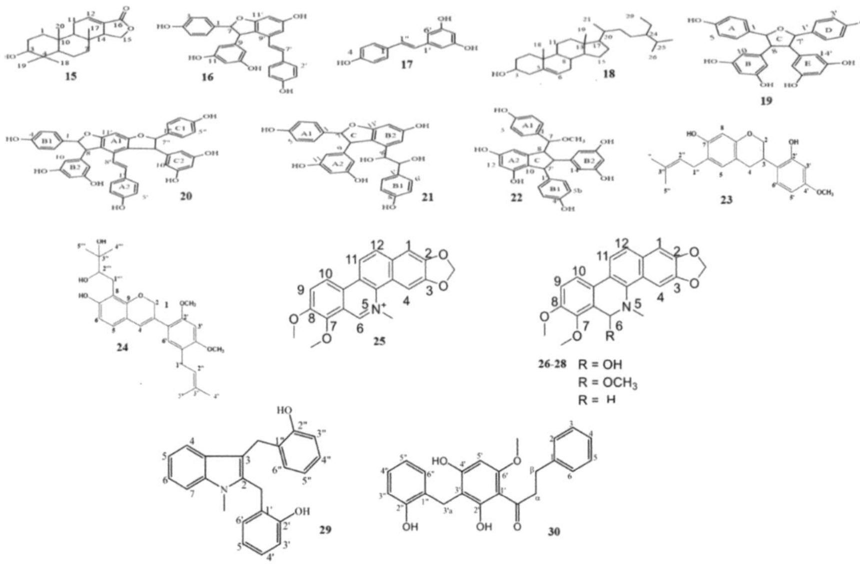

Fig. 6.1b Structure of compounds isolated from eight plants

eluted with polarity gradient of *n*-hexane/ethyl acetate, followed by mixtures of dichloromethane/methanol. One hundred forty-five fractions (50 mL each) were collected. Fractions 37–44, which were eluted with *n*-hexane/ethyl acetate (1:1), were further purified on a silica gel column (20 g) using isocratic elution with DCM/MeOH (9.5:0.5), yielding 3-hydroxyisoagatholactone (**15**, 53 mg, Fig. 6.1b). Compound **16** (45 mg, Fig. 6.1b) was isolated from fractions 17–36, eluted with *n*-hexane/ethyl acetate (3:1). Fractions 45–52, eluted with *n*-hexane/ethyl acetate (2:3), were further purified on a silica gel column (20 g) using 1–5% ethyl acetate in *n*-hexane as eluent to afford compound **17** (41 mg, Fig. 6.1b). For the methanol root extract (10 g), separation was performed using silica gel column chromatography with an increasing polarity gradient of *n*-hexane/ethyl acetate and DCM/MeOH mixtures, resulting in 98 fractions. Fractions 27–29, eluted with *n*-hexane/ethyl acetate (5:2), were further purified using isocratic elution with 5% methanol in dichloromethane to yield compounds **18** (11 mg, Fig. 6.1b), **19** (6 mg, Fig. 6.1b), and **20** (20 mg, Fig. 6.1b). Finally, fractions 53–57, eluted with *n*-hexane/ethyl acetate (3:1), were purified with isocratic elution using 10% methanol in dichloromethane, resulting in compounds **21** (40 mg, Fig. 6.1b) and **22** (7 mg, Fig. 6.1b).

The crude DCM/MeOH (1:1) root extract of *R. ferruginea* (14 g) was adsorbed onto 14 g of silica gel (mesh size 60–120) and subjected to column chromatography with 180 g of silica gel as the stationary phase. Elution was performed using an increasing gradient of ethyl acetate in *n*-hexane, followed by methanol in ethyl acetate. One hundred fifteen fractions (100 mL each) were collected and their purity was evaluated using TLC. Fractions with similar R_f values and colors on TLC (visualized under UV light) were combined. Fractions 35–38, which showed a single spot with an R_f of 0.52 in an ethyl acetate/*n*-hexane (1:1) system, yielded isoflavan (**23**, 120 mg, Fig. 6.1b). Fractions 43–45, exhibiting an R_f of 0.43 in the same solvent system, provided isoflavene (**24**, 50 mg, Fig. 6.1b).

The acid-base extract of *Z. chalybeum* roots (5.6 g) was subjected to silica gel column chromatography and eluted with a gradient of methanol in dichloromethane. Twenty-two fractions (50 mL each) were collected. Fractions 3–5, eluted with 100% dichloromethane, yielded compound **25** (60 mg, Fig. 6.1b) as a yellow powder. Fractions 7–10, combined and further purified with 2–5% methanol in dichloromethane, resulted in compound **26** (7.4 mg, Fig. 6.1b) as a yellow powder. Similarly, fractions 13–16, purified with 10–12% methanol in dichloromethane, provided compound **27** (11.7 mg, Fig. 6.1b). For the methanol extract of *Z. chalybeum* roots (30 g), silica gel column chromatography was performed with elution using an increasing gradient of ethyl acetate in *n*-hexane, followed by methanol in dichloromethane. Sixty-eight fractions (50 mL each) were collected. Fractions 8–10, eluted with 15% ethyl acetate in *n*-hexane, afforded compound **28** (81 mg, Fig. 6.1b).

The dried DCM/MeOH (1:1) root extract of *U. scheffleri* (30 g) was subjected to silica gel column chromatography and eluted with a gradient of ethyl acetate in *n*-hexane. Seventy-five fractions (50 mL each) were collected. Fraction 20, eluted with 45% ethyl acetate, yielded N-methyl-2,3-bis(2-hydroxybenzyl)-1H-indole (**29**, 12 mg, Fig. 6.1b) as pale gray crystals. Fractions 22–24, which exhibited identical spots on TLC, were combined and rechromatographed on a silica gel column using a gradient of ethyl acetate in *n*-hexane. From this, 22 fractions (25 mL each) were collected and subfraction 12, eluted with 40% ethyl acetate in *n*-hexane, provided uvaretin (**30**, 9 mg, Fig. 6.1b).

6.2.4 *Bioassay Analysis*

Crude extracts and isolated compounds from all plant sources were tested for their in vitro antibacterial activity using disc diffusion method. This evaluation focused on two Gram-positive bacteria (*Staphylococcus aureus* and *Bacillus subtilis*) and two Gram-negative bacteria (*Escherichia coli* and *Salmonella typhi*) (Hodge & Sterner, 2021; Kaur & Saini, 2020). Additionally, the in vitro antioxidant activity of crude extracts and isolated compounds from *B. antidysenterica* and root extract of *R. ferrugenia* were determined using the DPPH radical scavenging assay (El Sayed & El-Beltagi, 2018).

6.2.4.1 In Vitro Cytotoxicity Assay and Cell Cycle Study

Cell Lines and Culture Conditions

MCF-7 and MDA-MB-231 cells were grown in DMEM/F12 medium (Gibco, USA) supplemented with 10% (v/v) fetal bovine serum (FBS) (HyClone, USA) and maintained at 37 °C in a 5% CO_2 environment. Isolated compounds were dissolved in DMSO (Sigma, USA) to create a 10 mM stock solution, which was stored at − 20 °C until needed. For experimental use, the stock solutions were diluted in complete DMEM/F12 medium with 10% FBS under sterile conditions to obtain the required final concentrations.

Cell Viability Assay

Cell viability was evaluated using the MTS assay. Cells were plated at 5000 cells per well in 96-well plates with 50 μL of complete medium and allowed to adhere for 18 h. After adherence, the cells were treated with either complete medium containing 0.01–1.5% DMSO (as a vehicle control) or with the compounds at various concentrations for 72 h. Following treatment, 10 μL of freshly prepared MTS solution (2 mg/mL) mixed with phenazine methosulfate (PMS) at a 20:1 (v/v) ratio was added to each well. The plates were then incubated at 37 °C for an additional 2–3 h. Absorbance was measured at 490 nm using a CLARIOstar microplate reader (BMG Labtech). Untreated cells were used as the negative control, while doxorubicin served as the positive control for both cell lines (Wang et al., 2023).

Cell Cycle Assay

For cell cycle analysis, cells were plated at 30,000 cells per well in 24-well plates with 600 μL of complete medium and allowed to adhere for 18 h. Following adherence, the cells were treated with the compounds at their IC_{50} values for 72 h. The medium was then removed, and the cells were washed with PBS and detached using trypsin. The cell suspension was centrifuged at 400 g for 5 min, and the pellet was resuspended in 200 μL of hypotonic solution containing propidium iodide (5 μg/mL), trisodium phosphate, Triton X-100 (0.1% v/v), and RNase A (100 μg/mL). Cells were incubated at 4 °C for 12–24 h before analyzing DNA content by flow cytometry using a CytoFLEX Beckman-Coulter (Yang et al., 2024).

6.2.5 Molecular Docking Analysis

In this study, molecular docking simulations were performed using AutoDockTools (ADT) 1.5.2 and AutoDock 4.2. Ligand structures were optimized with HyperChem

7.0, and polar hydrogens were added while non-polar hydrogens were merged. Gasteiger charges were assigned, and all rotatable bonds in the ligands were allowed to rotate. Protein and ligand structures were prepared in PDBQT format, with docking conducted using a 46 × 46 × 46 $Å^3$ grid box and a 0.375 Å grid spacing. The Lamarckian Genetic Algorithm (LGA) was used for docking, with 200 runs per ligand and conformers clustered by RMSD tolerance of 2.0 Å. The conformation with the lowest binding energy was selected for further analysis (Morris et al., 2009; Trott & Olson, 2010).

6.3 Results and Discussion

6.3.1 Characterization of Isolated Compounds

Chromatographic separation of the roots of *C. anisata* led to the isolation of a carbazole alkaloid (**1**, Fig. 6.1a) and three coumarins (**2–4**, Fig. 6.1a) (Tamene et al., 2019). From the roots of *V. dainellii*, five alkaloids (**5–9**, Fig. 6.1a) were isolated (Anza et al., 2021b), while the roots of *E. schimperi* yielded two flavans (**10–11**, Fig. 6.1a) (Guyasa et al., 2018). The root extracts of *B. antidysenterica* provided three indole alkaloids (**12–14**, Fig. 6.1a) (Alemu et al., 2020), and the roots of *C. cyphopetalum* yielded eight compounds (**15–22**, Fig. 6.1b) (Degfie et al., 2022). Extracts from *R. ferruginea* roots produced two isoflavans (**23–24**, Fig. 6.1b) (Hussien et al., 2021), while *Z. chalybeum* gave four benzo[c] phenanthridine alkaloids (**25–28**, Fig. 6.1b) (Anza et al., 2021a). Additionally, *U. scheffleri* yielded an indole alkaloid (**29**, Fig. 6.1b) and a chalcone (**30**, Fig. 6.1b) (Anza et al., 2021c). The structures of these compounds were determined using UV-Vis, IR, NMR, and MS spectroscopic techniques and comparison with reported literature data. Detailed structural elucidation of these compounds has been provided in publications from our group between 2019 and 2022 (Tamene et al., 2019; Anza et al., 2021c; Alemu et al., 2020; Degfie et al., 2022; Hussien et al., 2021; Guyasa et al., 2018). For instance, the characterization of evodamine (**8**, Fig. 6.2) is illustrated below.

Compound **8** was isolated as a white powder with a melting point of 278–279 °C. The ESI-MS analysis revealed a molecular ion peak at m/z = 303. The ^{1}H NMR spectrum (300 MHz, DMSO-d_6, Fig. 6.3) indicated a singlet at δ 11.07, attributed to the N-H group, and eight aromatic methine protons at δ 7.81 (d, J = 7.8 Hz, 1H, H-4), 7.49 (m, 2H, H-2,3), 7.37 (d, J = 8.1 Hz, 1H, H-1), and 7.04 (m, 4H, H-9-12). Additionally, sp^3 methine signal was observed at δ 6.14 (s, 1H, H-14). Methylene protons bonded to sp^2 quaternary carbon and heteroatoms appeared at δ 2.72–2.85 (m, 2H, H-7), δ 4.62 (dd, J = 12.8, 1.6 Hz, 1H, H-7), and δ 3.25 (m, 1H, H-8). A singlet at δ 2.88 (3H, s) indicated a methyl group attached to nitrogen. The ^{13}C NMR spectrum (100 MHz, DMSO-d_6, Fig. 6.4) showed a carbonyl carbon at δ 164.7 (C-5), and six sp^2 quaternary carbons at δ 149.8 (C-1a), 136.3 (C-12a), 127.8 (C-13a), 125.2 (C-4a), 111.9 (C-9a), and 110.8 (C-8a). Additionally, eight sp^2

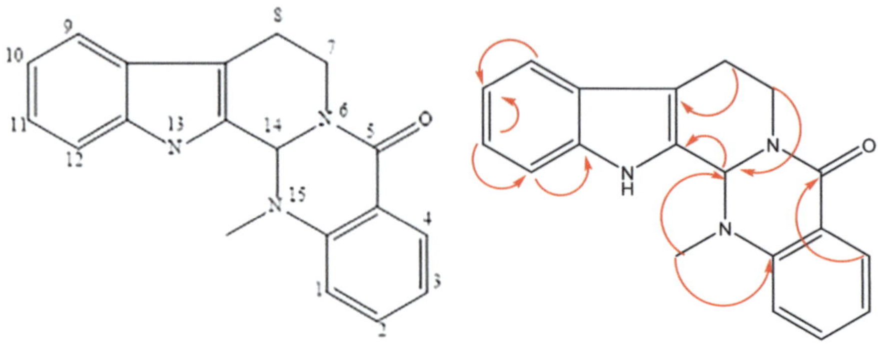

Fig. 6.2 Evodiamine (**8**) isolated from roots methanol extract of *V. dainellii*

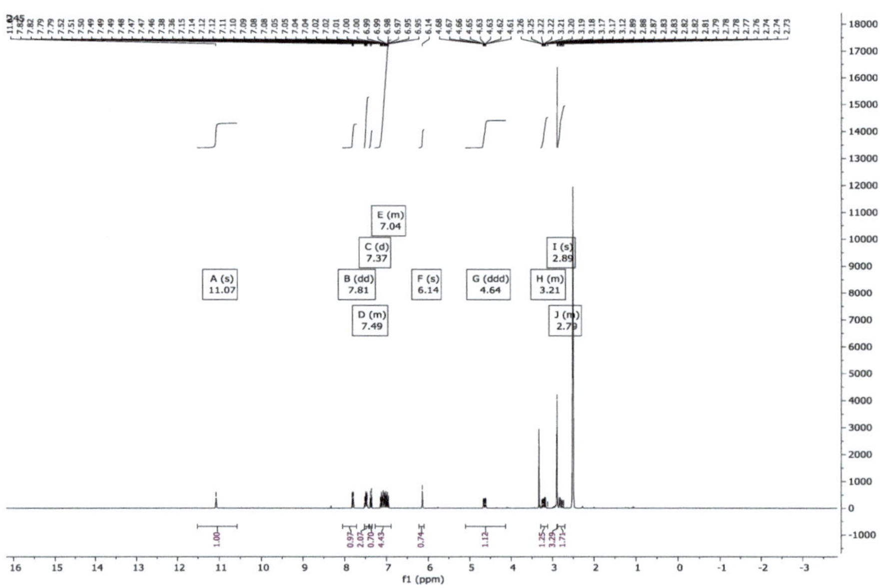

Fig. 6.3 ^{1}H NMR spectrum of evodamine (**8**), $CDCl_3/CD_3OD$, 300 MHz

methines were observed at δ 136.3 (C-2), 127.2 (C-3), 121.8 (C-4), 121.6 (C-9), 122.4 (C-10), 120.4 (C-11), 118.7 (C-1), and 117.8 (C-12) including one sp^3 methine at δ 68.7 (C-14), and a methyl group attached to nitrogen at δ 35.6. Two methylene signals appeared at δ 39.2 (C-7) and 19.7 (C-8), confirmed by DEPT-135 (Fig. 6.5). The HSQC experiment revealed (Fig. 6.6) a ^{1}J correlation between H-4 (7.81) and C-4 (121.8), H-2,3 (7.49) and C-2,3 (136.3, 127.2), H-1 (7.37) and C-1 (118.7), H-9,10,11,12 (7.04) and C-9,10,11,12 (121.9, 122.4, 120.4, 117.8), H-14 (6.14) and C-14 (68.7), H-7 (2.72–2.85) and C-7 (39.2), H-8 (4.62 and 3.25) and C-8 (19.7), and N-C$\underline{H}_3$ (2.88) and N-$\underline{C}H_3$ (35.6). The HMBC (Fig. 6.7) correlations demonstrated that the methyl group at δ 35.6 showed a ^{3}J correlation with C-1a (δ

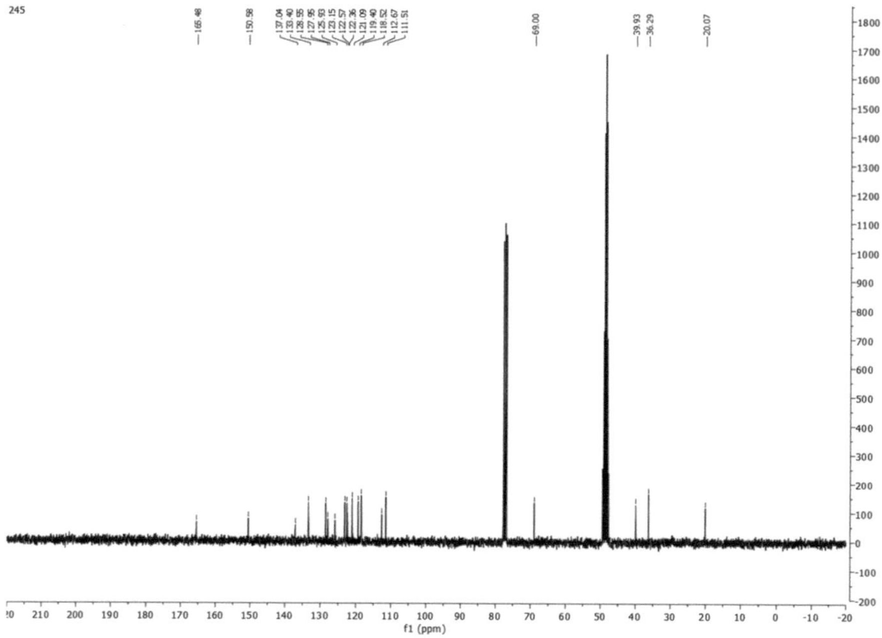

Fig. 6.4 ^{13}C NMR spectrum of evodamine (**8**), $CDCl_3/CD_3OD$, 75 MHz

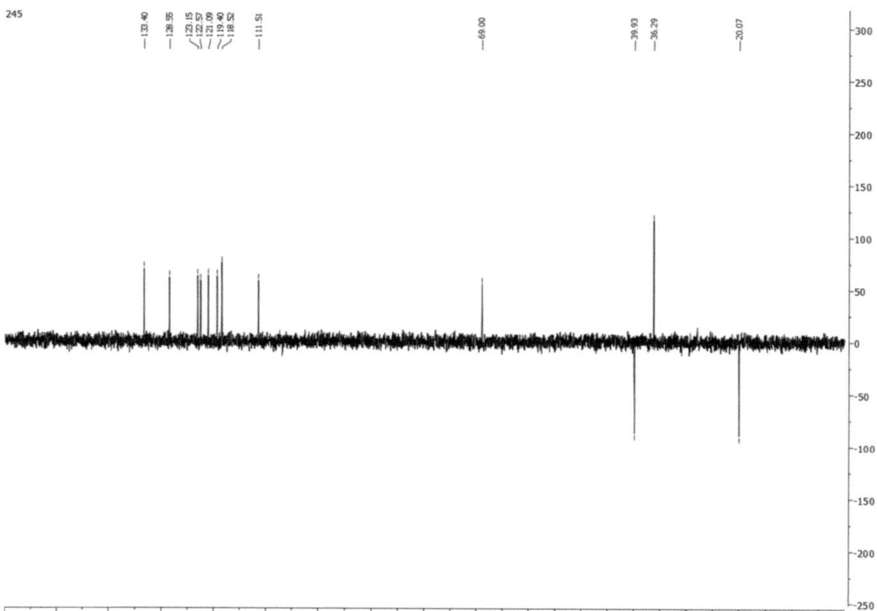

Fig. 6.5 DEPT-135 spectrum of evodiamine (8), 75 MHz, in $CDCl_3$/MeOD

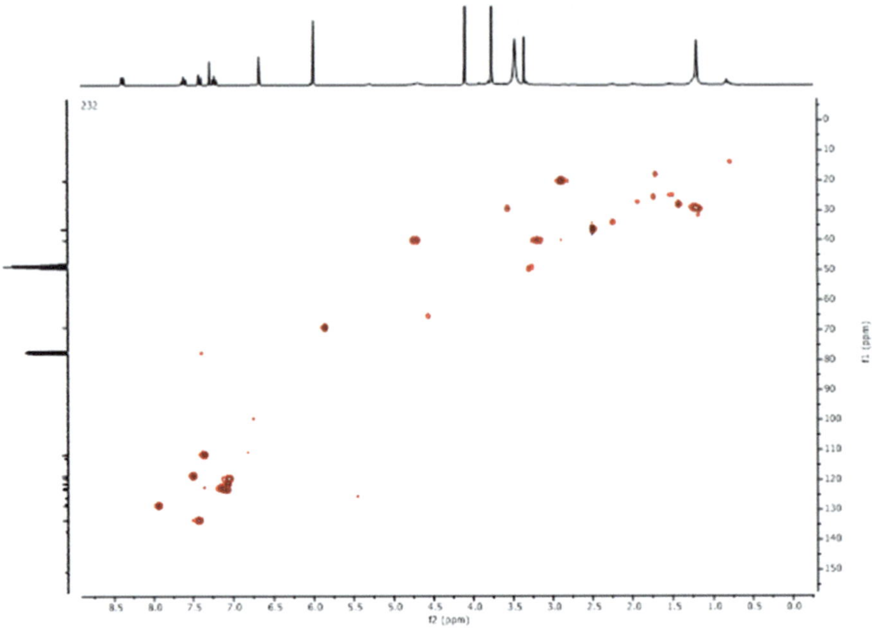

Fig. 6.6 HSQC spectrum of evodiamine (**8**), $CDCl_3$, 300 MHz

149.8) and C-14 (δ 68.7), and the methylene protons at H-7 and H-8 displayed ^{3}J and ^{1}J correlations with C-14 (δ 68.7) and C-8a (δ 110.8). These spectral data confirm that compound **8** is evodiamine (Fig. 6.2), previously isolated from *Zanthoxylum budrunga* (Joshi et al., 1991) but reported here for the first time from *Vepris* species (Table 6.1).

In a similar way, the structural elucidation of the remaining compounds was characterized and their structure has been presented here (Figs. 6.1a and 6.1b).

6.3.2 Antibacterial Activity

The root extracts of *C. anisata* and isolated compounds were tested at 20 μg/mL and showed significant antibacterial activity. Heptaphyline derivative (**1**) and imperatorin (**3**) had 14-mm inhibition zones against *S. aureus* and *B. subtilis*, slightly less than ciprofloxacin's 15-mm zone. Chalepin (**4**) was effective with a 14-mm zone against *S. aureus* and 12-mm against *P. aeruginosa*, and superior against *B. subtilis* (16 mm) compared to ciprofloxacin (15 mm) (Tamene et al., 2019). Epicatechin (**10**) showed a 15-mm zone against *S. aureus* and 12-mm against *E. coli*, comparable to gentamicin. Flavan (**11**) had 11-mm and 13-mm zones against *S. aureus* and *E. coli*, respectively, slightly less than gentamicin. Overall, chalepin (**4**) and epicatechin (**10**) were particularly effective (Guyasa et al., 2018).

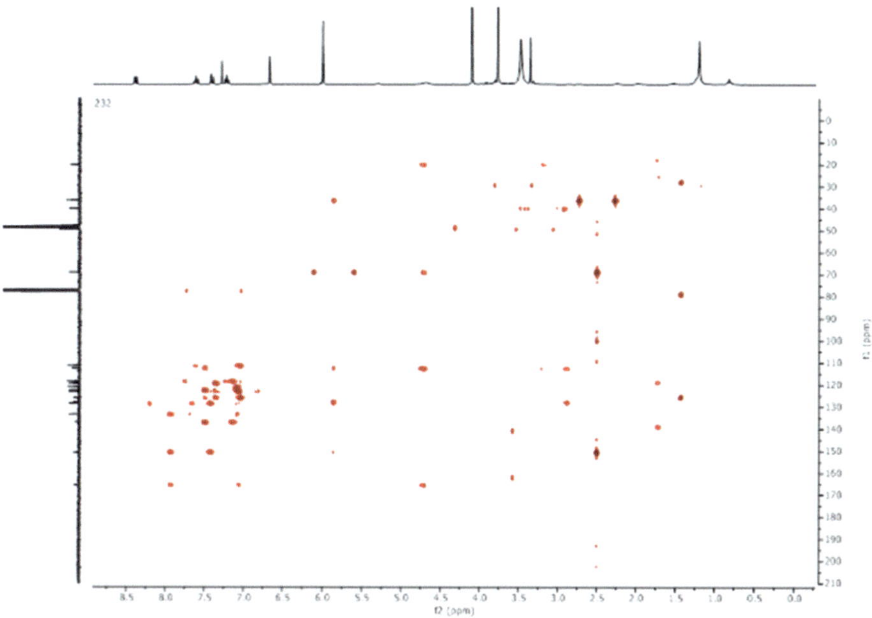

Fig. 6.7 HMBC spectrum of evodiamine (**8**), $CDCl_3$, 300 MHz

The antibacterial activity of *B. abyssinica* extracts and isolated alkaloids (**12**–**14**) revealed that canthine-6-one (**13**) and 1,11-dimethoxycanthin-6-one (**14**) exhibited moderate activity, with inhibition zones of 12.66 ± 0.6 mm and 12.5 ± 0.87 mm against *E. coli* and *S. typhimurium*, respectively. These results were notably less effective compared to ciprofloxacin, which had inhibition zones of 27.3 ± 2.52 mm and 29 ± 1.00 mm. The DCM/MeOH (1:1) and methanol extracts also showed moderate activity across various bacteria, with inhibition zones ranging from 11.6 to 13.6 mm (Alemu et al., 2020) (Table 6.2).

ε-Viniferin (**16**) demonstrated the highest inhibition against *E. coli* with a zone of 10.45 ± 0.75 mm at 1000 μg/mL, followed by gnetin H (**19**) at 10.02 ± 0.13 mm and parthenostilbenin B (**22**) at 9.52 ± 0.19 mm. Compounds **15**, **18**, **20**, and **21** exhibited moderate activity with inhibition zones between 7.32 ± 1.9 mm and 8.92 ± 0.04 mm. At 50 μg/mL, compound **17** had the highest inhibition against *E. coli* (8.55 ± 0.45 mm) and *S. aureus* (9.30 ± 1.39 mm). Gnetin H (**19**) showed the largest inhibition zone of 12.55 ± 0.65 mm against *P. aeruginosa*, surpassing chloramphenicol (11.76 ± 0.77 mm at 30 μg/mL). At 1000 μg/mL, all compounds were effective against *P. aeruginosa*, with gnetin H (**19**) being the most effective. ε-Viniferin diol (**21**) had an inhibition zone of 9.42 ± 0.42 mm at 50 μg/mL. For *S. aureus*, gnetin H (**19**) was the most effective at 12.45 ± 0.37 mm, followed by ε-viniferin (**16**) at 12.22 ± 0.74 mm and β-sitosterol (**18**) at 10.93 ± 1.54 mm. Compounds **15**–**22** showed dose-dependent inhibition against *P. aeruginosa*, except β-sitosterol (**18**) at lower concentrations (Degfie et al., 2022). Chelerythrine (**25**) had

Table 6.1 ^{1}H (300 MHz) and ^{13}C NMR (75 MHz), chemical shifts (ppm) in DMSO-d_6 of evodiamine (**8**)

	Evodamine (**8**)		Joshi et al. (1991)	
No	δ_H (*J* in Hz)	δC	δ_H (J in Hz)	^{13}C
1	7.37 (1H, *d*, *J* = 8.1)	118.7	6.92–7.82 (8, *m*, Ar-H)	119.7
1a	–	149.8	–	150.1
2	7.81 (1H, d, *J* = 7.8)	136.3		133.4
3	7.49 (2H = *m*)	127.2		127.1
4		121.8		122.6
4a	–	125.2	–	129.2
5	–	164.7	–	164.9
7	2.72–2.85 (*m*, 2H) 4.64 (*dd*, *J* = 12.8, 1.6, 1H), 3.25 (*m*, 1H)	39.2	2.85 (*m*, 2H)	40.6
8	2.72–2.85 (*m*, 2H)	19.7	4.64 (*m*, 1H) 3.20 (*m*, 1H)	20.4
8a	–	110.8	–	112.2
9	7.04 (4H, *m*)	121.6	6.92–7.82 (8, *m*, Ar-H)	122.8
9a	–	111.9	–	112.8
10	7.04 (4H, *m*)	122.4	6.92–7.82 (8, *m*, Ar-H)	122.9
11		120.4		119.2
12		117.8.		117.9
12a	–	136.3		138.1
13a		127.8		130.2
–	11.7 (*s*, N-H)		–	–
14	6.14 (1H, *s*)	68.7	5.98 (*s*)	69.9
NCH_3	2.88 (3H, *s*)	35.6	2.88 (*s*)	36.8

significant antimicrobial activity with IC_{50} values of 12.5 μg/mL against *S. aureus* and 50 μg/mL against *C. albicans*. Isoflavan (**23**) and isoflavene (**24**) showed inhibition zones of 9.67 ± 0.58 mm and 10.67 ± 0.58 mm against *E. coli*, and 10.33 ± 1.15 mm and 10 ± 1.00 mm against *S. aureus*, respectively. The dichloromethane/methanol extract from *U. scheffleri* had a MIC of 6.25 μg/mL against *S. aureus*, slightly less effective than gentamicin. N-methyl-2,3-bis (2-hydroxybenzyl)-1H-indole (**29**) was inactive, potentially due to synergistic effects or other bioactive constituents (Anza et al., 2021a, b).

6.3.3 Antioxidant Activity

The DPPH radical scavenging activities of the extracts and isolated compounds were recorded as 85.3%, 87.5%, and 78.4% for compounds **12**, **13**, and **14**, respectively.

Table 6.2 Zone of bacteria growth inhibition diameter (mm)

	Inhibition diameter (mm) ± SD			
Sample (0.5 mg/mL)	*E. coli*	*S. typhimurium*	*B. subtilis*	*S. aureus*
DCM/MeOH (1:1) extract of *B. antidysentrica*	11.6 ± 1.53	10.66 ± 1.15	11.33 ± 1.15	9.33 ± 0.58
Acid-base extract of *B. antidysentrica*	11.33 ± 2.08	11.0 ± 1.73	12.0 ± 3.74	11.33 ± 0.58
Compound **12**	11.33 ± 0.58	11.66 ± 3.06	11.33 ± 1.53	11.0 ± 1.00
Canthin-6-one (**13**)	12.66 ± 0.60	12.33 ± 2.31	11.0 ± 0.707	9.6 ± 1.16
1,11-dimethoxycanthin-6-one (**14**)	11.3 ± 0.58	12.5 ± 0.87	12.3 ± 1.65	10 ± 1.00
DCM: MeOH extracts of *B. abyssinica*	13 ± 0	11.6 ± 0.48	12.3 ± 1.25	13 ± 2
MeOH extracts of *B. abyssinica*	13.6 ± 0.55	12 ± 0	11.6 ± 0.48	12 ± 2
Ciprofloxacin	27.3 ± 2.52	29 ± 1.00	33.6 ± 3.22	23.5 ± 0.58

Alemu et al. (2020)
SD Standard deviation, each value represents the mean value from triplicate assays

In comparison, the DCM/MeOH (1:1) and acid-base crude extracts showed 83.3% and 80.00% inhibition, respectively, at a concentration of 100 μg/mL (Table 6.3). It was observed that DPPH scavenging activity increased with higher sample concentrations. Among the tested samples, canthine-6-one (**13**) achieved the highest percentage of DPPH radical inhibition, outperforming ascorbic acid, which reached its peak scavenging effect at a very low concentration (Alemu et al., 2020).

DPPH radical scavenging assay of the DCM:MeOH (1:1) roots extract of *R. ferruginea* and isoflavan (23) exhibited better radical scavenging activity with IC_{50} value of 17.7 and 32, respectively (Hussien et al., 2021).

6.3.4 Cytotoxicity and Cell Cycle Analysis

Evodiamine (**8**) exhibited the highest cytotoxicity and selectivity against the MDA-MB-231 breast cancer cell line, significantly inducing G2/M phase arrest. Evoxanthine (**6**) and arborinine (**7**) were more cytotoxic to MCF-7 cells than MDA-MB-231, with arborinine (**7**) being notably less active than evoxanthine (**6**) (Anza et al., 2021b) (Table 6.4).

Chelerythrine (**25**) and dihydrochelerythrine (**28**) both significantly reduced cell growth in MDA-MB-231 and MCF-7 breast cancer cell lines in a dose-dependent manner. Chelerythrine (**25**) showed higher potency against MDA-MB-231 with an IC_{50} of 3.616 ± 0.51 μM, compared to dihydrochelerythrine (**28**) with an IC_{50} of 24.14 ± 5.24 μM. Both compounds induced G2/M phase arrest in MDA-MB-231 cells, inhibiting mitosis of cells with damaged DNA (Anza et al., 2021b).

Table 6.3 Percentage of radical scavenging activity of the extracts and isolated compounds of *B. antidysenterica*

C (μg/mL)	Samples									
	DCM:MeOH(1:1)extract		Acid-base extract		**12**		**13**		**14**	
	A	% scavenging activity	A	% scavenging activity	A	% scavenging activity	A	% scavenging activity	A	% scavenging activity
100	0.2	83.3	0.26	80	0.15	85.3	0.1	87.5	0.28	78.4
50	0.3	75	0.27	79.2	0.23	77.4	0.2	83.3	0.31	76.1
25	0.35	70.8	0.3	76.9	0.3	70.5	0.3	75	0.34	73.8
12	0.4	66.6	0.4	69.2	0.4	60.7	0.38	68.3	0.45	65.4

Alemu et al. (2020)
A absorbance, *N.B* the experiment was done in triplicate

Table 6.4 Cytotoxicity of alkaloids (**6**–**8**) against breast cancer cell lines (IC_{50} in μM)

Alkaloids	MDA-MB-231	MCF-7
6	55.99 ± 12.17	17.44 ± 0.87
7	69.30 ± 5.34	40.63 ± 2.30
8	1.73 ± 0.48	26.47 ± 5.95
Doxorubicin	0.23	9.6

Anza et al. (2021b)

6.3.5 Molecular Docking Analysis Against Antibacterial and Anticancer Targets

In the docking study, the compounds exhibited binding energies ranging from −5.8 to −4.8 kJ/mol. Alkaloids (**12**–**14**) demonstrated specific binding energies of −5.8, −5.7, and −5.7 kcal/mol, respectively. DNA gyrase, which plays a crucial role in introducing negative supercoils into DNA, has 14 active site residues. Among these, only two residues, PHE198 and ALA51, interact directly with the ligands, while the others are hydrophobic and in close proximity. Flazin methyl ether (**3**) formed a hydrogen bond with PHE196 at 1.5 Å and established two hydrophobic interactions with TRY218 and GLU219. 1,11-Dimethoxycanthin-6-one (**14**) created a hydrogen bond with ALA51 at 1.5 Å and engaged in four hydrophobic interactions with ASN198, LEU52, VAL201, and LYS57. Canthin-6-one (**13**) demonstrated six hydrophobic interactions with VAL201, LEU52, LYS57, GLY200, ASN198, and ALA52. None of the compounds exhibited π-π or π-cation interactions with the protein. Given their high binding affinity and significant interactions with DNA gyrase, these alkaloids (**12**–**14**) are potential inhibitors of this essential bacterial enzyme.

The aromatase enzyme is critical in estrogen-positive breast cancers, which can be resistant to tamoxifen. Treatments for these cancers target estrogen production or action. In a docking study, active compounds **12**–**14** showed moderate to high binding scores ranging from −6.6 to −5.3 kcal/mol. Among them, compound **12** exhibited the highest binding affinity at −6.6 kcal/mol, followed by compound **13** at −5.5 kcal/mol and **14** at −5.3 kcal/mol. Compound **3** demonstrated strong interactions with aromatase, including hydrogen bonding with SER90, π-π interactions with PHE116, π-cation interactions with LYS376, and hydrophobic interactions with ASN136, ARG115, GLU92, and GLY117. Canthin-6-one (**13**) interacted via π-π interactions with TYR52 and hydrophobic interactions with TRP88 and PRO58. 1,11-Dimethoxycanthin-6-one (**14**) bound through a hydrogen bond with LYS376 and hydrophobic interactions with LYS119, GLU92, and GLY117.

Compounds **5**–**9** were docked against human DNA topoisomerase II α. Evodiamine (**8**) had the lowest binding affinity at −6.3 kcal/mol, interacting with residues Asp-479, Ser-480, Glu-477, Ala-481, Arg-503, Gly-478, Gly-504, and nucleic acids DT-8, DT-9, DG-10, DG-12, and DG-13. Norevoxanthine (**5**) and arborinine (**7**) showed moderate affinities similar to vosaroxin. These findings match in vitro cytotoxicity results, where alkaloids (**6**, **7**, and **8**) showed potential against MDA-MB-231 and MCF-7 cell lines. Compounds **15**–**20** had significant binding affinities: −6.6 to −7.6 kcal/mol for DNA gyrase B, −6.4 to −8.8 kcal/mol for PqsA, and − 4.4 to −5.8 kcal/mol for *S. aureus* PK. All tested compounds showed better binding affinity than chloramphenicol for DNA gyrase B, with compound **17** showing the highest affinity at −7.6 kcal/mol. For PqsA, compound **19** had the strongest affinity at −8.8 kcal/mol. For *S. aureus* PK, compounds **19** (−5.8 kcal/mol) and **3** (−5.4 kcal/mol) showed the best binding. These results align with in vitro activities, highlighting compounds **17** and **19** as promising candidates.

Compounds **25**–**28** showed strong binding affinities for topoisomerase II α, ranging from −6.3 to −6.1 kcal/mol, compared to vosaroxin's − 6.2 kcal/mol. Chelerythrine (**25**) and dihydrochelerythrine (**28**) had the highest affinities at −6.4 and −6.3 kcal/mol, respectively. They exhibited similar binding profiles to vosaroxin, interacting with amino acid residues Asp-479, Ser-480, Glu-477, Ala-481, Arg-503, Gly-478, Gly-504, and nucleic acid residues DT-8, DT-9, DG-10, DG-12, and DG-13.

6.4 Conclusion and Recommendation

This study examines the medicinal potential of Ethiopian plants, focusing on *C. anisata*, *U. scheffleri*, *Z. chalybeum*, *V. dainellii*, *B. antidysenterica*, *E. schimperi*, *C. cyphopetalum*, and *R. ferruginea*. Purification of crude extracts was done using silica gel column chromatography and yielded from *C. anisata*—a carbazole alkaloid (**1**) and three coumarins (**2**–**4**); from *V. dainellii*—five alkaloids (**5**–**9**); from *E. schimperi*—two flavans (**10**–**11**); from *B. antidysenterica*—three indole alkaloids (**12**–**14**); from *C. cyphopetalum*—eight compounds (**15**–**22**); from *R. ferruginea*—two isoflavans (**23**–**24**); and from *Z. chalybeum*—four benzo[c] phenanthridine alkaloids (**25**–**28**). Additionally, *U. scheffleri* provided an indole alkaloid (**29**) and a chalcone (**30**).

In antibacterial tests, Gnetin H (**19**) outperformed chloramphenicol against *P. aeruginosa* with an inhibition zone of 12.55 ± 0.65 mm. Evodiamine (**8**) was highly effective against MDA-MB-231 breast cancer cells, with an IC_{50} of 1.73 ± 0.48 μM, while evoxanthine (**6**) and arborinine (**7**) had IC_{50} values of 17.44 ± 0.87 μM and 40.63 ± 2.30 μM against MCF-7 cells, respectively. Chelerythrine (**25**) also showed significant activity against MDA-MB-231 cells (IC_{50} of 3.616 ± 0.51 μM) and effective antibacterial and antifungal properties. The *U. scheffleri* dichloromethane extract demonstrated strong antibacterial activity with a MIC of 6.25 μg/mL against *S. aureus*.

Heptaphylline (**1**) and imperatorin (**3**) exhibited comparable antibacterial activity, with inhibition zones of 14 mm and 13 mm, respectively. Epicatechin (**10**) had similar effects as gentamicin against *S. aureus* and *E. coli*. Isoflavan (**23**) showed a 10.67 ± 0.58 mm zone against *S. aureus*, slightly less than ciprofloxacin. DPPH radical scavenging assays showed compounds **12**–**14** had inhibition rates between 78.4% and 87.5%, indicating antioxidant potential. Molecular docking revealed binding affinities of −6.5 to −7.5 kcal/mol for *E. coli* DNA gyrase B and −6.1 to −6.4 kcal/mol for topoisomerase II α. Evodiamine (**8**) had the lowest binding affinity (−6.3 kcal/mol), while norevoxanthine (**5**) and arborinine (**7**) displayed moderate affinities, similar to vosaroxin's −6.2 kcal/mol.

Evoxanthine (**6**), evodiamine (**8**), chelerythrine (**25**), and dihydrochelerythrine (**28**) show promise as potent inhibitors of topoisomerase IIα, particularly for human breast cancer. Canthin-6-one (**13**) exhibited impressive antioxidant activity, with an 87.5% inhibition rate in the DPPH assay, surpassing ascorbic acid. Epicatechin (**10**) proved effective against *S. aureus*, while heptaphylline (**1**) and imperatorin (**3**) are potential antibacterial agents against *S. aureus* and *B. subtilis*. Chalepin (**4**) also shows potential against *B. subtilis*. The crude extracts from *U. scheffleri* and *C. anisata* demonstrated strong antibacterial properties against *S. aureus* and *B. subtilis*, respectively, and merit further study. In silico docking studies support the medicinal value of these plants, highlighting their antibacterial, antioxidant, and anticancer potentials. This research emphasizes the integration of traditional knowledge with modern science, reinforcing the therapeutic potential of these Ethiopian medicinal plants. Based on this study, further exploration of gnetin H (**19**), evodiamine (**8**), chelerythrine (**25**), and dihydrochelerythrine (**28**) is recommended due to their notable antibacterial, antioxidant, and anticancer activities. Gnetin H (**19**) and chelerythrine (**25**) showed promising antibacterial and antifungal effects, while evodiamine (**8**) exhibited significant activity against aggressive cancer cell lines. Compounds such as epicatechin (**10**), evoxanthine (**6**), and canthin-6-one (**13**) also demonstrated potential in antibacterial and antioxidant assays. Future research should include in vivo studies to confirm these compounds' therapeutic efficacy and safety. Additionally, the crude extracts from *U. scheffleri* and *C. anisata*, which displayed strong antibacterial activity, warrant further investigation. Integrating these findings with traditional knowledge could enhance the development of new therapeutic agents and support the use of Ethiopian medicinal plants in modern healthcare.

Acknowledgments The principal investigator ME acknowledges Adama Science and Technology University (ASTU), Adama, Ethiopia, for funding part of this research. We extend our gratitude to Prof. R. Eswaramoorthy from Saveetha Institute of Medical and Technical Sciences (SIMATS), Saveetha University, India, and Dr. Taye B. Demissie from the University of Botswana's Department of Chemistry for their invaluable contributions to the in silico computational analysis. We also thank Dr. Japheth Ombito from the University of Botswana and Prof. Maria L. Cardona from the University of Valencia, Spain, for their expertise in NMR analysis. Our appreciation goes to Prof. Diego Cortes from the University of Valencia for his assistance with anticancer and antibacterial tests. Finally, we acknowledge the dedication of MSc students Tewabech Alemu, Kalid Hussein, Dandena Tamene, and Babe Guyasa at ASTU, whose efforts were essential to this research.

References

Ajayi, I. A., Ajibade, O., & Oderinde, R. A. (2011). Preliminary phytochemical analysis of some plant seed. *Research Journal of Chemical Sciences., 1*(3), 58–62.

Alemu, T., Eswaramoorthy, R., Melaku, Y., & Endale, M. (2020). Molecular docking analysis of antibacterial indole alkaloids from roots of *Brucea antidysentrica*. *Ethiopian Journal of Science and Sustainable Development, 7*(2), 1–13.

Anza, M., Endale, M., Cardona, L., Cortes, D., Eswaramoorthy, R., Zueco, J., Rico, H., Trelis, M., & Abarca, B. (2021a). Antimicrobial activity, *In Silico* molecular docking, ADMET and DFT analysis of secondary metabolites from three Ethiopian medicinal plants. *Advances and Applications in Bioinformatics and Chemistry., 14*, 117–132.

Anza, M., Endale, M., Cardona, L., Cortes, D., Eswaramoorthy, R., Cabedo, N., Terlis, M. Z. J., & Rico, H. (2021b). Cytotoxicity, antimicrobial, molecular docking, drug likeness and DFT analysis of benzo[c]phenanthridine alkaloids from roots of *Zanthoxylum chalybeum*. *Biointerface Research in Applied Chemistry, 12*(2), 1569–1586.

Anza, M., Endale, M., Cardona, L., Cortes, D., Cabedo, N., Eswaramoorthy, R., Abarca, B., Domingo-Ortí, I., & Schätzlein, M. P. (2021c). Cytotoxicity, molecular docking, ADMET and DFT analysis of alkaloids from roots and fruits of *Vepris dainelli*. *Current Bioactive Compounds, 18*(10), 1–13.

Barbosa, L., Braz-Filho, R., & Vieira, I. (2011). Chemical constituents of plants from the genus Simaba (Simaroubaceae). *Chemistry and Biodiversity, 8*(12), 2163–2178.

Beerntje, H. J. (1994). *Kenya trees, shrubs and lianas*. National Museum of Kenya, Nairobi.

Cheikhyoussef, A., Shapi, M., Matengu, K., & Mu, H. (2011). Ethnobotanical study of indigenous knowledge on medicinal plant use by traditional healers in Oshikoto region. *Namibia, Journal of Ethnobiology and Ethnomedicine., 7*(10), 1–11.

Chekole, G. (2017). Ethnobotanical study of medicinal plants used against human ailments in Gubalafto. *Journal of Ethnobiology and Ethnomedicine., 13*(55), 1–29.

Chitra, M., Devi, C. S., & Sukumar, E. (2003). Antibacterial activity of Embelin. *Fitroterapia., 74*(4), 401–403.

Dagne, E., Yenesew, A., Waterman, P. G., & Gray, A. I. (1988). The chemical systematics of the Rutaceae, subfamily Toddalioideae, in Africa, biochemical systematic. *Ecology, 16*, 179–188.

Debebe, Y., Tefera, M., Mekonnen, W., Abebe, D., Woldekidan, S., Abebe, A., Belete, Y., Menberu, T., Belayneh, B., Tesfaye, B., Nasir, I., Yirsaw, K., Basha, H., Dawit, A., & Debella, A. (2015). Evaluation of anthelmintic potential of the Ethiopian medicinal plant Embelia schimperi *in vivo* and *in vitro* against some intestinal parasites. *BMC Complementry and Alternative Medicine., 15*, 187.

Degfie, T., Ombito, J. O., Demissie, T. B., Eswaramoorthy, R., Dekebo, A., & Endale, M. (2022). Biological activity, *In Silico* molecular docking, ADMET and DFT analysis of secondary metabolites isolated from roots of *Cyphostemma cyphpetalum*. *Advances and Applications in Bioinformatics and Chemistry., 15*, 1–19.

El Sayed, S. A., & El-Beltagi, H. S. (2018). Evaluation of antioxidant and antimicrobial activities of medicinal plants. *Antioxidants., 7*(3), 34.

Fouda, A., Aristide, F., Toze, A., Langat, M. K., Ngeufa, E., Leonard, L. M. M., Mulholland, D. A., Waffo, K. F. A., Sewald, N., & Duplex, J. (2017). Acridone alkaloids from Vepris verdoorniana (Excell & Mendonça) Mziray (Rutaceae). *Phytochemistry Letters, 19*, 191–195.

Guyasa, B., Melaku, Y., & Endale, M. (2018). Antibacterial activity of two flavans from the stem bark of *Embelia schimperi*. *Advances in Pharmacological Sciences, 9*, 2–8.

Hamza, O. J. M., Van den Bout-van, den Beukel, C. J. P., & Matee, M. I. N. (2006). Antifungal activity of some Tanzanian plants used traditionally for the treatment of fungal infections. *Journal of Ethnopharmacology, 108*(1), 124–132.

Haruna, A., & Yahaya, S. M. (2021). Recent advances in the chemistry of bioactive compounds from plants and soil microbes: A review. *Chemistry Africa., 4*(2), 231–248.

Hodge, J. E., & Sterner, G. (2021). A standardized method for the antibiotic susceptibility testing of clinical isolates: The disc diffusion method. *Journal of Clinical Microbiology., 59*(10), e01234–e01221.

Hussien, K., Melaku, Y., Eswaramoorthy, R., & Endale, M. (2021). Antibacterial and antioxidant activities of Isoflavans from the roots of *Rhynchosia ferruginea* and *In Silico* study on DNA gyrase and human peroxidoxin. *International Journal of Secondary Metabolites., 8*(4), 321–336.

Joshi, B. S., Moore, K. M., & Pelletier, S. W. (1991). Alkaloids of *Zanthoxylum budrunga* wall. : NMR assignments of Dihydrochelerythrine, Evodiamine and Zanthobungeanine. *Phytochemical Analysis., 2*(1), 20–25.

Kaur, P., & Saini, H. S. (2020). Antibacterial activity of medicinal plants and antibiotics: A review. *Journal of Pharmacognosy and Phytochemistry., 9*(3), 275–283.

Kebebew, M. R., & Mohammed, E. E. (2013). Distribution, abundance and population status of four indigenous threatened tree species in the Arba Minch Natural Forest, Southern Ethiopia. *International Journal of Natural Resource Ecology and Management, 2*(1), 1–8.

Kokwaro, J. (1993). *Medicinal plants of East Africa* (2nd ed., p. 164). E. A. Lit. Bureau, Nairobi.

Kuma, M., & Shibru, S. (2015). Floristic composition, vegetation structure and regeneration status of Kimphe Lafa Natural Forest. *Journal of Botany., 5*, 19–32.

Jima, T. T., & Megersa, M. (2018). Ethnobotanical study of medicinal plants used to treat human diseases in Berbere District, bale zone of Oromia regional state, South East Ethiopia. *Evidence-Based Complementary and Alternative Medicine., 2018*, 1–16.

Morris, G. M., Goodsell, D. S., Halliday, S. K., Huey, R., Hart, W. E., Belew, R. K., & Olson, A. J. (2009). AutoDock4 and AutoDockTools4: Automated docking with selective receptor flexibility. *Journal of Computational Chemistry, 30*(16), 2785–2791.

Muhammad, N., Uddin, N., & Ali, N. (2019). Ethnomedicinal uses and inter specific diversity encourage conservation of *Rhynchosia* species growing in hilly areas of swat. *Agricultural Research & Technology: Open Access Journal., 20*(2), 96–107.

Munda, S., & Bibhuti, B. K. (2017). Pharmacological evaluation of methanolic extract of *Zanthoxylum oxyphyllum. International Journal of Applied Biology and Pharmaceutical Technology., 8*, 5–10.

Patiño, O. J. L., & Cuca, L. E. S. (2010). Alkaloids of *Zanthoxylum quinduense* (Rutaceae). *Biochemical Systematic Ecology., 38*, 853–856.

Ross, S. A., Sultana, N. N., Burandt, C. L., ElSohly, M. A., Marais, P. J., & Ferreira, D. S. (2004). A new Antiplasmodial (+)-norepinephrine derivative from *Zanthoxylum syncarpum. Journal of Natural Product., 67*, 88–90.

Rungsung, W., Ratha, K., Dutta, S., Dixit, A., & Hazra, J. (2015). Secondary metabolites of plants in drugs discovery. *World Journal of Pharmaceutical Research., 4*(7), 604–613.

Sousa, F., Figueiredo, E., & Smith, G. F. (2011). The genus *Cyphostemma* (planch.) Alston (Vitaceae) in Angola the genus Cyphostemma (planch.) Alston (Vitaceae) in Angola. *Bradleya, 29*, 79–92.

Tamene, D., & Endale, M. (2019). Antibacterial activity of coumarins and carbazole alkaloid from roots of *Clausena anisata. Advances in Pharmacological Sciences, 10*, 1–9.

Teklehaymanot, T., Giday, M., Medhin, G., & Mekonnen, Y. (2007). Knowledge and use of medicinal plants by people around Debre Libanos monastery inEthiopia. *Journal of Ethnopharmacology, 111*(2), 271–283.

Tesfaye, A., & Sebsebe, D. (2009). *Ethnobotanical study of medicinal plants in Kafficho people, southwestern Ethiopia* (pp. 711–726). The 16th International Conference of Ethiopian Studies.

Trott, O., & Olson, A. J. (2010). AutoDock Vina: Improving the speed and accuracy of docking with a new scoring function, efficient optimization, and multithreading. *Journal of Computational Chemistry, 31*(2), 455–461.

Uwaifo, A. O. (1984). Mutagencities of seven coumarin derivatives and a furan derivative (nimbolide) isolated from three medicinal plants. *Journal of Toxicology and Environmental Health., 56*(3), 117–121.

Wang, Z., Zhang, X., & Xu, Y. (2023). Evaluation of cell viability assays: MTS, MTT, and XTT. *Biological Methods and Protocols., 25*(1), 58–72.

Wanzala, W., Marcellina, S. S., & Joseck, O. A. (2016). A survey of the applications and use of ethnomedicinal plants and plant products for healthcare from the Ukambani region in eastern Kenya. *Indian Journal of Ethnophyopharmaceutical., 2*, 6–58.

Yang, X., Zhao, L., & Liu, J. (2024). Optimized flow cytometry techniques for cell cycle analysis. *Journal of Flow Cytometry and Cell Analysis., 39*(2), 112–126.

Chapter 7
African Natural Products with Anti-malarial Activity

Esraa A. Elhawary, Nada M. Mostafa, Omayma A. Eldahshan, and Abdel Nasser Singab

Abstract Malaria is considered one of the deadliest and widespread endemic diseases in Africa. This disease does not fit in the category of Neglected Tropical Diseases (NTDs) as described by the World Health Organization (WHO), although it shares characteristics with NTDs, such as prevalence in tropical and subtropical regions and primarily affecting impoverished communities, malaria has received significant global attention and funding. NTDs generally refer to diseases that have historically received less attention and fewer resources compared to major diseases like malaria, HIV/AIDS, and tuberculosis. Malaria infection is usually located in tropical and subtropical regions where the transfer of the causative parasite, *Plasmodium malaria*, occurs through an intermediate host, the female *Anopheles* mosquito. Resistance is the main problem facing the efforts to combat malaria where the parasite and/or the mosquito can now resist most of the chemical agents utilized in the disease control. African plants and their natural products possess a lot of interesting treatment options against malaria. Most of these African plants may carry promising lead compounds that may act as an effective anti-malarial agent shortly. This chapter focuses on the African medicinal plants together with their active ingredients with potential anti-malarial activity either against the parasite itself or the mosquito malaria vector. Twenty plant families together with certain genera and species belonging to such families were detailed in this chapter. The main active phytoconstituents involved in the anti-malarial activity were flavonoids, phenolic acids, triterpenoids, fatty acids, chlorins, naphthoquinones, lignans, coumarins, and others. The data summarized in this chapter was accessed through many scientific databases such as Scopus®, Web of Science®, Google Scholar®, and PubMed®.

Keywords Natural products · Africa · Malaria · Neglected diseases · Tropical · Endemic

---

E. A. Elhawary · N. M. Mostafa · O. A. Eldahshan (✉) · A. N. Singab
Department of Pharmacognosy, Faculty of Pharmacy, Ain Shams University, Cairo, Egypt
e-mail: esraa.elhawary@pharma.asu.edu.eg; nadamostafa@pharma.asu.edu.eg; oeldahshan@pharma.asu.edu.eg; dean@pharma.asu.edu.eg

A. Dekebo (ed.), *Bioactive Secondary Metabolites from Medicinal Plants of Africa*, Interdisciplinary Biotechnological Advances,
https://doi.org/10.1007/978-981-96-8590-5_7

Abbreviations

NTDs	Neglected tropical diseases
P.	*Plasmodium*
mg	Milligram
kg	Kilogram
g	Gram
ml	Milliliter
bwt	Body weight
IC_{50}	Half maximal inhibitory concentration
LD_{50}	Lethal dose
LC_{50}	Half lethal concentration
MLD_{50}	Median lethal dose
p.o.	Per oral
IS	Inhibition score
MIC	Minimum inhibitory concentration
ppm	Part per million
DEET	N,N-diethyl-meta-toluamide
HPLC–UV	High-performance liquid chromatography coupled to ultraviolet detector
μg	Microgram

7.1 Introduction

Folk medicine has remained the cornerstone of many medical practices worldwide for hundreds of years. In Africa, traditional medicine and medical practices had been transferred verbally from one generation to another; thus, part of this folk knowledge is vulnerable to loss due to the lack of documentation and also due to obstacles regarding legislation of traditional practices during the colonial conquests. Nowadays, a lot of people still rely on traditional medical practices (Bouabdallah et al., 2022; El-Nashar et al., 2023). It was found that up to 75% of the African nation depends primarily on traditional medicine for treating their ailments and this is because of financial obstacles as well as the spiritual traditions related to some folk medical traditions (Abdullahi, 2011; James et al., 2018).

Malaria keeps being an overwhelming disease with prominent effects on people's health especially in Africa and other endemic regions worldwide. Through the last decade, the increase in malaria cases has reached an alarming point nevertheless in developing countries. According to recent reports 3.3 billion people are at risk of contracting the infection of which 1.2 billion are at high risk (Al-Huqail et al., 2023).

The number of recorded cases of malaria during 2013 reached 198 million cases accounting for 755,000 deaths, 90% of which occur in Africa were documented. In Africa, around three thousand children die because of malaria every day whereas

Nigeria, the largest African country, presented the highest number of malaria cases all around the year seasons throughout the whole country (El-Tabakh et al., 2023; Shehata et al., 2023).

Certain malaria-causing species have been nominated such as *Plasmodium vivax*, *P. ovale*, *P. malariae*, and *P. falciparum.* The control of such vectors through synthetic anti-malarial agents like primaquine and chloroquine has been tied to a wide range of drug resistance developed by the parasite in addition to severe and unwanted side effects to the patient. Thus, it was of great need to develop much safer options to undergo these tackles which necessitate the digging for new anti-malarial agents from natural resources (Elhawary et al., 2021). The literature is filled with many plant families which were reported for folk medicinal use against malaria from which about 160 plant families and 1200 plant species were documented either in folk use or even extended to testing in vitro and/or in vivo in order to validate their efficacy against malaria. Some of the traditional drugs used against malaria, namely, quinine and artemisinin, originated from plant extracts: *Cinchona calisaya* and *Artemisia annua*, respectively. This has enthused many researchers, especially in Africa to further intensify the search for anti-malarial agents (Lawal et al., 2015; Singab et al., 2022). This chapter was developed to highlight the main plant families and genera from African territories with documented anti-malarial activities to pave the way for new research on new anti-malarial drug leads from natural sources which will serve to combat such neglected tropical diseases in Africa. The data summarized in this chapter was accessed through many scientific databases such as Scopus®, Web of Science®, Google Scholar®, and PubMed® (Fig. 7.1a, b).

7.2 Neglected Tropical Diseases (NTDs)

Neglected tropical diseases (NTDs) can be defined as "a group of thirteen infectious diseases caused by parasitic worms, protozoa or bacteria." The WHO defines NTDs as "a diverse group of 20 communicable diseases caused by bacteria, viruses, helminths, or protozoa, cause 185,000 deaths a year." They usually spread among people in poor countries in Sub-Saharan Africa, Asia, and the Americas, causing life-long disability, disfigurement, reduced economic productivity, and social stigma. The global burden from such NTDs is huge, especially from malaria; thus, targeting this group of diseases is therefore widely promoted as a means to reaching some of the Millennium Development Goals (MDGs) (Kolaczinski et al., 2006).

The WHO set a roadmap for the management of NTDs from 2021 to 2030 to enhance the world's interest in such diseases in order to shift the focus from programs for each disease to more wide and sophisticated protocols that raise the patients' needs and to help communities suffering from such diseases. This roadmap also paves the way for multi-sectional coordination to provide suitable impact in affected areas. NTD management programs are also directed toward the reduction in the world financial cost paid to combat such NTDs. The draft roadmap clearly illustrates the protocols needed to achieve these targets through cross-cutting activities that intersect multiple diseases (Organization WH, 2020).

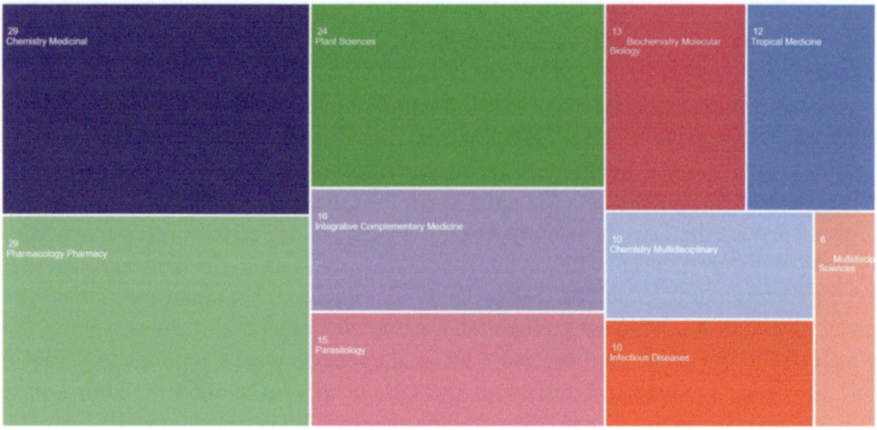

(a)

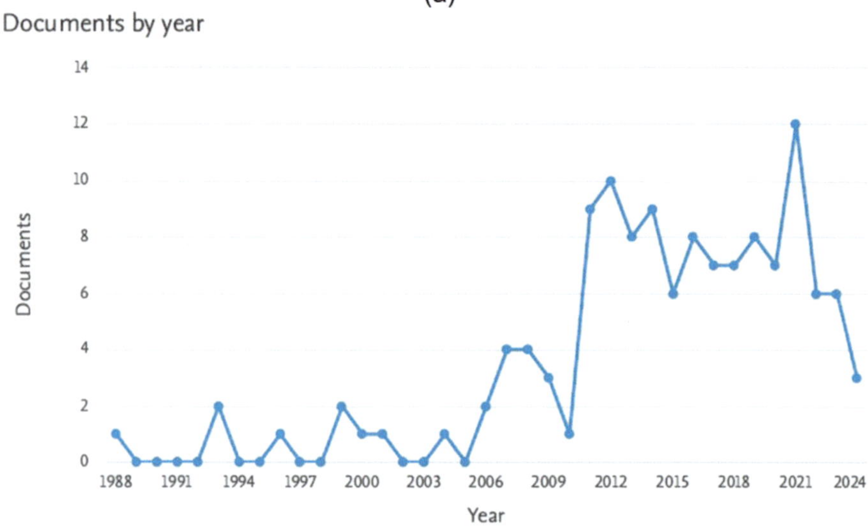

(b)

Fig. 7.1 Analysis of research data obtained from (**a**) Web of Science and (**b**) Scopus databases using relevant keywords

7.3 Health Burden of Neglected Tropical Diseases

In order to deliver more health facilities to infected zones, the global burden of drugs needed for such conditions has to be calculated. Unfortunately, a lot of statistics and work numbers do not give a clear image of the situation regarding the long-term effects and disabilities caused by such diseases on billions of people in poor affected areas. Policymakers, pharmaceutical companies, and other stakeholders require information about treatment success in addressing NTD epidemics over time to provide more support and to ensure the overall effects. The NTD treatment plans should be integrated in order to cover many diseases by one medication or a group of diseases through multiple medications at once. This led to the need for more models and platforms that cover the required drug-disease analysis. More sophisticated records are also required addressing the disability metrics, such as quality-adjusted life years or disability-adjusted life years (DALYs), to fully capture the disease burden of NTD infections that often have low mortality rates but high disability burdens (Hassoun & Cosler, 2023).

7.4 Malaria and Its Vectors

Mosquitos act as mechanical vectors for a number of tropical diseases such as malaria, filariasis, dengue fever, Japanese encephalitis, and yellow fever. Three thousand species of malaria are documented with one hundred of them being the vectors for human diseases. The regular control methods usually target one or a few numbers of such vectors either against the adult mosquito or its intermediate stages such as the larvae and pupae (Organization WH, 1997, 2014).

Mosquitos life cycle circles around four distinct stages starting from the egg and reaching the adult with two intermediate stages, namely, the larva and the pupa. Males and females mate together once in the female life cycle followed by the production of eggs by the female through intervals. In order for the females to lay eggs, they should suck blood which explains why males are not bloodsuckers rather they feed on plant juices. After sucking blood, the female requires two to three days to develop eggs in the tropics but longer in temperate zones. The gravid females search for suitable places to deposit their eggs, after which another blood meal is taken, and another batch of eggs is laid. This process is repeated until the mosquito dies (Msugupakulya et al., 2023; Organization WH, 1997).

According to the mosquito species, a female can lay a number of eggs ranging from 30 to 300 at a time. The females either lay their eggs on water surfaces singly (*Anopheles*) or stuck together in floating rafts (e.g., *Culex*). In the tropics, the eggs usually hatch within 2–3 days. Some species (e.g., *Aedes*) lay their eggs just above the water line or on wet mud; these eggs hatch only when flooded with water. If left dry, they can remain viable for many weeks. Once hatched, the larvae do not grow continuously but in four different stages (instars). The first instar measures about 1.5 mm in length, the fourth about 8–10 mm. Although they have no legs, they have a well-developed head and body covered with hair and swim with sweeping movements of the body. They feed on yeasts, bacteria, and small aquatic organisms. Most mosquito larvae have a siphon located at the tip of the abdomen through which air is taken in and comes to the water surface to breathe; they dive to the bottom for short periods to feed or escape danger. *Anopheles* larvae, which feed and breathe horizontally at the surface, have a rudimentary siphon (Ibrahim et al., 2024; Organization WH, 1997).

Female mosquitos acquire their blood meals either from animals or humans (Fig. 7.2). The mosquitos show certain selection criteria for certain animals or humans, certain body odors, CO_2, or even heat projecting from living bodies.

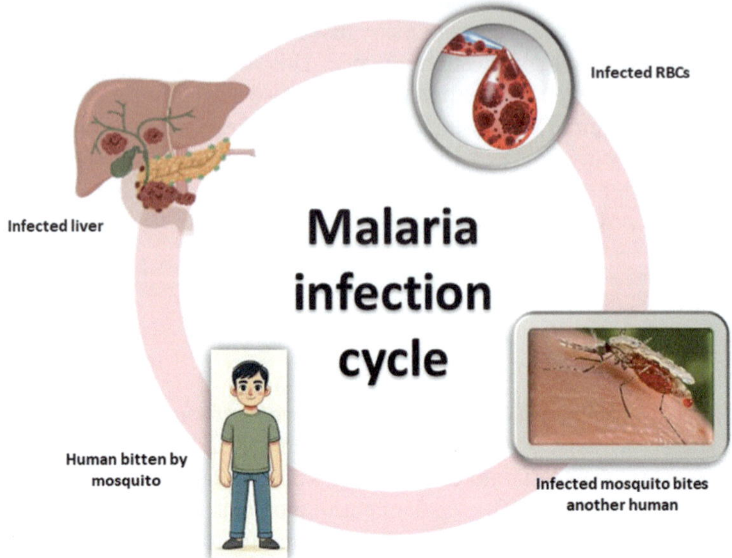

Fig. 7.2 Malaria infection cycle

Moreover, some species prefer to bite during certain times of the day during dusk and dawn or in the middle of the night. Feeding usually takes place during the night, but daytime biting also occurs. Some species prefer to feed in forests, some outside of houses, and others indoors (Ibrahim et al., 2024; Organization WH, 1997).

The process of blood digestion followed by egg laying requires 2–3 days as discussed above; thus, the egg-laying mosquito searches for a quiet place usually in the shade for protection. Some species prefer to rest in houses or cattle sheds, while others prefer to rest outdoors on vegetation, or at other natural sites. It is worth noting that, mosquitoes do not bite during the egg development stage which has an impact on the mosquito control measures by regarding them either nuisance insects or vectors of disease. In addition to that, mosquito species which bite animals are not disease transmitters among persons. Those that bite in the early evening may be more difficult to avoid than species that feed at night. Mosquitos that rest indoors are the easiest to control (Fikadu & Ashenafi, 2023; Organization WHO, 1997).

7.5 Treatment Options for Malaria Using Synthetic Drugs

Malaria is a treatable disease and through decades people utilized many natural resources for this purpose. One of the traditional and old medications is quinine which represented the treatment of choice for long; however, it showed many adverse effects combined with its high cost which provoked the need for a more safe and cost-effective replacement. Quinine is the principal alkaloid from *Cinchona* bark and it acted as a lead compound for the synthesis of many effective derivatives, namely, pamaquine, quinacrine (mepacrine; Atabrine®), sontaquine, and chloroquine. The loss of access by the allies to quinine supplies in World War II following the German conquest of the Netherlands and the subsequent occupation of the then Dutch East Indies by the Japanese in March 1942 were huge since these islands provided 90% of the world's quinine supplies. This was especially significant as much of the war would need to be fought in malarial areas of Southeast Asia and Southern Europe. Atabrine, the synthetic analog, helped in saving the lives of many of the allied forces in World War II. It was developed by German scientists in the 1930s. Other synthetic derivatives were further produced, including amodiaquine, mefloquine, and tafenoquine (Kingston & Cassera, 2022).

Artemisinin-based combination therapies (ACTs) are the most potent drug options available for malaria nowadays, especially effective against *P. falciparum* (the deadliest malaria species worldwide). These artemisinin-based therapies depend on the combination of two active ingredients working in two different ways of activity. This combination therapy consists of artemisinin-derived components from natural source (*Artemisia annua)* in addition to a partner drug (Dekebo et al., n.d.; Bukirwa et al., 2006; Eastman & Fidock, 2009; Kalbessa et al., 2019; van der Pluijm et al., 2021). Artemisinin acts by reducing the malaria parasite number through three days of treatment followed by the elimination of the rest of the parasites by the partner drug. WHO recommends that the treatment with artemisinin or its derivatives should be only initiated if the person tests positive for the malaria parasite in order to preserve the potency of such drug and its derivatives and to avoid the development of resistance against them. WHO does not support the promotion or use of *Artemisia* plant material (whether teas, tablets, or capsules) for the prevention or treatment of malaria. Over the last decade, parasite resistance to anti-malarial medicines has emerged as a threat in the fight against malaria, particularly in the Greater Mekong subregion. WHO is also concerned about more recent reports of drug-resistant malaria in Africa. Up till now, the reported drug-resistant malaria species are mainly *P. falciparum*, *P. vivax,* and *P. malariae*. However, nearly all patients infected with artemisinin-resistant parasites who are treated with an ACT are fully cured, provided the partner drug is highly efficacious (Organization WH, 2015). Anti-malarial drug resistance has emerged as one of the greatest challenges facing malaria control today. Plant resources remain the treasure source for developing countries in order to provide replacements for the drugs already resisted by malaria (Adeoye et al., 2017).

7.6 African Medicinal Plants with an Ethnopharmacological Role in Malaria

Malaria remains as one of the leading causes of morbidity and mortality in tropical countries. Drug resistance developed by the malaria parasite is increasing which provokes the need for new drug leads with potential anti-malarial activity (Zaruwa et al., 2018). Plants and their bioactive components can act as nontoxic mosquito treatment options as they contain a wide array of constituents that may act as alternative larvicidal, pupicidal, antifeedant, and repellent to known chemical insecticides (Kamaraj & Rahuman, 2010). The health hazards originating from the chemical insecticides have urged the need for safer, effective, and eco-friendly substitutes like botanicals and especially their volatile constituents that may be extracted either as essential oil or as *n*-hexane soluble volatiles (Zoubiri & Baaliouamer, 2014). The search for new plant-derived insecticidal agents is of great importance and was the focus of many research articles to find more effective natural agents that can combat insect vectors by many mechanisms (Ríos et al., 2017).

7.7 Review Method

The related information on African medicinal plants with anti-malarial activity was obtained from internationally recognized scientific databases and reputed publishers through the Internet [Web of Science®, PubMed®, American Chemical Society® (ACS), MDPI®, Springer Nature®, Royal Society of Chemistry®, Frontiers®, Wiley Online Library®, Elsevier®, and Egyptian Knowledge Bank® (EKB)]. The search term used was: "African" separated with the Boolean operator (AND) from the following terms: "plant", "extract", "malaria", "*Plasmodium*", "vector", "*Anopheles*", and "*Culex*." The names of the medicinal plant species were verified and retrieved from the International Plant Names Index® (IPNI) (International Plant Names Index [Internet], 2024) and the Plantlist® (Zhang, 2019), and all the available literature about them was reviewed up to June 2024 to obtain information about the African medicinal plants and their reported malaria-related activities. The inclusion criterion included articles where malaria was mentioned. Full-text articles were screened using the search terms previously mentioned in their titles, abstracts, or full text. The exclusion criteria were the gray literature and the articles addressing other diseases or vectors with no relation to malaria.

7.8 African Medicinal Plants with Anti-malarial Activity

Plants and plant compounds have been successfully used historically to cure a variety of illnesses, including malaria. Documents have even revealed that a few of the anti-malarial medications in use today have plant origins. For instance, the first anti-malarial medication was extracted from the stem bark of the *Cinchona* plant in 1820. The plant *Artemisia annua* is the source of even the well-known anti-malarial medication, artemisinin. Thus, these and numerous other achievements in the field of herbal medicine have prompted further studies on the different herbal plants that have historically been employed as anti-malarials. As a result, a number of plants have been traditionally used as anti-malarial agents in different African countries due to reports of their anti-malarial properties (Chinedu et al., 2014). In addition to protecting priceless indigenous knowledge and biodiversity, properly documenting traditional medicine and plants used in malaria prophylaxis and treatment is crucial for improving community access to and involvement in the advancement of malaria control interventions. Additionally, it is essential for advancing studies on the efficacy and safety of medicinal plants as well as the identification of chemical compounds that may one day be turned into new, standardized phytomedicines (Alebie et al., 2017). Many plant families (Fig. 7.3, Table 7.1) have been linked to either the prophylaxis or treatment of malaria as detailed below.

Fig. 7.3 Plant families with reported anti-malarial activity

Table 7.1 Examples of African medicinal plants with anti-malarial activity

No.	Plant name	English name	Common name(s)	Family name
1	*Achyranthes aspera* L. (*Achyranthes annua* Dinter)	Chaff-flower	Prickly chaff flower, devil's horsewhip	Amaranthaceae
2	*Aerva lanata* (L.) Schultes (*Ouret lanata*)	*Polpala*	Mountain knotgrass	
4	*Cussonia spicata*	*Cabbage tree*	Spiked, low veld cabbage tree or common cabbage tree	Araliaceae
5	*Adansonia digitata* L.	*The African baobab*	Baobab, monkey-bread tree, upside-down tree, cream of tartar tree.	Bombacaceae
6	*Cassia singueana*	*Scrambled egg*	Winter cassia, sticky pod	Caesalpiniaceae
7	*Cassia siamea* (*Senna siamea*)	Kassod Tree	Siamese cassia, Thailand shower, Johar, Jahor, Busuk-Busuk, Petai Belalang	
8	*Gynandropsis gynandra* (L.) Briq. (*Gynandropsis pentaphylla* DC.)	Spider Wisp	Shona cabbage, African cabbage, cat's whiskers, and stinkweed	Capparidaceae
9	*Maytenus senegalensis* (Lam.) Excell (*Gymnosporia montana* (Roth) Benth.)	Spike thorn	*Red spike-thorn, confetti tree*	Celastraceae
10	*Parinari excelsa* Sabine	Guinea plum	Gray plum, rough skin plum, rough-skinned plum, 'dawi' plum	Chrysobalanaceae
11	*Combretum molle*	Velvet bush willow	*Velvet leaf willow*	Combretaceae
12	*Combretum nigricans*	–	*Combretum, bush willow*	
13	*Combretum leprosum*	–	Mofumbo (Portuguese) and asucaró (Eastern Bolivian Spanish)	
14	*Mikania scandens*	Climbing hemp weed	American rope, mile-A-minute weed, bitter vine, Chinese creeper, climbing hemp vine	Compositae (Asteraceae)
15	*Tamarindus indica L.* (*T. officinalis* Hook.)	Tamarind	Tamarind, Asam, Asam Jawa, Tamarindo, Indian date	
16	*Momordica foetida* Schum. (*Momordica mannii* Hook.f.)	Bitter melon	Concombre sauvage (wild cucumber), nyanya-nua, sɔprɔpɔ, kakle, awoduan (snake food), cerassee, goya, bitter apple, bitter	Cucurbitaceae

(continued)

Table 7.1 (continued)

No.	Plant name	English name	Common name(s)	Family name
			gourd, bitter squash, balsam-pear, karavila	
17	*Luffa cylindrica* (L.) M.J. Roem. (*Luffa aegyptiaca* Mill.)	Loofah	Chishambo, vegetable sponge	
18	*Lagenaria siceraria*	Bottle gourd	Labu Ayer Puteh, calabash gourd	Ebenaceae
19	*Diospyros chamaethamnus*	Sand apple	–	
20	*Diospyros quaesita*	Calamander	*Calamander, kalu mediriya*	
21	*Diospyros conocarpa*	–	*Bolinda, Muika Ndombe*	
22	*Erythrina sacleuxii* Hua	–	Muungu	Fabaceae
23	*Erythrina schliebenii*	–	*Coral Tree*	
24	*Erythrina latissima*	Broad-leaved coral tree	*Lucky-bean tree*, breëblaarkoraalboom, grootblaar (Afr.)	
25	*Hoslundia opposita* Vahl (*Hoslundia oppositifolia* P.Beauv.)	Orange bird berry	*Bird gooseberry* (*Eng.*); *uyaweyawe* (*Zulu*)	Lamiaceae
26	*Leonotis ocymifolia*	Minaret flower	Rock lion's paw, minaret flower (Eng.); klipdagga, wildedagga (Afr.)	
27	*Cissampelos pariera*	Velvet leaf	False Pareira Brava, Abuta, Pereira root, Barbasco	Menispermaceae
28	*Cissampelos mucronata*	Hairy heart leaf	Heart-leaved vine	
29	*Acacia polyacantha* Willd	White thorn	*White catechu*. White cutch	Mimosaceae
30	*Albizia gummifera* J.F.G mel.	Peacock flower	Kmntia, Kisya, Musya, Mwethia, Kikuyu: Mukurue, Kipsigis, Seet, Se yet, Kisii: Omugonjoro, Luhya, Mukhonzu1i; Luhya (Bukusu), Kumulukhu; Maiakote: Motemwelo, Maasai: Onnoso, Olsamakupe, Osapukupes, Marakwet: Se, Set, Ste, Meru: Mukuruwe; Nandi: Seet, Seyet; Pokomo: Mchachampili, Pokot: Ses; Samburu: Sogore,	

(continued)

Table 7.1 (continued)

No.	Plant name	English name	Common name(s)	Family name
			Sogorogurri, Swahili: Mchani mbao, Taita: Msarawachi; Taveta: Kiririgiti, Kiririgwi; 'fugen: Se, Set, Seot; Turkana: Ekakwait, Ekeweit	
31	*Psidium guajava* L.	Guava	Guava, Jambu Batu, Jambu Biji, common guava, yellow guava, apple guava, Jambu Kampuchia	Myrtaceae
32	*Keetia leucantha*	–	*Sierra leone*	Rubiaceae
33	*Heinsia crinita* (Afzel.) G.Taylor subsp. *parviflora* (K.Schum. & K.Krause) Verde (*Gardenia pulchella* G.Don)	Bush apple	Jasmine-gardenia and small false gardenia	
34	*Pavetta crassipes* K. Schum. (*Pavetta barteri* Dawe)	Gland leaf brides bush	Mchoka (Kunda), likaka-lyanyombe, lilime-lyanyombe, manjatali, mpewu, mteruteru, matuwa-ngoma, mdyansefu, mtangoma (Nyanja)	
35	*Zanthoxylum chalybeum* Engl. (*Fagara ehalybea* (Engl.))	Lemon-scented knob wood	*Knob Wood, Kundanyoka, knob wood*	Rutaceae
36	*Paullinia pinnata* L.	Barbasco	Bread-and-cheese, bread and cheese, dragon weed, basket wiss, leyland cypress, timbo	Sapindaceae
37	*Clerodendrum myricoides (Rotheca myricoides)*	Butterfly bush	Bandamuchenene (Shona), blue cat's whiskers (English), Mudzimamuriro (Shona), Muparamhozva (Shona), Mutilingwa (Tonga: Zimbabwe), Nyamhepo (Shona), Zanzi (Shona)	Verbenaceae
38	*Cloredendrum inerme*	Indian privet	Seashore tube flower, bunga pawang, wild jasmine, sorcerers bush, seaside clerodendrum	

7.8.1 *Amaranthaceae*

7.8.1.1 *Achyranthes aspera* L. (*Achyranthes annua* Dinter)

A. aspera was rich with phenols, tannins, alkaloids, saponins, flavonoids (Fig. 7.4), and balsam. The extract was found safe in cytotoxic assay (dose 5000 mg/kg/bwt). The plant extract exerted a significant reduction ($p < 0.05$) in parasitemia (400 & 600 mg/kg/day) in Swiss albino mice. The plant decreased *P. berghei*-induced body weight loss and temperature reduction (Mankilik et al., 2021). Moreover, the methanol extract of *A. aspera* was tested against the malaria-causing vector (*P. berghei*) in vivo (50, 100, and 200 mg/Kg body weight) where the extract showed a significant reduction in schizonts (74%, 100%, and 87%, respectively) (Zaruwa et al., 2018).

7.8.1.2 *Aerva lanata* (L.) Schultes

The ethyl acetate and methanol extracts of *A. lanata* were evaluated for their potential anti-plasmodial effect. *A. lanata* (aerial parts-ethyl acetate extract) presented good anti-plasmodial activity against *Plasmodium* strain coded *Pf* 3D7 (IC_{50} 21–40 μg/ml) (Kaushik et al., 2015).

7.8.2 *Araliaceae*

7.8.2.1 *Cussonia* Species

Cussonia species were reported in African folk medicine to treat pain, inflammation, GIT ailments, malaria, and sexually transmitted diseases. The leaves of *Cussonia* species were evaluated against *P. falciparum* by the [3H]-hypoxanthine incorporation assay. The cytotoxicity was tested using T-cell leukemia (Jurkat) cell line (tetrazolium-based cellular toxicity assay). *C. arborea* methanol extract was effective against *P. falciparum* (13.68 g/mL) (De Villiers et al., 2010).

7.8.3 *Bombacaceae*

7.8.3.1 *Adansonia digitata* L. (Baobab)

A. digitata L. was reported for traditional use against malaria as well as other infections. *A. digitata* stem bark extract was investigated against *P. berghei*-infected mice. The plant was extracted with water and methanol then ethyl acetate and chloroform were fractionated from the methanol extract. The extracts and fractions

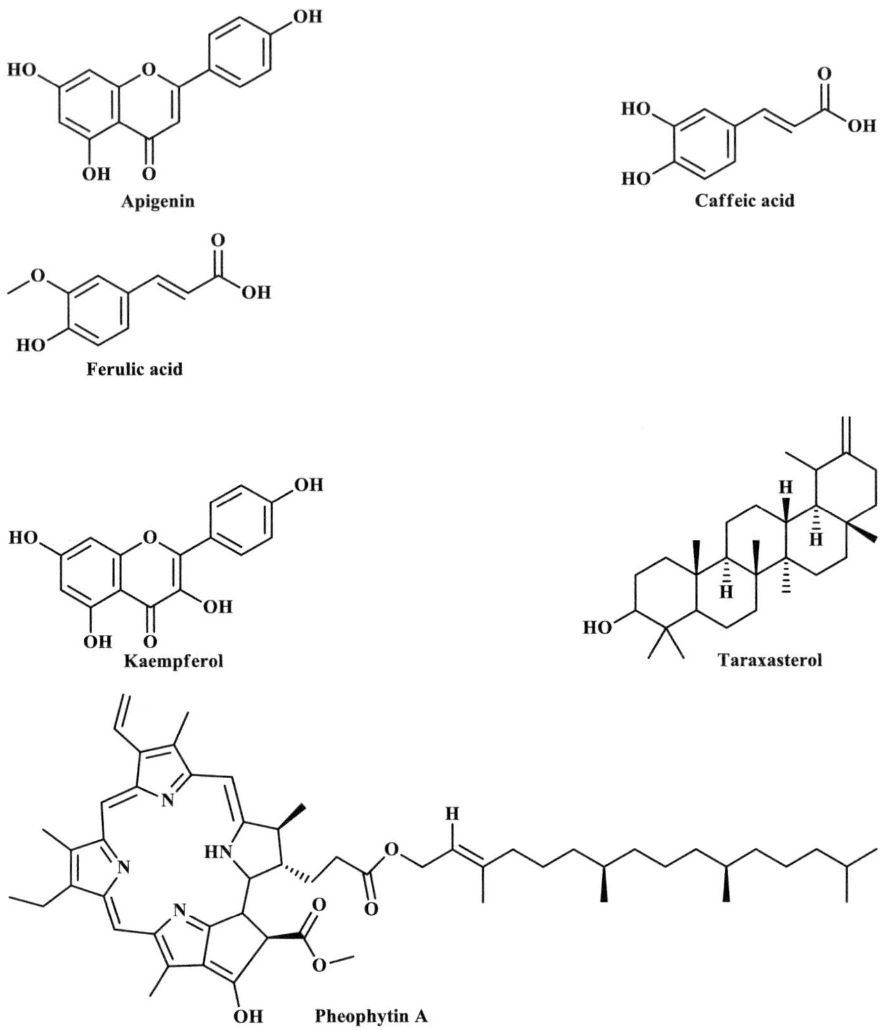

Fig. 7.4 Structures of the main identified phytoconstituents from African medicinal plants with anti-malarial activity

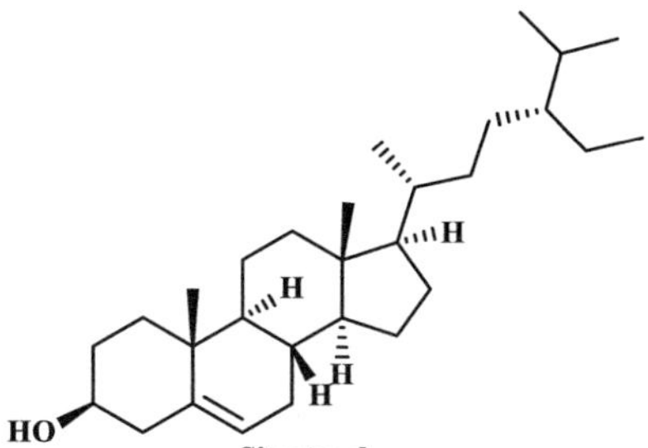

Sitosterol

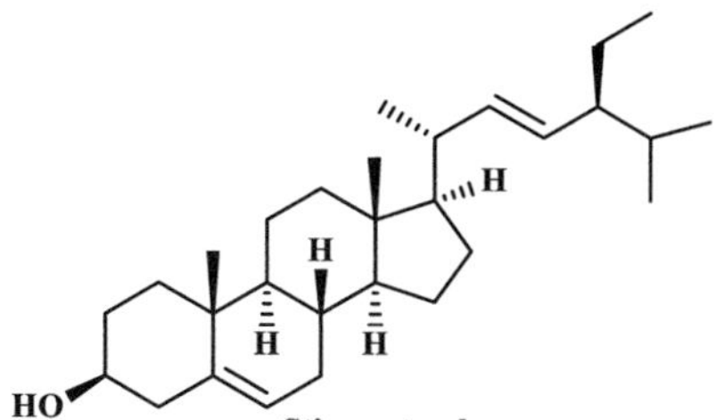

Stigmasterol

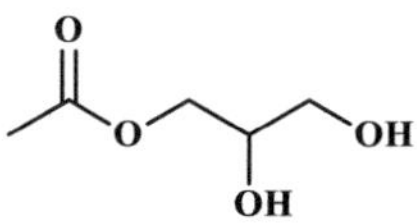

Monoacetyl glycerol

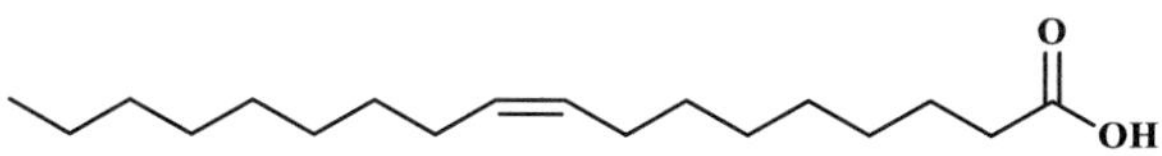

Oleic acid

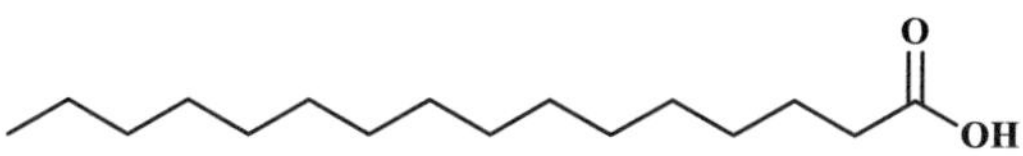

Hexadecanoic acid

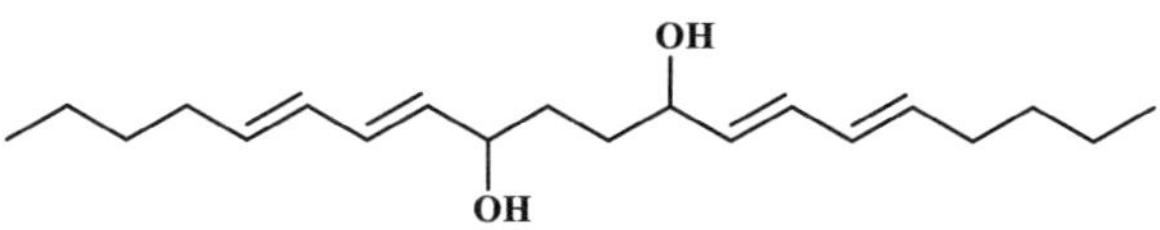

5,7,13,15-Eicosatetraen-9, 12-diol

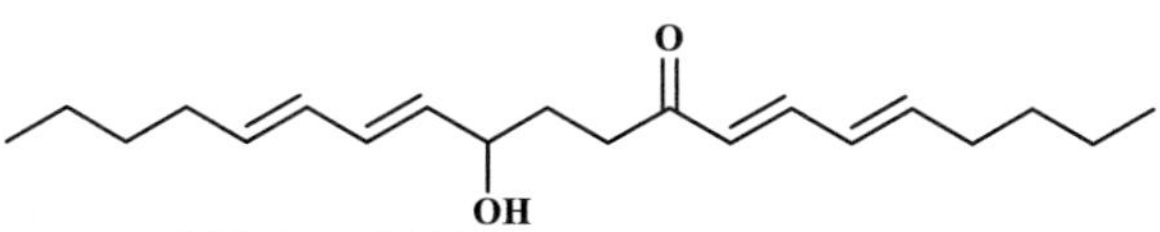

9-Hydroxy-5,7,13,15-eicosatetraen-12-one

Fig. 7.4 (continued)

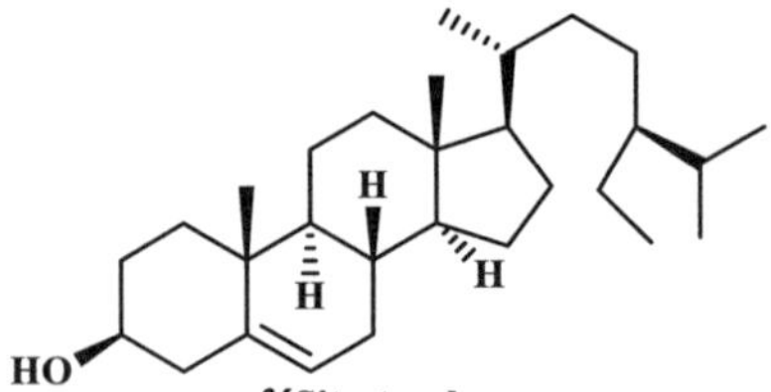

γ-Sitosterol

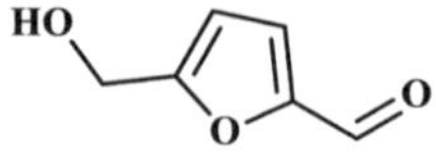

5-Hydroxymethyl furfural

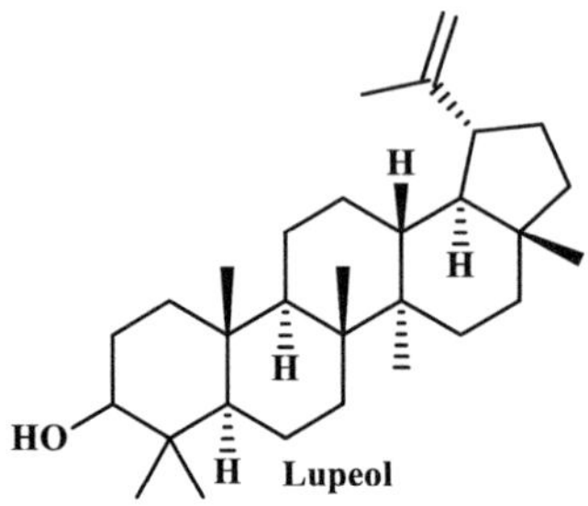

Lupeol

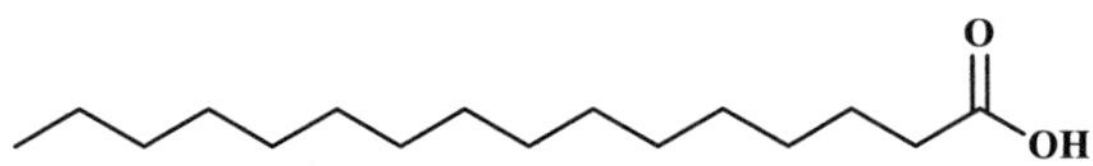

Palmitic acid

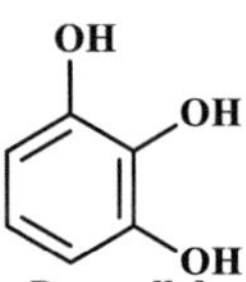

Pyrogallol

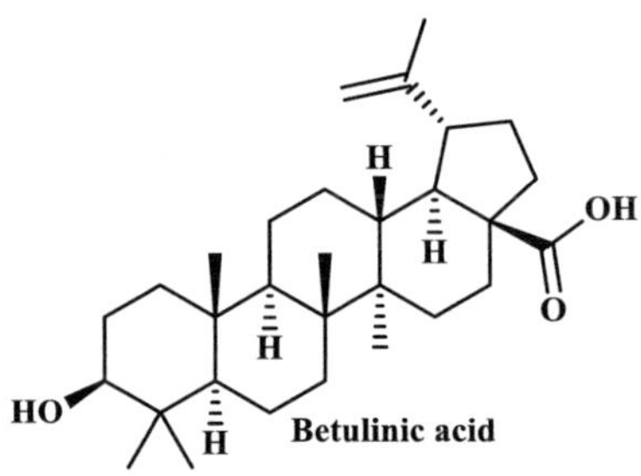

Betulinic acid

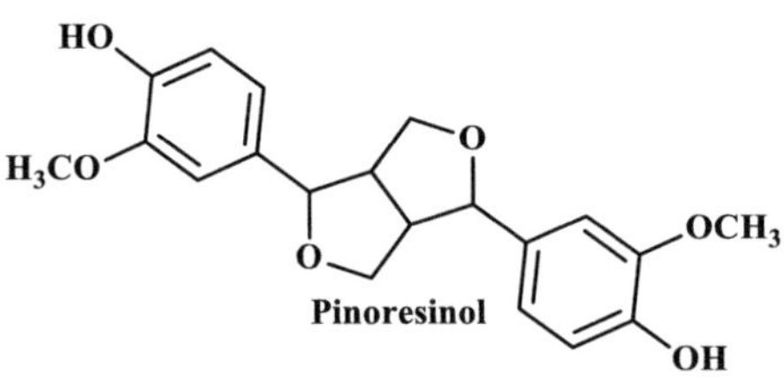

Pinoresinol

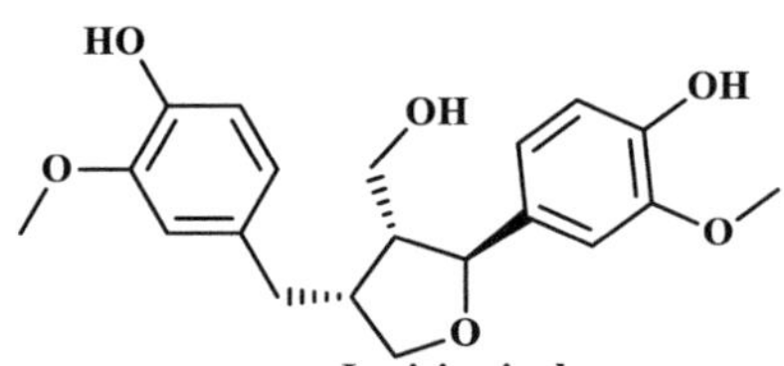

Lariciresinol

Fig. 7.4 (continued)

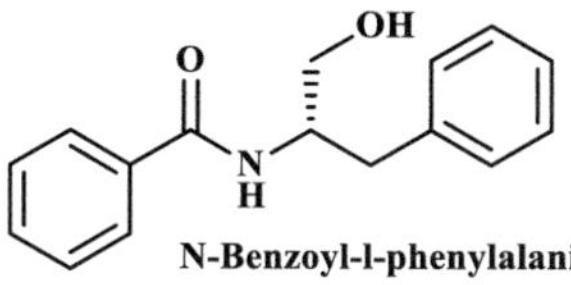

N-Benzoyl-l-phenylalaninol

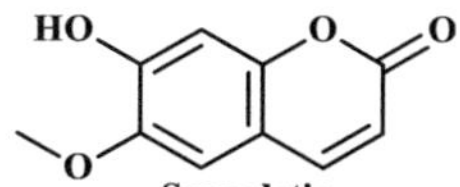

Scopoletin

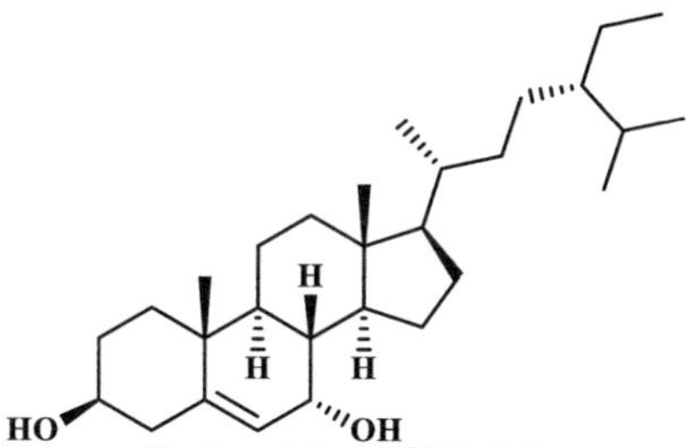

Poriferast-5-en-3β,7α-diol

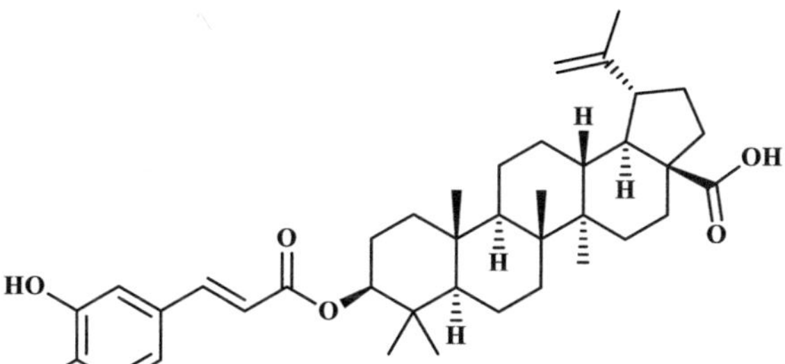

Betulinic acid-3- caffeate

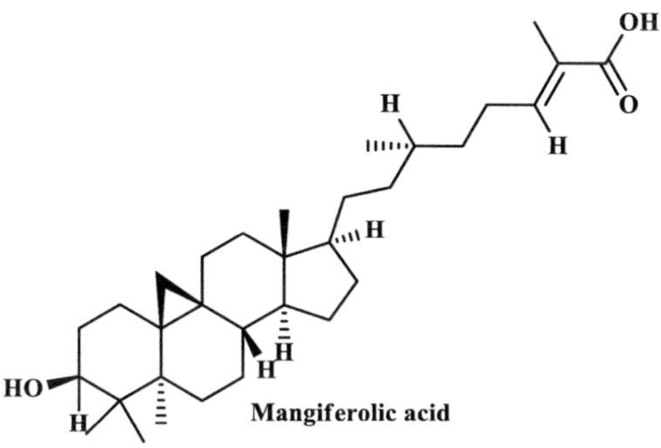

Mangiferolic acid

Fig. 7.4 (continued)

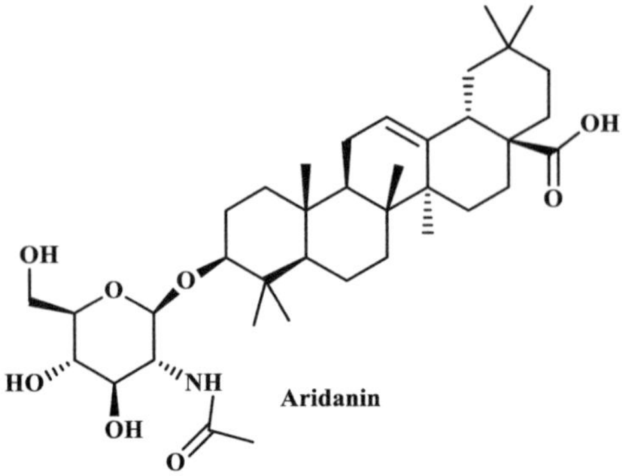

Aridanin

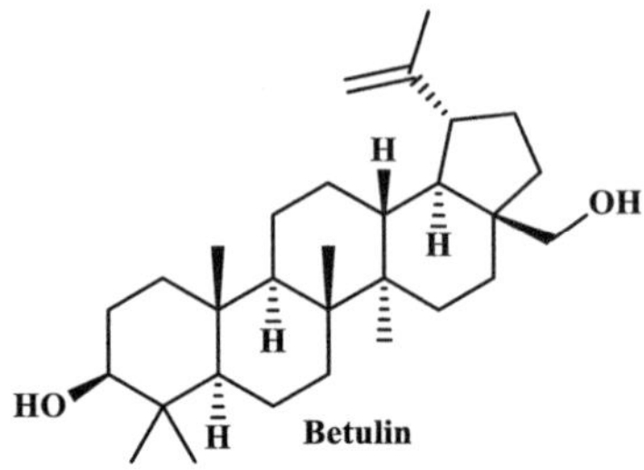

Betulin

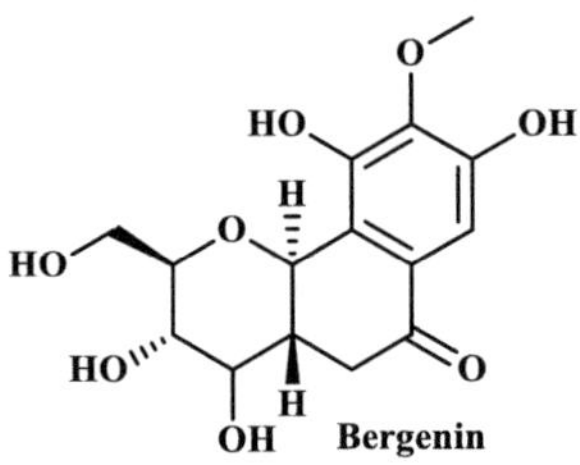

Bergenin

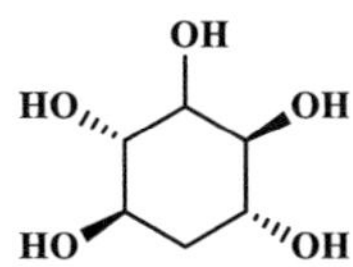

D-Quercitol

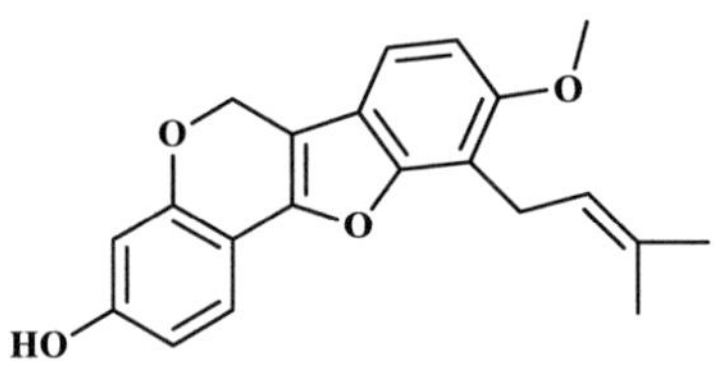

3-Hydroxy-9-methoxy-10-(3,3-dimethylallyl) pterocarpene

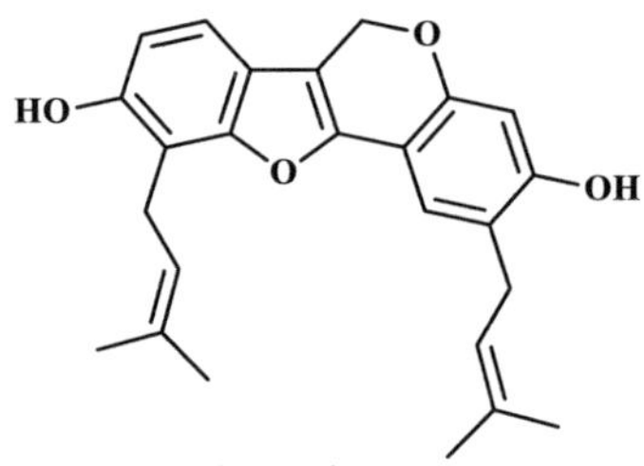

Erycristagallin

Fig. 7.4 (continued)

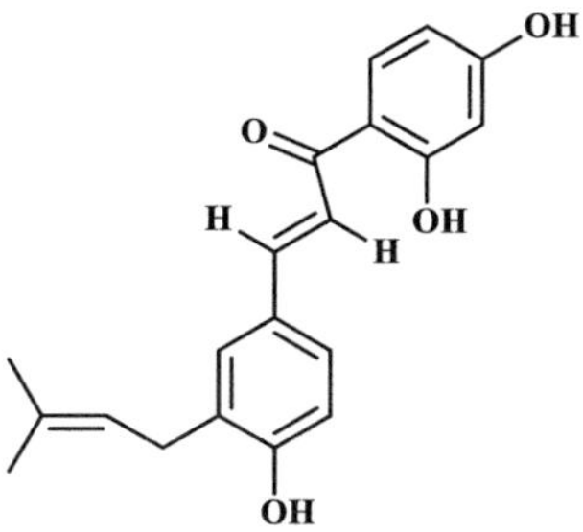

Licoagrochalcone A

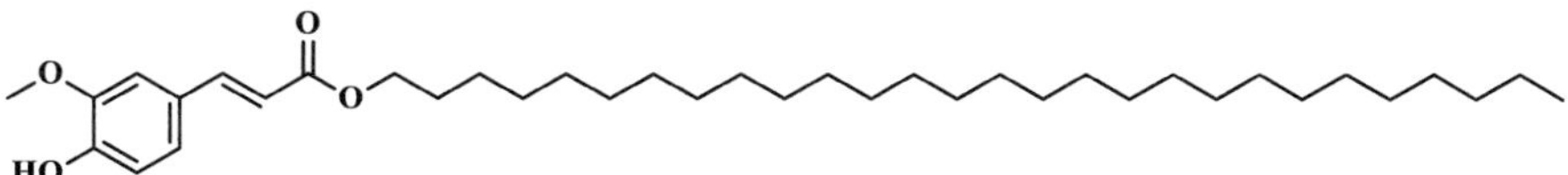

Octacosyl ferulate

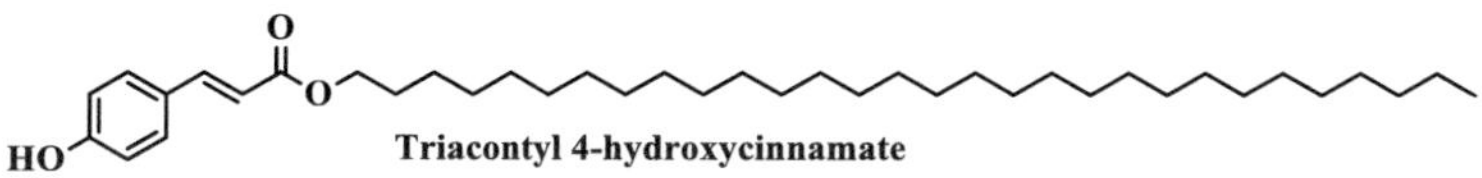

Triacontyl 4-hydroxycinnamate

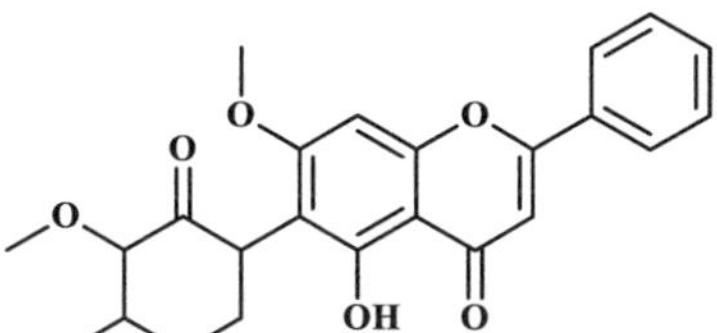

Hoslundin

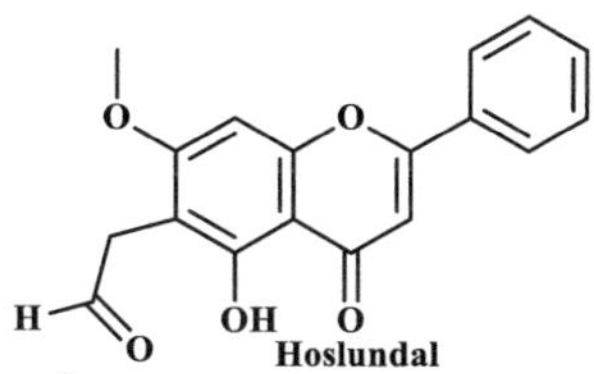

Hoslundal

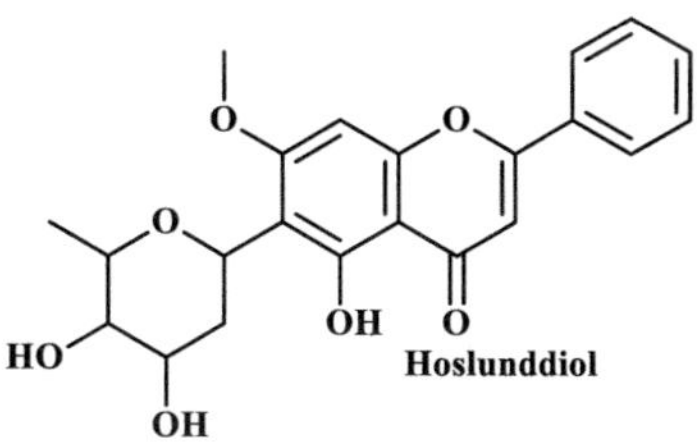

Hoslunddiol

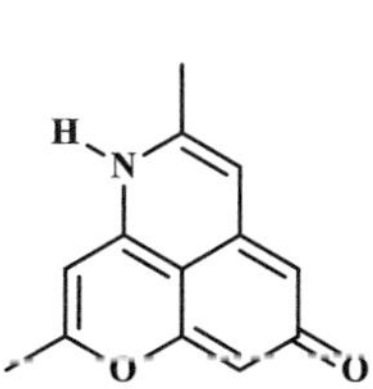

Cassiarin A

Fig. 7.4 (continued)

were administered to the albino mice (200 mg/kg body weight and 400 mg/kg body weight). Artemether-lumefantrine and chloroquine (5 mg/kg body weight/day) acted as the positive control. The extracts and fractions showed chemosuppression of the parasite which was dose-dependent. The highest dose (400 mg/kg body weight) showed the most potent activity in all the groups where the ethyl acetate fraction recorded the highest chemosuppression. Both the methanol extract and the ethyl acetate fraction showed a delay in parasitemia (Adeoye et al., 2017). Similarly, the stem bark extract of *A. digitata* was tested against experimental malaria. The extract significantly and in a dose-dependent manner elevated the packed cell volume and the % chemosuppression/clearance while significantly lowering the % parasitemia at different doses. In addition to that, the methanol extract at the dose of 400 mg/kg body weight significantly lowered tissue peroxidation and elevated the levels of reduced glutathione, catalase, and superoxide dismutase activity (Adeoye & Bewaji, 2018).

7.8.4 *Caesalpiniaceae*

7.8.4.1 *Cassia* Species

C. singueana is reported to be used in North Nigeria for the treatment of acute malarial attacks. The methanol extract of *C. singueana* root bark was investigated against *P. berghei* infection in mice and rats. The extract presented significant anti-plasmodial activity ($LD_{50} = 847+/-30$ mg/kg, i.p. in mice). Phytochemically the extract was rich in phenols, saponins, tannins, and some traces of anthraquinones which may be responsible for the deduced activity (Adzu et al., 2003). Another *Cassia* species called *C. siamea* was also tested as an anti-malarial agent. The plant leaves are known to be rich with Cassiarin A which is known for its anti-malarial potency. The anti-malarial activity test was done using *Plasmodium falciparum* 3D7. *C. siamea* leaf extract showed potent anti-malarial activity ($IC_{50} = 0.006$ μg/ml) (Tasiam et al., 2020).

7.8.5 *Capparidaceae*

7.8.5.1 *Gynandropsis gynandra* (L.) Briq. (*Gynandropsis pentaphylla* DC.)

Cleome gynandra is used widely in African folk medicine against malaria. Different extracts (hexane, ethyl acetate, and methanol) of this plant collected from three African countries, namely, Nigeria, Tanzania, and Uganda, were prepared. The extracts were rich with apigenin, caffeic acid, ferulic acid, kaempferol, taraxasterol, pheophytin A, sitosterol, stigmasterol, monoacetyl glycerol, 2-monoacetyl glycerol, oleic acid, and mixtures of fatty acids and triglycerides, hexadecanoic acid, and eicosatetraenes (5,7,13,15-eicosatetraen-9, 12-diol and 9-hydroxy-5,7,13,15-eicosatetraen-12-one). The methanol/ethyl acetate extract (12.5, 25, 50, 100, and 200 mg/kg) presented significant anti-malarial activity (73.2% clearance, dose 50 mg/kg of the extract) against *P. berghei* NK65 in vivo compared to 25 mg/kg of Halofantrine (control) for four days (Igoli et al., 2016).

7.8.6 *Celastraceae*

7.8.6.1 *Maytenus senegalensis* (Lam.) Excell (*Gymnosporia montana* (Roth) Benth)

Maytenus senegalensis is a widely used medicinal plant from Tanzania that has been evaluated in vivo as an anti-malarial agent. The ethanol root extract was tested in mice using the Peters 4-day in vivo anti-plasmodial effect against early rodent malaria infection in chloroquine-sensitive *P. berghei* NK 65 strain in mice. The reduction in parasitemia was the highest for sulfadoxine/pyrimethamine (25/1.25 mg/kg body weight) followed by 100 mg/kg of the extract, 75 mg/kg, 50 mg/kg, and 25 mg/kg body weight of the extract in a descending manner. Thus this plant can serve as a new and safe anti-plasmodial drug resource (Malebo et al., 2015). Likewise, several medicinal plants from Tanzania with documented anti-malarial activity had been tested through the 4-day suppressive test in *P. berghei*-infected mice and were assayed for cytotoxic activity in human carcinoma cell lines in vitro. The most potent among them was the root and stem bark of *M. senegalensis* (90% and 63% suppression of parasitemia, respectively, at 500 mg/kg body weight/day) (Gessler et al., 1995).

7.8.7 *Chrysobalanaceae*

7.8.7.1 *Parinari excelsa* Sabine

Parinari excelsa Sabine is an indigenous tree from West and Eastern Africa. This tree is used traditionally in Ivory Coast as an anti-malarial treatment. The plant's leaves and bark were investigated as anti-plasmodial (methylene chloride and methanol) in vitro. The anti-plasmodial activity of the extracts can be ordered as $IC_{50} = 3.41$ bark methylene chloride $< IC_{50} = 4.42$ leaf methanol against *P. falciparum*. The bark extract, being the most active, was fractionated and new ventiloquinone, five triterpenes, and one chlorogenic acid were identified. 16*β*-hydroxylupane-1,20(29)-dien-3-one presented the most potent activity ($IC_{50} = 28.3$ μM) (Attioua et al., 2012).

7.8.8 *Combretaceae*

7.8.8.1 *Combretum* Species

The *Combretum molle* seed extract was evaluated using the 4-day suppressive test against malaria. The plant was found safe in the cytotoxicity study, while in the anti-plasmodial testing, it resulted in 63.5% parasite suppression in *P. berghei*-infected mice (250 mg/kg). The extract-treated mice presented longer survival time against the parasite. The plant seed extract after four days of infection had significantly reduced body weight and temperature in addition to preserving the RBCs from hemolysis (at low dose). Using the optimum dose, the extract showed moderate chemosuppression and enhanced clinical picture in malaria-infected mice (Anato & Ketema, 2018). Similarly, *Combretum nigricans* is traditionally used as anti-malarial in North-Central Nigeria. *C. nigricans* aqueous/methanol leaf extract was tested against *P. berghei* in mice. The Peters' 4-day suppressive test against early malaria infection and Rane's curative test against established malaria were applied. The extract (200–800 mg/kg b.w. p.o.) gave significant ($P < 0.05$) chemosuppression against the plasmodial vector (400 mg/kg showed the best effect). All of the tested doses served to prolong the survival time against the parasite and prevented the decrease in body temperature in vivo (Enegide et al., 2021).

On the other hand, the flower extracts of different polarities were prepared from *Combretum leprosum* and assayed against the *Plasmodium* parasite. The most potent anti-malarial ($IC_{50} = 3.97$ µg/mL) and also highest cytotoxic activity with Median Lethal Dose ($MLD_{50} = 18.12$ µg/mL) was observed with the chloroform extract; however, this extract also presented higher cytotoxicity. The most suitable extract was the methanol extract [Inhibition score (IS) = 18.97] (Passarini et al., 2017). In another study, mice were infected with *P. berghei* and then were administered the methanol extract and solvent fractions of *C. molle* (100, 200, and 400 mg/kg/day). The chemosuppresive effect was methanol extract 27–68%, chloroform fraction 52.93–67.73%, *n*-hexane fraction 23.61–44.97% and aqueous fraction 29.48–51.13%. While the curative effect was methanol 25–49% and chloroform fraction exhibited 42.78–69.22%. The prophylactic effect was of 51–76.2% for the methanol extract and 46.57–71.96% for the chloroform fraction. Thus the methanol extract showed chemosuppresive, curative, and prophylactic effects, prevention of weight loss, temperature and packed cell volume, and an increase in mean survival time (Adigo Shibeshi et al., 2021).

7.8.9 *Compositae (Asteraceae)*

7.8.9.1 *Mikania* Species

The volatile oil of *Mikania scandens* was investigated in vitro against *Anopheles gambiae*. *Alpha*-bisabolol (39.34%) and stigmasterol (13.45%) were the major volatile components. The essential oil exhibited a maximum larval mortality rate at a dose of 1000 ppm. LC_{50} reached 488 ± 2.45 ppm, while enzyme inhibition assays resulted in a significant increase in CYP450, GST levels and a decrease α-β carboxylesterase level in both the third and fourth instar larvae ($p \leq 0.0001$). Upon histology of the mosquito gut, the mid-gut was severely damaged (Radhakrishnan et al., 2023).

7.8.9.2 *Tamarindus indica* L. (*T. officinalis* Hook)

T. indica L. is a beneficial tree in both medical and nutritive applications which is native to Sudan and grows in many other parts of the world. The fruits of *Tamarindus* are documented for their anti-malarial potency. The ethanol and aqueous extracts in addition to the *n*-hexane, chloroform, methanol, and water successive extracts were prepared to assess their anti-plasmodial activity. The chloroform extract presented the most potent activity (IC_{50} = 34.8 μg/mL) and was subjected to GC-MS analysis where aliphatic saturated alcohols (12.15%), aliphatic saturated fatty acids and their derivatives (17.41%), phenolic compounds (2.67%), aromatic acids (0.58%), γ-sitosterol (2.33%), and 5-hydroxymethyl furfural (30.74%) were the main components (Ahmed & Ayoub, 2015). Moreover, Tamarind (*T. indica* L.) anti-malarial activity was investigated. The methanol extract was evaluated against *P. falciparum* and *P. vivax* using the plate diffusion method (dose = 10 μl/plate) where inhibition zones were recorded. Clear zones of inhibition were observed using 5% and 10% extract, while the MIC values were 1.9 and 4.8 for *P. falciparum* and *P. vivax*, respectively (Usman & Sunil, n.d.).

7.8.10 *Cucurbitaceae*

7.8.10.1 *Momordica foetida* Schum. (*Momordica mannii* Hook.f.)

Momordica foetida had been folk medicinally utilized to treat malaria in East and Central Africa. Parasite lactate dehydrogenase was applied to test for *P. falciparum* drug sensitivity. Aqueous extract showed a weak in vitro effect ($IC_{50} > 28.00$ μg/ml), while the in vivo assay (dose 10, 100, 200, and 500 mg/kg 2 × day) elongated the survival of *P. berghei*-infected mice from 7.0 ± 1.8 to 17.9 ± 1.8 days. This study highlights the differences between the in vitro and in vivo experiments on plant extracts (Waako et al., 2005).

7.8.10.2 *Luffa cylindrica* (L.) M.J.Roem. (*Luffa aegyptiaca* Mill)

Luffa cylindrica is a Nigerian medicinal plant with a traditional history against malaria. Different extracts and fractions from this plant (hexane, ethyl acetate, methanol, and aqueous) were assayed in vitro against *P. falciparum*. The ring stage synchronized *P. falciparum* Pf3D7 strain, 1% parasitemia, and 2% hematocrit were incubated in a 96-well microplate for 96 h with different concentrations of plant extracts (0.05, 5, and 5 mg/ml) and fractions of the most potent extract (0.01, 0.1, and 1 mg/ml). Alkaloids, saponins, phenolics, and flavonoids were the main components of the extracts. The highest anti-plasmodial activity was observed with the leaf methanol extract ($IC_{50} = 2.75 \pm 0.21$ mg/ml) and then this active extract was fractionated, and the most potent fractions against the parasite growth were fractions F7–F10 with an $IC_{50} = 2.23 \pm 0.34$, 2.30 ± 0.45, 2.31 ± 0.64, and 2.20 ± 0.62 mg/ml, respectively (Tona et al., 2004).

7.8.11 Ebenaceae

7.8.11.1 Lagenaria Species

Zinc oxide nanoparticles were developed and incorporated with the peel water extract of *L. siceraria*. The extract and the prepared nanoparticles were tested against third instar larvae of *An. stephensi*. The aqueous extract of *L. siceraria* and its prepared ZnO nanoparticles were evaluated against *P. falciparum* chloroquine-sensitive strains. The extract and nanoparticles presented potent cytotoxicity ($IC_{50} = 62.5$ µg/mL, HeLa cell lines). Thus, *L. siceraria* extract and nanoparticles may act as a green source for the eradication of malaria (Kalpana et al., 2020). In another similar study, the aqueous peel extract of *L. siceraria* was applied to prepare ZnO nanoparticles which were then evaluated on different stages of *An. stephensi*. The peel extract and nanoparticles were found poisonous against the fourth instar larvae (LC_{50} 185.78 ppm, $LC_{90} = 389.05$ ppm) while the nanoparticles LC_{50} were 3.873 ppm and LC_{90} were = 8.83 ppm. The peel extract and the nanoparticles can serve as a green and safe option for future research on anti-malarial agents (Vellore Nagarajan & Vijayarangan, 2019).

On the other hand, extracts of different polarities were prepared from leaves and stems of *L. siceraria*. The prepared extracts were tested as a repellent against *Culex pipiens* L. adult. The petroleum ether leaf extract (3.33 mg/cm^2) was equally potent to DEET, the artificial insecticide, as a repellent, while the stem extract using the same solvent and dose showed less repellency (89.6%). The leaf and stem ethanol extracts were the least effective with repellency of 81.3% and 69.1% at 6.67 mg/cm^2, respectively. Thus the leaf and stem extracts of *L. siceraria* may work as an alternative option for mosquito control and may be commercialized in the future as repellent agents from natural sources (Hassan et al., 2014).

7.8.11.2 *Diospyros* Species

Diospyros chamaethamnus is a woody plant indigenous to different African countries, namely, Namibia, Botswana, Angola, and Zambia, with folk use against malaria. The dried root dichloromethane (DCM)-methanol (ME) and aqueous extracts of *D. chamaethamnus* were tested against *P. falciparum* (chloroquine-sensitive strain D10). Intermediate anti-malarial activity ($IC_{50} < 50$ mg/ml) was recorded in vitro (DCM: ME 19.51 mg/mL) and in (aqueous 18.30 mg/mL) extracts. Through GC/MS analysis, lupeol, palmitic acid, stigmasterol, and pyrogallol were identified (Du Preez, 2022). In another study, several triterpenes, namely, lupeol, *β*-sitosterol, stigmasterol, and diospyrin, were isolated from the wood extract of *D. glandulosa* and *D. rhodocalyx*, and then the identified compounds were evaluated in vitro against *P. falciparum* K1 giving IC_{50} values of 3.29–6.25 μg/mL (du Preez-Bruwer et al., 2022; Theerachayanan et al., 2007).

Similarly, aerial parts of *D. quaesita* were extracted and then subjected to fractionation. Betulinic acid, pinoresinol, lariciresinol, N-benzoyl-l-phenylalaninol, scopoletin, and poriferast-5-en-3b,7a-diol, betulinic acid-3-caffeate, and betulinic acid were identified. The latter compound had anti-plasmodial activity in vitro against *P. falciparum* strain D6 (chloroquine-sensitive, $IC_{50} = 1.40$ mm) and strain W2 (chloroquine-resistant, $IC_{50} = 0.98$ mm) (Ma et al., 2008).

Moreover, the Cameroonian species (*D. conocarpa)* was phytochemically evaluated where mangiferolic acid, 3*β*, 22(S)-dihydroxycycloart-24E-en-26-oic acid, lupeol, aridanin, betulin, betulinic acid, bergenin, and D-quercitol were isolated. Aridanin and the root extract of *D. conocarpa* showed moderate anti-plasmodial and anti-trypanosomal actions (Fouokeng et al., 2019).

7.8.12 Fabaceae

7.8.12.1 *Erythrina sacleuxii* Hua

E. sacleuxii is a native species from Kenya with documented anti-plasmodial, antimicrobial, and cytotoxic activities. The acetone extract of *E. sacleuxii* root and stem bark extracts exhibited anti-plasmodial activities against *P. falciparum* strains D6 (chloroquine-sensitive, $IC_{50} = 4.9$–22.0 μM) and W2 (chloroquine-resistant, $IC_{50} = 6.1$–24.9 μM) strains. The root bark extract yielded 5-deoxy-3′--prenylbiochanin A and other isoflavonoids (Andayi et al., 2006; Mwakumanya et al., 2022). From the root bark of *E. abyssinica*, 3-hydroxy-9-methoxy-10-(3,3-dimethylallyl) pterocarpene, 7,4′-dihydroxy-2′,5′-dimethoxyisoflav-3-ene, erycristagallin, licoagrochalcone A, octacosyl ferulate, and triacontyl 4-hydroxycinnamate were isolated. The extract and the isolated flavonoids and their derivatives presented anti-plasmodial effects (Yenesew et al., 2003).

E. schliebenii is previously reported against malaria in Tanzania. *The* anti-malarial activity was evaluated in vivo by using the 4-day suppression test. Mice were injected with RBCs infected with *P. berghei* ANKA and then were fed with the extract once daily. The extract showed suppression of *Plasmodium* which depended on the used dosage (Nondo et al., 2016). Similarly, the anti-plasmodial activity of the prenylated flavonoids from *E. latissima* from South Africa was evaluated using *P. falciparum* K1. Moderate activity was observed, and the IC_{50} was in the range of 1–30 μM (Tuenter et al., 2019).

7.8.13 *Lamiaceae*

7.8.13.1 *Hoslundia opposita* Vahl (*Hoslundia Oppositifolia* P.Beauv.)

The glycolytic pathway represents one of the important enzyme targets in the malaria cycle; *P. falciparum* lactate dehydrogenase is especially utilized as a target for anti-malarial drug development. The in silico anti-malarial activity of certain compounds isolated from *H. opposita* was evaluated. The isolated compounds were hoslundin, hoslundal, and hoslunddiol. Compound binding was promising and opened the way for future in vitro and in vivo studies (Shadrack et al., 2016). The in vitro chemosensitivity of the aqueous and hydroethanol extracts of *H. opposita* leaves on *P. falciparum* was tested using a maturation test. The maturation test of the extracts showed anti-malarial activity, with the aqueous extract being more potent (IC_{50} 12.66 μg/mL) while the hydro-ethanol extract showed moderate activity, $IC_{50} = 15–50$ μg/mL (Okou et al., 2019).

Abietane-type diterpenes such as 3-*O*-benzoylhosloppone, 3-*O*-cinnamoyl-hosloppone, 3-*O*-benzoylhinokiol, and 3-*O*-benzoylhosloquinone isolated from the root bark *H. opposita* were assessed against malaria. 3-*O*-Benzoylhosloppone inhibited the growth of *Plasmodium* strain K in vitro ($IC_{50} = 0.4$ μg/ml) (Achenbach et al., 1992; Akharaiyi et al., 2022).

7.8.13.2 *Leonotis* Species

The hydroalcohol extract of Leonotis ocymifolia (dose 100–800 mg/kg) was tested against *P. berghei*. The assay involved the recording of parasitemia, survival time, packed cell volume, temperature, and weight. The plant extract suppressed the parasite [41.4%, 800 mg/kg ($p < 0.001$)]. The suppression was succeeded by elongation in survival time for the doses of 400 and 800 mg/kg ($p < 0.05$) in a 4-day suppression assay and using 800 mg/kg ($p < 0.05$) in Rane's test. The plant extract has no effect on the packed cell volume or on the weight, although it inhibited the decrease in temperature using 400 and 800 mg/kg ($p < 0.05$) in a 4-day suppression test and at 800 mg/kg ($p < 0.05$) in Rane's assay (Teklu et al., 2020).

7.8.14 *Menispermaceae*

7.8.14.1 Genus *Cissampelos*

The rhizomes of *Cissampelos pariera* were rich with *bis*-benzylisoquinoline and isoliensinine. Malaria SYBR Green I fluorescence assay was utilized to assess the in vitro anti-plasmodial (strains D6, Dd2, and F32-ART5) and immediate ex vivo (IEV) susceptibility for 10 freshly collected *P. falciparum* isolates. IC_{50} speed assay and morphological analysis were executed to check the effectiveness of isoliensinine. Isoliensinine presented potent in vitro gametocytocidal activity ($IC_{50}{}^{gam}$ 0.41–0.69 μM) against *P. falciparum. bis*-benzyl-isoquinoline inhibited asexual replication ($IC_{50}{}^{Asexual}$ = 2.17 μM, 2.22 μM, and 2.39 μM for D6, Dd2, and F32-ART5, respectively) (Muema et al., 2023). Another *Cissampelos* species is called *C. mucronata* which is a climber plant from Northern Nigeria, and it is utilized by domestic people to cure illness and fever. The ethanol leaf extract of the plant was assessed (200 mg/kg/day intraperitoneally) in mice for four days. The extract inhibited the growth of *P. berghei* by 68.4%, 60.0%, and 73.7% compared with chloroquine (5 mg/kg/day, inhibition100%, 97.4%, and 89.5%) for suppressive, curative, and prophylactic tests, respectively (Byumanine et al., 2023).

7.8.15 *Mimosaceae*

7.8.15.1 *Acacia polyacantha* Willd

The methanol bark extract of *Acacia polyacantha* was evaluated against *Culex quinquefasciatus* larva. The extract was rich in tannins, saponins, steroids, flavonoids, alkaloids, and coumarins. The extract showed a superior dose-dependent effect against the tested larvae (100% mortality, 500 μL). The LC_{50} values were 26.55 μL and 27.11 μL at 24 h and 48 h, respectively, while the LC_{90} was 318.5 μL and 328.1 μL at 24 h and 48 h, respectively (Daffalla et al., 2018).

7.8.15.2 *Albizia gummifera* J.F.G mel

The methanol extract of *A. gummifera* was prepared and fractionated and then the fractions were evaluated against two strains of *P. falciparum*, namely, NF54, chloroquine sensitive, and ENT30, chloroquine resistant. The alkaloid-rich fraction showed potent activity (IC_{50} of 0.16+/−0.05, NF54 and 0.99+/−0.06 microg/ml, ENT30). Spermine alkaloids were isolated from the aforementioned fraction and were active against the two tested *Plasmodium* strains (IC_{50} 0.09+/−0.02 to 0.91+/−0.10 microg/ml). The alkaloids also were active in vivo in mice with chemosuppression of parasitemia of 43–72% (Rukunga et al., 2007). The anti-malarial activity of *A. gummifera* extract was tested against isolates of *P. falciparum* in vitro by radioisotopic uptake technique. The extract presented a potent anti-plasmodial effect ($IC_{50} = 3.5 + 1.6$ μg/ml) compared with chloroquine ($IC_{50} = 0.065 + 0.057$ μg/ml) (Ofulla et al., 1995).

7.8.16 *Myrtaceae*

7.8.16.1 *Psidium guajava* L.

Guava (*P. guajava*) aqueous and aqueous-ethanol leaf extracts were evaluated in vitro against *β*-hematin formation. The 35% ethanol extract of guava leaves (at concentrations 1 and 0.4 mg/ml) was the most effective hematin formation inhibitor. Through LCMS, quercetin 3-*O*-glycoside, quercetin-galloyl-hexoside, morin-3-*O*-lyxoside, morin-3-*O*-arabinoside, quercein-3-*O*-arabinoside (guaijaverin), and ellagic acid were identified (Akkawi et al., 2021). Moreover, the extract (dose = 100–500 mg/kg/day) and fractions (dose = 50–200 mg/kg/day) of *P. guajava* leaf were tested against *P. berghei*-infected mice. The crude extract significantly ($p < 0.05$) inhibited parasite growth as well as prevented body weight loss and packed cell volume reduction dose-dependently. The aqueous fraction was the most active with 54.26% inhibition of parasite growth at 200 mg/kg. The crude extract was rich with phenol, alkaloids, flavonoids, and terpenoids (Alozieuwa et al., 2022).

7.8.17 *Rubiaceae*

7.8.17.1 Genus *Keetia*

The triterpene esters isolated from *Keetia leucantha* were investigated in vivo as anti-malarial agents against *P. falciparum*. Ursolic and oleanolic acids were quantified in the twigs dichloromethane extract by validated HPLC–UV methods. They account for about 19% of this extract (16.9% for acids and 1.8% for esters). The esters were given intraperitoneally (50 mg/kg/day), and *Plasmodium*-infected mice presented a highly significant ($p < 0.01$) parasitemia inhibition (27.8 ± 5.4%) (Beaufay et al., 2017; Tuenter et al., 2019).

7.8.17.2 *Heinsia crinita* (Afzel.) G.Taylor subsp. *parviflora* (K.Schum. & K.Krause) Verde (*Gardenia pulchella* G.Don)

Heinsia crinita stem bark powder is marketed in Congo as a folk medicine for malaria. The crude extracts of the leaves, fruits, and stem bark of *H. crinita* were evaluated as anti-plasmodial agents. Different solvents were utilized in the extraction and new iridoids were identified from the dichloromethane extract. The extracts and isolated compounds were tested against *P. falciparum* 3D7 strain in vitro. The in vivo anti-plasmodial activity was assessed by the inhibition of *P. berghei* growth in infected mice treated with the ethanol extract of *H. crinita* stem bark (200 and 300 mg/Kg/day) being the most active. Moderate in vitro anti-malarial activity was recorded for the DCM extract (IC_{50} = 29.2 ± 1.39 μg/mL) and the isolated compounds; lamalbide-6, 7, 8-triacetate (IC_{50} = 16.39 ± 0.43 μg/mL) and lamiridosin-6, 7, 8-triacetate (IC_{50} = 0.44.56 ± 1.12 μg/mL). The ethanol stem bark extract (200 and 300 mg/kg/day, oral route) showed moderate in vivo anti-malarial activity (27.84 ± 2.75% and 48.54 ± 3.76% growth inhibition, respectively) (Tshibangu et al., 2017).

7.8.17.3 *Pavetta crassipes* K. Schum. (*Pavetta barteri* Dawe)

Pavetta crassipes is traditionally used against malaria in Burkina Faso; thus, the alkaloid extract of this plant was tested in vitro against two strains of *P. falciparum* W2 (chloroquine-resistant) and D6 (chloroquine-sensitive) where significant inhibitory activity was observed (IC_{50} = 1.23 microg/ml). The anti-malarial activity was also evaluated against six *P. falciparum* isolates from children (IC_{50} = 25–670 ng/ml) (Sanon et al., 2003).

7.8.18 Rutaceae

7.8.18.1 *Zanthoxylum chalybeum* Engl. (*Fagara ehalybea* (Engl.))

The root bark and the leaves of *Zanthoxylum chalybeum* are used traditionally in Uganda to treat malaria. The anti-plasmodial activity of the crude ether and methanol leaf extracts of *Z. chalybeum*. The anti-plasmodial activities of the ether and methanol extracts were EC_{50} 13.39 μg/ml (95% inhibition) and 8.10 μg/ml (95% inhibition), respectively (Bbosa et al., 2014).

7.8.19 Sapindaceae

7.8.19.1 *Paullinia pinnata* L.

The ethanol leaf extract of *P. pinnata* was investigated for activity against early, residual (repository), and established malaria infections using Swiss albino mice infected with *P. berghei* (doses = 12.5 mg/kg per day, 25 mg/kg per day, and 50 mg/kg per day). A dose-dependent chemosuppression of the parasites was not obtained at different dose levels of the extract tested. *P. pinnata* showed some level of suppression of parasitemia in the early and established infection stages at a dose of

12.5 mg/kg per day and 25 mg/kg per day (53% and 69%) and 25 mg/kg per day and 50 mg/kg per day (60% both), respectively (Maje et al., 2007).

7.8.20 Verbenaceae

7.8.20.1 *Clerodendrum* Species

The crude extract of *Clerodendrum myricoides* (200, 400, and 600 mg/kg/day) was administered orally to *Plasmodium*-infected mice. The crude leaf methanol extract of *C. myricoides* exhibited 82.50% suppression at the dose of 600 mg/kg (Deressa et al., 2010). Similarly, the anti-malarial effect of *Cloredendrum inerme* (all parts of the plants are used traditionally against malaria) was tested clinically in 30 patients using exclusion and inclusion criteria. The result showed that *C. inerme* stew water can reduce the density of malaria parasite from day 3 (Ileke & Adesina, 2018; Nindatu et al., 2020).

7.9 Mechanism of Malaria Treatment Using Natural Products

Quinine, being the oldest malaria remedy, is worthy of studying with concern of its mechanism of action to help future research for safer and more effective derivatives. The anti-malarial activity of quinine is mainly centered around its ability to inhibit the formation of hemozoin, thus helping in merozoites death by their toxic hematin product. X-ray diffraction showed the complexes formed between quinine and quinidine and iron(III) protoporphyrin IX (Fe(III)PPIX). These structures showed that both quinine and quinidine form five-coordinate complexes with the porphyrin through their benzylic C-9 hydroxy groups, as well as an intramolecular hydrogen bond between the propionate side chain of Fe(III)PPIX and the protonated quinuclidine nitrogen atom of the alkaloid. This complex then inhibits the formation of hemozoin. This finding is fully consistent with earlier synthetic work that showed that a hydroxy functionality and a rigid proximal N are required for forming a noncovalent quinine–heme complex, and this can be described as the heme theory. For the newer anti-malarial drugs like artemisinin and its derivatives, the redox theory was behind their mechanisms of action where it was proposed that such compounds develop synthetic peroxide analogs that disturb the redox system in malaria (Kingston & Cassera, 2022). Further research work may be helpful in this context by semi-synthesis or synthesis to create more effective and safe anti-malarial drug leads.

7.10 Conclusions and Future Perspectives

Malaria remains one of the deadliest and most challenging neglected tropical diseases which puts much burden on several countries due to its economic, environmental, humanitarian, and even political impacts. The search for new, safe, and cost-effective treatment options will play a key role in future research to overcome the multidrug resistance played by the *Plasmodium* itself and to fight against the mechanical vectors mainly the *Anopheles* mosquito by its many species and to a lesser extent the *Culex* mosquito. Future research should be conducted on the plant extracts that were listed in this chapter to pinpoint the main components responsible for combating malaria and for a better understanding of the mechanisms of action played by such natural products against the malaria parasite and its carrying vector.

References

Abdullahi, A. A. (2011). Trends and challenges of traditional medicine in Africa. *African Journal of Traditional, Complementary, and Alternative Medicines, 8*(5S).

Achenbach, H., Waibel, R., Nkunya, M. H., & Weenen, H. (1992). Antimalarial compounds from *Hoslundia opposita. Phytochemistry, 31*(11), 3781–3784.

Adeoye, A., & Bewaji, C. (2018). Chemopreventive and remediation effect of *Adansonia digitata* L. Baobab (Bombacaceae) stem bark extracts in mouse model malaria. *Journal of Ethnopharmacology, 210*, 31–38.

Adeoye, A., Bewaji, C., & Ademowo, G. (2017). *In vivo* antimalarial activity of *Adansonia digitata* stem bark extract and some fractions. *International Journal of Toxicological and Pharmacological Research, 9*, 26–32.

Adigo Shibeshi, M., Fentahun Enyew, E., Mequanint Adinew, G., & Jemere, A. T. (2021). Antimalarial activity of methanolic extracts and solvent fractions of *Combretum molle* leaves in *Plasmodium berghei* infected mice. *Journal of Experimental Pharmacology, Volume 13*, 69–89.

Adzu, B., Abbah, J., Vongtau, H., & Gamaniel, K. (2003). Studies on the use of *Cassia singueana* in malaria ethnopharmacy. *Journal of Ethnopharmacology, 88*(2–3), 261–267.

Ahmed, A. O. E. E., & Ayoub, S. M. H. (2015). Chemical composition and antimalarial activity of extracts of Sudanese *Tamarindus indica* L. (Fabaceae). *Pharmaceutical Innovation, 4(4, Part B)*, 90.

Akharaiyi, F. C., Imarhiagbe, O., Isunu, L. E., & Ajibola, A. T. (2022). Ethanol leaf extract of *Hoslundia opposita* in *in vivo* antioxidant and hepatoprotective activity using an animal model. *Biomedicine (Taipei)., 12*(3), 48–55.

Akkawi, M., Abu-Lafi, S., Abu-Remeleh, Q., & Lutgen, P. (2021). Screening of guava (*Psidium guajava*) leaves extracts against *β*-hematin formation. *Pharmacy and Pharmacology International Journal, 9*(1), 11–15.

Alebie, G., Urga, B., & Worku, A. (2017). Systematic review on traditional medicinal plants used for the treatment of malaria in Ethiopia: Trends and perspectives. *Malaria Journal, 16*, 1–13.

Al-Huqail, A. A., Bekhit, A. A., Ullah, H., Ayaz, M., & Mostafa, N. M. (2023). Antimalarial and antileishmanial flavonoids from *Calendula officinalis* flowers. *Agronomy, 13*(11), 2765.

Alozieuwa, U. B., Mann, A., Kabiru, A. Y., & Ogbadoyi, E. O. (2022). *In vivo* antimalarial efficacy of *Psidium guajava* leaf crude extract and fractions in *Plasmodium berghei* infected mice. *AROC in Natural Product Research, 2*(1), 28–37.

Anato, M., & Ketema, T. (2018). Anti-plasmodial activities of *Combretum molle* (Combretaceae) [Zwoo] seed extract in Swiss albino mice. *BMC Research Notes, 11*, 1–5.

Andayi, A. W., Yenesew, A., Derese, S., Midiwo, J. O., Gitu, P. M., Jondiko, O. J., et al. (2006). Antiplasmodial flavonoids from *Erythrina sacleuxii*. *Planta Medica, 72*(02), 187–189.

Attioua, B., Yeo, D., Lagnika, L., Harisolo, R., Antheaume, C., Weniger, B., et al. (2012). *In vitro* antileishmanial, antiplasmodial and cytotoxic activities of a new ventiloquinone and five known triterpenes from *Parinari excelsa*. *Pharmaceutical Biology, 50*(7), 801–806.

Bbosa, G. S., Mwebaza, N., Lubega, A., Musisi, N., Kyegombe, D. B., & Ntale, M. (2014). Antiplasmodial activity of leaf extracts of *Zanthoxylum chalybeum* Engl. *British Journal of Pharmaceutical Research, 4*(6), 705–713.

Beaufay, C., Hérent, M.-F., Quetin-Leclercq, J., & Bero, J. (2017). *In vivo* anti-malarial activity and toxicity studies of triterpenic esters isolated form *Keetia leucantha* and crude extracts. *Malaria Journal, 16*, 1–8.

Bouabdallah, S., Cianfaglione, K., Azzouz, M., Batiha, G. E.-S., Alkhuriji, A. F., Al-Megrin, W. A. I., et al. (2022). Sustainable extraction, chemical profile, cytotoxic and antileishmanial activities *in-vitro* of some *Citrus* species leaves essential oils. *Pharmaceuticals., 15(9)*, 1163.

Bukirwa, H., Yeka, A., Kamya, M. R., Talisuna, A., Banek, K., Bakyaita, N., et al. (2006). Artemisinin combination therapies for treatment of uncomplicated malaria in Uganda. *PLoS Clinical Trials, 1*(1), e7.

Byumanine, N. J. R., Mwaty, B. M., Irenge, M. H., Mbwine, M., Mayunga, D. K., Shungu, F. S., et al. (2023). Phytochemical study and *in vitro* test of the activity of total extracts of *Cissampelos mucronata* (Menispermaceae) leaves on *Plasmodia falciparum*. *The European Journal of Theoretical and Applied Sciences, 1*(4), 762–773.

Chinedu, E., Arome, D., & Ameh, S. (2014). African herbal plants used as anti-malarial agents—A review. *PharmaTutor., 2*(3), 47–53.

Daffalla, H. M., Ali, K. S. E., Tajelser, T., Hagr, T. E., Ahmed, N. S., & Ahmed, R. H. (2018). Larvicidal and antibacterial activities of methanol extract of *Acacia polyacantha* Willd. *Journal of Advanced Research in Pharmaceutical Sciences & Pharmacology Interventions, 2*, 7–11.

De Villiers, B., Van Vuuren, S., Van Zyl, R., & Van Wyk, B.-E. (2010). Antimicrobial and antimalarial activity of *Cussonia* species (Araliaceae). *Journal of Ethnopharmacology, 129*(2), 189–196.

Dekebo, A., Tesfaye, M., & Kumar, S. (n.d.). Molecular docking, chemical composition, antibacterial and antioxidant properties of *Ocimum lamiifolium* (Damakese) leaves: A common home treatment plant in Ethiopia.

Deressa, T., Mekonnen, Y., & Animut, A. (2010). *In vivo* anti-malarial activities of *Clerodendrum myricoides*, *Dodonea angustifolia* and *Aloe debrana* against *Plasmodium berghei*. *The Ethiopian Journal of Health Development, 24*(1).

Du Preez, I. (2022). *In vitro* antimalarial properties and chemical composition of *Diospyros chamaethamnus* extracts

du Preez-Bruwer, I., Mumbengegwi, D. R., & Louw, S. (2022). *In vitro* antimalarial properties and chemical composition of *Diospyros chamaethamnus* extracts. *South African Journal of Botany, 149*, 290–296.

Eastman, R. T., & Fidock, D. A. (2009). Artemisinin-based combination therapies: A vital tool in efforts to eliminate malaria. *Nature Reviews. Microbiology, 7*(12), 864–874.

Elhawary, E., Mostafa, N., Shehata, A., & Labib, R. (2021). Comparative study of selected *Rosa* varieties' metabolites through UPLC-ESI-MS/MS, chemometrics and investigation of their insecticidal activity against *Culex pipiens* L. *The Jordan Journal of Pharmaceutical Sciences, 14*(4).

El-Nashar, H. A., Sayed, A. M., El-Sherief, H. A., Rateb, M. E., Akil, L., Khadra, I., et al. (2023). Metabolomic profile, anti-trypanosomal potential and molecular docking studies of *Thunbergia grandifolia*. *Journal of Enzyme Inhibition and Medicinal Chemistry, 38*(1), 2199950.

El-Tabakh, M. A., Elhawary, E. A., Hwihy, H. M., Darweesh, K. F., Shaapan, R. M., Ghazala, E. A., et al. (2023). UPLC/ESI/MS profiling of red algae *Galaxaura rugosa* extracts and its activity against malaria mosquito vector, *Anopheles pharoensis*, with reference to *Danio rerio* and *Daphnia magna* as bioindicators. *Malaria Journal, 22*(1), 368.

Enegide, C., Akah, P., Ofili, C., Agatemor, U., Ameh, S., Dabum, J., et al. (2021). Evidence supporting the use of *Combretum nigricans* as an antimalarial agent in Ethnomedicine. *International Journal of Current Research in Physiology and Pharmacology*, 13–20.

Fikadu, M., & Ashenafi, E. (2023). Malaria: An overview. *Infection and Drug Resistance, 16*, 3339–3347.

Fouokeng, Y., Feusso, H. F., Teinkela, J. M., Noundou, X. S., Wintjens, R., Isaacs, M., et al. (2019). *In vitro* antimalarial, antitrypanosomal and HIV-1 integrase inhibitory activities of two Cameroonian medicinal plants: *Antrocaryon klaineanum* (Anacardiaceae) and *Diospyros conocarpa* (Ebenaceae). *South African Journal of Botany, 122*, 510–517.

Gessler, M., Tanner, M., Chollet, J., Nkunya, M., & Heinrich, M. (1995). Tanzanian medicinal plants used traditionally for the treatment of malaria: *In vivo* antimalarial and *in vitro* cytotoxic activities. *Phytotherapy Research, 9*(7), 504–508.

Hassan, M. I., Fouda, M. A., Hammad, K. M., Tanani, M. A., & Shehata, A. Z. (2014). Repellent effect of *Lagenaria siceraria* extracts against *Culex pipiens*. *Journal of the Egyptian Society of Parasitology, 44*(1), 243–248.

Hassoun, N., & Cosler, L. (2023). Global health impact: A model to alleviate the burden and expand access to treatment of neglected tropical diseases. *The American Journal of Tropical Medicine and Hygiene, 108*(4), 801.

Ibrahim, E. A., Wamalwa, M., Odindi, J., & Tonnang, H. E. Z. (2024). Insights and challenges of insecticide resistance modelling in malaria vectors: A review. *Parasites and Vectors, 17*(1), 174.

Igoli, J. O., Masao, C. A., Orkpeh, U., Nakamatte, E., Yakumbur, D. T., Nnadozie, T., et al. (2016). *In vivo* antimalarial activity of *Cleome gynandra* extracts. *J Nat Prod Res Updates, 2*, 19–29.

Ileke, K., & Adesina, F. (2018). Bioefficacy of larvicidal and pupicidal properties of *Clerodendrum capitatum* and *Bridelia machrantha* leaves extracts against malaria vector, *Anopheles gambiae* Giles (Diptera: Culicidae). *Journal of Biology and Medicine, 2*, 007–011.

International Plant Names Index [Internet]. The Royal Botanic Gardens, Kew, Harvard University Herbaria & Libraries and Australian National Herbarium. 2024 [cited May 2024]. Available from: http://www.ipni.org

James, P. B., Wardle, J., Steel, A., & Adams, J. (2018). Traditional, complementary and alternative medicine use in sub-Saharan Africa: A systematic review. *BMJ Global Health, 3*(5), e000895.

Kalbessa, A., Dekebo, A., Tesso, H., Abdo, T., Abdissa, N., & Melaku, Y. (2019). Chemical constituents of root barks of *Gnidia involucrata* and evaluation for antibacterial and antioxidant activities. *Journal of Tropical Medicine, 2019*(1), 8486214.

Kalpana, V., Alarjani, K. M., & Rajeswari, V. D. (2020). Enhancing malaria control using *Lagenaria siceraria* and its mediated zinc oxide nanoparticles against the vector *Anopheles stephensi* and its parasite *Plasmodium falciparum*. *Scientific Reports, 10*(1), 21568.

Kamaraj, C., & Rahuman, A. A. (2010). Larvicidal and adulticidal potential of medicinal plant extracts from South India against vectors. *Asian Pacific Journal of Tropical Medicine, 3*(12), 948–953.

Kaushik, N. K., Bagavan, A., Rahuman, A. A., Zahir, A. A., Kamaraj, C., Elango, G., et al. (2015). Evaluation of antiplasmodial activity of medicinal plants from North Indian Buchpora and South Indian Eastern Ghats. *Malaria Journal, 14*(1), 1–8.

Kingston, D. G., & Cassera, M. B. (2022). *Antimalarial natural products*. Springer.

Kolaczinski JH, Onapa AW, Kabatereine NB, Ndyomugyenyi R, Kakembo AS, Brooker S. (2006). Neglected tropical diseases and their control in Uganda

Lawal, B., Shittu, O. K., Kabiru, A. Y., Jigam, A. A., Umar, M. B., Berinyuy, E. B., et al. (2015). Potential antimalarials from African natural products: A review. *Journal of Intercultural Ethnopharmacology, 4*(4), 318.

Ma, C. Y., Musoke, S. F., Tan, G. T., Sydara, K., Bouamanivong, S., Southavong, B., et al. (2008). Study of antimalarial activity of chemical constituents from *Diospyros quaesita. Chemistry & Biodiversity, 5*(11), 2442–2448.

Maje, I., Anuka, J., Hussaini, I., Katsayal, U., Yaro, A., Magaji, M., et al. (2007). Evaluation of the anti-malarial activity of the ethanolic leaves extract of *Paullinia pinnata* (Linn.)(Sapindaceae). *Niger Journal of Pharmaceutical Sciences, 6*(2), 67–72.

Malebo, H. M., Wiketye, V., Katani, S. J., Kitufe, N. A., Nyigo, V. A., Imeda, C. P., et al. (2015). *In vivo* antiplasmodial and toxicological effect of *Maytenus senegalensis* traditionally used in the treatment of malaria in Tanzania. *Malaria Journal, 14*, 1–7.

Mankilik, M. M., Longdet, I. Y., & Luka, C. D. (2021). Evaluation of *Achyranthes aspera* shoot extract as an alternative therapy for malaria. *Journal of Basic and Applied Zoology, 82*, 1–9.

Msugupakulya, B. J., Urio, N. H., Jumanne, M., Ngowo, H. S., Selvaraj, P., Okumu, F. O., et al. (2023). Changes in contributions of different *Anopheles* vector species to malaria transmission in East and Southern Africa from 2000 to 2022. *Parasites and Vectors, 16*(1), 408.

Muema, J. M., Mutunga, J. M., Obonyo, M. A., Getahun, M. N., Mwakubambanya, R. S., Akala, H. M., et al. (2023). Isoliensinine from *Cissampelos pariera* rhizomes exhibits potential gametocytocidal and anti-malarial activities against *Plasmodium falciparum* clinical isolates. *Malaria Journal, 22*(1), 161.

Mwakumanya, M. O., Kyama, M. C., & Maina, E. N. (2022). Phytochemical analysis and safety evaluation of ethanol roots extract of *Erythrina sacleuxii* Hua in Wistar albino rats. *Journal of Medicinal Plant Research: Planta Medica, 16*(4), 126–140.

Nindatu, M., Titaley, C., Taihuttu, Y., & Kaihena, M. (Eds.). (2020). *Antimalarial effect of Lemburung meit (Clerondrum inerme Linn) stew on malarial infections at Piru Health Center, West Seram District, Maluku. J Physics: Conference Series*. IOP Publishing.

Nondo, R. S., Erasto, P., Moshi, M. J., Zacharia, A., Masimba, P. J., & Kidukuli, A. W. (2016). *In vivo* antimalarial activity of extracts of Tanzanian medicinal plants used for the treatment of malaria. *Journal of Advanced Pharmaceutical Technology & Research, 7*(2), 59–63.

Ofulla, A., Chege, G., Rukunga, G. M., Kiarie, F. K., Githure, J. I., & Kofi-Tsekpo, M. W. (1995). *In vitro* antimalarial activity of extracts of *Albizia gummifera, Aspilia mossambicensis, Melia azedarach* and *Azadirachta indica* against *Plasmodium falciparum. African Journal of Health Sciences, 2*(2), 309–311.

Okou, O. C., Beourou, S., Gnondjui, A. A., Akpoue, E. V., & Djaman, A. J. (2019). *Hoslundia opposita* Vahl (Lamiaceae) leaves, chemosensitivity evaluation on *Plasmodium falciparum. International Journal of Pharma and Bio Sciences, 10*(4), 97–104.

Organization WH. Vector control—Methods for use by individuals and communities. Organization WH, editor 1997. 425 p.

Organization WH. A global brief on vector-borne diseases. 2014.

Organization WH. (2015). *Guidelines for the treatment of malaria*. World Health Organization.

Organization WH. (2020). *Neglected tropical diseases: Draft road map for neglected tropical diseases 2021–2030*. WHO.

Passarini, G. M., de Medeiros, D. S., de Oliveira Meneguetti, D. U., Lima, R. A., Facundo, V. A., & de Medeiros, P. S. M. (2017). *In vitro* antiplasmodial activity of flower extracts from *Combretum leprosum* Mart. *(Mofumbo). Ciên Nat., 39*(1), 84–90.

Radhakrishnan, N., Karthi, S., Raghuraman, P., Ganesan, R., Srinivasan, K., Edwin, E.-S., et al. (2023). Chemical screening and mosquitocidal activity of essential oil derived from *Mikania scandens* (L.) Willd. Against *Anopheles gambiae* Giles and their non-toxicity on mosquito predators. All. *Life., 16(1)*, 2169959.

Ríos, N., Stashenko, E. E., & Duque, J. E. (2017). Evaluation of the insecticidal activity of essential oils and their mixtures against *Aedes aegypti* (Diptera: Culicidae). *Revista Brasileira de Entomologia, 61*(4), 307–311.

Rukunga, G., Muregi, F. W., Tolo, F., Omar, S., Mwitari, P., Muthaura, C., et al. (2007). The antiplasmodial activity of spermine alkaloids isolated from *Albizia gummifera*. *Fitoterapia, 78*(7–8), 455–459.

Sanon, S., Azas, N., Gasquet, M., Ollivier, E., Mahiou, V., Barro, N., et al. (2003). Antiplasmodial activity of alkaloid extracts from *Pavetta crassipes* (K. Schum) and *Acanthospermum hispidum* (DC), two plants used in traditional medicine in Burkina Faso. *Parasitology Research, 90*, 314–317.

Shadrack, D. M., Nyandoro, S. S., Munissi, J. J., & Mubofu, E. B. (2016). *In silico* evaluation of anti-malarial agents from *Hoslundia opposita* as inhibitors of *Plasmodium falciparum* lactate dehydrogenase (PfLDH) enzyme. *Computational Molecular Bioscience, 6*(2), 23–32.

Shehata, A. Z., Elhawary, E. A., Mokhtar, M. M., Waheeb, H. O., Fares, M., Emam, D. E., et al. (2023). Molecular docking and insecticidal activity of *Pyrus communis* L. extracts against disease vector, *Musca domestica* L. (Diptera: Muscidae). *Egyptian Journal of Basic and Applied Sciences, 10*(1), 800–811.

Singab, A. N. B., Mostafa, N. M., Elkhawas, Y. A., Al-Sayed, E., Bishr, M. M., Elissawy, A. M., et al. (2022). Cyclodepsipeptides: Isolation from endophytic fungi of *Sarcophyton ehrenbergi* and verification of their larvicidal activity *via in-vitro* and *in-silico* studies. *Marine Drugs, 20*(5), 331.

Tasiam, E., Primaharinastiti, R., & Ekasari, W. (2020). *In vitro* antimalarial activity and toxicity studies of Johar (*Cassia siamea*) leaves from three different locations. *African Journal of Infectious Diseases, 14*(2), 23–29.

Teklu, T., Engidawork, E., Nedi, T., Teklehaymanot, T., & Gebremeskel, L. (2020). *Evaluation of the antimalarial activity of the hydroalcoholic extract of leaf of Leonotis ocymifolia (Burm. f.) Iwarsson (Lamiaceae) against Plasmodium berghei in mice*. Evid-Based Complement Altern Med: eCAM. 2020: 5384804.

Theerachayanan, T., Sirithunyalug, B., & Piyamongkol, S. (2007). Antimalarial and antimycobacterial activities of dimeric naphthoquinone from *Diospyros glandulosa* and *Diospyros rhodocalyx*. *Chiang Mai University Journal of Natural Science, 6*(2), 253–258.

Tona, L., Cimanga, R., Mesia, K., Musuamba, C., De Bruyne, T., Apers, S., et al. (2004). *In vitro* antiplasmodial activity of extracts and fractions from seven medicinal plants used in the Democratic Republic of Congo. *Journal of Ethnopharmacology, 93*(1), 27–32.

Tshibangu, P. T., Kapepula, P. M., Kapinga, M. K., Mukuta, A. T., Kalenda, D. T., Tchinda, A. T., et al. (2017). Antiplasmodial activity of *Heinsia crinita* (Rubiaceae) and identification of new iridoids. *Journal of Ethnopharmacology, 196*, 261–266.

Tuenter, E., Zarev, Y., Matheeussen, A., Elgorashi, E., Pieters, L., & Foubert, K. (2019). Antiplasmodial prenylated flavonoids from stem bark of *Erythrina latissima*. *Phytochemistry Letters, 30*, 169–172.

Usman, M. R. M, & Sunil, B. P. (n.d.). Pharmacognostical and antimalarial studies of *Tamarindus indica* leaves.

van der Pluijm, R. W., Amaratunga, C., Dhorda, M., & Dondorp, A. M. (2021). Triple artemisinin-based combination therapies for malaria—A new paradigm? *Trends in Parasitology, 37*(1), 15–24.

Vellore Nagarajan, K., & Vijayarangan, D. R. (2019). *Lagenaria siceraria*–synthesised ZnO NPs—A valuable green route to control the malaria vector *Anopheles stephensi*. *IET Nanobiotechnology, 13*(2), 170–177.

Waako, P., Gumede, B., Smith, P., & Folb, P. (2005). The *in vitro* and *in vivo* antimalarial activity of *Cardiospermum halicacabum* L. and *Momordica foetida* Schumch. Et Thonn. *Journal of Ethnopharmacology, 99*(1), 137–143.

Yenesew, A., Derese, S., Irungu, B., Midiwo, J. O., Waters, N. C., Liyala, P., et al. (2003). Flavonoids and isoflavonoids with antiplasmodial activities from the root bark of *Erythrina abyssinica*. *Planta Medica, 69*(07), 658–661.

Zaruwa, M., Madu, J. O., Yusuf, A. A., & Muhammed, A. O. O. (2018). *In vivo* antiplasmodial and cytotoxic activity of *Achyranthes aspera* and *Ficus thoningii* in mice infected with *Plasmodium berghei*. *International Biological and Biomedical Journal, 4*(1), 48–53.

Zhang, J. (2019). *Plantlist: Looking up the status of plant scientific names based on the plant list database*. R package version 0.5. 3.

Zoubiri, S., & Baaliouamer, A. (2014). Potentiality of plants as source of insecticide principles. *Journal of Saudi Chemical Society, 18*(6), 925–938.

Chapter 8
African Natural Products with Antileishmanial Activity

Dina M. Bahgat, Muhammad Ajmal Shah, and Omayma A. Eldahshan

Abstract Leishmaniasis is an infectious parasitic disease located in tropical and subtropical regions of the world, including many African countries. Infection is caused by the bite of sand fly leading to cutaneous leishmaniasis (CL) that may develop into mucosal leishmaniasis (ML). Visceral leishmaniasis (VL) is a serious form of disease affecting the internal organs with a high mortality rate. There are several anti-parasitic medications used to treat leishmaniasis as amphotericin, miltefosine, antimonials, and paromomycin. Unfortunately, the long-term use of these drugs reduces their effectiveness and leads to multiple side effects. Thus, natural products may offer a convenient alternative or complementary treatment for leishmaniasis. This book chapter piles up the biological studies (2018–2024) describing the potential role of African medicinal plants and their secondary metabolites in the treatment of this disease, focusing on their efficacy and highlighting the mechanism of action. Results revealed that different 52 plant extracts showed variable in vitro antileishmanial activity while eight plant extracts treated cutaneous and visceral leishmaniasis in experimental animals significantly and could represent a starting point for clinical trials. Regarding pure isolated compounds, terpenes exhibited high antileishmanial activity compared to standard drugs. Additionally, most endophytic compounds exhibited leishmanicidal activity with IC_{50} values less than 50 μM.

Keywords Leishmaniasis · Natural products · Plant extracts · In vitro leishmanicidal · Endophytes

---

D. M. Bahgat · O. A. Eldahshan (✉)
Department of Pharmacognosy, Faculty of Pharmacy, Ain Shams University, Cairo, Egypt
e-mail: oeldahshan@pharma.asu.edu.eg

M. A. Shah
Department of Pharmacy, Hazara University, Mansehra, Pakistan

A. Dekebo (ed.), *Bioactive Secondary Metabolites from Medicinal Plants of Africa*, Interdisciplinary Biotechnological Advances,
https://doi.org/10.1007/978-981-96-8590-5_8

Abbreviations

VL	Visceral leishmaniasis
CL	Cutaneous leishmaniasis
ML	Mucosal leishmaniasis
MeOH	Methanol
DCM	Dichloromethane
IL	Interleukin
INF	Interferon

8.1 Introduction

Leishmaniasis is an infectious parasitic disease caused by different species of *Leishmania* protozoa. It is transmitted through the bite of an infected female sandfly. In Africa, these little hairy flies usually bite at night, transferring a flagellate form of the parasite known as promastigote, from its digestive tract to host blood. Promastigote are phagocytosed by host macrophages, then proliferate into intracellular flagellate form known as amastigote form (Fig. 8.1). The clinical manifestations take multiple forms and stages depending on the site of infection, the host immune response, and the parasite species. Disease severity ranges from being serologically positive but with no symptoms, up to cases of disfiguring and fatal manifestations (Rahmanipour et al., 2023).

Cutaneous leishmaniasis (CL) is caused by *L. tropica*, *L. major*, *L. amazonesis*, and *L. mexicana*. It is the most prevalent form of the disease appearing as bumps and ulcers a few weeks or months after sand fly bite. These lesions are mostly self-healing. Rarely, these bumps or ulcers are detected on many areas of the skin and can recur even with treatment. The resulting scars from CL affect social life of the patient, leading to anxiety, depression, and poor quality of life (Kahime et al., 2016).

Mucosal or mucocutaneous leishmaniasis (ML) is caused mainly by *L. vianniabraziliensis. L. vianniapanamensis* and *L. amazonesis* where the mucosal tissues of the nose, mouth, or throat are affected, leading to partial or complete damage of the mucosa and may cause facial disfigurement. The pathology of ML isn't fully understood, but it is assumed that it is developed due to the aggressive immune response of the patient (Carvalho et al., 2005; Strazzulla et al., 2013).

Visceral leishmaniasis (VL) is the most serious form of the disease with a fatality rate of more than 95% if left untreated. The causative species is *L. donovani* affecting the internal organs, spleen, and liver. It is characterized by intermittent fever attacks, weight loss, hepato-splenomegaly, and anemia. Patients commonly die from subsequent infections or severe anemia (Hasan et al., 2023; Lewis et al., 2020).

Despite the availability of many pharmaceutical compounds for the treatment of leishmaniasis, they are characterized by being expensive and toxic causing serious side effects on long-term use, and after a while, most *Leishmania* species develop

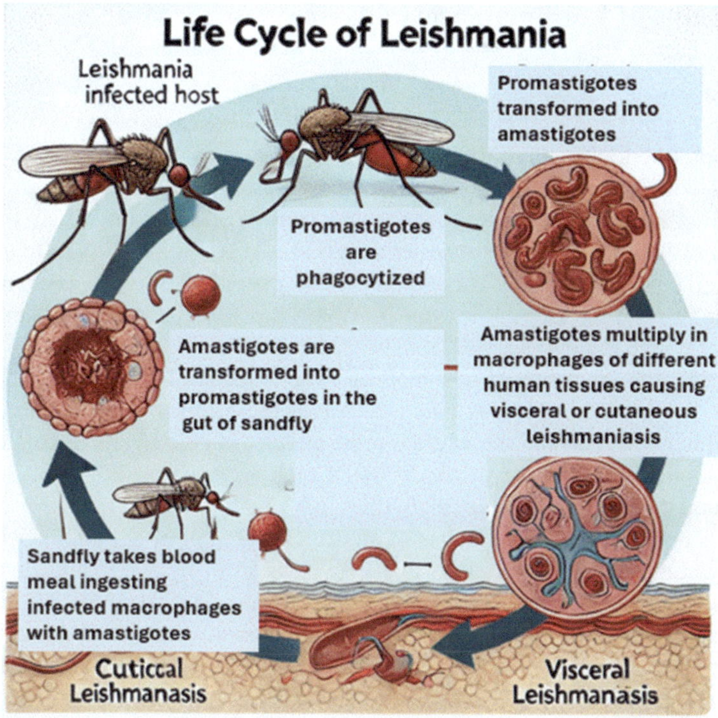

Fig. 8.1 Life cycle of leishmania

resistance to synthetic drugs. Thus, many research groups have investigated natural products as a more appropriate alternative or complementary therapy for curing leishmaniasis with fewer side effects and less expense in addition to being available in most endemic countries that already suffer from poverty and unavailable facilities for drug development.

8.2 Review Method

In this book chapter, we compile the results of studies performed in African countries on natural products exhibiting antileishmanial activity published in the period between 2018 and 2024. Different research engines were explored as: Google Scholar, PubMed, Scopus, SciFnder, Science Direct. The used keywords were: leishmaniasis, leishmania species, natural products, Africa, in vitro, in vivo, plant extract, endophytes, mechanism, flavonoids, terpenes, and phenolic compounds. The keywords were used in different combinations. Both reviews and original articles were reviewed. After excluding irrelevant publications, 88 references were included in this book chapter.

8.3 Status of Leishmaniasis in Africa

Leishmaniasis is categorized by the World Health Organization (WHO) as a neglected tropical disease hindering the attainment of the Sustainable Development Goals 2030 (Abdoul-Latif et al., 2024).

It is classified according to region of occurrence: Old World Leishmaniasis found in the Middle East, Europe, and Africa and New World Leishmaniasis found in Central and South America (Fischer et al., 2024). According to a WHO fact sheet issued in January 2023, there are about a billion people who live in endemic areas with a high risk of infection. It is estimated that from 700,000 to one million people get infected annually (WHO, 2022).

In Africa, the incidence of leishmaniasis varies from 14% to 50% in different countries. It is endemic in many countries across the continent, particularly in East Africa which is considered the world's most affected region by VL accounting for 45% of global cases (Alvar et al., 2021). Countries such as Ethiopia, Somalia, Kenya, and Sudan suffer from high rates of VL, while Algeria experiences a detectable percentage of CL (Abbas et al., 2022; Scarpini et al., 2022). Although data on leishmaniasis in Africa is available, it lacks accuracy due to undiagnosed or misdiagnosed cases with other diseases such as malaria. Consequently, it is assumed that the actual burden of leishmaniasis is notably higher than what is reported.

Massive human migration around African countries due to political conflicts or economic purposes leads to the expansion of endemic areas where non-immunized populations migrate to endemic foci or migrants can carry leishmaniasis to regions where the sand fly vector is present, yet the disease didn't spread. Thus, outbreaks of the disease have been frequently detected in countries such as Ethiopia, Kenya, and South Sudan (Argaw et al., 2013).

Another important factor leading to the exacerbation of leishmaniasis is the global climate changes, where the variation in ecological factors, temperature, rainfall patterns, and soil types, dramatically changed the number and distribution of the vector host. Additionally, vector control programs aren't well established or implemented by official authorities in endemic countries (Wamai et al., 2020).

Furthermore, socioeconomic conditions are a critical key factor triggering the outbreaks of the infection. Poverty leads to bad sanitary status favoring the reproduction of sand flies. Also, malnutrition and deficiency of protein and vitamins in daily diet is an important predisposing factor for disease complications and severity (Osorio et al., 2023).

8.4 Pharmaceutical Antileishmanial Drugs

There are few chemotherapeutics approved for the eradication of leishmaniasis that can be used solely or in combination (Fig. 8.2). These drugs have shown variable levels of clinical efficacy depending on the infectious species and site of infection.

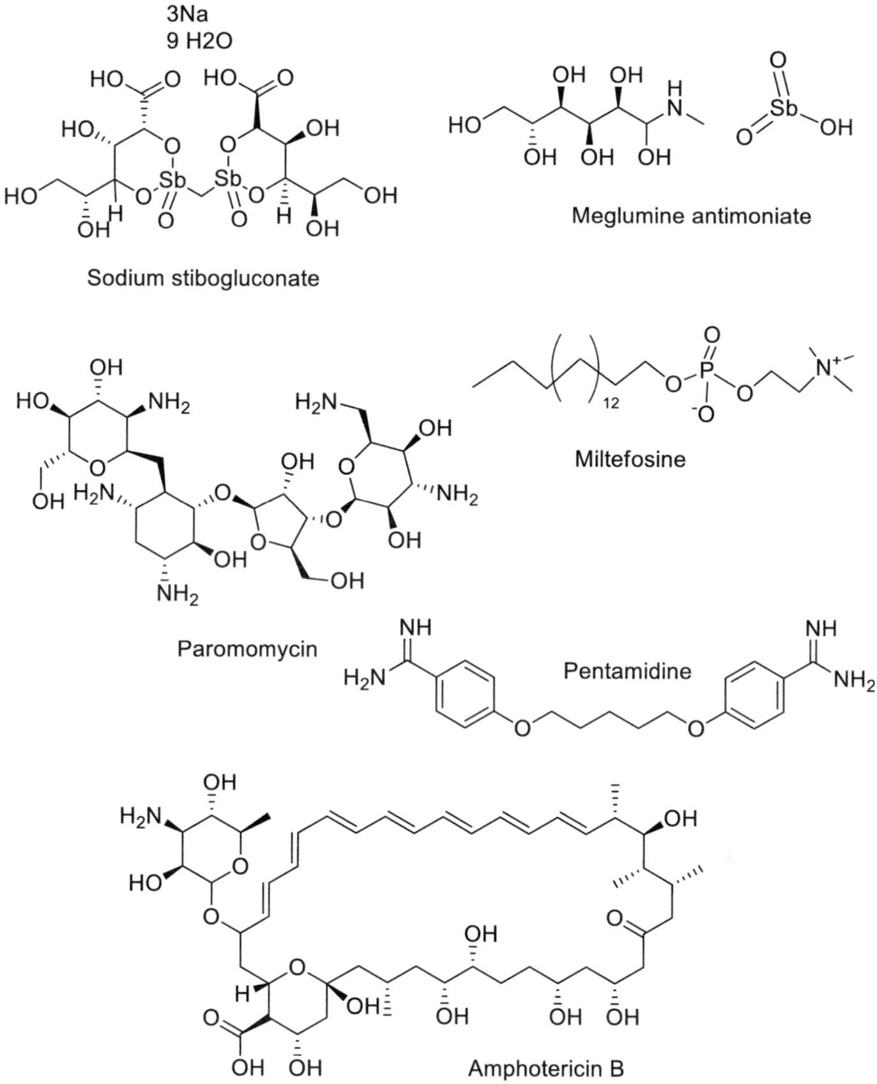

Fig. 8.2 Chemical structure of antileishmanial conventional therapy

8.4.1 Pentavalent Antimonial

Pentavalent antimonial is the oldest used treatment since 1945. It is available in two formulations: meglumine antimoniate and sodium stibogluconate which are administrated by injection. The precise mechanism isn't completely comprehended. It is assumed that pentavalent antimonials inhibit the parasite virulence factor: trypanothione. It also interrupts metabolic energy pathways inducing oxidative stress in the parasite cells leading to cellular apoptosis. Unfortunately, the long-term use of antimonials leads to serious side effects including renal failure and leucopenia (Berbert et al., 2018; Saad et al., 2022).

8.4.2 Miltefosine

Miltefosine is originally designed as an antitumor drug and then approved by the FDA as oral therapy for the treatment of VL. It causes apoptosis like the death of parasite cells. It is associated with multiple gastrointestinal side effects including nausea, vomiting, and diarrhea. It is prohibited to be used by pregnant women due to teratogenicity. Parasites usually develop resistance due to a long half-life of the drug reaching about 7 days (Hassan et al., 2022; Mathison & Bradley, 2023).

8.4.3 Pentamidine

Pentamidine is an aromatic diamidine exhibiting anti-fungal activity and was used in cases of resistance to antimonials; however, its use was restricted due to serious cardiac toxicity. It inhibits the active transport system and mitochondrial topoisomerase II leading to parasitic death (Kaur & Rajput, 2014).

8.4.4 Paromomycin

Paromomycin is an aminoglycoside antibiotic isolated showing antileishmanial effect antibacterial activity. It inhibits cellular protein synthesis of the parasite and alters the mitochondrial membrane potential. Its major drawback is poor absorption, so it is administrated through intramuscular injections (Ghorbani & Farhoudi, 2017).

8.4.5 Amphotericin B

Amphotericin B is an antifungal metabolite isolated from the bacteria *Streptomyces nodosus*. It causes membrane disruption in the parasite as it binds to membrane ergosterol. It is effective in unresponsive patients, but unfortunately, it also shows many side effects like fever, rigors, and renal failure. AmBisome is a liposomal formulation of amphotericin B that exhibits higher effectiveness and less toxicity than conventional amphotericin B making it a drug of choice in many endemic countries. However, its clinical importance is threatened by its high cost and development of resistance in clinical isolates (Jawed & Majumdar, 2018; Tiwari et al., 2019).

8.5 Natural Plant Products Exhibiting Antileishmanial Activity

Eco-diversity of Africa endowed the content with an exceptionally variable flora. This rich flora possesses high portion of remarkably important medicinal plants possessing valuable pharmacological activities including antileishmanial activity. In this section, we summarize the studies performed on the leishmanicidal effect of plant extracts or the secondary metabolites isolated from them and the associating endophytes.

8.5.1 Leishmanicidal Activity of Plant Extracts

In vitro leishmanicidal tests are performed on two forms of the parasite either the extracellular promastigotes or the intracellular, intra-macrophage amastigotes. Although extracellular assay is much simpler, technically it is more susceptible to false negative results where the tested drug may be active against promastigotes but inactive against cell-protected amastigotes. Thus, it is recommended to perform tests on amastigotes to investigate the mechanism of leishmanicidal activity in the host-mediated phase (De Muylder et al., 2011).

In vivo assay is performed after infecting the experimental animals with the proper species to induce cutaneous or visceral leishmaniasis then measuring certain parameters including the parasite burden, infection rate, and multiplication index beside other parameters to assess the effectiveness of the tested drug (Pinheiro et al., 2016). Toxicity of investigated extracts was measured first on experimental animals to determine the maximum safe dose. These in vivo studies can be considered as remarkable starting point to perform clinical trials on these plant extracts to find new available and safe treatment options for leishmaniasis.

The scientific research performed by African researchers on medicinal plants cultivated in different regions of Africa led to valuable results demonstrated in Tables 8.1 and 8.2. Plants belonging to the family Asteraceae and Fabaceae have been extensively subjected to antileishmanial investigation.

8.5.2 Leishmanicidal Activity of Secondary Metabolites Isolated from Plants

The advantage of testing pharmacological activities on pure isolated compounds is the ability to study the mechanism of action either by conducting further pharmacological assays or using in silico tests. Also, you can easily predict the toxicity and pharmacokinetic properties of the chemical compound. Using computer-aided drug design we can create a complete virtual library for possible pharmacophores, screen

Table 8.1 In vitro leishmanicidal activity of plant extracts growing in African countries

No.	Plant species	Extract/used part	Country	*Leishmania* species/stage	IC_{50} (μg/mL) / Effectiveness	IC_{50} (μg/mL) standard drug	Ref.
Family Anacardiaceae							
1.	*Pistacia lentiscus*	Essential oil/Leaves	Morocco	*L. infantum* Promastigotes	11.28 Moderate	gGlucantime < 50 μg/mL	Bouyahya et al. (2019)
				L. tropica Promastigotes	23.50 Moderate		
				L. major Promastigotes	17.52 Moderate		
		Essential oil/Flowers	Morocco	*L. infantum* Promastigotes	8.00 High		
				L. tropica Promastigotes	26.20 Moderate		
				L. major Promastigotes	21.42 Moderate		
2.	*Spondias mombin*	50% Aqu. EtOH/Leaves	Ghana	*L. donovani* Promastigotes	81.50 weak	Amphotericin B 3.20 μg/mL	Ohashi et al. (2018)
Family Annonaceae							
3.	*Annona senegalensis*	50% Aqu. EtOH/Leaves	Ghana	*L. donovani* Promastigotes	10.80 moderate	Amphotericin B 3.20 μg/mL	Ohashi et al. (2018)
		50% Aqu. EtOH/Stem bark			27.80 moderate		
4.	*Cleistopholis patens*	50% Aqu. EtOH/Stem bark	Ghana	*L. donovani* Promastigotes	60.20 weak		
Family Apiaceae							
5.	*Ferula communis*	80% MeOH/Whole plant	Ethiopia	*L. aethiopica* Promastigotes	11.38 Moderate	Amphotericin B 1.20 μg/mL	Nigatu et al. (2021)
				L. donovani Promastigotes	23.41 Moderate	Amphotericin B 1.30 μg/mL	

				L. aethiopica Amastigotes	14.32 Moderate	Amphotericin B 1.22 µg/mL	
				L. donovani Amastigotes	31.12 Moderate	Amphotericin B 1.46 µg/mL	
Family Apocynaceae							
6.	*Mondia whitei*	50% Aqu. EtOH/Root	Ghana	*L. donovani* Promastigotes	31.00 Moderate	Amphotericin B 3.20 µg/mL	Ohashi et al. (2018)
Family Asteraceae							
7.	*Acanthospermum hispidum*	50% Aqu. EtOH/Whole plant	Ghana	*L. donovani* Promastigotes	32.10 Moderate	Amphotericin B 3.20 µg/mL	Ohashi et al. (2018)
8.	*Artemisia absinthium*	Hydro-distilled essential oil/Leaves	Tunis	*L. major* Promastigotes	1.49 High	Amphotericin B 0.02 µg/mL	Mathlouthi et al. (2018)
9.	*Artemisia campestris*			*L. major* Promastigotes	2.20 High		
10.	*Artemisia herba-alba*				2.20 High		
		MeOH/Aerial parts	Algeria	*L. infantum* Amastigotes	68.25 Weak	Amphotericin B 0.20 µg/mL	Eddaikra et al. (2019)
				L. major Amastigotes	37.87 Moderate		
11.	*Bidens pilosa*	50% Aqu. EtOH/Whole plant	Ghana	*L. donovani* Promastigotes	28.90 moderate	Amphotericin B 3.20 µg/mL	Ohashi et al. (2018)
12.	*Silybum marianum*	MeOH/Rhizomes	Egypt	*L. major* Promastigotes	77.35 High	Amphotericin B 58.75 µg/mL	Darwish et al. (2019)
Family Balanophoraceae							
13.	*Thonningia sanguinea*	50% Aqu. EtOH/Whole plant	Ghana	*L. donovani* Promastigotes	18.60 moderate	Amphotericin B 3.20 µg/mL	Ohashi et al. (2018)

(continued)

Table 8.1 (continued)

No.	Plant species	Extract/used part	Country	*Leishmania* species/stage	IC_{50} (μg/mL) / Effectiveness	IC_{50} (μg/mL) standard drug	Ref.
Family Cistaceae							
14.	*Cistus crispus*	MeOH/Leaves	Morocco	*L. major* Promastigotes	84.29 Weak	Glucantime > 500 μg/mL	Bouyahya et al. (2018)
				L. infantum Promastigotes	82.39 Weak		
		n-Hexane/Leaves		*L. tropica* Promastigotes	96.82 Weak		
				L. major Promastigotes	47.29 Moderate		
Family Clusiaceae							
15	*Pentadesma butyracea*	DCM: MeOH (1:1)/Fruits	Cameroon	*L. donovani* Promastigotes	5.96 High	Amphotericin B 0.22 μg/mL	Garba et al. (2022)
		DCM: MeOH (1:1)/Stem bark			26.43 Moderate		
Family Colchicaceae							
16.	*Colchicum autumnale*	MeOH/Corms and bulbs	Egypt	*L. major* Promastigotes	60.09 High	Amphotericin B 58.75 μg/mL	Darwish et al. (2019)
Family Combretaceae							
17.	*Terminalia ivorensis*	50% Aqu. EtOH/Leaves	Ghana	*L. donovani* Promastigotes	24.90 Moderate	Amphotericin B 3.20 μg/mL	Ohashi et al. (2018)
Family Ericaceae							
18.	*Arbutus unedo*	*n*-Hexane/Leaves	Morocco	*L. infantum* Promastigotes	64.05 Weak	Glucantime > 500 μg/mL	Bouyahya et al. (2018)
				L. tropica Promastigotes	79.57 Weak		

19.	*Erica arborea*	MeOH/Flower	Algeria	*L. major* Promastigotes	43.98 Moderate	Amphotericin B 0.20 µg/mL	Eddaikra et al. (2019)
				L. infantum Promastigotes	61.27 Weak		
				L. major Amastigotes	36.00 Moderate		
				L. infantum Amastigotes	53.93 Weak		
Family Euphorbiaceae							
20.	*Croton caudatus*	EtOAc.: *n*-Hexane (9:1)/ Leaves	Morocco	*L. donovani* Promastigotes	10.00 High	Miltefosine 2.50 µg/mL	Dey et al. (2015)
				L. donovani Amastigotes	2.50 High	Miltefosine 5.00 µg/mL	
21.	*Jatropha pelargonifolia*	MeOH/Leaves	Egypt	*L. infantum* Promastigotes	3.2 High	Miltefosine 10.7 µg/mL	Okba et al. (2018)
Family Fabaceae							
22.	*Afzelia africana*	50% Aqu. EtOH/Bark	Ghana	*L. donovani* Promastigotes	77.10 weak	Amphotericin B 3.20 µg/mL	Ohashi et al. (2018)
23.	*Baphia nitida*	50% Aqu. EtOH/Bark	Ghana		34.40 moderate		
24.	*Cassia sieberiana*	50% Aqu. EtOH/Leaves	Ghana		62.90 Weak		
25.	*Erythrophleum ivorense*	MeOH/Root bark	Ghana	*L. donovani* Promastigotes	2.97 High	Amphotericin B 2.40 µg/mL	Armah et al. (2018)
26.	*Parkia clappertoniana*	50% Aqu. EtOH/leaves	Ghana	*L. donovani* Promastigotes	17.00 moderate	Amphotericin B 3.20 µg/mL	Ohashi et al. (2018)
		50% Aqu. EtOH/Bark			17.60 moderate		
27.	*Tamarindus indica*	50% Aqu. EtOH/Leaves	Ghana	*L. donovani* Promastigotes	58.12 weak		

(continued)

Table 8.1 (continued)

No.	Plant species	Extract/used part	Country	*Leishmania* species/stage	IC_{50} (µg/mL) / Effectiveness	IC_{50} (µg/mL) standard drug	Ref.
Family Gentianaceae							
28.	*Anthocleista nobilis*	50% Aqu. EtOH/leaves	Ghana	*L. donovani* Promastigotes	41.50 moderate	Amphotericin B 3.20 µg/mL	Ohashi et al. (2018)
		50% Aqu. EtOH/root			79.00 weak		
29.	*Centaurium erythraea*	*n*-Hexane/Flowers	Morocco	*L. tropica* Promastigotes	37.20 Moderate	Glucantime > 500 µg/mL	Hassan et al. (2022)
				L. major Promastigotes	64.52 Weak		
Family Lamiaceae							
30.	*Marrubium vulgare*	MeOH/Leaves	Algeria	*L. major* Promastigotes	45.84 Moderate	Amphotericin B 0.20 µg/mL	Eddaikra et al. (2019)
				L. infantum Promastigotes	35.63 Moderate		
				L. major Amastigotes	32.15 Moderate		
				L. infantum Amastigotes	18.64 Moderate		
31.	*Rosmarinus officinalis*	Essential oil/Leaves	Morocco	*L. infantum* Promastigotes	1.20 High	Glucantime > 10 µg/mL	Hassan et al. (2022)
				L. tropica Promastigotes	3.50 High		
				L. major Promastigotes	2.60 High		
32.	*Vitex fosteri*	50% Aqu. EtOH/Leaves	Ghana	*L. donovani* Promastigotes	72.40 weak	Amphotericin B 3.20 µg/mL	Ohashi et al. (2018)
		50% Aqu. EtOH/Stem bark			49.80 moderate		

Family Malvaceae							
33.	*Ceiba pentandra*	50% Aqu. EtOH/Stem bark	Ghana	*L. donovani* Promastigotes	31.10 moderate	Amphotericin B 3.20 µg/mL	Ohashi et al. (2018)
34.	*Cola acuminata*	50% Aqu. EtOH/Stem bark	Ghana	*L. donovani* Promastigotes	47.80 moderate		
35.	*Cola cordifolia*	50% Aqu. EtOH/Leaves	Ghana	*L. donovani* Promastigotes	18.20 moderate		
		50% Aqu. EtOH/Stem bark			25.10 moderate		
36.	*Glyphaea brevis*	50% Aqu. EtOH/Leaves	Ghana	*L. donovani* Promastigotes	43.40 Moderate		
Family Meliaceae							
37.	*Khaya grandifolia*	50% Aqu. EtOH/Stem bark	Ghana	*L. donovani* Promastigotes	43.20 Moderate	Amphotericin B 3.20 µg/mL	Ohashi et al. (2018)
Family Moraceae							
38.	*Ficus capensis*	50% Aqu. EtOH/Leaves	Ghana	*L. donovani* Promastigotes	88.90 Weak	Amphotericin B 3.20 µg/mL	Ohashi et al. (2018)
		50% Aqu. EtOH/Stem bark			37.00 Moderate		
39.	*Treculia africana*	50% Aqu. EtOH/Stem bark	Ghana	*L. donovani* Promastigotes	44.80 Moderate		
Family Moringaceae							
40.	*Moringa oleifera*	16% Aqu. EtOH/Leaves	Tunisia	*L. major* Promastigotes	6.87 High	Amphotericin B 0.97 µg/mL	Hammi et al. (2020)
				L. major Amastigotes	9.31 High		
Family Myrtaceae							
41.	*Eugenia uniflora*	50% Aqu. EtOH/Seeds	Ghana	*L. donovani* Promastigotes	26.60 moderate	Amphotericin B 3.20 µg/mL	Ohashi et al. (2018)

(continued)

Table 8.1 (continued)

No.	Plant species	Extract/used part	Country	*Leishmania* species/stage	IC_{50} (μg/mL) / Effectiveness	IC_{50} (μg/mL) standard drug	Ref.
Family Ochnaceae							
42.	*Lophira lanceolata*	50% Aqu. EtOH/Stem bark	Ghana	*L. donovani* Promastigotes	68.60 weak	Amphotericin B 3.20 μg/mL	Ohashi et al. (2018)
		50% Aqu. EtOH/Roots			66.00 weak		
Family Olacaceae							
43.	*Ximenia americana*	50% Aqu. EtOH/Stem and twigs	Ghana	*L. donovani* Promastigotes	36.10 moderate	Amphotericin B 3.20 μg/mL	Ohashi et al. (2018)
Family Phyllanthaceae							
44.	*Bridelia ferruginea*	50% Aqu. EtOH/Leaves	Ghana	*L. donovani* Promastigotes	16.50 moderate	Amphotericin B 3.20 μg/mL	Ohashi et al. (2018)
Family Rubiaceae							
45.	*Mitragyna inermis*	50% Aqu. EtOH/Leaves	Ghana	*L. donovani* Promastigotes	21.90 Moderate	Amphotericin B 3.20 μg/mL	Ohashi et al. (2018)
		50% Aqu. EtOH/Stem bark			28.00 Moderate		
Family Ruteacea							
46.	*Clausena anisata*	50% Aqu. EtOH/Root	Ghana	*L. donovani* Promastigotes	12.10 Moderate	Amphotericin B 3.20 μg/mL	Ohashi et al. (2018)
47.	*Zanthoxylum zanthoxyloides*	50% Aqu. EtOH/Root	Ghana	*L. donovani* Promastigotes	13.50 Moderate		
		50% Aqu. EtOH/Stem bark			45.20 Moderate		

Family Simaroubaceae							
48.	*Brucea antidysenterica*	EthAc fraction/Seeds	Ethiopia	*L. aethiopica* Promastigotes	6.53 High	Amphotericin B 0.17 µg/mL	Ketema et al. (2023)
				L. donovani Promastigotes	6.16 High	Amphotericin B 0.15 µg/mL	
				L. aethiopica Amastigotes	4.14 High	Amphotericin B 0.08 µg/mL	
				L. donovani Amastigotes	6.85 High	Amphotericin B 0.07 µg/mL	
Family Solanaceae							
49.	*Solanum nigrum*	MeOH/Leaves	Kenya	*L. major* Promastigotes	1.28 High	Amphotericin B 0.04 µg/mL	Mutoro et al. (2018)
		Aqu. /Leaves			1.18 High		
50.	*Solanum villosum*	MeOH/Leaves	Egypt	*L. infantum* Promastigotes	2.00 High	Miltefosine 10.7 µg/mL	Okba et al. (2018)
Family Verbenaceae							
51.	*Lippia adoensis*	70% Aqu. EtOH/Leaves	Cameroon	*L. donovani* Promastigotes	22.87 Moderate	Amphotericin B 1.11 µg/mL	Kemgne et al. (2024)
		70% Aqu. EtOH/Twigs			10.52 High		
Family Zingiberaceae							
52.	*Alpinia officinarum*	MeOH/Rhizomes	Egypt	*L. major* Promastigotes	65.16 High	Amphotericin B 58.75 µg/mL	Darwish et al. (2019)

Aqu. aqueous, *MeOH* methanol, *EtOH* ethanol, *DCM* dichloromethane, *EtOAc* ethyl acetate

Table 8.2 In vivo leishmanicidal activity of plant extracts growing in African countries

No.	Plant species	Country	Extract used/dose	Experimental animals/ administration route	*Leishmania* species/type of leishmaniasis	Effectiveness	Ref.
Family Annonaceae							
1.	*Annona muricata*	Kenya	Aqueous extract of fruits (250 mg/kg)	BALB/c mice Oral	*L. major* Cutaneous leishmaniasis	The extract effectively reduced lesion size and swelling compared to untreated group	Chimbevo (2019)
2.	*Annona squamosa*						
FamilyAsphodelaceae							
3.	*Aloe secundiflora* **A** combined with *Callistemon citrinus* **C** (Myrtaceae)	Kenya	Aqueous extract of leaves (2 mg/kg **A** + 5mg/kg **C**)	BALB/c mice Intraperitoneal	*L. major* Cutaneous leishmaniasis	The Infection rate decreased to 19%, compared to 21% for pentostam. The footpad lesion size was significantly reduced, and spleen burden of amastigotes was reduced by 73.46%	Ndeti (2022)
Family Apiaceae							
4.	*Foeniculum Vulgare*	Ethiopia	Methanol extract of seed (4500 μg/kg)	BALB/c mice Intralesional application	*L. major* Cutaneous leishmaniasis	Lesion size and parasite burden in spleen decreased significantly compared to the untreated group and glucantime treated group	Mostafa et al. (2021)
Family Canellaceae							
5.	*Warburgia ugandens*	Kenya	Methanol extract of leaves (20 mg/kg)	BALB/c mice Intraperitoneal	*L. donovani* Visceral leishmaniasis	Parasite burden in spleen decreased to 13% compared to 14 % in the sodium stibogluconate treated group. IgG serum level was inhibited	Mogaka et al. (2022)
Family Fabaceae							
6.	*Piliostigma thonningii*	Kenya	Methanol extract of leaves (20 mg/kg)	BALB/c mice Intraperitoneal	*L. donovani* Visceral leishmaniasis	Parasite burden in spleen decreased to 23% compared to 14 % in the sodium stibogluconate treated group. IgG serum level was inhibited	Mogaka et al. (2022)

7.	*Prosopis juliflora*	Kenya	Methanol extract of leaves (100 mg/kg)	BALB/c mice Intraperitoneal	*L. donovani* Visceral leishmaniasis	Extract exhibited a higher splenic anti-amastigote effect compared to sodium stibogluconate	Mutile et al. (2021)
Family Rosaceae							
8.	*Prunus africana*	Kenya	Methanol extract of leaves (20 mg/kg)	BALB/c mice Intraperitoneal	*L. donovani* Visceral leishmaniasis	Parasite burden in spleen decreased to 30% compared to 14 % in the sodium stibogluconate treated group. IgG serum level was inhibited	Mogaka et al. (2022)

it shortly to search for new entities, and predict structure-activity relationship of derivatives.

The leishmanicidal activity of natural metabolites (Figs. 8.3, 8.4 and 8.5) isolated from plants cultivated in Africa is summarized in Table 8.3.

8.5.3 Mechanism of Leishmanicidal Activity

The antileishmanial activity of plant extracts is attributed to the presence of certain bioactive phytoconstituents inhibiting the growth of this parasite by complex multiple mechanisms of action targeting various pathways in leishmania life cycle.

Terpenes are plant secondary metabolites exhibiting a wide array of pharmacological activities. They are considered as an effective treatment for visceral leishmaniasis in experimental rodents showing no or negligible toxicity. Terpenes can induce parasite apoptosis via damaging the plasma membrane and inhibiting DNA topoisomerases. In addition, they elevate the immune response profile while inhibiting the levels of inflammatory cytokines and significantly diminish the parasite burden (Raimundo et al., 2022). Concisely, the terpenes mentioned in Table 8.3 show strong leishmanicidal effect compared to the standard drug. Even compounds 10 and 11 were more active than pentamidine.

Flavonoids are the most abundant antioxidant phytoconstituents in the plant family. They can initiate some morphological changes in the parasite leading to the loss of cellular integrity and arresting cell cycle at G phase (Araújo et al., 2019). They can also damage mitochondria affecting the energy level of *Leishmania* (Inacio et al., 2019). Estela et al. reported that flavonoids demonstrate inhibitory effect against cysteine proteases which is essential for virulence and pathogenicity of amastigotes in mammalian host cell (Lourenço et al., 2023).

Coumarins are phenolic compounds that can induce parasite apoptosis in a dose-dependent manner. They represent promising candidate for the synthesis of potent antileishmanial medications as they act on two aspects. Firstly, as a treatment for leishmaniasis where they can induce mitochondrial membrane damage. Secondly, as a tool to control Leishmania vectors where they block the transmission of leishmaniasis via reducing parasite loads (Ferreira et al., 2017).

Alkaloids have been described as having biological activities. They are capable of reducing the viability of promastigotes via generating reactive oxygen species in these cells. The oxidative stress generated induces depolarization of mitochondrial membrane in the parasite (De Sarkar et al., 2019).

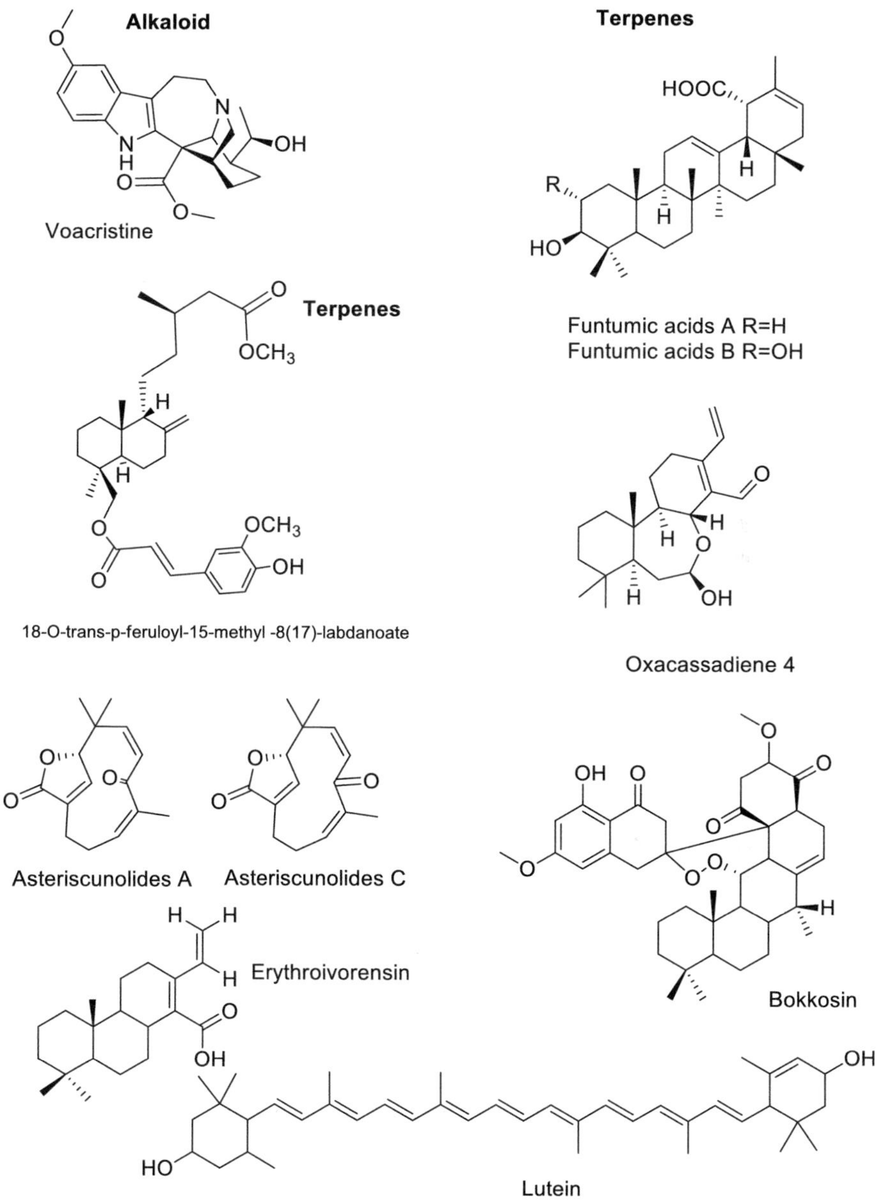

Fig. 8.3 Chemical structure of natural products with leishmanicidal activity (alkaloid and terpenes)

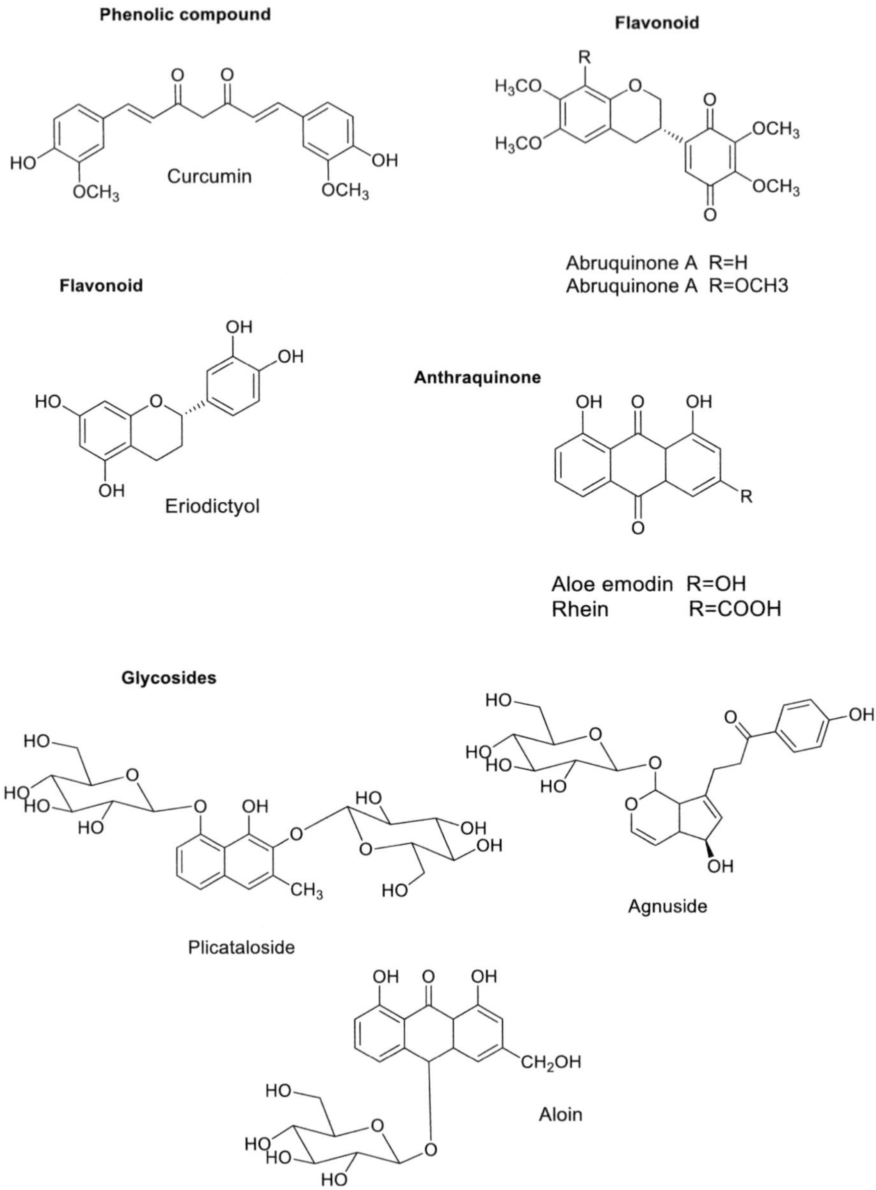

Fig. 8.4 Chemical structure of natural products with leishmanicidal activity (phenolic compound, flavonoids, and glycosides)

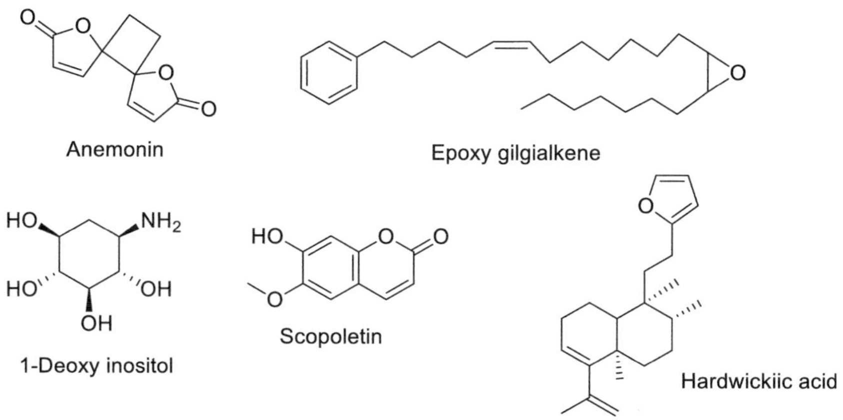

Fig. 8.5 Chemical structure of other natural products with leishmanicidal activity

8.6 Leishmanicidal Activity of Metabolites Isolated from Endophytes

Endophytes are microorganisms, fungi or bacteria, inhabiting all plant tissues, without causing any disease to plants. They represent a promising source of bioactive secondary metabolites characterized by high chemical diversity. Moreover, endophytes offer the possibility of growth by fermentation processes which are efficient, and reproducible, and allow the isolation of metabolites on a large and controllable scale. Note that endophytes usually biosynthesize multiple derivatives of the same compound, which facilitates the development of structure-activity relationship approaches (Hzounda Fokou et al., 2021). Scanning the FDA-approved new drug entities shows that natural products and their derivatives constitute about 33% of new drug entities; 25% of them are from endophytes (da Cunha et al., 2021).

Although an enormous number of studies were performed on endophytic metabolites as a source of leishmanicidal drugs, most of these studies were not performed in African countries on African plants. Thus, endophytic metabolites represent a rich untouched source for new molecules possessing antileishmanial activity.

The oxylipin (9*Z*,11*E*)-13-oxooctadeca-9,11-dienoic acid was isolated from *Penicillium herquei*; an endophytic fungus obtained from the leaves of *Laguncularia racemosa* (Combretaceae) a tree growing in Ghana. The compound exhibited a weak anti-parasite effect against *L. donovani* and *L. major* with IC_{50} values greater than 100 μM compared to amphotericin B, $IC_{50} = 0.31$ μM). These results strengthen the need for further manipulation of the compound structure by changing the conditions and medium for fungal growth (Hayibor et al., 2019).

In a review published in 2021, the authors reported that more than 250 plant species have been collected from African countries in the time interval from 1995 to 2019 mostly from Algeria, Egypt, Sudan, Cameroon, Nigeria, and

Table 8.3 Leishmanicidal activity of secondary metabolites isolated from African plants

No.	Compound name	Plant	Extract identification method	Country	Targeted species	IC_{50} compound	IC_{50} of standard drug	Ref.
Alkaloid								
1.	Voacristine	*Tabernaemontana ventricosa* (Apocynaceae)	Ethanol NMR	Uganda	*L. donovani* Promastigotes	11.00 μM	Miltefosine 5.5 μM	Andima et al. (2022)
Terpenes								
2.	Asteriscunolides A	*Asteriscus hierochunticus* (Asteraceae)	Methanol NMR, HRESIMS	Egypt	*L. donovani Promastigotes*	8.38 μg/mL	Pentamidine 4.4 μg/mL	Imieje et al. (2021)
3.	Asteriscunolides C					3.26 μg/mL	Pentamidine 29.3 μg/mL	
4.	Bokkosin	*Calliandra portoricensis* (Fabaceae)	Ethyl acetate NMR	Nigeria	*L. mexicana* Promastigotes	9.10 μM	Pentamidine 0.84 μM	Nvau et al. (2020)
5.	Erythroivorensin	*Erythrophleum ivorense* (Fabaceae)	Ethyl acetate NMR, qTOF-MS/MS	Ghana	*L. donovani* Promastigotes	29.10 μg/mL	Amphotericin B 2.40 μg/mL	Armah et al. (2018)
6.	Funtumic acids A	*Funtumia elastica* (Apocynaceae)	Methanol NMR, MS, IR, UV	Cameroon	*L. donovani* Promastigotes	10.49 μg/mL	Amphotericin B 0.05 μg/mL	Ma'mag et al. (2021)
7.	Funtumic acids B					13.21 μg/mL		
8.	Lutein	*Bridelia ferruginea* (Euphorbiaceae)	Methanol NMR, GC-MS	Nigeria	*L. donovani* Amastigote	9.30 μM	Not mentioned	Afolayan et al. (2018)
9.	Methylen-3β-hydroxy-15-methyl -8(17)-labdanoate	*Piliostigma thonningii* (Fabaceae)	Methanol NMR	Nigeria	*L. donovani* Amastigote	7.82 μM	Pentamidine 1.99 μM	Afolayan et al. (2018)

10.	18α-*O*-trans-*p*-feruloyl-15-methyl -8(17)-labdanoate	*Vachellia nilotica* (Fabaceae)	Hexane NMR	Nigeria	*L. major* Promastigotes	0.0154 μM	Pentamidine 0.63 μM	Al-Tannak et al. (2024)
11.	Oxacassadiene 4	*Acacia nilotica* (Fabaceae)	Hexane NMR	Nigeria	*L. mexicana* Promastigotes	0.28 μM	Pentamidine 0.78 μM	Anyam et al. (2021)
12.	Trigilgianin	*Trichilia gilgiana* (Meliaceae)	DCM:Methanol (1:1) NMR, EIMS	Cameroon	*L. donovani* Promastigotes	6.04 μg/mL	Amphotericin B 3.10 μg/mL	Kowa et al. (2020)
Flavonoid								
13.	Abruquinone A	*Abrus precatorius* (Fabaceae)	Ethyl acetate NMR, HR-ESIMS	Nigeria	*L. major* Promastigotes	6.35 μg/mL	Amphotericin B 0.39 μg/mL (*L. major*), 0.40 μg/mL (*L. tropica*) Pentamidine 3.15 μg/mL (*L. major*), 3.21 μg/mL (*L. tropica*)	Okoro et al. (2020)
					L. tropica Promastigotes	6.29 μg/mL		
14.	Abruquinone B				*L. major* Promastigotes	6.32 μg/mL		
					L. tropica Promastigotes	6.31 μg/mL		
15.	Eriodictyol	*Erythrophleum Ivorense* (Fabaceae)	Ethyl acetate UPLC-QTOF-MS/MS	Ghana	*L. donovani* Promastigotes	103.80 μg/mL	Amphotericin B 2.40 μg/mL	Armah et al. (2018)
Phenolic compounds								
16.	Curcumin	*Curcuma longa* (Zingiberaceae)	Standard drug with high purity	Egypt	*L. major* Promastigotes	35 μM	Not mentioned	Elamin et al. (2021)
Anthraquinone								
17.	Aloe-emodin	*Aloe macrocarpa* (Asphodelaceae)	Latex ESI-MS	Ethiopia	*L. aethiopica* Amastigotes	2.23 μg/mL	Amphotericin B 2.45 μM	

(continued)

Table 8.3 (continued)

No.	Compound name	Plant	Extract identification method	Country	Targeted species	IC_{50} compound	IC_{50} of standard drug	Ref.
					L. donovani Amastigotes	2.75 μg/mL		Tewabe et al. (2019)
18.	Rhein				*L. aethiopica* Amastigotes	1.01 μg/mL		
					L. donovani Amastigotes	1.17 μg/mL		
Glycosides								
19.	Agnuside	*Vitex grandifolia* (Lamiaceae)	Methanol NMR	Nigeria	*L. donovani* Amastigotes	16.98 μg/mL	Amphotericin B 1.14 μg/mL	Bello et al. (2018)
20.	Aloin A/B	*Aloe macrocarpa* (Asphodelaceae)	Latex ESI-MS	Ethiopia	*L. aethiopica* Amastigotes	2.80 μg/mL	Amphotericin B 2.45 μM	Tewabe et al. (2019)
					L. donovani Amastigotes	2.92 μg/mL		
21.	Plicataloside	*Aloe rugosifolia* (Asphodelaceae)	Methanol NMR, HR-ESIMS	Ethiopia	*L. aethopica* Promastigotes	14.22 μg/mL	Amphotericin B 8.10 μg/mL	Chemeda et al. (2022)
					L. donovani Promastigotes	18.86 μg/mL	Amphotericin B 7.20 μg/mL	
Others								
22.	Anemonin (Spirolactone)	*Ranunculus multifidus* (Ranunculaceae)	80% methanol NMR, HR-ESIMS	Ethiopia	*L. aethiopica* Amastigotes	0.24 μg/mL	Amphotericin B 0.015 μg/mL	Sirak et al. (2021)
					L. donovani Amastigotes	0.37 μg/mL	Amphotericin B 0.006 μg/mL	
23.	1-Deoxy inositol (Sugar alcohol)	*Diospyros gracilescens* (Ebenaceae)	80% ethanol NMR, HR-ESIMS	Cameroon	*L. donovani* Promastigotes	241.71 μM	Amphotericin B 0.34 μg/mL	Njanpa et al. (2021)
						120.00 μM		

					L. donovani Amastigotes		Amphotericin B 0.12 μg/mL	
24.	Epoxy gilgialkene (Phenyl alkene)	*Trichilia gilgiana* (Meliaceae)	DCM:Methanol (1:1) NMR, EIMS	Cameroon	*L. donovani* Promastigotes	56.81 μg/mL	Amphotericin B 3.10 μg/mL	Kowa et al. (2020)
25.	Scopoletin (Coumarin)					6.80 μg/mL		
26.	Hardwickiic acid (Furan derivative)	*Croton sylvaticus* (Euphorbiaceae)	Petroleum ether NMR	Ghana	*L. donovani* Promastigotes	31.57 μM	Amphotericin B 3.35 μM	Crentsil et al. (2020)

South Africa leading to isolation and identification of almost 700 endophytes (fungi and bacteria). Searching different databases about these endophytes, mainly fungi, revealed that leishmanicidal activity was evaluated for some of them in the last five years in non-African countries. The results are summarized in Table 8.4, Fig. 8.6.

Regarding endophytic bacteria, the lipopeptide biosurfactant LPB produced by the bacterium *Acinetobacter junii* B6 was found to ameliorate *L. tropica* infection and modulate the immune response via affecting cytokine gene expression in infected macrophages. LPB was evaluated for antileishmanial effect alone and in combination with meglumine antimoniate MA. Results showed that MA/LPB combination reduced the parasite burden more than each alone. Also, the combination therapy elevated the levels of IL-12p40 and IFN-γ and inhibited the level of IL-10. Thus MA/LPB can be considered as a possible efficient treatment for CL (Sharifi et al., 2023).

8.7 Limitations of Using Natural Products-Based Therapy and Attempts to Overcome Those Limitations

Despite all the advantages of using natural products for curing leishmaniasis, still, the process of developing a pharmaceutical entity based on natural products faces a lot of burdens. So logically there has been a steep decline in the number of new natural product-based entities reaching clinical over the past decades (Wright, 2019). The isolation process is very complex and tedious and yields small amounts of pure compounds that are usually not enough for in vivo studies. In addition, most compounds isolated from plants are already previously reported, lacking novelty makes it less likely to lead to academic publications. This motivates scientists to explore new pharmacological activities for ready-known metabolites (Walsh & Tang, 2017).

Another important challenge is the undesirable physical properties of natural products. Drug absorption can be predicted from certain equations as Lipinski's rule of five. Unfortunately, most natural metabolites do not obey this rule. This is in addition to the poor aqueous solubility. The absence of specifically authorized legislation and international agreements regulating the access to plant resources makes the documentation of using herbal products a difficult mission. Finally, climate change has a highly negative impact on the environment and flora. Environmental problems such as desertification and global warming led to the extinction of many valuable plants including important medicinal species (Ungogo et al., 2020).

Researchers are attempting to overcome these limitations through different approaches including the use of recent biotechnological technologies such as plant cell cultures or microbial fermentation in order to provide larger quantities of natural compounds in a more controlled and reproducible manner compared to traditional plant-based extraction. Additionally, the use of advanced analytical techniques facilitates the automated purification and could help in the identification process of

Table 8.4 Leishmanicidal activity of endophytic fungi

No.	Endophyte name	Compound or extract	Targeted species	Effectiveness/ IC_{50} or LC_{50}	Ref.
Fungi					
1.	*Alternaria alternata*	Alternarlactone A (**1**) Verrulactone B (**2**) Altenuisol (**3**) 5-*O*-Methylether-3-hydroxy alternariol (**4**)	*L. donovani* promastigotes	(**1**) $IC_{50} = 4.7$ µM (**2**) $IC_{50} = 2.4$ µM (**3**) $IC_{50} = 7.1$ µM (**4**) $IC_{50} = 7.5$ µM	Shi et al. (2019)
2.	*Aspergillus* spp	Monomethylsulochrin (**5**)	*L. amazonensis* promastigotes	(**5**) $IC_{50} = 18.04$ µM	Silva-Silva et al. (2022)
3.	*Botryosphaeria rhodina*	Botryosphaeran (**6**)	*L. amazonensis* promastigotes	(**6**) $IC_{50} = 7.0$ µg/mL	Carloto et al. (2022)
			L. Amazonensis amastigote	(**6**) IC_{50} 9.3 µg/mL	
4.	*Colletotrichum* spp.	DCM extract	*L. amazonensis* amastigote	$IC_{50} = 36$ µg/mL	de Azevedo et al. (2021)
5.	*Diaporthe* spp.	EtAcO extract	*L. infantum* promastigotes	A dose of 4000 µg/mL caused 80% mortality in 24 h	Moreira et al. (2020)
		DCM extract	*L. amazonensis* amastigote	$IC_{50} = 21$ µg/mL	de Azevedo et al. (2021)
6.	*Emericella nidulans*	Fungal enzyme extract (lipase)	*L. amazonensis* amastigote	A dose of 5 mg/mL reduced viability of amastigotes by 39.65%	Alves et al. (2018)
7.	*Fusarium graminarum*	Silver nanoparticles of fungal aqu. Extract	*L. tropica* promastigotes	Significant reduction in parasite burden after 24 h treatment with 100 µg/ml	Mohammed et al. (2019)
			L. tropica amastigotes		
8.	*Geosmithia langdonii*	(1*S*,2*R*,3*R*,4*R*,5*R*)-2,3,4-trihydroxy-5-methylcyclohexyl-2′,5′-dihydroxybenzoate (**7**) (1*S*,2*S*,3*S*,4*R*,5*R*)-4-	*L. donovani* Promastigotes	(**7**) $IC_{50} = 100.00$ µM (**8**) $IC_{50} = 57.00$ µM	Malak et al. (2018)

(continued)

Table 8.4 (continued)

No.	Endophyte name	Compound or extract	Targeted species	Effectiveness/ IC_{50} or LC_{50}	Ref.
		[(2′,5′-dihydroxybenzyl)oxy]-5-methylcyclohexane-1,2,3-triol (**8**)			
9.	*Nectria pseudotrichia*	10-acetyl trichoderonic acid A (**9**) 6′-Acetoxy-piliformic acid (**10**) Hydroheptelidic acid (**11**)	*L. braziliensis* Promastigotes	(**9**) $IC_{50} = 21.4$ µM (**10**) $IC_{50} = 28.3$ µM (**11**) $IC_{50} = 24.8$ µM	Cota et al. (2018)
10.	*Paecilomyces* spp.	Harzialactone A (**12**)	*L. amazonensis* promastigotes	$IC_{50} = 5.25$ µg/mL	Braun et al. (2021)
			L. amazonensis Amastigotes	$IC_{50} = 18.18$ µg/mL	
11.	*Penicillium citrinum*	Cyclo(L-pro-L-Leu) (**13**) Cyclo-(L-pro-L-Phe) (**14**) Tryprostatin B (**15**)	*L. amazonensis* Amastigotes	All the compounds reduced parasite growth by 50% at a dose of 100 µM. Compound (**15**) is the most active	GuimarÃ et al. (2019)
			L. infantum Amastigotes		
12.	*Trichoderma asperelloides*	EtOH extract	*L. amazonensis* promastigotes	Parasite viability was reduced by 30–90% at conc. 0.31–4.95 ng/µl	Lopes et al. (2020)

DCM dichloromethane, *EtOH* ethanol, *EthAcO* ethyl acetate, *Aqu* aqueous

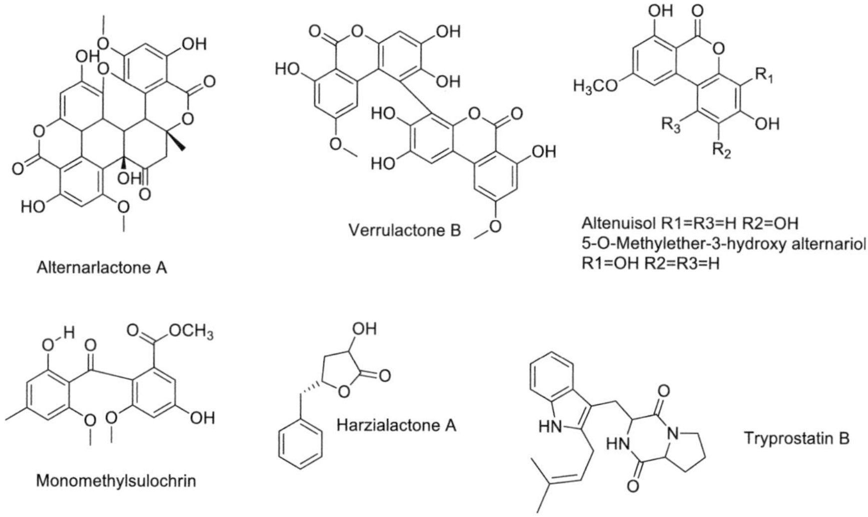

Fig. 8.6 Chemical structure of some endophytic products with leishmanicidal activity

compounds. Combinatorial chemistry could also help synthesize derivatives of known compounds to create novel molecules possessing better bioactivity and bioavailability. The undesirable physical properties or poor bioavailability can be handled using nanotechnology. Encapsulation of natural products in nanoparticles or liposomes can improve solubility and bioavailability. Nanotechnology can also enhance targeted delivery and reduce side effects, making natural products more effective. Implementing sustainable harvesting techniques and cultivating endangered medicinal plants in controlled environments (e.g., greenhouses, botanical gardens) can protect these species from climate-related threats. Also, developing in vitro techniques for plant propagation as tissue culture can ensure the continuous supply of medicinal plants regardless of environmental challenges (Chaachouay & Zidane, 2024; Dzobo, 2022).

8.8 Conclusion and Future Prospects

African flora is characterized by high diversity offering human a valuable source of metabolites covering a wide array of pharmacological activities, one of which is the antileishmanial activity. In the past five years, there has been an intensive scientific research in the field of natural metabolites isolated from plants and their associated endophytes and their biological activity. A total of 60 plants belonging to different families mainly Fabaceae and Asteraceae were evaluated either in vivo or in vitro for leishmanicidal activity against different strains. The plant extracts combated extracellular and intracellular forms of *leishmania* in the in vitro assays with effectiveness ranging from weak to very effective. The extracts were mainly polar (aqueous, methanol, and ethyl acetate). Interestingly, most of the tested essential oils (*Pistacia lentiscus*, *Rosmarinus officinalis*, *Artemisia absinthium*) were highly active. The eight plant extracts that were assisted in vivo treated visceral and cutaneous leishmaniasis and showed no toxicity. In addition, 26 pure isolated compounds exhibited antileishmanial effect. The most effective compounds compared to the standard drugs were terpenes. The mechanism of action was explained in some studies. Regarding endophytic metabolites, they represent a rich untouched mine for new therapeutic molecules. The IC_{50} of most of the tested molecules was less than 50 μM. To sum up, African plants offer a promising future cure for leishmaniasis. Many approaches can be adopted with the use of computer-aided drug design tools not only to investigate the effectiveness of natural compounds but also to understand the mechanism of action via simulating different targets in the parasite and also to get better assumptions about the bioavailability of effective molecules. This eventually facilitates the transfer to the phase of clinical trials aiming to find better therapeutic alternatives to a serious pandemic disease.

References

Abbas, M. A. S., Lachheb, J., Chelbi, I., Louati, D., Dachraoui, K., Ben Miled, S., & Zhioua, E. (2022). Independent circulation of Leishmania major and Leishmania tropica in their respective sandfly vectors for transmission of zoonotic and chronic cutaneous leishmaniasis co-existing in a mixed focus of Central Tunisia. *Journal of Pathogens, 11*(8), 855–867.

Abdoul-Latif, F. M., Oumaskour, K., Abdallah, N., Ainane, A., Houmed Aboubaker, I., Merito, A., et al. (2024). Overview of research on Leishmaniasis in Africa: Current status, diagnosis, therapeutics, and recent advances using by-products of the Sargassaceae family. *Pharmaceuticals, 17*(4), 53–544.

Afolayan, M., Srivedavyasasri, R., Asekun, O. T., Familoni, O. B., Orishadipe, A., Zulfiqar, F., et al. (2018). Phytochemical study of Piliostigma thonningii, a medicinal plant grown in Nigeria. *Journal of Medicinal Chemistry and Research, 27*, 2325–2330.

Al-Tannak, N. F., Anyam, J. V., Santali, E. Y., Gray, A. I., Ibeji, C. U., & Igoli, J. O. (2024). Antiparasitic activity and computational studies on a novel labdane diterpene from the roots of Vachellia nilotica. *Open Chemistry, 22*(1), 2024–2035.

Alvar, J., den Boer, M., & Dagne, D. A. J. T. L. G. H. (2021). Towards the elimination of visceral leishmaniasis as a public health problem in East Africa: Reflections on an enhanced control strategy and a call for action. *The Lancet Global Health, 9*(12), 1763–1769.

Alves, D. R., Morais, S. M., Tomiotto-Pellissier, F., Vasconcelos, F. R., Freire, F. d. C. O., Silva, I. N. G. d., … Conchon-Costa, I. (2018). Leishmanicidal and fungicidal activity of lipases obtained from endophytic fungi extracts. PLoS One, 13(6), 1967–1996.

Andima, M., Ndakala, A., Derese, S., Biswajyoti, S., Hussain, A., Yang, L. J., et al. (2022). Antileishmanial and cytotoxic activity of secondary metabolites from Taberneamontana ventricosa and two aloe species. *Natural Product Research, 36*(5), 1365–1369.

Anyam, J. V., Daikwo, P. E., Ungogo, M. A., Nweze, N. E., Igoli, N. P., Gray, A. I., et al. (2021). Two new antiprotozoal diterpenes from the roots of Acacia nilotica. *Frontiers in Chemistry, 9*, 6247–6255.

Araújo, M. V., Queiroz, A. C., Silva, J. F., Silva, A. E., Silva, J. K., Silva, G. R., et al. (2019). Flavonoids induce cell death in Leishmania amazonensis: In vitro characterization by flow cytometry and Raman spectroscopy. *Analyst, 144*(17), 5232–5244.

Argaw, D., Mulugeta, A., Herrero, M., Nombela, N., Teklu, T., Tefera, T., et al. (2013). Risk factors for visceral leishmaniasis among residents and migrants in Kafta-Humera, Ethiopia. *PLOS Neglected Tropical Diseases, 7*(11), 2543–2550.

Armah, F. A., Amponsah, I. K., Mensah, A. Y., Dickson, R. A., Steenkamp, P. A., Madala, N. E., & Adokoh, C. K. (2018). Leishmanicidal activity of the root bark of Erythrophleum Ivorense (Fabaceae) and identification of some of its compounds by ultra-performance liquid chromatography quadrupole time of flight mass spectrometry (UPLC-QTOF-MS/MS). *Journal of Ethnopharmacology, 211*, 207–216.

Bello, O. M., Ogbesejana, A. B., & Osibemhe, M. (2018). Iridoid glucosides from Vitex grandifolia displayed anti-inflammatory and antileishmania effects and structure activity relationship. *Journal of Applied Sciences and Environmental Management, 22*(3), 373–378.

Berbert, T. R. N., Mello, T. F. P.d., Wolf Nassif, P., Mota, C. A., Silveira, A. V., Duarte, G. C., … practice. (2018). Pentavalent antimonials combined with other therapeutic alternatives for the treatment of cutaneous and mucocutaneous leishmaniasis: A systematic review. *Dermatology Research,* 2018(1), 9014726–9014734.

Bouyahya, A., Assemian, I. C. C., Mouzount, H., Bourais, I., Et-Touys, A., Fellah, H., et al. (2019). Could volatile compounds from leaves and fruits of Pistacia lentiscus constitute a novel source of anticancer, antioxidant, antiparasitic and antibacterial drugs? *Journal of Industrial Crops and Products, 128*, 62–69.

Bouyahya, A., Et-Touys, A., Dakka, N., Fellah, H., Abrini, J., & Bakri, Y. (2018). Antileishmanial potential of medicinal plant extracts from the North–West of Morocco. *Beni-Suef University Journal of Basic and Applied Sciences, 7*(1), 50–54.

Braun, G. H., Ramos, H. P., Candido, A. C., Pedroso, R. C., Siqueira, K. A., Soares, M. A., et al. (2021). Evaluation of antileishmanial activity of harzialactone a isolated from the marine-derived fungus Paecilomyces sp. *Natural Product Research, 35*(10), 1644–1647.

Carloto, A. C. M., da Silva Bortoleti, B. T., Rodrigues, A. C. J., Silva, T. F., Tomiotto-Pellissier, F., Bidóia, D. L., et al. (2022). Botryosphaeran,[(1→ 3)(1→ 6)-β-D-glucan], induces apoptosis-like death in promastigotes of Leishmania amazonensis, and exerts a leishmanicidal effect on infected macrophages by activating NF-kB and producing pro-inflammatory molecules. *Chemico-Biological Interactions, 351*, 109713–109721.

Carvalho, L. P., Passos, S., Dutra, W. O., Soto, M., Alonso, C., Gollob, K., et al. (2005). Effect of LACK and KMP11 on IFN-γ production by peripheral blood mononuclear cells from cutaneous and mucosal Leishmaniasis patients. *Scandinavian Journal of Immunology, 61*(4), 337–342.

Chaachouay, N., & Zidane, L. J. D. (2024). Plant-derived natural products: A source for drug discovery and development. *Drugs Candidates, 3*(1), 184–207.

Chemeda, G., Bisrat, D., Yeshak, M. Y., & Asres, K. (2022). In vitro antileishmanial and antitrypanosomal activities of plicataloside isolated from the leaf latex of Aloe rugosifolia Gilbert & Sebsebe (Asphodelaceae). *Molecules, 27*(4), 1400–1421.

Chimbevo, L. M. (2019). Efficacy and safety, nutritional andAntioxidant Activityof Kenyan Annona muricata (L.) and Annona squamosa (L.). In *Fruits extracts in BALB/c mice model of L.* major Leshmaniasis.

Cota, B. B., Tunes, L. G., Maia, D. N. B., Ramos, J. P., Oliveira, D. M.d., Kohlhoff, M., et al. (2018). Leishmanicidal compounds of Nectria pseudotrichia, an endophytic fungus isolated from the plant Caesalpinia echinata (Brazilwood). *Memórias Inst Oswaldo Cruz, 113*, 102–110.

Crentsil, J. A., Yamthe, L. R. T., Anibea, B. Z., Broni, E., Kwofie, S. K., Tetteh, J. K. A., & Osei-Safo, D. (2020). Leishmanicidal potential of hardwickiic acid isolated from Croton sylvaticus. *Frontiers in Pharmacology, 11*, 753–761.

da Cunha, A. F., Di, C., Oliveira, Y. L., Dolabella, S. S., Scher, R., Souto, E. B., Lopez, J. A., & Jain, S. (2021). Anti-leishmanial compounds from microbial metabolites: A promising source. *Applied Microbiology and Biotechnology, 105*, 1–14.

Darwish, A., Samy, M., Sugimoto, S., Matsunami, K., & Otsuka, H. (2019). In vitro antileishmanial activity of methanolic extracts for some selected medicinal plants. *Pharmacognosy Magazine, 15*(62), 62–71.

de Azevedo, R. N., Ferreira, M. C., de Assis, J. C. S., Junior, P. A. S., Maia, D. N. B., de Almeida Alves, T. M., et al. (2021). *Bioprospecting of bioactive secondary metabolites of Endophytic fungi of Carapichea ipecacuanha (Rubiaceae), a Neotropical medicinal species* (pp. 321–336). Neotrop Endophyt Fungi Diver Eco Biotech App.

De Muylder, G., Ang, K. K., Chen, S., Arkin, M. R., Engel, J. C., McKerrow, J. H. J. P., & n. t. d. (2011). A screen against Leishmania intracellular amastigotes: Comparison to a promastigote screen and identification of a host cell-specific hit. *PLOS Neglected Tropical Diseases, 5*(7), 1253–1260.

De Sarkar, S., Sarkar, D., Sarkar, A., Dighal, A., Staniek, K., Gille, L., & Chatterjee, M. (2019). Berberine chloride mediates its antileishmanial activity by inhibiting Leishmania mitochondria. *Parasitology Research, 118*, 335–345.

Dey, S., Mukherjee, D., Chakraborty, S., Mallick, S., Dutta, A., Ghosh, J., et al. (2015). Protective effect of Croton caudatus Geisel leaf extract against experimental visceral leishmaniasis induces proinflammatory cytokines in vitro and in vivo. *Experimental Parasitology, 151*, 84–95.

Dzobo, K. (2022). The role of natural products as sources of therapeutic agents for innovative drug discovery. *Comprehensive Pharmacology*, 408–418.

Eddaikra, N., Boudjelal, A., Amine Sbabdji, M., Eddaikra, A., Boudrissa, A., Mounir Bouhenna, M., et al. (2019). Leishmanicidal and cytotoxic activity of Algerian medicinal plants on Leishmania major and Leishmania infantum. *Journal of Medical Microbiology and Infectious Diseases, 7*(3), 66–71.

Elamin, M., Al-Olayan, E., Abdel-Gaber, R., & Yehia, R. (2021). Anti-proliferative and apoptosis induction activities of curcumin on Leishmania major. *Revista Argentina de Microbiología, 53*(3), 240–247.

Ferreira, C., Passos, C., Soares, D., Costa, K., Rezende, M., Lobão, A., et al. (2017). Leishmanicidal activity of the alkaloid-rich fraction from Guatteria latifolia. *Experimental Parasitology, 172*, 51–60.

Fischer, T., Fischer, M., Schliemann, S., & Elsner, P. (2024). Treatment of mucocutaneous leishmaniasis–A systematic review. *Journal der Deutschen Dermatologischen Gesellschaft, 22*, 763.

Garba, J. K., Nguengang, R. T., Youmbi, G. T., Menatche, J. N., Ngansop, C. A. N., Bankeu, J. J. K., et al. (2022). Antileishmanial, antibacterial and cytotoxicity activity of the extracts, fractions, and compounds from the fruits and stem bark extracts of Pentadesma butyracea Sabine. *Zeitschrift für Naturforschung B, 77*(1), 9–15.

Ghorbani, M., & Farhoudi, R. J. D.d., development. (2017). Leishmaniasis in humans: Drug or vaccine therapy? *Drug Design, Development and Therapy, 12*, 25–40.

GuimarÃ, L., Bueno, D. N., Zani, C. L., Rosa, L. H., & Cota, B. B. (2019). Diketopiperazine alkaloids produced by the endophytic fungus Penicillium citrinum and evaluation of their antileishmanial activity. *African Journal of Microbiology Research, 13*(28), 562–567.

Hammi, K., Essid, R., Tabbene, O., Elkahoui, S., Majdoub, H., Ksouri, R. J. S. A. J., & o. B. (2020). Antileishmanial activity of Moringa oleifera leaf extracts and potential synergy with amphotericin B. *South African Journal of Botany, 129*, 67–73.

Hasan, M. M., Proma, S. B., Hossain, M. S., Arifuzzaman, M., Islam, N., Siddique, M. A. B., & Amiruzzaman. (2023). A case report on Para-kala-azar dermal leishmaniasis: An unresolved mystery. *BMC Infectious Diseases, 23*(1), 885.

Hassan, A. A., Khalid, H. E., Abdalla, A. H., Mukhtar, M. M., Osman, W. J., & Efferth, T. (2022). Antileishmanial activities of medicinal herbs and phytochemicals in vitro and in vivo: An update for the years 2015 to 2021. *Molecules, 27*(21), 75–79.

Hayibor, K., Kwain, S., Osei, E., Nartey, A. P., Tetevi, G. M., Owusu, K. B.-A., et al. (2019). Ghanaian mangrove wetland endophytic fungus, Penicillium herquei strain BRS2A-AR produces (9Z, 11E)-13-oxooctadeca-9, 11-dienoic acid with activity against trichomonas mobilensis. *International Journal of Biological and Chemical Sciences, 13*(4), 1918–1937.

Hzounda Fokou, J. B., Dize, D., Etame Loe, G. M., Nko'o, M. H. J., Ngene, J. P., Ngoule, C. C., & Boyom, F. F. (2021). Anti-leishmanial and anti-trypanosomal natural products from endophytes. *Journal of Parasitology Research, 120*, 785–796.

Imieje, V. O., Zaki, A. A., Metwaly, A. M., Eissa, I. H., Elkaeed, E. B., Ali, Z., et al. (2021). Antileishmanial derivatives of humulene from Asteriscus hierochunticus with in silico tubulin inhibition potential. *Journal of Records of Natural Products, 16*, 150–171.

Inacio, J. D., Fonseca, M. S., & Almeida-Amaral, E. E. (2019). (−)-epigallocatechin 3-O-gallate as a new approach for the treatment of visceral Leishmaniasis. *Journal of Natural Products, 82*(9), 2664–2667.

Jawed, J. J., & Majumdar, S. (2018). Recent trends in Leishmania research: A therapeutic perspective. *Journal of Infectiology and Epidemiology, 1*(3), 34–45.

Kahime, K., Boussaa, S., Laamrani-El Idrissi, A., Nhammi, H., & Boumezzough, A. (2016). Epidemiological study on acute cutaneous leishmaniasis in Morocco. *Journal of Acute Disease, 5*(1), 41–45.

Kaur, G., & Rajput, B. (2014). Comparative analysis of the omics technologies used to study antimonial, amphotericin B, and pentamidine resistance in leishmania. *Journal of Parasitology Research, 2014*(1), 726328–726334.

Kemgne, E. A. M., Tali, M. B. T., Dize, D., Ngansop, C. A. N., Kamdem, B. P., & Boyom, F. F. (2024). Crude extracts from Lippia adoensis (Hochst.). In *Inhibit the growth of plasmodium, Leishmania and Trypanosoma parasites*. Sear Life-Sci Lit.

Ketema, T., Tadele, M., Gebrie, Z., Makonnen, E., Hailu, A., & Abay, S. M. (2023). In vitro anti-Leishmanial activities of methanol extract of Brucea antidysenterica JF mill seeds and its solvent fractions. *Journal of Experimental Pharmacology, 15*, 123–135.

Kowa, T. K., Tchokouaha, L. R., Cieckiewicz, E., Philips, T. J., Dotse, E., Wabo, H. K., et al. (2020). Antileishmanial and cytotoxic activities of a new limonoid and a new phenyl alkene from the stem bark of Trichilia gilgiana (Meliaceae). *Natural Product Research, 34*(22), 3182–3188.

Lewis, M. D., Paun, A., Romano, A., Langston, H., Langner, C. A., Moore, I. N., et al. (2020). Fatal progression of experimental visceral leishmaniasis is associated with intestinal parasitism and secondary infection by commensal bacteria, and is delayed by antibiotic prophylaxis. *PLoS Pathogens, 16*(4), 1008456–1008463.

Lopes, D. D. S., Santos, U. R. D., Dos Anjos, D. O., Silva Júnior, L. J. C. D., Paula, V. F. D., Vannier-Santos, M. A., et al. (2020). Ethanolic extract of the fungus Trichoderma asperelloides induces ultrastructural effects and death on Leishmania amazonensis. *Frontiers in Cellular and Infection Microbiology, 10*, 306.

Lourenço, E. M., da Silva, F., das Neves, A. R., Bonfá, I. S., Ferreira, A. M. T., Menezes, A. C., et al. (2023). Investigation of the potential targets behind the promising and highly selective antileishmanial action of synthetic flavonoid derivatives. *Journal of ACS Infectious Diseases, 9*(10), 2048–2061.

Ma'mag, L. K., Zintchem, A. A. A., Théodora, K. K., de Théodore Atchadé, A., Lauve, T. Y., Frédérich, M., et al. (2021). Antiplasmodial and antileishmanial inhibitory activity of triterpenes and steroidal alkaloid from the leaves of Funtumia elastica (Preuss) Stapf (Apocynaceae). *Fitoterapia, 151*, 104869–104877.

Malak, L. G., Ibrahim, M. A., Moharram, A. M., Pandey, P., Tekwani, B., Doerksen, R. J., et al. (2018). Antileishmanial carbasugars from Geosmithia langdonii. *Journal of Natural Products, 81*(10), 2222–2227.

Mathison, B. A., & Bradley, B. T. (2023). Review of the clinical presentation, pathology, diagnosis, and treatment of leishmaniasis. *Laboratoriums Medizin, 54*(4), 363–371.

Mathlouthi, A., Belkessam, M., Sdiri, M., Diouani, M. F., Souli, A., El-Bok, S., & Ben Attia, M. J. J. O. E. O. B. (2018). Chemical composition and anti-leishmania major activity of essential oils from Artemesia spp. grown in Central Tunisia. *Journal of Essential Oil Bearing Plants, 21*(5), 1186–1198.

Mogaka, M. D., Mutiso, J. M., Macharia, J. C., Ayako, R. M., Osero, B., Gicheru, M. M. J. J., & o. Z. R. (2022). Evaluation of the therapeutic potential of Warburgia ugandensis, Prunus africana, and Piliostigma thonningii against Leishmania donovani in vitro and in Balb/c mice. *Journal of Zoological Research, 4*(4).

Mohammed, O. T., Abdulkhaliq, R. J., & Mohammed, S. T. (2019). *The effects of Fusarium graminarum silver nanoparticles on leishmania tropica*. Paper presented at the J Physic: Conference Series.

Moreira, C. C., Luna, G. L. F., Soriano, B., Cavicchioli, R., de Sousa, C. P., Anibal, F. F., Lacava, P. T. J. A. J., & o. M. R. (2020). Leishmanicidal, cytotoxic, antimicrobial and enzymatic activities of Diaporthe species, a mangrove-isolated endophytic fungus. *African Journal of Microbiology Research, 14*(9), 516–524.

Mostafa, G., Nahid, J., Alireza, D., & Ebrahim, S. S. (2021). Effect of Foeniculum Vulgare aqueous and alcoholic seed extract against zoonotic cutaneous Leishmaniasis. *Ethiopian Journal of Health Sciences, 31*(2), 22–34.

Mutile, M. C., Muli, M. J., & Muita, G. M. (2021). Safety and efficacy of Prosopis juliflora leaf extract as a potential treatment against visceral Leishmaniasis in Balb/c mice. *Iranian Journal of Parasitology, 16*(4), 652–665.

Mutoro, C. N., Kinyua, J. K., Ng'ang'a, J. K., Kariuki, D. W., Ingonga, J. M., & Anjili, C. O. (2018). In vitro study of the efficacy of Solanum nigrum against Leishmania major. *Journal of F1000Research, 7*, 1329–1337.

Ndeti, C. M. (2022). *Synergistic effects of extracts of Aloe secundiflora Eng L. and Callistemon citrinus William C. In Leishmania major infected BALB/c mice*. JKUAT-COHES.

Nigatu, H., Belay, A., Ayalew, H., Abebe, B., Tadesse, A., Tewabe, Y., Degu, A. J. J., & o. E. P. (2021). In vitro antileishmanial activity of some Ethiopian medicinal plants. *Journal of Experimental Pharmacology, Volume 13*, 15 22.

Njanpa, C. A. N., Wouamba, S. C. N., Yamthe, L. R. T., Dize, D., Tchatat, B. M. T., Tsouh, P. V. F., et al. (2021). Bio-guided isolation of anti-leishmanial natural products from Diospyros gracilescens L.(Ebenaceae). *BMC Complementary Medicine and Therapies, 21*, 1–12.

Nvau, J. B., Alenezi, S., Ungogo, M. A., Alfayez, I. A., Natto, M. J., Gray, A. I., et al. (2020). Antiparasitic and cytotoxic activity of bokkosin, a novel diterpene-substituted chromanyl benzoquinone from Calliandra portoricensis. *Journal of Frontiers in Chemistry, 8*, 574103–574112.

Ohashi, M., Amoa-Bosompem, M., Kwofie, K. D., Agyapong, J., Adegle, R., Sakyiamah, M. M., et al. (2018). In vitro antiprotozoan activity and mechanisms of action of selected G hanaian medicinal plants against trypanosoma, leishmania, and plasmodium parasites. *Journal of Phytotherapy Research, 32*(8), 1617–1630.

Okba, M. M., Sabry, O. M., Matheeussen, A., & Abdel-Sattar, E. (2018). In vitro antiprotozoal activity of some medicinal plants against sleeping sickness, Chagas disease and leishmaniasis. *Journal of Future Medicinal Chemistry, 10*(22), 2607–2617.

Okoro, E. E., Ahmad, M. S., Osoniyi, O. R., & Onajobi, F. D. (2020). Antifungal and antileishmanial activities of fractions and isolated isoflavanquinones from the roots of Abrus precatorius. *Journal of Comparative Clinical Pathology, 29*, 391–396.

Osorio, E. Y., Uscanga-Palomeque, A., Patterson, G. T., Cordova, E., Travi, B. L., Soong, L., Melby, P. C. J. P., & n. t. d. (2023). Malnutrition-related parasite dissemination from the skin in visceral leishmaniasis is driven by PGE2-mediated amplification of CCR7-related trafficking of infected inflammatory monocytes. *PLOS Neglected Tropical Diseases, 17*(1), 11040–11050.

Pinheiro, I. M., Carvalho, I. P., de Carvalho, C. E. S., Brito, L. M., da Silva, A. B. S., Júnior, A. M. C., et al. (2016). Evaluation of the in vivo leishmanicidal activity of amphotericin B emulgel: An alternative for the treatment of skin leishmaniasis. *Journal of Experimental Parasitology, 164*, 49–55.

Rahmanipour, M., Mohebali, M., Koosha, M., Kazemirad, E., Yasami-Khiabani, S., Mirjalali, H., & Hajjaran, H. J. E. P. (2023). Effect of Leishmania RNA virus 2 on virulence factors and cytokines gene expression in a human macrophage infected with Leishmania major: A preliminary study. *Journal of Experimental Parasitology, 246*, 108459–108467.

Raimundo, V. D., Carvalho, R. P. R., Machado-Neves, M., & de Almeida Marques-da-Silva, E. J. P. R. (2022). Effects of terpenes in the treatment of visceral leishmaniasis: A systematic review of preclinical evidence. *Journal of Pharmacology Research, 177*, 106117–106125.

Saad, K. A., Abdalla, I. O., Alkailani, H. A., Elbakush, A. M., & Zreiba, Z. A. (2022). Evaluation of hepatotoxicity effect of sodium Stibogluconate (Pentostam) in mice model. *Al-Mukhtar Journal of Sciences*, 22–28.

Scarpini, S., Dondi, A., Totaro, C., Biagi, C., Melchionda, F., Zama, D., et al. (2022). Visceral leishmaniasis: Epidemiology, diagnosis, and treatment regimens in different geographical areas with a focus on pediatrics. *Journal of Microbiology, 10*(10), 1887–1895.

Sharifi, F., Sharifi, I., Ohadi, M., Mohamadi, N., Salarkia, E., Banat, I. M., Dehghannoudeh, G. J. I. J., & o. P. R. (2023). The potential role of lipopeptide biosurfactant generated by Acinetobacter junii B6 on Leishmania tropica: The synergy of lipopeptide biosurfactant and Glucantime. *International Journal of Peptide Research and Therapeutics, 29*(4), 57–68.

Shi, Y.-N., Pusch, S., Shi, Y.-M., Richter, C., Maciá-Vicente, J. G., Schwalbe, H., et al. (2019). (±)-Alternarlactones A and B, two Antiparasitic Alternariol-like dimers from the fungus Alternaria alternata P1210 isolated from the halophyte Salicornia sp. *The Journal of Organic Chemistry, 84*(17), 11203–11209.

Silva-Silva, J. V., Moreira, R. F., Watanabe, L. A., de Souza, C.d. S. F., Hardoim, D. D. J., Taniwaki, N. N., et al. (2022). Monomethylsulochrin isolated from biomass extract of Aspergillus sp. against Leishmania amazonensis: In vitro biological evaluation and molecular docking. *Journal of Frontiers in Cellular and Infection Microbiology, 12*, 9749–9755.

Sirak, B., Asres, K., Hailu, A., Dube, M., Arnold, N., Häberli, C., et al. (2021). In vitro Antileishmanial and antischistosomal activities of anemonin isolated from the fresh leaves of Ranunculus multifidus Forsk. *Molecules, 26*(24), 7473–7480.

Strazzulla, A., Cocuzza, S., Pinzone, M. R., Postorino, M. C., Cosentino, S., Serra, A., et al. (2013). Mucosal leishmaniasis: An underestimated presentation of a neglected disease. *Journal of BioMed Research International, 2013*(1), 805108–805114.

Tewabe, Y., Kefarge, B., Belay, H., Bisrat, D., Hailu, A., & Asres, K. (2019). Antileishmanial evaluation of the leaf latex of Aloe macrocarpa, Aloin A/B, and its semisynthetic derivatives against two Leishmania species. *Journal of Evidence-Based Complementary & Alternative Medicine, 2019*(1), 4736181–4736189.

Tiwari, N., Kumar, A., Singh, A. K., Bajpai, S., Agrahari, A. K., Kishore, D., et al. (2019). Leishmaniasis control: Limitations of current drugs and prospects of natural products. In *Discovery and development of therapeutics from natural products against neglected tropical diseases* (pp. 293–350). Elsevier.

Ungogo, M. A., Ebiloma, G. U., Ichoron, N., Igoli, J. O., De Koning, H. P., Balogun, E. O. J. F., & i. c. (2020). A review of the antimalarial, antitrypanosomal, and antileishmanial activities of natural compounds isolated from Nigerian flora. *Journal of Frontiers in Chemistry, 8*, 617448–617452.

Walsh, C. T., & Tang, Y. (2017). *Natural product biosynthesis*. Royal Society of Chemistry.

Wamai, R. G., Kahn, J., McGloin, J., Ziaggi, G. J. J., & o. G. H. S. (2020). Visceral leishmaniasis: A global overview. *Journal of Global Health and Science, 2*(1), 21–30.

WHO. (2022). Leishmaniasis Fact Sheet. *World Health Organization*. https://www.who.int/news-room/fact-sheets/detail/leishmaniasis. Last accessed June, 2024

Wright, G. D. J. M.b. (2019). Unlocking the potential of natural products in drug discovery. *Journal of Microbiology and Biotechnology, 12*(1), 55–57.

Chapter 9
A Review on Antimicrobial Compounds from Selected African Medicinal Plants

Namukobe Jane, Sekandi Peter, Imalingat Betty, Kiprotich Sammy, and Byamukama Robert

Abstract Medicinal plants have been used in the treatment of several diseases, and many plant-derived therapeutic agents are being used to combat microbial infection. However, antimicrobial resistance is still a threat to treatment of microbial diseases. The discovery and development of novel antimicrobial drugs is extremely challenging in Africa. Nevertheless, the role of natural products from plants with diverse and complex structures cannot be underestimated, and as such, continuous efforts towards searching for antimicrobial agents from plants to combat antimicrobial resistance are crucial. Due to Africa's rich biodiversity and historical use of medicinal plants in traditional medicine, Africa could be a potential source of effective antimicrobial agents to overcome multidrug resistance. In this review, we discuss the natural sources of antimicrobial compounds and summarize the isolated compounds from African medicinal plants as potential novel candidates in the development of alternative therapeutic antimicrobial drugs. The study included 290 articles identified by searching articles published using ScienceDirect, PubMed and Web of Science databases. Novel drug discovery proves challenging due to the frequent re-discovery of known antimicrobial agents, and the yield of unknown antimicrobial agents is low for further research to be conducted; thus, there is a need to synthesize analogues of the isolated compounds which have exhibited significant antimicrobial activity so as to optimize the activity and ascertain structure-activity relationships.

Keywords African medicinal plants · Drug resistance · Antimicrobial phytochemicals · Molecular docking

---

N. Jane (✉) · S. Peter (✉) · I. Betty · B. Robert
Department of Chemistry, College of Natural Sciences, Makerere University, Kampala, Uganda

K. Sammy
Department of Chemistry, Ben-Gurion University of the Negev, Be'er Sheva, Israel

A. Dekebo (ed.), *Bioactive Secondary Metabolites from Medicinal Plants of Africa*, Interdisciplinary Biotechnological Advances,
https://doi.org/10.1007/978-981-96-8590-5_9

Abbreviations

AACs	Aminoglycoside N-acetyltransferases
ABC	Adenosine triphosphate-binding cassette
ADMET	Absorption, distribution, metabolism, excretion and toxicity
AMR	Antimicrobial resistance
CATs	Chloramphenicol acetyltransferases
DNA	Deoxyribonucleic acid
ERM	Erythromycin ribosyltransferases
IC_{50}	Half-maximal inhibitory concentration
MBC	Minimum bactericidal concentration
$MBIC_{50}$	Minimum biofilm inhibition concentration 50%
MefA	Macrolide efflux protein A
MefE	Macrolide efflux protein E
MFS	Major facilitator superfamily
MIC	Minimum inhibitory concentration
PBP	Penicillin-binding proteins
RND	Resistance nodulation division
SMR	Small multidrug resistance
Tet	Tetracycline resistance proteins
TtgA	Toluene tolerance gene A
TtgB	Toluene tolerance gene B

9.1 Introduction

Antimicrobial compounds are compounds used to treat or prevent infections caused by bacteria, fungi, parasites and viruses. This categorizes them into antibiotic, antiviral, antifungal, antiprotozoal and disinfectant agents for bacterial, viral, fungal, protozoal and non-specific infections, respectively (Bondaryk et al., 2017; Clardy et al., 2013; Ndhlala et al., 2013). Many antimicrobial compounds on market today are synthetic and have scaffolds of natural origin (Bbosa et al., 2014). For example, penicillin (1) was isolated from *Penicillin notatum* in 1928 by Fleming Alexander and formed the basis for antimicrobial therapy (Laskin et al., 2023; Canzani & Aldeek, 2017; Gaynes, 2017; Luc, 2015). A number of antimicrobials such as penicillin G (**2**), amoxicillin (**3**), ticarcillin (**4**), piperacillin (**5**) (Fig. 9.1) have been synthesized, leading to effective treatment of many microbial infections (Canzani & Aldeek, 2017).

The global health concern to date is the emerging antimicrobial resistance (AMR). AMR has led to treatment failures causing the loss of many lives. Over 1.2 million people died in 2019, and this number will increase to 10 million deaths per year by 2050 if no interventions are put in place to control AMR (Shallcross & Davies, 2014; Tang et al., 2023). Some of the major causes of AMR include misuse,

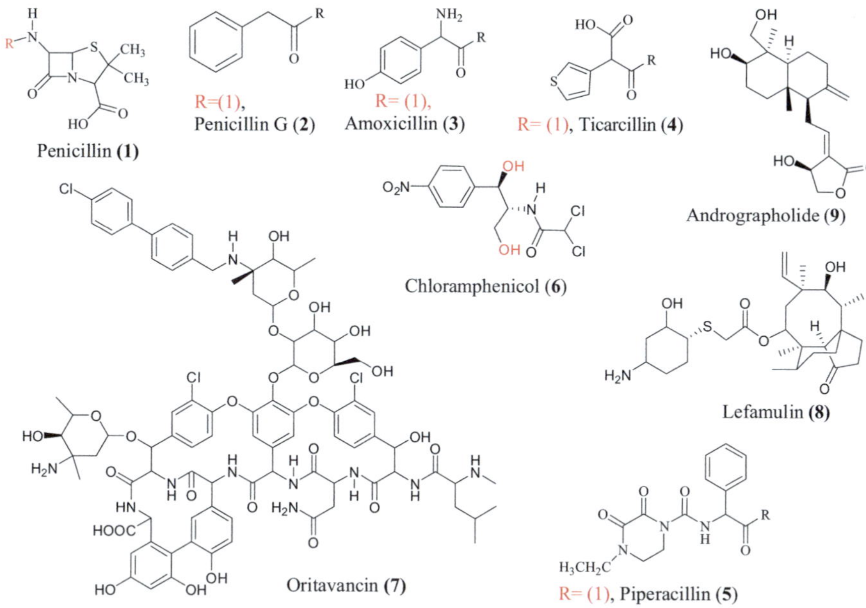

Fig. 9.1 Chemical structures of penicillin derivative drugs (**1–5**) and selected antimicrobial compounds (Laskin et al., 2023; Canzani & Aldeek, 2017; Luc, 2015)

overuse and lack of antimicrobial agents for humans and agriculture (Abushaheen et al., 2020; Jadimurthy et al., 2023).

Efforts have been made to fight antimicrobial resistance through exploring alternative sources of antimicrobial agents, like plants. Therefore, plants could be a source due to their abundant bioactive compounds, which could be used to manage infections caused by resistant organisms (Moiketsi et al., 2023). Phytochemicals from plants possess a wide range of activity, including antibacterial, antifungal, antiviral and antiprotozoal activities. Several extracts from plants are already being tested in humans, and those identified with potential antimicrobial activity have been recommended by both medical and traditional doctors (Mokoka et al., 2014; Okoye et al., 2014). Africa presents a rich cultural heritage and a vast array of plant species, many of which have been used in traditional medicine since the dawn of humankind (Agyare et al., 2016).

Africa has a unique biodiversity that has led to the evolution of plants with remarkable bioactive properties, including antimicrobial compounds (Anand et al., 2019; Radol et al., 2016). Compounds of plant origin with antimicrobial potential are used as an alternative agent to reduce the prevalence of drug-resistant infections. This plays an important role in the healthcare systems by using effective remedies against various diseases (Anand et al., 2019; Owusu et al., 2021). Consequently, scientists have begun to unravel the secrets of plants through isolating and characterizing antimicrobial compounds (Namuga et al., 2024; Sekandi et al., 2023).

There is a growing concern about developing new antimicrobial drugs from various sources. Much of the focus has been on isolating and identifying antimicrobial agents from plants. However, less effort has been put to develop the identified compounds into antimicrobial drugs. This chapter highlights an exhaustive list of selected African medicinal plants, detailed information on the antimicrobial activity of crude extracts and isolated compounds as an alternative source of antimicrobial compounds. The chapter also reports why many of the isolated compounds from African plants are not commercialized.

9.2 Antimicrobial Resistance and Mechanism of Action of Antimicrobial Agents

Antimicrobial resistance is a global burden predicted to cause ten million deaths annually by 2050 (Achilonu & Udensi, 2022; Mikecz et al., 2020). There are three fundamental mechanisms through which resistance can be understood; they include enzymatic degradation, alteration of bacterial proteins and changes in the membrane permeability to antimicrobial agents (Abreu et al., 2012; Todar, 2014; Tyagi et al., 2015). Enzymatic degradation (modification of antimicrobials) is the widespread antibiotic inactivation that involves enzymes. The bacteria are resistant to β-lactams mediated by the overproduction of β-lactamases by Gram-negative bacteria that hydrolyse methicillin and related penicillin antibiotics (Arora et al., 2014; Munoz-Price et al., 2013). The enzymes break the β-lactam ring rendering the antibiotic inactive (Canzani & Aldeek, 2017). Thus, looking for compounds that bind to β-lactamases is an important target to reduce microbial resistance since the inhibitors of β-lactamases protect the antibiotic from degradation. This enables β-lactam drugs interact with the penicillin-binding proteins (PBPs) (Dever et al., 2018). According to Abreu et al., administration of antimicrobials with β-lactam inhibitors has been proven the most effective antibiotic combinations.

The second key mechanism of antimicrobial resistance involves alterating the bacteria protein that are antimicrobial targets with the help of target-modifying enzymes, once the target site is modified, the antibiotic agent does not recognise the bacteria protein resulting into resistance (Bbosa et al., 2014; Moiketsi et al., 2023). There are ways through which the bacteria protein target could be modified, by adding either acetyl, phosphate, adenyl, ribosyl or methyl group on the target site. Examples of target modifying enzymes include aminoglycoside N-acetyltransferases (AACs), chloramphenicol acetyltransferases (CATs), erythromycin ribosyl-transferases (ERM), tetracycline resistance proteins (Tet) and β-lactamase inhibitor such as TEM-1 (Abushaheen et al., 2020; Luthra et al., 2018; Manchester et al., 2012; Singh et al., 2021).

Fungi and bacteria are protected from being killed by the antimicrobial agents by resistant proteins found on the DNA. Antibiotics with cationic aminoglycoside interact with the aminoglycoside site of ribosomes using the hydroxyl and amine groups; this impairs the mechanism of codon–anticodon decoding and results in the synthesis of aberrant proteins (Gall et al., 2022; Zimmermann et al., 2021). Genes encoding AACs have been reported in both clinical and environmental bacteria, such as *A. baumannii* (Moiketsi et al., 2023). CATs have contributed to resistance by deprotonating the hydroxyl group of chloramphenicol (**6**) as shown in Fig. 9.1. The latter binds to 50S ribosomal subunit and inhibits the synthesis of bacterial protein. There are two types of CATs, class A from *E. coli* and class B from *Pseudomonas aeruginosa* (Dever et al., 2018).

Resistance may result from decrease in the membrane's permeability to inhibitors. Antibiotics face resistance when the organism lacks channels or carriers as transport routes for the antibiotic. This reduces the concentration of effective intracellular antibiotic due to decreased permeability. For example, Gram-negative bacteria are covered with lipopolysaccharide, an outer membrane that is a barrier to drugs. Others form biofilms for protecting themselves against antibiotics. Thus, identifying agents that disrupt the already-formed biofilm may mitigate the high prevalence of resistance (Todar, 2014; Wang et al., 2019).

Resistance may result from biofilm formation where microbes organize themselves in a media composed of DNA, polysaccharides and proteins (Banerjee et al., 2017). This formulation makes microbes less susceptible to antibiotics than those growing planktonically. Biofilm resistance depends on several causes, including extracellular structure medium, difference in the physiological state of biofilm-growing and planktonic bacteria. This difference results from different nutrients and oxygen supplied inside and on the surface. Extracellular medium may change antibiotic activity by either diminishing the diffusion or seizing the antibiotic through its binding media, thus causing resistance to the drug (Abebe et al., 2016; Borges et al., 2013).

Efflux pumps are an energy-dependent resistance method that eliminates aminoglycosides from the cell (Basler et al., 2018). Compounds that are not related to the structure or chemical composition of the bacteria are removed without altering or degrading the compound; thus, the bacteria can survive in toxic environments (Basler et al., 2018). Categories of efflux systems include major facilitators (MFS) associated with mefA/mefE genes of *Streptococcus pyogenes*, adenosine triphosphate-binding cassette (ABC), small multidrug resistance (SMR), and resistance nodulation cell division (RND), which is common in *Pseudomonas* species. For example, in Spain, *Pseudomonas putida* was reported to possess RND efflux consisting of proteins in the inner (TtgB), fusion (TtgA), and outer (TtgC) membrane channels (Abebe et al., 2016; Basler et al., 2018). Another common efflux system is the toxic and multidrug compound extrusion (MATE) type consisting of NorA and MsrA genes in strains of *S. aureus*. Compounds such as andrographolide (**9**) can improve the antibiotic susceptibility by inhibiting the expression of bacterial efflux pumps and competitively inhibiting antibiotic resistance sites in bacteria (Abushaheen et al., 2020; Ye et al., 2020).

According to Gareth (2003), antimicrobials can make bacteria membrane permeable to specific essential small molecules, a phenomenon known as ionophoric antibiotic action, where ionophores penetrate and increase membrane permeability to ions. This reduces ionic concentration until it is the same on both membrane sides (Ekinci et al., 2023; Roy & Talukdar, 2021). The reduction is often sufficient to cause destruction of the bacterial cell envelope, which is irreversible and leads to cell damage and death. Carrier ionophores like sterols increase the permeability of the membrane, allowing cellular contents to leak from the cell. Ionophores like salinomycin, lasalocid and monesin, among others, are toxic to bacteria and fungi and thus could be exploited as antimicrobial agents (Adaszynska-Skwirzynska & Szczerbinska, 2018; Novilla, 2018; Roy & Talukdar, 2021).

In order to reduce the burden of antimicrobial resistance, various antimicrobial agents of plant origin have various modes of antimicrobial action. The mechanism of action of antimicrobial compounds depends on the structure of the antimicrobial agent. Polyphenols provide a basis for their antimicrobial activity through altering the cell membranes, especially in Gram-negative bacteria. Phenolic compounds have a high affinity for the lipids in the bacteria cell membrane; thus, they interact and dissolve in the cell membrane exerting a lethal effect onto the target bacteria; this mode of action is known as lipophilicity (Allah et al., 2023; Alshwyeh, 2020). Computational studies indicated that 10 out of the 26 compounds identified from *Capparis decidua* Edgew (Forssk) showed acceptable lipophilicity with binding free energy range of −43.09 to −62.2 kcal/mol (Allah et al., 2023). Therefore, compounds from fatty acids (essential oils), resin acids (phenolics), acyl glycerides, and sterol esters could possess antibacterial activity due to their lipophilicity (Alshwyeh, 2020; Stefano et al., 2021).

Glycopeptides such as oritavancin (**7**) exhibit antibacterial activity through lipophilic action mode. Glycopeptides inhibit *trans*-glycosylation and *trans*-peptidation in the cell membrane. The alkyl and fatty acyl lipophilic side chains of the drug cause effective membrane anchoring, increase dimerization and prolong the half-life of the drug, thus inhibiting the bacterial growth (Devasahayam et al., 2010). Similarly, flavonoids complex to soluble extracellular proteins through hydrogen bonding and hydrophobic effects, then bind to the cell wall to disrupt cell membranes through lipophilic reactions (Ferdes, 2018). This inactivates enzymes and disrupts cell envelope transport proteins (Moiketsi et al., 2023; Shashank & Abhay, 2013). Compounds from quinones inhibit protein synthesis in the microorganism based on their redox properties of the carbonyl group (Ferdes, 2018; Veve & Wagner, 2018). For instance, lefamulin (**8**) binds to the peptidyl transferase centre of the bacterial ribosome and prevents the binding of tRNA for peptide transfer (Veve & Wagner, 2018).

Phenolic compounds and essential oils have shown diverse antimicrobial mechanisms against different bacterial strains from synergistic activity with conventional antibiotics via inhibition of efflux pumps. Several studies suggest that phenolic compounds interact with and disrupt the lipid bilayer of bacterial cell membranes, leading to leakage of cellular contents and finally cell death (Borges et al., 2013; Moiketsi et al., 2023; Osuntokun & Cristina, 2019). According to Borges et al., ferulic and gallic acid had antimicrobial activity against *P. aeruginosa, E. coli, S. aureus* and *L. monocytogenes* by significantly changing surface hydrophobicity, charge, induced propidium iodide uptake and potassium ion leakage in the cell.

Ferulic and gallic acids on the surface of the cell membrane rupture it or induce pore formation in the cell membranes of Gram-negative and positive bacteria. Phenolic compounds can bind to and inhibit essential enzymes such as urease, sortase A and dihydrofolate reductase relevant for energy metabolism, DNA, replication and cell wall biosynthesis (Dever et al., 2018; Ye et al., 2020). The observed activities make phenolic compounds a good candidate for future in vivo studies and even clinical trials (Akpalo et al., 2020). Saponins penetrate into the lipid bilayer, accumulate in the lipid raft microdomain and bind to cholesterol in the lipid membrane. Saponins block cholesterols from interacting with sphingomyelin. The saponin-cholesterol micelles destabilize the lipid raft structure and disrupt the lipid bilayer, leading to cell haemolysis (Kundishora et al., 2020).

9.3 African Medicinal Plants with Antimicrobial Potential

More than 5400 plant species are used in traditional medicine in Africa, of which 10% of the species with antimicrobial potential have been regularly traded on international markets. Africa largely depended on the use of medicinal plants to manage the outbreak on COVID-19 pandemic (Adebisi et al., 2021). From West Africa, *Citrus aurantifolia* (Christm), *Cucumis metuliferus* E.Mey, *Loranthus micranthus* Linn and *Macaranga barteri* Mull were documented to manage the COVID-19 virus (Popoola et al., 2022). In Nigeria, *Macaranga kilimandscharica* Pax, *Guizotia scabra* (Vis.) Chiov, *Pavetta ternifolia* Hiern, *Eriosema montanum* Baker, *Entada abyssinica* (EnE) Steud and *Syzygium brazzavillense* Aubrév have been reported to be active on poliomyelitis, coxsackie, measles and vesicular stomatitis virus (Attah et al., 2021). In Ethiopia, *Acokanthera schimperi* (A.DC.) Benth & Hook.f. ex Schweinf., *Euclea schimperi* (A. DC.) Dandy, *Plumbago zeylanica* L. and *Inula confertiflora* A. Rich., among others, are used in the treatment of viral skin diseases, and their methanolic extracts showed antiviral activity against coxsackie virus B3, influenza A virus and herpes simplex virus type 1 Kupka. (Dhale & Markandeya, 2011). African medicinal plants have shown a wide range of antimicrobial activity (Table 9.1, Figs. 9.2 and 9.3). *Adansonia digitata* L., *Harpagophytum procumbens* DC., *Harungana Madagascariensis* Poir, *Kigelia africana* (Lam) and *Prunus africana* (Hook. f.) have exhibited potent antibacterial activity against *Bacillus* and *Staphylococcus* species with minimum inhibitory concentration (MIC) of 100 μg/mL (Ndhlala et al., 2013). Table 9.1 and Figs. 9.2 and 9.3 show some African medicinal plants with antimicrobial activity.

Namukobe et al. (2021b) found that extracts of *P. febrifugum, P. caespitosus* and *E. tomentosa* showed good antibacterial and antifungal activity against *Pseudomonas aeruginosa, Staphylococcus aureus, Escherichia coli, Klebsiella pneumoniae, Streptococcus pyogenes* and *Salmonella typhi* with MIC ranging between 3.12 and 12.5 mg/mL. *Hoslundia opposita Vahl* extracts inhibited the growth of *P. aeruginosa* and *S. aureus* biofilms with $MBIC_{50}$ of 28.37 mg/mL and 10 mg/mL, respectively (Namuga et al., 2024). *Plectranthus barbatus* Andrews extract showed antibacterial

Table 9.1 Some of the African medicinal plants with antimicrobial potential

Plant name	Plant part used	Local/trade name (country)	Antimicrobial activity
Harpagophytum procumbens (Burch.)	Seeds, stem bark	*Duiwelsklou*, *Otjihangatene* (Southern Africa)	Antibacterial, antifungal and antiviral (Khumalo et al., 2022; Ndhlala et al., 2013)
Hibiscus sabdariffa Linn.	Leaves	*Musayi* (Uganda), *Olugbu* (Nigeria, Ghana), *Mundele* (Lingala-Congo), *Bitterblaar* (South Africa)	Antiviral activity, antibacterial (Erhabor et al., 2019; Eto, 2013; Ndhlala et al., 2013)
Psorospermum febrifugum Spach	Stem bark	*Kanzironziro*, *Umukúbagwá* (Uganda, Tanzania, Rwanda)	Antibacterial, anti-acne (Félicienne et al., 2019; Larrosa et al., 2011; Namukobe et al., 2021a)
Coffea Arabica Linn	Seeds	*Mwanyi* (Uganda)	HIV/AIDS and related conditions (Lamorde et al., 2010)
Pausinystalia johimbe (K. Schum.)	Leaves	(Cameroon, Madagascar)	Antifungal (Khumalo et al., 2022)
Ricinus communis Linn	Leaves	*Ensogasoga* (Uganda), *Uhlakuva, Umhlakuva* (South Africa)	Antimalarial, antibacterial, antifungal activity (Nalule et al., 2011; Van-Wyk, 2015)
Aloe vera L (Burm)	Leaves	*Kigagi, Rukaka, Kikakarubamba* (Uganda)	Retroviral, malaria, bacterial and fungal infection (Komlaga et al., 2015; Namukobe et al., 2021a)
Plectranthus caespitosus Lukhoba	Leaves	*Ewe aje* (Nigeria), *Tikil gish* (Ethiopia), *Umhlonishwa* (Zulu)	Antibacterial activity (Kipkoech et al., 2019; Namukobe et al., 2021b)
Erlangea tomentosa S. Moore	Leaves, flowers	*Ekyoganyanja/ Ekijjo/ Ekitokotoko, Ettwatwa* (Uganda), *Lengalenga* (Swahili), Tikil guracha (Amharic-Ethiopia)	Opportunistic infections, antibacterial activity (Mugisha et al., 2014)
Hypoxis hemerocallidea Fisch	Corns	African potato (South Africa), *Ekibondo* (Uganda), *Ekigongoni* (Kiswahili), *Omakipundi* (Namibia)	Antibacterial, antifungal, antiviral activity (Matotoka et al., 2019; Ndhlala et al., 2013; Sikhakhane, 2013)
Catharanthus roseus Linn	Leaves	*Magalela, Isikhokhokho* (South Africa), *Okoro-akapa, Ewe, Aso, Yoruba* (Nigeria), periwnkle (Madagascar)	Antimalarial, antiviral, antiulcer (Anand et al., 2019; Nyamukuru et al., 2017)
Khaya anthotheca (Welw.)	Seeds, stem bark	*Mvule* (Kenya, Tanzania), *Muvule* (Uganda), Red *Mahogany* (South Africa), *Kambala* (Lingala-Congo)	Antiplasmodial (Obbo et al., 2024)
Vernonia amygdalina Del.	Leaves	*Omululuza* (Uganda), *Umubiriizi* (Rwanda), bitter leaf (English), *Ndole* (Ghana)	Antibacterial, antifungal, antiplasmodial (Adeola et al., 2015; Holmes, 2017)

(continued)

Table 9.1 (continued)

Plant name	Plant part used	Local/trade name (country)	Antimicrobial activity
Ocimum gratisimum L.	Leaves	African/clove basil (English), *Efirin, Nchuanwu, Daidoya* (Nigeria), *Akokoro* (Ghana), *Njansang* (Cameroon)	Anthelmintic (Adeola et al., 2015)
Nicotiana tabacum Linn.	Leaves	*Taaba* (Uganda), Tobacco (English), *Tabak* (South Africa), *Uba, Tabu, Taba* (Nigeria)	Anti-venom, anti-stress, anthelmintic (Sirama, 2014; Valluru & Van-den, 2011)
Talinum triangulare (Jacq.)	Leaves	Jewels of Opar/fameflower (English), *Akan* (Ghana), *Efo lyana* (Nigeria)	Anthelmintic (Ebenezer, 2011)
Stomatostemma monteiroae (Oliv.)	Seeds, stem bark, root tubers	*Wildehoendervy* (South Africa), *Omunguvi* (Namibia), *Sekopane* (Botswana)	Antibacterial (Matotoka et al., 2019)
Acokanthera schimperi (A. DC)	Leaves	*Chaka* (Amharic-Ethiopia), *Mporokoyo* (Kiswahili)	Antiviral activity (Pascaline et al., 2011)
Euclea schimperi (A. DC.) F White	Leaves, roots	Bush/red Guarri (English), *Rooigwarrie* (South Africa), *Muhugu* (Kiswahili), *Umushikiri* (Rwanda), *Mdaa* (Tanzania)	Antiviral, antibacterial, anti-fungal activity (Holmes, 2017)
Inula confertiflora	Leaves	*Geelbos* (South Africa), *Khara-om* (Namibia), *Sekgophana* (Botswana)	Antiviral, antibacterial activity (Habtamu et al., 2024)
Melilotus elegans	Leaves	French melilot (English), *Khayt-el-baqar* (Morocco), *Witheuning* (South Africa)	Antiviral activity (Sowa-borowiec et al., 2025)
Plumbago zeylanica	Leaves	Ceylon leadwort (English), *Blousyselbos, Isicakathi* (South Africa), *Oju-owo* (Nigeria)	Antimicrobial activity (Dhale & Markandeya, 2011)
Artemisia afra (Jacq.)	Roots, stems and leaves	*Wildeals, Umhlonyane,* (South Africa), *Lengana* (Zimbabwe), *Nyumba* (Kiswahili), *Ariti* (Ethiopia)	Antibacterial, antifungal, coughs, haemorrhoids, fevers, malaria, anti-COVID (Achilonu & Udensi, 2022; Adeleye et al., 2021; Okoye et al., 2014)
Strychnos henningsii (Gilg.)	Stem bark	*Doringklapper, Umvithi* (South Africa), *Omukithi* (Namibia), *Motlhware* (Botswana)	Anti-venom, antimalarial (Karunamoorthi et al., 2013)
Tropaeolum majus Linn.	Seeds	Nasturium (English), Groot Kappisser (South Africa)	Urinary tract infections, antibacterial (Okoye et al., 2014; van Vuuren & Frank, 2020)

(continued)

Table 9.1 (continued)

Plant name	Plant part used	Local/trade name (country)	Antimicrobial activity
Urena lobata Linn.	Roots	*Chemulmeswo* (Kenya), *Aramina, Uvovovo* (South Africa), *Ewe-erin* (Nigeria)	Skin disease, cough, poisonous bites (Jeruto et al., 2015)
Ximenia americana Linn	Leaves, stem bark	*Gbani* (Guinea), *Moretologa wa pudi* (South Africa), *Mogwale* (Botswana), *Damara* (Namibia)	Stomach ulcers, antibacterial, antifungal, pesticidal, antineoplastic (Moiketsi et al., 2023; Okoye et al., 2014)
Sorghum bicolor (L.) Pers.	Seeds	*Amabele* (South Africa), *Okababa* (Nigeria), *Tekel* (Ethiopia)	Anti-poison (Okoye et al., 2014)
Solanum aethiopicum L.	Fruits, leaves	*Ekitulatula, Nakati* (Uganda), *Osuga* (Kiswahili), *Gomen zinj* (Ethiopia)	Cough (Uwimbabazi et al., 2023)
Psidium guajava L.	Leaves, stem bark	*Mapera* (Uganda), Guava (English), *Igwavava* (South Africa, *Gwaaba* (Nigeria), *Peri* (Kiswahili)	Cough, antibacterial, antileishmanial (Owusu et al., 2021)
Pittosporum viridiflorum Sims	Whole plant	Cheesewood (English), *Kasuurboom, Umgqalutye* (South Africa)	Fungal infections, antidote for insect bites, antiplasmodial, antibacterial (van Vuuren & Frank, 2020)
Piptadeniastum africana Pip (Hook f.)	Leaves and roots	*Ekouyi/Kokowa* (Gabon), *Onyaa* (Ghana), *Afara* (Nigeria), *Muadi* (DRC), *Ebele/Ebaye* (Cameroon)	Anti-ulcers, anti-helmintic (Eto, 2013)
Phyllanthus muellerianus (Kuntze)	Leaves	Tropical leaf-flower (English), *Nyamekye* (Ghana), *Irawo ile* (Nigeria)	Wound dressing, antiplasmodial (Owusu et al., 2021)
Parkia biglobosa (Jacq.)	Leaves	*Dorowa/Iru* (Nigeria), *Dawadawa* (Ghana), *Netetou* (Senegal), *Kapseu* (Cameroon)	Antimalarial, antiviral (Popoola et al., 2022)
Nigella sativa L	Seeds	Black seed (English), *Habbatu* (Hausa), *Habbat al-barakah* (Egypt)	Anticonvulsant, antileishmanial (Rokni et al., 2021)
Newbouldia laevis	Seeds, leaves	*Ogilisi* (Nigeria) *Senkwen* (Ghana), *Ebanda, Mvulu* (Cameroon), *Sesemesa* (Ghana)	Antibacterial, antimalarial (Komlaga et al., 2015)
Momordica charantia L.	Leaves, seeds	Bitter melon (English), *Akara-aje* (Nigeria), *Akonta* (Ghana)	Anthelmintic, anti-ulcer, antiviral (Jia et al., 2017)
Moringa oleifera Lam.	Seeds, leaves	*Moringa* (Uganda), *Ewelle* (Nigeria), *Morogo* (South Africa)	Antispasmodic, antibacterial, antihyperglycaemia and antiviral (Akindele et al., 2022)
Hyptis suaveolens Poit	Roots, stem back	(Curry leaf) Nigeria	Fever, eczema antibacterial, antifungal (Mishra et al., 2021)

(continued)

Table 9.1 (continued)

Plant name	Plant part used	Local/trade name (country)	Antimicrobial activity
Hybanthus enneaspermus (L.)	Leaves, roots	*Abiwere* (Nigeria)	Antimalarial, antiplasmodial (Patel et al., 2013)
Alstonia boonei De Wild	Leaves, Stem bark	*Ahum, Egbe* (Nigeria), *Kyenkyen* (Ghana), *Ekpili* (Cameroon), Cheese wood (English)	Antimalarial, antibacterial (Unuigbe & Okhankhuele, 2023)
Enantia chlorantha Oliv	Roots and stem bark	*Awogba* (Nigeria)	Antimalarial
Plectranthus barbatus Andrews	Leaves	*Ekicuncu, Kibwankulata* (Uganda)	Colic pain, antiviral (Gumisiriza et al., 2020; Namukobe et al., 2021a)
Hoslundia opposita Vahl	Leaves	*Kamunye* (Uganda)	Antibacterial (Nambejja et al., 2019; Namuga et al., 2024)
Cucumis metuliferus E. Mey	Roots	Mozambique	Stomach pain (Ribeiro et al., 2010)
Dilochopentas decora (S. Moore)	Leaves	*Kabyakyasha* (Uganda)	Antifungal, antibacterial (Ahumuza & Kirimuhuzya, 2011; Sekandi et al., 2024)
Capparis decidua Edgew	Roots, seed	*Mlonge* (Tanzania), *Esho* (Kenya)	Anticancer, antibacterial (Allah et al., 2023; Savoia, 2015)
Spermacoce princeae	Leaves	*Kishakimwe* (Uganda)	Antibacterial, antifungal (Sekandi et al., 2023).

activity against *S. aureus* (3.12 mg/mL), *S. epidermidis*, *Streptococcus pneumoniae* and *E. coli* with an MIC of 6.25 mg/mL. In South Africa, 34 *Aloe* species possessed antimalarial activity, where 50% of the parasite growth was inhibited with IC_{50} ranging from 32 to 77 μgml^{-1} (Sacan et al., 2017). This information shows that Africa is a rich source of antimicrobial agents from medicinal plants.

9.4 Classes of Phytochemicals with Antimicrobial Activity

Antimicrobial agents of plant origin are mainly primary and secondary metabolites, which play a vital role in their survival. Primary metabolites include carbohydrates, proteins and fats, whereas secondary metabolites include phenolics and polyphenols, terpenes, quinones and alkaloids. Primary metabolites serve as food reserves for both plants and animals (Sharma et al., 2018; Stirk et al., 2019). In many cases, the

Fig. 9.2 Pictures of some African plants with antimicrobial agents (Khumalo et al., 2022; Van-Wyk, 2015)

secondary metabolites serve as plant defence mechanisms against predation by microorganisms, insects and herbivores (Prasad & Prasad, 2019). Some give plants their odours (Asikin et al., 2018); others are responsible for plant pigment and flavour (Bharali & Gamo, 2022; Himesh et al., 2012). Researchers have made efforts to synthesise primary metabolites, but less has been carried out to synthesize

Fig. 9.3 Pictures of some African plants with antimicrobial agents (Sikhakhane, 2013)

secondary metabolites. This is attributed to the fact that secondary metabolites are either present only in certain species, can only be identified at a certain stage of growth and development within a specific species, may be produced during periods of stress caused by the attack of microorganisms, or the soil on to which the plant grows. In addition, they have complex stereo structures with many chiral centres that are relevant in biochemical reactions, weather conditions and stage of maturity (Frimpong et al., 2023; Migabo et al., 2023). The different classes of phytochemicals are discussed in detail below.

9.4.1 Phenolic Compounds

Phenolic compounds are secondary metabolites which contain an aromatic ring with either one or many hydroxyl groups (polyphenols) located in the cell vacuole. Flavonoids form the largest group of phenolic compounds, where over 8000 phenolic compounds have been identified in various forms. Only 4000 belong to the flavonoids and are mainly found in fruits, vegetables, beverages and cereals (Azmir et al., 2013). Phenolic compounds have multiple functions for plants such as defence against ultraviolet radiation, oxidizing agents or the aggression of some microorganism agents, and they enable the plant adapt biotic and abiotic stress.

Polyphenols are heterocyclic molecules that belong to different groups of phenolic compounds with many chemical structures that could alter their biological activities. They are categorized into different classes such as phenolic acids, flavonoids, lignans and stilbenes (Amber et al., 2014; Kakkar & Bais, 2014). The presence of a carboxylic group is a unique structural similarity with phenolic acids. Polyphenols are derived from phenylalanine via the shikimic and malonic acid pathway. Lignans consist of two phenylpropane units linked by a $\beta - \beta'$ carbon atoms of their side chains (Decembrino et al., 2020). Phenolic metabolites have exhibited a wide range of antimicrobial activity, indicating their potential use as key elements for human health (Altemimi et al., 2017; Krishnaiah et al., 2011; Shin et al., 2019). The antioxidant properties of polyphenols provide a basis for antimicrobial effects (Othman et al., 2019). The antioxidant action of phenolic compounds on bacterial cells to inhibit virulence factors such as enzymes, toxins and suppression of bacterial biofilm formation enhances antibacterial activity (Shouping et al., 2020).

Phenolic acids such as *p*-coumaric acid (**10**), caffeic acid (**11**), ferulic acid (**12**), sinapic acid (**13**), gallic acid (**14**), coumarin (**15**), catechol (**16**), resveratrol (**17**) and protocatechuic acid (**18**) have shown antibacterial activity against *Xylella fastidiosa, E. coli, Bacillus subtilis, P. aeruginosa* and *S. aureus* strains, with the MIC ranging from 64 to 1024 μg/mL (Byamukama et al., 2015; Maddox et al., 2010; Tyagi et al., 2015). Studies also indicated that ferulic (**12**) and gallic (**14**) acids had antimicrobial activity with MIC of 500 mg/mL against *P. aeruginosa*, 1500 mg/mL against *E. coli* and 1750 mg/mL against *S. aureus* (Borges et al., 2013). In South Africa, resveratrol has been identified in the leaves of *Reynoutria japonica* Houtt, a plant used to manage COVID-19 (Attah et al., 2021) and possess antibacterial activity against *Klebsiella pneumoniae* KP55 strain with MIC of 16 μg/mL (Maddox et al., 2010; Seukep et al., 2016b). Catechol is a phenolic compound identified in the roots of *Diospyros kaki* Thunb and *Anthospermum perrieri* Homolle ex Puff, among others, with antibacterial activity (5 mM) against *Pseudomonas putida*, *P. pyocyanea* and *Corynebacterium xerosis* (Ferdes, 2018; Ralambonirina et al., 2018). The chemical structures of the aforementioned phenolic acids are shown in Fig. 9.4.

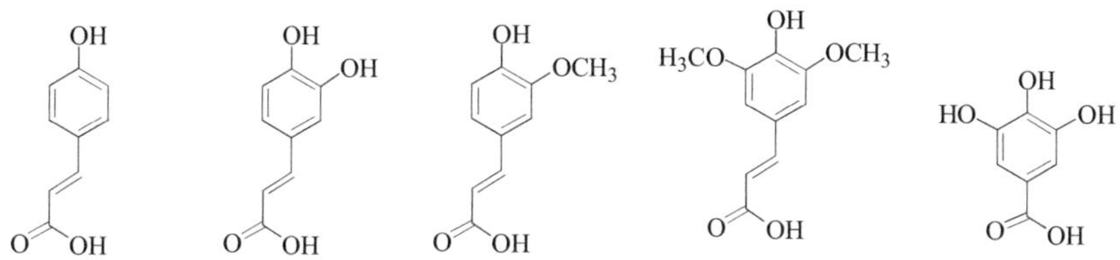

p-Coumaric acid (**10**) Caffeic acid (**11**) Ferulic acid (**12**) Sinapic acid (**13**) Gallic acid (**14**)

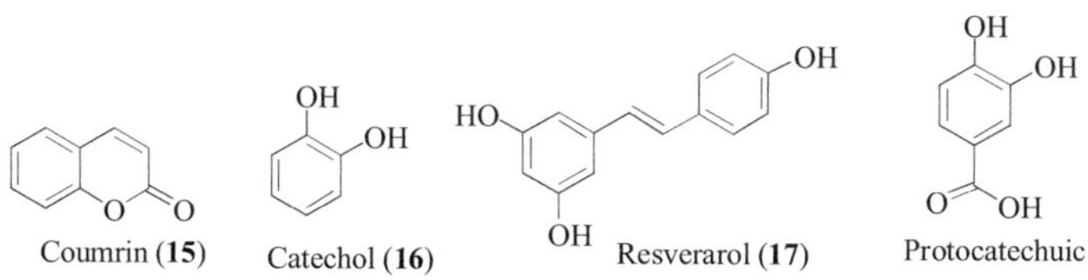

Coumrin (**15**) Catechol (**16**) Resverarol (**17**) Protocatechuic acid (**18**)

Fig. 9.4 Some phenolic compounds (**10–18**) with potential antimicrobial activity (Borges et al., 2013; Maria et al., 2015; Quinn et al., 2017)

9.4.2 Quinones

Quinones (**26**) are composed of aromatic hexacyclic saturated ring with two ketone substitutions (Fig. 9.5), synthesized by plants in response to browning reactions in fruits and vegetables. The main classes of quinones include benzoquinones, hydroquinones, naphthoquinones, phenanthrenequinones, polycyclic quinones and anthraquinones, characterized according to the number of benzene rings in their structure (Eyong et al., 2013). Quinones are synthesized in plants via the shikimate (*o*-succinoyl benzoic acid) or polyketide pathways (Klopčic & Dolenc, 2019). Quinones are mainly from families such as Bignoniaceae, Ebenaceae, Droseraceae, Rhamnaceae, Juglandaceae, Plumbaginaceae, Polygonaceae, Boraginaceae, Lythraceae, Proteaceae, Verbenaceae, Rubiaceae, Leguminosae and Labiatae (Gurib-Fakim, 2006; Kumar et al., 2017). *Aloe* species have been reported to accumulate bianthraquinoids, preanthraquinones and anthraquinones (van Zyl et al., 2002).

A large number of quinones possess significant antimicrobial activities (Kumar et al., 2017). For example, aloinoside (**19**), aloin (**20**) and microdontin (**21**) quinones from the leaf latex of *Aloe calidophila* Reynolds in Ethiopia possess antileishmanial activities with IC_{50} values ranging from 3.12 to 10.92 μM against *Leishmania aethiopica* and from 3.70 to 15.26 μM against *Leishmania major* (Simoben et al., 2018). Methanolic root extract of *Pentas micrantha* (Backer) from Kenya and 5,6-dihydroxylucidin-11-*O*-methyl ether (**22**) tested in the same study have been reported to show moderate antiplasmodial activity against the W2 and D6 *Plasmodium falciparum* strains with IC_{50} values of 3.37 μg/mL and 4.00 μg/mL, respectively (Endale et al., 2013). Adamabianthrone (**23**) is a quinone commonly isolated from plants belonging to the genus *Psorospermum*. The compound has shown antibacterial activity against *Citrobacter freundii*, *E. coli*, *P. aeruginosa*,

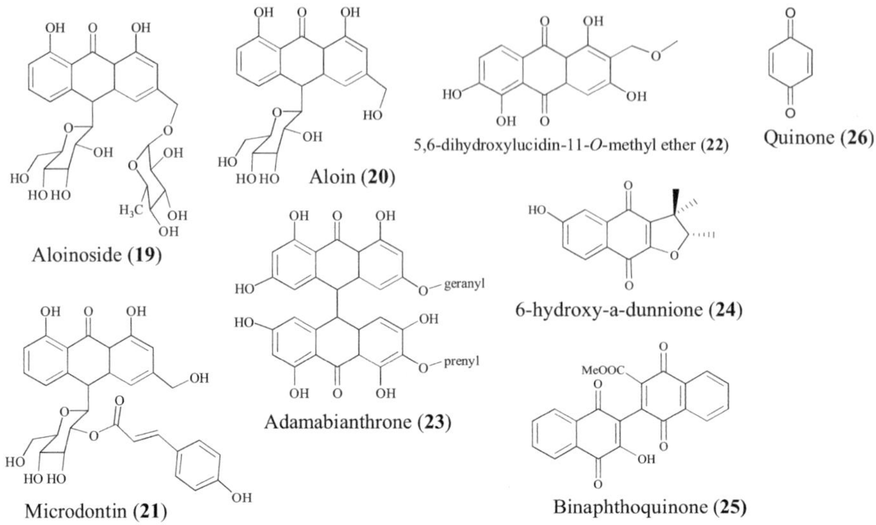

Fig. 9.5 Structures of some quinones (**19–26**) with antiplasmodial and antibacterial activities (Abeje et al., 2014; Endale et al., 2013; Xiao et al., 2011)

K. pneumoniae and *S. typhi* with MIC ranging between 27.34 μM and 218.74 μM (Epifano et al., 2013).

In spite of strong bioactivity, the medical use of naphthoquinones is limited. This is attributed to the toxic nature of quinones for most living organism. For example, a naphthoquinone; 6-hydroxy-a-dunnione (**24**) and binaphthoquinone; methyl 1,1′,4,4′-tetrahydro-3-hydroxy-1,1′,4,4′-tetraoxo[2,2′-binaphthalene]-3′-carboxylate (**25**) from *Didymocarpus hedyotideus* Chun in Madagascar were reported to be cytotoxic with IC_{50} values of less than 10 μM and 12.6 μM, respectively (Xiao et al., 2011). The chemical structures of the aforementioned quinone compounds are shown in Fig. 9.5.

9.4.3 Flavonoids

Flavonoids are phenolic structures with one carbonyl group, synthesized by plants in response to microbial infection (Shashank & Abhay, 2013). Biosynthesis of flavonoids involves a combination of shikimic acid and three acetate units. They possess a basic aromatic (Ar) C_6 unit structure, with a C_3 unit attached to the aromatic ring, i.e. an Ar–C_3–Ar skeleton forms the basic structure of flavonoids (Gurib-Fakim, 2006; Prasad & Prasad, 2019). There are several classes of flavonoids (Fig. 9.6) found in higher plants that include flavones, flavonols, flavanones, flavanonol, isoflavones, flavanols and catechins (Khater et al., 2020; Kucharska et al., 2017).

Flavones are found especially in vascular plants in which they form glycosides and aglycones. They have an unsaturated 3-C chain with a C=C double bond, and

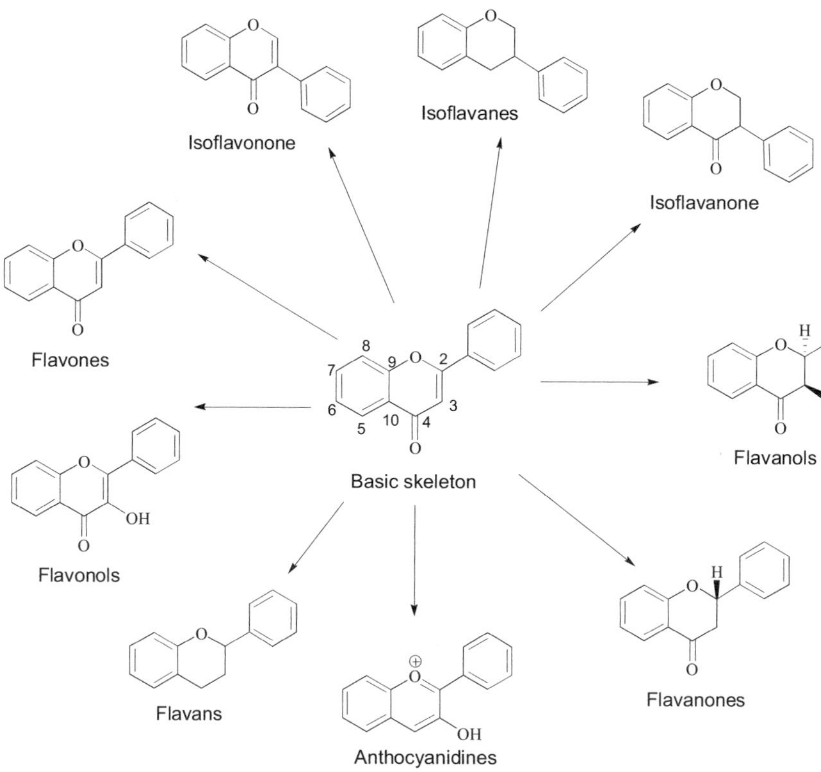

Fig. 9.6 The general structures of the different classes of flavonoids (Ferdes, 2018)

they differ from flavonols with the absence of a hydroxyl group at position C-3. Flavanols (catechins) are the most reduced form of the C-3 unit in flavonoid compounds and have been extensively researched due to their presence in green tea and other types of tea (Ferdes, 2018; Gurib-Fakim, 2006; Rihacek & Roubal, 2017).

Several flavonoids (Fig. 9.7) have been shown to possess potent antimicrobial activity (Attah et al., 2021; Kemboi et al., 2022). For example, rutin (**26**) and kaempferol-3-*O*- rutinoside (**27**), isolated from *Spermacoce princeae* (K. Schum) and *Dolichopentas decora* (S. Moore) (Sekandi et al., 2025), have been reported for their antibacterial and antifungal activity against *P. aeruginosa*, *E. coli*, *B. subtilis* and *S. aureus* (Asogwa & Okoye, 2019; Reddy et al., 2016; Sekandi et al., 2023). In 2016, Reddy et al. reported that rutin (**26**) at 50 μg/mL had significant antibacterial activity against *P. aeruginosa*, *E. coli*, *B. subtilis* and *S. aureus*. A study conducted by Sekandi et al. (2023) observed a good antifungal activity of kaempferol-3-*O*-rutinoside (**27**) against *C. albicans* (zone of inhibition, 23.0 ± 0.1 mm) isolated from *S. princeae*; a plant used in Uganda for treatment of skin infections.

Additionally, epigallocatechin (**28**) has been reported to inhibit DNA synthesis in *P. vulgaris* and galangin (**29**) has shown antibacterial activity (6.3 μg/mL) against

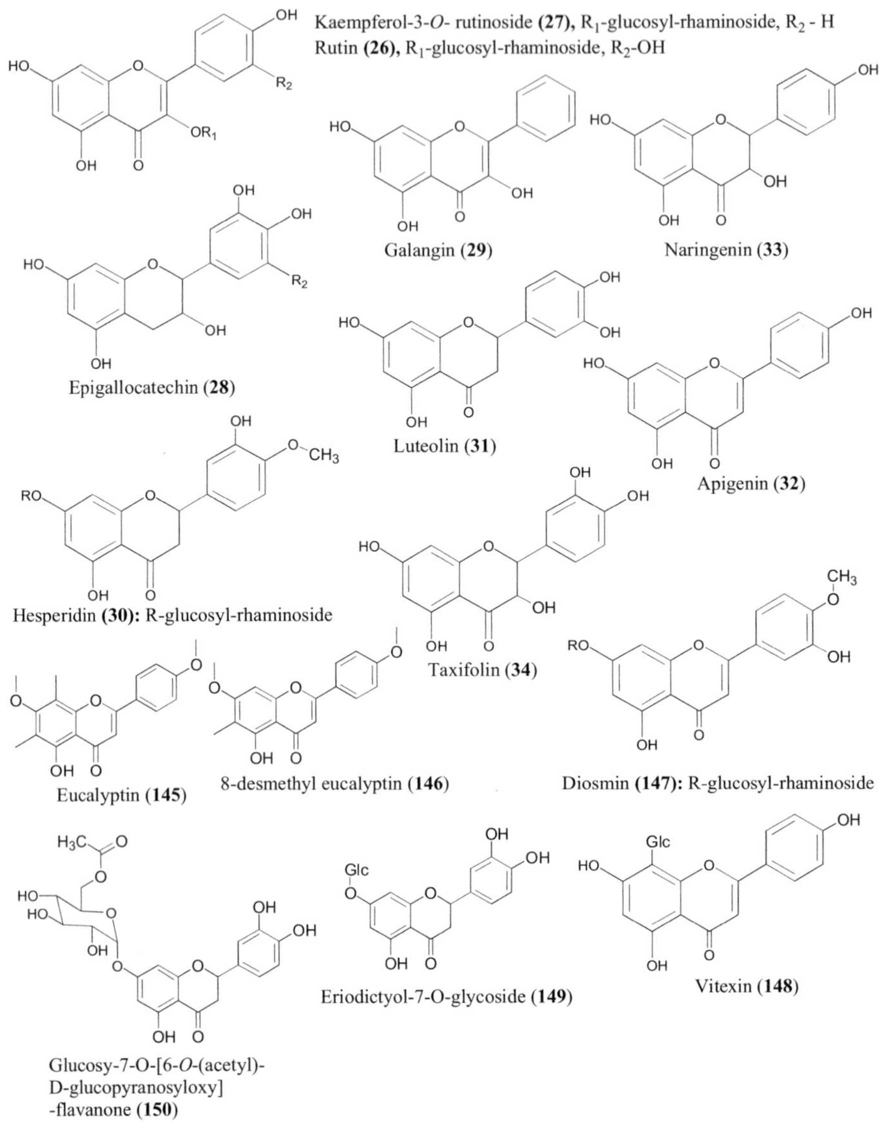

Fig. 9.7 Chemical structures of different flavonoids (**26–34, 145–150**) with potential antimicrobial activity (Namukobe et al., 2024; Shashank & Abhay, 2013)

Staphylococcus epidermidis (Fyhrquist et al., 2014; Shashank & Abhay, 2013). Hesperidin (**30**) from the stem bark of *Zanthoxylum leprieurii* Guill is an antimycobacterial agent against $H_{37}Rv$ and MDR-Tuberculosis strains with MIC of 73.4 μg/mL (Oloya et al., 2021). In Cameroon, *Zanthoxylum heitzii* (Aubrev & Pellegr) P.G. Waterman and *Capsicum annuum* L. were used to manage COVID-19 pandemic and have been reported to contain luteolin (**31**), thus possessing antiviral activity (Fedoung et al., 2023). Apigenin (**32**) from *Moringa oleifera* Lam., naringenin (**33**) and taxifolin (**34**) have shown significant antiangiogenic, antiviral and antibacterial activity (Adeleye et al., 2021; Fedoung et al., 2023; Khater et al., 2020; Maddox et al., 2010). In Zimbabwe, alkaloid extracts from *Callistemon citrinus* Skeels were found active against *S. aureus* and *P. aeruginosa* at 0.0025 mg/mL (MIC) and 0.835 mg/mL (MBC) (Mabhiza et al., 2016).

Phytochemical analysis of *C. citrus* showed presence of eucalyptin (**145**) and 8-desmethyl eucalyptin (**146**) with potent anti-biofilm activity up to 71% and 62.3% against *S. aureus* (Khanh et al., 2016; Shehabeldine et al., 2020). *Persicaria salicifolia* Brouss. ex Willd, *Zanthoxylum bungeanum* Maxim. and *Erlangea tomentosa* S. Moore are some of the medicinal plants used to manage bacterial skin infections in Egypt and Uganda. Phytochemical analysis of Z. *bungeanum, P. salicifolia* and *E. tomentosa* led to identification of diosmin (**147**), vitexin (**148**), eriodictyol-7-O-glycoside (**149**) and 5,7,3′,4′-tetrahydroxy-7-O-[6″-O-(acetyl)-β-D-glucopyranosyloxy]-flavanone (**150**) (Hesham et al., 2018; Namukobe et al., 2024; Zhang et al., 2014).

9.4.4 Alkaloids

Alkaloids are nitrogen-containing organic compounds with the exception of amino acids, peptides, purines and derivatives of amino sugars. There are several types of alkaloids, namely, pyrrolizidine alkaloids, tropane alkaloids, vinca alkaloids, among others (Kuete, 2014). Due to their structural biodiversity, each class of alkaloids possess a specific biosynthetic pathway. For instance, monoterpene indole alkaloids are synthesized from strictosidine (**35**) which is derived from tryptamine and secologanin through a Pictet-Spengler condensation reaction (Xiao-Feng et al., 2022).

Alkaloids are widely distributed in angiosperms from families such as Apocyanaceae, Ranunculaceae, Amaryllidaceae, Papaveraceae, Solanaceae, Rutaceae, Rubiaceae, Camptothecin, Solanaceae, Annonaceae, Loganiaceae and Fabaceae (Masi et al., 2019; Xiao-Feng et al., 2022). Lower plants, insects, marine organisms and microorganisms have been also reported to be a source of alkaloids. Alkaloids are mainly produced by dicotyledonous plants and to a lesser extent the cryptogams such as ferns, mosses, lichens, except bryophytes and algae (Ferdes, 2018).

Alkaloids have demonstrated a wide range of antimicrobial activity. For example, lysicamine (**36**), trivalvone (**37**), palmatine (**38**) and jatrorrhizine (**39**), isolated from *Annickia kummeriae* (Engl. & Diels) of Tanzania, showed strong antiplasmodial activity (IC_{50} values of 0.08–2.4 μg/mL), moderate anti-trypanosomal activity (IC_{50} values of 2.80–14.3 ± 0.001 μg/mL) and antileishmanial activity (IC_{50} values of 2.7–20.4 μg/mL) (Malebo et al., 2013). Camptothecin (**40**)-producing plants such as spices of *Albizia* are known to contain coccinine (**41**) and montanine (**42**) (Figs. 9.8 and 9.9) alkaloids, which have shown significant antiplasmodial activity (Masi et al.,

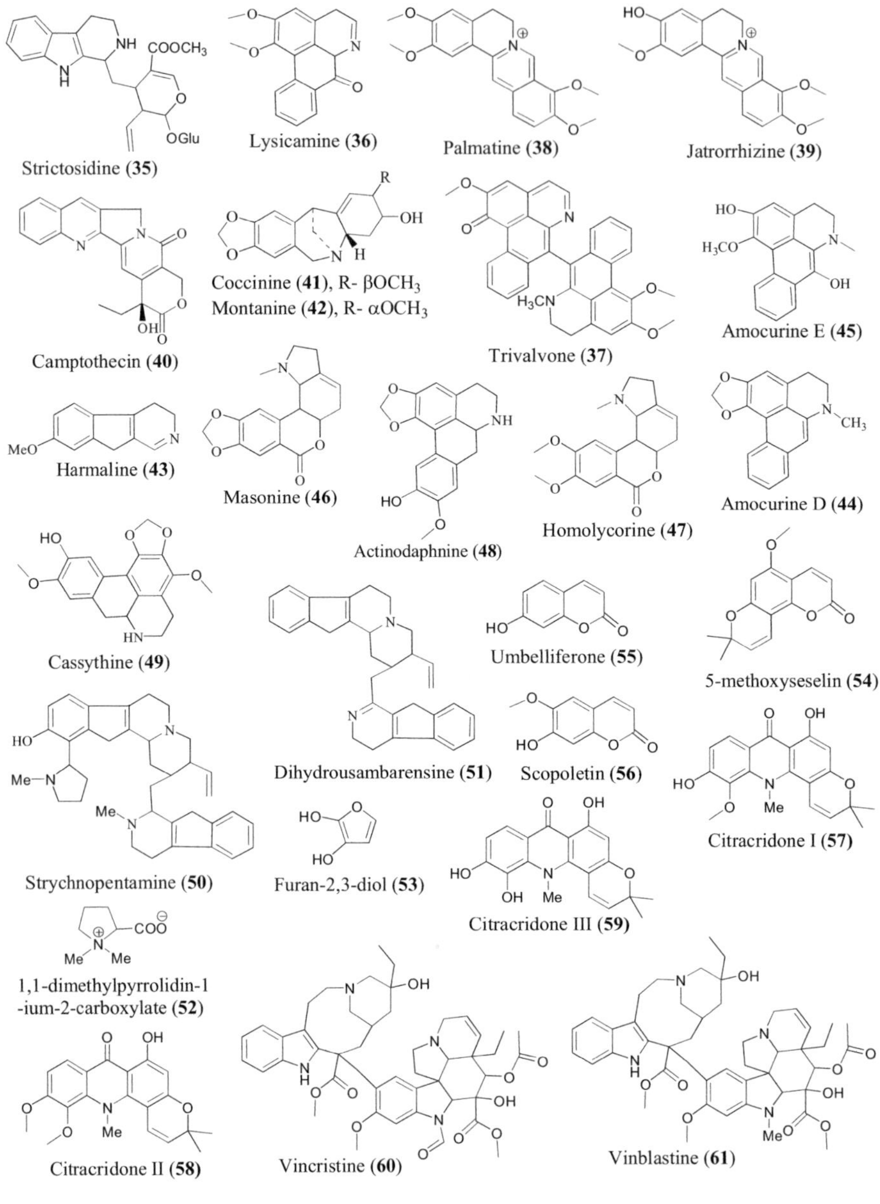

Fig. 9.8 Structures of alkaloids (**35–61**) identified from African plants with antimicrobial properties (Chumkaew et al., 2021; Malebo et al., 2013)

2019; Nair & Van, 2019). Among the indole alkaloids, harmaline (**43**) has been reported to possess significant antileishmanial activity with an IC_{50} value of 116.8 μM (Khanum et al., 2024). Amocurine D (**44**) and E (**45**) isolated from *Amoora cucullata* Roxb. showed antibacterial activities against *B. subtilis*,

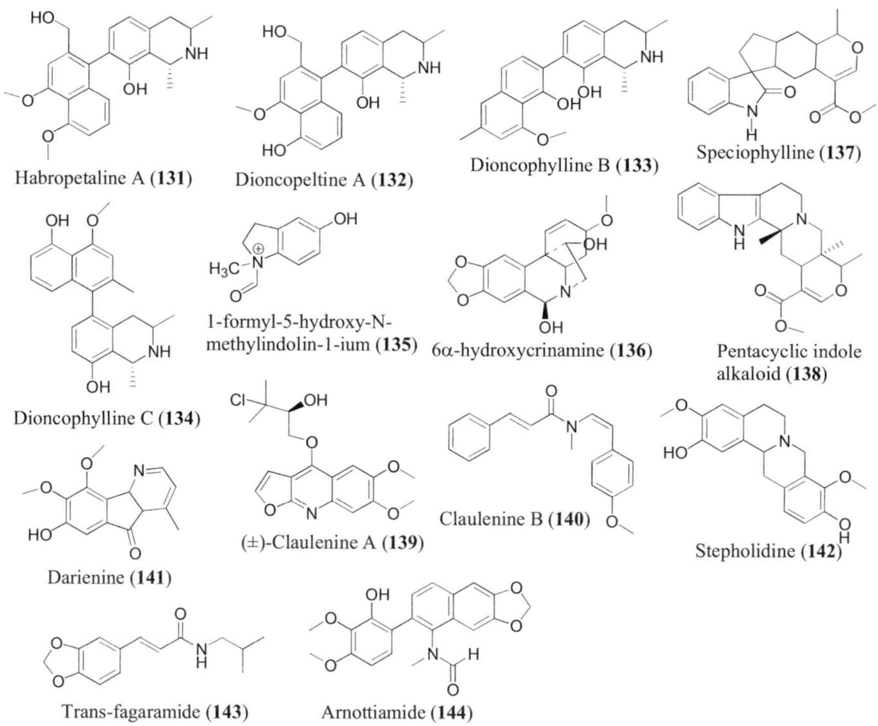

Fig. 9.9 Structures of alkaloids (**131**–**144**) identified from African plants with antimicrobial properties (Adewusi et al., 2022; Bringmann et al., 2003; Gbedema et al., 2015; Kamurthy et al., 2014; Zhu et al., 2022)

S. aureus, *P. aeruginosa* and *E. coli* with an MIC value of 3.1 μg/mL (Chumkaew et al., 2021). Masonine (**46**), homolycorine (**47**), aporphine such as actinodaphnine (**48**) and cassythine (**49**) isolated from *Cassytha filiformis* L. of Burundi showed antitrypanosomal activities with IC_{50} values of 10.29 and 17.60 μM, respectively (Simoben et al., 2018).

Indole alkaloids have shown more potency against a wide range of plasmodium species. For example, strychnopentamine (**50**) and dihydrousambarensine (**51**) have shown antiplasmodial activity against *P. falciparum* with IC_{50} values of 150 nM and 23 nM, respectively (Fanny-Aimee et al., 2023; Ogunwande et al., 2010). In Cameroon, 1,1-dimethylpyrrolidin-1-ium-2-carboxylate (**52**), furan-2,3-diol (**53**), 5-methoxyseselin (**54**), umbelliferone (**55**), scopoletin (**56**) and citracridones (I-III) (**57–59**) isolated from roots of the *Citrus Paradisi* Macfad were shown to have antimalarial activity with IC_{50} values ranging between 99.2 and 250.2 μM (Fanny-Aimee et al., 2023).

In Liberia and Ivory Coast, *Triphyophyllum peltatum* (Hutch. & Dalziel) is used to manage malaria. The plant has been reported to contain habropetaline A (**131**), dioncopeltine A (**132**), dioncophylline B (**133**) and dioncophylline C (**134**) as shown in Fig. 9.9. The naphthylisoquinolines (131–134) showed a very good antiplasmodium activity against *P. falciparum* with IC_{50} ranging between 5.0 and 2.3 ng/mL (Simoben et al., 2018). In Cameroon, *Citrus* × *paradisi*, a hybrid of *Citrus maxima* (Burm.) Merr. and *Citrus sinensis* (L.) Osbeck, has been used in folk medicine as an antibacterial, antifungal, antiviral and preservative drug due to the presence of citracridone (**57–59**) and 1-formyl-5-hydroxy-N-methylindolin-1-ium (**135**) (Fanny-Aimee et al., 2023).

Boophane disticha Dinter & G.M. Schulze is used traditionally in Southern Africa in the treatment of several neurological disorders due to the presence of 6β-hydroxycrinamine (**136**). However, the compounds were cytotoxic with IC_{50} values of 54.5 and 61.7 μM (Adewusi et al., 2022). The leaves of *Pentas lanceolata* K. Schum. plant common in East Africa have been reported to contain speciophylline (**137**) and pentacyclic indole alkaloid (**138**) (Kamurthy et al., 2014), *Clausena lenis* Drake in Cameroon, Angola and Ghana has been used as a folk medicine for the treatment of influenza, colds, bronchitis and malaria. The plant has been reported to contain (±)-claulenine A (**139**) and claulenine B (**140**) that contribute to its anti-neuroinflammatory effects with IC_{50} values in the range of 17.6–40.9 μM (Zhu et al., 2022). Darienine (**141**) and stepholidine (**142**) isolated from *Polyalthia longifolia* (Sonn) in Ghana exhibited antiplasmodial activity at IC_{50} 105 mg/mL (Gbedema et al., 2015). Trans-fagaramide (**143**) and arnottianamide (**144**) isolated from the stem bark of *Zanthoxylum leprieurii* Guill in Uganda possess antimycobacterial activity against the susceptible tuberculosis strain at 12.2 ug/mL (Oloya et al., 2021).

Despite the wide range of bioactivity, alkaloids remain largely unused due to their toxicity to humans and complexity in structure (Kuete, 2014). Masonine, homolycorine, coccinine and montanine from *Haemanthus humilis* Jacq in South Africa were reported to exhibit antibacterial and antiplasmodial activity at a concentration less than 50 μM (Coz & Ducombs, 2017; Masi et al., 2019). Vincristine (**60**) and vinblastine (**61**) isolated from *Catharanthus roseus* (L.) G.Don in Madagascar are antibacterial agents with high toxicity. They both have neurotoxic effects such as neuralgia, myalgia and paresthesia, loss of the tendon reflexes, depression and headache. It is reported that no attempts have been made to synthesize these compounds in the laboratories (Mahomoodally, 2013).

9.4.5 Terpenes and Terpenoids

Terpenes are hydrocarbons formed from isoprene (isopentane) units and can be synthesized via mevalonic pathway. These are the largest and most diverse groups of secondary metabolites in plants, animals and some insects such as termites or swallowtail butterflies. Terpenes are classified basing on the number of isoprene units from which they are biogenetically derived. They include hemiterpenes (C_5), monoterpenes (C_{10}), sesquiterpenes (C_{15}), diterpenes (C_{20}) and triterpenes (C_{30}) (Armijos et al., 2020; Zhang et al., 2024). The most widely distributed terpenoids

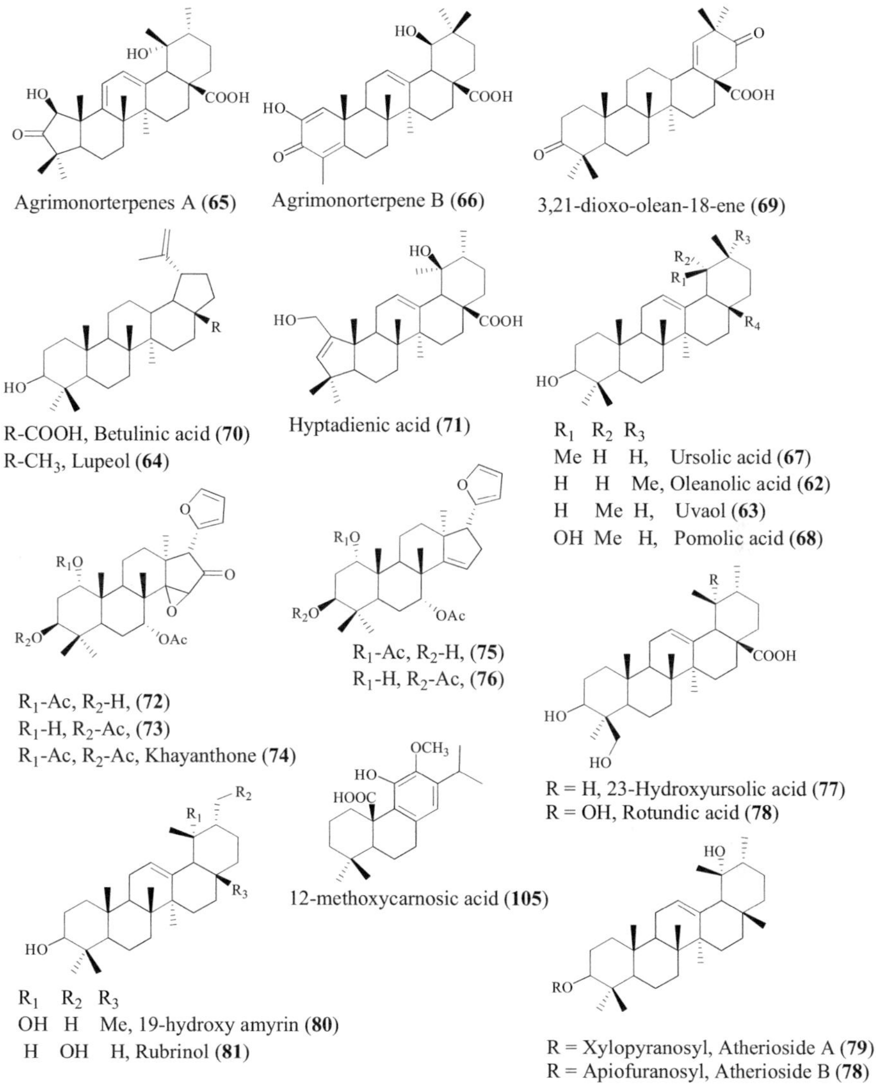

Fig. 9.10 Chemical structures of some triterpenoids (**62–81**,**105**) with antimicrobial activities (Osorio et al., 2012; Zhang et al., 2022)

and abundant in plants are the pentacyclic triterpenoids (Fig. 9.10), namely, oleananes (**62**), ursanes (**63, 67**) and lupanes (**64, 70**) (Hui et al., 2017; Lakhal et al., 2014).

Pentacyclic triterpenoids such as agrimonorterpenes A (**65**) and B (**66**) isolated from *Agrimonia pilosa* Ledeb possess antimalarial activity with selective cytotoxicity against HCT116, BGC823 and HepG2 cell lines at a concentration of 16.31 μM, 21.94 μM and 23.40 μM, respectively (Zhang et al., 2022). *Agrimonia* *pilosa*, also known as hairy agrimony, is native to Asia and Europe. However, due to its medicine values, the plant has been naturalized in Algeria, Egypt, Ethiopia, Morocco and South Africa (Attallah et al., 2021; Zhang et al., 2022). Ursolic acid (**67**), rotundic acid (**78**), 3-hydroxyursolic acid (**77**) and pomolic acid from *Rytigynia kigeziensis* Verdc (**68**) inhibited the growth of *Providencia stuartii, P. aeruginosa, K. pneumoniae, E. aerogenes, E. cloacae* and *E. coli* with MICs ranging between 32 and 256 μg/mL (Nascimento et al., 2014; Sekandi et al., 2023; Seukep et al., 2016a). 3,21-Dioxo-olean-18-ene (**69**) obtained from *Cassine xylocarpa* Vent in Rwanda displayed potent antiviral activity (Osorio et al., 2012).

Betulinic acid (**70**), lupeol (**64**) and hyptadienic acid (**71**), isolated from *Eriobotrya japonica* (Thunb.) Lindl. and *Ocimum basilicum* L., showed antibacterial activity against *S. aureus* and *B. subtilis* (Hui et al., 2017; Nguyen et al., 2017; Tshilanda et al., 2014). The crude extract of *Khaya anthotheca* C. DC. shows antiplasmodial activity ranging between 0.96 and 1.59 μg/mL. Limonoids (**72–76**) from *Khaya anthotheca* have potential antiplasmodial activity at 1.32 μg/mL (Obbo et al., 2024). Oleanolic acid (**62**) from *Pentas lanceolata* K. Schum. (Sweelam et al., 2018) and uvaol (**63**), Atherioside A (**79**), atherioside B (**78**), 9-hydroxy amyrin (**80**), rubrinol (**81**) from *Tricalysia atherura* N. have been reported to exhibit moderate antimalarial activity against chloroquine-resistant (Dd2) strains at IC_{50} between 64.99 and 92.29 μg/mL (Jikam-Sime et al., 2023). 3-Hydroxyursolic acid (**77**) is known to induce cancer cell apoptosis (Won et al., 2018), and used for the treatment of oedema, appendicitis, endometriosis and inflammation at 121.0 μM (Won et al., 2018). In South Africa, 12-methoxycarnosic acid (**105**) from *Salvia repens* Burch showed antiprotozoal activity against axenically grown *Leishmania donovani* amastigotes with an IC_{50} of 0.75 μM, with marginal cytotoxicity against the L6-cells (IC_{50}, 17.3 μM) (Mokoka et al., 2014).

Essential oils also consist of different classes of terpenes such as monoterpenes, sesquiterpenoids and coumarins having various antimicrobial activities. Examples of monoterpenes (Fig. 9.11) found in essential oils include α-pinene (**82**), β-pinene, (**83**), linalool (**84**), menthol (**85**), borneol (**86**) and 1,8-cineole (**87**), geraniol (**88**), perillyl alcohol (**89**), isopulegol (**90**) and citronellol (**91**) (Akakpo et al., 2023; Steenhuisen et al., 2010).

According to the International Organization for Standardization, essential oils are products obtained from a natural raw material of plant origin, by either steam distillation, mechanical processes from the epicarp of citrus fruits or by dry distillation after separation of the aqueous phase. Essential oils can also be obtained from flowers, stem barks, leaves and roots (Ferdes, 2018; Gurib-Fakim, 2006). According to Ferdes (2018), the antimicrobial activity of essential oils depends on their composition.

The most active compounds against bacterial and fungal strains include limonene (**92**), thymol (**93**), carvacrol (**94**), eugenol (**95**), vanillin (**96**), allicin (**97**) and squalene (**98**), among other compounds (Fig. 9.11) (Ferdes, 2018; Gurib-Fakim, 2006). Monoterpenes **82–91** showed moderate antiplasmodial activity with IC_{50} values ranging from 52 to 519 μM (Turcheniuk et al., 2015). In Uganda, *P. febrifugum* is one of the most used plants to manage skin infections due to its

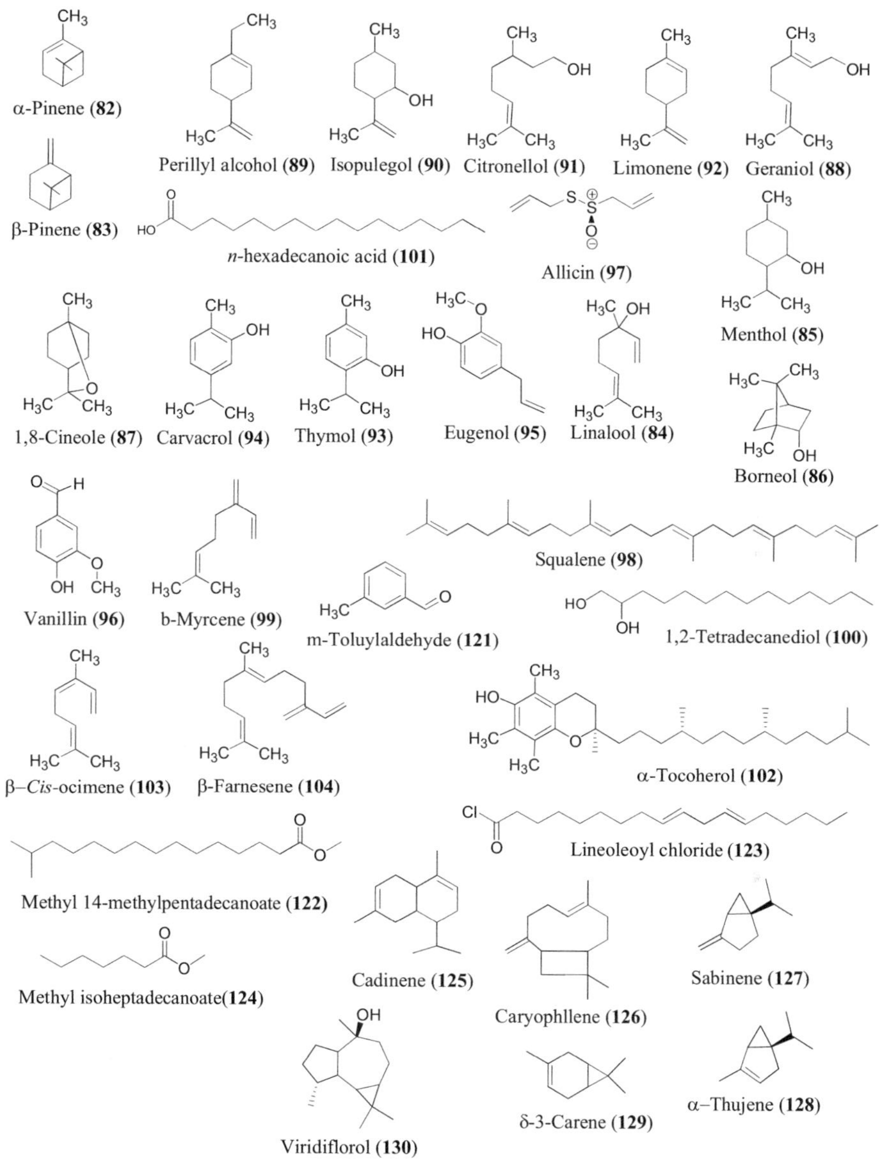

Fig. 9.11 Some common monoterpenes (**82–104, 121–130**) found in essential oils with antibacterial and antifungal activities (Ferdes, 2018; Gurib-Fakim, 2006; Nagawa et al., 2015)

abundant oils in the stem bark. GC-MS analysis of *P. febrifugum* extracts revealed the presence of β-myrcene (**99**), 1,2-tetradecanediol (**100**), *n*-hexadecanoic acid (**101**), α-tocoherol (**102**), β-*Cis*-ocimene (**103**) and β-farnesene (**104**) (Nagawa et al., 2015). Hexane extracts of *P. febrifugum* showed antibacterial activity against *S. aureus*, *E. coli*, *K. pneumoniae* and *P. aeruginosa* with an MIC of 6.25 mg/mL.

This activity was attributed to the presence of compounds **92–98**. *Aglaophenia cupressina* Lamouroux has antibacterial agents against *Salmonella typhi* (Johannes & Litaay, 2016). Aberu et al.'s (2012) review indicates that allicin (**97**) isolated from garlic (*Allium sativum*) possesses antibacterial activity against *Staphylococcus* spp. and *P. aeruginosa.*

Crude essential oils obtained from *Piper nigrum* L., *Melaleuca alternifolia* Cheel, *Copaifera officinalis* L., *Cinnamomum cassia* (L.), *Syzygium aromaticum* (L.) Merr. & L.M.Perry were found active against *S. aureus*, *E. coli* and *P. aeruginosa* strains with MIC value of 8.29 mg/mL (Murbach-Teles et al., 2014). Essential oils from Lemon grass (*Cymbopogon citratus*) showed antimicrobial activity with MIC ranging from 1 μg/mL to 32 μg/mL for sensitive species which included *Escherichia*, *Enterococci*, *Serratia*, *Erwinia, Pragia*, *Staphylococci* and *Klebsiella* (Bhoj et al., 2013; Shanjun et al., 2020). Squalene (**94**) detected in *Pentas decora* S.Moore (Sekandi et al., 2025) and *Hoslundia opposita* Vahl has exhibited antifungal and antibacterial properties against *Trichophyton mentagrophytes* and is currently used for numerous vaccine and drug delivery emulsions due to its stability-enhancing properties and biocompatibility (Kong et al., 2021; Namuga et al., 2024). Despite extensive investigation on African medicinal plants, the biological activity and/or chemical composition of many species still remain unidentified (Yazdi et al., 2019).

In Sudan, *Azadirachta indica* A. Juss. (Neem) have been reported to possess anti-cutaneous Leishmaniasis at LC_{50} of 10.2, 4.94 and 89.38 μg/ml. GC-MS analysis of *Azadirachta indica* showed several compounds including; m-Toluylaldehyde (**121**), Methyl 14-methylpentadecanoate (**122**), lineoleoyl chloride (**123**), Methyl isoheptadecanoate(**124**) in the methanol extract (Sarkar et al., 2021). In South Africa, over 31 essential oils displayed antimicrobial activity (MIC ≤1.00 mg/ mL) against *Brevibacillus agri, Brevibacillus epidermidis and Brevibacillus linens*. Among these, essential oils from *Plectranthus grandidentatus* Gurke and *Salvia africana-lutea* displayed antimicrobial activity (MIC = 1.00 mg/ mL) against *B. linens* (Van Vuuren et al., 2019). Phytochemical analysis of *P. grandidentatus* and *S. africana-lutea* revealed the presence of antimicrobial oils such as cadinene (**125**), caryophyllene (**126**), sabinene (**127**), α-thujene (**128**), δ-3-carene (**129**), viridiflorol (**130**), among others (Anthony et al., 2024; Mota et al., 2013).

9.5 Plant-Derived Approved Antimicrobial Drugs

The prevalence of infectious diseases is increasing, and this has led to high morbidity and mortality especially in this era of increased antimicrobial resistance. This creates a dire need for new and effective antimicrobial agents (Sanchez Armengol et al., 2021). Plants have continuously been used in treatment of infections, but few of them are commercialized. For example, many plant-derived compounds (Table 9.1) show antifungal properties, but a few of them reach the market. Therefore, well-designed clinical trials are needed to validate their use (Simonetti et al., 2018).

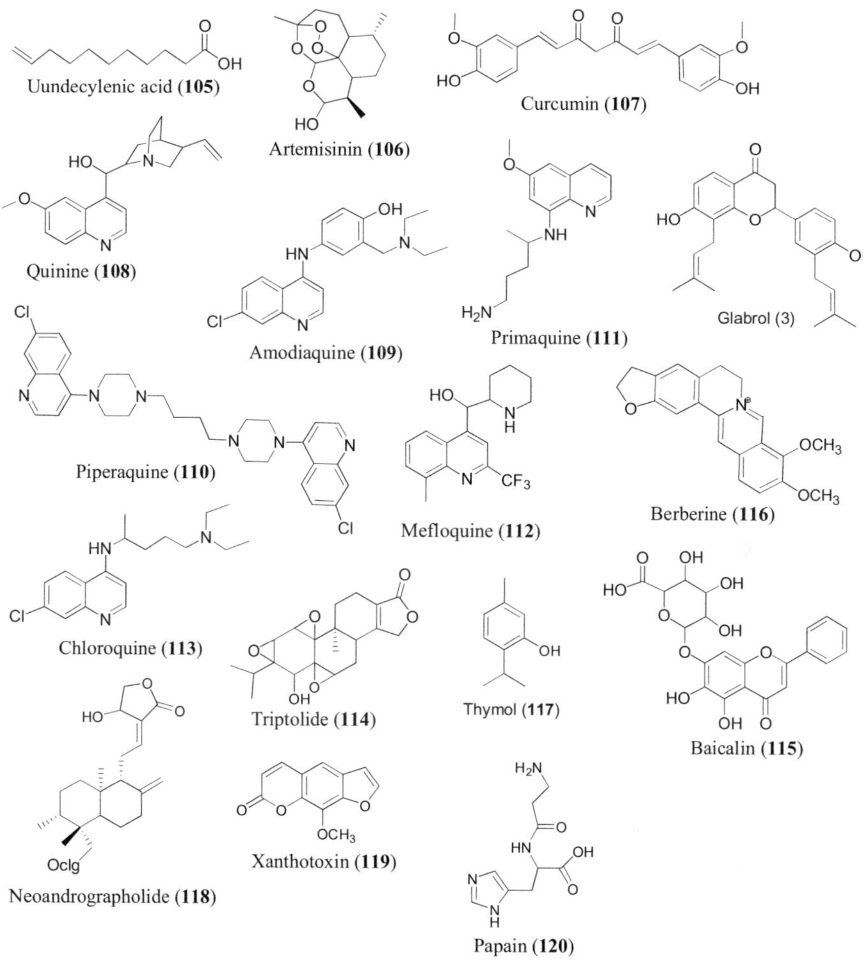

Fig. 9.12 Some of the plant-derived antimicrobial compounds (**105–120**) (Babalola et al., 2023; Van-Wyk, 2015; Ye et al., 2020)

Presently, azoles, polyenes and echinocandins are the major structural classes of approved antifungal agents (Bondaryk et al., 2017). Other classes include pyrimidine analogues and allylamines, most of which are not plant-derived (Campoy & Adrio, 2017; Vanreppelen et al., 2023). A number of plant-derived antifungal agents (Fig. 9.12) have been approved, and some are in clinical trials. These include undecylenic acid (**105**) from *R. communis* and thymol (**117**) from *Thymus vulgaris*, which are important compounds used as a drug against fungal infections (Abreu et al., 2012).

Artemisinin (**106**), a sesquiterpene lactone endoperoxide, was isolated from *Artemisia annua* (Cheong et al., 2020; Ye et al., 2020) and is now cultivated in African countries like Kenya, Uganda and Madagascar. Artemisinin (**106**) acted as a model for synthesis of a number of derivatives. These include artesunate, artemether, arteether, dihydroartemisinin, artemiside and artemisone (Ansari et al., 2013). Artemisinin and derivatives like artesunate, artemether, arteether, dihydroartemisinin have been key and first line remedy for the treatment of uncomplicated malaria caused by *Plasmodium falciparum* (Loo et al., 2017). Artemisinin combination therapies (ACTs) have been designed to reduce the burden of antimicrobial resistance. Here, artemisinin is simultaneously administered with one or more drugs. For example, artemether/lumefantrine, piperaquine, artesunate/amodiaquine and artesunate/mefloquine (Karunamoorthi et al., 2013; WHO, 2010). Apart from antimalarial activity, artemisinin and its analogues have been found to have activity against virus, fungi and protozoa like trypanosomes (Naß & Efferth, 2018).

Andrographolide (**9**), isolated from *Andrographis paniculata* (Burm.f.) (king of bitters) in 1951, is a labdane diterpenoid used as a Chinese drug (Daniel et al., 2017). Andrographolide (**9**) and its derivatives are known for antibacterial, antimalarial, antihepatitic and anti-HIV activities (Kumar et al., 2020; Ye et al., 2020). It is used against symptoms of inflammation that could be caused by bacterial infections (Daniel et al., 2017; Lee et al., 2011). Andrographolide (**9**) and other compounds synthesized from it have recommendable antibacterial activity against various strains of bacteria (Arifullah et al., 2013; Banerjee et al., 2017). This implies that the natural product derived from the plant source is a potential source of antibacterial drugs. Andrographolide (**9**) has properties that inhibit virus replication and virus pathogenesis. This enhances its antiviral properties against a variety of viral infections, including SARS-CoV-2 (Adiguna et al., 2021; Gupta et al., 2017). It can therefore be used as a potential natural agent for the synthesis of drugs against SARS-CoV-2 (Das et al., 2022). Its analogues are reported to have activity against influenza virus (Yuan et al., 2016).

It was discovered that analogue of andrographolide (**9**) has activity against influenza A virus, including the deadly H5N1 strain (Kumar et al., 2020). Human enterovirus A71 (EV-A71) infections have been treated in China using andrographolide sulphonate, a synthesized derivative of the compound. Therefore, this can cure hand, foot and mouth disease (Xu et al., 2022). When tested in combination with curcumin and artesunate against erythrocytic stages of *Plasmodium falciparum* and *P. berghei*, a potent antiplasmodial activity was recorded. It can be a potential protease inhibitor against malaria parasites (Dey et al., 2011; Megantara et al., 2015). Neoandrographolide (**118**), another compound isolated from *Andrographis paniculata*, is used in the treatment of dysentery (Gupta et al., 2017; Kumar et al., 2020).

Curcumin (**107**) isolated from *Curcuma longa* L. is another important commercial compound with antimicrobial properties. When *S. aureus, E. faecalis, E. coli* and *P. aeruginosa* were exposed to 25 μg, 50 μg and 100 μg of curcumin concentrations for 2 hours and plating the aliquots at selected time intervals on brain heart infusion plates, curcumin showed strong killing potential against the bacteria (Tyagi et al., 2015).

Quinine (**108**), which is in a class of quinolines, was first isolated from the bark of Cinchona tree and has been widely used for treatment of complicated malaria. The quinoline moiety is the active site for the antimalarial activity of quinine. Therefore, quinine and other quinoline compounds have antimalarial activity (Eyong et al., 2013; O'Donnell et al., 2010). Other antimalarial agents in the quinoline family include amodiaquine (**109**), piperaquine (**110**), primaquine (**111**) and mefloquine (**112**) (Kaur et al., 2010). Chloroquine (**113**) is another antimalarial agent that was used over time. However, it has become less effective due to evolution of resistant strains (Yenesew et al., 2012). Xanthotoxin (**119**) and papain (**120**) used in the USA are derived from *Ruta montana* L. Xanthotoxin and papain were tested against different strains of bacteria and fungi; the results showed that at 50 mg/mL, the compounds were effective against *S. aureus, Bacillus subtilis* and *C. albicans*. At 80 mg/mL, **119** and **120** were effective against *P. aeruginosa* and *Aspergillus fungi*. Papain, a proteolytic enzyme that can be extracted from the latex of the fruit of *Carica papaya* Linn., shows antibacterial and antifungal potential against different strains of bacteria and fungi (Babalola et al., 2023; Van-Wyk, 2015).

Triptolide (**114**), a diterpenoid epoxide isolated from *Tripterygium wilfordii* Hook.f (Huangteng), is a Chinese medicine marketed as *Tripterygium wilfordii* glycoside tablets (Ye et al., 2020). It is reported to have anticancer, anti-inflammatory, antidiabetic and antiobesity activities. A study done on the drug suggested that it has antiviral activity against HIV-1 (Wan & Chen, 2014). Baicalin (**115**), derived from *Scutellaria radix* SR, is another plant-derived drug used in Chinese medicine. It is a glycosidic flavone possessing antimicrobial properties, including antibacterial and antiviral abilities (Huang et al., 2019; Nayak et al., 2014). Studies have shown that Baicalin is a potent antibacterial compound against several strains of bacteria like multidrug resistant *S. aureus* and *E. coli*. Although the mechanism of action has not been extensively studied, Baicalin offers protection against influenza virus (Shouping et al., 2020; Zhao et al., 2018).

Berberine (**116**), an isoquinoline alkaloid, is derived from the root, stem bark or rhizome of *Biberis, Hydrastis canadensis* and *Coptidis rhizome* from Canada (Ye et al., 2020). Berberine-based derivatives have antimicrobial activity. Synthesized derivatives showed antibacterial and antiplasmodial activity (Jeyakkumar et al., 2016; Samosorn, 2016). Berberine and its combination with clindamycin and rifampicin separately have excellent anti-*S. aureus* activity (Xia et al., 2022). A number of essential oils from plants have been reported to have antimicrobial activity and have been used against multidrug-resistant strains. These include citrus essential oil, rosemary oil, oregano oil, basil oil and mentha oil. Mentha oil is known to have antimicrobial activity, and it is used for the treatment of respiratory diseases like sinusitis, tuberculosis and bronchitis. Menthol, one of the major components of mentha oil, is an important antiseptic used in dermatology. It inhibits biofilm formation (Murbach-Teles et al., 2014; Tyagi & Malik, 2011).

9.6 Scalability and Commercialization of Plant-Derived Products

Considering the large diversity of identified compounds and their therapeutic potential from African medicinal plants, thousands of them remain untested and developed due to the small quantity of the isolated compound usually enough for spectroscopic analysis and cannot be tested on the intended microorganism. Therefore, synthesizing enough quantities of a target molecule using scalable routes has emerged as important in natural product synthesis (Van-Wyk, 2015, 2020). These routes have yielded enough that enable further investigations and eventually leads to commercial production. Metabolic engineering using plant tissues or cell cultures have served as alternatives for the overproduction of high-value plant pharmaceuticals that can be conveniently scaled up in a controlled environment (Tang et al., 2021). Progress in biology systems has aided a better understanding of genome-wide interconnectivity in either plants or microorganisms. In this section, we discuss some compounds industrially synthesized using plant tissues or cell cultures.

Several plant cell lines have been developed to synthesize important pharmaceuticals such as taxol, shikonin and vinflunine (Fig. 9.13) at industrial levels. Concentrations of taxol (**151**), as high as 0.5% of dry weight in Pacific yew, were achieved in plant cell culture with methyl jasmonate elicitation (Yazdiniapour et al., 2022). Shikonin (**152**), used in cosmetic, is a naphthoquinone pigment that is derived from the roots of *Lithospermum erythrorhizon* cell suspension cultures (Nasim et al., 2022). Vinflunine (**153**) is one of the fluorinated tropane alkaloids used to treat cancer. It is produced by feeding fluorinated phenyllactic acid analogues to *Datura stramonium root cultures* (Lombardi et al., 2017) and scopolamine (**154**) from root cultures of *Datura metel* (Shakeran et al., 2017).

Microorganisms have sparked significant interest in their engineering for the synthesis of high-value plant metabolites due to fewer cell organelles. Microorganisms such as *E. coli* and yeast *S. cerevisiae* have been explored to produce a number of plant alkaloids. For example, reticuline (**155**) is a pharmacophore for antimalarial agents, and was successfully biosynthesized in *E. coli* by introducing monoamine oxidase in *Coptis japonica* plant (Herraiz & Guillén, 2018; Hong et al., 2022). The outcomes of chosen metabolic engineering strategies in plant tissue and cell cultures are often difficult to predict or control. These challenges could be resolved with the application of computer-aided techniques, namely, molecular docking, drug-likeness and ADMET profiling (Aboagye et al., 2023; Ameji et al., 2023; Amer et al., 2021).

Molecular docking involves using computer algorithms to examine the extent and mode of interaction of particular proteins with the active sites of a target pharmacophore. Computer-aided ADMET profiling is a component of modern drug design that involves predicting absorption, distribution, metabolism, excretion and toxicity of a therapeutic target compound (Ameji et al., 2023; Osafo et al., 2023). Docking is used at early stages of development to select the best structure of a ligand by using a score function in a structure established by the target-binding site, a process where novel and potent compounds are identified. Molecular docking also

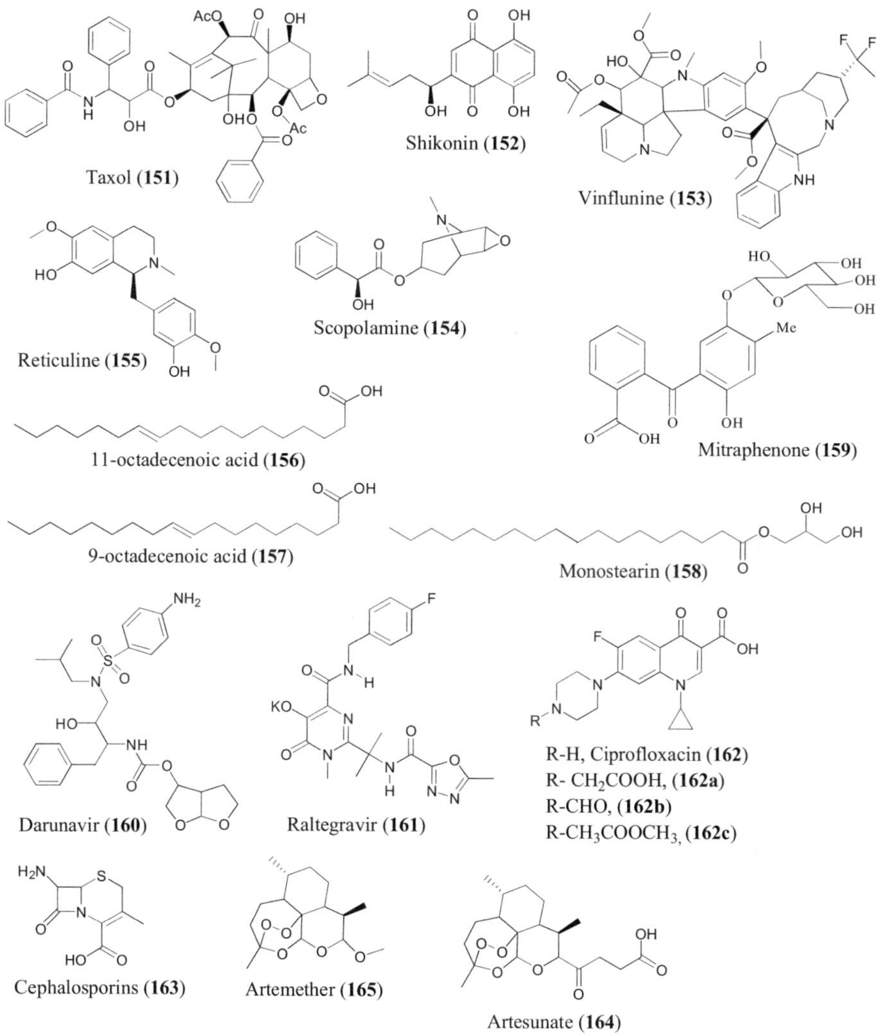

Fig. 9.13 Antimicrobial compounds (**151–165**) biosynthesized using plant tissues and cell cultures (Nasim et al., 2022; Shakeran et al., 2017; Yazdiniapour et al., 2022)

explains the mechanism of action of a target pharmacophore on different types of enzyme systems (Aboagye et al., 2023). This section brings you some notable drugs on market that have benefited from docking studies.

In Sudan, 11-octadecenoic acid (**156**), hexadecanoic acid (**101**) and 9-octadecenoic acid (**157**) in *Capparis decidua* (Forssk) had good binding affinity for Mcl-1 protein widely expressed in various cancer cells. The compounds have good docking scores with binding free energy ranging from −43.09 to −62.2 kcal/mol. This molecular evidence supports the anticancer potential of *C. decidua* (Allah et al., 2023). Molecular docking studies of monostearin (**158**) and lupeol **(64)** from *Parkia biglobosa* in Ghana, with trypanothione reductase and pteridine reductase 1, showed that both compounds exhibited binding affinities of −7.1 kcal/mol and −10.2 kcal/mol thus are potential antileishmanial agents (Osafo et al., 2023). Mitraphenone **(159)** from *Dolichopentas decora* (S. Moore) in Uganda binds better to lysine through the hydroxyl group of C-6″ with a binding affinity value ($\Delta G = -3.7$ kcal/mol). These compounds are predicted to be potential antimicrobial agents (Sekandi et al., 2025).

Darunavir (**160**) and raltegravir (**161**) are second-generation protease/integrase inhibitors developed to optimize their binding to HIV-1 protease and replication cycle. Docking was used to ensure that **160** could block the replication of HIV more effectively (Eto, 2013; Pala et al., 2020). Darunavir has a stereochemically and conformationally constrained bicyclic *bis*-tetrahydrofuran heterocycle ring that forms strong hydrogen bonds with HIV-1 protease enzyme (Ghosh et al., 2019; Ruela Corrêa et al., 2012). Docking simulations have been used to study the interactions between ciprofloxacins (**162**), penicillins (Fig. 9.1), cephalosporins (**163**) and bacterial DNA and identify numerous derivatives (such as **162a-c**) with better activity against resistant strains (Bbosa et al., 2014; Sharma et al., 2010; Zhang et al., 2018).

The discovery of artemisinin (**106**) from *Artemisia annua* gave the plant popularity in African countries such as Uganda because of its antimalarial property (Ellman, 2015). Efforts to increase the yield of **106** in the crop and other plants of the same genus gave a momentum to be grown in a wide range of sub-tropical and temperate environments. Artemisinin scaffolds such as artesunate (**164**) and artemether (**165**) on market have been developed using computational docking techniques (Ansari et al., 2013; Gurjar et al., 2023). These examples highlight that docking has become an indispensable tool in modern drug discovery process along with other computational methods.

Plant-based NMR metabolomics is another way of metabolite profiling that ensures quality and reproducibility of herbal medicine. The technique generates enough knowledge on composition of the herbal products and their quality (Heyman & Meyer, 2012). This will enable herbal medicine industry to scale up the production of herbal products. This will create opportunities in harnessing these natural products into modern medicine, with a focus on drug discovery, development, and commercialization.

9.7 Conclusion

This review discussed medicinally significant plant species from Africa and showed that many of their crude extracts have antimicrobial activity but lack detailed phytochemical analyses involving the characterization of their chemical constituents. This review highlights *Psorospermum febrifugum* Spach, *Vernonia amygdalina* Del, *Moringa oleifera* Lam. *Plectranthus barbatus* Andrews,

Plectranthus caespitosus Lukhoba as some of the most important African plants with rutin, quercetin, ursolic acid and oleanolic as the major antimicrobial compounds. For many compounds such as mitraphenone, 12-methoxycarnosic acid, uvaol, atheriosides, agrimonorterpenes, vincristine, vinblastine, trans-fagaramide, arnottianamide, darienine, stepholidine, no further evaluation of their efficacy was found due to low yield. There is a need to create a natural product database that will allow the exploitation of a larger fraction of the identified compounds, and eventually expand the antimicrobial repertoire.

Acknowledgement We acknowledge all authors, especially Namukobe Jane, for the idea to write a review on antimicrobial compounds from African medicinal plants and their guidance throughout the process.

Competing Interests The authors have no conflict of interests.

References

Abebe, E., Tegegne, B., & Tibebu, S. (2016). A review on molecular mechanisms of bacterial resistance to antibiotics. *European Journal of Applied Sciences, 8*(5), 301–310. https://doi.org/10.5829/idosi.ejas.2016.8.5.1143

Abeje, F., Bisrat, D., Hailu, A., & Asres, K. (2014). Phytochemistry and antileishmanial activity of the leaf latex of Aloe calidophila reynolds. *Phytotherapy Research, 28*(12), 1801–1805. https://doi.org/10.1002/ptr.5204

Aboagye, C. I., Ampomah, G. B., Mensah, J. O., Mensah, C. N., Nartey, D., Gasu, E. N., & Borquaye, L. S. (2023). N-benzylimidazoles as potential antimicrobial and antibiofilm agents – Syntheses, bioactivity and molecular docking studies. *Scientific African, 19*, e01529. https://doi.org/10.1016/j.sciaf.2022.e01529

Abreu, A. C., McBain, A. J., & Simões, M. (2012). Plants as sources of new antimicrobials and resistance-modifying agents. *Natural Product Reports, 29*(9), 1007–1021. https://doi.org/10.1039/c2np20035j

Abushaheen, M. A., Fatani, A. J., Alosaimi, M., Mansy, W., George, M., Acharya, S., Rathod, S., Divakar, D. D., Jhugroo, C., Vellappally, S., Khan, A. A., Shaik, J., & Jhugroo, P. (2020). Antimicrobial resistance, mechanisms and its clinical significance. *Disease-a-Month, 66*(6), 1–21. https://doi.org/10.1016/j.disamonth.2020.100971

Achilonu, C. C., & Udensi, O. U. (2022). Systematic review of antimicrobial activities of medicinal plants against bacterial and fungal microbes: A research trend in Africa from 1980 to 2019. *Research Square, 2022*(April), 1–22. https://doi.org/10.21203/rs.3.rs-1529001/v1

Adaszynska-Skwirzynska, M., & Szczerbinska, D. (2018). The antimicrobial activity of lavender essential oil (Lavandula angustifolia) and its influence on the production performance of broiler chickens. *Journal of Animal Physiology and Animal Nutrition, 102*(4), 1020–1025.

Adebisi, Y. A., Jimoh, N. D., Ogunkola, I. O., Uwizeyimana, T., Olayemi, A. H., Ukor, N. A., & Lucero-Prisno, D. E. (2021). The uşe of antibiotics in COVID-19 management: A rapid review of national treatment guidelines in 10 African countries. *Tropical Medicine and Health, 49*(1). https://doi.org/10.1186/s41182-021-00344-w

Adeleye, O. A., Femi-oyewo, M. N., Bamiro, O. A., Bakre, L. G., Alabi, A., Ashidi, J. S., Balogunagbaje, O. A., & Hassan, O. M. (2021). Ethnomedicinal herbs in African traditional medicine with potential activity for the prevention, treatment, and management of coronavirus disease 2019. *Future Journal of Pharmaceutical Sciences, 7*(72), 1–14

Adeola, A. O., Adewole, S. O., & Olofintoye, L. K. (2015). Phytochemical screening of four different medicinal plants. *Unique Research Journal of Agricultural Sciences Full Length Research Paper, 3*(1), 1–6.

Adewusi, E. A., Fouche, G., & Steenkamp, V. (2022). Cytotoxicity and acetylcholinesterase inhibitory activity of an isolated crinine alkaloid from Boophane disticha (Amaryllidaceae). *Journal of Ethnopharmacology, 143*(2), 572–578.

Adiguna, S. P., Panggabean, J. A., Atikana, A., Untari, F., Izzati, F., Bayu, A., Rosyidah, A., Rahmawati, S. I., & Putra, M. Y. (2021). Antiviral activities of andrographolide and its derivatives: Mechanism of action and delivery system. *Pharmaceuticals, 14*(11), 1–20. https://doi.org/10.3390/ph14111102

Agyare, C., Boakye, Y. D., Bekoe, E. O., Hensel, A., Dapaah, S. O., & Appiah, T. (2016). Review: African medicinal plants with wound healing properties. *Journal of Ethnopharmacology, 177*, 85–100. https://doi.org/10.1016/j.jep.2015.11.008

Ahumuza, T., & Kirimuhuzya, C. (2011). Qualitative (phytochemical) analysis and antifungal activity of Pentas decora (De wild), a plant used traditionally to treat skin fungal infections in Western Uganda. *Research In Pharmaceutical Biotechnology, 3*(7), 75–84.

Akakpo, A. Y., Kabore, D., Somda, M. K., Sawadogo, A., Mihin, H. B., Semde, Z., & Ouattara, A. S. (2023). Potential antifungal activity of essential oil from Cymbopogon citratus (DC) Stapf and its performances as biostabilization of tiger nut milk (Cyperus exculentus). *Scientific African, 19*, e01497. https://doi.org/10.1016/j.sciaf.2022.e01497

Akindele, A. J., Sowemimo, A., Agunbiade, F. O., So, M. O., Awodele, O., Ade-ademilua, O., Orabueze, I., Ishola, I. O., Ayolabi, C. I., Salu, O. B., Akinleye, M. O., Oreagba, I. A., Unilag, A., & Team, C.-R. (2022). *Bioprospecting for anti-COVID-19 interventions from African medicinal plants: A review*. https://doi.org/10.1177/1934578X221096968

Akpalo, A. E., Saloufou, I. K., Eloh, K., & Kpegba, K. (2020). Wound healing biomolecules present in four proposed soft aqueous extractions of Ageratum conyzoides Linn. *International Journal of Biological and Chemical Sciences, 14*(2), 638–651.

Allah, A. A. D., Yousif, H. A., Hasaballa, N. O., Elkhawad, E. A., Abdallah, R. B., Ahmed, H. M., Abdelrahman, A. M., Hago, S., Abdelgadir, A. A., Alzain, A. A., & Ahmed, E. M. (2023). Identification of phytochemicals from Tundub Capparis decidua (Forssk) Edgew seed oil as potential anticancer agents using gas chromatography-mass spectroscopy analysis, molecular docking, and molecular dynamics studies. *Scientific African, 19*, e01517.

Alshwyeh, H. A. (2020). Phenolic profiling and antibacterial potential of Saudi Arabian native date palm (Phoenix dactylifera) cultivars. *International Journal of Food Properties, 23*(1), 627–638. https://doi.org/10.1080/10942912.2020.1751196

Altemimi, A., Lakhssassi, N., Baharlouei, A., Watson, D. G., & Lightfoot, D. A. (2017). Phytochemicals: Extraction, isolation, and identification of bioactive compounds from plant extracts. *Plants, 6*(4), 42. https://doi.org/10.3390/plants6040042

Amber, K. T., Shiman, M. I., & Badiavas, E. V. (2014). The use of antioxidants in radiotherapy-induced skin toxicity. *Integrative Cancer Therapies, 13*(1), 38–45.

Ameji, P. J., Uzairu, A., Shallangwa, G. A., & Uba, S. (2023). Molecular docking simulation, drug-likeness assessment, and pharmacokinetic study of some cephalosporin analogues against a penicillin-binding protein of Salmonella typhimurium. *Journal of Antibiotics, 76*(4), 211–224.

Amer, H. H., Eldrehmy, E. H., Abdel-Hafez, S. M., Alghamdi, Y. S., Hassan, M. Y., & Alotaibi, S. H. (2021). Antibacterial and molecular docking studies of newly synthesized nucleosides and Schiff bases derived from sulfadimidines. *Scientific Reports, 11*(1), 1–14.

Anand, U., Jacobo-Herrera, N., Altemimi, A., & Lakhssassi, N. (2019). A comprehensive review on medicinal plants as antimicrobial therapeutics: Potential avenues of biocompatible drug discovery. *Metabolites, 9*(11), 1–13. https://doi.org/10.3390/metabo9110258

Ansari, M., Saify, Z., Sultana, N., Ahmad, I., Saeed-Ul-Hassan, S., Tariq, I., & Khanum, M. (2013). Malaria and artemisinin derivatives: An updated review. *Mini-Reviews in Medicinal Chemistry, 13*(13), 1879–1902. https://doi.org/10.2174/13895575113136660097

Anthony, C., Rita, E., Aguchem, N., Chigozie, E., Favour, W., Ezeorba, C., Uzochukwu, I., Timothy, O., & Chidike, P. (2024). Salvia africana – lutea L.: A review of ethnobotany, phytochemistry, pharmacology applications and future prospects. *Advances in Traditional Medicine, 24*(3), 703–724. https://doi.org/10.1007/s13596-023-00726-x

Arifullah, M., Namsa, N. D., Mandal, M., Chiruvella, K. K., Vikrama, P., & Gopal, G. R. (2013). Evaluation of anti-bacterial and anti-oxidant potential of andrographolide and echiodinin isolated from callus culture of Andrographis paniculata Nees. *Asian Pacific Journal of Tropical Biomedicine, 3*(8), 604–610. https://doi.org/10.1016/S2221-1691(13)60123-9

Armijos, C., Matailo, A., Bec, N., Salinas, M., Aguilar, G., Solano, N., Calva, J., Ludeña, C., Larroque, C., & Vidari, G. (2020). Chemical composition and selective BuChE inhibitory activity of the essential oils from aromatic plants used to prepare the traditional Ecuadorian beverage horchata lojana. *Journal of Ethnopharmacology, 263*, 113162. https://doi.org/10.1016/j.jep.2020.113162

Arora, R., Vig, A. P., & Arora, S. (2014). Glucosinolates: Transposing trends of identification methods from paper. *International Journal of Life Sciences Biotechnology and Pharma Research, 3*(1), 42–61.

Asikin, Y., Kusumiyati, Shikanai, T., & Wada, K. (2018). Volatile aroma components and MS-based electronic nose profiles of dogfruit (Pithecellobium jiringa) and stink bean (Parkia speciosa). *Journal of Advanced Research, 9*, 79–85. https://doi.org/10.1016/j.jare.2017.11.003

Asogwa, F. C., & Okoye, C. O. B. (2019). Anti–psoriatic activity of Psorospermum febrifugum stem bark extract using the rat – dinitrofluorobenzene induced model. *International Journal of Scientific Engineering and Science, 3*(2), 1–5.

Attah, A. F., Fagbemi, A. A., Olubiyi, O., Dada-Adegbola, H., Oluwadotun, A., Elujoba, A., & Babalola, C. P. (2021). Therapeutic potentials of antiviral plants used in traditional African medicine with COVID-19 in focus: A Nigerian perspective. *Frontiers in Pharmacology, 12*, 1–38.

Attallah, N. G. M., El-Kadem, A. H., Negm, W. A., Elekhnawy, E., El-Masry, T. A., Elmongy, E. I., Altwaijry, N., Alanazi, A. S., Al-Hamoud, G. A., & Ragab, A. E. (2021). Promising antiviral activity of Agrimonia pilosa phytochemicals against severe acute respiratory syndrome coronavirus 2 supported with in vivo mice study. *Pharmaceuticals, 14*, 1313. https://doi.org/10.3390/ph14121313

Azmir, J., Zaidul, I. S. M., Rahman, M. M., Sharif, K. M., Mohamed, A., Sahena, F., Jahurul, M. H. A., Ghafoor, K., Norulaini, N. A. N., & Omar, A. K. M. (2013). Techniques for extraction of bioactive compounds from plant materials: A review. *Journal of Food Engineering, 117*(4), 426–436.

Babalola, B. A., Akinwande, A. I., Gboyega, A. E., & Otunba, A. A. (2023). Extraction, purification and characterization of papain cysteine-proteases from the leaves of Carica papaya. *Scientific African, 19*, 1–13.

Banerjee, M., Parai, D., Chattopadhyay, S., & Mukherjee, S. K. (2017). Andrographolide: Antibacterial activity against common bacteria of human health concern and possible mechanism of action. *Folia Microbiologica, 62*(3), 237–244. https://doi.org/10.1007/s12223-017-0496-9

Basler, G., Thompson, M., Tullman-Ercek, D., & Keasling, J. (2018). A pseudomonas putida efflux pump acts on short-chain alcohols. *Biotechnology for Biofuels, 11*, 136. https://doi.org/10.1186/s13068-018-1133-9

Bbosa, G., Mwebaza, N., Odda, J., Kyegombe, D. B., & Ntale, M. (2014). Antibiotics/antibacterial drug use, their marketing and promotion during the post-antibiotic golden age and their role in emergence of bacterial resistance. *Health, 06*(05), 410–425.

Bharali, P., & Gamo, Y. (2022). *Current Science Association Phytochemical and biochemical study of four legume plants with detergent and anti-lice properties from the Eastern Himalayan region of India author (s): Pankaj Bharali, Yabid Gamo, Arup Kumar Das, Hui Tag, Ananta Madhab B.*

Bhoj, R.-S., Singh, V., Singh, K. R., & Ebibeni, N. (2013). Cymbopogon citrates oil showing antimicrobial activity against microbes of environmental, clinical and food origin. *Asian Journal of Pharmacy and Technology, 3*(2), 67–72.

Bondaryk, M., Staniszewska, M., Zielińska, P., & Urbańczyk-Lipkowska, Z. (2017). Natural antimicrobial peptides as inspiration for design of a new generation antifungal compounds. *Journal of Fungi, 3*(3), 46 https://doi.org/10.3390/jof3030046

Borges, A., Ferreira, C., Saavedra, M. J., & Simões, M. (2013). Antibacterial activity and mode of action of ferulic and gallic acids against pathogenic bacteria. *Microbial Drug Resistance, 19*(4), 256–265. https://doi.org/10.1089/mdr.2012.0244

Bringmann, G., Messer, K., Schwöbel, B., Brun, R., & Aké Assi, L. (2003). Habropetaline a, an antimalarial naphthylisoquinoline alkaloid from triphyophyllum peltatum. *Phytochemistry, 62*(3), 345–349. https://doi.org/10.1016/S0031-9422(02)00547-2

Byamukama, R., Barbara, G., Namukobe, J., Heydenreich, M., & Kiremire, B. T. (2015). Bioactive compounds in the stem bark of Albizia coriaria (Welw. ex Oliver)*, 9*(April), 1013–1024.

Campoy, S., & Adrio, J. L. (2017). Antifungals. *Biochemical Pharmacology, 133*, 86–96. https://doi.org/10.1016/j.bcp.2016.11.019

Canzani, D., & Aldeek, F. (2017). Penicillin G's function, metabolites, allergy, and resistance. *Journal of Nutrition and Human Health, 01*(01), 28-40. https://doi.org/10.35841/nutrition-human-health.1000105

Cheong, D. H. J., Tan, D. W. S., Wong, F. W. S., & Tran, T. (2020). Anti-malarial drug, artemisinin and its derivatives for the treatment of respiratory diseases. *Pharmacological Research, 158*, 104901. https://doi.org/10.1016/j.phrs.2020.104901

Chumkaew, P., Teerapongpisan, P., & Pechwang, J. (2021). Two New Aporphine Alkaloids from Amoora cucullata and their antibacterial activity. *Chemistry of Natural Compounds, 57*(5), 907–910.

Clardy, J., Fischbach, M. A., & Currie, C. R. (2013). The natural history of antibiotics. *Current Biology, 19*(11), R440. https://doi.org/10.1039/9781849735483-00273

Daniel, T., Liao, W., Zhou, S., & Wong, W. S. F. (2017). Is there a future for andrographolide to be an anti-inflammatory drug? Deciphering its major mechanisms of action. *Biochemical Pharmacology, 139*, 71–81.

Das, B. S., Das, N. C., Swain, S. S., Mukherjee, S., & Bhattacharya, D. (2022). Andrographolide induces anti-SARS-CoV-2 response through host-directed mechanism: An in silico study. *Future Virology, 17*(9), 651–673. https://doi.org/10.2217/fvl-2021-0171

Decembrino, D., Ricklefs, E., Wohlgemuth, S., Girhard, M., Schullehner, K., Jach, G., & Urlacher, V. B. (2020). Assembly of plant enzymes in E. coli for the production of the valuable (−)-Podophyllotoxin precursor (−)-Pluviatolide. *SyntheticBiology, 9*(11), 3091–3103.

Devasahayam, G., Scheld, W. M., & Hoffman, P. S. (2010). Newer antibacterial drugs for a new century. *Expert Opinion on Investigational Drugs, 19*(2), 215–234.

Dever, L. A., et al. (2018). Mechanisms of bacterial multiresistance to antibiotics. *Ecological Genetics, 16*(3), 4–17. https://doi.org/10.17816/ecogenl634-17

Dey, N., Mishra, K., & Dash, A. P. (2011). Andrographolide: A novel antimalarial diterpene lactone compound from Andrographis paniculata and its interaction with curcumin and artesunate. *Journal of Tropical Medicine, 2011*(365645), 1–6. https://doi.org/10.1155/2011/579518

Dhale, D. A., & Markandeya, S. K. (2011). Antimicrobial and phytochemical screening of Plumbago zeylanica Linn. (Plumbaginaceae) Leaf. *Journal of Experimental Sciences, 2*(3), 04–06. file:///C:/Users/Usuario/Downloads/Exp Sci 6028-16279-1-PB[1] (Dhale).pdf.

Ebenezer, I. O. (2011). Preliminary investigations on the ethnomedicinal plants of Akoko Division, South West Nigeria. *Global Journal of Health Science, 3*(2), 84. https://doi.org/10.5539/gjhs.v3n2p84

Ekinci, İ. B., Chłodowska, A., & Olejnik, M. (2023). Ionophore toxicity in animals: A review of clinical and molecular aspects. *International Journal of Molecular Sciences, 24*(2). https://doi.org/10.3390/ijms24021696

Ellman, A. (2015). Cultivation of Artemisia annua in Africa and Asia. *Research Information Limited*, 84–88. https://doi.org/10.1564/21apr08

Endale, M., Ekberg, A., Alao, J. P., Akala, H. M., Ndakala, A., Sunnerhagen, P., Erdélyi, M., & Yenesew, A. (2013). Anthraquinones of the roots of Pentas micrantha. *Molecules, 18*(1), 311–321.

Epifano, F., Fiorito, S., & Genovese, S. (2013). Phytochemistry and pharmacognosy of the genus Psorospermum. *Phytochemistry Reviews, 12*, 673–684.

Erhabor, C. R., Erhabor, J. O., & Mcgaw, L. J. (2019). South African Journal of Botany The potential of South African medicinal plants against microbial bio fi lm and quorum sensing of foodborne pathogens: A review. *South African Journal of Botany, 126*, 214–231. https://doi.org/10.1016/j.sajb.2019.07.024

Eto, B. (2013). Research in clinical phytopharmacology to develop health care in developing countries: State of the art and perspectives. *Phytopharmacology, 4*(2), 149–205.

Eyong, K. O., Kuete, V., & Efferth, T. (2013). Quinones and benzophenones from the medicinal plants of Africa. In *Medicinal plant research in Africa*. Elsevier.

Fanny-Aimee, E., Kouam, A. D. K., Nyobe, J. C. N., Ngah, L., Frese, M., Ndom, J. C., Langat, M. K., Lenta, B. N., Mulholland, D. A., Sewald, N., & Wansi, J. D. (2023). Coumarinolignoid and indole alkaloids from the roots of the hybrid plant citrus × paradisi Macfad (Rutaceae). *Molecules, 28*(3), 2–9.

Fedoung, E. F., Bernard, A., Christine, B., Nyangono, F., Tounkam, M. N., Akono, P., Véronique, N., Nguiamba, P., Marie, D., Preasious, E., Funwi, F., Merlin, G., Kemeuze, V., Jean, D., Nole, S., Sandrine, I., Zra, B., Alexandre, T., Boum, T., et al. (2023). A review of Cameroonian medicinal plants with potentials for the management of the COVID – 19 pandemic. *Advances in Traditional Medicine, 23*(1), 59–84. https://doi.org/10.1007/s13596-021-00567-6

Félicienne, A., Isabelle, S. T., Pascal, T. A., Espérance, M., Rose, K. E., Eugène, A., Pascal, A. D. C., Frédéric, L., Anatole, L., Jean-Marc, A., Maximin, S., & Alphonse, S. (2019). The aqueous extract of the root bark of *Psorospermum febrifugum* Spach effectively corrects anaemia. Experimental study on *Wistar* rats. *Journal of Applied Biosciences, 139*(1), 14137. https://doi.org/10.4314/jab.v139i1.1

Ferdes, M. (2018). Antimicrobial compounds from plants. IAPC-OBP, Zagreb, Croatia. In *Chapter 10* (pp. 245–269).

Frimpong, Y. O., Antwi, C., Agbenyega, O., Oppong, S. K., Anim-Jnr, A. S., Burimuah, V., Folitse, R., & Emikpe, B. O. (2023). Nutrient profiling of ruminant feed resources in Ghana. *Scientific African, 19*. https://doi.org/10.1016/j.sciaf.2022.e01495

Fyhrquist, P., Laakso, I., Garcia Marco, S., Julkunen-Tiitto, R., & Hiltunen, R. (2014). Antimycobacterial activity of ellagitannin and ellagic acid derivate rich crude extracts and fractions of five selected species of Terminalia used for treatment of infectious diseases in African traditional medicine. *South African Journal of Botany, 90*, 1–16. https://doi.org/10.1016/j.sajb.2013.08.018

Gaynes, R. (2017). The discovery of penicillin—New insights after more than 75 years of clinical use. *Emerging Infectious Diseases, 23*(5), 849–853. https://doi.org/10.3201/eid2305.161556

Gbedema, S. Y., Bayor, M. T., Annan, K., & Wright, C. W. (2015). Clerodane diterpenes from Polyalthia longifolia (Sonn) Thw. var. pendula: Potential antimalarial agents for drug resistant Plasmodium falciparum infection. *Journal of Ethnopharmacology, 169*, 176–182. https://doi.org/10.1016/j.jep.2015.04.014

Ghosh, A. K., Weber, I. T., & Mitsuya, H. (2019). Beyond Darunavir: Recent development of next generation HIV-1 protease inhibitors to combat drug resistance. *ChemComm, 2–21*. https://doi.org/10.1039/x0xx00000x

Gumisiriza, H., Sesaazi, C. D., Olet, E. A., & Birungi, G. (2020). Medicinal plants used to treat African diseases by the local communities of Bwambara Sub-county in Rukungiri, Western Uganda. *Journal of Ethnopharmacology, 20*, 1–43.

Gupta, S., Mishra, K. P., & Ganju, L. (2017). Broad-spectrum antiviral properties of andrographolide. *Archives of Virology, 162*(3), 611–623. https://doi.org/10.1007/s00705-016-3166-3

Gurib-Fakim, A. (2006). Medicinal plants: Traditions of yesterday and drugs of tomorrow. *Molecular Aspects of Medicine, 27*(1), 1–93.

Gurjar, S. P., Gupta, A., & Roy, A. (2023). Molecular docking studies of phytocompounds from Artemisia monosperma against ERK2 kinase in lung cancer. *Journal of Molecular Chemistry, 3*(2 SE-Medicinal Chemistry), 591. https://pubs.thesciencein.org/journal/index.php/jmc/article/view/a591

Habtamu, A., Meku, E., & Leshe, S. (2024). Investigation of the preliminary compounds; Assessing the antioxidant and in- vitro antibacterial activities of Inula confertiflora leaf extracts. *Journal of Narural Product Research and Application, 4*(2), 19–31. https://doi.org/10.46325/jnpra.v4i02.75

Herraiz, T., & Guillén, H. (2018). Monoamine oxidase-A inhibition and associated antioxidant activity in plant extracts with potential antidepressant actions. *BioMed Research International*, 1–10. https://doi.org/10.1155/2018/4810394

Hesham, S., Atef, A. E., & Nevein, M. A. (2018). Utility of silver nanoparticles for the analysis of diosmin and rutin in Persicaria salicifolia extract, authentic and pharmaceutical dosage forms monitored with their haemostatic activity. *African Journal of Pharmacy and Pharmacology, 12*(20), 248–262. https://doi.org/10.5897/ajpp2018.4910

Heyman, H. M., & Meyer, J. J. M. (2012). NMR-based metabolomics as a quality control tool for herbal products. *South African Journal of Botany, 82*, 21–32. https://doi.org/10.1016/j.sajb.2012.04.001

Himesh, S., Nanda, S., Singhai, A. K., & Jitender, M. (2012). Radical scavenging activities and natural indicator activity of aqueous and ethanolic extract of Rosa damascena. *International Journal of Pharmacy and Pharmaceutical Sciences, 4*, 581–586.

Holmes, T. (2017). Shifts in African traditional herbal medicine (THM): Relevance for HIV/AIDS, as foremost among new diseases, and impacts of stigma and culture change: A Review. *ResearchGate, May*, 258–280.

Hong, J., Xie, Z., Yang, F., Jiang, L., Jian, T., Wang, S., Guo, Y., & Huang, X. (2022). Erianin suppresses proliferation and migration of cancer cells in a pyruvate carboxylase-dependent manner. *Fitoterapia, 157*(2021), 105136. https://doi.org/10.1016/j.fitote.2022.105136

Huang, T., Liu, Y., & Zhang, C. (2019). Pharmacokinetics and bioavailability enhancement of Baicalin: A review. *European Journal of Drug Metabolism and Pharmacokinetics, 44*(2), 159–168. https://doi.org/10.1007/s13318-018-0509-3

Hui, T., Sonam, T., & Shimizu, K. (2017). The potential of triterpenoids from loquat leaves (Eriobotrya japonica) for prevention and treatment of skin disorder. *International Journal of Molecular Sciences, 18*(5), 1030.

Jadimurthy, R., Jagadish, S., Nayak, S. C., Kumar, S., Mohan, C. D., & Rangappa, K. S. (2023). Phytochemicals as invaluable sources of potent antimicrobial agents to combat antibiotic resistance. *Life, 13*(4), 1–34. https://doi.org/10.3390/life13040948

Jeruto, P., Too, E., Mwamburi, L., & Omari, A. (2015). An inventory of medicinal plants used to treat gynaecological-obstetric-urino-genital disorders in South Nandi Sub County in Kenya quality standards determination of some selected herbal medicinal products in the Kenyan market view project PLANT TAXONOMY. *Journal of Natural Sciences Research, 5*(18), 136–152. https://www.researchgate.net/publication/282696097

Jeyakkumar, P., Zhang, L., Avula, S. R., & Zhou, C. H. (2016). Design, synthesis and biological evaluation of berberine-benzimidazole hybrids as new type of potentially DNA-targeting antimicrobial agents. *European Journal of Medicinal Chemistry, 122*, 205–215. https://doi.org/10.1016/j.ejmech.2016.06.031

Jia, S., Shen, M., Zhang, F., & Xie, J. (2017). Recent advances in momordica charantia: Functional components and biological activities. *International Journal of Molecular Sciences, 18*(12). https://doi.org/10.3390/ijms18122555

Jikam-Sime, G., Mbabi Nyemeck, N., Zintchem, A. A. A., October, N., Missi, M. B., Farooq, R., Khan, K. M., Ngono Bikobo, D. S., Choudhary, M. I., & Pegnyemb, D. E. (2023). Secondary metabolites of the leaves of Tricalysia atherura N. Hallé (Rubiaceae) and their potential antiplasmodial activity. *Natural Product Research, 37*(17), 2830–2840.

Johannes, E., & Litaay, M. (2016). The bioactivity of hexadecanoic acid compound isolated from hydriod Aglaophenia ccupressinaLarmoureoux as antibacterialagentagainst Salmonella typhi. *International Journal of Biological & Medical Research, 6*(2), 5469–5472.

Kakkar, S., & Bais, S. (2014). A review on protocatechuic acid and its pharmacological potential. *ISRN Pharmacology, 2014*, 1–9.

Kamurthy, H., Dontha, S., & Duggi, S. (2014). Phytochemical screening on Pentas lanceolata leaves – isolation of saponin and anthracene glycosides and alkaloids. *American Journal of Ethnomedicine, 1*(4), 206–215.

Karunamoorthi, K., Sabesan, S., Jegajeevanram, K., & Vijayalakshmi, J. (2013). Role of traditional antimalarial plants in the battle against the global malaria burden. *Vector-Borne and Zoonotic Diseases, 13*(8), 521–544. https://doi.org/10.1089/vbz.2011.0946

Kaur, K., Jain, M., Reddy, R. P., & Jain, R. (2010). Quinolines and structurally related heterocycles as antimalarials. *European Journal of Medicinal Chemistry, 45*(8), 3245–3264. https://doi.org/10.1016/j.ejmech.2010.04.011

Kemboi, D., Langat, M. K., Siwe-Noundou, X., Krause, R. W. M., Isaacs, M. L., & Tembu, V. J. (2022). In vitro antibacterial and cytotoxic effects of Euphorbia grandicornis Blanc chemical constituents. *BMC Complementary Medicine and Therapies, 22*(1), 1–10.

Khanh, P. N., Duc, H. V., Huong, T. T., Ha, V. T., Van, D. T., Son, N. T., Kim, Y. H., Viet, D. Q., Cuong, N. M., Giay, C., & Giay, C. (2016). Phenolic compounds from Callistemon citrinus. *Journal of School Health, 54*(2), 190–197. https://doi.org/10.15625/0866-708X/54/2/6469

Khanum, A., Bibi, Y., Khan, I., Mustafa, G., Attia, K. A., Mohammed, A. A., Yang, S. H., & Qayyum, A. (2024). Molecular docking of bioactive compounds extracted and purified from selected medicinal plant species against covid-19 proteins and in vitro evaluation. *Scientific Reports, 14*(1), 1–19. https://doi.org/10.1038/s41598-024-54470-6

Khater, M., Greco, F., & Osborn, H. M. I. (2020). Antiangiogenic activity of flavonoids: A systematic review and meta-analysis. 1–31.

Khumalo, G. P., Van Wyk, B. E., Feng, Y., & Cock, I. E. (2022). A review of the traditional use of southern African medicinal plants for the treatment of inflammation and inflammatory pain. *Journal of Ethnopharmacology, 283*(June), 114436. https://doi.org/10.1016/j.jep.2021.114436

Kipkoech, S., Melly, D. K., Mwema, B. W., Mwachala, G., Musili, P. M., Hu, G., & Wang, Q. (2019). Conservation priorities and distribution patterns of vascular plant species along environmental gradients in Aberdare ranges forest. *PhytoKeys, 131*, 91–113. https://doi.org/10.3897/phytokeys.131.38124

Klopčic, I., & Dolenc, M. S. (2019). Chemicals and drugs formin reactive quinone and quinone-imine metabolites. *Chemical Research in Toxicology, 31*, 1–34.

Komlaga, G., Agyare, C., Dickson, R. A., Mensah, M. L. K., Annan, K., Loiseau, P. M., & Champy, P. (2015). Medicinal plants and finished marketed herbal products used in the treatment of malaria in the Ashanti region, Ghana. *Journal of Ethnopharmacology, 172*, 333–346.

Kong, X., Tang, C., Singh, A., Ahmed, S. A., Al-Hatmi, A. M. S., Chowdhary, A., Nenoff, P., Gräser, Y., Hainsworth, S., Zhan, P., Meis, J. F., Verweij, P. E., Liu, W., & de Hoog, G. S. (2021). Antifungal susceptibility and mutations in the squalene epoxidase gene in dermatophytes of the trichophyton mentagrophytes species complex. *Antimicrobial Agents and Chemotherapy, 65*(8), 10-1128. https://doi.org/10.1128/AAC.00056-21

Krishnaiah, D., Sarbatly, R., & Nithyanandam, R. (2011). A review of the antioxidant potential of medicinal plant species. *Food and Bioproducts Processing, 89*, 217–233.

Kucharska, A. Z., Sokól-Łętowska, A., Oszmiánski, J., Piórecki, N., & Fecka, I. (2017). Iridoids, phenolic compounds and antioxidant activity of edible honeysuckle berries (Lonicera caerulea var. kamtschatica Sevast.). *Molecules, 22*(3), 1–20.

Kuete, V. (2014). Health effects of alkaloids from African medicinal plants. In *Toxicological survey of African medicinal plants*. Elsevier Inc.

Kumar, P., Sharma, A. K., & Shukla, P. (2017). Physiochemical screening of marker Anthraquinones and derivatives from the Pentas Lanceolata (Forss. K.) Deflers Leaves. *European Journal of Pharmaceutical and Medicinal Research, 4*(5), 439–446.

Kumar, G., Singh, D., Tali, J. A., Dheer, D., & Shankar, R. (2020). Andrographolide: Chemical modification and its effect on biological activities. *Bioorganic Chemistry, 95*, 103511. https://doi.org/10.1016/j.bioorg.2019.103511

Kundishora, A., Sithole, S., & Mukanganyama, S. (2020). Determination of the cytotoxic effect of different leaf extracts from Parinari curatellifolia (Chrysobalanaceae). *Journal of Toxicology, 2020*. https://doi.org/10.1155/2020/8831545

Lakhal, H., Kabouche, A., Alabdul Magid, A., Voutquenne-Nazabadioko, L., Harakat, D., & Kabouche, Z. (2014). Triterpenoids from Salvia argentea var. aurasiaca (Pomel) Batt. & Trab. and their chemotaxonomic significance. *Phytochemistry, 102*, 145–151.

Lamorde, M., Tabuti, J. R. S., Obua, C., Kukunda-Byobona, C., Lanyero, H., Byakika-Kibwika, P., Bbosa, G. S., Lubega, A., Ogwal-Okeng, J., Ryan, M., Waako, P. J., & Merry, C. (2010). Medicinal plants used by traditional medicine practitioners for the treatment of HIV/AIDS and related conditions in Uganda. *Journal of Ethnopharmacology, 130*(1), 43–53. https://doi.org/10.1016/j.jep.2010.04.004

Larrosa, F. H., Maune, J. F., Erazzú, L. E., & Camadro, E. L. (2011). In vitro antimicrobial assay of plants used in traditional medicine in Bukoba rural district, Tanzania. *Plant Biology, 4*(4), no-no. https://doi.org/10.1111/j.1438-8677.2011.00470.x

Allen I. Laskin, Bennett, J. W., & Gadd, G. M. (2023). Advances in applied microbiology. https://doi.org/10.1016/s0065-2164(23)00035-7.

Coz, C. J. Le, & Ducombs, G. (2017). Plants and plant products. Springer-Verlag (Springer Berlin Heidelberg). In *Chapter 41* (pp. 751–796).

Le Gall, T., Berchel, M., Davies, L., Mottais, A., Ghanem, R., Fautrel, A., Gill, D., Hyde, S., Lehn, P., Lehn, J. M., Lemiègre, L., Benvegnu, T., Jaffrès, P. A., Pitard, B., & Montier, T. (2022). Aerosol-mediated non-viral lung gene therapy: The potential of aminoglycoside-based cationic liposomes. *Pharmaceutics, 14*, 25. https://doi.org/10.3390/pharmaceutics14010025

Lee, K. C., Chang, H. H., Chung, Y. H., & Lee, T. Y. (2011). Andrographolide acts as an anti-inflammatory agent in LPS-stimulated RAW264.7 macrophages by inhibiting STAT3-mediated suppression of the NF-κB pathway. *Journal of Ethnopharmacology, 135*(3), 678–684. https://doi.org/10.1016/j.jep.2011.03.068

Lombardi, V. R. M., Carrera, I., & Cacabelos, R. (2017). In vitro screening for cytotoxic activity of herbal extracts. *Evidence-Based Complementary and Alternative Medicine, 2017*.

Loo, C. S. N., Lam, N. S. K., Yu, D., Su, X. Z., & Lu, F. (2017). Artemisinin and its derivatives in treating protozoan infections beyond malaria. *Pharmacological Research, 117*, 192–217. https://doi.org/10.1016/j.phrs.2016.11.012

Luc, M. (2015). *A comparison of disc diffusion and microbroth dilution methods for the detection of antibiotic resistant subpopulations in Gram negative bacilli*.

Luthra, S., Rominski, A., & Sander, P. (2018). The role of antibiotic-target-modifying and antibiotic-modifying enzymes in mycobacterium abscessusdrug resistance. *Frontiers in Microbiology, 9*, 2179. https://doi.org/10.3389/fmicb.2018.02179

Mabhiza, D., Chitemerere, T., & Mukanganyama, S. (2016). Antibacterial properties of alkaloid extracts from Callistemon citrinus and Vernonia adoensis against Staphylococcus aureus and Pseudomonas aeruginosa. *International Journal of Medicinal Chemistry, 2016*, 1–7.

Maddox, C. E., Laur, L. M., & Tian, L. (2010). Antibacterial activity of phenolic compounds against the phytopathogen Xylella fastidiosa. *Current Microbiology, 60*(1), 53–58.

Mahomoodally, M. F. (2013). Traditional medicines in Africa: An appraisal of ten potent African medicinal plants. *Evidence-based Complementary and Alternative Medicine, 2013*, 1–14. https://doi.org/10.1155/2013/617459

Malebo, H. M., Wenzler, T., Cal, M., Swaleh, S. M., Omolo, M. O., Hassanali, A., Séquin, U., Häussinger, D., Dalsgaard, P., Hamburger, M., Brun, R., & Ndiege, I. O. (2013). Anti-protozoal activity of aporphine and protoberberine alkaloids from Annickia kummeriae. *BMC Complementary and Alternative Medicine, 13*(48), 1–10.

Manchester, J. I., Dussault, D. D., Rose, J. A., Boriack-Sjodin, P. A., Uria-Nickelsen, M., Ioannidis, G., Bist, S., Fleming, P., & Hull, K. G. (2012). Discovery of a novel azaindole class of antibacterial agents targeting the ATPase domains of DNA gyrase and topoisomerase IV. *Bioorganic and Medicinal Chemistry Letters, 22*(15), 5150–5156. https://doi.org/10.1016/j.bmcl.2012.05.128

Maria, M., Kępa, M., Wojtyczka, R. D., Idzik, D., Zdebik, A., Orlewska, K., & Tomasz, J. W. (2015). Antibacterial activity of protocatechuic acid ethyl ester on. *Molecules, 20*, 13536–13549.

Masi, M., Van Slambrouck, S., Gunawardana, S., van Rensburg, M. J., James, P. C., Mochel, J. G., Heliso, P. S., Albalawi, A. S., Cimmino, A., van Otterlo, W. A. L., Kornienko, A., Green, I. R., & Evidente, A. (2019). Alkaloids isolated from Haemanthus humilis Jacq., an indigenous South African Amaryllidaceae: Anticancer activity of coccinine and montanine. *South African Journal of Botany, 126*, 277–281.

Matotoka, M. M., Ndhlala, A. R., & Masoko, P. (2019). South African Journal of Botany In vitro inhibition of HIV-1 reverse transcriptase and anti-in fl ammatory activities of some herbal concoctions sold in the Limpopo Province. *South African Journal of Botany, 126*, 65–69. https://doi.org/10.1016/j.sajb.2019.07.023

Megantara, S., Levita, J., & Surantaatmadja, S. I. (2015). In silico study of andrographolide as protease inhibitors for antimalarial drug discovery. In *Proceedings of the 3rd international conference on computation for science and technology* (Vol. 5, pp. 36–39). https://doi.org/10.2991/iccst-15.2015.8

Migabo, H., Izere, C., Habyarimana, T., Karemera, U., & Josiane, N. (2023). Phytochemistry analysis and anti-bacterial properties of Rwandese medicinal plants. *NeuroQuantology, 20*(17), 965–978. https://doi.org/10.48047/nq.2022.20.17.nq880124

Mikecz, O., Pica-Ciamarra, U., Felis, A., Nizeyimana, G., Okello, P., & Brunelli, C. (2020). Data on antimicrobial use in livestock: Lessons from Uganda. *One Health, 10*(June), 100165. https://doi.org/10.1016/j.onehlt.2020.100165

Mishra, P., Sohrab, S., & Mishra, S. K. (2021). A review on the phytochemical and pharmacological properties of Hyptis suaveolens (L.) Poit. *Future Journal of Pharmaceutical Sciences, 7*(1), 1–11. https://doi.org/10.1186/s43094-021-00219-1

Moiketsi, B. N., Makale, K. P. P., Rantong, G., Rahube, T. O., & Makhzoum, A. (2023). Potential of selected African medicinal plants as alternative therapeutics against multi-drug-resistant bacteria. *Biomedicine, 11*, 2605. https://doi.org/10.3390/biomedicines11102605

Mokoka, T. A., Peter, X. K., Fouche, G., Moodley, N., Adams, M., Hamburger, M., Kaiser, M., Brun, R., Maharaj, V., & Koorbanally, N. (2014). Antileishmanial activity of 12-methoxycarnosic acid from Salvia repens Burch. ex Benth. (Lamiaceae). *South African Journal of Botany, 90*, 93–95. https://doi.org/10.1016/j.sajb.2013.10.014

Mota, L., Figueiredo, A. C., Pedro, L. G., Barroso, J. G., & Ascensão, L. (2013). *Glandular trichomes, histochemical localization of secretion, and essential oil composition in Plectranthus grandidentatus growing in Portugal†. May.* https://doi.org/10.1002/ffj.3175

Mugisha, M. K., Asiimwe, S., Namutebi, A., Borg-Karlson, A. K., & Kakudidi, E. K. (2014). Ethnobotanical study of indigenous knowledge on medicinal and nutritious plants used to manage opportunistic infections associated with HIV/AIDS in western Uganda. *Journal of Ethnopharmacology, 155*(1), 194–202.

Munoz-Price, L. S., Poirel, L., Bonomo, R. A., Schwaber, M. J., Daikos, G. L., Cormican, M., Cornaglia, G., Garau, J., Gniadkowski, M., Hayden, M. K., Kumarasamy, K., Livermore, D. M., Maya, J. J., Nordmann, P., Patel, J. B., Paterson, D. L., Pitout, J., Villegas, M. V., Wang, H., et al. (2013). Clinical epidemiology of the global expansion of Klebsiella pneumoniae carbapenemases. *The Lancet Infectious Diseases, 13*(9), 785–796.

Murbach-Teles, A. B., Nunes Barbosa, L., Da Silva Probst, I., & Fernandes Júnior, A. (2014). Antimicrobial activity of essential oils. *Journal of Essential Oil Research, 26*(1), 34–40.

Nagawa, C. B., Böhmdorfer, S., & Rosenau, T. (2015). Chemical composition of volatiles extracted from indigenous tree species of Uganda: Composition of bark extracts from Psorospermum febrifugum and Milicia excels. *Holzforschung, 69*(6), 815–821.

Nair, J. J., & Van, S. (2019). Antiplasmodial constituents in the minor alkaloid groups of the Amaryllidaceae. *South African Journal of Botany, 126*, 362–370.

Nalule, A. S., Mbaria, J. M., Olila, D., & Kimenju, J. W. (2011). Ethnopharmacological practices in management of livestock helminthes by pastoral communities in the drylands of Uganda. *Livestock Research for Rural Development, 23*(2).

Nambejja, C., Tugume, P., Nyakoojo, C., & Kamatenesi-mugisha, M. (2019). Medicinal plant species used in the treatment of skin diseases in Katabi Sub- County, Wakiso District, Uganda. *Ethnobotany Research and Applications, 18*(20), 1–17.

Namuga, C., Muwonge, H., Nasifu, K., Sekandi, P., Sekulima, T., & Kirabira, J. B. (2024). Hoslundia opposita vahl; a potential source of bioactive compounds with antioxidant and antibiofilm activity for wound healing. *BMC Complementary Medicine and Therapies, 24*(1), 1–16. https://doi.org/10.1186/s12906-024-04540-z

Namukobe, J., Lutaaya, A., Asiimwe, S., & Byamukama, R. (2021a). An ethnobotanical study of medicinal plants used in the management of dermatological disorders in Buyende and Kayunga Districts, Uganda. *European Journal of Medicinal Plants, 32*(2), 15–40.

Namukobe, J., Sekandi, P., Byamukama, R., Murungi, M., Nambooze, J., Ekyibetenga, Y., Nagawa, C. B., & Asiimwe, S. (2021b). Antibacterial, antioxidant, and sun protection potential of selected ethno medicinal plants used for skin infections in Uganda. *Tropical Medicine and Health, 49*, 1–10.

Namukobe, J., Ekyibetenga, Y., Oyedeji, A. O., Byamukama, R., & Heydenrech, M. (2024). Bioactive phytochemicals from leaf extract of Erlangea tomentosa S. Moore (Bothriocline longipes (Oliv. & Hiern.)) used to treat bacterial infections in Uganda. *Future Journal of Pharmaceutical Sciences, 10*(176), 2–11. https://doi.org/10.1186/s43094-024-00749-4

Nascimento, P. G., Lemos, T., Bizerra, A., Arriaga, Â., Ferreira, D., Santiago, G., Braz-filho, R., & Costa, J. G. (2014). Antibacterial and antioxidant activities of ursolic acid and its derivatives. *Molecules, 19*, 1317–1327.

Nasim, N., Sriram, I., & Sujata, S. (2022). Plant – derived natural products for drug discovery: Current approaches and prospects. *The Nucleus, 65*(3), 399–411. https://doi.org/10.1007/s13237-022-00405-3

Naß, J., & Efferth, T. (2018). The activity of Artemisia spp. and their constituents against Trypanosomiasis. *Phytomedicine, 47*, 184–191. https://doi.org/10.1016/j.phymed.2018.06.002

Nayak, M. K., Agrawal, A. S., Bose, S., Naskar, S., Bhowmick, R., Chakrabarti, S., Sarkar, S., & Chawla-Sarkar, M. (2014). Antiviral activity of baicalin against influenza virus H1N1-pdm09 is due to modulation of NS1-mediated cellular innate immune responses. *Journal of Antimicrobial Chemotherapy, 69*(5), 1298–1310. https://doi.org/10.1093/jac/dkt534

Ndhlala, A. R., Amoo, S. O., Ncube, B., Moyo, M., Nair, J. J., & Van Staden, J. (2013). Antibacterial, antifungal, and antiviral activities of African medicinal plants. *Medicinal Plant Research in Africa: Pharmacology and Chemistry. Elsevier Inc.* https://doi.org/10.1016/B978-0-12-405927-6.00016-3

Nguyen, H. T., Ho, D. V., Vo, H. Q., Le, A. T., Nguyen, H. M., Kodama, T., Ito, T., Morita, H., & Raal, A. (2017). Antibacterial activities of chemical constituents from the aerial parts of Hedyotis pilulifera. *Pharmaceutical Biology, 55*(1), 787–791.

Novilla, M. N. (2018). Ionophores. In *Veterinary toxicology: Basic and clinical principles: Third edition*. Elsevier Inc. https://doi.org/10.1016/B978-0-12-811410-0.00078-7

Nyamukuru, A., Tabuti, J. R. S., Lamorde, M., Kato, B., Sekagya, Y., & Aduma, P. R. (2017). Medicinal plants and traditional treatment practices used in the management of HIV/AIDS clients in Mpigi District, Uganda. *Journal of Herbal Medicine, 7*, 51–58. https://doi.org/10.1016/j.hermed.2016.10.001

O'Donnell, F., Smyth, T. J. P., Ramachandran, V. N., & Smyth, W. F. (2010). A study of the antimicrobial activity of selected synthetic and naturally occurring quinolines. *International Journal of Antimicrobial Agents, 35*(1), 30–38. https://doi.org/10.1016/j.ijantimicag.2009.06.031

Obbo, C. J. D., Mwangi, E. M., Kariuki, S. T., Gathirwa, J. W., Fairhurst, K. J., Burkhard, A. Y., Fidock, D. A., Sadgrove, N., Langat, M. K., Mas-Claret, E., & Mulholland, D. A. (2024). Antiplasmodial limonoids from Khaya anthotheca (Welw.) C.DC. *Phytochemistry Letters, 61*-(October 2023), 1–6. https://doi.org/10.1016/j.phytol.2024.03.007

Ogunwande, I. A., Walker, T. M., Bansal, A., Setzer, W. N., & Essien, E. E. (2010). Essential oil constituents and biological activities of Peristrophe bicalyculata and Borreria verticillata *#. NPC Natural Product Communications, 5*(11).

Okoye, T. C., Uzor, P. F., Onyeto, C. A., & Okereke, E. K. (2014). Safe African medicinal plants for clinical studies. In *Toxicological survey of African medicinal plants*. Elsevier. https://doi.org/10.1016/B978-0-12-800018-2.00018-2

Oloya, B., Namukobe, J., Heydenreich, M., Ssengooba, W., Schmidt, B., & Byamukama, R. (2021). Antimycobacterial activity of the extract and isolated compounds from the stem bark of Zanthoxylum leprieurii Guill. and Perr. *Natural Product Communications, 16*(8), 2–9.

Osafo, T., Philips, T. J., Adomako, A. K., Borquaye, L. S., Ekuadzi, E., Appiah-Opong, R., & Dickson, R. A. (2023). In vitro antileishmanial activity and molecular docking studies of lupeol and monostearin, isolated from Parkia biglobosa. *Scientific African, 19*, 1–8. https://doi.org/10.1016/j.sciaf.2022.e01464

Osorio, A. A., Muñóz, A., Torres-Romero, D., Bedoya, L. M., Perestelo, N. R., Jiménez, I. A., Alcamí, J., & Bazzocchi, I. L. (2012). Olean-18-ene triterpenoids from Celastraceae species inhibit HIV replication targeting NF-kB and Sp1 dependent transcription. *European Journal of Medicinal Chemistry, 52*, 295–303.

Osuntokun, O. T., & Cristina, G. M. (2019). Bio isolation, chemical purification, identification, antimicrobial and synergistic efficacy of extracted essential oils from stem bark extract of Spondias mombin(Linn). *International Journal of Molecular Biology, 4*(4), 135–143.

Othman, L., Sleiman, A., & Abdel-Massih, R. M. (2019). Antimicrobial activity of polyphenols and alkaloids in middle eastern plants. *Frontiers in Microbiology, 10*, 49–56.

Owusu, E., Ahorlu, M. M., Afutu, E., Akumwena, A., & Asare, G. A. (2021). Antimicrobial activity of selected medicinal plants from a sub-Saharan African country against bacterial pathogens from post-operative wound infections. *Medical Sciences (Basel, Switzerland), 9*(2), 1–16. https://doi.org/10.3390/medsci9020023

Pala, N., Esposito, F., Tramontano, E., Singh, P. K., Sanna, V., Carcelli, M., Haigh, L. D., Satta, S., & Sechi, M. (2020). Development of a Raltegravir-based photoaffinity-labeled probe for human immunodeficiency virus-1 integrase capture. *ACS Medicinal Chemistry Letters, 11*(10), 1986–1992. https://doi.org/10.1021/acsmedchemlett.0c00009

Pascaline, J., Charles, M., Lukhoba, C., & George, O. (2011). Phytochemical constituents of some medicinal plants used by the Nandis of south Nandi district, Kenya. *Journal of Animal & Plant Sciences, 9*(3), 1201–1210. http://www.biosciences.elewa.org/JAPS;ISSN2071-7024

Patel, D. K., Kumar, R., Sairam, K., & Hemalatha, S. (2013). Hybanthus enneaspermus (L.) F. Muell: A concise report on its phytopharmacological aspects. *Chinese Journal of Natural Medicines, 11*(3), 199–206. https://doi.org/10.1016/S1875-5364(13)60017-5

Popoola, T. D., Segun, P. A., Ekuadzi, E., Dickson, R. A., Awotona, O. R., Nahar, L., Sarker, S. D., & Fatokun, A. A. (2022). West African medicinal plants and their constituent compounds as treatments for viral infections, including SARS-CoV-2/COVID-19. *DARU, Journal of Pharmaceutical Sciences, 30*(1), 191–210. https://doi.org/10.1007/s40199-022-00437-9

Prasad, R., & Prasad, S. B. (2019). A review on the chemistry and biological properties of Rutin, a promising nutraceutical agent. *Asian Journal of Pharmacy and Pharmacology, 5*, 1–20.

Quinn, L., Gray, S. G., Meaney, S., Finn, S., Kenny, O., & Hayes, M. (2017). Sinapinic and protocatechuic acids found in rapeseed: Isolation, characterisation and potential benefits for human health as functional food ingredients. *Irish Journal of Agricultural and Food Research, 56*(1), 104–119. https://doi.org/10.1515/IJAFR-2017-0012

Radol, A., Kiptoo, M., Makokha, A., & Tolo, F. (2016). Types of herbal medicine used for HIV conditions in Vihiga County, Kenya. *European Journal of Medicinal Plants, 13*(2), 1–23.

Ralambonirina, T. S., Grougnet, R., Michel, S., Rakotobe Guillou, C., & Deguin, B. (2018). Chemical constituents of Anthospermum perrieri (Rubiaceae). *Biochemical Systematics and Ecology, 80*, 29–31.

Reddy, V., Swetha, K., & Avinash, G. (2016). Evaluation of antibacterial and antifungal activity of Rutin and Berberine. *Indo American Journal of Pharmaceutical Research, 6*(08), 6465–6470.

Ribeiro, A., Romeiras, M. M., Tavares, J., & Faria, M. T. (2010). Ethnobotanical survey in Canhane village, district of Massingir, Mozambique: Medicinal plants and traditional knowledge. *Journal of Ethnobiology and Ethnomedicine, 6*. https://doi.org/10.1186/1746-4269-6-33

Rihacek, T., & Roubal, J. (2017). The proportion of integrationists among czech psychotherapists and counselors: A comparison of multiple criteria. *Journal of Psychotherapy Integration, 27*.

Rokni, N., Faridnia, R., Esboei, B. R., Eslami, S., Fakhar, M., Youssefi, M. R., Kalani, H., & Keighobadi, M. (2021). Peganum harmala and Nigella sativa: Anti-leishmanial activity against Leishmania major promastigotes and amastigotes: In vitro and ex vivo experiment. *Annals of Parasitology, 67*(2), 313–319. https://doi.org/10.17420/ap6702.344

Roy, A., & Talukdar, P. (2021). Recent advances in bioactive artificial ionophores. *Chembiochem, 22*(20), 2925–2940. https://doi.org/10.1002/cbic.202100112

Ruela Corrêa, J. C., D'Arcy, D. M., Reis Serra, C. H., & Nunes Salgado, H. R. (2012). Darunavir: A critical review of its properties, use and drug interactions. *Pharmacology, 90*(1–2, 102), –109. https://doi.org/10.1159/000339862

Sacan, O., Akev, N., & Yanardag, R. (2017). In vitro inhibitory effect of Aloe vera (L.) Burm. f. Leaf extracts on the activity of some enzymes and antioxidant activity. *Indian Journal of Biochemistry and Biophysics, 54*(1–2), 82–89.

Samosorn, S. (2016). *AS NOVEL ANTIMICROBIAL AGENTS A thesis submitted in fulfillment of the requirements of the award of from Department of Chemistry University of Wollongong*.

Sanchez Armengol, E., Harmanci, M., & Laffleur, F. (2021). Current strategies to determine antifungal and antimicrobial activity of natural compounds. *Microbiological Research, 252*(July), 126867. https://doi.org/10.1016/j.micres.2021.126867

Sarkar, S., Singh, R. P., & Bhattacharya, G. (2021). Exploring the role of Azadirachta indica (neem) and its active compounds in the regulation of biological pathways: An update on molecular approach. *3 Biotech, 11*(4), 1–12. https://doi.org/10.1007/s13205-021-02745-4

Savoia, D. (2015). Plant derived antimicrobial compounds: Alternatives to Antibiotics. *Future Microbiology*, 7(8), 979–990.

Sekandi, P., Namukobe, J., Byamukama, R., Nagawa, C. B., Barbini, S., Bacher, M., Böhmdorfer, S., & Rosenau, T. (2023). Antimicrobial, antioxidant, and sun protection potential of the isolated compounds from Spermacoce princeae (K. Schum). *BMC Complementary Medicine and Therapies, 23*(201), 1–14.

Sekandi, P., Namukobe, J., Byamukama, R., Nagawa, C. B., Rosenau, T., Langat, L., Mulholland, D., Nagawa, C. B., Bacher, M., Rosenau, T., Langat, L., Mas-claret, E., & Mulholland, D. (2024). The antibacterial, antioxidant and sun protection potential of a benzophenone from Dolichopentas decora (S. Moore). *Pharmacological Research - Natural Products*, 100135. https://doi.org/10.1016/j.prenap.2024.100135

Sekandi, P., Namukobe, J., Byamukama, R., Betty, C., Bacher, M., Rosenau, T., Langat, L., Mas-claret, E., & Mulholland, D. (2025). The antibacterial, antioxidant and sun protection potential of a benzophenone from Dolichopentas decora (S . Moore). *Pharmacological Research - Natural Products, 6*, 100135. https://doi.org/10.1016/j.prenap.2024.100135

Seukep, J. A., Sandjo, L. P., Ngadjui, B. T., & Kuete, V. (2016a). Antibacterial activities of the methanol extracts and compounds from Uapaca togoensis against Gram-negative multi-drug resistant phenotypes. *South African Journal of Botany, 103*, 1–5.

Seukep, J. A., Sandjo, L. P., Ngadjui, B. T., & Kuete, V. (2016b). Antibacterial and antibiotic-resistance modifying activity of the extracts and compounds from Nauclea pobeguinii against Gram-negative multi-drug resistant phenotypes. *BMC Complementary and Alternative Medicine, 16*(1). https://doi.org/10.1186/s12906-016-1173-2

Shakeran, Z., Keyhanfar, M., & Ghanadian, M. (2017). Biotic elicitation for scopolamine production by hairy root cultures of. *Molecular Biology Research Communications, 6*(4), 169–176. https://doi.org/10.22099/mbrc.2017.25776.1275

Shallcross, L. J., & Davies, S. C. (2014). The world health assembly resolution on antimicrobial resistance. *Journal of Antimicrobial Chemotherapy, 69*(11), 2883–2885. https://doi.org/10.1093/jac/dku346

Shanjun, G., Liu, G., Li, J., Chen, J., Li, L., Li, Z., Zhang, X., Zhang, S., Thorne, R. F., & Zhang, S. (2020). Antimicrobial activity of lemongrass essential oil (Cymbopogon flexuosus) and its active component citral against dual-species biofilms of Staphylococcus aureus and Candida species. *Frontiers in Cellular and Infection Microbiology, 10*, 1–14.

Sharma, P. C., Jain, A., Jain, S., Pahwa, R., & Yar, M. S. (2010). Ciprofloxacin: Review on developments in synthetic, analytical, and medicinal aspects. *Journal of Enzyme Inhibition and Medicinal Chemistry, 25*(4), 577–589. https://doi.org/10.3109/14756360903373350

Sharma, A., Rai, P. K., & Prasad, S. (2018). GC–MS detection and determination of major volatile compounds in Brassica juncea L. leaves and seeds. *Microchemical Journal, 138*(2017), 488–493.

Shashank, K., & Abhay, K. P. (2013). Chemistry and biological activities of flavonoids: An overview. *The Scientific World Journal, 2013*, 1–16.

Shehabeldine, A. M., Ashour, R. M., Okba, M. M., & Saber, F. R. (2020). Callistemon citrinus bioactive metabolites as new inhibitors of methicillin- resistant Staphylococcus aureus biofilm formation. *Journal of Ethnopharmacology, 254*(October 2019), 112669. https://doi.org/10.1016/j.jep.2020.112669

Shin, S., Cho, S. H., Park, D., & Jung, E. (2019). Anti-skin aging properties of protocatechuic acid in vitro and in vivo. *Journal of Cosmetic Dermatology, 19*, 977–084.

Shouping, Z., Hu, B., Xu, J., Ren, Q., Wang, Z., Wang, S., Dong, Y., & Yang, G. (2020). Baicalin suppress growth and virulence-related factors of methicillin-resistant Staphylococcus aureus in vitro and vivo. *Microbial Pathogenesis, 139*, 103899.

Sikhakhane, X. (2013). *Evaluating the anticancer and antimicrobial properties of extracts from Hypoxis hemerocallidea (African potato)*. University of Johannesburg (South Africa).

Simoben, C. V., Ntie-Kang, F., Akone, S. H., & Sippl, W. (2018). Compounds from African medicinal plants with activities against selected parasitic diseases: Schistosomiasis, Trypanosomiasis and Leishmaniasis. *Natural Products and Bioprospecting, 8*(3), 151–169.

Simonetti, G., Valletta, A., Kolesova, O., & Pasqua, G. (2018). *Natural products as source of molecules with therapeutic potential*. https://doi.org/10.1007/978-3-030-00545-0

Singh, A. K., Bilal, M., Iqbal, H. M. N., Meyer, A. S., & Raj, A. (2021). Bioremediation of lignin derivatives and phenolics in wastewater with lignin modifying enzymes: Status, opportunities and challenges. *Science of the Total Environment, 777*, 145988. https://doi.org/10.1016/j.scitotenv.2021.145988

Sirama, V. O. (2014). Ethnobotanical, bioactivity and phytochemical evaluation of anthelmintic herbal remedies of Migori county, Kenya. University of Nairobi.

Sowa-borowiec, P., Czernicka, M., & Jarecki, W. (2025). Sweet Clover (Melilotus spp.) as a Source of Biologically Active Compounds. *Molecules, 30*, 526.

Steenhuisen, S. L., Raguso, R. A., Jürgens, A., & Johnson, S. D. (2010). Variation in scent emission among floral parts and inflorescence developmental stages in beetle-pollinated Protea species (Proteaceae). *South African Journal of Botany, 76*(4), 779–787. https://doi.org/10.1016/j.sajb.2010.08.008

Stefano, B., Jaxel, J., Karlström, K., Rosenau, T., & Potthast, A. (2021). Multistage fractionation of pine bark by liquid and supercritical carbon dioxide. *Bioresource Technology, 341*, 125862.

Stirk, W. A., Tarkowská, D., Gruz, J., Strnad, M., Ördög, V., & van Staden, J. (2019). Effect of gibberellins on growth and biochemical constituents in Chlorella minutissima (Trebouxiophyceae). *South African Journal of Botany, 126*, 92–98.

Sweelam, H., Abd-Alla, H. I., Abdelwahab, A. B., Gabr, M. M., & Kirsch, G. (2018). Secondary metabolites and biological activity of Pentas species: A minireview. *Journal of Advanced Research, 10*, 21–30.

Tang, W. Y., Wang, D. P., Tian, Y., Fan, X., Wang, C., Lu, X.-Y., Li, P.-W., Ji, X.-J., & Liu, H.-H. (2021). Metabolic engineering of Yarrowia lipolytica for improving squalene production. *Bioresource Technology, 323*, 124652.

Tang, K. W. K., Millar, B. C., & Moore, J. E. (2023). Antimicrobial Resistance (AMR). *British Journal of Biomedical Science, 80*, 11387. https://doi.org/10.3389/bjbs.2023.11387

Todar, K. (2014). Bacterial mechanisms of antibiotic resistance. *Bacterial Resistance to Antibiotics*, 1–4. https://doi.org/10.1002/9780470720837.ch13

Tshilanda, D. D., Mpiana, P. T., Onyamboko, D. N. V., Mbala, B. M., Ngbolua, K. t. N., Tshibangu, D. S. T., Bokolo, M. K., Taba, K. M., & Kasonga, T. K. (2014). Antisickling activity of butyl stearate isolated from Ocimum basilicum (Lamiaceae). *Asian Pacific Journal of Tropical Biomedicine, 4*(5), 393–398. https://doi.org/10.12980/APJTB.4.2014C1329

Turcheniuk, V., Raks, V., Issa, R., Cooper, I. R., Cragg, P. J., Jijie, R., Dumitrascu, N., Mikhalovska, L. I., Barras, A., Zaitsev, V., Boukherroub, R., & Szunerits, S. (2015). Antimicrobial activity of menthol modified nanodiamond particles. *Diamond and Related Materials, 57*, 2–8. https://doi.org/10.1016/j.diamond.2014.12.002

Tyagi, A. K., & Malik, A. (2011). Antimicrobial potential and chemical composition of Mentha piperita oil in liquid and vapour phase against food spoiling microorganisms. *Food Control, 22*(11), 1707–1714.

Tyagi, B., Dubey, A., Verma, A. K., & Tiwari, S. (2015). Antibacterial activity of phenolics compounds against pathogenic bacteria. *International Journal of Pharmaceutical Sciences Review and Research, 35*(1), 16–18.

Unuigbe, C. A., & Okhankhuele, O. N. (2023). Phytochemical screening, mineral content, antioxidant potential and antibacterial activity of the leaves extract of Alstonia boonei. *ChemSearch Journal, 14*(1), 8–13.

Uwimbabazi, M., Kabonesa, B., Ongarep, S. V., Omujal, F., & Agaba, H. (2023). Ethnobotanical survey of medicinal plants used for the treatment of diabetes in Uganda. *Ethnobotany Research and Applications, 26*. https://doi.org/10.32859/era.26.56.1-14

Valluru, R., & Van-den, W. (2011). Myo-inositol and beyond – Emerging networks under stress. *Plant Science, 181*(4), 387–400.

van Vuuren, S., & Frank, L. (2020). Review: Southern African medicinal plants used as blood purifiers. *Journal of Ethnopharmacology, 249*, 112434. https://doi.org/10.1016/j.jep.2019.112434

Van Vuuren, S., Ramburrun, S., Kamatou, G., & Viljoen, A. (2019). Indigenous South African essential oils as potential antimicrobials to treat foot odour (bromodosis). *South African Journal of Botany, 126*, 354–361. https://doi.org/10.1016/j.sajb.2019.06.019

Van Wyk, B. E. (2020). A family-level floristic inventory and analysis of medicinal plants used in Traditional African Medicine. *Journal of Ethnopharmacology, 249*(May 2019), 112351. https://doi.org/10.1016/j.jep.2019.112351

van Zyl, R. L., Viljoen, A. M., & Jäger, A. K. (2002). In vitro activity of aloe extracts against Plasmodium falciparum. *South African Journal of Botany, 68*(1), 106–110. https://doi.org/10.1016/s0254-6299(16)30465-3

Vanreppelen, G., Wuyts, J., Van Dijck, P., & Vandecruys, P. (2023). Sources of antifungal drugs. *Journal of Fungi, 9*(2), 1–20. https://doi.org/10.3390/jof9020171

Van-Wyk, B. E. (2015). A review of commercially important African medicinal plants. *Journal of Ethnopharmacology, 176*, 118–134.

Veve, M. P., & Wagner, J. L. (2018). Lefamulin: Review of a promising novel pleuromutilin antibiotic. *Pharmacotherapy, 38*(9), 935–946. https://doi.org/10.1002/phar.2166

Wan, Z., & Chen, X. (2014). Triptolide inhibits human immunodeficiency virus type 1 replication by promoting proteasomal degradation of Tat protein. *Retrovirology, 11*(1), 1–13. https://doi.org/10.1186/s12977-014-0088-6

Wang, J., Jiao, H., Meng, J., Qiao, M., Du, H., He, M., Ming, K., Liu, J., Wang, D., & Wu, Y. (2019). Baicalin inhibits biofilm formation and the quorum-sensing system by regulating the MsrA drug efflux pump in Staphylococcus saprophyticus. *Frontiers in Microbiology, 10-*(December), 1–11. https://doi.org/10.3389/fmicb.2019.02800

WHO. (2010). *Guidelines for the treatment of malaria*. World Health Organization.

Won, J. H., Chung, K. S., Park, E. Y., Lee, J. H., Choi, J. H., Tapondjou, L. A., Park, H. J., Nomura, M., Hassan, A. H. E., & Lee, K. T. (2018). 23-Hydroxyursolic acid isolated from the stem bark of Cussonia bancoensis induces apoptosis through fas/caspase-8-dependent pathway in HL-60 human promyelocytic leukemia cells. *Molecules, 23*(12), 2–13.

Xia, S., Ma, L., Wang, G., Yang, J., Zhang, M., Wang, X., Su, J., & Xie, M. (2022). In vitro antimicrobial activity and the mechanism of Berberine against methicillin-resistant Staphylococcus aureus isolated from bloodstream infection patients. *Infection and Drug Resistance, 15*, 1933–1944. https://doi.org/10.2147/IDR.S357077

Xiao, X., Lin, Y., Xu, G., Gong, X., Gu, Y., & Tong, J. (2011). Two new cytotoxic naphthoquinones from Didymocarpus hedyotideus. *Helvetica Chimica, 94*, 404–409.

Xiao-Feng, W., Ren-Jie, W., Qiang-Qiang, F., He, W.-Y., Jiang, K., Qu, S.-J., Tan, J.-J., & Tan, C.-H. (2022). Monoterpene indole alkaloids from Melodinus cochinchinensis (Lour.) Merr. *Fitoterapia, 157*, 105118.

Xu, Z., Tang, Q., Xu, T., Cai, Y., Lei, P., Chen, Y., Zou, W., Dong, C., Lan, K., Wu, S., & Zhou, H. B. (2022). Discovery of aminothiazole derivatives as novel human enterovirus A71 capsid protein inhibitors. *Bioorganic Chemistry, 122*(1), 1–22. https://doi.org/10.1016/j.bioorg.2022.105683

Yazdi, S. E., Prinsloo, G., Heyman, H. M., Oosthuizen, C. B., Klimkait, T., & Meyer, J. J. M. (2019). Anti-HIV-1 activity of quinic acid isolated from Helichrysum mimetes using NMR-based metabolomics and computational analysis. *South African Journal of Botany, 126*, 328–339.

Yazdiniapour, Z., Mirian, M., Zolfaghari, B., Mehdifar, P., Ghanadian, M., & Lanzotti, V. (2022). Myrsinane-type diterpenes from Euphorbia gedrosiaca with cell growth inhibitory activity and apoptotic effects on melanoma cancer cells. *Fitoterapia, 157*(January), 105138.

Ye, L., Zhang, J., Xiao, W., & Liu, S. (2020). Efficacy and mechanism of actions of natural antimicrobial drugs. *Pharmacology and Therapeutics, 216*, 107671.

Yenesew, A., Akala, H. M., Twinomuhwezi, H., Chepkirui, C., Irungu, B. N., Eyase, F. L., Kamatenesi-Mugisha, M., Kiremire, B. T., Johnson, J. D., & Waters, N. C. (2012). The antiplasmodial and radical scavenging activities of flavonoids of Erythrina burttii. *Acta Tropica, 123*(2), 123–127. https://doi.org/10.1016/j.actatropica.2012.04.011

Yuan, L., Zhang, C., Sun, H., Liu, Q., Huang, J., Sheng, L., Lin, B., Wang, J., & Chen, L. (2016). The semi-synthesis of novel andrographolide analogues and anti-influenza virus activity evaluation of their derivatives. *Bioorganic and Medicinal Chemistry Letters, 26*(3), 769–773. https://doi.org/10.1016/j.bmcl.2015.12.100

Zhang, Y., Wang, D., Yang, L., Zhou, D., & Zhang, J. (2014). Purification and characterization of flavonoids from the leaves of Zanthoxylum bungeanum and correlation between their structure and antioxidant activity. *PLoS One, 9*(8). https://doi.org/10.1371/journal.pone.0105725

Zhang, G. F., Liu, X., Zhang, S., Pan, B., & Liu, M. L. (2018). Ciprofloxacin derivatives and their antibacterial activities. *European Journal of Medicinal Chemistry, 146*, 599–612. https://doi.org/10.1016/j.ejmech.2018.01.078

Zhang, J., Yang, Y. N., Feng, Z. M., Yuan, X., Zhang, X., Jiang, J. S., & Zhang, P. C. (2022). The triterpenoids and sesquiterpenoids from the plant of Agrimonia pilosa. *Fitoterapia, 157*, 105104.

Zhang, J.-X., Lv, T. M., Huang, X. X., & Song, S. J. (2024). Phytochemical investigation on the leaves of Viburnum chingii Hsu (Adoxaceae family) and the chemotaxonomic significance. *Biochemical Systematics and Ecology, 112*(December 2023), 104765.

Zhao, Q. Y., Yuan, F. W., Liang, T., Liang, X. C., Luo, Y. R., Jiang, M., Qing, S. Z., & Zhang, W. M. (2018). Baicalin inhibits Escherichia coli isolates in bovine mastitic milk and reduces antimicrobial resistance. *Journal of Dairy Science, 101*(3), 2415–2422. https://doi.org/10.3168/jds.2017-13349

Zhu, S. S., Zhang, Y. F., Ding, M., Zeng, K. W., Tu, P. F., & Jiang, Y. (2022). Anti-neuroinflammatory components from Clausena lenis Drake. *Molecules, 27*, 1971.

Zimmermann, A., Jaber, Q. Z., Koch, J., Riebe, S., Vallet, C., Loza, K., Hayduk, M., Steinbuch, K. B., Knauer, S. K., Fridman, M., & Voskuhl, J. (2021). Luminescent amphiphilic aminoglycoside probes to study transfection. *Chembiochem, 22*(9), 1563–1567. https://doi.org/10.1002/cbic.202000725

Chapter 10
Antiviral Medicinal Plant Natural Products of Africa

Yonatan Alebachew

Abstract Natural products possess immense potential as a source of antiviral drugs. Numerous antiviral natural products have been discovered, many of which are promising antiviral drug candidates currently undergoing various stages of clinical and preclinical investigation. However, the status of antiviral natural products isolated from African traditional medicinal plants remains largely unexplored. This chapter therefore provides a comprehensive review of the chemistry and antiviral potential of various classes of natural products as well as extracts derived from African medicinal plants and discusses their future prospects. A total of 81 antiviral terpenoids, alkaloids, and phenolics, along with 2 anti-HIV proteins isolated from African medicinal plants, are described. Notably, 53% of these compounds are novel and potent antivirals. In addition, there are 101 African medicinal plants belonging to 42 families and 79 genera with extracts and fractions active against various pathogenic viruses. Nevertheless, these natural products are at early stages of antiviral drug development; hence, further studies and approaches needed to enhance their potency and selectivity and to translate them into clinical candidates are discussed. Furthermore, the discovery of novel antivirals from untapped African traditional plant resources using different innovative techniques is briefly highlighted.

Keywords Africa · Antiviral natural products · Medicinal plants · Terpenoids · Alkaloids · Phenols · Anti-HIV activity · Cell-based assays · Enzyme assays · In vivo

Abbreviations

ACV	Acyclovir
AIDS	Acquired immunodeficiency syndrome

Y. Alebachew (✉)
Department of Pharmaceutical Chemistry and Pharmacognosy, School of Pharmacy, College of Health Sciences, Addis Ababa University, Addis Ababa, Ethiopia
e-mail: yonathan.alebachew@aau.edu.et

A. Dekebo (ed.), *Bioactive Secondary Metabolites from Medicinal Plants of Africa*, Interdisciplinary Biotechnological Advances,
https://doi.org/10.1007/978-981-96-8590-5_10

COVID-19	Coronavirus disease
CPE	Cytopathic effect
CVB3	Coxsackievirus B3
DNA	Deoxyribonucleic acid
ELISA	Enzyme-linked immunosorbent assay
GFP	Green fluorescent protein
HBV	Hepatitis B virus
HIV	Human immunodeficiency virus
HSV	Herpes simplex virus
HSV-1	Herpes simplex virus type 1
HSV-2	Herpes simplex virus type 2
LRA	Latency reversal agents
MTT	3-(4,5-dimethylthiazol-2-yl)-2,5-diphenyl tetrazolium bromide
NA	Neuraminidase
PBMC	Peripheral blood mononuclear cells
PKC	Protein kinase C
RdRp	RNA-dependent RNA polymerase
RNA	Ribonucleic acid
RSV	Respiratory syncytial virus
SAR	Structure-activity relationships
SARS-CoV-2	Severe acute respiratory syndrome coronavirus 2
TM	Traditional medicine
XTT	2,3-bis-[2-methoxy-4-nitro-5-sulfophenyl] -2H-tetrazolium-5-carboxanilide inner salt

10.1 Introduction

Viruses are microscopic organisms that contain few proteins and genetic material, all enclosed within a protective protein shell known as a capsid. They are unable to reproduce outside of their host. Viruses carry their genetic material in the form of either ribonucleic acid (RNA) or deoxyribonucleic acid (DNA), which can be single-stranded or double-stranded. Accordingly, they are classified based on the type of nucleic acid content, as well as the size and shape of the protein capsid and the presence of a surrounding lipoprotein envelope. The size and structure of viruses are determined by the amount and organization of their proteins and nucleic acids. Most viruses are either filamentous or spherical (polygonal) in shape; however, larger and more intricate bacteriophages exhibit a blend of both filamentous and polygonal features (Mansfield & King, 1998).

The global prevalence of human viral diseases is estimated to exceed 4.07 billion (Naghavi et al., 2024; WHO, 2024). Some of the most common viral infections include influenza, coronavirus disease (COVID-19), human immunodeficiency virus/acquired immunodeficiency syndrome (HIV/AIDS), hepatitis (A, B, C), measles, polio, and dengue fever. To address these challenges, various mono- and combination-antiviral drug regimens, along with multiple vaccines, are in use. Different classes of antiviral drugs act through distinct mechanisms. Based on their structure and target or function, they can be grouped into several categories: acyclic nucleoside analogs, nucleoside analogs, non-nucleoside analogs, nucleoside reverse transcriptase inhibitors, non-nucleoside reverse transcriptase inhibitors, integrase inhibitors, protease inhibitors, entry inhibitors, neuraminidase inhibitors, immunomodulators, antimitotic inhibitors, and oligonucleotides (Tompa et al., 2021). However, the emergence of new pandemics, such as COVID-19, alongside the complex biochemistry and pathogenesis of viral diseases and issues related to drug resistance, creates an urgent need for the discovery and development of novel antiviral drugs and vaccines, while maintaining continuous public health monitoring and surveillance. In the antiviral drug discovery, natural products, in particular, have played a significant role as lead compounds. Numerous promising antiviral natural products, including those derived from medicinal plants, are currently under different stages of preclinical and clinical investigations (Louis-Félix et al., 2015; Prinsloo et al., 2018; Darshani et al., 2022; González-Cardenete et al., 2022; Sopjani et al., 2024). Consequently, there exists a wealth of excellent reviews and book chapters that discuss the role of phytochemical natural products—their identity, antiviral activity, and development—in antiviral drug discovery and therapy (Asres et al., 2005; Singh et al., 2005; Mukhtar et al., 2008; Liu & Du, 2012; Fekadu et al., 2024). However, the status of African natural products and their antiviral potential remain relatively unexplored in this regard. Therefore, this chapter presents a comprehensive review of the chemistry and antiviral activities of pure natural products and extracts derived from African traditional medicinal plants and discusses their future prospects.

10.2 Review Method

To conduct the comprehensive review, an electronic database (Google Scholar, Science Direct, PubMed, Springer Link) literature search was done using keywords such as the scientific names of the plants or their synonyms, antiviral natural products, antiviral activity, antiviral plant secondary metabolite, antiviral medicinal plants, Africa, and antiviral compounds from African medicinal plants. A total of 118 African medicinal plants were reviewed, and 89 articles from 1991 to 2025 were included. The review solely included antiviral natural products derived from plants collected from Africa; antiviral natural products from plants collected outside of Africa were excluded. Additionally, inactive antiviral natural products from plants collected from Africa were also excluded.

10.3 Antiviral Activities of African Medicinal Plants

10.3.1 Extracts and Fractions

According to the systemic data in Table 10.1, there are 101 African medicinal plants with extracts and fractions active against various pathogenic viruses. The plants belong to 42 families and 79 genera (Fig. 10.1). The Asteraceae family, one of the largest flowering plant families known for their bioactive sesquiterpene lactones, has the highest representation, comprising 22 species (21.35%). This is followed by Euphorbiaceae (6 species, 5.8%), Convolvulaceae (6, 5.8%), Apiaceae (4, 3.88%), Bignoniaceae (4, 3.88%), and Leguminosae (4, 3.88%), along with 36 other families that collectively account for 53 species (51.45%). The majority of the plants (84 species, 80.5%) were collected from a few regions of Africa, mostly from South Africa, Kenya, Ethiopia, Rwanda, Nigeria, Malawi, Tanzania, and Togo, as a part of broad-scale antiviral plant extract screening projects. The plants were selected based on their history of traditional use, previous ethnobotanical and pharmacological reports, strong traditional claims, published traditional medicine (TM) monographs, books, historical manuscripts, and religious TM manuscripts for treating viral diseases or viral-like symptoms. In addition, some of the studies involved ethnobotanical or ethno-medicinal surveys to select the plants.

In terms of plant parts used for antiviral activity, leaves are the most frequently reported parts (38.39%), preceded by stem barks (16.9%), areal parts (16%), and roots (14.28%) (Fig. 10.1). Other plant parts, including root barks (4.4%), twigs, shoots, flowers, bulbs, and seeds, account for the remaining 12.5%. Leaves are commonly investigated due to their higher traditional use and ease of collection throughout the growing seasons. From a bio-conservation perspective, they also pose less ecological concern. Most of these plant parts were extracted with alcoholic, hydroalcoholic, or aqueous solvents by cold maceration in line with traditional preparation procedures and for the reason that these solvents are capable of extracting all polar, medium polar, and non-polar components of the plants. A large proportion of extracts or fractions were tested against the retroviral RNA virus HIV (82, 57.74%) (Fig. 10.1). Other evaluated RNA viruses include echovirus (15, 10.56%), poliovirus (7, 4.93%), rhinovirus type 2 (7, 4.93%), respiratory syncytial virus (RSV) (6 count, 4.22%), influenza A virus (4 count, 2.81%), measles virus (4, 2.81%), severe acute respiratory syndrome coronavirus 2 (SARS-CoV-2) (4, 2.81%), and coxsackievirus B3(CVB3) (2, 1.4%). Meanwhile, the only DNA virus evaluated was herpes simplex virus (HSV), with 11 (10%) count. Apart from *Croton dichogamu* and *Tithonia diversifolia,* further fractionation of the extracts has not been carried out in these antiviral screening studies of medicinal plants. However, novel antiviral pure compounds have been discovered from African medicinal plants, including some listed in Table 10.1, in separate phytochemical studies, which are discussed in the next section.

Cell-based phenotypic assays are the primary in vitro antiviral screening methods (76.85%) utilized for the medicinal plant extracts and fractions. Among them, cell-

Table 10.1 Antiviral activities of extracts and fractions from African medicinal plants

No	Plant species	Family	Plant part	Extract/fraction	Virus	Model	IC_{50} /EC_{50}	Positive control	References
1	*Acacia mellifera*(Vahl)Beuth	Leguminosae	Stem bark	Hot water extract	HIV-1	HIV-1 RT radioactive assay.	$IC_{50} = 3.5$ µg/mL	Forscanet	Kanyara and Njagi (2005)
2	*Acokanthera schimperi*	Apocynaceae	Leaves	80% methanol extract	CVB3	Cell-based colorimetric assay	$IC_{50} = 1.34$ µg/ml	Guanidine hydrochloride, amantadine, phosphonoformic acid	Gebre-Mariam et al. (2006)
						Plaque reduction antiviral assay	$IC_{50} = 1.4$ µg/mL		
3	*Adansonia digitata*	Bombacaceae	Roots Leaves	Methanol extract	Herpes simplex virus	Cell-based colorimetric assay	$IC_{50} = 62.5$ µg/mL	None	Hudson et al. (2000)
4	*Ageratum conyzoides* L.	Asteraceae	Leaves	Methanol extract	Echovirus (E7)	Cell-based MTT assay	$IC_{50} = 0.00208$ µg/mL	None	Ogbole et al. (2018)
					Echovirus (E19)		$IC_{50} = 0.000564$ µg/mL		
5	*Alepidea amatymbica*	Apiaceae	Aerial parts and roots	Aqueous extract	HIV-1 strains	Cell-based colorimetric assay	$EC_{50} = 22$ µg/mL	None	Louvel et al. (2013)
6	*Aloe vera,*	Aloaceae	–	Aqueous extract	HIV-1	Cell-based MTT assay	$EC_{50} = 4.4$ µg/mL	Zidovudine	Muriuki et al. (2017)
7	*Anthocleista grandiflora* Gilg	Loganiaceae	Stem bark	Water extract	RSV	Plaque reduction antiviral assay	$IC_{50} = 13$ µg/mL	None	Mollel et al. (2022)
8	*Artemisia afra*	Asteraceae	Leaves	Aqueous extract	SARS-CoV-2	Plaque reduction antiviral assay	$EC_{50} = 120$ µg/mL	None	Nie et al. (2021)
9	*Artemisia annua*	Asteraceae	Leaves	Aqueous extract	SARS-CoV-2	Plaque reduction antiviral assay	$EC_{50} = 280$ µg/mL	None	Nie et al. (2021)
10	*Aspilia pluriseta*	Asteraceae	Leaves	80% ethanol extract	HIV-1	Cell-based MTT assay	$EC_{50} = 16.13$ µg/mL	None	Cos et al. (2002)

(continued)

Table 10.1 (continued)

No	Plant species	Family	Plant part	Extract/fraction	Virus	Model	IC_{50} /EC_{50}	Positive control	References
11	*Badula insularis*	Myrsinaceae	Leaves	Dichloromethane extract	Rhinovirus type 2	Plaque reduction antiviral assay	$EC_{50} = 25$ μg/mL	None	Beuscher et al. (1994)
12	*Bersama abyssinica* Fresen	Melianthaceae	Leaves and stem bark	Dichloromethane extract	SARS-CoV-2 delta variant	Plaque reduction antiviral assay	$IC_{50} = 50$ μg/mL	Remdesivir	Zekeya et al. (2022)
				Ethyl acetate extract			$IC_{50} = 15.8$ μg/mL		
				Ethanol extract					
				Infusion extract					
				Decoction extract					
			Leaves	Hexane extract	HIV-1	Cell-based MTT assay	$EC_{50} = 2.3$ μg/mL	None	Asres et al., 2001
					HIV-2		$EC_{50} = 2.3$ μg/mL		
				Hexane extract	HIV-1		$EC_{50} = 2.7$ μg/mL		
					HIV-2		$EC_{50} = 2.7$ μg/mL		
			Root bark	Acetone extract	HIV-1		$EC_{50} = 2.8$ μg/mL		
					HIV-2		$EC_{50} = 2.8$ μg/mL		
				Methanol extract	HIV-1		$EC_{50} = 3.1$ μg/mL		
					HIV-2		$EC_{50} = 12$ μg/mL		
13	*Bridelia micrantha*	Euphorbiaceae	Roots	Methanol extract	HIV-1	HIV-1 RT RNA-dependent-DNA polymerase (RDDP) inhibition radiometric assay	$IC_{50} = 10.5$ μg/mL	AZT-TP	Bessong et al. (2005)
				Aqueous extract		HIV-1 RT RNA-dependent-DNA polymerase (RDDP) inhibition radiometric assay	$IC_{50} = 18.5$ μg/mL		

				Methanol extract		RNase H inhibition radiometric assay	$IC_{50} = 10.3$ µg/mL		
				Aqueous extract		RNase H inhibition radiometric assay	$IC_{50} = 11$ µg/mL		
14	*Bryophyllum pinnatum* (lam.) Oken	Crassulaceae	Leaves	Methanol extract	Echovirus (E7)	Cell-based MTT assay	$IC_{50} = 3.13$ µg/mL	None	Ogbole et al. (2018)
					Echovirus (E19)		$IC_{50} = 2.03$ µg/mL		
15	*Bulbine alooides*	Asphodelaceae	Roots	Aqueous extract	HIV-1	HIV-RT colorimetric assay (ELISA)	$IC_{50} = 200$ µg/mL	Nevirapine Ritonavir	Klos et al. (2009)
				Ethanol extract	HIV-1	HIV-1 PR fluorescence assay	$IC_{50} = 94$ µg/mL		
16	*Cassia sieberiana*	Leguminosae	Roots	Dichloromethane/ methanol 1:1 extract	HIV–1	Cell-based flow cytometric assay	$IC_{50} = 84.8$ µg/mL	Indinavir, Efavirenz	Tietjen et al. (2016)
17	*Caucalis incognita*	Apiaceae	Leaves	80% ethanol extract	HIV-1	Cell-based MTT assay	$EC_{50} = 10.1$ µg/mL	None	Cos et al. (2002)
18	*Chironia krebsii*	Gentianaceae	Roots	Dichloromethane extract	Herpes simplex virus	Plaque reduction antiviral assay	$EC_{50} = 12.5$ µg/mL	None	Beuscher et al. (1994)
19	*Clematis hirsuta* Guill. & Perr.	Ranunculaceae	Whole plant	Water extract	RSV	Plaque reduction antiviral assay	$IC_{50} = 59$ µg/mL	None	Mollel et al. (2022)
20	*Combretum molle*	Combretaceae	Roots	Methanol extract	HIV-1	HIV-1 RT RNA-dependent-DNA polymerase (RDDP) inhibition radiometric assay	$IC_{50} = 9.5$ µg/mL	AZT-TP	Bessong et al. (2005)
				Aqueous extract		HIV-1 RT RNA-dependent-DNA polymerase (RDDP) inhibition radiometric assay	$IC_{50} = 37.5$ µg/mL		
				Methanol extract		RNase H inhibition radiometric assay	$IC_{50} = 9.7$ µg/mL		
				Aqueous extract		RNase H inhibition radiometric assay	$IC_{50} = 13.7$ µg/mL		

(continued)

Table 10.1 (continued)

No	Plant species	Family	Plant part	Extract/fraction	Virus	Model	IC_{50} /EC_{50}	Positive control	References
21	*Combretum paniculatum* vent.	Combretaceae	Leaves	Acetone extract	HIV-1	Cell-based MTT assay	EC_{50} = 15 μg/mL	None	Asres et al. (2001)
					HIV-2		EC_{50} = 3 μg/mL		
				Methanol extract	HIV-1		EC_{50} = 5.2 μg/mL		
					HIV-2		EC_{50} = 24.6 μg/mL		
22	*Crassula swaziensis*	Crassulaceae	Aerial parts	Dichloromethane extract	Poliovirus	Plaque reduction antiviral assay	EC_{50} = 25 μg/mL	None	Beuscher et al. (1994)
					Rhinovirus type 2		EC_{50} = 25 μg/mL		
23	*Crinum jagus* (J. Thomps.) Dandy	Amaryllidaceae	Bulb	Methanol extract	Echovirus (E7)	Cell-based MTT assay	IC_{50} = 0.21 μg/mL	None	Ogbole et al. (2018)
					Echovirus (E19)		IC_{50} = 1.94 μg/mL		
24	*Croton dichogamus* Pax	Euphorbiaceae	Twigs	Hexane fraction	HIV-1 IIIB	Cell-based MTT assay	IC_{50} = 0.03 μg/ml	Zidovudine, Abacavir, Nevirapine	Terefe et al. (2022a)
				Dichloromethane fraction			IC_{50} = 0.03 μg/mL		
				Methanol fraction			IC_{50} = 0.06 μg/mL		
25	*Croton megalobotrys* Müll Arg.	Euphorbiaceae	Bark	Dichloromethane/ methanol 1:1 extract	HIV–1	Cell-based flow cytometric assay	IC_{50} = 1.61 μg/mL	Indinavir, Efavirenz	Tietjen et al. (2016)
26	*Dodonaea angustifolia* L.f.	Sapindaceae	Leaves	Petroleum ether extract	HIV-1	Cell-based MTT assay	EC_{50} = 0.6 μg/mL	None	Asres et al. (2001)
					HIV-2		EC_{50} = 0.6 μg/mL		
27	*Dodonaea viscosa* (L.) Jacq.	Sapindaceae	Seeds	Hexane extract	HIV-1	Cell-based MTT assay	EC_{50} = 4.9 μg/mL	None	Asres et al. (2001)
					HIV-2		EC_{50} = 4.9 μg/mL		
				Dichloromethane extract	HIV-1		EC_{50} = 4.7 μg/mL		
					HIV-2		EC_{50} = 4.7 μg/mL		
28	*Dovyalis abyssinica* (Rich.) Warburg	Flacourtiaceae	Leaves	Dichloromethane extract	HIV-1	Cell-based MTT assay	EC_{50} = 3.8 μg/mL	None	Asres et al. (2001)
					HIV-2		EC_{50} = 3.8 μg/mL		
				Acetone extract	HIV-1		EC_{50} = 4.5 μg/mL		
					HIV-2		EC_{50} = 4.5 μg/mL		

29	*Erythrina abyssinica* D.C.	Leguminosae	Stem bark	Water extract	RSV	Plaque reduction antiviral assay	$IC_{50} = 15$ μg/mL	None	Mollel et al. (2022)
					HSV-2		$IC_{50} = 27$ μg/mL		
30	*Euclea schimperi*	Ebenaceae	Leaves	80% methanol extract	CVB3	Cell-based colorimetric assay	$IC_{50} = 33.6$ μg/mL	Guanidine hydrochloride, amantadine, phosphonoformic acid	Gebre-Mariam et al. (2006)
					Influenza A virus		$IC_{50} = 6.22$ μg/mL		
31	*Ficus ovata*	Moraceae	Roots	Methanol extract	Herpes simplex virus	Cell-based colorimetric assay	$IC_{50} = 62.5$ μg/mL	None	Hudson et al. (2000)
			Leaves				$IC_{50} = 62.5$ μg/mL		
32	*Guiera senegalensis*	Combretaceae	Galls	Aqueous acetone extract	Fowlpox virus (FPV)	Plaque reduction antiviral assay	$EC_{50} = 1.1$ and 1.6 μg/mL	None	Lamien et al. (2005)
33	*Gynostemma pentaphyllum*	Cucurbitaceae	Leaves	Ethyl ether extract	HIV-1	Cell-based flow cytometric assay	$IC_{50} = 6.21$ μg/mL	None	Okoye et al. (2012)
				Methanol extract			$IC_{50} = 8.32$ μg/mL		
				Water extract			$IC_{50} = 5.8$ μg/mL		
34	*Helichrysum acutatum*	Asteraceae	Aerial parts	Dichloromethane extract	HIV-1	Cell-based colorimetric assay	$IC_{50} = 4$ μg/mL	Efavirenz	Heyman et al. (2015)
35	*Helichrysum anomalum*	Asteraceae	Aerial parts	Dichloromethane extract	HIV-1	Cell-based colorimetric assay	$IC_{50} = 23$ μg/mL	Efavirenz	Heyman et al. (2015)
36	*Helichrysum appendiculatum*	Asteraceae	Aerial parts	Methanol/water extract	HIV-1	Cell-based colorimetric assay	$IC_{50} = 17$ μg/mL	Efavirenz	Heyman et al. (2015)
37	*Helichrysum aureonitens*	Asteraceae	Shoots	Aqueous extract	HSV-1	Cell-based (cytopathic effect) assay	MIC = 1350 μg/mL	None	Meyer et al. (1996)
			Aerial parts	Dichloromethane extract	HIV-1	Cell-based colorimetric assay	$IC_{50} = 9$ μg/mL	Efavirenz	Heyman et al. (2015)
38	*Helichrysum cephaloideum*	Asteraceae	Aerial parts	Dichloromethane extract	HIV-1	Cell-based colorimetric assay	$IC_{50} = 19$ μg/mL	Efavirenz	Heyman et al. (2015)
39	*Helichrysum chionosphaerum*	Asteraceae	Aerial parts	Dichloromethane extract	HIV-1	Cell-based colorimetric assay	$IC_{50} = 16$ μg/mL	Efavirenz	Heyman et al. (2015)

(continued)

Table 10.1 (continued)

No	Plant species	Family	Plant part	Extract/fraction	Virus	Model	IC_{50} /EC_{50}	Positive control	References
40	*Helichrysum cymosum*	Asteraceae	Aerial parts	Dichloromethane extract	HIV-1	Cell-based colorimetric assay	$IC_{50} = 10$ μg/mL	Efavirenz	Heyman et al. (2015)
41	*Helichrysum melanacme*	Asteraceae	Aerial parts	Dichloromethane extract	HIV-1	Cell-based colorimetric assay	$IC_{50} = 10$ μg/mL	Efavirenz	Heyman et al. (2015)
42	*Helichrysum oxyphyllum*	Asteraceae	Aerial parts	Methanol/water extract	HIV-1	Cell-based colorimetric assay	$IC_{50} = 19$ μg/mL	Efavirenz	Heyman et al. (2015)
43	*Helichrysum pallidum*	Asteraceae	Aerial parts	Dichloromethane extract	HIV-1	Cell-based colorimetric assay	$IC_{50} = 31$ μg/mL	Efavirenz	Heyman et al. (2015)
44	*Helichrysum pilosellum*	Asteraceae	Aerial parts	Dichloromethane extract	HIV-1	Cell-based colorimetric assay	$IC_{50} = 20$ μg/mL	Efavirenz	Heyman et al. (2015)
45	*Helichrysum populifolium*	Asteraceae	Aerial parts	Dichloromethane extract	HIV-1	Cell-based colorimetric assay	$IC_{50} = 33$ μg/mL	Efavirenz	Heyman et al. (2015)
46	*Helichrysum splendidum*	Asteraceae	Aerial parts	Dichloromethane extract	HIV-1	Cell-based colorimetric assay	$IC_{50} = 7$ μg/mL	Efavirenz	Heyman et al. (2015)
47	*Helichrysum umbraculigeru*	Asteraceae	Aerial parts	Dichloromethane extract	HIV-1	Cell-based colorimetric assay	$IC_{50} = 17$ μg/mL	Efavirenz	Heyman et al. (2015)
48	*Heteromorpha trifoliata*	Apiaceae	Root bark	Methanol extract	Poliovirus	Plaque reduction antiviral assay	$EC_{50} = 25$ μg/mL	None	Beuscher et al. (1994)
49	*Hibiscus fuscus*	Malvaceae	Roots	80% ethanol extract	HIV-1	Cell-based MTT assay	$EC_{50} = 1.19$ μg/mL	None	Cos et al. (2002)
50	*Holarrhena pubescens*	Apocynaceae	Stem bark	Dichloromethane extract	Rhinovirus type 2	Plaque reduction antiviral assay	$EC_{50} = 25$ μg/mL	None	Beuscher et al. (1994)
51	*Hoslundia opposita* Vahl.	Lamiaceae	Leaves	Water extract	RSV	Plaque reduction antiviral assay	$IC_{50} = 25$ μg/mL	None	Mollel et al. (2022)
					HSV-2		$IC_{50} = 45$ μg/mL		
52	*Hypericum revolutum*	Clusiaceae	Roots	80% ethanol extract	HIV-1	Cell-based MTT assay	$EC_{50} = 11.9$ μg/mL	None	Cos et al. (2002)

53	*Hypoxis hemerocallidea*	Hypoxidaceae	–	Aqueous extract	HIV-1	Cell-based MTT assay	EC_{50} = 0.002–0.2 μg/mL	Zidovudine	Muriuki et al. (2017)
54	*Hypoxis sobolifera*	Hypoxidaceae	Corms	Aqueous extract	HIV-1	HIV-RT colorimetric assay (ELISA)	IC_{50} = 200 μg/mL	Nevirapine	Klos et al. (2009)
55	*Inula confertiflora*	Asteraceae	Leaves	80% methanol extract	Influenza A virus	Cell-based colorimetric assay	IC_{50} = 6.50 μg/mL	Guanidine hydrochloride, amantadine, phosphonoformic acid	Gebre-Mariam et al. (2006)
56	*Ipomoea asarifolia* (Desr.) Roem. & Schult.	Convolvulaceae	Leaves	Methanol extract	Echovirus (E7)	Cell-based MTT assay	IC_{50} = 0.00252 μg/mL	None	Ogbole et al. (2018)
57	*Justicia secunda* Vahl.	Acanthaceae	Aerial parts	Methanol extract	Herpesvirus type-1 (HHV-1)	Cell-based MTT assay	IC_{50} = 2.16 μg/mL	Acyclovir	Świątek et al. (2023)
58	*Kalanchoe crenata*	Crassulaceae	Stem	80% ethanol extract	HIV-1	Cell-based MTT assay	EC_{50} = 11.1 μg/mL	None	Cos et al. (2002)
59	*Labourdonnaisia glauca*	Sapotaceae	Leaves	Ethanol extract	Poliovirus	Plaque reduction antiviral assay	EC_{50} = 25 μg/mL	None	Beuscher et al. (1994)
60	*Lantana camara* L. cultivars	Verbenaceae	Fresh flowers and leaves	95% ethanol extract	SARS-CoV-2	Plaque reduction antiviral assay	IC_{50} = 3.181 to 15.05 μg/mL	None	Darwish et al. (2022)
61	*Lantana trifolia*	Verbenaceae	Leaves	80% ethanol extract	HIV-1	Cell-based MTT assay	EC_{50} = 4.01 μg/mL	None	Cos et al. (2002)
62	*Leonotis leonurus*	Lamiaceae	Leaves	Aqueous extract	HIV-1	HIV-RT colorimetric assay (ELISA)	IC_{50} = 200 μg/mL	Nevirapine	Klos et al. (2009)
				Ethanol extract	HIV-1	HIV-1 PR fluorescence assay	IC_{50} = 120 μg/mL	Ritonavir	
63	*Lippia multiflora* Moldenke	Verbenaceae	Leaves	Methanol extract	Echovirus (E7)	Cell-based MTT assay	IC_{50} = 1.532 μg/mL	None	Ogbole et al. (2018)

(continued)

Table 10.1 (continued)

No	Plant species	Family	Plant part	Extract/fraction	Virus	Model	IC_{50} /EC_{50}	Positive control	References
64	*Lobostemon trigonus* (Thunb.) H. Buek	Boraginaceae	Leaves	Aqueous extract	HIV	HIV-RT colorimetric assay (ELISA)	IC_{50} = 15.8 μg/mL	None	Harnett et al. (2005)
			Aerial parts	Aqueous extract	HIV-1 subtype A, B, and C	Single-cycle neutralization assay	IC_{50} = 0.10–2 μg/mL		Chawuke et al. (2021)
					HIV-1 subtype B virus—$HIV\text{-}1_{RF}$	Cell-based XTT assay	EC_{50} = 8.9 μg/mL		
					HIV-1 subtype B envelope CAAN5342.12 and HIV-1 subtype C envelope COT6.15	TZM-bl cell neutralization assay	IC_{50} = 1.24 and 2.19 μg/mL	Maraviroc, cyanovirin-N, Griffithsin	
					HIV–1IIIB	HIV-1 neutralization assay in peripheral blood mononuclear cells	IC_{50} = 0.97 μg/mL	Zidovudine, Nevirapine	
					$HIV\text{-}1_{BAL}$	Cell (macrophages)-based assay and XTT assay	EC_{50} = 4.4 μg/mL	Zidovudine	
65	*Macaranga barteri* Mull. Arg.	Euphorbiaceae	Leaves	Methanol extract	Echovirus (E7)	Cell-based MTT assay	IC_{50} = 0.0000276 μg/mL	None	Ogbole et al. (2018)
					Echovirus (E19)		IC_{50} = 0.00000169 μg/mL		
66	*Macaranga kilimandscharica*	Euphorbiaceae	Leaves	80% ethanol extract	HIV-1	Cell-based MTT assay	EC_{50} = 5.23 μg/mL	None	Cos et al. (2002)
			Stem				EC_{50} = 8.2 μg/mL		
67	*Maytenus buchananii* (Loes)Wilczer	Celastraceae	Stem bark	Hot water extract	HIV-1	HIV-1 RT radioactive assay.	IC_{50} = 6 μg/mL	Forscanet	Kanyara and Njagi (2005)
68	*Maytenus senegalensis* (lam.) Excell.	Celastraceae	Roots	Hot water extract	HIV-1	HIV-1 RT radioactive assay.	IC_{50} = 4 μg/mL	Forscanet	Kanyara and Njagi (2005)

69	*Melia azedarach*	Meliaceae	Leaves	Warm water extract	HIV-1	HIV-1 RT radioactive assay.	$IC_{50} = 7$ μg/mL	Forscanet	Kanyara and Njagi (2005)
70	*Mitragyna inermis*	Rubiaceae	Leaves	Dichloromethane extract	Poliovirus	Plaque reduction antiviral assay	$EC_{50} = 25$ μg/mL	None	Beuscher et al. (1994)
					Rhinovirus type 2		$EC_{50} = 25$ μg/mL		
71	*Mondia whitei* (Hook.f.) Skeels	Asclepiadaceae	Leaves	Methanol extract	Echovirus (E7)	Cell-based MTT assay	$IC_{50} = 0.04$ μg/mL	None	Ogbole et al. (2018)
					Echovirus (E19)		$IC_{50} = 0.(0005)35$ μg/mL		
72	*Myrothamnus flabellifolia*	*Myrothamnaceae*	Leaves	Polyphenol fraction	HIV-1	M-MLV RT fluorescence assay	$IC_{50} = 3.24$ μg/mL	None	Kamng'ona et al. (2011)
						HIV-1 RT fluorescence assay	$IC_{50} = 22.04$ μg/mL		
73	*Olinia rochetiana*	Oliniaceae	Stem bark	Methanol extract	Measles virus (MV)	Plaque neutralization assay	$IC_{50} = 19$ μg/mL	None	Parker et al. (2007)
74	*Palisota hirsuta*	Commelinaceae	Roots Leaves	Methanol extract	Herpes simplex virus	Cell-based colorimetric assay	$IC_{50} = 62.5$ μg/mL	None	Hudson et al. (2000)
75	*Peltophorum africanum*	Leguminosae	Roots	Methanol extract	HIV-1	HIV-1 RT RNA-dependent-DNA polymerase (RDDP) inhibition radiometric assay	$IC_{50} = 8$ μg/mL	AZT-TP	Bessong et al. (2005)
				Aqueous extract		HIV-1 RT RNA-dependent-DNA polymerase (RDDP) inhibition radiometric assay	$IC_{50} = 38.3$ μg/mL		
				Methanol extract		RNase H inhibition radiometric assay	$IC_{50} = 13.7$ μg/mL		
			Stem bark	Aqueous extract		RNase H inhibition radiometric assay	$IC_{50} = 49$ μg/mL		
				Methanol extract		HIV-1 RT RNA-dependent-DNA polymerase (RDDP) inhibition radiometric assay	$IC_{50} = 3.5$ μg/mL		

(continued)

Table 10.1 (continued)

No	Plant species	Family	Plant part	Extract/fraction	Virus	Model	IC_{50} /EC_{50}	Positive control	References
				Methanol extract		RNase H inhibition radio-metric assay	$IC_{50} = 10.6$ µg/ml		
76	*Pittosporum viridiflorum*	Pittosporaceae	Leaves	Methanol extract	Influenza A virus (IAV)	Cell-based MTT assay	$EC_{50} = 3.6$ µg/mL	None	Mehrbod et al. (2018)
				Ethanol extract (100%)			$EC_{50} = 3.4$ µg/mL		
				Ethanol extract (30%)			$EC_{50} = 19.2$ µg/mL		
				Acetone extract			$EC_{50} = 82.3$ µg/mL		
77	*Polygala virgata*	Polygalaceae		Dichloromethane extract	Rhinovirus type 2	Plaque reduction antiviral assay	$EC_{50} = 25$ µg/mL	None	Beuscher et al. (1994)
78	*Prunus africana* (Hook.f.) Kalkm.	Rosaceae	Stem bark	Hot water extract	HIV-1	HIV-1 RT radioactive assay.	$IC_{50} = 20$ µg/mL	Forscanet	Kanyara and Njagi (2005)
79	*Psorospermum febrifugum* Spach	Clusiaceae	Leaves	Water extract	RSV	Plaque reduction antiviral assay	$IC_{50} = 21.5$ µg/mL	None	Mollel et al. (2022)
80	*Pterocarpus angolensis*	Fabaceae	Stem bark	Ethanol extract	Rhinovirus type 2	Plaque reduction antiviral assay	$EC_{50} = 25$ µg/mL	None	Beuscher et al. (1994)
81	*Rapanea melanophloeos*	Myrsinaceae	Leaves	Methanol extract	Influenza A virus (IAV)	Cell-based MTT assay	$EC_{50} = 113.3$ µg/mL	None	Mehrbod et al. (2018)
82	*Rhus natalensis* (Kraus)	Anacardiaceae	Leaves	Warm water extract	HIV-1	HIV-1 RT radioactive assay.	$IC_{50} = 5.5$ µg/mL	Forscanet	Kanyara and Njagi (2005)
83	*Securidaca Iongepedunculata*	*Polygalaceae*	Roots	Dichloromethane extract	Poliovirus	Plaque reduction antiviral assay	$EC_{50} = 10$ µg/mL	None	Beuscher et al. (1994)
				Methanol extract	Poliovirus		$EC_{50} = 10$ µg/mL		
				Methanol extract	Herpes simplex Virus		$EC_{50} = 25$ µg/mL		
				Ethanol extract	Herpes simplex virus		$EC_{50} = 6$ µg/mL		

84	*Securidaca longepedunculata* Fresen	Polygalaceae	Root bark	Hexane extract	HIV-1	Cell-based MTT assay	$EC_{50} = 8.3$ µg/mL	None	Asres et al. (2001)
					HIV-2		$EC_{50} = 27.1$ µg/mL		
85	*Spathodea campanulata*	Bignoniaceae	Stem bark	Decoction polar fractions	HIV-1	Cell-based MTT assay	$EC_{50} = 6.16$ µg/mL	Zidovudine	Niyonzima et al. (1999)
86	*Spondias mombin* L.	Anacardiaceae	Bark	Methanol extract	Echovirus (E7)	Cell-based MTT assay	$IC_{50} = 0.1$ µg/mL	None	Ogbole et al. (2018)
87	*Steganotaenia araliacea*	Apiaceae	Roots	Ethanol extract	Rhinovirus type 2	Plaque reduction antiviral assay	$EC_{50} = 10$ µg/mL	None	Beuscher et al. (1994)
88	*Stephania abyssinica* (Quart. Dill.ex Rich.) Walp.	Menispermaceae	Leaves	Alkaloid extract	HIV-1	Cell-based MTT assay	$EC_{50} = 12.7$ µg/mL	None	Asres et al. (2001)
			Twigs	Alkaloid extract	HIV-2		$EC_{50} = 12.7$ µg/mL		
					HIV-1		$EC_{50} = 12.5$ µg/mL		
					HIV-2		$EC_{50} = 12.5$ µg/mL		
89	*Terminalia ivorensis* A. Chev.	Combretaceae	Bark	Methanol extract	Echovirus (E7)	Cell-based MTT assay	$IC_{50} = 0.01$ µg/mL	None	Ogbole et al. (2018)
90	*Tetracera alnifolia* Willd.	Dilleniaceae	Leaves	Methanol extract	Echovirus (E7)	Cell-based MTT assay	$IC_{50} = 1.61$ µg/mL	None	Ogbole et al. (2018)
91	*Tithonia diversifolia*	Asteraceae	Leaves	80% ethanol extract	HIV-1	Cell-based MTT assay	$EC_{50} = 1.60$ µg/mL	None	Cos et al. (2002)
				Petroleum ether fraction			$EC_{50} = 18.3$ µg/mL		
				Ethylacetate fraction			$EC_{50} = 0.2$ µg/mL		
				Water fraction			$EC_{50} = 0.04$ µg/mL		
92	*Tragia furialis*	Euphorbiaceae	Leaves, roots	Water extract	RSV	Plaque reduction antiviral assay	$IC_{50} = 14$ µg/mL	None	Mollel et al. (2022)
					HSV-2		$IC_{50} = 19$ µg/mL		
93	*Triumfetta rhomboïdea*	Tiliaceae	Leaves	80% ethanol extract	HIV-1	Cell-based MTT assay	$EC_{50} = 0.03$ µg/mL	None	Cos et al. (2002)

(continued)

Table 10.1 (continued)

No	Plant species	Family	Plant part	Extract/fraction	Virus	Model	IC_{50} /EC_{50}	Positive control	References
94	*Uvaria chamae*	Annonaceae	Root bark	Dimethyl sulfoxide (DMSO) extract	Measles virus	Cell-based MTT assay	$IC_{50} = 2.5$ μg/mL	None	Oluremi and Adeniji (2015)
			Stem bark				$IC_{50} = 2.5$ μg/mL		
95	*Vernonia jugalis*(-Olive& Hiern)	Asteraceae	Leaves	Warm water extract	HIV-1	HIV-1 RT radioactive assay.	$IC_{50} = 18$ μg/mL	Forscanet	Kanyara and Njagi (2005)
96	*Vernonia miombicola*	Asteraceae	Leaves	80% ethanol extract	HIV-1	Cell-based MTT assay	$EC_{50} = 14.62$ μg/mL	None	Cos et al. (2002)
97	*Vitex doniana*	Lamiaceae	Roots	Dichloromethane/ methanol 1:1 extract	HIV–1	Cell-based flow cytometric assay	$IC_{50} = 62.5$ μg/mL	Indinavir, Efavirenz	Tietjen et al. (2016)
98	*Warburgia ugandensis*	Canellaceae	Stem bark	Methanol extract	Measles virus (MV)	Plaque neutralization assay	$IC_{50} = 19$ μg/mL	None	Parker et al. (2007)
99	*Ximenia americana* L.	Olacaceae	Stem bark	Hexane extract	HIV-1	Cell-based MTT assay	$EC_{50} = 8.3$ μg/mL	None	Asres et al. (2001)
					HIV-2		$EC_{50} = 27.1$ μg/mL		
100	*Xylopia aethiopica*	Annonaceae	Leaves	Dimethyl sulfoxide (DMSO) extract	Measles virus	Cell-based MTT assay	$IC_{50} = 2.5$ μg/mL	None	Oluremi and Adeniji (2015)
101	*Zanha africana*	Sapindaceae	Root bark	Dichloromethane extract	Poliovirus	Plaque reduction antiviral assay	$EC_{50} = 25$ μg/mL	None	Beuscher et al. (1994)

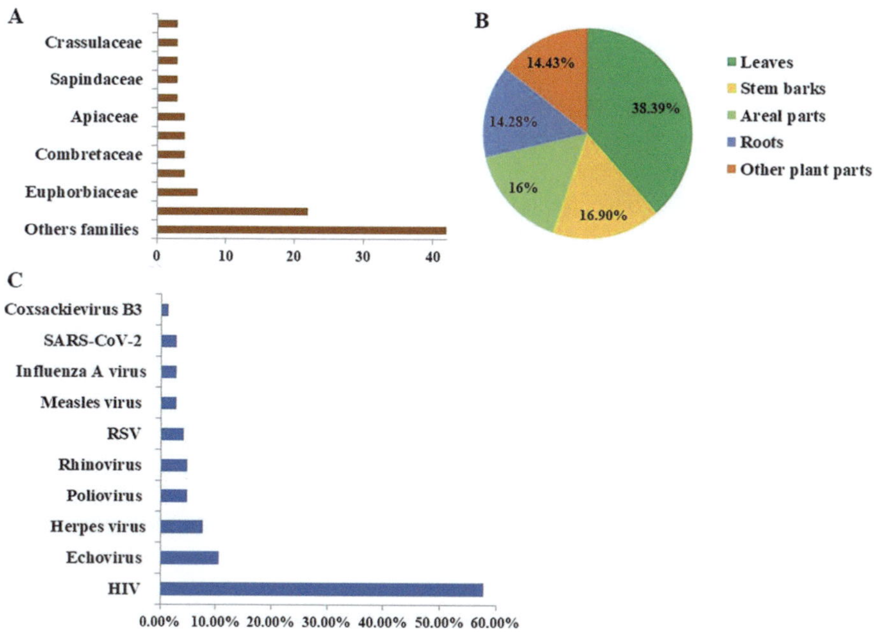

Fig. 10.1 Antiviral activities of African medicinal plants. (**a**). Plant families with active extracts and fractions. (**b**). Plant parts extracted for antiviral activities. (**c**). Viruses or viral components tested

based MTT (3-(4,5-dimethylthiazol-2-yl)-2,5-diphenyl tetrazolium bromide) assay is highly reputed (30.57%). Various cell lines—such as Vero cells, CEM-GXR25 cells, MT-4 cells, HeLa Ohio (human cervix carcinoma) cells, GMK (green monkey kidney) cells, and human rhabdomyosarcoma cells—are employed for these assays. Generally, the cell lines are infected with the viral strains and incubated with test substances and standard drugs at 37 °C for 2 to 5 days or until negative control shows the maximum cytopathic effect (CPE), depending on the method used. After that, cell viability, CPE, or cell infection signals are measured using different techniques. In the MTT assay, cell viability (and hence antiviral activity) is determined by spectrophotometric measurement of the reduction of yellow-colored MTT to blue formazan by mitochondrial dehydrogenase in metabolically active cells. In other words, the antiviral activities of the extracts or fractions are calculated or determined by comparing the absorbance of the treated cells (with the test substance) to that of the negative controls (untreated-infected and untreated-uninfected cells). Likewise, in the XTT assay, the metabolic reduction of 2,3-bis-[2-methoxy-4-nitro-5-sulfophenyl]-2H-tetrazolium-5-carboxanilide inner salt (XTT)-tetrazolium into a water-soluble dark-colored formazan in viable cells is spectrophotometrically measured. Moreover, enzyme-linked immunosorbent assay (ELISA) technique is employed to measure cell viability in cell-based colorimetric assays, neutralization assays, and macrophage-based assays. In neutralization or macrophage-based assays, ELISA measures the amount of p24 antigen in infected cells. Furthermore, in a cell-based flow cytometric assay, the number of infected cells expressing the green fluorescent protein (GFP) is determined by the flow cytometer; in a single-cycle neutralization assay, infected cells are detected by luminescence; in the plaque reduction assay, the number of infected cells (plaques) is counted manually to determine their viral concentration. On the other hand, enzyme (molecular) assays account for 23.15% of the in vitro anti-viral screening methods. The enzymatic activity of the extracts/fractions is detected by incorporating radioactive or fluorescent substrates and measuring their radioactivity or spectroscopic absorbance.

Although almost all the plant extracts and fractions were tested in vitro, five plant extracts—*Carissa edulis*, *Chrysanthemum cinerariaefolium, Justicia schimperiana*, *Phytolacca dodecandra*, and *Croton macrostachyus*—have been evaluated in vivo against herpes simplex and rabies viruses. Tolo et al. (2006) investigated the in vitro and in vivo anti-HSV activity of the hot water extract of a Kenyan medicinal plant *C. edulis*. The extract was tested against both wild-type (HSV-1 and HSV-2) and resistant strains of HSV (APr HSV-1 and TK − HSV-1). In the plaque reduction assay, it inhibited all HSV strains—HSV-1 (EC_{50}, 15.1 μg/mL), HSV-2 (EC_{50}, 6.9 μg/mL), APr HSV-1 (EC_{50}, 8.1 μg/mL), and TK − HSV-1 (EC_{50}, 11.1 μg/mL)—with wide selectivity (therapeutic) index (SI, > 50). More importantly, the extract exhibited in vivo anti-HSV-1 effect at 250 mg/kg dose. In HSV-1-infected mice, it significantly delayed the onset of infection, prolonged the mean survival time (>6 days), and had a lower mortality rate (30%) compared to the negative control, which had a 5-day mean survival time and 100% mortality rate. In comparison, the standard drug acyclovir (ACV) showed better efficacy with a 10% mortality rate. Interestingly, the extract showed an equal survival time to ACV and an even lower mortality rate (20%) in the HSV-2-infected mice model compared to ACV (40%). It was also found to be active in APr HSV-1-infected mice. Similarly, the methanol and aqueous extracts of another Kenyan plant, *C. cinerariaefolium,* displayed in vivo anti-HSV-1 activity, significantly delaying the onset of local lesions and progression of infection in HSV-1-infected mice (Simon et al., 2016). Moreover, Zewde et al. (2019) reported the in vivo anti-rabies activities of the 80% ethanol extracts from the leaves of *J. schimperiana*, the roots of *P. dodecandra*, and the root barks of *C. macrostachyus*, as well as their combination. The extracts and their combination at doses of 5000 mg/kg and 2000 mg/kg significantly prolonged the mean survival time compared to untreated groups.

As the only mechanistic study of the plants, Darwish et al. (2022) attempted to elucidate the in vitro anti-COVID-19 (viral plaque inhibitory activity) mechanism of action of the extracts of *L. camara* cultivars. The mechanism was found to be inhibition of the expression of the RNA-dependent RNA polymerase (RdRp) gene. Thus, the extracts exhibited % downregulation of the RdRp gene in the range of 87.48 to 99.99%. In addition, eight plants—including *Artemisia afra, Adansonia digitata, Elaeodendron croceum, Elaeodendron ransvaalense,* and *Euclea natalensis*—displayed in vitro antiviral effects against the rift valley fever virus (RVFV), a mosquito-borne virus endemic to sub-Saharan Africa for which there is currently no effective antiviral treatment (More et al., 2021). Moreover, the most potent extracts or fractions (in vitro), with IC_{50} values in the range of 0.00169 ng/mL to 0.044 μg/mL, are from the following list of plants: *Acokanthera schimperi, Ageratum conyzoides, Crinum jagus, Croton dichogamus, Croton megalobotrys, Dodonaea angustifolia, Guiera senegalensis, Hibiscus fuscus, Hypoxis hemerocallidea, Ipomoea asarifolia, Macaranga barteri, Mondia whitei, Spondias mombin, Terminalia ivorensis, Tetracera alnifolia, Tithonia diversifolia,* and *Triumfetta rhomboïdea.*

10.3.2 Pure Natural Products

10.3.2.1 Terpenoids

Currently, more than 30 antiviral terpenoids have been discovered from African traditional medicinal plants. Terpenoids are a large class of natural products biosynthetically derived from the 5-carbon isoprene units, isopentenyl diphosphate (IPP) and dimethylallyl diphosphate (DMAPP). They are structurally diverse and generally classified as hemiterpenoids, monoterpenoids, sesquiterpenoids, diterpenoids, triterpenoids, and tetraterpenoids based on the number of isoprene units and the functional groups they possess. Most of the antiviral terpenoids isolated from African medicinal plants are triterpenoids (Fig. 10.2),

Some examples of terpenoids include carvacrol (**1**), an aromatic monoterpenoid isolated from a Tunisian *Thymus capitatus,* which showed anti-HSV activity. Carvacrol, along with two other compounds, was isolated from the ethanol extract of the plant. This is because among the three extracts of *T. capitatus* tested, the ethanol extract had the highest activity and chemo-selectivity in the plaque reduction assay against herpes simplex virus type 1 and 2 (HSV-1 and HSV-2) with an EC_{50} (and SI) value of 16.6 μg/mL (3.5) and 2.3 μg/mL (206.8) μg/mL, respectively. The compound displayed moderate activity against HSV-2 (EC_{50} = 51.9 μg/mL, SI = 12.1) and an acyclovir-resistant HSV-2 strain (HSV-2 ACV-r) (EC_{50} = 82.2 μg/mL, SI = 7.6). In addition, it was shown to have a virucidal effect at 90% effective concentration in the virus inactivation assay Toujani et al. (2018).

Even more potent novel antiviral sesquiterpenoids and diterpenoids were isolated from a couple of African medicinal plants. For example, three new anti-HIV eudesmane-type sesquiterpene lactones named ermiasolide A, B, and C (**2–4**) were reported from the stem bark of *Corton megalocarpus*. In a cell-based MTT assay, the compounds showed potent and profoundly selective antiviral activity against the HIV-1IIIB strain. Notably, all three had equal or greater efficacy compared to the clinically used anti-HIV drugs zidovudine, tenofovir, abacavir, and nevirapine. Ermiasolide A exhibited the highest activity with an IC_{50} value of 0.002 μg/mL and a viral inhibition rate of 93.4%, while ermiasolide B and C inhibited viral replication with IC_{50} values of 0.002 and 0.044 μg/mL, respectively (Terefe et al., 2022b).

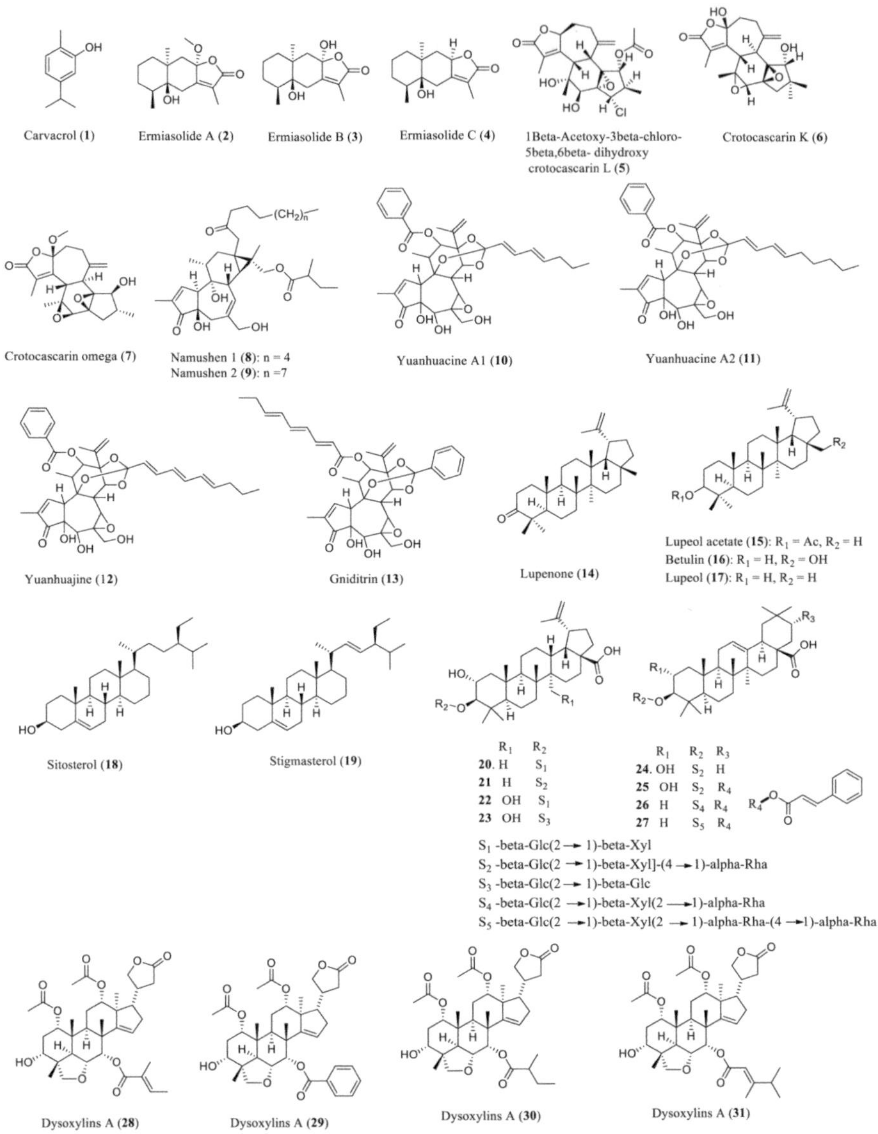

Fig. 10.2 Antiviral terpenoids isolated from African medicinal plants

Similarly, species of *C. megalocarpus, Croton dichogamus, Croton megalobotrys, Gnidia sericocephala,* and *Suregada zanzibariensis* were reported to have novel antiviral diterpenoids. Accordingly, novel crotofolane-type anti-HIV diterpenoids, 1β-Acetoxy-3β-chloro-5α,6α-dihydroxycrotocascarin L (**5**), crotocascarin K (**6**), and crotocascarin *ω* (**7**) were isolated from the stem bark of *C. megalocarpus* and twigs of *C. dichogamus*, respectively (Terefe et al., 2022d,

2023). In MT-4 cell-based MTT assay, the compounds exhibited potent and selective antiretroviral activity against HIV-1IIIB replication, with an IC_{50} value of 0.012 μg/mL for compound **5** and 0.002 μg/mL for compounds **6** and **7**. They had better potency and maximum antiviral effect than the clinically used anti-HIV drugs tenofovir, abacavir, and nevirapine. In addition, enzymatic assay and molecular docking analysis of the compounds showed that they inhibit HIV-1 protease (IC_{50}, 10 μg/mL) and interact with key amino acid residues in the enzyme's binding site. Moreover, diterpenoid phorbol esters, namushen 1 (**8**) and 2 (**9**), were identified as promising HIV latency reversal agents (LRA) following a case study-based evaluation of a traditional three-step medicinal plant regimen used for HIV/AIDS management in Botswana. The three-step medicinal plant regimen comprises *Cassia sieberiana* root, *Vitex doniana* root, and *C. megalobotrys* bark. The crude extracts of the three plants were prepared guided by a focus group discussion with HIV/AIDS patients who successfully utilized the treatment regimen. To prove the efficacy of these extracts, the researchers first tested their anti-HIV activity in a cell-based cytoflometric assay and found that the *C. megalobotrys* extract was approximately 500- and 1700-fold more effective than *V. doniana* and *C. sieberiana* extracts, respectively. In the HIV latency reversal assay as well, only *C. megalobotrys* extract displayed activity inducing HIV proviral GFP reporter gene expression through activation of protein kinase C (PKC) signaling. Its effectiveness and toxicity were comparable to the known LRA, prostratin. Lastly, the phorbol esters namushen 1 (**8**) and 2 (**9**) were found to be the major active components of the extract. Both compounds exhibit more potent HIV latency reversal activity than the extract. They activated GFP expression at lower concentrations (0.03 μg/mL) and are also PKC activators (Tietjen et al., 2018). Furthermore, the roots of *Gnidia sericocephala*, a traditional medicinal plant in South Africa, gave daphnane-type antiviral diterpenoids, namely, yuanhuacine A**1** (**10**), yuanhuacine A**2** (**11**), yuanhuajine (**12**), and gniditrin (**13**). The compounds suppressed HIV replication and, to varying degrees, reversed its latency in vitro (Tembeni et al., 2022). The anti-HIV screening was carried out using a dual enhancement of cell infection to phenotypic resistance assay (deCIPhR), and the HIV latency reversal activity was done by applying J-Lat T cell-based cytofluorometric assay for GFP expression. Yuanhuacine A**1** (**10**) and yuanhuacine A**2** (**11**) (EC_{50}, 0.08 μg/mL) showed more potent inhibition of HIV replication compared to yuanhuajine (**12**) and gniditrin (**13**) (EC_{50}, 4 μg/mL). More interestingly, **11** and **12** exhibited latency reversal activities of 65.5% and 58.2% at 0.15 μM, respectively, which is 16 times higher than prostratin PKC activator positive control. Moreover, they acted via the PKC pathway and were found to be less toxic than prostratin.

Triterpenoids are isoprenoids containing 30 carbons (C_{30}). They are biosynthetically derived from squalene, a precursor molecule formed by tail-to-tail joining of two farnesyl PP (C_{15}) molecules. Typically, they occur as alcohol, esters, or glycosides. Thus, six anti-HIV triterpenoid alcohols and esters—lupenone (**14**), lupeol acetate (**15**), betulin (**16**), lupeol (**17**), sitosterol (**18**), and stigmasterol (**19**)—were isolated from the leaves of *Croton macrostachyus*, a traditional Kenyan plant known for its use in managing HIV symptoms through leaf and root decoctions (Terefe et al., 2022c). In MT-4 cell-based MTT assay, compounds (**14–16**) exhibited strong anti-HIV activity, with an IC_{50} of 0.002 μg/mL, inhibiting virus-induced CPE. Compounds (**17–19**) also showed pronounced activity, with IC_{50} values of 0.047 μg g/mL, 5.58 μg g/mL, and 0.14 μg/mL, respectively. Moreover, all the compounds except **18** had a better effectiveness and cytotoxicity safety profile than standard antiretroviral drugs (zidovudine, tenofovir, abacavir, and nevirapine). Compound **18**, isolated from *T. capitatus,* was found to possess strong anti-HSV-2 activity ($EC_{50,}$ 2.7 μg/mL) (Toujani et al., 2018).

In contrast to the above triterpenoids, Mair et al. (2018) discovered new polar antiviral triterpenoid glycoside saponins (**20–27**) from the ethanolic extract of the stem bark of *Burkea africana*. The compounds were isolated by polyamide column fractionation of the ethanolic extract, specifically targeting the most active tannin-free triterpenoid saponin fractions. Compounds **20** to **23** are lupane-type pentacyclic triterpene saponins, while compounds **24** to **27** are oleanane-type triterpene saponins. In cell-based assays, compounds **21**, **26**, and **27** showed the highest anti-influenza virus activity against the HK/68 and/or Jena/8178 viral strains, with IC_{50} values of 1.1 μM and 1.9 μM, 0.05 μM and 0.27 μM, or 0.17 μM and 0.16 μM, respectively. On the other hand, compounds **20**, **22–25** exhibit weak anti-influenza virus activity with IC_{50} values ranging from 1.7 to 43 μM (Mair et al., 2018).

Furthermore, Chen et al. (2007) identified four new classes of antiviral triterpenoids called dysoxylins A-D (**28–31**) from the bark of *Dysoxylum gaudichaudianum*, a Papua New Guinea medicinal plant. Dysoxylins are tetranortriterpenoids—modified triterpenoids in which four terminal carbons are removed from the side chain of their triterpenoid skeleton during biosynthesis. Compounds **28**, **29**, and **31** displayed potent and selective anti-RSV activity that was even higher than standard drug ribavirin (EC_{50}, 3.6 μg/mL) in cell-based screening. The compounds (**28**, **29**, and **31**) had EC_{50} values of 2, 4, and 1 μg/mL, respectively. Additionally, all the compounds **28–31** were also found to be active in plaque reduction assays, with EC_{50} values in the range of 40 to 150 μg/mL (Chen et al., 2007).

10.3.2.2 Phenols

Phenols are a large class of secondary metabolites found ubiquitously in plants. Structurally, they are diverse, characterized by aromatic moieties that can range from simple aromatic rings to complex polymers. These compounds are also present as glycosides. The biosynthesis of phenols occurs through the shikimate pathway, the acetate pathway, or a combination of both. Known for their numerous biological activities, phenols possess antioxidant, anticancer, antimicrobial, and anti-inflammatory properties, among others. Notably, the antiviral phenolic classes reported from traditional African medicinal plants include simple phenols, lignans, anthraquinones, benzophenones (and their glycosides), xanthones, flavonoids (and their glycosides), rotenoids, isoflavones, and tannins (Fig. 10.3).

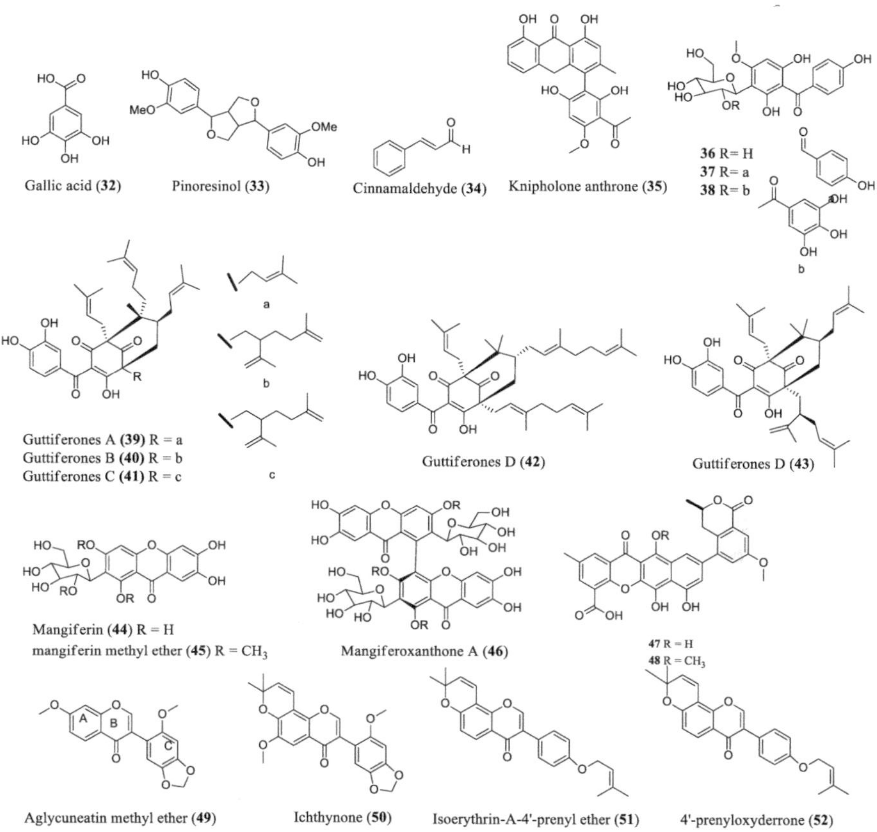

Fig. 10.3 Antiviral phenolic compounds isolated from African medicinal plants

Gallic acid (**32**), a simple phenol, and pinoresinol (**33**), a lignin, which are oxidative coupling dimers of phenylpropanoids, have been isolated from the shoot extract of *Limonium densiflorum*. **32** showed strong virucidal effects (100% inhibition) against HSV-1 in plaque reduction assay, while **33** exhibited weak activity (26% inhibition). Similarly, the phenylpropene phenol cinnamaldehyde (**34**), isolated from *T. capitatus,* showed moderate anti-HSV-2 activity, with an $EC_{50} = 39.7$ µg/mL (Medini et al., 2016).

Knipholone anthrone (**35**), a phenyl anthraquinone derivative isolated from the Ethiopian endemic medicinal plant *Kniphofia foliosa*, possessed strong anti-HIV activity. In primary peripheral blood mononuclear cell (PBMC)-based assay (ELISA), the compound displayed antiviral effect against HIV-1c(MJ4) with an EC_{50} value of 1.8 µg/mL. However, it was also found to be substantially cytotoxic to primary PBMCs in an MTT assay (Feilcke et al., 2019). Furthermore, Richard et al. (2020) discovered **35** as a promising HIV-1 latency reversal lead. The study revealed that, among 216 or more compounds screened from the pan-African Natural Product

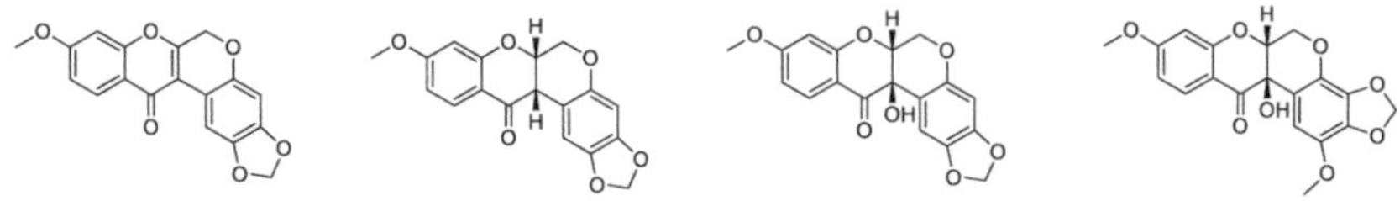

Oblarotenoid G (**53**) Oblarotenoid C (**54**) Oblarotenoid A (**55**) Oblarotenoid D (**56**)

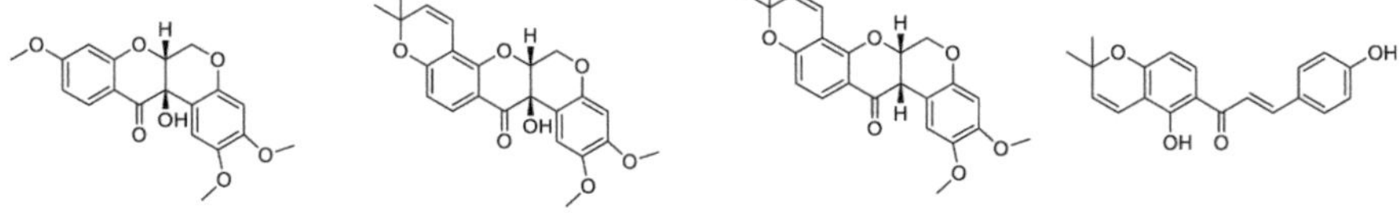

12a-hydroxymunduserone (**57**) Tephrosin (**58**) Deguelin (**59**) Isobavachromene (**60**)

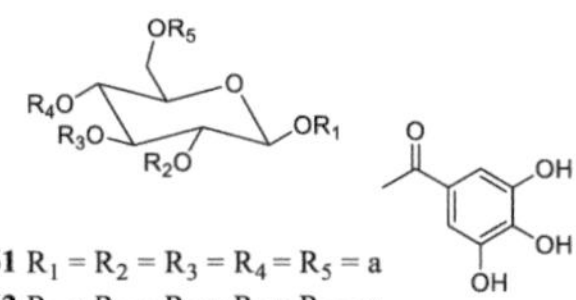

61 $R_1 = R_2 = R_3 = R_4 = R_5 = a$
62 $R_1 = R_2 = R_3 = R_4 = R_5 = a$
63 $R_1 = R_2 = R_3 = R_4 = R_5 = H$

a

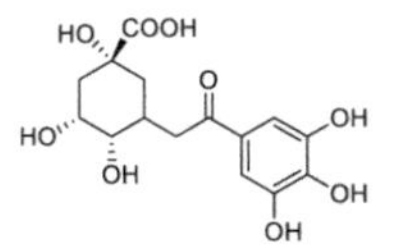

3-O-galloyl quinic acid (**64**)

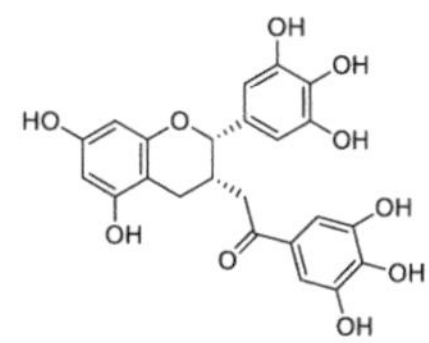

Epigallocatechine gallate (**65**)

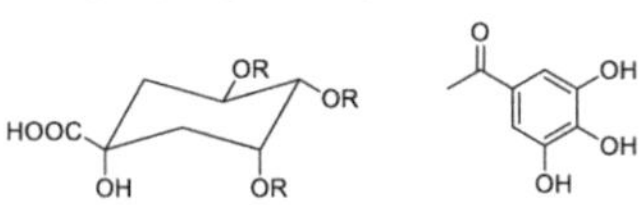

a

66 3,4,5 tri-O-galloylquinic acid R= a

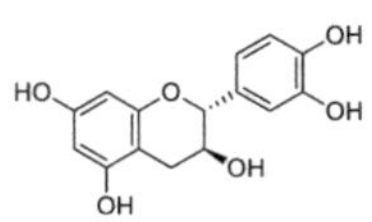

Catechin (**67**)

Fig. 10.3 (continued)

Library in a cell-based latency reversal assay, **35** had the highest activity, inducing GFP expression in 6.1% of cells. It also induced GFP expression across multiple cell lines and proviral integration sites. Nonetheless, it was found to be equally cytotoxic at its latency reversal concentrations, making it poorly selective. It also induced T cell activation by increasing the expression of CD69 and CD25 markers. In addition, metal ions were found to regulate or inhibit the latency reversal properties of the compound. Moreover, in vitro global gene expression analysis of J-Lat 10.6 cells suggested that **35** has **a** distinct mechanistic pathway from the standard LRAs. Finally, it showed a synergistic effect with positive control LRAs, enhancing both HIV RNA and protein production in primary cells from HIV-infected donors (Richard et al., 2020).

Another phenolic group, benzophenone *C*-glycosides (**36–38**) isolated from the active *n*-butanol fraction of the stem bark of *Mangifera indica*, displayed antiviral activity against influenza coxsackie viruses in neuraminidase (NA) and protease inhibitory assays. Compound **38** was relatively the most active one, with an IC_{50} values of 100 μM (Abdel-Mageed et al., 2014). Similarly, Gustafson et al. (1992) discovered novel anti-HIV polyisoprenylated benzophenones called guttiferones

A-E (**39–43**) from the tropical species *Symphonia globulifera* and *Garcinia livingstonei*, which were collected from the Central African Republic and Tanzania, respectively. In the cell-based XXT assay, all the compounds exhibit selective anti-HIV activity, with EC_{50} values of 1–10 mg/mL (Gustafson et al., 1992).

Xanthones are oxygenated heterocyclic compounds having a dibenzo-γ-pyrone moiety. They are biosynthesized by the oxidative coupling of hydroxylated benzophenones. Three xanthone glucosides, mangiferin (**44**), mangiferin methyl ether (**45**), and mangiferoxanthone A (**46**), isolated from the stem bark of *M. indica*, showed modest anti-influenza and anti-Coxsackie virus activity in NA and protease inhibitory assays (IC_{50}, 100 *μM*) (Abdel-Mageed et al., 2014). Likewise, using bioassay-guided fractionation, Omolo et al. (2012) isolated two new xanthones (**47, 48**) as the anti-HIV active principle of the tubers of *Pyrenacantha kaurabassana*. *P. kaurabassana* was one of the four plants utilized by traditional healers in Tanzania for the preparation of an effective herbal remedy. The two compounds (**47, 48**) displayed anti-HIV activity in the cell-based colorimetric replication assay, with an IC_{50} of 21 and 2 μg/mL, respectively; however, they showed poor selectivity (Omolo et al., 2012).

Flavonoids are a class of phenolics that consist of two aromatic rings (A and B) linked through a 3-carbon chain oxygenated heterocyclic ring (C) (**49**). They are biosynthesis products of a phenylpropane unit (cinnamoyl-CoA) and three acetate units (malonyl-CoA). They are further classified as chalcones, flavanones, flavones, anthocyanins, and isoflavones. From the leaves and root wood extracts of *Millettia oblata* antiviral isoflavones, aglycuneatin methyl ether (**49**), ichthynone (**50**), isoerythrin-A-4′-prenyl ether (**51**), and 4′-prenyloxyderrone (**52**) were reported by Kiganda et al. (2024). Compounds **50–52** are prenylated isoflavones via dimethylallylation in the form dimethylpyrano heterocyclic ring (D). The compounds exhibited substantial antiviral activity against RSV in the viral plaque reduction assay using HEp-2 cell cultures, with an IC_{50} range of 2.1 to 10 μM. Although **51** was the most potent (IC_{50}, 2.1 *μ*M), it lacks selectivity, as it was found to be toxic to both HEp-2 cell lines and primary human embryonic lung fibroblast cells (HELFs) at lower concentrations, 4 and 9 *μ*M, respectively. In addition, another group of isoflavone derivative antivirals known as rotenoids (**53–59**) (some of which contain C_5 isoprene units in the form a dimethylpyrano system (ring D) (**58, 59**) were also isolated from the leaves of *M. oblate*. All the rotenoids—oblarotenoid G (**53**), oblarotenoid C (**54**), oblarotenoid A (**55**), oblarotenoid D (**56**), 12a-hydroxymunduserone (**57**), tephrosin (**58**), and deguelin (**59**)—showed anti-RSV activity (IC_{50} 0.4–53 *μ*M), with tephrosin (**58**) being most active (IC_{50}, 0.4 *μ*M). Isobavachromene (**60**), a potent antiviral chalcone with anti-RSV activity and an IC_{50} value of 0.8 *μ*M, was also isolated from root wood extract of the same plant (Kiganda et al., 2024).

Tannins are high-molecular-weight polyphenolic compounds of plant origin capable of forming complexes with proteins and other macromolecules. They consist of a diverse group of oligomers and polymers, classified into two groups: hydrolyzable tannins, which include gallotannins and ellagitannin complex tannins, and condensed tannins (proanthocyanidins), which are not easily hydrolysable. Four gallotannins with anti-influenza and anti-coxsackie activity were reported from the stem bark of *M. indica*: penta-*O*-galloyl-glucose (**61**), tetra-*O*-galloyl-glucose (**62**), gallic glucoside (**63**), and 3-*O*-galloyl quinic acid (**64**). All showed antiviral activity, with **60** and **61** being the most potent against the influenza virus in the NA inhibitory assay, with IC_{50} values of 11.9 and 9.2 μM, respectively (Abdel-Mageed et al., 2014). Another gallotannin, epigallocatechin gallate (**65**), isolated from the shoots of *L. densiflorum* through bioassay-guided fractionation, exhibited strong inhibitory activity (100% inhibition) against HSV-1 while remaining non-toxic to Vero cells (Medini et al., 2016). Similarly, the gallotannin 3,4,5 tri-*O*-galloylquinic acid (**66**) was found to be the major antiviral component of a phenol-rich extract of *Myrothamnus flabellifolia.* In an ethidium bromide-based fluorescent assay, the compound displayed non-competitive inhibition of M-MLV and HIV-1 reverse transcriptases, with IC_{50} values of 5 μM and 34 μM, respectively (Kamng'ona et al., 2011). Furthermore, catechin (**67**), a flavonoid and condensed tannin monomer, along with unidentified gallotannin, isolated from the methanolic extract of *Peltophorum africanum*, displayed anti-HIV activity in various in vitro enzymatic models, including the newly developed protocols. The compounds were characterized through anti-HIV enzymatic screening of the aqueous and methanol extracts from nine South African medicinal plants used by traditional healers in the Limpopo Province for the treatment of HIV/AIDS, from which *P. africanum* was found to be the most active. The gallotannin inhibited the RDDP activity of HIV-1 RT with an IC_{50} of 6 μM and similarly inhibited the RNase H activity with an IC_{50} of 5 μM. Additionally, it possessed an inhibitory effect against the 3′-end processing activity of HIV-1 IN (IC_{50}, 6 μM). In a newly developed HIV-1 subtype C (HIV-1-C) pseudovirus assay, catechin demonstrated strong antiviral activity against two pseudoviruses (HR 4199 and HR 4012) and two reverse transcriptase inhibitor-resistant (HR 4018 and HR 4073) viruses, with IC_{50} values of 0.49, 2.3, 47.3, and 102 μM, respectively. Similarly, gallotannin exhibited moderate inhibition against these pseudoviruses (HR 4199 and HR 4012) and transcriptase inhibitor-resistant viruses (HR 4018 and HR 4073) with IC_{50} values of 3.86, 87.47, 15.22, and 13.03 μM, respectively (Bessong et al., 2005; Mavhandu et al., 2020).

10.3.2.3 Alkaloids

Alkaloids are a class of basic, naturally occurring plant constituents that contain at least one nitrogen atom, usually in a heterocyclic rings such as pyrrolidine, piperidine, quinoline, isoquinoline, indole, quinolizidine, indolizidine, imidazole, purine, and so on. These compounds exhibit diverse physiological and pharmacological effects and are important active components of many clinically used drugs. Amino acids, which serve as the source of the nitrogen atoms in alkaloids, are key precursors in their biosynthetic pathways. Approximately 12,000 alkaloids have been discovered from plants. However, only a small number of michellamine-type dimeric naphthylisoquinoline alkaloids, isolated from two species of Ancistrocladus collected in Cameroon and the Democratic Republic of the Congo, are identified as antiviral alkaloids of African origin (Fig. 10.4).

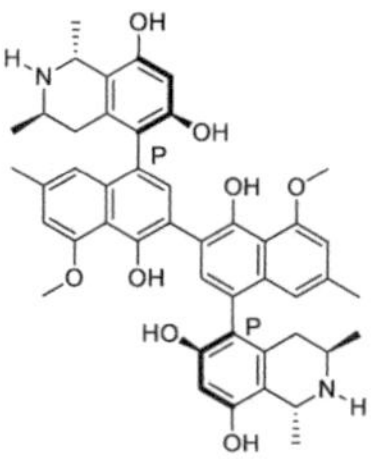

Michellamines A (**68**)

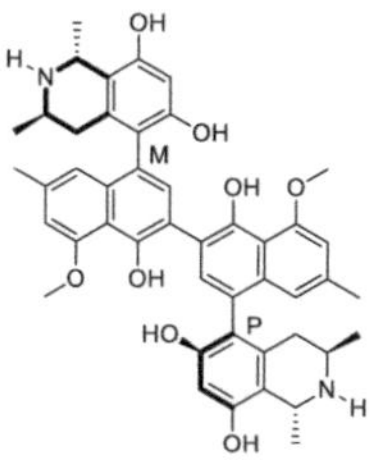

Michellamines B (**69**)

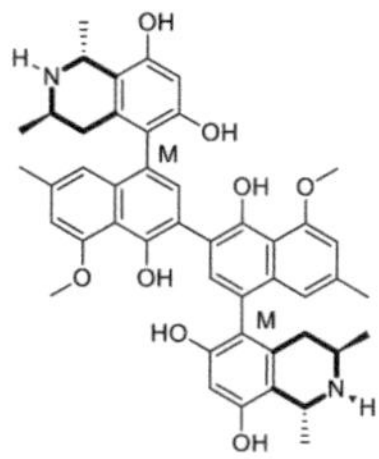

Michellamines C (**70**)

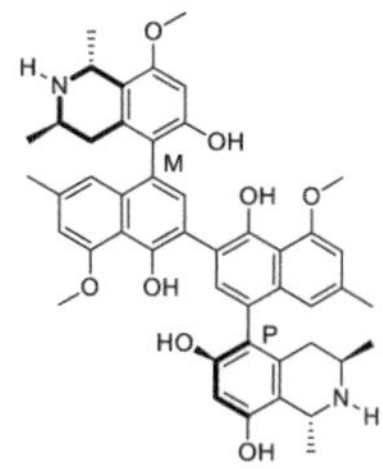

Michellamines D (**71**)

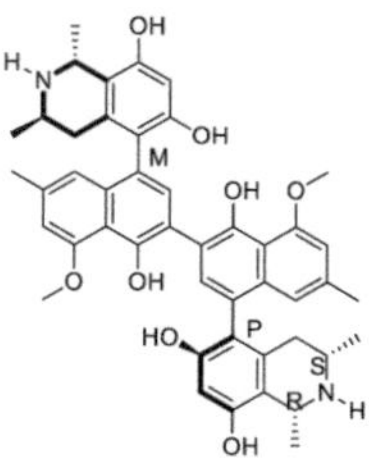

Michellamines E (**72**)

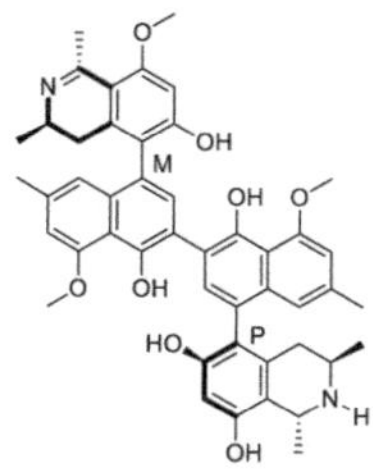

Michellamines F (**73**)

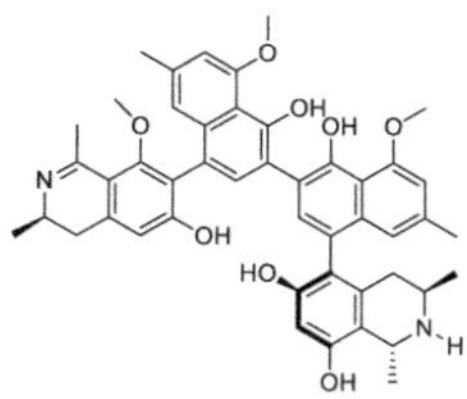

korundamine A (**74**)

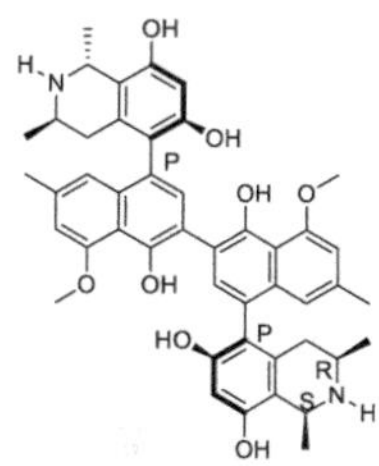

Michellamines A_2 (**75**)

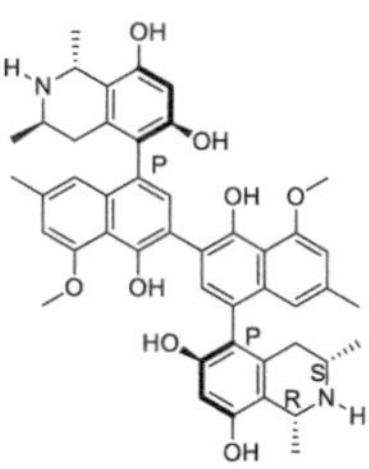

Michellamines A_3 (**76**)

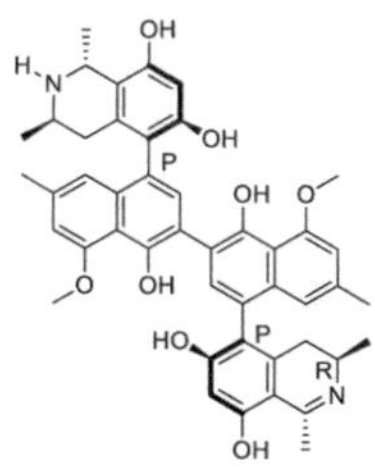

Michellamines A_4 (**77**)

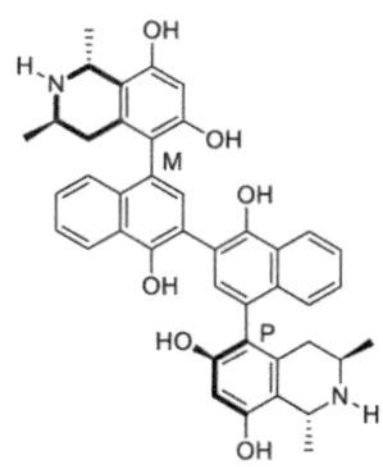

78

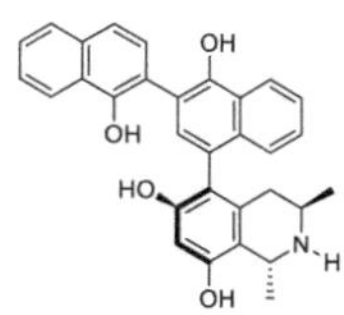

79

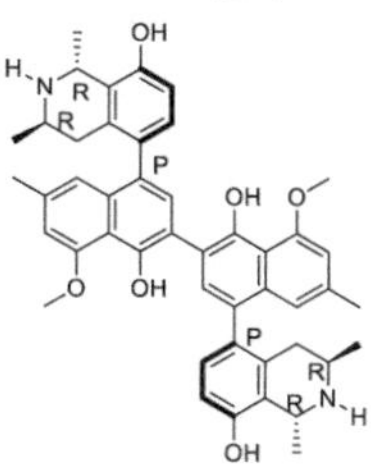

Jozimine C (**80**),

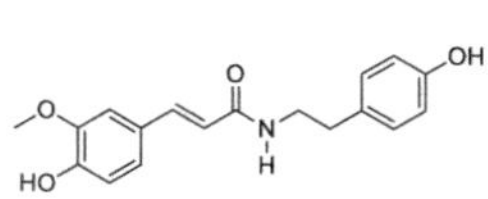

N-trans-ferulolyl tyramine (**81**)

Fig. 10.4 Antiviral alkaloids isolated from African medicinal plants

Michellamines A (**68**), B (**69**), and C (**70**) are novel anti-HIV atropisomeric naphthylisoquinoline alkaloid dimers, having an axial chirality due to hindered rotation about single bonds. These compounds were discovered from the leaves and stems of a Cameroonian liana *Ancistrocladus korupensis.* The pairs michellamines A (**68**) and B (**69**), as well as B (**69**) and C (**70**) are atropodiastereomers, while A (**68**) and C (**70**) are atropoenantiomers. In a cell based XTT assay, all the compounds completely inhibited the replication and cytopathic effects of HIV-1 and HIV-2, with EC_{50} value in the range of 2 to13 μM. Notably, michellamine **B** displayed broad-spectrum selective antiviral activity against diverse laboratory and clinical strains of HIV-1 and HIV-2 across a wide range of cell lines, showing its most potent activity is against HIV-1 laboratory strain in CEM-SS cell lines (EC_{50}, 1 μM). The compound acts by inhibiting reverse transcription at an early stage of the HIV life cycle and blocking cellular fusion and syncytium formation at later stages (Manfredi et al., 1991; Boyd et al., 1994; McMahon et al., 1995). Additionally, Hallock et al. (1997) discovered three more new anti-HIV michellamines D-F (**71**–**73**), dimeric naphthylisoquinoline alkaloids, from the leaves of the same plant, *A. korupensis*. In a cell-based screening against several strains of HIV-1, the compounds displayed comparable antiviral activity to michellamine B, with the highest EC_{50} values in the range of 2 to 6 μM (Hallock et al., 1997). Likewise, Hallock et al. (1998) isolated a novel heterodimeric naphthylisoquinoline anti-HIV alkaloid korundamine A (**74**) from *A. korupensis*. Unlike michellamines, korundamine A is biosynthesized from two dissimilar monomeric halves along the biaryl axis. The compound inhibited the cytopathic effect and replication of HIV-l as potently as michellamines, with EC_{50} value of 2 μM. It was also found to be active against several resistant strains (EC_{50}, 6-10 μM) (Hallock et al., 1998).

From the root bark of another *Ancistrocladus* species, Central African liana *Ancistrocladus congolensis*, Bringmann et al. (2016) further isolated new antiviral michellamines A_2, A_3, and A_4 (**75**–**77**), which are dimeric naphthylisoquinoline alkaloids. All the compounds (**75**–**77**) inhibited HIV replication in a cell-based XXT assay, with corresponding IC_{50} values of 29.6, 15.2, and 35.9 μM, respectively. However, their anti-HIV activity was lower compared to that of the clinical antiviral drugs zidovudine, tenofovir, and nevirapine (IC_{50}, 5–500 nM) (Bringmann et al., 2016). Simplified analogs of (**78, 79**) of michellamine B (**69**) were synthesized and evaluated for their HIV RT inhibitory activity to rationalize the pharmacophoric role of the main functional moieties—methoxy, methyl, dihydroxybinaphthalene, tetrahydroisoquinoline groups—present in **69**. Nevertheless, while **78** had comparable activity to **69** (IC_{50}, 33 μ*M*) with an IC_{50} value of 62 μ*M*, **79** was found to be poorly active (IC_{50},1000 μ*M*) (Upender et al., 1996). Similarly, another close analog of **69,** jozimine C (**80**), a dimer of the naphthylisoquinoline alkaloid dioncophylline C, was totally synthesized using atropo-enantioselective methodology. The compound demonstrated antiviral activity against HIV-1 (EC_{50}, 27 μg/mL), comparable to that of michellamine B (EC_{50}, 14 μg/mL) in a cell-based assay (Bringmann et al., 1998).

As the only non-dimeric naphthylisoquinoline alkaloids, N-trans-ferulolyl tyramine (**81**) was isolated from the shoot extract of *L. densiflorum* through bioassay-guided fractionation. This non-heterocyclic amide-alkaloid showed a moderate virucidal effect against HSV-1, having 26% inhibition in the plaque reduction assay (Medini et al., 2016).

10.3.2.4 Proteins

Charan et al. (2000) reported the isolation and characterization of MHL, a 9284-Da cysteine-rich protein, as the active component of the aqueous extract of the rhizomes *Myrianthus holstii.* In a cell-based XXT assay, the protein exhibited anti-HIV activity, with an EC_{50} value of 150 nM (Charan et al. 2000). Likewise, Khan et al. (2023) identified a 30 kDa protein, termed MoMo30-plant, as the active constituent of the aqueous extract of the leaves of *Momordica balsamina.* The protein selectively inhibited the replication of various HIV-1 strains, exhibiting IC_{50} values ranging from 2.8 to 9.0 nM in a multicistronic anti-genetic inhibition (MAGI) cell-based assay (Khan et al., 2023).

10.4 Conclusions and Future Perspectives

The present review found 81 pure antiviral natural products, 2 anti-HIV proteins, and more than 100 active antiviral extracts and fractions from African medicinal plants. Interestingly, 53.08% of the pure compounds were novel or new antivirals: 22.22% terpenoids, 16.04% alkaloids, and 14.81% phenolics. Their antiviral potency ranges from 0.00747 μM to 200 μM. However, as discussed in previous sections, for all the compounds and 95% of the plant extracts and fractions, only in vitro antiviral activity data are available. This shows that these natural products are at early stages of antiviral drug development even though their discovery time varies. For this reason, further studies are needed to enhance the potency and selectivity of these compounds, establish structure-activity relationships (SAR), and improve their pharmacokinetics and target interactions—especially for the most potent leads, such as ermiasolide A, B, and C (**2–4**), crotocascarin K (**6**), crotocascarin ω (**7**), lupenone (**14**), lupeol acetate (**15**), betulin (**16**), saponin 26, michellamines A (**68**) and B (**69**), and korundamine A (**74**). In this regard, there are a plethora of examples of antiviral drug candidates developed from plant natural product leads, from which valuable insights for drug discovery and development can be gained. For example, Cheng et al. (2001), Yeo et al. (2005), and Janmanchi et al. (2013) synthesized more than 100 analogs of helioxanthin, a lignan lactone isolated from *Taiwania cryptomerioides* and *Heliopsis scabra*, to develop more selective and potent anti-hepatitis B virus (HBV) agents and to establish SARs. As a result, the lactam and cyclichydrazide derivatives of helioxanthin were reported to have 125 and 300 times more anti-HBV potency than the parent compound. Similarly, Wu et al. (2013) reported the synthesis and SAR of potent in vitro and in vivo active antiviral analogs of antofine, a phenanthroindolizidine alkaloid isolated from *Cynanchum komarovii*; likewise, González-Cardenete et al. (2022) discovered ferruginol and its analog as broad-spectrum antivirals through semisynthetic modification of abietane diterpenoids. Furthermore, recent reviews by Khwaza et al. (2018), Mattio et al. (2020), Popović-Djordjević et al. (2022), and La Monica et al. (2024) have discussed the synthesis and SAR of analogs of antiviral leads, oleanolic acid (a triterpenoid), stilbenoids, flavonoids, and other classes of natural products.

In addition to improving the efficacy and the pharmacokinetics of the existing antiviral natural products, the exploration and discovery of novel antivirals from Africa's untapped traditional medicinal plant resources should be pursued. Currently, only a few natural products, particularly antivirals, have been discovered from African flora. Innovative technologies in the separation and identification of natural products could markedly accelerate the discovery process. Some of the major advances in the field include metabolomics-guided discovery, genome mining, artificial intelligence, machine learning, dereplication, molecular networking, highly sensitive structural elucidation techniques, and modern high-throughput screening and bioassay technologies (Atanasov et al., 2021; Bauman et al., 2021; Gangwal & Lavecchia, 2025). By utilizing one or a combination of these approaches, the discovery of novel bioactive natural products and clinical candidates is on the rise (Atanasov et al., 2021; Bauman et al., 2021; Domingo-Fernández et al., 2024). However, so far, none of the antiviral compounds isolated from African traditional plants have been discovered following these approaches. Instead, they were isolated through phytochemical investigations and antiviral evaluation of both traditionally claimed and randomly selected plant extracts, or by using the classical bioassay-guided fractionation methods.

In conclusion, natural products, with their vast chemical space, structural diversity, and evolutionarily selected bioactive scaffolds, have provided—and will continue to provide—novel therapeutics for viral and other diseases. A significant milestone has been achieved in the discovery of novel antiviral leads from African medicinal plants. However, further research is essential to translate them into clinics. Equally important is the discovery of novel antivirals from unexplored African traditional plant species, employing both classic bioassay-guided fractionation and the innovative techniques mentioned earlier.

References

Abdel-Mageed, W. M., Bayoumi, S. A., Chen, C., Vavricka, C. J., Li, L., Malik, A., et al. (2014). Benzophenone C-glucosides and gallotannins from mango tree stem bark with broad-spectrum anti-viral activity. *Bioorganic & Medicinal Chemistry, 22*(7), 2236–2243.

Asres, K., Bucar, F., Kartnig, T., Witvrouw, M., Pannecouque, C., & De Clercq, E. (2001). Antiviral activity against human immunodeficiency virus type 1 (HIV-1) and type 2 (HIV-2) of ethnobotanically selected Ethiopian medicinal plants. *Phytotherapy Research, 15*(1), 62–69.

Asres, K., Seyoum, A., Veeresham, C., Bucar, F., & Gibbons, S. (2005). Naturally derived anti-HIV agents. *Phytotherapy Research, 19*(7), 557–581.

Atanasov, A. G., Zotchev, S. B., Dirsch, V. M., & Supuran, C. T. (2021). Natural products in drug discovery: Advances and opportunities. *Nature Reviews Drug Discovery, 20*(3), 200–216.

Bauman, K. D., Butler, K. S., Moore, B. S., & Chekan, J. R. (2021). Genome mining methods to discover bioactive natural products. *Natural Product Reports, 38*(11), 2100–2129.

Bessong, P. O., Obi, C. L., Andréola, M. L., Rojas, L. B., Pouységu, L., Igumbor, E., et al. (2005). Evaluation of selected South African medicinal plants for inhibitory properties against human immunodeficiency virus type 1 reverse transcriptase and integrase. *Journal of Ethnopharmacology, 99*(1), 83–91.

Beuscher, N., Bodinet, C., Neumann-Haefelin, D., Marston, A., & Hostettmann, K. (1994). Antiviral activity of African medicinal plants. *Journal of Ethnopharmacology, 42*(2), 101–109.
Boyd, M. R., Hallock, Y. F., Cardellina, J. H., Manfredi, K. P., Blunt, J. W., McMahon, J. B., et al. (1994). Anti-HIV michellamines from *Ancistrocladus korupensis*. *Journal of Medicinal Chemistry, 37*(12), 1740–1745.
Bringmann, G., Holenz, J., Weirich, R., Rübenacker, M., Funke, C., Boyd, M. R., et al. (1998). First synthesis of the antimalarial naphthylisoquinoline alkaloid dioncophylline C, and its unnatural anti-HIV dimer, jozimine C. *Tetrahedron, 54*(3–4), 497–512.
Bringmann, G., Steinert, C., Feineis, D., Mudogo, V., Betzin, J., & Scheller, C. (2016). HIV-inhibitory michellamine-type dimeric naphthylisoquinoline alkaloids from the Central African liana *Ancistrocladus congolensis*. *Phytochemistry, 128*, 71–81.
Charan, R. D., Munro, M. H., O'Keefe, B. R., Sowder, R. C., McKee, T. C., Currens, M. J., et al. (2000). Isolation and characterization of *Myrianthus holstii* lectin, a potent HIV-1 inhibitory protein from the plant *Myrianthus holstii*. *Journal of Natural Products, 63*(8), 1170–1174. https://doi.org/10.1021/np000039h.
Chawuke, P., van den Berg, N., Fouche, G., Maharaj, V., Shoko, T., van der Westhuizen, C. J., et al. (2021). *Lobostemon trigonus* (Thunb.) H. Buek, a medicinal plant from South Africa as a potential natural microbicide against HIV-1. *Journal of Ethnopharmacology, 277*, 114,222.
Chen, J. L., Kernan, M. R., Jolad, S. D., Stoddart, C. A., Bogan, M., & Cooper, R. (2007). Dysoxylins A– D, Tetranortriterpenoids with potent anti-RSV activity from *Dysoxylum gaudichaudianum*. *Journal of Natural Products, 70*(2), 312–315.
Cheng, Y. C., Chou, C. K, Fu, L., Kuo, Y. H, Yeh, S. F, & Zhu, J. (2001) Inhibition and treatment of Hepatitis B virus and Flavivirus by Helioxanthin and its analogs. U.S. Patent No. 6,306,899. Washington, DC: U.S. Patent and Trademark Office. https://patents.google.com/patent/US6306899B1/en
Cos, P., Hermans, N., De Bruyne, T., Apers, S., Sindambiwe, J. B., Witvrouw, M., et al. (2002). Antiviral activity of Rwandan medicinal plants against human immunodeficiency virus type-1 (HIV-1). *Phytomedicine, 9*(1), 62–68.
Darshani, P., Sen Sarma, S., Srivastava, A. K., Baishya, R., & Kumar, D. (2022). Anti-viral triterpenes: A review. *Phytochemistry Reviews, 21*(6), 1761–1842.
Darwish, R. S., El-Banna, A. A., Ghareeb, D. A., El-Hosseny, M. F., Seadawy, M. G., & Dawood, H. M. (2022). Chemical profiling and unraveling of anti-COVID-19 biomarkers of red sage (*Lantana camara* L.) cultivars using UPLC-MS/MS coupled to chemometric analysis, in vitro study and molecular docking. *Journal of Ethnopharmacology, 291*, 115,038.
Domingo-Fernández, D., Gadiya, Y., Preto, A. J., Krettler, C. A., Mubeen, S., Allen, A., et al. (2024). Natural products have increased rates of clinical trial success throughout the drug development process. *Journal of Natural Products, 87*(7), 1844–1851.
Feilcke, R., Arnouk, G., Raphane, B., Richard, K., Tietjen, I., Andrae-Marobela, K., et al. (2019). Biological activity and stability analyses of knipholone anthrone, a phenyl anthraquinone derivative isolated from *Kniphofia foliosa* Hochst. *Journal of Pharmaceutical and Biomedical Analysis, 174*, 277–285.
Fekadu, M., Lulekal, E., Tesfaye, S., Ruelle, M., Asfaw, N., Awas, T., et al. (2024). The potential of Ethiopian medicinal plants to treat emergent viral diseases. *Phytotherapy Research, 38*(2), 925–938.
Gangwal, A., & Lavecchia, A. (2025). Artificial intelligence in natural product drug discovery: Current applications and future perspectives. *Journal of Medicinal Chemistry, 68*(4), 3948–3869.
Gebre-Mariam, T., Neubert, R., Schmidt, P. C., Wutzler, P., & Schmidtke, M. (2006). Antiviral activities of some Ethiopian medicinal plants used for the treatment of dermatological disorders. *Journal of Ethnopharmacology, 104*(1–2), 182–187.
González-Cardenete, M. A., Hamulic, D., Miquel-Leal, F. J., González-Zapata, N., Jimenez-Jarava, O. J., Brand, Y. M., et al. (2022). Antiviral profiling of C-18-or C-19-functionalized semisynthetic Abietane diterpenoids. *Journal of Natural Products, 85*(8), 2044–2051.

Gustafson, K. R., Blunt, J. W., Munro, M. H., Fuller, R. W., McKee, T. C., Cardellina, J. H., II, et al. (1992). The guttiferones, HIV-inhibitory benzophenones from *Symphonia globulifera, Garcinia livingstonei, Garcinia ovalifolia* and *Clusia rosea*. *Tetrahedron, 48*(46), 10,093–10,102.

Hallock, Y. F., Manfredi, K. P., Dai, J. R., Cardellina, J. H., Gulakowski, R. J., McMahon, J. B., et al. (1997). Michellamines D– F, new HIV-inhibitory dimeric naphthylisoquinoline alkaloids, and korupensamine E, a new antimalarial monomer, from *Ancistrocladus korupensis*. *Journal of Natural Products, 60*(7), 677–683.

Hallock, Y. F., Cardellina, J. H., II, Schäffer, M., Bringmann, G., François, G., & Boyd, M. R. (1998). Korundamine A, a novel HIV-inhibitory and antimalarial "hybrid" naphthylisoquinoline alkaloid heterodimer from *Ancistrocladus korupensis*. *Bioorganic & Medicinal Chemistry Letters, 8*(13), 1729–1734.

Harnett, S. M., Oosthuizen, V. D. V., & Van de Venter, M. (2005). Anti-HIV activities of organic and aqueous extracts of *Sutherlandia frutescens* and *Lobostemon trigonus*. *Journal of Ethnopharmacology, 96*(1–2), 113–119.

Heyman, H. M., Senejoux, F., Seibert, I., Klimkait, T., Maharaj, V. J., & Meyer, J. J. M. (2015). Identification of anti-HIV active dicaffeoylquinic-and tricaffeoylquinic acids in *Helichrysum populifolium* by NMR-based metabolomic guided fractionation. *Fitoterapia, 103*, 155–164.

Hudson, J. B., Anani, K., Lee, M. K., De Souza, C., Arnason, J. T., & Gbeassor, M. (2000). Further iinvestigations on the antiviral activities of medicinal plants of Togo. *Pharmaceutical Biology, 38*(1), 46–50.

Janmanchi, D., Lin, C. H., Hsieh, J. Y., Tseng, Y. P., Chen, T. A., & Jhuang, H. J. (2013) Synthesis and biological evaluation of helioxanthin analogues. *Bioorganic & Medicinal Chemistry, 21*(7), 2163–2176. https://doi.org/10.1016/j.bmc.2012.11.037.

Kamng'ona, A., Moore, J. P., Lindsey, G., & Brandt, W. (2011). Inhibition of HIV-1 and M-MLV reverse transcriptases by a major polyphenol (3, 4, 5 tri-O-galloylquinic acid) present in the leaves of the South African resurrection plant, *Myrothamnus flabellifolia*. *Journal of Enzyme Inhibition and Medicinal Chemistry, 26*(6), 843–853.

Kanyara, J. N., & Njagi, E. N. M. (2005). Anti-HIV-1 activities in extracts from some medicinal plants as assessed in an in vitro biochemical HIV-1 reverse transcriptase assay. *Phytotherapy Research, 19*(4), 287–290.

Khan, M., Diop, A., Gbodossou, E., Xiao, P., Coleman, M., De Barros, K., et al. (2023). Anti-human immunodeficiency virus-1 activity of MoMo30 protein isolated from the traditional African medicinal plant *Momordica balsamina*. *Virology Journal, 20*(1), 50.

Khwaza, V., Oyedeji, O. O., & Aderibigbe, B. A. (2018). Antiviral activities of oleanolic acid and its analogues. *Molecules, 23*(9), 2300.

Kiganda, I., Bogaerts, J., Wieske, L. H., Deyou, T., Atilaw, Y., Uwamariya, C., et al. (2024). Antiviral Rotenoids and Isoflavones Isolated from *Millettia oblata* ssp. teitensis. *Journal of Natural Products, 87*(4), 1003–1012.

Klos, M., Van de Venter, M., Milne, P. J., Traore, H. N., Meyer, D., & Oosthuizen, V. (2009). In vitro anti-HIV activity of five selected South African medicinal plant extracts. *Journal of Ethnopharmacology, 124*(2), 182–188.

La Monica, G., Bono, A., Alamia, F., Lauria, A., & Martorana, A. (2024). Bioisosteric heterocyclic analogues of natural bioactive flavonoids by scaffold-hopping approaches: State-of-the-art and perspectives in medicinal chemistry. *Bioorganic & Medicinal Chemistry, 117*, 791.

Lamien, C. E., Meda, A., Mans, J., Romito, M., Nacoulma, O. G., & Viljoen, G. J. (2005). Inhibition of fowlpox virus by an aqueous acetone extract from galls of *Guiera senegalensis* JF Gmel (Combretaceae). *Journal of Ethnopharmacology, 96*(1–2), 249–253.

Lihana, R. W., & Mwau, M. (2017). An in-vitro evaluation of a triple herbal therapy used in Kenya for HIV/AIDS management. *Journal of Dental and Medical Sciences, 16*(8), 74–78.

Liu, A. L., & Du, G. H. (2012). Antiviral properties of phytochemicals. In *Dietary phytochemicals and microbes* (pp. 93–126). Springer.

Louis-Félix, N. S., Christophhe, P., Franck, R., Leen, D., Johan, N., Fanny, R., et al. (2015). Antiviral activity of Diterpene Esters on Chikungunya Virus and HIV replication. *Journal of Natural Products, 78*(6), 1277–1283.

Louvel, S., Moodley, N., Seibert, I., Steenkamp, P., Nthambeleni, R., Vidal, V., et al. (2013). Identification of compounds from the plant species *Alepidea amatymbica* active against HIV. *South African Journal of Botany, 86*, 9–14.

Mair, C. E., Grienke, U., Wilhelm, A., Urban, E., Zehl, M., Schmidtke, M., & Rollinger, J. M. (2018). Anti-influenza triterpene saponins from the bark of *Burkea africana*. *Journal of Natural Products, 81*(3), 515–523.

Manfredi, K. P., Blunt, J. W., Cardellina, J. H., McMahon, J. B., Pannell, L. L., Cragg, G. M., & Boyd, M. R. (1991). Novel alkaloids from the tropical plant *Ancistrocladus abbreviatus* inhibit cell killing by HIV-1 and HIV-2. *Journal of Medicinal Chemistry, 34*(12), 3402–3405.

Mansfield, K., & King, N. (1998). Viral diseases. In *Nonhuman primates in biomedical research* (pp. 1–57). Academic.

Mattio, L. M., Catinella, G., Pinto, A., & Dallavalle, S. (2020). Natural and nature-inspired stilbenoids as antiviral agents. *European Journal of Medicinal Chemistry, 202*(112), 541.

Mavhandu, L. G., Cheng, H., Bor, Y. C., Tebit, D. M., Hammarskjold, M. L., Rekosh, D., & Bessong, P. O. (2020). Development of a pseudovirus assay and evaluation to screen natural products for inhibition of HIV-1 subtype C reverse transcriptase. *Journal of Ethnopharmacology, 258*(112), 931.

McMahon, J. B., Currens, M. J., Gulakowski, R. J., Buckheit, R. W., Jr., Lackman-Smith, C., Hallock, Y. F., & Boyd, M. R. (1995). Michellamine B, a novel plant alkaloid, inhibits human immunodeficiency virus-induced cell killing by at least two distinct mechanisms. *Antimicrobial Agents and Chemotherapy, 39*(2), 484–488.

Medini, F., Megdiche, W., Mshvildadze, V., Pichette, A., Legault, J., St-Gelais, A., & Ksouri, R. (2016). Antiviral-guided fractionation and isolation of phenolic compounds from *Limonium densiflorum* hydroalcoholic extract. *Comptes Rendus. Chimie, 19*(6), 726–732.

Mehrbod, P., Abdalla, M. A., Njoya, E. M., Ahmed, A. S., Fotouhi, F., Farahmand, B., et al. (2018). South African medicinal plant extracts active against influenza A virus. *BMC Complementary and Alternative Medicine, 18*, 1–10.

Meyer, J. J. M., Afolayan, A. J., Taylor, M. B., & Engelbrecht, L. (1996). Inhibition of herpes simplex virus type 1 by aqueous extracts from shoots of *Helichrysum aureonitens* (Asteraceae). *Journal of Ethnopharmacology, 52*(1), 41–43.

Mollel, J. T., Said, J. S., Masalu, R. J., Hannoun, C., Mbunde, M. V., Nondo, R. S., et al. (2022). Anti-respiratory syncytial virus and anti-herpes simplex virus activity of six Tanzanian medicinal plants with extended studies of *Erythrina abyssinica* stem bark. *Journal of Ethnopharmacology, 292*(115), 204.

More, G. K., Makola, R. T., & Prinsloo, G. (2021). In vitro evaluation of anti-Rift Valley Fever Virus, antioxidant and anti-inflammatory activity of South African medicinal plant extracts. *Viruses, 13*(2), 221.

Mukhtar, M., Arshad, M., Ahmad, M., Pomerantz, R. J., Wigdahl, B., & Parveen, Z. (2008). Antiviral potentials of medicinal plants. *Virus Research, 131*(2), 111–120.

Muriuki, JK., Ng'ang'a, ZW., Lwembe, RM., Lihana, RW., & Mwau, M. (2017). An In-Vitro Evaluation of A Triple Herbal Therapy Used in Kenya For HIV/AIDS Management. *Journal of Dental and Medical Sciences, 16*(8), 74–78. https://doi.org/10.9790/0853-1608027478.

Naghavi, M., Mestrovic, T., Gray, A., Hayoon, A. G., Swetschinski, L. R., Aguilar, G. R., et al. (2024). Global burden associated with 85 pathogens in 2019: A systematic analysis for the Global Burden of Disease Study 2019. *The Lancet Infectious Diseases, 24*(8), 868-895.

Nie, C., Trimpert, J., Moon, S., Haag, R., Gilmore, K., Kaufer, B. B., & Seeberger, P. H. (2021). In vitro efficacy of Artemisia extracts against SARS-CoV-2. *Virology Journal, 18*, 182.

Niyonzima, G., Laekeman, G., Witvrouw, M., Van Poel, B., Pieters, L., Paper, D., et al. (1999). Hypoglycemic, anticomplement and anti-HIV activities of *Spathodea campanulata* stem bark. *Phytomedicine, 6*(1), 45–49.

Ogbole, O. O., Akinleye, T. E., Segun, P. A., Faleye, T. C., & Adeniji, A. J. (2018). In vitro antiviral activity of twenty-seven medicinal plant extracts from Southwest Nigeria against three serotypes of echoviruses. *Virology Journal, 15*, 1–8.

Okoye, E. L., Nworu, C. S., Ezeifeka, G. O., & Esimone, C. O. (2012). Inhibition of HIV-1 lentiviral particles infectivity by *Gynostemma pentaphyllum* extracts in a viral vector based assay. *African Journal of Biotechnology, 11*(7), 1782–1788.

Oluremi, B. B., & Adeniji, J. A. (2015). Anti-viral activity evaluation of selected medicinal plants of Nigeria against measles virus. *British Microbiology Research Journal, 7*(5), 218–225.

Omolo, J. J., Maharaj, V., Naidoo, D., Klimkait, T., Malebo, H. M., Mtullu, S., et al. (2012). Bioassay-guided investigation of the Tanzanian plant *Pyrenacantha kaurabassana* for potential anti-HIV-active compounds. *Journal of Natural Products, 75*(10), 1712–1716.

Parker, M. E., Chabot, S., Ward, B. J., & Johns, T. (2007). Traditional dietary additives of the Maasai are antiviral against the measles virus. *Journal of Ethnopharmacology, 114*(2), 146–152.

Popović-Djordjević, J., Quispe, C., Giordo, R., Kostić, A., Stanković, J. S. K., & Fokou, P. V. T. (2022) Natural products and synthetic analogues against HIV: A perspective to develop new potential anti-HIV drugs. *European journal of medicinal chemistry, 233*,114217. https://doi.org/10.1016/j.ejmech.2022.114217.

Prinsloo, G., Marokane, C. K., & Street, R. A. (2018). Anti-HIV activity of southern African plants: Current developments, phytochemistry and future research. *Journal of Ethnopharmacology, 210*, 133–155.

Richard, K., Schonhofer, C., Giron, L. B., Rivera-Ortiz, J., Read, S., Kannan, T., et al. (2020). The African natural product knipholone anthrone and its analogue anthralin (dithranol) enhance HIV-1 latency reversal. *Journal of Biological Chemistry, 295*(41), 14,084–14,099.

Simon, A. O., Festus, M. T., & Makokha, A. O. (2016). Preclinical anti-HSV-1 activity of aqueous and methanol extracts of Kenya grown Pyrethrum (*Chrysanthemum cinerariaefolium*). *Journal of Medicinal Plants Research, 10*(44), 811–817.

Singh, I. P., Bharate, S. B., & Bhutani, K. K. (2005). Anti-HIV natural products. *Current Science*, 269–290.

Sopjani, M., Falco, F., Impellitteri, F., Guarrasi, V., Nguyen Thi, X., Dërmaku-Sopjani, M., & Faggio, C. (2024). Flavonoids derived from medicinal plants as a COVID-19 treatment. *Phytotherapy Research, 38*(3), 1589–1609.

Świątek, Ł., Sieniawska, E., Sinan, K. I., Zengin, G., Boguszewska, A., Hryć, B., et al. (2023). Chemical characterization of different extracts of *Justicia secunda* vahl and determination of their anti-oxidant, anti-enzymatic, anti-viral, and cytotoxic properties. *Antioxidants, 12*(2), 509.

Tembeni, B., Sciorillo, A., Invernizzi, L., Klimkait, T., Urda, L., Moyo, P., et al. (2022). HPLC-based purification and isolation of potent Anti-HIV and latency reversing Daphnane Diterpenes from the medicinal plant *Gnidia sericocephala* (Thymelaeaceae). *Viruses, 14*(7), 1437.

Terefe, E. M., Okalebo, F. A., Derese, S., Batiha, G. E. S., Youssef, A., Alorabi, M., & Muriuki, J. (2022a). Cytotoxicity and anti-HIV activities of extracts of the twigs of *Croton dichogamus* Pax. *BMC Complementary Medicine and Therapies, 22*(1), 49.

Terefe, E. M., Okalebo, F. A., Derese, S., Langat, M. K., Mas-Claret, E., Qureshi, K. A., et al. (2022b). Anti-HIV ermiasolides from *Croton megalocarpus*. *Molecules, 27*(20), 7040.

Terefe, E. M., Okalebo, F. A., Derese, S., Langat, M. K., Mas-Claret, E., Aljarba, N. H., et al. (2022c). In vitro anti-HIV and cytotoxic effects of pure compounds isolated from *Croton macrostachyus* Hochst. Ex Delile. *BMC Complementary Medicine and Therapies, 22*(1), 159.

Terefe, E. M., Okalebo, F. A., Derese, S., Muriuki, J., Rotich, W., Mas-Claret, E., et al. (2022d). Constituents of *Croton megalocarpus* with potential Anti-HIV activity. *Journal of Natural Products, 85*(7), 1861–1866.

Terefe, E. M., Okalebo, F. A., Derese, S., Muriuki, J., Mas-Claret, E., & Langat, M. K. (2023). Anti-HIV crotocascarin ω from Kenyan *Croton dichogamus*. *Natural Product Research, 37*(17), 2809–2816.

Tietjen, I., Gatonye, T., Ngwenya, B. N., Namushe, A., Simonambanga, S., Muzila, M., et al. (2016). *Croton megalobotrys* Müll Arg. and *Vitex doniana* (Sweet): Traditional medicinal plants in a three-step treatment regimen that inhibit in vitro replication of HIV-1. *Journal of Ethnopharmacology, 191*, 331–340.

Tietjen, I., Ngwenya, B. N., Fotso, G., Williams, D. E., Simonambango, S., Ngadjui, B. T., et al. (2018). The *Croton megalobotrys* Müll Arg. traditional medicine in HIV/AIDS management: Documentation of patient use, in vitro activation of latent HIV-1 provirus, and isolation of active phorbol esters. *Journal of Ethnopharmacology, 211*, 267–277.

Tolo, F. M., Rukunga, G. M., Muli, F. W., Njagi, E. N., Njue, W., Kumon, K., et al. (2006). Antiviral activity of the extracts of a Kenyan medicinal plant *Carissa edulis* against herpes simplex virus. *Journal of Ethnopharmacology, 104*(1–2), 92–99.

Tompa, D. R., Immanuel, A., Srikanth, S., & Kadhirvel, S. (2021). Trends and strategies to combat viral infections: A review on FDA approved antiviral drugs. *International Journal of Biological Macromolecules, 172*, 524–541.

Toujani, M. M., Rittà, M., Civra, A., Genovese, S., Epifano, F., Ghram, A., et al. (2018). Inhibition of HSV-2 infection by pure compounds from *Thymus capitatus* extract in vitro. *Phytotherapy Research, 32*(8), 1555–1563.

Upender, V., Pollart, D. J., Liu, J., Hobbs, P. D., Olsen, C., Chao, W. R., et al. (1996). The synthesis and biological activity of two analogs of the anti-HIV alkaloid michellamine B. *Journal of Heterocyclic Chemistry, 33*(4), 1371–1384.

WHO. (2024). World health statistics: Monitoring health for the SDGs, Sustainable Development Goals.

Wu, M., Han, G., Wang, Z., Liu, Y. & Wang, Q, (2013). Synthesis and antiviral activities of antofine analogues with different C-6 substituent groups. *Journal of agricultural and food chemistry, 61*(5), 1030–1035. https://doi.org/10.1021/jf304905k.

Yeo, H., Li, Y., Fu, L., Zhu, J. L., Gullen, E. A., & Dutschman, G. E. (2005) Synthesis and antiviral activity of helioxanthin analogues. *Journal of medicinal chemistry, 48*(2), 534–546. https://doi.org/10.1021/jm034265a.

Zekeya, N., Mamiro, B., Ndossi, H., Mallya, R. C., Kilonzo, M., Kisingo, A., et al. (2022). Screening and evaluation of cytotoxicity and antiviral effects of secondary metabolites from water extracts of *Bersama abyssinica* against SARS-CoV-2 Delta. *BMC Complementary Medicine and Therapies, 22*(1), 280.

Zewde, D., Dawo, F., Hurisa, B., Mengesha, A., & Tadele, A. (2019). Determination of anti-rabies virus activities of crude extracts from some traditionally used medicinal plants in East Wollega, Ethiopia. *International Journal of Basic and Applied Virology, 8*(2), 28–37.

Chapter 11
Medicinal Plants Used for Managing Erectile Dysfunction in Africa

Emmanuel Kademeteme, Elliot Nyagumbo, Ian Mutasa, William Pote, Herbet Zirima, and Mazuru Gundidza

Abstract The use of indigenous medicinal plants in Africa to address erectile dysfunction has gained significant attention, highlighting the efficacy of diverse plant species. This chapter presents a detailed documentation of medicinal plants utilized across Africa for managing erectile dysfunction, drawing insights from sources such as Scopus, Science Direct, Google Scholar, and grey literature. Factors like cultural preferences, geographical locations, and traditional beliefs influence the diagnosis, selection of potent plants, preparation methods, and administration for managing erectile dysfunction. The exact estimates of the prevalence of erectile dysfunction in Africa are difficult to ascertain, and estimates suggest rates exceeding 80% among hypertensive patients in public hospitals and 60% among diabetic patients. Erectile dysfunction is more prevalent in diabetic and hypertensive patients compared to those without these conditions. The documented medicinal plants are grouped into three categories: category 1 ($n = 126$) comprises plants lacking experimental research, category 2 ($n = 44$) includes plants studied experimentally, and category 3 ($n = 8$) covers plants investigated in clinical research. Studies exploring the use of medicinal plants from Africa, including ethnobotanical surveys, preclinical trials, and clinical investigations, underscore the potential of these plants

E. Kademeteme (✉) · I. Mutasa · W. Pote
Department of Biomedical Sciences, Physiology Unit, School of Medical and Health Sciences, Great Zimbabwe University, Masvingo, Zimbabwe
e-mail: ekademeteme@gzu.ac.zw

E. Nyagumbo
National Pathology Research and Diagnostic Centre, Midlands State University, Gweru, Zimbabwe

H. Zirima
Department of Psychiatry and Behavioural Sciences, School of Medical and Health Sciences, Great Zimbabwe University, Masvingo, Zimbabwe

M. Gundidza
Department of Biomedical Sciences, Herbal Medicine Unit, School of Medical and Health Sciences, Great Zimbabwe University, Masvingo, Zimbabwe

A. Dekebo (ed.), *Bioactive Secondary Metabolites from Medicinal Plants of Africa*, Interdisciplinary Biotechnological Advances,
https://doi.org/10.1007/978-981-96-8590-5_11

in managing erectile dysfunction. These findings highlight the significant role of medicinal plants in addressing this global health concern.

Keywords Medicinal plants · Erectile dysfunction · African countries

11.1 Introduction

Medicinal plants have been used for centuries to enhance virility and elevate social experiences. Ancient Kings, who sought to satisfy multiple wives, used herbs (Chifamba & Zengeni, 2024). Much of the knowledge about the use of medicinal plants to manage different conditions, which has been kept secret, began to be documented ever since ethnobotanical surveys being conducted across many regions in Africa (Iwu & Maurice, 2014; Liengme, 1983; Mahwasane et al., 2013; Terashima et al., 1991).

The purpose of this chapter is to compile a comprehensive record of medicinal plants utilized in Africa for the management of erectile dysfunction (ED). This inventory establishes a foundation for conducting extensive research in the field of andrology. Researchers interested in this area can employ this information to identify the most potent plant species available in Africa and validate the claims made by custodians of traditional knowledge.

Ethnobotanical information on the use of plants to address sexual dysfunction concerns in many regions of Africa is limited compared to other conditions due to the sensitive nature of the topic and is considered taboo in many cultures (Kacholi & Amir, 2023). The use of medicinal plants to restore men's sexual health has remained shrouded in secrecy, making it difficult to identify the most effective herbs.

The medicinal plants are often regarded as aphrodisiacs that possess a wide range of applications in managing diverse sexual health issues in both men and women. They are used to enhance libido, heighten sexual sensations, prevent premature ejaculation, or facilitate orgasm. Experiments have been conducted on some of the medicinal plants to investigate the mechanism of action of phytochemical compounds and revealed their ability to increase nitric oxide synthesis and phosphodiesterase-5 (PDE5) inhibition to maintain an erection (Opuwari & Moundipa, 2021). These plants exhibit the capacity to boost testosterone levels and modulate other key andrological proteins, thereby reinvigorating male sexual health.

Medicinal plants have exhibited therapeutic outcomes in managing underlying conditions such as stress, depression, hypertension, diabetes, and Parkinson's disease, all of which are factors contributing to erectile dysfunction. Instead of relying solely on PDE5 inhibitors, the focus of managing erectile dysfunction has shifted toward addressing its underlying causes (Saikia et al., 2022).

Medicinal plants used in Africa are claimed to offer a holistic approach in managing ED by targeting involved pathophysiological aspects and other underlying conditions (Owaba et al., 2021). Medicinal plants, for managing erectile dysfunction, hold promise in restoring long-term erectile function, surpassing the need for temporary boosts in every sexual encounter, as provided by PDE5 inhibitors (Abdillahi & Van Staden, 2012a).

11.2 Erectile Dysfunction Trends in Africa

The prevalence of ED in Africa is a concern, with studies showing high rates of affected men across different regions. The pooled prevalence of ED in patients with diabetes mellitus in Africa ranges from 61.62% to 71.45% (Shiferaw et al., 2020). Factors such as body mass index (BMI) and glycated hemoglobin $\geq$7% show potential links to ED development (Shiferaw et al., 2020). The West African countries, Ghana and Nigeria, have an ED prevalence of 66% and 57.4%, respectively, followed by a 20% magnitude in the Sub-Saharan region (Afolayan & Yakubu, 2009). In the Eastern African country of Kenya, the prevalence of hypertensive patients attending public hospitals was 95% (Mkonyi et al., 2024). The prevalence of diabetic patients is 60% (Gobena et al., 2023). Hypertension and diabetes are the most risk factors for ED due to various causes. These two chronic non-communicable diseases (NCDs) contribute to nerve damages, hormonal imbalances, reduced nitric oxide production, atherosclerosis, and blood vessel damages, therefore increasing the risk of ED. However, not all hypertensive or diabetes mellitus patients experience ED. This is seemingly due to the individual variability and the severity of the conditions. As evidenced by Nkafu et al. (2023), 5% of hypertensive people and 40% of diabetic patients did not experience ED. Egypt showed a mild prevalence of erectile dysfunction of 17.2% in the population-based study (Nkafu et al., 2023).

The prevalence statistics of ED are varied across studies but remain a significant health concern (Yafi et al., 2017). There is a growing concern of ED affecting young men due to sedentary lifestyles, drugs, and substance abuse. Drug use, including recreational drugs, can interfere with normal sexual function. Substance abuse, such as excessive alcohol consumption and smoking, damages blood vessels, resulting in erectile dysfunction. All these concerns can be addressed by providing sexual health services and support, hence promoting healthy lifestyles and education, pivotal for boosting sexual health. The cases of ED in Africa are often underreported due to cultural taboos and stigmatization, which surround sexual health issues (Afolayan & Yakubu, 2009). This hinders accurate assessments and appropriate interventions tailored for affected individuals. Many African countries are facing economic hardships, which have an impact on health care services. The limited resources and developed hospital infrastructure affect the availability and accessibility of specialized treatment (Fig. 11.1) for ED. Remote areas in Africa are the most affected by the economic hardships (Mkonyi et al., 2024), contributing to difficulties in diagnosing and treating ED effectively. Primarily, medicinal plants have been used to address many cases of ED due to the limited availability of advanced medical treatments for this condition (Josephine Ozioma & Antoinette Nwamaka Chinwe, 2019). In a typical standard set up, the treatment algorithm has to be employed to guide

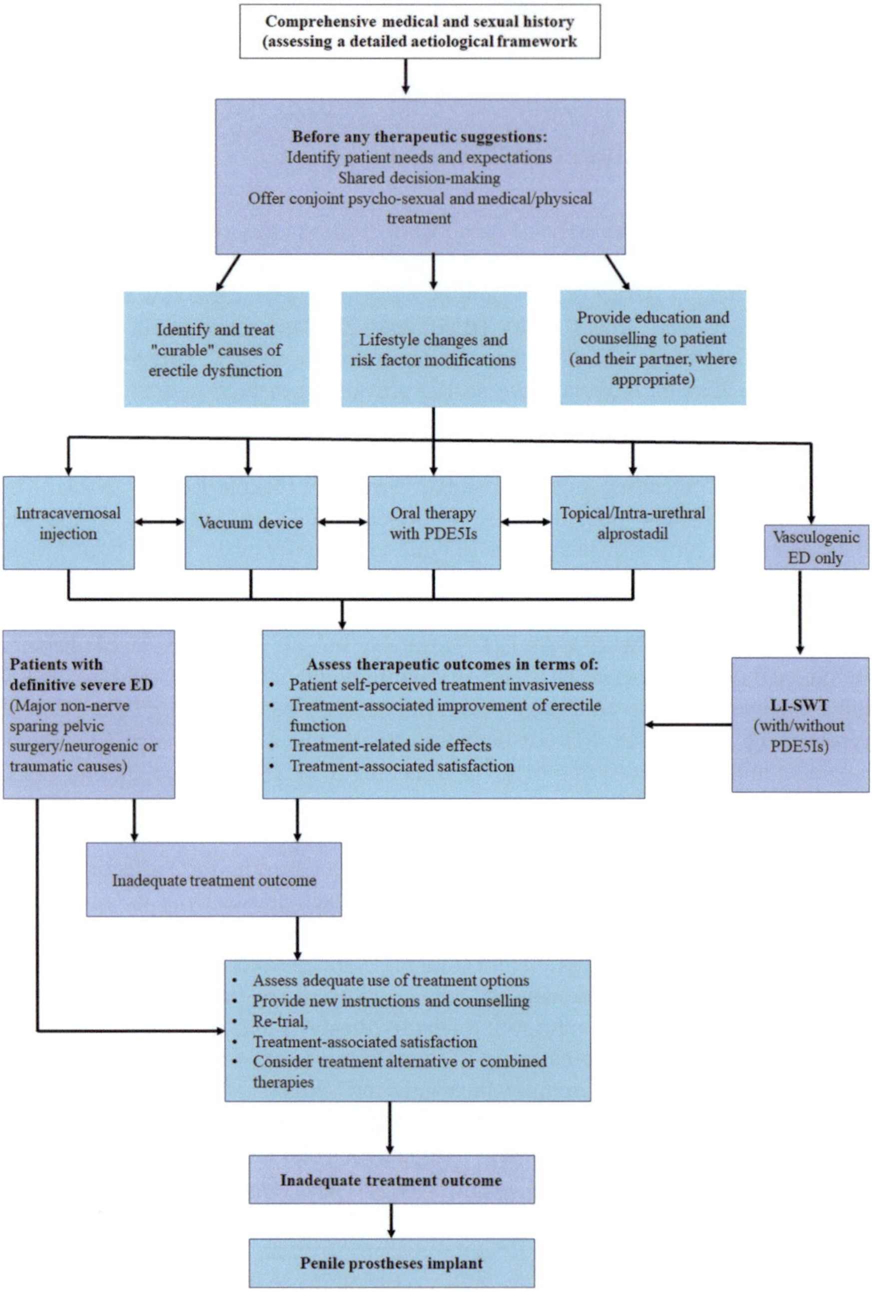

Fig. 11.1 Treatment algorithm for ED. *BPH* Benign prostatic hyperplasia, *LUTS* Lower urinary tract infections, *PDE5* Phosphodiesterase 5. (Adopted from Josephine Ozioma and Antoinette Nwamaka Chinwe (2019))

healthcare providers in a systematic approach for managing ED. The treatment algorithm takes into account the severity of the condition (Fig. 11.1) before incorporating the herbal medicines. Reliance reflects the effectiveness of medicinal plants in restoring erectile function to the affected individuals.

11.3 African Traditional Medicine

African traditional medicine (ATM) is one of the oldest and most practiced forms of traditional medicine. ATM utilizes natural resources as medicines, which include animals, minerals, and plants. Medicinal plants are mostly used due to their availability. The ATM offers a holistic healthcare system organized in three levels, which include divination, spiritualism, and herbalism (Mothibe & Sibanda, 2019). A traditional healer often practices divination, and spiritual healers are affiliated with Christian denominations, specifically the African apostolic movement. Herbalism is a practice that is often passed down through generations, with traditional healers playing a significant role in educating herbalists. Many herbalists acquire their knowledge and skills from experienced traditional healers, subsequently becoming herbal vendors. In some cases, community elders, possessing knowledge of therapeutic plants, also practice herbalism. It is worth noting that herbalism is an accessible field, and anyone with the necessary knowledge and understanding of medicinal plants can potentially become a herbalist. The way the system is set up allows it to generate income and sustain livelihoods by creating connections. Traditional healers also educate interested individuals in learning about specific plants and their uses. It is worth noting that the information is closely guarded, and the traditional healers share this knowledge with close relatives or chosen individuals in order to inherit the profession (Ouma, 2022). Eventually, these individuals often go on to become herbalists and sometimes vendors. Usually, herbalists and vendors have a go-to traditional healer whom they consult on plant species and applications. The journey of some herbalists began as assistants to traditional healers. The herbalists will be acting on instructions, location directions, and proper harvesting techniques, therefore preparing the plants accordingly. Nowadays, there are fewer traditional healers, contributed to by the use of modern medicine, religion, and other social factors. The banning of traditional practices under Witchcraft Suppression Act in many Southern African countries has been a key factor. (Abdullahi, 2011). The decrease in traditional healers has led to a decline in therapeutic plants' knowledge. The belief is, this knowledge comes from a higher source with unlimited resources. To establish a connection with this higher source, one typically needs to practice traditional healing. Some herbalists even use literature as a source of information on exotic plants, incorporating them within their practices.

The integration of ATM into mainstream healthcare has the potential to improve the health situation in low-income countries. According to the World Health Organization (WHO, 2012), approximately 80% of the African population relies on

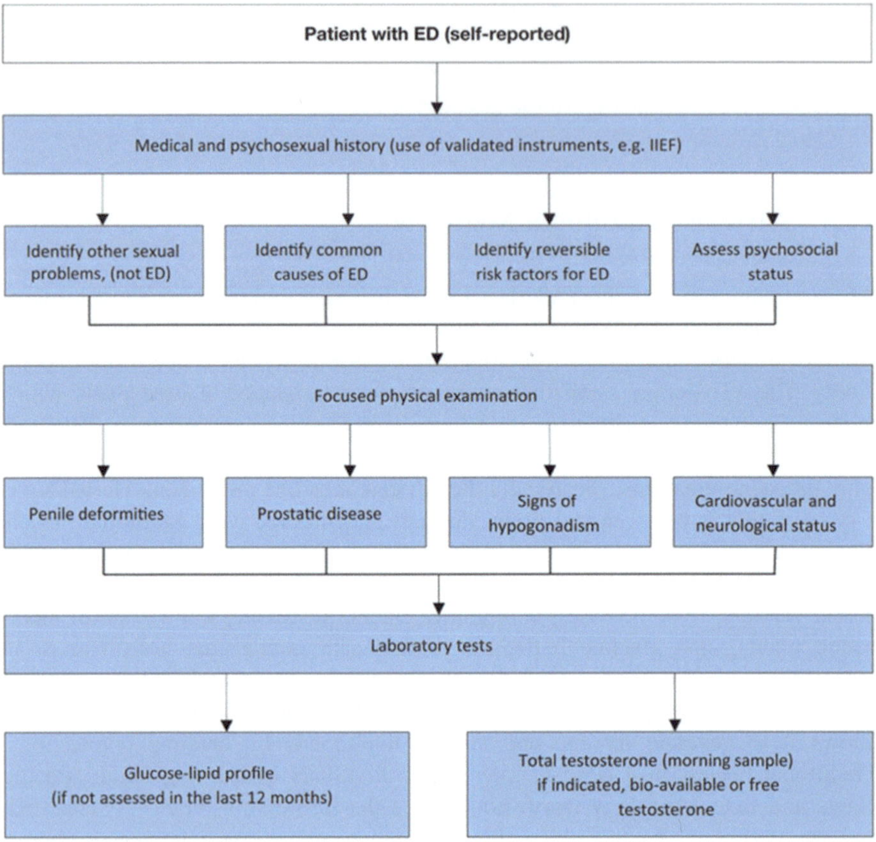

Fig. 11.2 *ED* Erectile dysfunction; *IIEF* International index of erectile function adopted from (Josephine Ozioma & Antoinette Nwamaka Chinwe, 2019)

traditional medicine to address their primary health concerns. ATM encompasses various forms of treatment, including spiritual and physical healing.

Many traditional healers in the ATM attribute ED to witchcraft. Due to the lack of proper diagnostic techniques (Fig. 11.2) in ATM, traditional healers often perform rituals to determine the causes of the disease. The choice of treatment is determined through consultation with spirit mediums. In the case of ED, the traditional healer prescribes medicinal plants to improve erectile function if a patient complains about a lack of erection. The psychological aspect is not often considered as one of the factors causing ED in ATM, since there are no proper diagnoistic procedures, and the belief is that natural remedies can offer a solution regardless of the other comorbidities.

Incorporating ATM into mainstream medicine can help address the healthcare needs of low-income countries. This helps them embrace traditional healing practices, therefore integrating them with modern diagnostic techniques and evidence-based treatments. This integration has the potential to enhance healthcare accessibility and provide comprehensive care that considers both physical and spiritual aspects of health.

11.3.1 Role of Traditional Medicine in Managing ED

In many African countries, erectile dysfunction (ED) is one of the prevalent conditions that often drives people to seek help from traditional healers and explore herbal markets. The utilization of traditional medicine for managing male sexual health has become increasingly popular, and herbal vendors often prioritize selling aphrodisiac herbs due to their higher income potential compared to alternative herbs.

The treatment of ED in traditional medicine uses various herbal concoctions. One form of treatment is the detoxification polyherbal infusion. This concoction is prepared by combining different plant parts. Of mention would be various plants such as *Pterocarpus angolensis* DC, *Cassia abbreviata* Oliv, and *Elephantorrhiza elephantina* (Burch.) Skeels, among others. Further research is needed to explore the potential of detoxification herbal remedies in addressing erectile dysfunction (van Vuuren & Frank, 2020). Upon consuming the detoxification herb, patients may experience frequent urination and bowel movements. They are advised to stay at home during the treatment period, or in remote areas, they may be advised to be in the bush to facilitate defecation whenever necessary. Traditional healers strongly believe that the detoxification process helps cleanse the urinary tract system and blood vessels, leading to improved blood circulation. This, in turn, is believed to be beneficial for addressing ED.

Traditional healers claim that if a man engages in sexual intercourse with a menstruating woman, he may be at risk of contracting certain illnesses that can affect the urinary tract system, hindering his erectile capabilities. A detoxification herb is prescribed to address this infection, detoxifying the body and eliminating toxins. Traditional healers assert if a man suspects having copulated with a person having a sexually transmitted infection (STI), the detoxification herb is given within 72 h, as a prophylactic measure. The traditional medical practitioners believe that specific types of STIs can affect erectile function.

Another category of herbal medicines prescribed in traditional medicine is those claimed to enhance sperm count and semen production. These herbs are believed to play a crucial role in addressing fertility issues. In African culture, there is a strong emphasis on men being virile, and the use of these herbs to increase semen production is seen as a source of satisfaction, indicating their ability to produce an ample amount of semen. Experiments conducted in this area have indicated that these herbs may have a mechanism of action that boosts testosterone levels, which in turn can contribute to improved erectile function (Moyo, 2013).

Another herb commonly prescribed in traditional medicine is one that is believed to promote firm erections. Some of these herbs are also favored for their potential to prolong sexual intercourse. Having a strong erection is highly valued in African culture, and while it may occur naturally, consuming these herbs is often seen as a means to achieve and maintain a healthy erection. These three types of herbal medicines can be taken together, although caution is advised with the detoxification herb due to potential adverse effects that could be fatal. Many men are encouraged to regularly consume these herbs even if they do not have clinically diagnosed ED, as a way to maintain a healthy erection.

In certain cultures, particularly in southern parts of Africa, men would gather at traditional ceremonies where traditional beer, infused with herbs, would be consumed. Alternatively, a traditional meal, such as meat soup infused with herbs, would be prepared and drunk. These practices were carried out regularly to prevent any complaints in ED. Young boys would sometimes be given herbs by their elders to prevent general weakness, which was believed to improve their sexual health. There are anecdotal reports, though not widely documented, that suggest giving these young boys herbs could potentially lead to increased penis size and improved fertility.

The beliefs and practices are rooted in cultural traditions and may not have scientific evidence to support their effectiveness or safety.

11.3.2 Case Reports

Famous herbal medicines have emerged from Southern Africa, which are called *vuka vuka* (wake up wake up) and *muchemedzambuya* (make a grandmother cry). The deeper meaning about *muchemedzambuya* is that the herb can bring pleasure to older women during intercourse, as it is believed that older women may have difficulty reaching climax. Consuming this herb is said to enhance sexual pleasure for older women, allowing them to experience climax more easily. These are purported to be strong aphrodisiac herbs to improve erectile problems and boost libido in men experiencing weak erections. *Muchemedzambuya* (*Pittopsorum viridiflorum* Sims var. *viridiflorum*) (Fig. 11.3a) implied by history to have been used by a King from

Fig. 11.3 (**a**) *Eriosema kraussianum pauciflorum* Klotzsch picture taken by Bart Wursten at Haka Game Park, Cleveland Dam on 23 September 2007. (**b**) *Ximenia caffra* Sond. var. caffra pictures taken by Bart Wursten at Valley of Kings, Mutare on 27 Sep 2004. (**c**) Sonchus oleraceus L pictures taken by Bart Wursten at Ndundu Lodge, Vumba, on 19 May 2004. (**d**) *Ozoroa insignis reticulata* (Baker f.) R. Fern. & A. Fern pictures taken by MA Hyde and BT Wursten at Cleveland Dam on 5 November 2004

the Great Zimbabwe who had 200 wives. The plant is widely consumed in Zimbabwe and South Africa for its many healing properties, and it is famous for its aphrodisiac properties (Odmell et al., 2018).

11.3.3 Medicinal Plants Used for Managing ED in Africa

The information on medicinal plants used in Africa has primarily been obtained from literature sources, with most of them from gray literature. Only a limited amount of information has been sourced from journals indexed in databases such as Scopus, Google Scholar, PubMed, and Science Direct. Many ethnobotanical surveys conducted in Africa have been published in low-impact factor journals, making the information less accessible, despite its value.

Tables 11.1, 11.2, and 11.3 shown in their categories provide information about medicinal plants that are used to manage ED in Africa. The information was obtained from various ethnobotanical surveys conducted in many regions of Africa.

The ethnobotanical data typically lacks some important details such as dosage forms and specific preparation methods. This raises significant concerns among researchers and consumers regarding the appropriate dosage for individuals. The common method of preparation indicated in literature is that a handful of plant material can be taken in the form of decoction, infusion, or taken with traditional dishes as medicine (Madani et al., 2017). The plant parts used for making medicines depend on the type of plant being used, according to the ancient knowledge which has been passed down from generation to generation in which the most favored plant parts are the barks and the roots. Decoctions and infusions (Table 11.1) are the most commonly used method of preparation where a decoction is prepared by boiling or immersing the hardest plant parts, which are the barks or roots (Odongo et al., 2018). An infusion is prepared by soaking medicinal plant extracts in water until the water absorbs all the plant materials, and it is often indicated by a color change (Faleyimu & Akinyemi, 2010). Obtaining detailed information on the exact methods from traditional healers can be challenging, as some may be reluctant to disclose all details as requested by researchers.

The medicinal plants in the chapter have been categorized into following three categories.

11.3.3.1 Category 1

Plants in this category (Table 11.1) are those for which studies validating their safety, efficacy, and mechanism of action in managing erectile dysfunction (ED) have not yet been conducted. Although these medicinal plants may have a long history of traditional use, there is a lack of scientific evidence supporting their use. Below are some of the important plant species in each category, which can be

Table 11.1 Medicinal plants used for managing ED in Africa categorized as plants without scientific validation

Family	Scientific name	Description of medicinal plants	Country	Method of preparation and administration	References
Acanthaceae	*Thomandersia hensii* De wild. & T. Durand	Shrub or small tree	DR Congo	Root bark shredded and inserted anally	(Mola et al., 2022)
Amaryllidaceae	*Ammocharis coranica* (Ker Gawl.)	Herb	South Africa	Root is boiled for 10 min and taken orally	(Semenya & Potgieter, 2013)
Anacardiaceae	*Lannea schweinfurthii* (Engl.) Engl.	Small to medium-sized tree with drooping branches	Zimbabwe	Half cup of root decoction is taken orally once a day	(Shopo et al., 2022a)
	Ozoroa sphaerocarpa R. Fern. & A. Fern.	Small tree with whitish flowers up to 7 m	South Africa	Bark pounded and taken with porridge	(Semenya & Potgieter, 2013)
	Ozoroa reticulata (Baker f.) R. Fern. & A. Fern.	A small, much-branched deciduous tree	Tanzania Somalia South Africa	Root decoction is taken orally	(Kacholi & Amir, 2023)
	Rhus vulgaris Meikle	Shrub or small tree	Uganda[1] Kenya[2]	Raw fruits are eaten[1] Whole plant decoction is taken orally[2]	(Kamatenesi-Mugisha & Oryem-Origa, 2005; Kyarimpa et al., 2023)
Annonaceae	*Annona senegalensis* Pers.	Large shrub or small tree	Mali	Oral 1 teacup of 75 ml once a day	(Togola et al., 2020)
	Monanthotaxis glauococarpa (Baill.) Verdc.	Herbaceous flowering plant	Comoros	Root decoction is taken orally	(Saive et al., 2020)
Apiaceae	*Heteromorpha arborescens* (Spreng.) Cham. & Schltdl	Small to medium-sized deciduous tree	South Africa Zimbabwe	1 teaspoon of powdered roots in milk and boiled, and is drunk twice a day	(Kamatenesi-Mugisha & Oryem-Origa, 2005)

Apocynaceae	*Ceropegia purpurascens* K. Schum	Perennial twinning herb with annual stems	South Africa	Roots boiled for 10 min	(Semenya & Potgieter, 2013)
	Gomphocarpus fruticosus (L.) Aiton f	Herbaceous, perennial, spindly shrub	South Africa	Root decoction is taken orally	(Semenya & Potgieter, 2013)
	Gymnosporia senegalensis (Lam.) Loes.	Shrub or multi-stemmed tree	South Africa	Roots and leaves decoction is taken orally	(Semenya & Potgieter, 2013)
	Tabernaemontana crassa Benth	Fragrant flowering small tree	South Africa	Stem bark decoction is inserted anally	(Mola et al., 2022)
	Holarrhena africana A. DC	Shrub or small tree with milky latex	Sierra Leone Zimbabwe[2]	Roots are chewed or powder is cooked with meat soup and taken orally[2]	(Johnny et al., 2022)
	Pergularia daemia (Forssk.) Chiov.	Perennial twinning herb with milky sap	Namibia	No detailed information	(Cheikhyoussef et al., 2011)
	Dregea macrantha (Klotzsch) Schltr.	Robust perennial climber	Namibia	No detailed information	(Cheikhyoussef et al., 2011)
Arecaceae	*Cocos nucifera* L.	Tree up to 30 m	Ghana	Whole plant No detailed information	(Asigbaase et al., 2024)
	Artemisia annua L	Annual short-day plant	South Africa	Root decoction is taken orally	(Semenya & Potgieter, 2013)
	Callilepsis laureola DC	Perennial herb with tufts of stem	South Africa	Root decoction is taken orally	(Semenya & Potgieter, 2013)
	Sonchus oleraceus L.	Annual herb up to 1.2 m	Kenya Zimbabwe	Leaves overnight infusion is taken three times a day till recovery	(Jeruto et al., 2015)
	Vernonia cinerea (L.)	Less erect annual to 1.3 m tall	Uganda	Leaves and root decoction are taken orally	(Kamatenesi-Mugisha & Oryem-Origa, 2005)

(continued)

Table 11.1 (continued)

Family	Scientific name	Description of medicinal plants	Country	Method of preparation and administration	References
	Borassus aethiopum Mart.	Large palm tree	Benin	Fruit or root decoction is taken orally	(Batcho, 2020)
Asteraceae	*Microglossa pyrifolia*	Climber or small shrub	Comoros	Seeds are taken orally by chewing them	(Saive et al., 2020)
	Kleinia longifora DC	Succulent shrub	Namibia	No detailed information	(Cheikhyoussef et al., 2011)
Balsaminaceae	*Impetiens* sp.	Erect annual herbs up to 60 cm	Uganda	Whole plant can be chewed or boiled and taken orally	(Kamatenesi-Mugisha & Oryem-Origa, 2005)
Burseraceae	*Commiphora Africana* (A. Rich.) Engl.	Shrub or small tree	Zimbabwe	A handful of chopped bark is added to 2 L boiling water ½ cup is taken 3 times a day	(Shopo et al., 2022a)
Canellaceae	*Warburgia ugandensis* Sprague	Spreading evergreen tree up to 30 m	Uganda Kenya	Bark, leaf, and root are pounded and taken with tea or porridge	(Kamatenesi-Mugisha & Oryem-Origa, 2005; Kyarimpa et al., 2023)
Cannabaceae	*Cannabis sativa* L.	Aromatic annual herb, up to 3 m	Uganda	Leaves and roots decoction is drunk or inhaled as smoke	(Mola et al., 2022)
Capparaceae	*Boscia coriaceae*	Twiggy evergreen, usually multi-stemmed shrub or small tree, 2–6 m	Kenya	Root decoction with *U. Leptacladon* and *C. Hereroense* roots is taken orally	(Kyarimpa et al., 2023)
	Capparis tomentosa Lam.	Spiny, much-branched shrub or small tree	South Africa Sudan Benin	Root is taken with *Cyperus* sp. roots and decoction are taken orally	(Abdillahi & Van Staden, 2012b; Saloufou et al., 2018)
	Cleome gynandra L.	Annual, herbaceous, erect, and branched plant that can grow	Uganda	Leaf, roots, and flowers are chewed or boiled	(Kamatenesi-Mugisha & Oryem-Origa, 2005)
Caricaceae	*Carica papaya*	Tree with palm-like trunk up to 8 m	South Africa	Roots pounded and taken with porridge	(Semenya & Potgieter, 2013)

Clusiaceae	*Garcinia epunctata* Stapf	Tall tropical rainforest tree	Dr Congo	Seed is chewed	(Mola et al., 2022)
Combretaceae	*Terminalia mollis* M.A, Lawson	Medium-sized to large deciduous tree.	DR Congo	Root decoction taken orally 2x a day	(Valentin et al., 2022)
	Guiera senegalensis JF Gmel	Evergreen flowering shrub	Mali	Roots can be chewed	(Togola et al., 2020)
Costaceae	*Costus afer* Ker Gawl.	Tall perennial herb	DR Congo	Stem sap extract is taken orally	(Mola et al., 2022)
Cynomoriaeae	*Cynomorium coccineum* L.	Distinctive plant with blackish red inflorescence from an underground rhizome	Libya	No detailed information	(El-mokasabi et al., 2018)
Dipterocarpaceae	*Monotes engleri* Gilg.	Small to medium-sized tree	Zimbabwe	Root and bark decoction, and half cup is taken three times a day.	(Shopo et al., 2022a)
Ebenaceae	*Diospyros heudelotii* Hiern	Evergreen shrub or tree	Sierra Leone	Root and leaf decoction is taken orally	(Johnny et al., 2022)
Eriospermaceae	*Eriospermum Aabyssinica* Baker	Perennial herb growing from a depressed globose tuber	Cameroon Zimbabwe	Pulverized bulbs mixed with red oil are taken orally Bulb is chewed orally	(Tacham, 2015)
Euphorbiaceae	*Acalypha villicaulis* Hochst. ex A. Rich.	A bushy perennial, up to 2 m	Uganda Kenya	Infusion is taken orally. The plant is also an aphrodisiac	(Valentin et al., 2022)
	Alchornea hirtella Benth.	Climber, a spindly shrub or a small tree	DR Congo	Stem bark decoction is inserted in the anus	(Mola et al., 2022)
	Euphorbia tirucalli L. Manny	Branched succulent tree up to 3.5 m	South Africa Tanzania	No detailed information	(Abdillahi & Van Staden, 2012b; Ajao et al., 2019)
	Jatropha curcas L.	Glabrous shrub or small tree	South Africa[1] Sudan[2] Côte d' Ivoire	Root decoction is taken orally.[1] Seed is chewed immediately before act[2]	(Semenya & Potgieter, 2013; Saloufou et al., 2018)(Denis-esdras et al., 2017)

(continued)

Table 11.1 (continued)

Family	Scientific name	Description of medicinal plants	Country	Method of preparation and administration	References
	Macaranga capensis (Baill.). Benth. Ex Sim	Medium to large tree	Zimbabwe	Root decoction is taken orally	(Maroyi, 2013)
	Tragia brevipes Pax	Climbing perennial herb with stinging hairs	Uganda Ethiopia	Leaves are pounded, boiled, and taken orally	(Kamatenesi-Mugisha & Oryem-Origa, 2005)
	Manihot esculenta Crantz	Herb or shrub with elongated tubers	Mali	Tuber can be chewed once a day	(Togola et al., 2020)
	Mallotus oppositifolius (Geiseler) Müll. Arg	Shrub or a tree with elongated branches	Benin	No detailed information	(Madge, 1998)
Fabaceae	*Acacia mellifera* (Vahl) Benth	*Acacia mellifera* (Vahl) Benth shrub or small tree	Ethiopia	The roots are taken with the root and barks of *Amaranthus cruentus*, and decoction is drunk orally	(Asmerom et al., 2021)
	Acacia abysinica Hochst. ex Benth	Flat-topped tree, to 16 m	Uganda	Decoction of root or stem bark is taken orally	(Kyarimpa et al., 2023)
	Albizia adianthifolia W. Wight	Large flat-crowned tree.	DR Congo	Root decoction is taken orally three times a day	(Valentin et al., 2022)
	Albizia antunesiana Harms	Small to medium-sized tree	Angola Zimbabwe	Leaf decoction is taken orally	(Novotna et al., 2020; Maroyi, 2014)
	Albizia versicolor Welw. ex	Tree	Zimbabwe	Root decoction and ¼ cup is taken three times a day	(Shopo et al., 2022a)
	Arachis hypogaea L.	Annual herb, erect or straggling up to 30 cm	Uganda Mali	Seeds are roasted and taken as food	(Kamatenesi-Mugisha & Oryem-Origa, 2005)

Fabaceae	*Cassia sieberiana* DC	Perennial legume shrub tree	Mali	Root or fruit macerated and one tea cup is taken orally once a day	(Togola et al., 2020)
	Prosopis Africana (Guill. & Perr.) Taub.	Small to medium-sized tree	Mali[1] Western Sudan[2] Benin	Root or stem is chewed in the evening[1] Bark decoction is taken orally[2]	(Togola et al., 2020; Doka & Yagi, 2009) (Batcho, 2020)
	Bobgunnia madagascariensis	Small deciduous tree 3–4 m in height	DR Congo	Root and bark powder is soaked in palm oil and applied 15 min before the act	(Valentin et al., 2022)
	Elephantorrhiza elephantine (Burch.) Skeels	Low-growing suffrutex, arising from massive underground tuberous roots.	Zimbabwe South Africa	Pounded root decoction. Hot milk infusion	(Abdillahi & Van Staden, 2012a)
	Eriosema salignum E. Mey.	Perennial with slender erect stems up to 30 cm	South Africa	Hot milk infusion taken in small doses morning and night	(Abdillahi & Van Staden, 2012a)
	Piptadeniastrum africanum (Hook. f.)	Tall deciduous tree up to 50 m	DR Congo Cameroon	Root macerated is inserted anally. Root bark decoction is taken orally	(Mola et al., 2022; Natacha et al., 2022)
	Senegalia mellifera (Vahl) Seigler & Ebinger	Shrub or small tree	Zimbabwe Tanzania	Cooked with meat and taken orally Handful roots are chopped and added to boiling water	(Kacholi & Amir, 2023; Shopo et al., 2022a)
	Senna petersiana (Bolle) Lock	Shrub or small deciduous tree	DR Congo Zimbabwe	Root infusion and ½ glass is taken 2 times a day	(Valentin et al., 2022)
Fabaceae	*Drimia elata* Jacq.	Perennial herb up to 1.2 m with overlapping reddish scales on the bulb	South Africa	Tuber is boiled for 5 min	(Semenya & Potgieter, 2013)
	Eucomis pallidiflora Baker	Perennial growing from a large bulb	South Africa	Bulb decoction is taken orally	(Semenya & Potgieter, 2013)

(continued)

Table 11.1 (continued)

Family	Scientific name	Description of medicinal plants	Country	Method of preparation and administration	References
	Baphia massaiensis Taub.	Shrub or small tree	Namibia	No detailed information	(Cheikhyoussef et al., 2011)
	Desmodium ramosissimum G Don	Perennial erect semi-woody herb	Comoros	No detailed information	(Saive et al., 2020)
	Tetrapleura tetraptera (Schum & Thonn). Taubert.	Flowering herbaceous plant	Côte d' Ivoire	Bark infusion is taken orally	(Malan et al., 2015)
Haloragaceae	*Gunnera perpensa* L.	Herbaceous perennial	South Africa	No detailed information	(Van Wyk, 2015)
Hypericaceae	*Hypericum monogynum* L.	Bushy shrub	Mauritius	Leaf decoction is taken orally	(Suroowan et al., 2019)
Hypoxidaceae	*Hypoxis colchicifolia* Bak.	Herbaceous perennial plant	South Africa	No detailed information	(Abdillahi & Van Staden, 2012b)
	Hypoxis hemerocallidea Fisch., C.A. Mey. & Ave- Lall	Robust perennial herb, up to 40 cm tall	South Africa Mozambique	Tuber decoction is taken orally	(Semenya & Potgieter, 2013
	Hypoxis obtusa Burch. Ex Ker Gawl.	Similar to *H. hemerocallidea* but differing in its leaves being stiff, twisting with age	South Africa	Pounded and mixed with *E. pallidiflora* (bulb) taken with warm water	(Semenya & Potgieter, 2013)
Icacenaceae	*Pyrenacantha scandens* Planch. Ex. Harv	Climber up to 3 m	South Africa	Details not provided	(Abdillahi & Van Staden, 2012b)
Icacenaceae	*Icacina oliviformis* (Poir.) J. Raynal	Drought resistant sub shrub	Benin	No detailed information	(Batcho, 2020)

Irvingiaceae	*Irvingia smithii* Hook.f.	Tree up to 30 m	Dr Congo	Stem bark decoction is taken orally	(Mola et al., 2022)
Lamiaceae	*Ocimum basilicum* L.	Erect perennial herb	DR Congo	Root is chewed	(Mola et al., 2022)
	Clerodendrum myricoides (Hochst.) R. ex Vatke	Small shrub up to 3.5 m	South Africa, Kenya	No detailed information	(Abdillahi & Van Staden, 2012b)
Loganiaceae	*Strychnos cocculoides* Baker	Small deciduous tree	DR Congo[1] Zambia[2]	Roots and leaves decoction is taken orally[1] Unripe fruit powder is mixed with tea taken orally before bedtime[2]	(Valentin et al., 2022)
	Strychnos stuhlmannii L.f.	Small to medium-sized tree with dense foliage	DR Congo	Root powder decoction is taken orally 1 cup each evening for 1 week	(Valentin et al., 2022)
Lythraceae	*Woodfordia fruiticosa* (L.) Kurz	Branched tropical shrub	Comoros	Root decoction is taken orally	(Saive et al., 2020)
Malvaceae	*Grewia flavescens* Juss.	Scrambing shrub	Zimbabwe	No detailed information	(Maroyi, 2013)
	Sida tenuicarpa Vollesen	Small shrub	Uganda	Leaves are pounded and boiled taken orally	(Odongo et al., 2018; Kamatenesi-Mugisha & Oryem-Origa, 2005)
	Cola acuminata Schott & Endl.	Evergreen tree Small to medium-sized tree	Kenya Uganda Benin	Seeds are chewed Fruit is eaten, seeds can be taken in tea or milk as beverage	(Kamatenesi-Mugisha & Oryem-Origa, 2005; Kyarimpa et al., 2023)
Meliaceae	*Pterorhachis zenkeri* Harms.	Shrub	Cameroon	Infusion of 2.5 g of roots is taken orally or chewed	(Tacham, 2015)
	Khaya senegalensis (Desr.) A. Juss.	Deciduous tree	Mali	Root decoction is taken once a day	(Togola et al., 2020)

(continued)

Table 11.1 (continued)

Family	Scientific name	Description of medicinal plants	Country	Method of preparation and administration	References
	Neobeguea mahafaliensis J.-F. Leroy	Small to medium-sized tree	Madagascar	No detailed information is used	(Van Wyk, 2015)
Melianthaceae	*Bersama lucens* (Hochst.)	Small to medium-sized tree of 3 to 10 m high	South Africa	Bark is used	(Abdillahi & Van Staden, 2012b)
	Bersama stayneri Phill.	Shrub or small tree 3 to 10 m high.	South Africa	Bark decoction is taken orally	(Abdillahi & Van Staden, 2012b)
	Bersama swinnyi E. Phillips	Small to medium-sized tree	South Africa	Bark	(Abdillahi & Van Staden, 2012b)
Meliceae	*Ekebergia benguelensis* Welw. ex. C. DC	Small to medium-sized deciduous tree.	DR Congo Tanzania Zimbabwe	Root powder decoction is taken orally ½ cup three times a day	(Kacholi & Amir, 2023; Valentin et al., 2022)
Mimosaceae	*Dichrostachys cinerea* (L.) Wight & Arn	Shrub or small tree, sometimes suckering and thicket forming	Uganda[1] Eritrea[1] South Sudan[2]	Bark is pounded and boiled in water taken orally[1] Roots powder decoction taken orally[2]	(Musa et al., 2011; Chinsembu, 2016; Yemane et al., 2017)
Myricaceae	*Myrica salicifolia* Hochst. Ex A. Rich.	Dioecious tree or shrub	Uganda	Root and bark decoction taken orally	(Kamatenesi-Mugisha & Oryem-Origa, 2005)
	Macrotyloma axillare (E.Mey.) Verdc	Climbing herb up to 3.5 m	Uganda	Leaf roots and tuber are pounded and taken orally	(Kamatenesi-Mugisha & Oryem-Origa, 2005)
Myrothamnaceae	*Myrothamnus flabellifolius* Welw.	Prostate, ascending or erect subshrub	South Africa	Whole plant is taken as tea	(Semenya & Potgieter, 2013)
Oleaceae	*Schrebera trichoclada* Welw.	Shrub or bushy tree	DR Congo	Decoction of a handful of roots for 45 min in 2 L of water. Cold drink ½ cup 3x / day	(Valentin et al., 2022)

	Ximenia caffra Sond.	Dioecious, sparsely branched shrub or small tree up to 6 m.	Namibia Zimbabwe Botswana South Africa	Leaf and root decoction taken orally Root powder is taken with beer in Botswana	(Ajao et al., 2019; Maroyi & Cheikhyoussef, 2015)
	Ximenia Americana L	Small or large shrub or small tree	Mali Namibia	Bark or root decoction is taken once a day	(Cheikhyoussef et al., 2011)
Orchidaceae	*Himantoglossum hircinum* (L.) Spreng.	Tuberous geophyte	Algeria	Root dry powder decoction is taken orally	(Zatout et al., 2021)
Phytolaccaceae	*Phytolacca dodecandra* L'Her.	Robust perennial climber or scrambling shrub	Uganda Ethiopia	Roots are taken with mashed bananas	(Kamatenesi-Mugisha & Oryem-Origa, 2005)
Plumbaginaceae	*Plumbago zeylanica* L.	Scrambling or climbing perennial up to 1.5 m	Zimbabwe[1] South Africa[1] Uganda[1] Ethiopia[2]	Root decoction is taken orally[1] Fresh leaf infusion is taken orally[2]	(Shopo et al., 2022a; Kyarimpa et al., 2023)
Poaceae	*Eragrostis curvula* (Schrad.) Nees.	Densely caespitose perennial without rhizomes or stolons	Lesotho	Root infusion taken orally	(Seleteng Kose et al., 2015)
	Penniselum typhoides (L.) Morrone	Annual grass up to 3 m high with leaves up to 1 m long and up to 7 cm wide. Inflorescence dense, globose or long linear and spike-like.	Mali	Seed is pounded, and material is taken with warm water	(Rakuambo, 2002)
	Albemoschus esculentus (L.)	Erect annual plant slightly woody	Mauritius	Fruits are chewed	
Polygonaceae	*Rumex abyssinicus* Thunb.	Perennial scrambling or climbing dioecious herb	Uganda	Leaf and stem can be chewed	(Odongo et al., 2018; Kamatenesi-Mugisha & Oryem-Origa, 2005)
Rosaceae	*Prunus persica* (L.) Bartsch	Tree-cultivated plant	South Africa	Root pounded and taken with soft porridge	(Jeruto et al., 2015; Semenya & Potgieter, 2013)

(continued)

Table 11.1 (continued)

Family	Scientific name	Description of medicinal plants	Country	Method of preparation and administration	References
	Prunus africana (Hook. f.) Kalkman	Medium to large evergreen tree	Ethiopia	Fresh roots are crushed and soaked in water and then one cup is drunk	(Denis-esdras et al., 2017; Rakuambo et al., 2006)
	Rubus pinnatus Willd.	Climber shrub under 2 m	Kenya	Freshly collected root decoction is taken three times a day	(Jeruto et al., 2015)
	Rubus steudner Schweinf	Hairy, scrambling shrub	Kenya	Root decoction is taken orally three times a day	(Jeruto et al., 2015)
	Crossopteryx febrifuga (Afzel. Ex G. Don) Benth	Shrub or small deciduous tree	DR Congo	Root powder decoction is taken orally three times a day	(Valentin et al., 2022)
Rubiaceae	*Coffea Arabica* L.	Shrub or small tree.	Uganda	Seeds are roasted and taken as tea. They can be chewed	(Kamatenesi-Mugisha & Oryem-Origa, 2005)
Rubiaceae	*Gardenia resiniflua* Hiern	Shrub or small tree	Zimbabwe	Bark fruits or roots added to 2 L of boiling water ½ cup is taken orally three times a day	(Shopo et al., 2022a)
	Gardenia ternifolia Schumach. & Thonn	Shrub or small tree	Mali Zimbabwe	Root decoction is taken once a day	(Togola et al., 2020)
	Gardenia erubescens Stapf & Hutch	Tree	Benin	No detailed information	(Cheikhyoussef et al., 2011; Lin et al., 2022)
	Tarenna graveolens (S. Moore)	Evergreen shrub or tree up to 14 m	Uganda Kenya	Bark, leaf, and root are pounded, boiled, and taken orally	(Odongo et al., 2018; Kamatenesi-Mugisha & Oryem-Origa, 2005; Kyarimpa et al., 2023)

Rutaceae	*Citrus aurantium* L.	Evergreen tree up to 5 m-	DR Congo Comoros	Roots are chewed	(Mola et al., 2022; Saive et al., 2020)
	Zanthoxylum capense (Thunb). Have.	Small to medium-sized tree	South Africa	Roots are mixed with roots of a *Cyperus* species	(Abdillahi & Van Staden, 2012b)
	Zanthoxylum humile (E.A. Bruce) P.G. Waterman	A dioecious plant sufrutex	South Africa	Roots pounded and taken with soft porridge or warm water	(Semenya & Potgieter, 2013)
Santalaceae	*Osyris lanceolate* Hochst. & Steud. Ex A. DC	Large evergreen shrub	South Africa	Root decoction is taken orally	(Semenya & Potgieter, 2013)
Sapindaceae	*Eriocoelum microspermum* Radlk. ex De Wild.	Tree up to 6 m	DR Congo	Stem decoction taken orally	(Mola et al., 2022)
Solanaceae	*Capsicum frutescens*	Herb	Uganda Mali	Fruit can be chewed or boiled	(Kacholi & Amir, 2023; Kamatenesi-Mugisha & Oryem-Origa, 2005)
	Solanum incanum Dunal	Perennial herb or herbaceous small shrub	Tanzania	Root decoction taken orally	(Kacholi & Amir, 2023)
Tiliaceae	*Grewia similis* K. Schum.	Shrub, up to 3(−4) m tall	Uganda Tanzania	Leaf or bark pounded/ boiled taken orally	(Shopo et al., 2022b; Giesen et al., 2007) (Shopo et al., 2022b)
Urticeae	*Urtica massaica* Mildbr.	Rhizomatous perennial herb up to 2 m	Uganda Kenya	Leaves infused overnight are taken three times a day	(Kamatenesi-Mugisha & Oryem-Origa, 2005)
Pandanaceae	*Pandanus mayotteensis* H. St. John	Dioecies palm like tree	Comoros	Root decoction is taken orally	(Saive et al., 2020)

(continued)

Table 11.1 (continued)

Family	Scientific name	Description of medicinal plants	Country	Method of preparation and administration	References
Rubiaceae	*Pentas lanceolate* (Forssk.) Deflers	Many branched sprawling plant	Comoros	No detailed information	(Saive et al., 2020)
Zingiberaceae	*Alpinia officinarum* Hance	Herbaceous plant	Morocco	Root powder decoction is taken orally	(Abouri et al., 2012; Hyde et al., 2025)
	Aframomum melegueta K. Schum.	Perennial herb with a palm like appearance	Benin	No detailed information	(Batcho, 2020)

The numerical superscripts indicate the country of origin of the method of preparation for plant remedies for Erectile Dysfunction, with superscript 1 representing method of preparation 1 and superscript 2 representing method of preparation 2.

Table 11.2 Medicinal plants used for managing ED in Africa categorized as plants with scientific validation

Family	Scientific name	Description of medicinal plant	Traditional method of administration	Parts and extracts used	Pharmacology	References
Areraceae	*Borrasus aethiopium* Mart.	Large palm tree	Root decoction is taken three times a day in Mali	Seed methanolic extract	Increased sexual behavior	(Abdullahi, 2011)
Apiaceae	*Conium maculatum* L.	Biennial herbaceous plant up to 2.4 m. Toxic plant	No detailed information	Arial parts ethanolic extract	Low-dose PDE5 inhibition	(Anand Ganapathy et al., 2021)
Amaryllidaceae	*Tulbaghia* sp. cf. *violacea* Harv	Rare Perennial herb, smelling strongly of garlic.	No detailed information	Leaves and rhizomes ethanolic extract	Extract significantly increased LH-induced testosterone production in testicular cell cultures	(Ebrahim & Pool, 2010)
Apocynaceae	*Hunteria umbellate* (K. Schum.) Hallier f.	Shrub or small tree	No detailed information	Seed hot water extract	PDE5 inhibition, ACE, and AChE at 100mg/kg. Increase in serum levels of LH, FSH, and testosterone	(Masuku et al., 2020)
	Mondia whitei (Hook. F.) Skeels	Robust perennial climber, growing from a large tuberous rootstock	It is ingested orally after grinding, chewing, or boiling and can be consumed with food, water, or tea- Cameroon Uganda, Kenya	Root aqueous and hexane extract	Nitric oxide synthesis activity increased by 7% following a 6-week administration of *Mondia* extract at a dosage of 200 mg/kg	(Tacham, 2015; Babalola et al., 2023)
	Wrightia natalensis Stapf.	Small to medium-sized deciduous tree	Root bark powder and portions are mixed with beer and water. The mixture is consumed immediately or within two days-South Africa	Root ethanolic extract	Dose-dependent relaxation in the corpus cavernosumsmooth muscle at final concentrations of 13 and 6.5 mg/ml	(Abdillahi & Van Staden, 2012a)

(continued)

Table 11.2 (continued)

Family	Scientific name	Description of medicinal plant	Traditional method of administration	Parts and extracts used	Pharmacology	References
Asphodelaceae	*Bulbine latifolia* (L.f.) Roem. & Schult. f.		Root: erectile dysfunction. South Africa	Roots ethanolic extract	Increased sexual function Androgenic potential Penile reflexes enhancement	(Abouri et al., 2012; GamalEl Din et al., 2019)
Myristicaceae	*Myristica fragrans* Houtt.	Small aromatic evergreen tree	Seed powder mixed with *Ammodaucus leucotrichus* and honey is taken orally	Seed ethanolic extract	Increased sexual behavior	(Abdullahi, 2011)
Fabaceae	*Pilostigma reticulum* DC	Evergreen shrub	Fruit is pounded, and a teaspoon is taken once a day in Mali	Leaf ethanolic extract	Increased sexual behavior of male rats and increased testosterone	(Dasofunjo et al., 2013)
	Tamarindus indica L.	Medium to large evergreen tree	Root or stem is chewed in the evening	Fruit aqueous extract	Increased sperm production, increased sexual behavior in male rats	(Dasofunjo et al., 2013)
Orobanchaceae	*Cistanche tubulosa* (Schenk) R.	Branchless pyramid shaped plant	No detailed information	Details not provided	Showed recovery of the sex hormone in diabetic male rats. Showed anti-oxidant, anti-inflammatory, and steroidogenesis effects	(Kong et al., 2018)
Arecaceae	*Cocos nucifera* L.	Tree up to 30 m	Fruits are consumed by men in Ghana	Ethanolic fruit contents extract	Improves sexual behavior in rats	(Ajao et al., 2019; Singh et al., 2012)

Asparagaceae	*Asparagus africanus* Lam	Slender scrambling shrub or climber	The infusion of the powdered root is taken two tablespoonfuls with half a teaspoonful of honey in Nigeria	Root aqueous extract	The plant's potential aphrodisiac properties may be attributed to its capacity to augment the synthesis of hormones, including androgens, estrogens, and thyroxines	(Babalola et al., 2023)
Asphodelaceae	*Aloe excelsa* A. Berger	Single-stemmed small tree up to 6 m	Details not provided	Orange sap freeze-dried to fine powder	Increase sexual behavior	(Gundidza et al., 2009)
Asteraceae	*Elephantopus scaber* subsp. Plurisetus (O. Hoffm.) Philipson	Erect perennial herb up to 1.3 m	No detailed information	Whole plant ethanolic extract	In vitro PDE5 inhibition	(Anand Ganapathy et al., 2021)
Asteraceae	*Inula glomerata* Oliv. & Hiern	Robust perennial herb up to 2 m	No detailed information	Leaves methanolic	Lowered ACE activities facilitated smooth muscle relaxation in vitro	(Ojo et al., 2022)
Celastraceae	*Catha edulis*	Shrub or small to medium-sized unarmed tree	Leaves or stems are chewed orally in Uganda	Leaf aqueous extract	Increase in plasma testosterone Levels by more than twofold	(Mali et al., 2022)
	Salacia kraussii (Harv.) Harv.	*suffrutex* with annual shoots up to 3 m	Fruits are eaten, and root decoction is taken orally in South Africa	Root methanolic extract	Increased testosterone and nitric oxide levels and simultaneously elevated the activities of arginase and acetylcholinesterase in rats	(Ojo et al., 2022)
Clusiaceae	*Allanblackia floribunda* Oliv.	A medium-sized fruit tree native to sub-Saharan Africa	Details not provided	Stem bark aqueous and ethanolic extracts	The crude ethanolic extract (CEE) induced a relaxation on aorta rings with EC50¼1172 μg/mL	(Ajao et al., 2019; Brusotti et al., 2016)

(continued)

Table 11.2 (continued)

Family	Scientific name	Description of medicinal plant	Traditional method of administration	Parts and extracts used	Pharmacology	References
Clusiaceae	*Garcinia kola* Heckel.	Medium-sized evergreen tree of lowland tropical rainforest up to 15 m tall	Seed is chewed in DR Congo	Seed extract	The seed was shown to increase sexual behavior parameters such as mount frequency, number of ejaculations, and intromission frequency, with the low dose (200 mg/kg)	(Opuwari & Moundipa, 2021)
Combretaceae	*Anogeissus leiocarpus* (DC.) Guill. & Perr.	Tree up to 18 m	Leaves powder taken with soft drink or meals everyday in Nigeria and used as an aphrodisiac in Chad	Stem bark aqueous extract	Stem bark extract (50 and 100 mg/kg) on antidepressant drug (paroxetine)-induced sexual dysfunction in male adult Wistar rats for a period of 21 days	(Faleyimu & Akinyemi, 2010; Masuku et al., 2020)
Combretaceae	*Terminalia catappa* L.	Tree to 25 m tall	An infusion of the kernel is taken orally some few min before sexual intercourse in Nigeria	Seed aqueous extract	Increase in mount and intromission frequencies, as well as a prolonged ejaculatory latency in male rats. Effectively suppressed arginase activity	(Babalola et al., 2023)
Convolvulaceae	*Ipomoea batatas (L.) Lam.*	Perennial herb, widely cultivated for its edible tubers	Roots are shredded and inserted anally in DR Congo	Tuber and aerial part aqueous extract	Extract improved semen quality, increased levels of testosterone, FSH, LH, and estradiol.	(Mola et al., 2022; Majid et al., 2019)

Euphorbiaceae	*Alchornea cordifolia*	A shrub or small tree	Roots are mixed with the root of *Tabernanthe iboga Bail* and taken orally with water in Gabon	Methanolic leaf extract	Increase in testicular weight, sperm count, and motility, and serum testosterone levels at amounts lower than 400 mg/kg	(Ajao et al., 2019)
Fabaceae	*Cylicodiscus gabunensis*	Large tree up to 61 m	No detailed information	Root bark aqueous extraction	Extracts inhibit PDE-5, arginase, and ACE	(Anand Ganapathy et al., 2021)
Fabaceae	*Eriosemakraussianum* Klotzsch	Ascending to suberect pyrophytic herb growing from a woody rootstock	Hot milk infusions. Pounded roots decoctions are taken in small doses- South Africa	Hot milk infusion root stock extract	Smooth muscle relaxation activity	(Owaba et al., 2021; Babalola et al., 2023)
Fabaceae	*Mucuna pruriens* (L.) DC.	Trailing or climbing herb with annual stems growing from perennial rootstock	No detailed information	Seed ethanolic extract	Showed a significant increase in MF, IF, and EL and decreased the ML, IL, PEI, and inter intromission interval	(Mali et al., 2022)
Geraniaceae	*Monsonia angustifolia* E.Mey. ex. A. Rich.	Decumbent or sub-erect annual herb.	Decoction prepared from crushed dried leaves and stems is taken orally- South Africa	Aerial parts aqueous extract	Enhancement of sexual activity by increasing the testosterone level. Nitric oxide activity	(Ajao et al., 2019)
Liliaceae	*Gloriosa superba* L.	Erect or scandent herb Toxic to humans	Powdered corms taken for impotency and as aphrodisiacs in South Africa The root powder is taken with *tej* in Ethiopia	Extracts of water, chloroform, and alcohol of *G. superba* stem and leaves	Increased blood testosterone levels and copulating performance in male albino rats	(Abdillahi & Van Staden, 2012b; Asmerom et al., 2021; Masuku et al., 2020)

(continued)

Table 11.2 (continued)

Family	Scientific name	Description of medicinal plant	Traditional method of administration	Parts and extracts used	Pharmacology	References
Meliceae	*Ekebergia capensis* C. Presl ex C. DC.	Medium-sized evergreen tree	Stem bark decoction is taken orally in Uganda	100 g of *M. whitei* and *E. capensis* ethanolic extract	Improve sexual behavior in rats	(Kyarimpa et al., 2023; Gundidza et al., 2009)
Moringaceae	*Moringa oleifera* Lour.	Small tree, cultivated for its edible roots and leaves	Bark and roots are ground together and taken with meals in Nigeria Three seeds are chewed once a day in Mali	Leaf aqueous and ethanolic extract	In vitro PDE5-inhibition. Improves blood testosterone concentration and mating performance	(Anand Ganapathy et al., 2021)
Musaceae	*Musa paradisiaca* L.	Introduced (cultivated) hybrid species	Infusion of the root and stem as juice or juice from the fruit or decoction of the unripe fruit is taken orally in Nigeria	Aqueous extract	Stimulation of the normal functioning of the testes through androgenic activity. Increase testosterone levels	(Ajao et al., 2019; Chauhan et al., 2014; Gupta et al., 2019)
	Morella serrata Lam.	Small tree or shrub up to 12 m	Decoction of the root is taken orally in South Africa	Aqueous root extract	Increase in sexual behavior, such as erection, quick flips, long flips, and total penile reflexes	(Ajao et al., 2019)
Myricaceae						
Piperaceae	*Pentadiplandra brazzeana*	Evergreen shrub or liane	Root is chewed in Uganda	Root aqueous extract	Increases serum and testicular testosterone levels.	(Mola et al., 2022; Chauhan et al., 2014)
Piperaceae	*Piper guineense* Schumach. &Thonn.	A climbing perennial	Leaf decoction is inserted anally or root is chewed in DR Congo	Dried fruits aqueous extract	Increased penile erection index	(Gupta et al., 2019; Kamtchouing et al., 2002)

Pittosporaceae	*Pittosporum viridiflorum* Sims var.	Small to medium-sized tree	Root powder is taken with ox hooves soup in Zimbabwe and South Africa	Leaves and roots ethanolic extract	Improved sexual behavior in male rats	(Chifamba & Zengeni, 2024; Abdillahi & Van Staden, 2012b)
Polygalaceae	*Securidaca longepedunculata* Fresen.	Deciduous shrub or small tree	Infusion, decoction, or maceration of roots are taken orally. It can be mixed with traditional beverage in Zimbabwe and South Africa	Chloroform and ethanolic root extract	Smooth muscle relaxation in vivo	(Abdillahi & Van Staden, 2012b; Babalola et al., 2023)
Rubiaceae	*Fadogia agrestis*	*Fadogia agrestis* Shrub up to 6 m	The stem is boiled, powdered, or mixed with spices to make a drink. A stem infusion taken orally half an hour before sexual intercourse in Nigeria	Stem aqueous extract	The ejaculatory latency was prolonged, and there was also a significant increase in serum testosterone concentrations	(Ajao et al., 2019; Babalola et al., 2023)
Rutaceae	*Citropsis articulate*	Evergreen shrub or small tree from 2.5 to 7 m	Root and bark chewed and swallowed	Crude ethanolic leaf extract	Increase the level of serum testosterone	(Owaba et al., 2021)
Smilacaceae	*Smilax kraussiana* Willd.	A vigorous scrambling vine.	A decoction of the root is made either with boiling water or local gin before oral consumption in Nigeria	Root methanolic extract	Improved sexual behavior in rats	(Babalola et al., 2023)
Verbenaceae	*Clerodendrum capitatum*	Shrub, scrambling to 6 m	Root decoction is taken orally in Sudan	Roots methanolic extract	PDE5 enzyme inhibitory activity at a value of 0.161 mg/mL	(Anand Ganapathy et al., 2021)

(continued)

Table 11.2 (continued)

Family	Scientific name	Description of medicinal plant	Traditional method of administration	Parts and extracts used	Pharmacology	References
Vitaceae	*Rhoicissus tridentata* (L.f.) Wild & R.B. Drum	Climber with tendrils but occasionally a shrub or very small tree	Details not provided have been used traditionally by *Vhavenda* people to treat erectile dysfunction for centuries	Root bark chloroform ethanolic extract	In vitro smooth muscle relaxation activity of New Zealand rabbit penis at low concentrations	(Abdillahi & Van Staden, 2012b; Rakuambo, 2002; Rakuambo et al., 2006)
Zingiberaceae	*Aframomum danielli* (Hook.f) K.Schum	A perennial flowering plant up to 4 m	Details not provided	Seed alkaloidal fraction	Hepatoprotective. The plant has pro-ejaculatory androgenic aphrodisiac effects	(Anand Ganapathy et al., 2021)

Table 11.3 Medicinal plants used for managing ED categorized based on clinical trial stages or being widely consumed as part of diet

Family	Scientific name	Description of botanical plant	TMPA	Extraction method	Pharmacology	Reference
Alliaceae	*Allium sativum* L.	A herb growing from a strongly aromatic, rounded bulb	Details not provided. Traditionally used for treating ED in many African countries	Bulb and seed aqueous, petroleum, and ethanolic extract	Increased sexual activity or behavior through the activity of sulphated compounds	(Owaba et al., 2021; Kyarimpa et al., 2023)
Amaryllidaceae	*Allium cepa* L.	Biennial widely cultivated plant	Infusion of bulb leaf and root is taken orally in Uganda, Mauritius, and Comoros	Onion peel	Enhances copulatory behavior in male rats. Generally safe for human consumption	(Saive et al., 2020; Ajao et al., 2019; Suroowan et al., 2019)
Cucurbitaceae	*Cucurbita pepo L.*	Annual, monoecious climber cultivated worldwide for its edible fruits	Roasted pumpkin seeds are chewed for their nutritional value	Seed powder aqueous extract	Improve sexual behavior in rats	(Gundidza et al., 2009)
Malveceae	*Sida acuta*	Erect branched suffrutex	Root decoction is taken orally in Zimbabwe, Benin, Ethiopia, and Nigeria	Polyherbal formulation of *Boesenbergia rotunda* (L.), *Dactyloctenium aegyptium* (L.) Willd, *Oryza sativa* L	Double-blind randomized trial showed an increase IIEF-5 scores	(Lin et al., 2022)
Punicaceae	*Punica granatum* L.	Multi-stemmed deciduous tree- cultivated plant	Fresh juice is made from fruit and taken orally in Tanzania and Kenya	Fruit juice extract	Enhanced sexual behavior in rats Safe for dietary consumption in humans	(Kacholi & Amir, 2023; Lydia et al., 2019)

(continued)

Table 11.3 (continued)

Family	Scientific name	Description of botanical plant	TMPA	Extraction method	Pharmacology	Reference
Rubiaceae	*Pausinystalia johimbe* Pierre. Ex. Bielle	Tree up to 30 m tall	Stem bark decoction is prepared and administered at a dosage of one wineglassful, three times per day in Nigeria	Isolated yohimbine compound	Displace epinephrine from alpha-2 adrenergic receptors in the pelvic areas, and it increases the proneness to arousal by supplying epinephrine from the alpha-2 receptors to the central nervous system	(Opuwari & Moundipa, 2021; Babalola et al., 2023)
Zingiberaceae	*Zingiber officinale* Roscoe	Herbaceous perennial plant	The exudate is chewed and swallowed		Widely consumed by humans as spices	(Kamatenesi-Mugisha & Oryem-Origa, 2005)
Zygophyllaceae	*Tribulus terrestris*	Prostrate annual herb. Poisonous to livestock	The decoction of the whole plant is taken at a dose of three wineglassful a day in Nigeria	Herbal drug	Increase endogenous testosterone levels in vivo. Clinical trials show androgenous activity in humans, but the efficacy is not well established	(Babalola et al., 2023; GamalEl Din et al., 2019)

TMPA Traditional method of preparation and administration.

considered potent African traditional remedies that can be prioritized for the development of scientifically acceptable natural remedies or to develop drugs.

Eriosema kraussianum (Fig. 11.3a) is a used to make a famous remedy in South Africa known as Bangalala. Hot milk infusions of *Eriosema* roots and/or pounded boiled root decoctions of the plant are taken in small doses in the morning and at night for impotence (Rayise et al., 2024).

Ximenia caffra (Fig. 11.3b) is used in various regions of Africa for its numerous medicinal properties. The roots are used as a remedy for ED. The seeds from the fruit are used to produce oils that are used for skincare, promoting blood circulation and preventing skin aging.

Sonchus oleraceus L (Fig. 11.3c) has been indicated in the ethnobotanical information that the consumption of this herb for consecutive days has been associated with complete recovery, thereby eliminating the need for continued medicinal use (Jeruto et al., 2015). To substantiate these claims, an experiment has to be conducted to validate the efficacy of this herb.

Ozoroa insgnis (Fig. 11.3d) is a plant with many medicinal properties that is used for managing ED. It has been widely studied for its efficacy in managing cancer (Rea et al., 2003) and no other efficacy studies on managing ED. The plant has been used to make a famous aphrodisiac remedy known as *vuka vuka*, which is made by a polyherbal formulation of root bark of *Mondia whiteii*, roots of *Albizzia antunesiana*, and stem bark of *Ozoroa insignis* (Hostettmann et al., 1998). Both the roots and bark of *Ozoroa insignis* can be utilized either in a polyherbal concoction or as a standalone formulation for managing ED.

11.3.3.2 Category 2

Medicinal plants in this category (Table 11.2) have been scientifically explored for their efficacy in managing ED. Many of them have been studied for their mechanism of action, which often involves inhibiting PDE5 (phosphodiesterase-5) and improving nitric oxide synthesis. Some plants in this category also have an impact on the central nervous system, which is crucial for sexual stimulation and achieving erections. Studies have been carried out outside of Africa, indicating the need for further research, development, and investment in equipment within African countries. The choice of research methods to test the efficacy of these medicinal plants is still limited, and certain methods, such as sexual behavioral studies, are considered subjective by some researchers. Implementing improved research designs is essential to produce reliable results and determine the status of the medicinal plants discussed in this chapter.

The roots of *Mondia whitei* (Fig. 11.4a) are used to make a famous aphrodisiac remedy in most parts of Africa. It is among the plants that are used to make a polyherbal *vuka vuka* in Zimbabwe and South Africa. The roots when dried produce a grayish white powder often sold in the streets as an aphrodisiac to help men alleviate problems with erection and libido.

Fig. 11.4 (**a**) *Mondia whitei* (Hook. f.) Skeels picture taken by Bart Wursten on 23 July 2004 in Cleveland dam, Zimbabwe. (**b**) *Rhoicissus tridentata* (L.f.) Wild & R.B. Drumm. picture taken by Sisanda Alakhe Velembo in Kirstenbosch National Botanical Garden, South Africa on 17 September 2017 (**c**) *Pittosporum viridiflorum* Sims var. picture taken by Bart Wursten in Ndundu Lodge, Vumba, on *5 January 2006*. (**d**) *Securidaca longepedunculata* Fresen. (Picture taken by BT Wursten at Rd. to Osborne Dam on 31 October 2004)

Rhoicissus tridentata (Fig. 11.4b) has been widely used by the *Vhenda* people from South Africa to cure ED. The plant has been pharmacologically validated and has been shown to cause rabbit smooth muscle relaxation.

Pittosporum viridiflorum (Fig. 11.4c) is a popular plant in Zimbabwe given the name *muchemedzambuya* with a sexual connotation meaning mourning of an old woman. The roots are cooked with meat soup, and the soup is drunk; roots can be put in a traditional beer as a traditional remedy for ED.

Securidaca longipendunculata (Fig. 11.4d) has been pharmacologically validated as a medicinal plant that is widely used in South Africa for managing erectile dysfunction. The plant root extract is used to make *Mpesu*, a famous traditional beverage as a medicine for ED (Rayise et al., 2024).

11.3.3.3 Category 3

Medicinal plants in this category (Table 11.3) have been involved in clinical trials with human subjects. These medicinal plants are used to make drugs and supplementary remedies found on the market for treating and managing ED. While these plants have been involved in clinical trials, it is important to note that they may not have official approval as evidence-based treatments. Other products available in the market that have not undergone testing, and some have even been banned, as they were found to contain synthetic PDE5 inhibitors, which should only be prescribed. Herbal drugs are often marketed as nutraceuticals and supplements, claiming to be organic with fewer side effects. Plants that are part of the regular diet are also included in this category, implying that they are generally considered safe for consumption and may have additional health benefits in improving erectile function.

Sida acuta (Fig. 11.5a) is used mainly for making outdoor sweeping brooms by indigenous people of Southern Africa, Nigeria, Zambia, and Zimbabwe. The whole plant is used to make a remedy for ED in most parts of Africa. In Nigeria, the traditional healers claim that if you chew the fresh roots of the plant will give a hard rock erection.

Fig. 11.5 (**a**) *Sida acuta* Burm. f picture taken by Bart Wursten at Lake Kyle Recreational Park, Zimbabwe on 7 January 2012. (**b**) *Tribulus terrestris* picture taken by Bart Wursten on 7 Jan 2010 at Guvulala Pan, Hwange NP, Zimbabwe. (**c**) *Pausinystalia johimbe* picture taken by Ehoarn Bidault at World Agroforestry Centre, Kenya

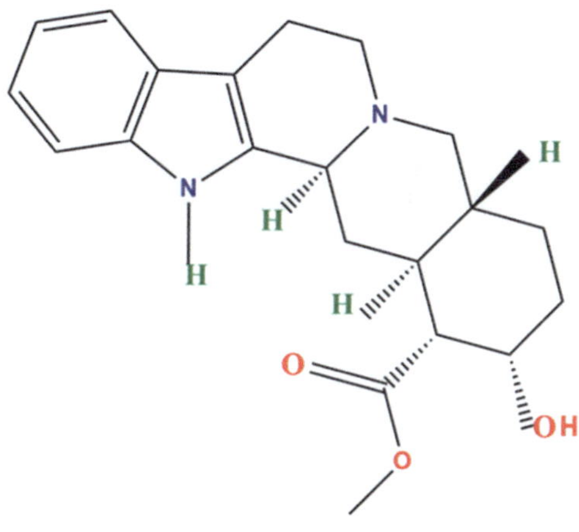

Fig. 11.6 2D chemical structure of Yohimbine indole alkaloid found in *Pausinystalia johimbe*. (Image was drawn using ChemDraw Ultra version 12.0)

Tribulus terrestris (Fig. 11.5b) is one of the most important plants widely used to manage ED in most parts of the world. Clinical trials have been conducted and show positive results, but there is no conclusive evidence on the safety and efficacy profile.

Pausinystalia johimbe (Fig. 11.5c) is an important plant species in managing ED, which has a worldwide attention and has been proven to be an effective remedy for ED. It grows in evergreen closed-canopy forest native to DR Congo, Cameroon, Equatorial Guinea, Gabon, and Nigeria.

An alkaloid isolated called yohimbine (Fig. 11.6) is a clinically proven cure for ED in low doses. Yohimbine chemical formula $C_{21}H_{26}N_2O_3$, molecular weight 354.4 g/mol. Yohimbine is an indole alkaloid with alpha-2 adrenoceptor antagonist activity.

The chemical name for yohimbine is methyl (1S,15R,18S,19R,20S)-18-hydroxy-1,3,11,12,14,15,16,17,18,19,20,21-dodecahydroyohimban-19-carboxylate (PubChem, accessed on 4 February 2025).

The compound serves as an active ingredient in nine pharmaceutical products made according to drug databank and is the first drug derived from plants used for the treatment of ED.

11.4 Future Perspective

Medicinal plants utilized for managing erectile dysfunction and other sexual dysfunctions are among the most commonly used and heavily sold herbal remedies. However, their unregulated use poses a significant threat to public health. There is a need for rigorous scientific validation and standardization of these medicinal plants. Regulatory authorities and governing bodies in Africa often do not give adequate attention to these issues (Bhagavathula et al., 2015).

Collaborations between biomedical scientists, botanists, and traditional healers are crucial in researching and developing drugs sourced from plant medicine. These partnerships will facilitate the dissemination of information from traditional healers, particularly details such as the conservation, cultivation, timing of harvesting, preparation of medicinal plants, and administration techniques, which are often scarce. This shared knowledge will guide future research in the development of natural medicines.

Medicinal plants make a significant contribution to the healthcare of men suffering from ED. The chapter highlighted that about one-third of the plants have been scientifically validated, and a few have undergone rigorous clinical trials. The increasing reliance on these natural remedies is driven by the limitations of conventional treatments, such as high costs and side effects. Research in the use of medicinal plants for selection of the most potent, non-toxic herbal remedies is the most unavoidable part of life since they form an integral part of our daily living.

Further research will assist in pharmacovigilance and educating people on the safe and toxic medicinal plants, given that numerous products are being manufactured without information on pharmacodynamics and pharmacokinetics information. Africa serves as the hub for unregulated herbal medicinal products. Research into herbal medicines in Africa will aid in mitigating the production of unregulated medicines.

Medicinal plants can be a huge economic value, particularly in rural communities. The conservation of these medicines can be beneficial to the community environment. Cultivation and sustainable of medicinal plants can provide income, support community resilience and improve research, making the raw materials easily accessible.

11.5 Conclusion

Throughout history, the management of erectile dysfunction using medicinal plants in Africa has remained prevalent. Even today, many men consult traditional medical practitioners for these valuable aphrodisiac medicinal plants. Ethnobotanical surveys have been conducted to document the knowledge and usage of these plants in various regions of Africa.

Some of these plants have been selected for further study to assess their efficacy in improving erectile function, often utilizing animal models. These investigations have yielded promising results, demonstrating desirable effects. Additionally, a subset of these plants has undergone clinical trials involving human subjects, further substantiating their potential effectiveness.

By addressing the existing gaps in research, African medicinal plants have the potential to make significant contributions to addressing this global health concern. Continued scientific exploration, documentation, and rigorous research will not only enhance our understanding of these plants' mechanisms of action but also provide valuable insights to the development of evidence-based therapies. The chapter highlights the importance of further research and the utilization of traditional medicinal knowledge in Africa to unlock the full potential of these medicinal plants. By bridging the research gaps, these plants can play a vital role in addressing the global challenge of managing erectile dysfunction.

References

Abdillahi, H. S., & Van Staden, J. (2012a). South African plants and male reproductive healthcare: Conception and contraception. *Journal of Ethnopharmacology, 143*(2), 475–480.

Abdillahi, H. S., & Van Staden, J. (2012b). South African plants and male reproductive healthcare: Conception and contraception. *Journal of Ethnopharmacology [Internet], 143*(2), 475–480. Available from: https://doi.org/10.1016/j.jep.2012.06.047

Abdullahi, A. A. (2011). Trends and challenges of traditional medicine in Africa. *African Journal of Traditional, Complementary and Alternative Medicines, 8*(5 SUPPL), 115–123.

Abouri, M., El, M. A., Msanda, F., Boubaker, H., Saadi, B., & Cherifi, K. (2012). An ethnobotanical survey of medicinal plants used in the Tata Province, Morocco. *International Journal of Medicinal Plants Research, 1*(7), 99–123.

Afolayan, A. J., & Yakubu, M. T. (2009). Erectile dysfunction management options in Nigeria. *The Journal of Sexual Medicine, 6*(4), 1090–1102.

Ajao, A. A., Sibiya, N. P., & Moteetee, A. N. (2019). Sexual prowess from nature: A systematic review of medicinal plants used as aphrodisiacs and sexual dysfunction in sub-Saharan Africa. *South African Joutnal of Botany [Internet], 122*, 342–359. Available from: https://doi.org/10.1016/j.sajb.2018.08.011

Anand Ganapathy, A., Hari Priya, V. M., & Kumaran, A. (2021). Medicinal plants as a potential source of Phosphodiesterase-5 inhibitors: A review. *Journal of Ethnopharmacology [Internet], 267*(August 2020), 113536. Available from: https://doi.org/10.1016/j.jep.2020.113536

Asigbaase, M., Adusu, D., Musah, A. A., Anaba, L., Nsor, C. A., Abugre, S., et al. (2024). Ethnobotanical and ethnopharmacological survey of medicinal tree species used in the treatment of diseases by forest-fringe communities of Southwestern Ghana. *Heliyon [Internet]., 10*(1), e23645. Available from: https://doi.org/10.1016/j.heliyon.2023.e23645

Asmerom, D., Kalay, T. H., Araya, T. Y., Desta, D. M., Wondafrash, D. Z., & Tafere, G. G. (2021). Medicinal plants used for the treatment of erectile dysfunction in Ethiopia: A systematic review. *BioMed Research International, 2021*, 6656406.

Babalola, O. O., Iwaloye, O., Ottu, P. O., Aturamu, P. O., & Olawale, F. (2023). Biological activities of African medicinal plants in the treatment of erectile dysfunction: A mechanistic perspective. *Hormone Molecular Biology and Clinical Investigation [Internet], 44*(4), 357–370. Available from: https://doi.org/10.1515/hmbci-2022-0090

Batcho, I.A. (2020). Diversity and Endogenous Knowledge of Aphrodisiac Plants in South and Central Benin Methods. Preprint, Research Square (pp. 1–32).

Bhagavathula, A. S., Elnour, A. A., Shehab, A., & Supplements, H. (2015). Pharmacovigilance on Sexual Enhancing Herbal Supplements Department of Clinical Pharmacy, College of Medicine and Health Sciences, University Associate Professor, Pharmacology Department; College of Medicine and Health.

Brusotti, G., Papetti, A., Serra, M., Temporini, C., Marini, E., Orlandini, S., et al. (2016). Allanblackia floribunda Oliv.: An aphrodisiac plant with vasorelaxant properties. *Journal of Ethnopharmacology [Internet], 192*, 480–485. Available from: https://pubmed.ncbi.nlm.nih.gov/27647010/

Chauhan, N. S., Sharma, V., Dixit, V. K., & Thakur, M. (2014). A review on plants used for improvement of sexual performance and virility. *BioMed Research International, 2014.*

Cheikhyoussef, A., Shapi, M., Matengu, K., & Mu, A. H. (2011). Ethnobotanical study of indigenous knowledge on medicinal plant use by traditional healers in Oshikoto region, Namibia. *Journal of Ethnobiology and Ethnomedicine [Internet], 7*(1), 10. Available from: http://www.ethnobiomed.com/content/7/1/10

Chifamba, J., & Zengeni, S. (2024). From Zimbabwean Indigenous Knowledge: Scientific Substantiation of the Aphrodisiac Efficacy and Safety of Pittosporum viridiflorum in Male Rat Models, *International Journal of Biochemistry Research & Review, 33*(5), 18–32.

Chinsembu, K. C. (2016). Ethnobotanical study of plants used in the management of HIV/AIDS-related diseases in Livingstone, Southern Province, Zambia. *Evidence-Based Complementary and Alternative Medicine, 2016*, 4238625.

Dasofunjo, K., Asuk, A. A., Ezugwu, H. C., Nwodo, O. F. C., & Olatunji, T. L. (2013). Aphrodisiac effect of ethanol extract of Piliostigma thonningii leaf on male albino wistar rats. *Journal of Applied Pharmaceutical Science, 3*(10), 130–135.

Denis-esdras, A. A., Fofana, S., & Kafana, S. (2017). Ethnobotany study of Loranthaceae, hemiparasitic plants used in traditional medicine by population, in the Sud-Comoé region (Côte d ' Ivoire). *Journal of Medicinal Plants, 5*(5), 215–222.

Doka, I. G., & Yagi, S. M. (2009). Ethnobotanical survey of medicinal plants in West Kordofan (Western Sudan). *Ethnobotanical Leaflets, 2009*(11), 8.

Ebrahim, M., & Pool, E. J. (2010). The effect of Tulbaghia violacea extracts on testosterone secretion by testicular cell cultures. *Journal of Ethnopharmacology [Internet], 132*(1), 359–361. Available from: https://doi.org/10.1016/j.jep.2010.08.018

El-mokasabi, F. M., Al-sanousi, M. F., & El-mabrouk, R. M. (2018). Taxonomy and Ethnobotany of Medicinal Plants in Eastern Region of Libya. *Journal of Environmental Science Toxicology and Food Technology, 12*(8), 14–23.

Faleyimu, O. I., & Akinyemi, O. (2010). Herbal approaches to the problem of erectile dysfunction in Kaduna state, Nigeria. *Egyptian Journal of Biology, 12*, 103–107.

GamalEl Din, S. F., Abdel Salam, M. A., Mohamed, M. S., Ahmed, A. R., Motawaa, A. T., Saadeldin, O. A., et al. (2019). Tribulus terrestris versus placebo in the treatment of erectile dysfunction and lower urinary tract symptoms in patients with late-onset hypogonadism: A placebo-controlled study. *Urology Journal, 86*(2), 74–78.

Giesen, W., Giesen, P., & Giesen, K. (2007, February 6). Lewa wildlife conservancy habitat changes at lewa wildlife conservancy, Kenya from cattle ranch to conservation area: Effects of changing management on habitat changes at Lewa wildlife conservancy, Kenya from cattle ranch to conservation area: Effects of changing management on habitat from 1962–2006. Available from: http://rgdoi.net/10.13140/RG.2.2.12661.73442

Gobena, M. B., Abdosh, T., & Dheresa, M. (2023, June). *E*rectile dysfunction and associated factors among patients with diabetes attending follow-up at a public hospital, Harar, Eastern Ethiopia. A cross-sectional study design. *Frontiers in Endocrinology,* (pp. 1–8).

Gundidza, G. M., Mmbengwa, V. M., Magwa, M. L., Ramalivhana, N. J., & Mukwevho, N. T. (2009). Aphrodisiac properties of some Zimbabwean medicinal plants formulations. *African Journal of Biotechnology, 8*(22), 6402–6407.

Gupta, A. K., Jain, A., & Singh, R. (2019). Aphrodisiac: An overview in present context of ailment. *Asian Pacific Journal of Health Sciences., 6*(2), 184–193.

Hostettmann, K., Potterat, O., & Wolfender, J. L. (1998). The potential of higher plants as a source of new drugs. *Chimia (Aarau), 52*(1–2), 10–17.

Hyde, M. A., Wursten, B. T., Ballings, P., & Coates, P. M. (2025). Flora of Zimbabwe: Species information: Cenchrus americanus [Internet]. Flora of Zimbabwe. Available from: https://www.zimbabweflora.co.zw/speciesdata/species.php?species_id=107810

Iwu, M. M., & Maurice, M. (2014). *Iwu African Medicinal Plants [Internet].* 183–184 p. Available from: https://www.routledge.com/Handbook-of-African-Medicinal-Plants/Iwu/p/book/9781466571976

Jeruto, P., Too, E., Mwamburia, L. A., & Amuka, O. (2015). An inventory of medicinal plants used to treat gynaecological-obstetric-urino-genital disorders in South Nandi Sub-County in Kenya. *Journal of Natural Sciences Research [Internet], 5*(18), 136–152. Available from: https://www.researchgate.net/publication/282696097

Johnny, J., Lebbie, A., & Wadsworth, R. (2022). Ethnobotanical survey of medicinal plants utilized by forest edge communities in southern Sierra Leone. *Journal of Medicinal Plants Research, 16*(1), 11–25.

Josephine Ozioma, E. O., & Antoinette Nwamaka Chinwe, O. (2019). Herbal medicines in African traditional medicine. *Herbal Medicine, IntechOpen*.

Kacholi, D. S., & Amir, H. M. (2023). Ethnobotanical study of medicinal plants traditionally used against erectile dysfunction in Tabora region, Tanzania. *Ethnobotany Research and Applications*, 25, 1-12.

Kamatenesi-Mugisha, M., & Oryem-Origa, H. (2005). Traditional herbal remedies used in the management of sexual impotence and erectile dysfunction in western Uganda. *African Health Sciences, 5*(1), 40–49.

Kamtchouing, P., Mbongue, G. Y. F., Dimo, T., Watcho, P., Jatsa, H. B., & Sokeng, S. D. (2002). SHORT REPORT E ¡ ects of Aframomum melegueta and Piper guineense on sexual behaviour of male rats, *13*(3), 243–247.

Kong, Z. L., Johnson, A., Ko, F. C., He, J. L., & Cheng, S. C. (2018). Effect of Cistanche tubulosa extracts on male reproductive function in streptozotocin–nicotinamide-induced diabetic rats. *Nutrients, 10*(10), 1562.

Kyarimpa, C., Nagawa, C. B., Omara, T., Odongo, S., Ssebugere, P., Lugasi, S. O., et al. (2023). Medicinal plants used in the management of sexual dysfunction, infertility and improving virility in the east African community: A systematic review. *Evidence-Based Complementary and Alternative Medicine, 2023*, 1–28.

Liengme, C. A. (1983). A survey of ethnobotanical research in southern Africa. *Bothalia, 14*(3/4), 621–629.

Lin, C., Pattraraachachai, J., Pawa, K. K., & Wongyai, S. (2022). A preliminary study of the efficacy of the polyherbal preparation Sao Thong Tai for erectile dysfunction among elderly men: A double-blind, randomized controlled trial. *Clinical Phytoscience [Internet], 8*(1), 9. Available from: https://doi.org/10.1186/s40816-022-00341-4

Lydia, K. K., Charles, I. M., Caleb, O. O., Collins, K. K., Benson, G. M., & Paul, N. W. (2019). Effectiveness of pomegranate (Punica Granatum L.) fruit extract on the sexual function in rats. *Journal of Microbial and Biochemical Technology, 11*, 424. https://doi.org/10.35248/1948-5948.19.11.424

Madani, S., Amel, B., Noui, H., & Djamel, S. (2017). An ethnobotanical survey of galactogenic plants of the Berhoum District (M ' sila, Algeria). *Journal of Intercultural Ethnopharmacology, 6*, 311–315.

Madge, C. (1998). Therapeutic landscapes of the Jola, The Gambia, West Africa. *Health & Place, 4*(4), 293–311.

Mahwasane, S. T., Middleton, L., & Boaduo, N. (2013). An ethnobotanical survey of indigenous knowledge on medicinal plants used by the traditional healers of the Lwamondo area, Limpopo province, South Africa. *South African Journal of Botany [Internet], 88*, 69–75. Available from: https://doi.org/10.1016/j.sajb.2013.05.004

Majid, M., Ijaz, F., Baig, M. W., Nasir, B., Khan, M. R., & Haq, I. U. (2019). Scientific validation of ethnomedicinal use of Ipomoea batatas L. Lam. as aphrodisiac and gonadoprotective agent against Bisphenol A induced testicular toxicity in male Sprague Dawley rats. *BioMed Research International, 2019(1)*. 8939854

Malan, D. F., Neuba, D. F. R., & Kouakou, K. L. (2015). Medicinal plants and traditional healing practices in ehotile people , around the aby lagoon (eastern littoral of Côte d ' Ivoire). *Journal of Ethnobiology and Ethnomedicine, 11*, 1–18.

Mali, V. A., Chanda, M. A. A., Ladda, D. P. L., Naikwade, N. S., & Tamboli, S. A. (2022). An overview: Some medicinal plants as aphrodisiac agents. *International Journal of Pharmaceutical Sciences Review and Research, 75*(21), 116–123.

Maroyi, A. (2013). Traditional use of medicinal plants in south-central Zimbabwe: Review and perspectives. *Journal of Ethnobiology and Ethnomedicine, 9*, 1–18.

Maroyi, A. (2014). Traditional use of medicinal plants in south-central Zimbabwe: Review and perspectives. *Journal of Ethnobiology and Ethnomedicine,* 9:31.

Maroyi, A., & Cheikhyoussef, A. (2015). A comparative study of medicinal plants used in rural areas of Namibia and Zimbabwe. *Indian Journal of Traditional Knowledge, 14*(3), 401–406.

Masuku, N. P., Unuofin, J. O., & Lebelo, S. L. (2020). Promising role of medicinal plants in the regulation and management of male erectile dysfunction. *Biomedicine & Pharmacotherapy [Internet], 130*(March), 110555. Available from: https://doi.org/10.1016/j.biopha.2020.110555

Mkonyi, E., Trent, M., Mwakawanga, D. L., Massae, A. F., Ross, M. W., Bonilla, Z. E., et al. (2024). Male erectile dysfunction problems in healthcare facilities in Dar-es-Salaam, Tanzania. *East African Journal of Health Science, 7*(1), 46–59.

Mola, A. M., Idrissa, A. Z., Mwange, R. K., Mukeba, F. B., Cakupewa, M. F., Mangambu, J. D. D., et al. (2022). Ethnobotanical survey of plants used against erectile dysfunction in the commune of Ngaba in Kinshasa/DR Congo. *World Journal of Advanced Research and Reviews, 13*, 193–200.

Mothibe, M. E., & Sibanda, M. (2019). African traditional medicine: South African perspective. *Traditional and Complementary Medicine*. IntechOpen. Available from: https://doi.org/10.5772/intechopen.83790

Moyo, S. (2013). Indigenous knowledge systems and attitudes towards male infertility in Mhondoro-Ngezi, Zimbabwe. *Culture, Health & Sexuality, 15*(6), 667–679.

Musa, M. S., Abdelrasool, F. E., Elsheikh, E. A., Ahmed, L. A. M. N., Mahmoud, A. L. E., & Yagi, S. M. (2011). Ethnobotanical study of medicinal plants in the Blue Nile State, South-eastern Sudan. *Journal of Medicinal Plants Research, 5*(17), 4287–4297.

Natacha, N. A., Jean, L. B., Evariste, F. F., Mathurin, T., Pascal, B. F., Armand, N. B., et al. (2022). Ethno botanical study of medicinal plants used by Baka people in the treatment of erectile dysfunction. *Journal of Medicinal Plants Research, 16*(8), 245–257.

Nkafu, T. S., Wambo, S., Gaetan, A., Eyong, E., & Divine, E. (2023). Prevalence of sexual dysfunction, factors, and psychological effects on adult males in the Buea Health*, 15*-(December), 304–314.

Novotna, B., Polesny, Z., Pinto-Basto, M. F., Van Damme, P., Pudil, P., Mazancova, J., et al. (2020). Medicinal plants used by 'root doctors', local traditional healers in Bié province, Angola. *Journal of Ethnopharmacology [Internet], 260*, 112662. Available from: https://doi.org/10.1016/j.jep.2020.112662

Odmell, C., Mamimine, P. W., & Kudakwashe, C. (2018). A situational analysis of health tourism for appropriation of indigenous herbal sexual stimulants and fertility enhancers in Zimbabwe: A situational analysis paper. *Cogent Social Sciences [Internet],* 4(1),*1440498*. Available from: https://doi.org/10.1080/23311886.2018.1440498

Odongo, E., Mungai, N., Mutai, P., Karumi, E., Mwangi, J., & Omale, J. (2018). Ethnobotanical survey of medicinal plants used in Kakamega County, western Kenya. *Applied Medical Research, 4*(1), 22.

Ojo, M. C., Osunsanmi, F. O., Cele, N. D., Zharare, G. E., Mosa, R. A., Opoku, A. R., et al. (2022). Effects of Methanolic crude extracts Inula glomerata and Salacia kraussii on erectile dysfunction. *African Journal of Biomedical Research, 25*(3), 425–430.

Opuwari, C. S., & Moundipa, P. F. (2021). Herbal medicine used to treat andrological problems: Africa [Internet]. Herbal Medicine in Andrology. Elsevier Inc.; 83–91 p. Available from: https://doi.org/10.1016/B978-0-12-815565-3.00014-X

Ouma, A. (2022). Intergenerational learning processes of traditional medicinal knowledge and socio-spatial transformation dynamics, *7*(July), 1–10.

Owaba, A. D., Etim, E. I., Johnson, E. C., & Umoh, U. F. (2021). Aphrodisiac agents used in traditional medicine and their mechanism of action – A review. *Journal of Pharmacognosy and Phytochemistry, 10*(3), 126–153.

Rakuambo, N. C. (2002). *The traditional use of medicinal plants to treat erectile dysfunction and the isolation of their bioactive compounds*. University of Pretoria.

Rakuambo, N. C., Meyer, J. J. M., Hussein, A., Huyser, C., Mdlalose, S. P., & Raidani, T. G. (2006). In vitro effect of medicinal plants used to treat erectile dysfunction on smooth muscle relaxation and human sperm. *Journal of Ethnopharmacology, 105*, 84–88.

Rayise, S. S., Chukwuneme, C. F., Takaidza, S., & Ssemakalu, C. (2024). A holistic view on erectile dysfunction: South African medicinal plants as sustainable alternative therapies. *Journal of Natural Remedies, 24*(9), 1863–1876.

Rea, A. I., Schmidt, J. M., Setzer, W. N., Sibanda, S., Taylor, C., & Gwebu, E. T. (2003). Cytotoxic activity of Ozoroa insignis from Zimbabwe. *Fitoterapia, 74*(970704), 732–735.

Saikia, Q., Hazarika, A., & Mishra, R. (2022). A review on the pharmacological importance of PDE5 and its inhibition to manage biomedical conditions. *Journal of Pharmacology and Pharmacotherapeutics, 13*(3), 246–257.

Saive, M., Frederich, M., Fauconnier, M., & laure. (2020). Plants used in traditional medicine in the Comoros archipelago. A review. *Biotechnologie, Agronomie, Société et Environnement, 24*(2), 117–141.

Saloufou, K. I., Boyode, P., Simalou, O., Eloh, K., Idoh, K., Melila, M., et al. (2018). Chemical composition and antioxidant activities of different parts of Ficus sur. *Journal of Herbmed Pharmacology, 7*(3), 185–192.

Seleteng Kose, L., Moteetee, A., & Van Vuuren, S. (2015). Ethnobotanical survey of medicinal plants used in the Maseru district of Lesotho. *Journal of Ethnopharmacology [Internet], 170*, 184–200. Available from: https://doi.org/10.1016/j.jep.2015.04.047

Semenya, S. S., & Potgieter, M. J. (2013). Ethnobotanical survey of medicinal plants used by Bapedi traditional healers to treat erectile dysfunction in the Limpopo Province, South Africa. *Journal of Medicinal Plants Research [Internet], 7*(7), 349–357. Available from: http://www.academicjournals.org/JMPR

Shiferaw, W. S., Akalu, T. Y., & Aynalem, Y. A. (2020). Prevalence of erectile dysfunction in patients with diabetes mellitus and its association with body mass index and glycated hemoglobin in Africa: A systematic review and meta-analysis. *International Journal of Endocrinology, 2020*, 1–24.

Shopo, B., Mapaya, R. J., & Maroyi, A. (2022a). Ethnobotanical study of medicinal plants traditionally used in Gokwe South District, Zimbabwe. *South African Journal of Botony [Internet], 149*, 29–48. Available from: https://doi.org/10.1016/j.sajb.2022.05.052

Shopo, B., Mapaya, R. J., & Maroyi, A. (2022b). The traditional uses of plant diversity in Gokwe South District, Zimbabwe: Timber & construction, ethnoveterinary medicine, firewood & charcoal, food, tools and handicraft, religious ceremonies & rituals & other uses. *South African Journal of Botany*, 29–48.

Singh, R., Singh, S., & Jeyabalan, A. A. (2012). An overview on traditional medicinal plants as aphrodisiac agent. *Journal of Pharmacognosy and Phytochemistry, 1*(4), 43–56.

Suroowan, S., Pynee, K. B., & Mahomoodally, M. F. (2019). South African Journal of Botany A comprehensive review of ethnopharmacologically important medicinal plant species from

Mauritius. *South African Journal of Botany [Internet]., 122*, 189–213. Available from: https://doi.org/10.1016/j.sajb.2019.03.024

Tacham, W. N. (2015). Medicinal plants of Aguambu – Bamumbu in the Lebialem highlands, southwest province of Cameroon; (January 2009).

Terashima, H., Kalala, S., & Malasi, N. (1991). Ethnobotany of the lega tropical rain forest of eastern Zaire. Part 1 Zone de Mwenga. *African Study Monograph, 1*(15), 1–62.

Togola, I., Dembélé, J., Daou, C., Dénou, A., Diarra, N., Badiaga, M., et al. (2020). Ethnobotanical survey and phytochemical screening of some plants used in the management of Erectile Dysfunction in Bwatun (Mali). *Journal of Natural Product and Plant Resources, 9*(1), 1–8.

Valentin, B. C., Gracia, A. A., Arlette, K. M., Salvius, B. A., & Ndjolo, O. (2022, November). *World Journal of Advanced Research and Reviews.*

van Vuuren, S., & Frank, L. (2020). Review: Southern African medicinal plants used as blood purifiers. *Journal of Ethnopharmacology [Internet], 249*, 112434. Available from: https://doi.org/10.1016/j.jep.2019.112434

Van Wyk, B. E. (2015). A review of commercially important African medicinal plants. *Journal of Ethnopharmacology [Internet], 176*, 118–134. Available from: https://doi.org/10.1016/j.jep.2015.10.031

Yafi, F. A., Jenkins, L., Albersen, M., Corona, G., Andrea, M., Goldfarb, S., et al. (2017). Erectile dysfunction. *Nature Reviews Disease Primers,* (pp. 1–47).

Yemane, B., Andebrhan, M., & Reddy, K. S. (2017). *Traditional medicinal plants used by Tigrigna Ethnic Group in Central Region of Eritrea, 12*(3), 40–46.

Zatout, F., Benarba, B., Bouazza, A., Babali, B., Bey, N. N., & Morsli, A. (2021). Ethnobotanical investigation on medicinal plants used by local populations in tlemcen national park (extreme North West Algeria). *Mediterranean Botany, 42*, 1–12.

Chapter 12
Overview of Medicinal Plants-Mediated Bio-nanoparticles: Synthesis, Their Characterization, and Applications as Therapeutic Agents

Dennis Nzilu, David Makhanu, Edwin Madivoli, Gideon Kirui, Vincent Mwangi, and Patrick Kareru

Abstract Nanotechnology shows revolutionary attempts to address humanity's challenges, especially in addressing the severe disease burden of the world's population. The use of conventional medicines is reported to be limited due to the development of antimicrobial resistance coupled with little research on new drug development, which continues to strain healthcare in contemporary times. However, using medicinal plants as remedies for diseases has increasingly garnered attention in research and is promising to combat the diseases that threaten human life. As a developing research area, researchers are exploiting the possibility of developing bio-nanoparticles from medicinal plants, which were earlier reported to possess therapeutic potential and enhance their effectiveness by incorporating a nanotechnology approach. Using medicinal plant extracts as reducing, capping, and stabilizing agents to prepare therapeutic metallic nanoparticles has shown promising results in advancing pharmaceutical nanotechnology. The plant-mediated nanoparticles exhibit enhanced drug delivery to the site of action, improved bioavailability, and better drug solubility. In this way, synthesizing bio-nanoparticles for therapeutic purposes provides a new approach to curbing the global disease burden emanating from antimicrobial resistance to conventional antibiotics and reducing millions of morbidities and mortalities. This chapter highlights an overview of nanoparticles and the different synthesizing methods, primarily focusing on African medicinal plants, their characterization, and their applications as therapeutic agents.

D. Nzilu (✉) · E. Madivoli · G. Kirui · V. Mwangi · P. Kareru
Chemistry Department, Jomo Kenyatta University of Agriculture and Technology, Nairobi, Kenya
e-mail: dennis.nzilu@students.jkuat.ac.ke

D. Makhanu
Department of Biological and Physical Sciences, Karatina University, Karatina, Kenya

A. Dekebo (ed.), *Bioactive Secondary Metabolites from Medicinal Plants of Africa*, Interdisciplinary Biotechnological Advances,
https://doi.org/10.1007/978-981-96-8590-5_12

Keywords Medicinal plants · Green nanotechnology · Bio-nanoparticles · Green chemistry

12.1 Introduction

The emergence of infectious diseases threatens public health, leading to the declaration of pandemics, the recent being COVID-19. While the efforts to curb new diseases using conventional antibiotics have been limited due to the propagation of antimicrobial resistance among pathogens, new efforts to develop new drugs remain vibrant to address this challenge (Prestinaci et al., 2015; Selvarajan et al., 2022). The World Health Organization (WHO) declared antimicrobial resistance a public health threat and called for concerted efforts to develop new drugs to manage the ever-emerging resistance (World Health Organization, 2023). As a consequence, in 2019, over 1.2 million deaths were reported to occur due to drug-resistance-related infections (World Health Organization, 2023), and the death statistics are expected to rise to ten million by 2050, according to the United Nations estimates (Anderson, 2023). The Centre for Disease Control (CDC) further documents that over 2.8 million antibiotic-resistant infections occur annually in the United States (CDC, 2021).

Despite the health menace attributed to antimicrobial resistance-related infections, the use of medicinal plants remains a vibrant area as herbal remedies for diseases and a cornerstone area relying on natural products to develop new drugs for therapeutic purposes in contemporary times (Aschale et al., 2021; Najmi et al., 2022). Several African-based traditional medicinal plants have been reported to offer therapeutic remedies for fungal and viral infections, such as those Fabaceae, Lamiaceae, Caesalpiniaceae, and Asteraceae plant families (Aschale et al., 2021). These natural plant-based medicinal remedies are either administered as raw plants or their modified forms as extracts and powders (Gabriel Kigen et al., 2020). However, because medicinal plants are traditionally administered as raw products, their bioavailability in the human body lowers their efficacy in managing targeted ailments. To date, researchers are developing interests in combining medicinal chemistry and nanotechnology knowledge to solve the multi-drug resistance exhibited by many pathogens (Hetta et al., 2023; Mba & Nweze, 2021).

The revolution of the nanotechnology field dates back many years. The initial scientific discovery started with Richard Feynman through a lecture he gave in 1959 at Caltech. It emphasized the need to advance technologies and miniaturization with a famous quote, "There's plenty of room at the bottom" (Bayda et al., 2019). It was not until 1974 when Norio Taniguchi from Japan's University of Tokyo was recognized as the father of nanotechnology, describing it as the technology of manipulating materials at the nanometer scale (Bayda et al., 2019). The United States' National Nanotechnology Initiative (NNI) defined nanotechnology as "a science, engineering, and technology conducted at the nanoscale" (1–100 nm) (Bayda et al., 2019). The nanomaterials, therefore, were described as any material with a size dimension from 0 to 100 nm, which can exist as particles, carbon dots, nanofibers, nanowires, and carbon nanotubes, among others (Gebre & Sendeku,

2019; Mekuye & Abera, 2023). Essentially, the nanomaterials have been demonstrated to have enhanced properties than their bulk counterparts as they have improved stability, smaller surface area to volume ratio, and the ability to form homogenous particle size and shape, which makes them suitable for different applications (Mekuye & Abera, 2023; Mourdikoudis et al., 2018; Baig et al., 2021). More importantly, nanotechnology has been explored in diverse applications, including biomedicine, electronics, energy, food packaging, automobiles, information technology, and environmental remediation (Jain & Mehata, 2017; Malik & Muhammad, 2023). In the therapeutic area, nanosystems enhance the transportation of bioactive compounds at proper concentrations during the treatment time, ensure safe delivery to sites of action, improve their overall potency, and minimize drugs' side effects (Chen & Hu., 2025; Yusuf et al., 2023).

To this end, this chapter highlights the need to exploit nanotechnology and medicinal chemistry to solve the increasing disease menace worldwide by discussing synthesis routes and characterization techniques for synthesized metallic nanoparticles. This chapter also reviews medicinal plant-mediated metallic and metal oxide nanoparticles and their reported application as therapeutic agents. The reviewed articles were searched using the key terms green synthesis, African medicinal plants, and nanoparticles from MPDI, Springer, and Elsevier databases published within the last 10 years (2015–2024).

12.2 Synthesis Methods for Metallic and Metal Oxide Nanoparticles

Generally, metallic and metal oxide nanoparticles can be produced using the top-down approach or the bottom-up approach (Bayda et al., 2019; Gebre & Sendeku, 2019). The top-down approach involves breaking bulk materials into their nanosized scale material, while the bottom-up approach involves assembling small particles to form nanosized counterparts (Kumar et al., 2017; Xulu et al., 2022). These approaches can further be classified into three broad methods of synthesizing nanoparticles: physical, chemical, and biological, each capable of preparing nanoparticles with different sizes and morphology (Baig et al., 2021). Figure 12.1 presents a summarized classification method for synthesizing metallic and metal oxide nanoparticles.

12.2.1 Chemical Methods

Chemically synthesizing metallic and metal oxide nanoparticles relies on chemical reagents as starting materials, specifically reducing solutions or capping agents, which reduce metal ion salts to their nanosized particles (Gebre & Sendeku,

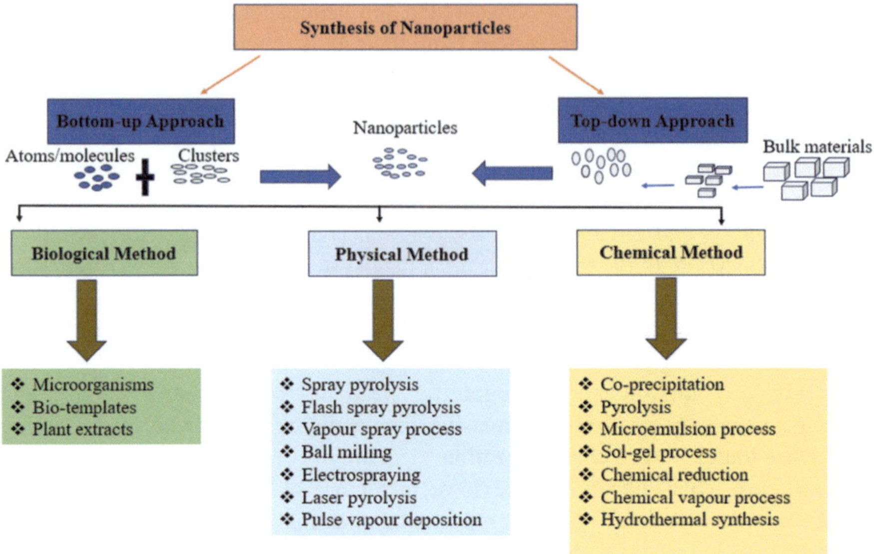

Fig. 12.1 A scheme showing approaches and methods of synthesizing metallic and metal oxide nanoparticles

2019). The capacity to manipulate particle size, shape, and dispersion led to a surge of interest in chemical techniques for producing metallic and metal oxide nanoparticles. Silver (Ag) nanoparticles with consistent sizes and excellent stability were produced by chemical reduction of silver nitrate using sodium borohydride and ethanol (Thiruvengadam & Bansod, 2020). In another study, Hussain et al. (2020) synthesized gold (Au) nanoparticles by chemical reduction of chloroauric acid (HAuCl4) using ethanol, L-ascorbic acid, and polyvinylpyrrolidone.

Pure zinc (Zn) nanoparticles were also synthesized via chemical reduction of zinc nitrate hexahydrate using sodium hydroxide, as depicted in the study by Kalia et al. (2021). Elsewhere, copper (Cu) nanoparticles were prepared by chemical reduction of copper sulphate pentahydrate salt with ascorbic acid or sodium hydroxide as reducing agents (Khan et al., 2016; Tyagi et al., 2023). Titanium dioxide (TiO_2) nanoparticles were also synthesized from the sol-gel method in which titanium tetra isopropoxide, glacial acetic, and sodium hydroxide were primary reagents (Ahmad et al., 2021). A study by Dhoke (2023) reported the synthesis of zinc oxide (ZnO) nanoparticles by chemical reduction of zinc (II) acetate and potassium hydroxide. Cobalt oxide (Co_3O_4) nanoparticles were also prepared from the chemical reduction of cobalt nitrate via the precipitation method (Lakra et al., 2021).

Despite their ability to produce nanoparticles for various applications, chemical methods have been classified to possess drawbacks that limit their wide-scale use in the production of nanoparticles because of their expensive nature to procure chemicals, the presence of toxic precursors, and the possibility of production of harmful byproducts (Kumari et al., 2023).

12.2.2 Physical Methods

Physical methods for producing metallic and metal oxide nanoparticles are advantageous over chemical methods as the technique does not utilize chemical reagents, thus eliminating potential contamination. Rahmah et al. (2024) report the synthesis of Ag nanoparticles of precisely regulated size distribution via laser ablation. In this method, no toxic chemical was utilized, and the nanoparticle size was controlled through the intensity of the laser. In another study, Wang et al. (2021) synthesized vanadium dioxide nanoparticles by using ball milling at ambient conditions. Synthesis of Au nanoparticles was also reported using ultrasonic spray pyrolysis (Shariq et al., 2018). Poostforooshan et al. (2023) report the synthesis of pure and Fe-doped TiO_2 in which the research employed electrospray-assisted flame spray pyrolysis. Elsewhere, the successful synthesis of ZnO nanoparticles was achieved using pulse laser deposition (Yudasari et al., 2019). In these methods, the variation of synthesis parameter conditions results in the production of nanoparticles of regulated shape, size, and fit for various functionalities.

Although physical methods can produce metallic and metal oxide nanoparticles for various applications, it is costly because the processes consume large quantities of power and energy (Mekuye & Abera, 2023). However, physical methods are environmentally friendly compared to chemical methods as they do not use chemical solvents and can produce homogenous nanoparticles (Al et al., 2024).

12.2.3 Biological Methods

Compared to physical and chemical methods, biological methods promote environmentally green chemistry and provide an alternative route to produce metallic and metal oxide nanoparticles for different applications without threatening environmental contamination. These methods are further classified based on the materials they use as a source of reducing agents: microorganisms, bio templates, or plant extracts. Microorganisms such as *Verticillium* sp., *Fusarium oxysporum*, and *Aspergillus flavus*, among others, have been exploited and found to be rich in reducing materials for the bioproduction of nanoparticles (Koul et al., 2021; Mwangi et al., 2024; Vincent et al., 2024).

For instance, Au nanoparticles for catalytic degradation of organic pollutants were achieved using *Lactobacillus casei* bacteria as a reducing agent (Miran et al., 2023). In another study, Zn nanoparticles were produced from *Aspergillus niger* fungal strain biomass, which demonstrated to possess antibacterial properties (Abdelkader et al., 2022). Mohamed (2022) employed *Cystoseira Myrica* aqueous extract to synthesize CuO nanoparticles for cancer treatment studies. Madhav et al. (2023) report the synthesis of Ag nanoparticles from *Azadirachta indica*. In addition, selenium nanoparticles were synthesized using *Bacillus subtilis* as an economically viable biosynthetic technique (Ullah et al., 2021).

Although microorganism-mediated synthesis of nanoparticles shows promising advantages in producing green nanoparticles, they are associated with safety concerns, mainly when target nanoparticles are to be used in health or environmental remediation. These concerns include contaminations, toxic final products, and health risks while handling the microorganisms (Nzilu et al., 2024).

Due to these shortcomings and the need to develop green and pollution-free metallic and metal oxide nanoparticles, researchers are using natural plant extracts as sources of reducing agents in the production of nanoparticles. Specifically, using plant extracts as reducing agents for metal ion solutions has been characterized as a non-toxic, economical, and promising nanoparticle synthesis route that promotes sustainable, environmentally friendly chemistry (Ahmed & Mustafa, 2020; Kumari et al., 2023). The plant extracts have been established to possess rich natural products comprising secondary metabolites such as terpenoids, polyphenols, phenolic acids, and alkaloids, among others, which help in the bioreduction of metal ions into their nanoparticles (Lung & Stegarescu, 2023; Nzilu et al., 2023; Ying et al., 2022). Some common metallic and metal oxide nanoparticles synthesized from plant extracts are depicted in Table 12.1, where emphasis is placed on African medicinal plants. As mentioned, chemical and physical methods are considered cost-intensive and can potentially produce toxic byproducts that do not promote environmentally friendly chemistry. As a result of these discoveries, green synthesis of nanoparticles employing medicinal plant extracts for therapeutic purposes creates an avenue for optimizing nanotechnology and plant biotechnology to push for sustainable solutions for treating diseases, which is a burden worldwide (Puja et al., 2023).

12.3 Characterization of Bio-nanoparticles

The effectiveness of bio-nanoparticles for their intended purpose depends on their physicochemical properties. These properties also determine the ability of the nanoparticles to perform their specific functions, biodistribution, and safety concerns (Mourdikoudis et al., 2018; Xulu et al., 2022). Therefore, the characterization of nanoparticles provides an avenue for assessing the functional properties of the bio-synthesized nanoparticles as medicinal remedies to treat diseases. Standard analytical techniques used to characterize nanoparticles include UV-Vis spectroscopy, Fourier-transform infrared spectroscopy, X-ray diffractometry, dynamic light scattering, scanning electron microscopy, transmission electron microscopy, and energy-dispersive X-ray spectroscopy (Puja et al., 2023; Xulu et al., 2022). Although different articles have been published on the principles of operation of these analytical characterization methods, this chapter elaborates explicitly on the use of these methods for characterizing bio-nanoparticles from medicinal plant extracts for therapeutic applications.

Table 12.1 A summary of African medicinal plants-mediated bio-nanoparticles and their efficacy in therapeutic applications

African medicinal plants used as extracts	Bio-nanoparticles type	Particle size (nm)	Application as a therapeutic agent	References
Lantana trifolia	Silver	35–70	Ag NPs showed antimicrobial activity against *S. aureus, E. coli, P. aeruginosa, B. subtilis, C. albicans*	Madivoli et al. (2020)
Aspilia pluriseta	Silver	6–12	Ag NPs showed antimicrobial activity against *S. aureus, E. coli, P. aeruginosa, B. subtilis, C. albicans*	Nyabola et al. (2020)
Annona muricata	Silver	60	Potential for use in medical areas	Gavamukulya et al. (2019)
Ageratum conyzoides	Iron	85.98	Showed antimicrobial activity against *S. aureus, E. coli, P. aeruginosa, B. subtilis, C. albicans*	Madivoli et al. (2019)
Hypoxis hemerocallidea	Gold	26	Au NPs showed immunomodulatory effects on microphages and natural killer cells	Elbagory et al. (2019)
Galenia africana	Gold	2–16	Showed antibacterial activity against *P. aeruginosa*, effective in wound healing	Elbagory et al. (2017)
Aloe vera	Selenium	7–48	Antioxidant properties (free radical inhibition on ABTS, FRAP, and DPPH assay), chemoprevention in cancer diagnosis	Vyas and Rana (2017)
Plumbago auriculata	Silver	15–30	Antibacterial activity against *E. coli, S. aureus, K. pneumoniae*, and *S. typhimurium*	Singh et al. (2018b)
Erythrina abyssinica	Silver	8–10	Antibacterial against *Salmonella* and *E. coli*	Mukaratirwa-Muchanyereyi et al. (2022)
Detarium microcarpum	Silver	81–84	Inhibitory effect against cervical and pancreatic cancer	Adebayo et al. (2020)
Morinda lucida	Silver	11	Antimicrobial activity against Citrobacter, *E. coli, P. vulgaris, S. typhi, V. cholerae,* and *E. faecalis* and antioxidant activity	Labulo et al. (2022)
Eucomis autumnalis	Silver	112	Antimicrobial activity against *E. faecalis*	Lediga et al. (2018)
Sclerocarya birrea	Silver	56	Antimicrobial activity against *A. baumannii*	Lediga et al. (2018)

(continued)

Table 12.1 (continued)

African medicinal plants used as extracts	Bio-nanoparticles type	Particle size (nm)	Application as a therapeutic agent	References
Physalis peruviana	Gold	10	Antimicrobial activity against *S. aureus* and *B. subtilis*, *E. coli*, and *P. aeruginosa*, and *C. albicans*, potential in wound healing	Odongo et al. (2022)
Crassocephalum rubens	Silver	15–25	Ag NPs exhibited DPPH radical scavenging activity	Adewale et al. (2020)
Crassocephalum rubens	Gold	10–20	Antioxidant activity observed against DPPH radical	Adewale et al. (2020)
Zingiber officinale	Gold	5–53	Inhibition against prokaryotic and eukaryotic microorganisms	Fouda et al. (2022)
Pistacia chinensis	Gold	1–100	Enzymatic inhibition against urease and carbonic anhydrase	Alhumaydhi (2022)
Krameria sp.	Copper	5.2–7.7	Scavenging DPPH and its potential as an antimicrobial agent and antioxidant therapy	Alshammari et al. (2023)
Alstonia scholaris	Copper		Antimicrobial activity against *S. aureus* and *E. coli*	Labaran et al. (2024)
Ocimum sanctum	Titanium dioxide	75–123	Demonstrated efficacy in wound healing	Ahmad et al. (2022)
Cassia fistula	Zinc oxide	68.1	Inhibition against clinical pathogens *E. coli* and *S. aureus*	Naseer et al. (2020)
Melia azadarach	Zinc oxide	3.62	Inhibition against clinical pathogens *E. coli* and *S. aureus*	Naseer et al. (2020)

12.3.1 UV-Vis Spectroscopy

UV-Vis spectroscopy (UV-Vis) is an essential method for monitoring the synthesis of nanoparticles and their stability over the production period. It is the primary characterization method that researchers depend on to establish the formation of specific nanoparticles, as the technique allows visualization of intrinsic optical properties of the nanoparticles displayed as surface plasmonic resonance spectrum (SPR) when solution interacts with UV light at a specific wavelength (Shard et al., 2020). Several nanoparticles from medicinal plant-mediated synthesis were characterized by UV-Vis, and the SPR associated with their formation was reported. Ajaykumar et al. (2023) synthesized silver nanoparticles (Ag NPs) using the *Uvaria narum* medicinal plant leaf extracts. These researchers found the SPR peak associated with Ag NPs to be observed at 427 nm. In another study, Giri et al. (2022) used leaves extracts of a medicinal plant, *Eugenia roxburghii* D.C., to synthesize Ag NPs with an SPR peak observed at 417 nm. Gavamukulya et al. (2019) synthesized Ag NPs from *Annona muricata* ethanolic extracts and obtained an SPR band associated with the synthesized nanoparticles at 427 nm. In our previous study, we synthesized

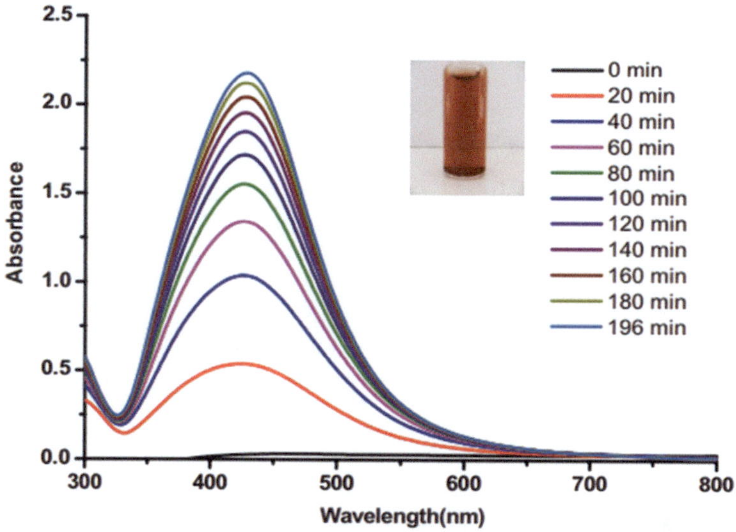

Fig. 12.2 UV-Vis spectra of Ag Nps from *Aspilia pluriseta* medicinal plant (Nyabola et al., 2020)

Ag NPs from *Lantana trifolia* aqueous extracts, and the UV-Vis absorption peak showing a consistent SPR band of the Ag NPs was centered at 443 nm (Madivoli et al., 2020). Additionally, a study by Nyabola et al. (2020) synthesized Ag NPs from *Aspilia pluriseta* extract with the SPR band being observed at 427 nm, as depicted in Fig. 12.2.

Other UV-Vis results for other nanoparticles were also reported. Cu nanoparticles were synthesized using *Alstonia scholaris* leaves aqueous extract with UV-Vis results showing a peak associated with Cu nanoparticles at 243 nm (Labaran et al., 2024). Elsewhere, Cu nanoparticles were synthesized from *Krameria* sp. (Rhatany medicinal plant), with the UV-Vis absorption peak appearing at 422–430 nm (Alshammari et al., 2023). Additionally, Cu NPs were synthesized using leaves extract of *Celastrus paniculatus* with the UV-Vis absorption peak occurring at 269 nm, as shown in Fig. 12.3 (Mali et al., 2020). Au nanoparticles were synthesized using *Pistacia chinensis* extracts, and the UV-Vis absorption peak associated with their formation occurred from 530 to 550 nm (Alhumaydhi, 2022). Additionally, Au nanoparticles were synthesized from *Zingiber officinale* aqueous extract with the SPR band observed at 530 nm, as depicted in Fig. 12.4 (Fouda et al., 2022). The leaf extracts of *Ocimum sanctum* were used to synthesize TiO_2 nanoparticles, which showed that the UV-Vis absorption spectra occurred at 235 and 320 nm (Ahmad et al., 2022). ZnO nanoparticles were synthesized from leaf extracts of *Cassia fistula* and *Melia azadarach* with the SPR band occurring at 320 nm and 324 nm, respectively (Naseer et al., 2020).

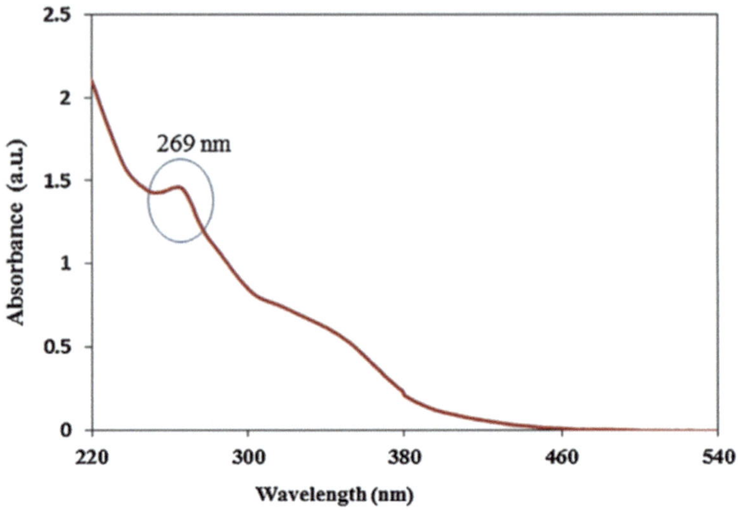

Fig. 12.3 UV-Vis spectrum of Cu nanoparticles synthesized using leaves extracts of *Celastrus paniculatus* (Mali et al., 2020)

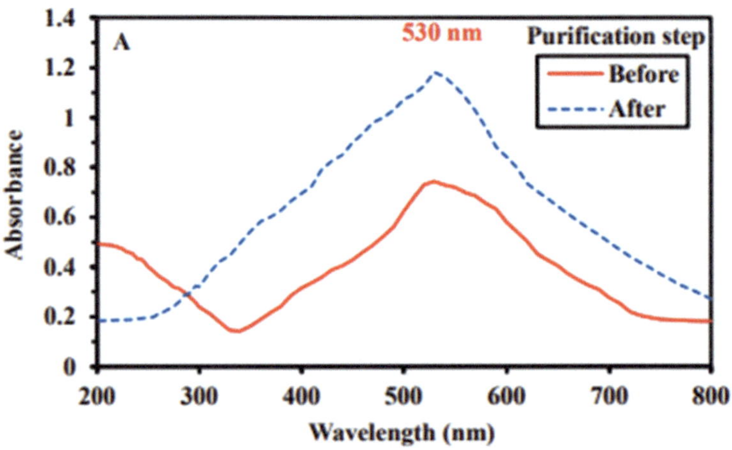

Fig. 12.4 UV-Vis spectrum of Au nanoparticles synthesized from *Zingiber officinale* aqueous extract (Fouda et al., 2022)

Worth noting, absorption peaks of the nanoparticles under UV-Vis analysis depend on several factors, such as the concentration of reducing agents, making the SPR absorption peak to be observed within the UV-Vis wavelength range (Gavamukulya et al., 2019).

12.3.2 Fourier-Transform Infrared Spectroscopy

The Fourier-transform infrared spectroscopy (FTIR) is a valuable technique that provides information on the functional groups present during nanoparticle synthesis. It gives an idea of the available secondary metabolites, such as phenols, flavonoids, terpenoids, and tannins, among others, which facilitate the reduction of metal salts to nanoparticles and stabilize the formed nanoparticles (Mourdikoudis et al., 2018). FTIR analysis provides information on the bonding of the biomolecules with the respective metal nanoparticles (Xulu et al., 2022). Another way researchers can monitor the formation of the nanoparticles is by comparing the FTIR spectrum of the medicinal plant extract with that of the synthesized nanoparticles. Figure 12.5 compares the FTIR spectra of *Parthenium hysterophorus* aqueous extract with that of zinc oxide nanoparticles with the differences attributed to the bonding interactions between the secondary metabolites with the zinc metal (Nzilu et al., 2024).

In another study, TiO_2 nanoparticles were synthesized from *Ocimum sanctum* leaf extracts, demonstrating the secondary metabolites' role in reducing titanium salt and forming the corresponding nanoparticles, as depicted in Fig. 12.6 (Ahmad et al., 2022). The IR spectrum also proved the synthesis of Au nanoparticles using *Zingiber*

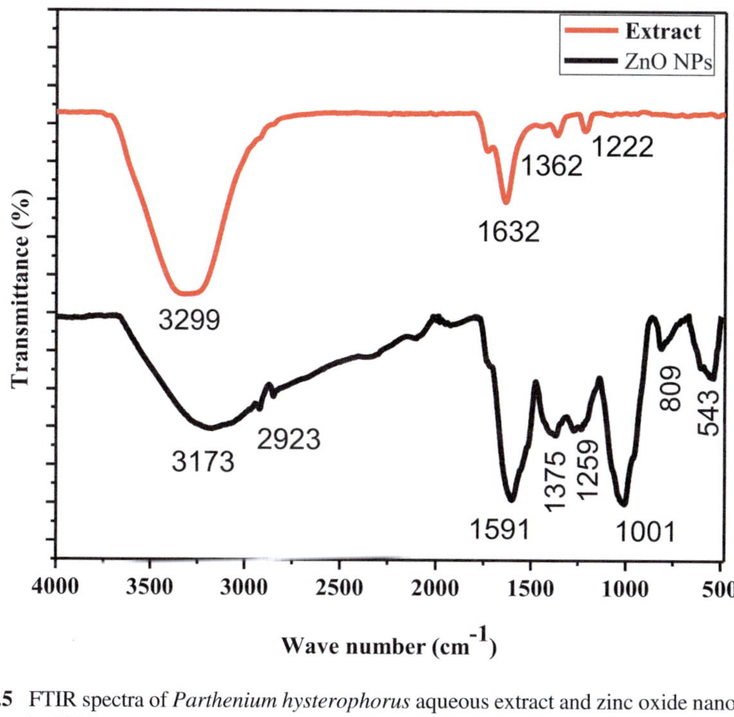

Fig. 12.5 FTIR spectra of *Parthenium hysterophorus* aqueous extract and zinc oxide nanoparticles (Nzilu et al., 2024)

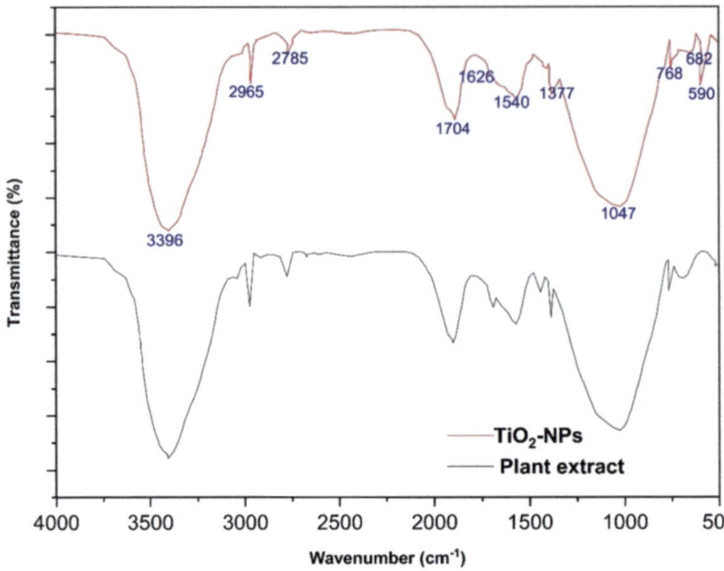

Fig. 12.6 FTIR spectra of TiO_2 nanoparticles produced from *Ocimum sanctum* leaf extracts (Ahmad et al., 2022)

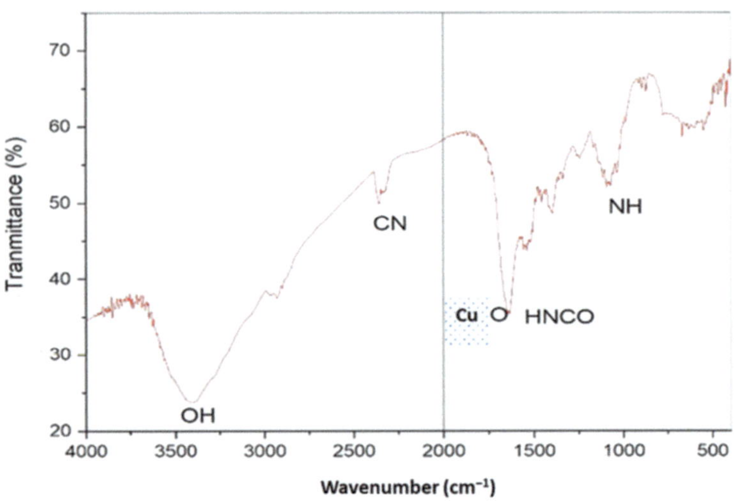

Fig. 12.7 FTIR spectrum of Cu nanoparticles synthesized from *Krameria* sp. root extract (Alshammari et al., 2023)

officinale aqueous extracts, showing the secondary metabolites' role in nanoparticle formation (Fouda et al., 2022). Figure 12.7 further shows the IR spectrum of Cu nanoparticles synthesized from *Krameria* sp. root extract (Alshammari et al., 2023).

12.3.3 X-Ray Diffractometry

X-ray diffractometry (XRD) is a valuable nanoparticle characterization technique that provides qualitative information about nanoparticles' crystallinity and particle size (Mourdikoudis et al., 2018). Each metal nanoparticle has a specific configuration of diffraction peaks associated with its structural characteristics (Xulu et al., 2022). In addition, the XRD can be used to calculate the particle size of the nanoparticles using the Debye-Scherrer formula, $D = (\frac{k\lambda}{\beta \cos\theta})$, where D is the average crystalline size in nm, k is the constant (0.94), λ is the wavelength of X-ray radiation, β is the full-width at half maximum of the peak in radians, and θ is the diffraction angle in degrees (Nzilu et al., 2023). For instance, Giri et al. (2022) obtained an XRD pattern of Ag NPs from *Eugenia roxburghii* D.C. leaf extracts and calculated the average crystalline size to be 35 nm. Furthermore, the study by Nyabola et al. (2020) obtained the XRD diffractograms synthesized from *Aspilia pluriseta* extracts, with crystallite size calculated and found to be 25.83 nm (Nyabola et al., 2020).

The XRD pattern of Au nanoparticles is depicted in Fig. 12.8, where the crystallographic planes correspond to (111), (200), (220), and (311) (Fouda et al., 2022). In another study, the XRD pattern of Cu nanoparticles is shown in Fig. 12.9 with the crystallographic planes (100), (111), and (200) (Alshammari et al., 2023). Elsewhere, Co_3O_4 nanoparticles synthesized from *Psidium guajava* leaves extracts, and their XRD diffractograms showed that the pattern corresponds to crystallographic planes at (220), (311), (222), (400), (511), and (440), as depicted in Fig. 12.10 (Govindasamy et al., 2022).

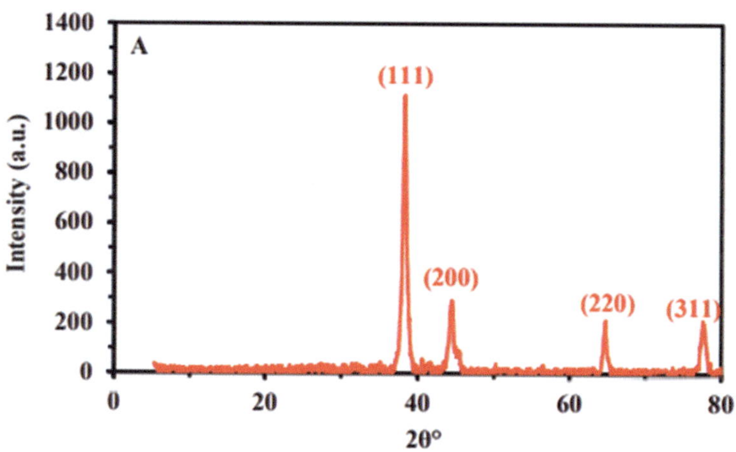

Fig. 12.8 The XRD pattern of gold nanoparticles (Fouda et al., 2022)

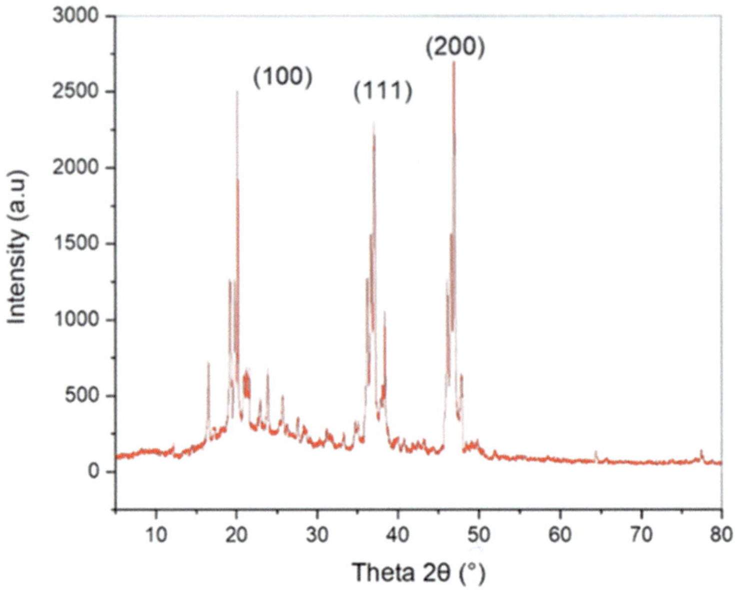

Fig. 12.9 XRD pattern of Cu NPs (Alshammari et al., 2023)

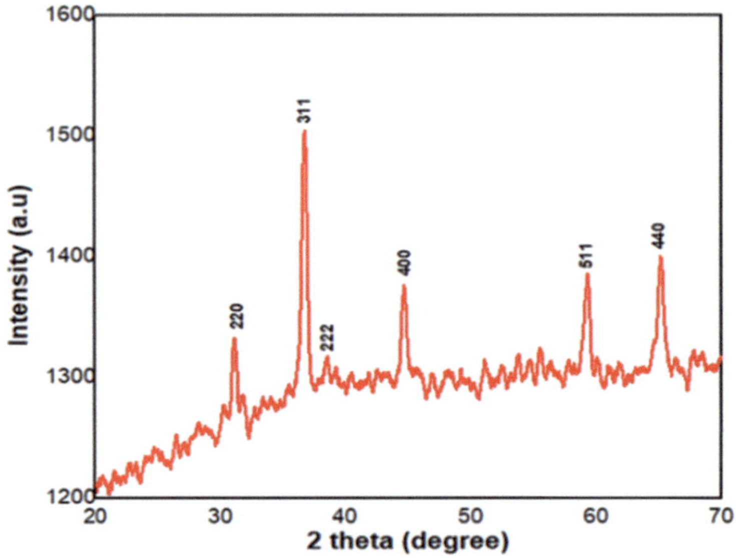

Fig. 12.10 XRD pattern cobalt oxide nanoparticles (Govindasamy et al., 2022)

12.3.4 Dynamic Light Scattering

Dynamic light scattering (DLS) is a valuable technique in determining the size distribution of the nanoparticles and zeta potential, which gives clues about the stability of the nanoparticles. Uniquely, DLS is a number-dependent technique that tends to measure larger particle sizes than those determined using SEM and TEM due to Brownian motion (Xulu et al., 2022). For instance, the study by Gavamukulya et al. (2019) in which Ag NPs were synthesized using *Annona muricata* ethanolic extracts showed that the particles were large as DLS revealed were 103.5 nm. Farooq et al. (2023) synthesized Ag NPs from *Ammi visnaga* plant extract with size distribution from DLS analysis revealing larger particles than those obtained from SEM and TEM. Our previous study calculated the hydrodynamic diameter of Ag NPs synthesized using *Lantana trifolia* aqueous extracts. The obtained size distribution ranged from 35 to 65 nm, as depicted in Fig. 12.11 (Madivoli et al., 2020).

In another study, *Ocimum sanctum* leaf extracts utilized in the synthesis of TiO_2, the DLS analysis showed that the average particle size was 130.3 nm with a polydispersity index of 0.237, as depicted in Fig. 12.12 (Ahmad et al., 2022). The DLS results obtained in a study by Fouda et al. (2022) showed that the average particle size of gold nanoparticles was 93 nm, as shown in Fig. 12.13.

As noted earlier, DLS techniques tend to be sensitive to large particle sizes, which makes the method record larger particle sizes than what is reported in SEM, TEM, and XRD (Gavamukulya et al., 2019; Madivoli et al., 2020; Nyabola et al., 2020).

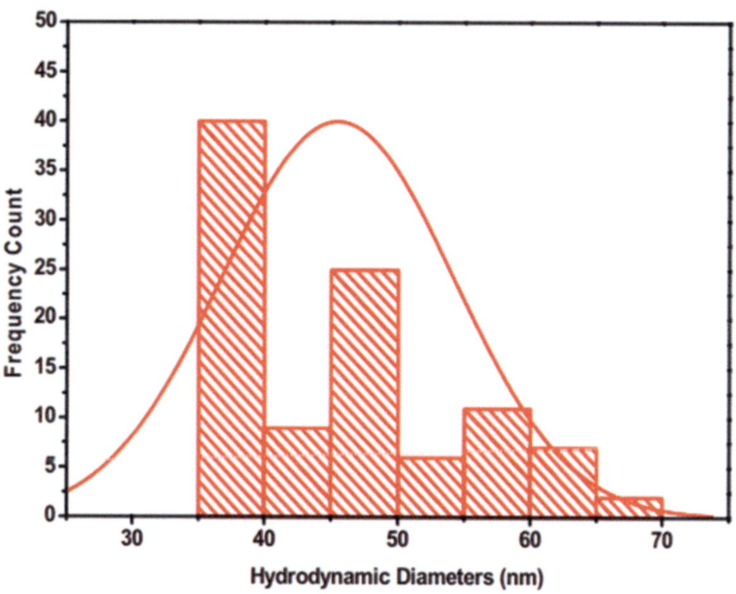

Fig. 12.11 Particle size distribution of Ag NPs synthesized using *Lantana trifolia* aqueous extracts DLS analysis (Madivoli et al., 2020)

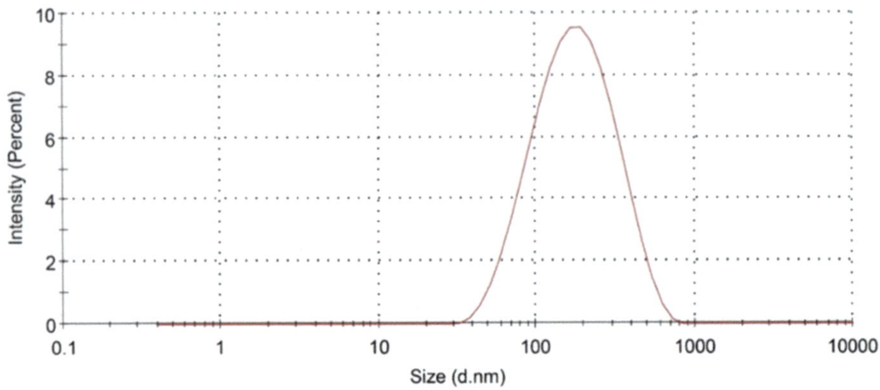

Fig. 12.12 Particle size distribution of TiO_2 nanoparticles synthesized from *Ocimum sanctum* leave extracts (Ahmad et al., 2022)

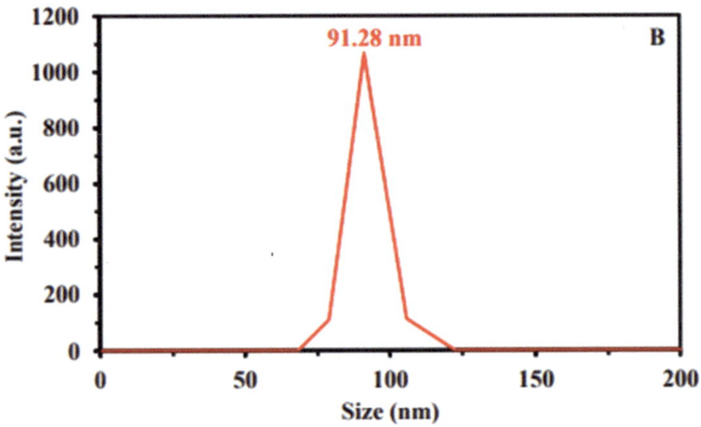

Fig. 12.13 DLS analysis of gold nanoparticles (Fouda et al., 2022)

12.3.5 Scanning Electron Microscopy

Scanning electron microscopy (SEM) is a valuable characterization technique that gives information on the morphology of the nanoparticles (Vladár & Hodoroaba, 2020). As noted earlier, particle morphology determines their suitability for specific applications, and observing the surface morphology of the bio-nanoparticles for therapeutic purposes becomes essential. In most instances, SEM is combined with energy-dispersive X-ray spectroscopy (EDX), providing information on surface morphology and elemental composition (Xulu et al., 2022). More importantly, SEM includes information on the extent of accumulation of the bio-nanoparticles, which offers an idea of the effectiveness of the synthesis methodology. A study by Ajaykumar et al. (2023) obtained Ag NPs from *Uvaria narum* medicinal plant leaf extract with spherical surface morphology particles. In our previous study, in which we synthesized Ag NPs from *Lantana trifolia* extracts, we observed via SEM that the nanoparticles were amorphous, as shown in Fig. 12.14 (Madivoli et al., 2020). The evidence of agglomeration of the nanoparticles can be attributed to the secondary metabolites in the *Lantana trifolia* extracts, which support the bonding of the metal ions with the functional groups of the secondary metabolites (Madivoli et al., 2020).

The SEM images of Cu nanoparticles synthesized from *Alstonia scholaris* leaves are shown in Fig. 12.15, revealing that the particles were irregular in shape, polygons, and spherical (Labaran et al., 2024). In another study, the SEM micrographs of ZnO nanoparticles synthesized from the leaf extract of *Cassia fistula* at different magnifications were less spherical with some evidence of aggregations, as shown in Fig. 12.16 (Naseer et al., 2020).

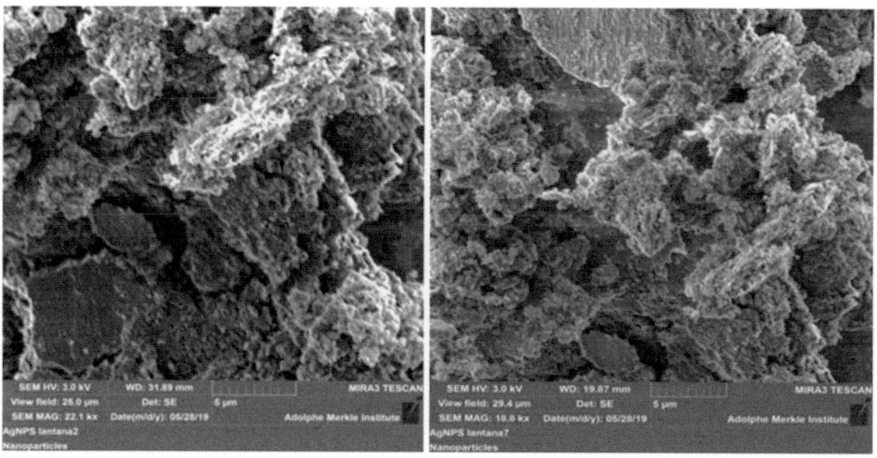

Fig. 12.14 SEM images of Ag NPs synthesized from *Lantana trifolia* extracts medicinal plant leaf extract (Madivoli et al., 2020)

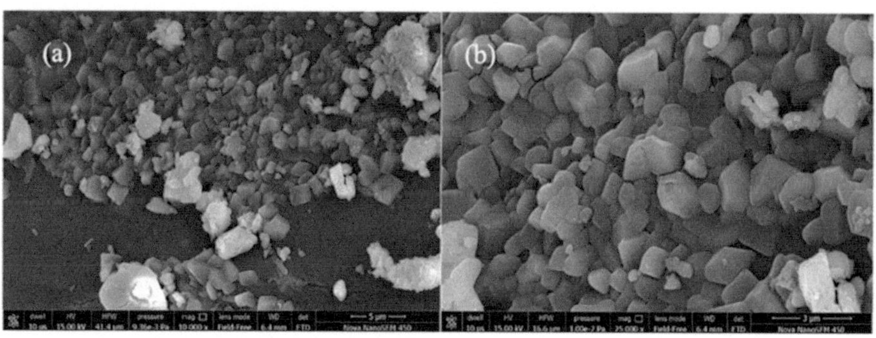

Fig. 12.15 SEM images of Cu NPs synthesized from *Alstonia scholaris* leaves at 5 μm (**a**) and 3 μm (**b**) maginifications (Labaran et al., 2024)

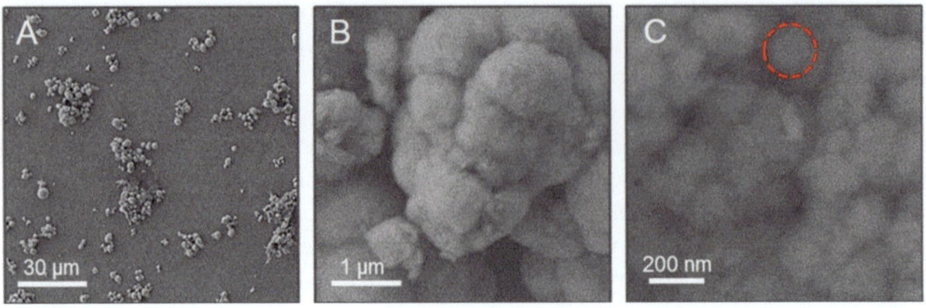

Fig. 12.16 SEM micrographs of zinc oxide nanoparticles synthesized from leaf extract of *Cassia fistula* at 30 µm, 1 µm, and 200 nm magnifications (Naseer et al., 2020)

12.3.6 Transmission Electron Microscopy

Transmission electron microscopy (TEM) provides essential information about nanoparticles' surface morphology, size, and shape. Characterization information from SEM and TEM gives a high degree of confirmation of the morphology and shape of the bio nanoparticles. The TEM images of Ag NPs were synthesized from a medicinal plant-mediated route in which the particles were found to be spherical with an average particle size of 16 nm (Ajaykumar et al., 2023). Additionally, we obtained a TEM micrograph (Fig. 12.17) of Ag NPs synthesized using *Aspilia pluriseta* extracts to be spherical and crystalline with an average diameter of 6 nm (Nyabola et al., 2020). The TEM images of Ag NPs, according to the study by Gavamukulya et al. (2019), revealed that the nanoparticles were spherical and had an average diameter of 51 nm.

Amaliyah et al. (2020) performed the TEM and selected area electron diffraction (SAED) analysis, revealing that Cu nanoparticles synthesized from *Piper retrofractum vahl* extracts had an amorph phase (Fig. 12.18). Elsewhere, Ag nanoparticles were synthesized using *Uvaria narum* extracts, and their TEM and SAED images were obtained in which SAED revealed the particles were bright circular rings, as shown in Fig. 12.19 (Ajaykumar et al., 2023).

Combining these characterization techniques enables researchers to effectively confirm the synthesis of bio-nanoparticles from medicinal plant-mediated routes as they give exclusive information on nanoparticle size, shape, composition, and morphology. Hence, a single characterization method cannot substantially prove nanoparticles' formation; instead, it combines the techniques.

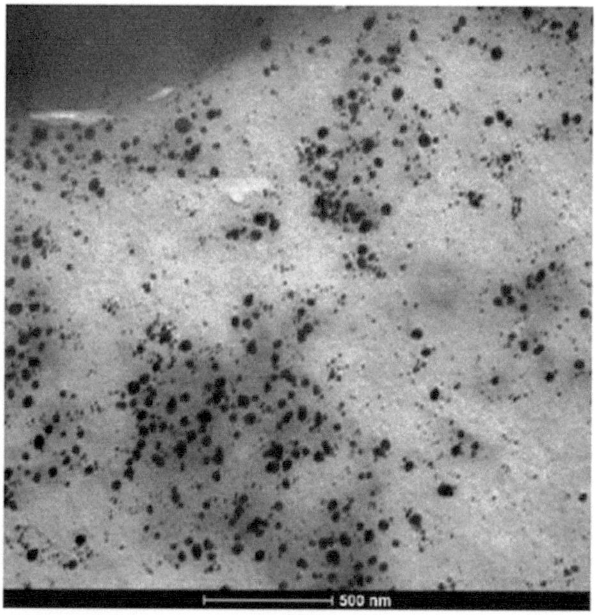

Fig. 12.17 TEM images of Ag NPs obtained from *Aspilia pluriseta* extracts (Nyabola et al., 2020)

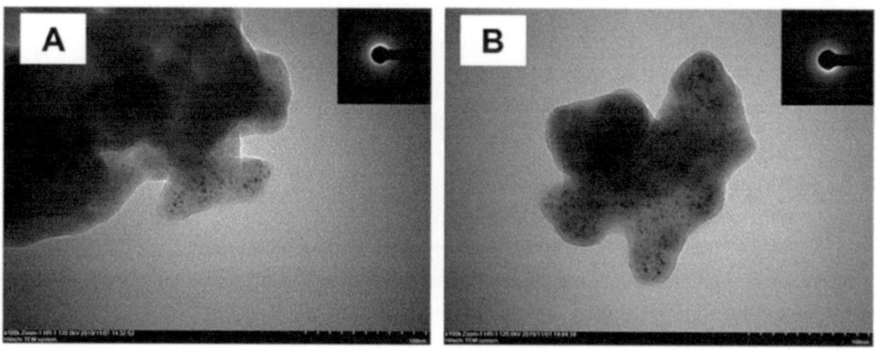

Fig. 12.18 TEM and SAED images for the stirred sample (**a**) and the sonicated sample (**b**) of Cu nanoparticles synthesized from *Piper retrofractum* vahl extracts (Amaliyah et al., 2020)

12.4 Application of African Medicinal Plants-Mediated Bio-nanoparticles as Therapeutic Agents

Researchers have embraced the green synthesis of metallic and metal oxide nanoparticles using extracts from medicinal plants' origins to improve strategies for addressing diseases. Table 12.1 identifies some bio-nanoparticles produced from African medicinal plant extracts and their role as therapeutic agents.

Fig. 12.19 TEM and SAED image (E) of silver nanoparticles synthesized from *Uvaria narum* extracts (Ajaykumar et al., 2023)

12.5 Mechanism of Action of Synthesis of Nanoparticles from Plant Extracts

Several studies have postulated the mechanism of synthesis of metallic and metal oxide nanoparticles from plant extracts (Hosseingholian et al., 2023; Singh et al., 2023; Abuzeid et al., 2023). The synthesis process involves three primary steps, namely, the activation step, where the metal ions are reduced, followed by nucleation of the metal ions; the growth step, where the smaller nanoparticles interact and cluster to form large-sized nanoparticles; and the termination step, which determines the resulting shape of the nanoparticles, as depicted in Fig. 12.20 (Hosseingholian et al., 2023). As mentioned earlier, the plant extracts possess rich phytochemicals, which reduce metal ions and stabilize them into nanoparticles. Primarily, phenolics, terpenoids, tannins, saponins, and flavonoids help in the metal ions reduction leading to the formation of the nanoparticles as they have functional groups capable of reducing the metal ions (Singh et al., 2018a, b, 2023).

12.6 Future Perspectives

Medicinal plants have been used as herbal remedies to treat numerous diseases, and their incorporation into nanoparticles has been shown to be fruitful in solving myriad diseases. This chapter identifies a significant research knowledge gap from these

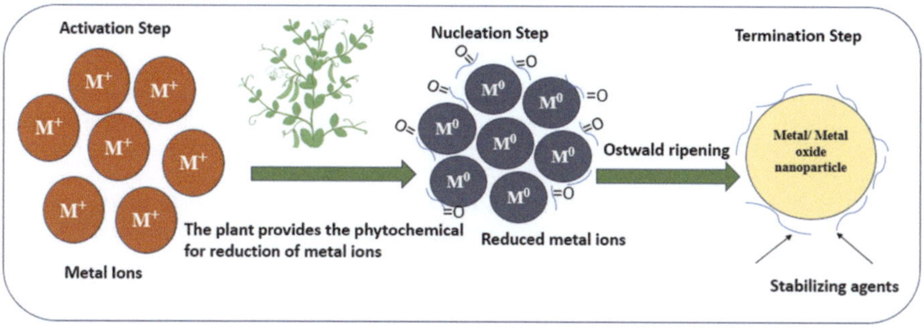

Fig. 12.20 A scheme showing the mechanism of synthesis of metal and metal oxide nanoparticles facilitated by phytochemicals from plant extracts

establishments in studies that have combined medicinal chemistry and nanotechnology to develop new drugs for therapeutic purposes. Therefore, this chapter recommends future work to exploit the potential of proven African medicinal plants as herbal drugs, incorporate them as nanoparticles to improve the effectiveness of addressing diseases that have been a problem due to multi-drug resistance, and create avenues for developing better medications for emerging diseases. The newly developed drugs incorporating nanotechnology should undergo clinical trials for approval, assessing the suitability of the developed nano-drugs with their appropriate dosage for optimal efficacy.

12.7 Conclusion

This chapter highlighted the need to develop medicinal-plant mediated bio-nanoparticles to improve the treatment of disease menace facing many of the world's populations amid existing multi-faced drug resistance. As new diseases emerge, a promising alternative exists to exploit nanotechnology and develop drug modalities from herbal options that have improved bioavailability at the target site of action and enhanced potency. Medicinal plant-mediated nanoparticles intended for use as therapeutic agents have demonstrated potential in addressing numerous diseases. However, more research is needed to exploit the effectiveness of medicinal plants and nanotechnology in new drug development. By exploiting this research area, a solution to the disease menace can be achieved, and help African countries and the world treat diseases experienced in modern healthcare systems.

Acknowledgments The authors thank the Department of Chemistry, JKUAT, for their support.

Competing Interests The authors have no conflicts of interest to declare that are relevant to the content of this chapter.

References

Abdelkader, D. H., Negm, W. A., Elekhnawy, E., Eliwa, D., Aldosari, B. N., & Almurshedi, A. S. (2022). Zinc oxide nanoparticles as potential delivery carrier: Green synthesis by aspergillus Niger endophytic fungus, characterization, and in vitro/in vivo antibacterial activity. *Pharmaceuticals, 15*(9), 1057. https://doi.org/10.3390/ph15091057

Abuzeid, H. M., Julien, C. M., Zhu, L., & Hashem, A. M. (2023). Green synthesis of nanoparticles and their energy storage, environmental, and biomedical applications. *Crystals, 13*(11),1576. https://doi.org/10.3390/cryst13111576

Adebayo, I. A., Arsad, H., Gagman, H. A., Ismail, N. Z., & Samian, M. R. (2020). Inhibitory effect of eco-friendly naturally synthesized silver nanoparticles from the leaf extract of medicinal Detarium microcarpum plant on pancreatic and cervical cancer cells. *Asian Pacific Journal of Cancer Prevention, 21*(5), 1247–1252. https://doi.org/10.31557/APJCP.2020.21.5.1247

Adewale, O. B., Egbeyemi, K. A., Onwuelu, J. O., Potts-Johnson, S. S., Anadozie, S. O., Fadaka, A. O., Osukoya, O. A., Aluko, B. T., Johnson, J., Obafemi, T. O., & Onasanya, A. (2020). Biological synthesis of gold and silver nanoparticles using leaf extracts of Crassocephalum rubens and their comparative in vitro antioxidant activities. *Heliyon, 6*(11), e05501. https://doi.org/10.1016/j.heliyon.2020.e05501

Ahmad, M. M., Mushtaq, S., Al Qahtani, H. S., Sedky, A., & Alam, M. W. (2021). Investigation of TiO2 nanoparticles synthesized by sol-gel method for effectual photodegradation, oxidation and reduction reaction. *Crystals, 11*(12), 1456. https://doi.org/10.3390/cryst11121456

Ahmad, M. Z., Alasiri, A. S., Ahmad, J., Alqahtani, A. A., Abdullah, M. M., Abdel-Wahab, B. A., Pathak, K., Saikia, R., Das, A., Sarma, H., & Alzahrani, S. A. (2022). Green synthesis of titanium dioxide nanoparticles using Ocimum sanctum leaf extract: In vitro characterization and its healing efficacy in diabetic wounds. *Molecules, 27*(22), 1–17. https://doi.org/10.3390/molecules27227712

Ahmed, R. H., & Mustafa, D. E. (2020). Green synthesis of silver nanoparticles mediated by traditionally used medicinal plants in Sudan. *International Nano Letters, 10*(1), 1–14. https://doi.org/10.1007/s40089-019-00291-9

Ajaykumar, A. P., Mathew, A., Chandni, A. P., Varma, S. R., Jayaraj, K. N., Sabira, O., Rasheed, V. A., Binitha, V. S., Swaminathan, T. R., & Basheer, V. S. (2023). Green Synthesis of silver nanoparticles using the leaf extract of the medicinal plant, Uvaria narum and its antibacterial, antiangiogenic, anticancer and catalytic properties. *Antibiotics, 12*(3), 564.

Al-Harbi, N., & Abd-Elrahman, N.K. (2024). Physical methods for preparation of nanomaterials their characterization and applications: A review. *Journal of Umm Al-Qura University for Applied Sciences, 1-22*. https://doi.org/10.1007/s43994-024-00165-7

Alhumaydhi, F. A. (2022). Green synthesis of gold nanoparticles using extract of Pistacia chinensis and their in vitro and in vivo biological activities. *Journal of Nanomaterials, 2022*, 23–27. https://doi.org/10.1155/2022/5544475

Alshammari, S. O., Mahmoud, S. Y., & Farrag, E. S. (2023). Synthesis of green copper nanoparticles using medicinal plant Krameria sp. root extract and its applications. *Molecules, 28*(12), 4629. https://doi.org/10.3390/molecules28124629

Amaliyah, S., Pangesti, D. P., Masruri, M., Sabarudin, A., & Sumitro, S. B. (2020). Green synthesis and characterization of copper nanoparticles using Piper retrofractum Vahl extract as bioreductor and capping agent. *Heliyon, 6*(8), e04636. https://doi.org/10.1016/j.heliyon.2020.e04636

Anderson, S. (2023). *Antimicrobial resistance death toll could catch up to cancer by 2050, and pollution is fuelling its spread* (pp. 1–5). Health Policy Watch.

Aschale, Y., Wubetu, M., Abebaw, A., Yirga, T., Minwuyelet, A., & Toru, M. (2021). A systematic review on traditional medicinal plants used for the treatment of viral and fungal infections in Ethiopia. *Journal of Experimental Pharmacology, 13*(August), 807–815. https://doi.org/10.2147/JEP.S316007

Baig, N., Kammakakam, I., & Falath, W. (2021). Nanomaterials: A review of synthesis methods, properties, recent progress, and challenges. *Materials Advances, 2*, 1821. https://doi.org/10.1039/d0ma00807a

Bayda, S., Adeel, M., Tuccinardi, T., Cordani, M., & Rizzolio, F. (2019). The history of nanoscience and nanotechnology: From chemical-physical applications to nanomedicine. *Molecules, 25*(1), 1–15. https://doi.org/10.3390/molecules25010112

CDC. (2021). *2019 Antibiotic resistance threats report | CDC*. Center for Disease Control, July, 1–113. https://www.hhs.gov/sites/default/files/michael-craig-cdc-talk-thursday-am-508.pdf

Chen, L. H., & Hu, J. N. (2025). Development of nano-delivery systems for loaded bioactive compounds: using molecular dynamics simulations. *Critical Reviews in Food Science and Nutrition, 65*(10), 1811–1832. https://doi.org/10.1080/10408398.2023.2301427

Dhoke, S. K. (2023). Synthesis of nano-ZnO by chemical method and its characterization. *Results in Chemistry, 5*(August 2022), 100771. https://doi.org/10.1016/j.rechem.2023.100771

Elbagory, A. M., Hussein, A. A., & Meyer, M. (2019). The in vitro immunomodulatory effects of gold nanoparticles synthesized from hypoxis hemerocallidea aqueous extract and hypoxoside on macrophage and natural killer cells. *International Journal of Nanomedicine,* 9007–9018.

Elbagory, A., Meyer, M., Cupido, C., & Hussein, A. A. (2017). Inhibition of bacteria associated with wound infection by biocompatible green synthesized gold nanoparticles from South African plant extracts. *Nanomaterials, 7*(12). https://doi.org/10.3390/nano7120417

Farooq, U., Qureshi, A. K., Noor, H., Farhan, M., Khan, M. E., Hamed, O. A., Bashiri, A. H., & Zakri, W. (2023). Plant extract-based fabrication of silver nanoparticles and their effective role in antibacterial, anticancer, and water treatment applications. *Plants, 12*, 2337.

Fouda, A., Eid, A. M., Guibal, E., Hamza, M. F., Hassan, S. E. D., Alkhalifah, D. H. M., & El-Hossary, D. (2022). Green synthesis of gold nanoparticles by aqueous extract of Zingiber officinale: Characterization and insight into antimicrobial, antioxidant, and in vitro cytotoxic activities. *Applied Sciences, 12*(24). https://doi.org/10.3390/app122412879

Gabriel Kigen, W. K., Wanjohi, B., & Haruki, B. (2020). Medicinal plants used by traditional healers in Sangurur, Elgeyo Marakwet County, Kenya. *Pharmacognosy Research, 10*(October), 24–30. https://doi.org/10.4103/pr.pr

Gavamukulya, Y., Maina, E. N., Meroka, A. M., Madivoli, E. S., El-Shemy, H. A., Wamunyokoli, F., & Magoma, G. (2019). Green synthesis and characterization of highly stable silver nanoparticles from ethanolic extracts of fruits of Annona muricata. *Journal of Inorganic and Organometallic Polymers and Materials, 30*, 1231. https://doi.org/10.1007/s10904-019-01262-5

Gebre, S. H., & Sendeku, M. G. (2019). New frontiers in the biosynthesis of metal oxide nanoparticles and their environmental applications: An overview. *SN Applied Sciences, 1*(8), 928. https://doi.org/10.1007/s42452-019-0931-4

Giri, A. K., Jena, B., Biswal, B., Pradhan, A. K., Arakha, M., Acharya, S., & Acharya, L. (2022). Green synthesis and characterization of silver nanoparticles using Eugenia roxburghii DC . extract and activity against biofilm – producing bacteria Full Width at Half Maximum. *Scientific Reports, 12*, 1–9. https://doi.org/10.1038/s41598-022-12484-y

Govindasamy, R., Raja, V., Singh, S., Govindarasu, M., Sabura, S., Rekha, K., Rajeswari, V. D., Alharthi, S. S., Vaiyapuri, M., Sudarmani, R., Jesurani, S., Venkidasamy, B., & Thiruvengadam, M. (2022). Green synthesis and characterization of cobalt oxide nanoparticles using Psidium guajava leaves extracts and their photocatalytic and biological activities. *Molecules, 27*(17), 5646. https://doi.org/10.3390/molecules27175646

Hetta, H. F., Ramadan, Y. N., Al-harbi, A. I., Ahmed, E. A., Battah, B., Ellah, N. H. A., Zanetti, S., & Donadu, M. G. (2023). Nanotechnology as a promising approach to combat multidrug resistant bacteria: A comprehensive review and future perspectives. *Biomedicine, 11*(2), 413. https://www.mdpi.com/2227-9059/11/2/413#

Hosseingholian, A., Gohari, S. D., Feirahi, F., Moammeri, F., Mesbahian, G., Moghaddam, Z. S., & Ren, Q. (2023). Recent advances in green synthesized nanoparticles: From production to application. *Materials Today Sustainability, 24*, 100500. https://doi.org/10.1016/j.mtsust.2023.100500

Hussain, M. H., Fitrah, N., Bakar, A., Mustapa, A. N., Low, K., Othman, N. H., & Adam, F. (2020). Synthesis of various size gold nanoparticles by chemical reduction method with different solvent polarity. *Nanoscale Research Letters, 15*, 140–150.

Jain, S., & Mehata, M. S. (2017). Medicinal plant leaf extract and pure flavonoid mediated green synthesis of silver nanoparticles and their enhanced antibacterial property. *Scientific Reports, 7*(1), 1–13. https://doi.org/10.1038/s41598-017-15724-8

Kalia, A., Kaur, J., Tondey, M., Manchanda, P., Bindra, P., Alghuthaymi, M. A., Shami, A., & Abd-Elsalam, K. A. (2021). Differential antimycotic and antioxidant potentials of chemically synthesized zinc-based nanoparticles derived from different reducing/complexing agents against pathogenic fungi of maize crop. *Journal of Fungi, 7*(3), 223. https://doi.org/10.3390/jof7030223

Khan, A., Rashid, A., Younas, R., & Chong, R. (2016). A chemical reduction approach to the synthesis of copper nanoparticles. *International Nano Letters, 6*(1), 21–26. https://doi.org/10.1007/s40089-015-0163-6

Koul, B., Poonia, A. K., & Yadav, D. (2021). Microbe-mediated biosynthesis of nanoparticles: Applications and future prospects. *Biomolecules, 11*, 886.

Kumar, S., Bhushan, P., & Bhattacharya, S. (2017). Fabrication of nanostructures with bottom-up approach and their utility in diagnostics, therapeutics, and others. In *Environmental, chemical and medical sensors* (pp. 67–198). Springer Singapore.

Kumari, S., Raturi, S., Kulshrestha, S., Chauhan, K., Dhingra, S., András, K., Thu, K., Khargotra, R., & Singh, T. (2023). A comprehensive review on various techniques used for synthesizing nanoparticles. *Journal of Materials Research and Technology, 27*, 1739–1763. https://doi.org/10.1016/j.jmrt.2023.09.291

Labaran, A. N., Zango, Z. U., Tailor, G., Alsadig, A., Usman, F., Mukhtar, M. T., Garba, A. M., Alhathlool, R., Ibnaouf, K. H., & Aldaghri, O. A. (2024). Biosynthesis of copper nanoparticles using Alstonia scholaris leaves and its antimicrobial studies. *Scientific Reports, 14*(1), 1–10. https://doi.org/10.1038/s41598-024-56052-y

Labulo, A. H., David, O. A., & Terna, A. D. (2022). Green synthesis and characterization of silver nanoparticles using Morinda lucida leaf extract and evaluation of its antioxidant and antimicrobial activity. *Chemical Papers, 76*(12), 7313–7325. https://doi.org/10.1007/s11696-022-02392-w

Lakra, R., Rahul Kumar, D. N. T., Sahoo, P. K., & Soam, A. (2021). Synthesis and characterization of cobalt oxide (Co3O4) nanoparticles. *Materials Today: Proceedings, 41*(2), 269–271. https://doi.org/10.1007/s10948-012-1618-x

Lediga, M. E., Malatjie, T. S., Olivier, D. K., Ndinteh, D. T., & van Vuuren, S. F. (2018). Biosynthesis and characterisation of antimicrobial silver nanoparticles from a selection of fever-reducing medicinal plants of South Africa. *South African Journal of Botany, 119*, 172–180. https://doi.org/10.1016/j.sajb.2018.08.022

Lung, I., & Stegarescu, A. (2023). Biosynthesis of nanoparticles using plant extracts and essential oils. *Molecules, 28*, 3060.

Madhav, K., Anand, N., & Pandita, A. (2023). Antimicrobial and Antibiofilm activity of Azadirachta Indica silver nanoparticles against enteric pathogens isolated from effluent samples. *International Journal of Pharmaceutical Sciences and Research, 14*(11), 5493. https://doi.org/10.13040/IJPSR.0975-8232.14(11).5493-03

Madivoli, E. S., Gachoki, P. K., Gachui, E. M., Augustine, N. O., Indire, S., Wanakai, S. I., & Nyang'au, J. O. (2019). Biosynthesis of iron nanoparticles using Ageratum conyzoides extracts, their antimicrobial and photocatalytic activity. *SN Applied Sciences, 1*(5), 1–11. https://doi.org/10.1007/s42452-019-0511-7

Madivoli, E. S., Kareru, P. G., Gachanja, A. N., Mugo, S. M., Makhanu, D. S., Wanakai, S. I., & Gavamukulya, Y. (2020). Facile synthesis of silver nanoparticles using Lantana trifolia aqueous extracts and their antibacterial activity. *Journal of Inorganic and Organometallic Polymers and Materials, 30*, 2842. https://doi.org/10.1007/s10904-019-01432-5

Mali, S. C., Dhaka, A., Githala, C. K., & Trivedi, R. (2020). Green synthesis of copper nanoparticles using Celastrus paniculatus Willd. leaf extract and their photocatalytic and antifungal properties. *Biotechnology Reports, 27*, e00518. https://doi.org/10.1016/j.btre.2020.e00518

Malik, S., & Muhammad, K. (2023). Nanotechnology: A revolution in modern industry. *Molecules, 28*, 661.

Mba, I. E., & Nweze, E. I. (2021). Nanoparticles as therapeutic options for treating multidrug – Resistant bacteria: Research progress, challenges, and prospects. *World Journal of Microbiology and Biotechnology, 37*(6), 1–30. https://doi.org/10.1007/s11274-021-03070-x

Mekuye, B., & Abera, B. (2023). Nanomaterials: An overview of synthesis, classification, characterization, and applications. *Nano Select, 4*(8), 486–501. https://doi.org/10.1002/nano.202300038

Miran, S. B., & Ali, F. A. (2023). Biosynthesis of gold nanoparticles with four different Lactobacillus species. *Eurasian Journal of Science and Engineering, 9*(3), 47–56. https://doi.org/10.23918/eajse.v9i3p05

Mohamed, R. M. (2022). Green biosynthesis, structural characterization and anticancer activity of copper oxide nanoparticles from the brown alga Cystoseira myrica Green biosynthesis, structural characterization and anticancer activity of copper oxide nanoparticles from the br. *Egyptian Journal of Aquatic Biology & Fisheries, 25*, 341–358.

Mourdikoudis, S., Pallares, R. M., & Thanh, N. T. K. (2018). Characterization techniques for nanoparticles: Comparison and complementarity upon studying nanoparticle properties. *Nanoscale, 10*(27), 12871–12934. https://doi.org/10.1039/c8nr02278j

Mukaratirwa-Muchanyereyi, N., Gusha, C., Mujuru, M., Guyo, U., & Nyoni, S. (2022). Synthesis of silver nanoparticles using plant extracts from Erythrina abyssinica aerial parts and assessment of their anti-bacterial and anti-oxidant activities. *Results in Chemistry, 4*(March), 100402. https://doi.org/10.1016/j.rechem.2022.100402

Mwangi, V. N., Madivoli, E. S., Kangogo, M., Wanakai, S. I., Waudo, W., & Nzilu, D. M. (2024). Antimicrobial surface coating as a pathway to curb resistance: Preparation, mode of action and future perspective. *Journal of Coatings Technology and Research, 21*(3), 799–810. https://doi.org/10.1007/s11998-023-00879-z

Najmi, A., Javed, S. A., Al Bratty, M., & Alhazmi, H. A. (2022). Modern approaches in the discovery and development of plant-based natural products and their analogues as potential therapeutic agents. *Molecules, 27*, 349.

Naseer, M., Aslam, U., Khalid, B., & Chen, B. (2020). Green route to synthesize zinc oxide nanoparticles using leaf extracts of Cassia fistula and Melia azadarach and their antibacterial potential. *Scientific Reports, 10*(1), 1–10. https://doi.org/10.1038/s41598-020-65949-3

Nyabola, A. O., Kareru, P. G., Wanakai, I. S., Madivoli, E. S., & Gachui, E. M. (2020). Formation of silver nanoparticles via Aspilia pluriseta extracts their antimicrobial and catalytic activity. *Journal of Inorganic and Organometallic Polymers and Materials, 30*, 3493.

Nzilu, D. M., Madivoli, E. S., Makhanu, D. S., Wanakai, S. I., Kiprono, G. K., & Kareru, P. G. (2023). Green synthesis of copper oxide nanoparticles and its efficiency in degradation of rifampicin antibiotic. *Scientific Reports, 13*(1), 1–18. https://doi.org/10.1038/s41598-023-41119-z

Nzilu, D., Madivoli, E., Makhanu, D., Wanakai, S., Kirui, G., Mwangi, V., & Kareru, P. (2024). Synthesis and characterization of Parthenium hysterophorus – Mediated ZnO nanoparticles for methylene blue dye degradation. *Journal of Chemistry, 2024*(1), 1088430.

Odongo, A. S., Oluoch Okumu, F., Omwoma Lugasi, S., Opiyo Onani, M., & Gaya Agong, S. (2022). Biogenic synthesis of gold nanoparticles from Physalis peruviana and application in wound healing. *Journal of Chemistry, 2022*(1), 9034840. https://doi.org/10.1155/2022/9034840

Poostforooshan, J., Belbekhouche, S., Olszok, V., Stodt, M. F. B., Simmler, M., Bierwirth, M., Nirschl, H., Kiefer, J., Fritsching, U., & Weber, A. P. (2023). Synthesis of pure and Fe-doped TiO2 nanoparticles via electrospray-assisted flame spray pyrolysis for antimicrobial applications. *ACS Applied Nano Materials, 6*(24), 22660–22672. https://doi.org/10.1021/acsanm.3c03107

Prestinaci, F., Pezzotti, P., & Pantosti, A. (2015). Antimicrobial resistance: A global multifaceted phenomenon. *Pathogens and Global Health, 109*(7), 309–318. https://doi.org/10.1179/2047773215Y.0000000030

Puja, A. M., Rupa, E. J., Kim, Y. J., & Yang, D. (2023). Medicinal plant enriched metal nanoparticles and nanoemulsion for inflammation treatment: A narrative review on current status and future perspective UV-vis focus nanoparticles. *Immuno, 3*, 182–194.

Rahmah, M. I., Ahmed, A. M., Rashid, T. M., & Qasim, A. J. (2024). Preparation of silver nanoparticles using laser ablation for in vitro treatment of MCF-7 cancer cells with antibacterial activity. *Plasmonics, 19*(4), 2097–2105. https://doi.org/10.1007/s11468-023-02150-y

Selvarajan, R., Obize, C., Sibanda, T., Abia, A. L. K., & Long, H. (2022). Evolution and emergence of antibiotic resistance in given ecosystems: Possible strategies for addressing the challenge of antibiotic resistance. *Antibiotics, 12*(1), 28. https://doi.org/10.3390/antibiotics12010028

Shard, A. G., Schofield, R. C., & Minelli, C. (2020). Characterization of nanoparticles chapter 3. 2. 3 – Ultraviolet – visible spectrophotometry. In *Micro and nano technologies* (pp. 10–12). Elsevier.

Shariq, M., Friedrich, B., Budic, B., Hodnik, N., Ruiz-Zepeda, F., Majerič, P., & Rudolf, R. (2018). Successful synthesis of gold nanoparticles through ultrasonic spray pyrolysis from a gold(III) nitrate precursor and their interaction with a high electron beam. *ChemistryOpen, 7*(7), 533–542. https://doi.org/10.1002/open.201800101

Singh, J., Dutta, T., Kim, K. H., Rawat, M., Samddar, P., & Kumar, P. (2018a). "Green" synthesis of metals and their oxide nanoparticles: Applications for environmental remediation. *Journal of Nanobiotechnology, 16*(1), 84. https://doi.org/10.1186/S12951-018-0408-4

Singh, K., Naidoo, Y., Mocktar, C., & Baijnath, H. (2018b). Biosynthesis of silver nanoparticles using Plumbago auriculata leaf and calyx extracts and evaluation of their antimicrobial activities. *Advances in Natural Sciences: Nanoscience and Nanotechnology, 9*(3), 035004. https://doi.org/10.1088/2043-6254/aad1a3

Singh, H., Desimone, M. F., Pandya, S., Jasani, S., George, N., Adnan, M., Aldarhami, A., Bazaid, A. S., & Alderhami, S. A. (2023). Revisiting the green synthesis of nanoparticles: Uncovering influences of plant extracts as reducing agents for enhanced synthesis efficiency and its biomedical applications. *International Journal of Nanomedicine, 18*(August), 4727–4750. https://doi.org/10.2147/IJN.S419369

Thiruvengadam, V., & Bansod, A. V. (2020). Characterization of silver nanoparticles synthesized using chemical method and its antibacterial property. *Biointerface Research in Applied Chemistry, 10*(6), 7257–7264. https://doi.org/10.33263/BRIAC106.72577264

Tyagi, P. K., Kumar, A., Mazumder, A., & Tyagi, S. (2023). Development of copper nanoparticles and their prospective uses as antioxidants, antimicrobials, and anti-cancer agents in the pharmaceutical sector. *Precision Nanomedicine, 6*(2), 1048–1065. https://doi.org/10.33218/001c.83932

Ullah, A., Yin, X., Wang, F., Xu, B., Mirani, Z. A., Xu, B., Chan, M. W. H., Ali, A., Usman, M., Ali, N., & Naveed, M. (2021). Biosynthesis of selenium nanoparticles (via bacillus subtilis bsn313), and their isolation, characterization, and bioactivities. *Molecules, 26*(18), 5559. https://doi.org/10.3390/molecules26185559

Vincent, N., Shigwenya, M., Madivoli, E., Kangogo, M., Clement, M., Sammy, I., Mwanza, W., Nzilu, D., & Waudo, W. (2024). An evaluation of antimicrobial potency of silver nanoparticles synthesised from Fusarium sp. *Discover Applied Sciences, 6*(4), 201. https://doi.org/10.1007/s42452-024-05870-w

Vladár, A. E., & Hodoroaba, V.-D. (2020). Characterization of nanoparticles chapter 2. 1. 1 – Characterization of nanoparticles by scanning electron microscopy. In *Micro and nano technologies* (pp. 23–24). Elsevier.

Vyas, J., & Rana, S. (2017). Antioxidant activity and green synthesis of selenium nanoparticles using allium sativum extract. *International Journal of Phytomedicine, 9*(4), 634. https://doi.org/10.5138/09750185.2185

Wang, C., Xu, H., Liu, T., Yang, S., Nie, Y., Wang, C., Guo, X., Wang, B., Ma, X., & Jiang, X. (2021). One-step ball milling synthesis of VO2 (M) nanoparticles with exemplary thermochromic performance. *SN Applied Sciences, 3*(4), 1–10. https://doi.org/10.1007/s42452-021-04154-x

World Health Organization. (2023). *Antimicrobial resistance*. https://www.who.int/news-room/fact-sheets/detail/antimicrobial-resistance

Xulu, J. H., Ndongwe, T., Ezealisiji, K. M., Tembu, V. J., Mncwangi, N. P., Witika, B. A., & Siwe-noundou, X. (2022). The use of medicinal plant-derived metallic nanoparticles in theranostics. *Pharmaceutics, 14*, 2437.

Ying, S., Guan, Z., Ofoegbu, P. C., & Clubb, P. (2022). Environmental technology & innovation green synthesis of nanoparticles: Current developments and limitations. *Environmental Technology & Innovation, 26*, 102336. https://doi.org/10.1016/j.eti.2022.102336

Yudasari, N., Kennedy, D. S., & Suliyanti, M. M. (2019). Pulse laser deposition (PLD) technique for ZnO photocatalyst fabrication. *Journal of Physics: Conference Series, 1191*(1), 012009. https://doi.org/10.1088/1742-6596/1191/1/012009

Yusuf, A., Almotairy, A. R. Z., Henidi, H., Alshehri, O. Y., & Aldughaim, M. S. (2023). Nanoparticles as drug delivery systems: A review of the implication of nanoparticles ' physicochemical properties on responses in biological systems. *Polymers, 15*(7), 1596.

Chapter 13
Protocol for the Extraction of Alkaloids from Medicinal Plant Extracts

Albert Fouda Atangana, Isaac Silvere Gade, Gaetan Bayiha Ba Njock, and Moïse Diguir

Abstract Man has always found food and medicines in plants, which justify his particular interest. With the development of modern medicine, interest in plants has increased due to the bioactive molecules found there and used as active ingredients. Among these molecules, alkaloïds are highly sought-after nitrogen compounds because of the interesting pharmacological properties provided by these types of molecules. They are found in living organisms, particularly in plants. Alkaloïds of natural origin, which are used in modern medicine, are highly appreciated, such as morphine, quinine, caffeine, and codeine. However, their extraction from plants should attract particular attention. The extraction of alkaloïds is based on their basic character and solubility profile. Generally, alkaloïds are extracted using mainly three methods: extraction with non-polar organic solvents in alkaline media, extraction with polar organic solvents, and extraction with acidified water. This chapter aims to present and describe the different procedures for extracting alkaloïds from medicinal plants.

Keywords Extraction · Alkaloïds · Medicinal plant · Extraction procedures

Abbreviations

HBAs	Hydrogen acceptors
HBDs	Hydrogen-bond donors
HPTLC	High-performance thin layer chromatography
NDES	Natural deep eutectic solvent
p-DMAB	p-Dimethylaminobenzaldehyde
SDS	Sodium dodecyl sulphate
SME	Surfactant-mediated extraction
TLC	Thin layer chromatography
UAMME	Ultrasonic-assisted micelle-mediated extraction

A. Fouda Atangana (✉) · I. S. Gade · G. Bayiha Ba Njock · M. Diguir
Department of Chemistry, Faculty of Science, University of Maroua, Maroua, Cameroon

A. Dekebo (ed.), *Bioactive Secondary Metabolites from Medicinal Plants of Africa*, Interdisciplinary Biotechnological Advances,
https://doi.org/10.1007/978-981-96-8590-5_13

13.1 Introduction

Extracts from medicinal plants have been used since ancient times as an alternative source of medicines in the treatment of various ailments. These extracts are composed of secondary metabolites produced by plants during their biosynthetic process and are responsible for the beneficial health effects observed. In fact, secondary metabolites of plants constitute a large and varied group of organic compounds that are synthesized in small quantities and have no direct function in essential processes of the plant like photosynthesis (Mera et al., 2019). They are often involved in plant protection against biotic and abiotic stresses (Pagare et al., 2015).

Alkaloïds are one of the three main groups of secondary metabolites, in addition to terpenoïds and phenolic compounds. They are found in living organisms, essentially in plants. Narcotine and morphine are the first alkaloïds to be extracted and isolated from plants, respectively, in 1803 and 1806 (Bruneton, 2015). Today, more than 20,000 alkaloïds have been discovered (Wangchuk, 2019). Alkaloïds are substances with a basic character carrying one or more nitrogen atoms in the ring and possess important pharmacological effects such as depressants (morphine, scopolamine), stimulants (strychine, caffeine), antimalarial (quinine), anticancer (vinblastine, vincristine, camptothecine), cholinesterase inhibitors (pilocarpine, galantamine), and anticholinergic (atropine, hyosciamine) (Bruneton, 2015). Alkaloïds are often toxic and have many dramatic physiological effects, hence their widespread use in medicine (Guttiérrez-Grijalva et al., 2020).

The extractions of alkaloïds from plant material as part of phytochemical or biological investigations present specific challenges for researchers. A successful extraction starts with careful selection and preparation of plant samples. The extraction of alkaloïds is also based on the use of their physico-chemical properties. Several techniques are proposed and based on the differential solubility of bases and salts in water on one side, and in organic solvents on the other side. This chapter proposes some protocols currently used for the extraction of alkaloïds: extraction with non-polar organic solvents in alkaline media, extraction with polar organic solvents, and extraction with acidified water. This chapter aims to summarize the different protocols of extraction of alkaloïds from medicinal plants.

13.2 Distribution/Occurrence

Alkaloïds are found in living organisms such as plants, microorganisms, and animals (Bérdy, 2005). In plants, alkaloïds generally occur in all parts of the plant (leaves, barks, roots, stems, seeds, etc.), but sometimes accumulate only in a particular organ. For a given plant, the content of alkaloïd may be different depending on the organs; at the same time, other organs are free from alkaloïds. Example: Quinine, an antiplasmodial compound, accumulates in the bark of *Cinchona* but is absent from the leaves; in the same way, conessine accumulates in the seeds and bark of Kurchi (*Holarrhena pubescens* Will. ex G. Don) but is absent from the leaves. The edible tubers of the potato plant are devoid of alkaloïds, whereas the green parts contain poisonous alkaloïd solanine. We can also note that the organs in which alkaloïds accumulate are not always the site of their synthesis. Example: in tobacco, nicotine is produced in the roots and translocated to the leaves where it accumulates (Bruneton, 2015).

In this section, we also want to briefly summarize the botanical families of plants which are sources of alkaloïds, particularly those of pharmacological importance. Plants are considered as sources of alkaloïds when the species contain more than 0.01% of alkaloïds (Yang and Stöckigt, 2010; Wang et al., 2019).

Alkaloïds are essentially present in plants, particularly in some botanical families. The abundance of alkaloïds can vary from one plant family to another, or in the same family, they can vary between genera or species, and others are free from it. We can therefore cite several families of plants source of alkaloïds:

The Amaryllidaceae family: This family is composed of about 1100 species and 75 genera. Some species are *Crinum poweli*, *Hippeastrum puniceum*, and *vitanum*, *Lycoris radiata*.

The Apocynaceae family: This family is composed of around 2000 species and about 200 genera. We can cite some species like *Alstonia angustiloba* and *scholaris*, *Catharanthus roseus*, *Cynanchum paniculatum*, *Ochrosia elliptica*, and *Rauvolfia caffra*.

The Papaveraceae family: This family is composed of about 775 species distributed into 42 genera. Some species are *Papaver somniferum*, *Papaver rhoeas*, *Glaucium flavum*, and *Argemone Mexicana*.

The Asteraceae family: It is made up of approximately 1600 genera and over 23,000 species. As species, we have *Senecio cineraria*, *Centaurea montana*, and *Ageratum conyzoides*.

The Solanaceae family: We distinguish more than 2000 species and 90 genera. Example of some species: *Datura innoxia*, *Duboisia myoporoides*, *Nicotiana glauca*, *Solanum nigrum*, and *Solanum torvum*.

The Rutaceae family: This family is distributed across tropical and subtropical areas. Some species are *Dictamnus albus*, *Skimmia japonica*, and *Acronychia baueri*.

The Fabaceae family: This family is composed of about 18,000 species and 650 genera. Some species are *Albizia gummifera*, *Erythrina variegata*, and *Sophora flavescens*.

The Rubiaceae family: The family contains 13,500 species and 611 genera. Example of some species are *Cinchona officinalis*, *Mitragyna speciosa*, *Nauclea orientalis*, and *Uncaria tomentosa*.

13.3 Definition and Classification

The term alkaloïd was introduced by W. Meissner in the early nineteenth century to designate natural products reacting as bases, like alkali (from Arabic *al kaly*, sodium hydroxide; and from Greek, *eidos*, appearance) (Bruneton, 2015). There is no simple and precise definition of alkaloïd, and it is difficult to make a difference between alkaloïds and other natural metabolites containing nitrogen. However, we can notice the interesting definition of Pelletier (1983): "An alkaloid is a cyclic organic compound containing nitrogen in a negative oxidation state which is of limited distribution among living organism." Alkaloïds are also defined as basic compounds synthesized by living organisms containing one or more heterocyclic nitrogen atoms, derived from amino acids (with some exceptions), moreover they are pharmacologically active compounds (Brielmann et al., 2006).

Alkaloïds are a large family of more than 20,000 nitrogen-containing secondary metabolites (Wangchuk, 2019). Their classification is based on several criteria such as taxonomical origin, biosynthetic origin, pharmacological properties, and chemical structure. Among this criteria, the biosynthetic origin is the most accepted way to specify the alkaloïds. This classification based on the biosynthetic origin and the type of nitrogen is also called HEGNAUER's classification. According to this criterion, alkaloïds are classified into three different groups: true alkaloids (Fig. 13.1), protoalkaloïds (Fig. 13.2), and pseudoalkaloïds (Fig. 13.3) (Ranjitha & Sudha, 2015).

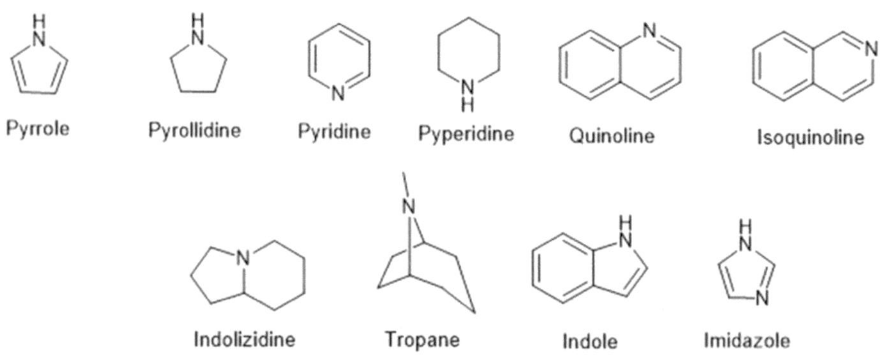

Fig. 13.1 Skeletal structure of true alkaloïds

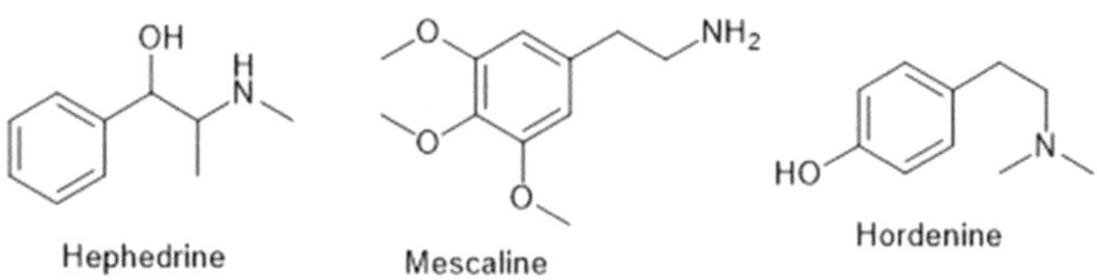

Fig. 13.2 Structure of some protoalkaloïds

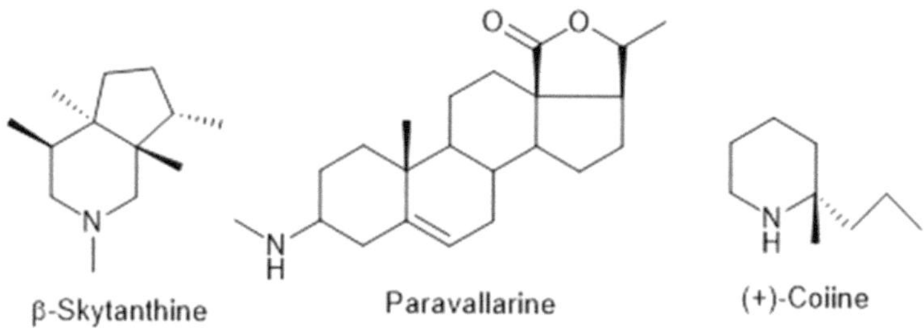

Fig. 13.3 Structure of some pseudoalkaloïds

True alkaloïds: They have a heterocyclic ring with nitrogen and are derived from amino acids. True alkaloïds are divided into pyrrole, pyrrolidine, pyrrolizidine, pyridine, piperidine, tropane, quinolone, isoquinoline, aporphine, quinolizidine, indole, indolizidine, and imidazole (Tidjani & Rhouati, 2016).

Protoalkaloïds: Like true alkaloïds, they are derived from amino acids or biogenetic amines, but they do not contain any heterocyclic system.

Pseudoalkaloïds: They are apparently unrelated to amino acids. They are nitrogen-containing molecules, but they have carbon skeletons derived from monoterpenes and other acetate derivatives and aliphatic polyketoacids.

13.4 Physico-Chemical Properties

Alkaloïds have complex and varied chemical structures, and they are compounds with molecular weight varying from 100 to 900. Most of non-oxygenated alkaloïds are liquids at room temperature (nicotine, sparteine, coniine), while those with oxygen atoms are normally crystallizable solids and rarely colored. Crystallized bases have a precise melting point without decomposition, especially above 200 °C. They are almost always able to deflect polarized light (Bruneton, 2015). Alkaloïds are derived from plants and have a bitter taste. The hydrogen potential greatly influences the soluble character of alkaloïds (Vercauteren, 2011).

As a general rule, basic alkaloïds are insoluble or partially soluble in water. However, they are soluble in non-polar or less polar organic solvents such as benzene, ether, and dichloromethane; as well as in high-strength alcohols. The basic character of alkaloïds allows them to form salts with mineral or organic acids according to the availability of the doublet of the nitrogen atom which brings them closer to amines. These salts are soluble in water and dilute alcohols.

Alkaloïd salts have an acidic character. They are insoluble in non-polar organic solvents, moderately soluble in polar organic solvents, and very soluble in water.

Alkaloïds are sensitive to heat, light, and oxygen (Bruneton, 2015).

13.5 Method of Extraction of Alkaloïds from Natural Substances

The extraction of alkaloïds is based on their differential solubility in acid medium and alkaline medium on the one hand, and the nature of the solvent on the other hand (Bruneton, 2009). Depending on the pH of the medium, the alkaloïds exist in the form of bases or salts. This property has an influence on the type of solvent to be used during the extraction: the alkaloïds in the base form (basic pH) are soluble in an organic solvent of low polarity, while in their salt form (acidic pH), they are soluble in an aqueous alcoholic solution (Amor, 2008a).

In general, there are four types of alkaloid extraction methods, including extraction by non-polar organic solvent, extraction by polar organic solvent, extraction by acidified water (Razafindrambao, 1973), and an extraction method developed in 2006, carried out using a sonic solution containing a surfactant as an extractant (Djilani et al., 2006). Recently, some methods of alkaloïds extractions have also been reported, such as non-ionic surfactant-mediated extraction (SME), and non-ionic-sulfants and natural deep eutectic solvent (NDES) (Takla et al., 2018).

13.5.1 Extraction with Non-Polar Organic Solvent

The dried material or sample (barks, leaves, roots, etc.) in powder form is treated with an alkali (ammonia, sodium carbonate, etc.) (Amor, 2008a). Indeed, alkaloïds are soluble in a low-polar medium, in a basic medium, and in water in acid medium (Sharad & Swati, 2012). The extraction can be done by maceration, percolation, or using a Soxhlet (Amor, 2008a). The organic solution obtained contains alkaloïds as described in Fig. 13.4.

13.5.2 Extraction with Polar Organic Solvent

This extraction with polar organic solvents is the most used method. The dried material is treated with an alcoholic solution, and this extract can be defatted with a non-polar solvent. Then, it is diluted in acid (with pH of 5) and with an organic solvent. The organic phase is free from alkaloïds. The next step is to alkalinize the medium (pH 11) and partition with an organic solvent (Sharad & Swati, 2012; Suau et al., 2002; Jones & Kinghorn, 2012). All this is summarized in Fig. 13.5.

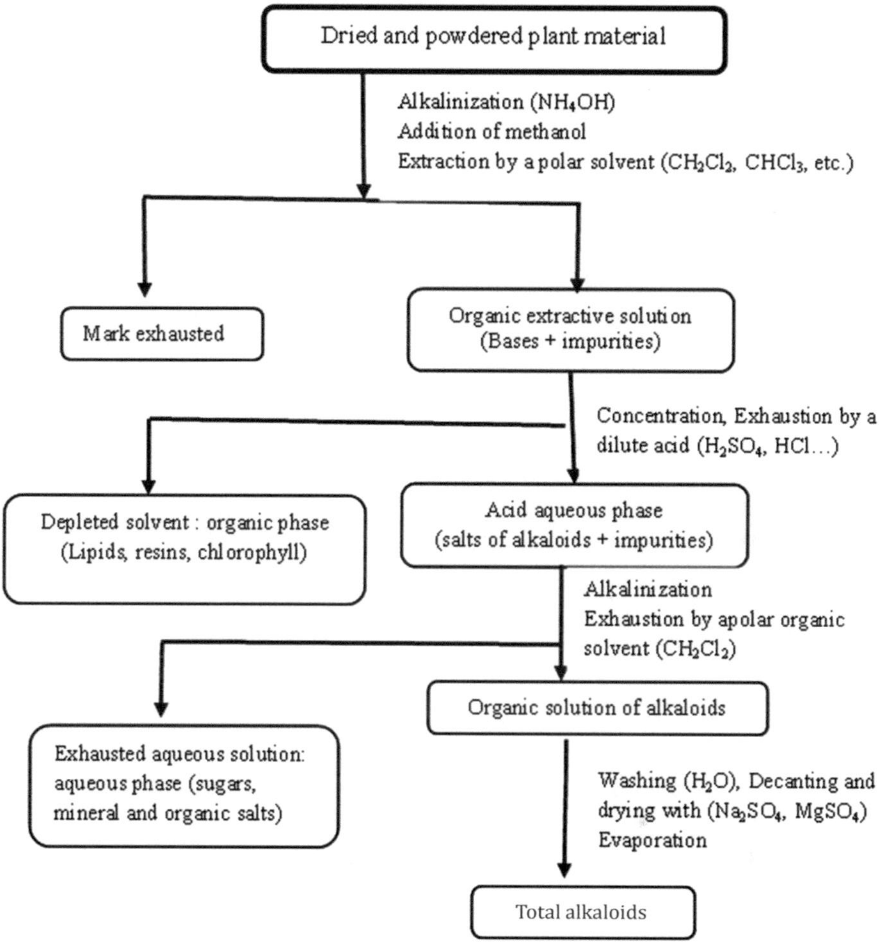

Fig. 13.4 Scheme of the extraction of alkaloïds in an alkaline medium (Bruneton, 2009)

13.5.3 Extraction with Acidified Water

Water is a solvent widely used in the field of traditional medicine for extractions at variable temperatures, for decoctions or infusions. However, many researchers prefer not to use this method because of the difficulty in obtaining the aqueous extract by evaporating water in the rotary evaporator due to its high boiling temperature (Jones & Kinghorn, 2012). Nevertheless, one can obtain total alkaloïds by macerating the powder of a dried part of the plant with acidulated water. The acidified aqueous solution of alkaloid salts obtained is alkalinized and leads to the precipitation of alkaloïds, which are finally extracted with a non-polar organic solvent (Fig. 13.6) (Vercauteren, 2011; Amor, 2008a).

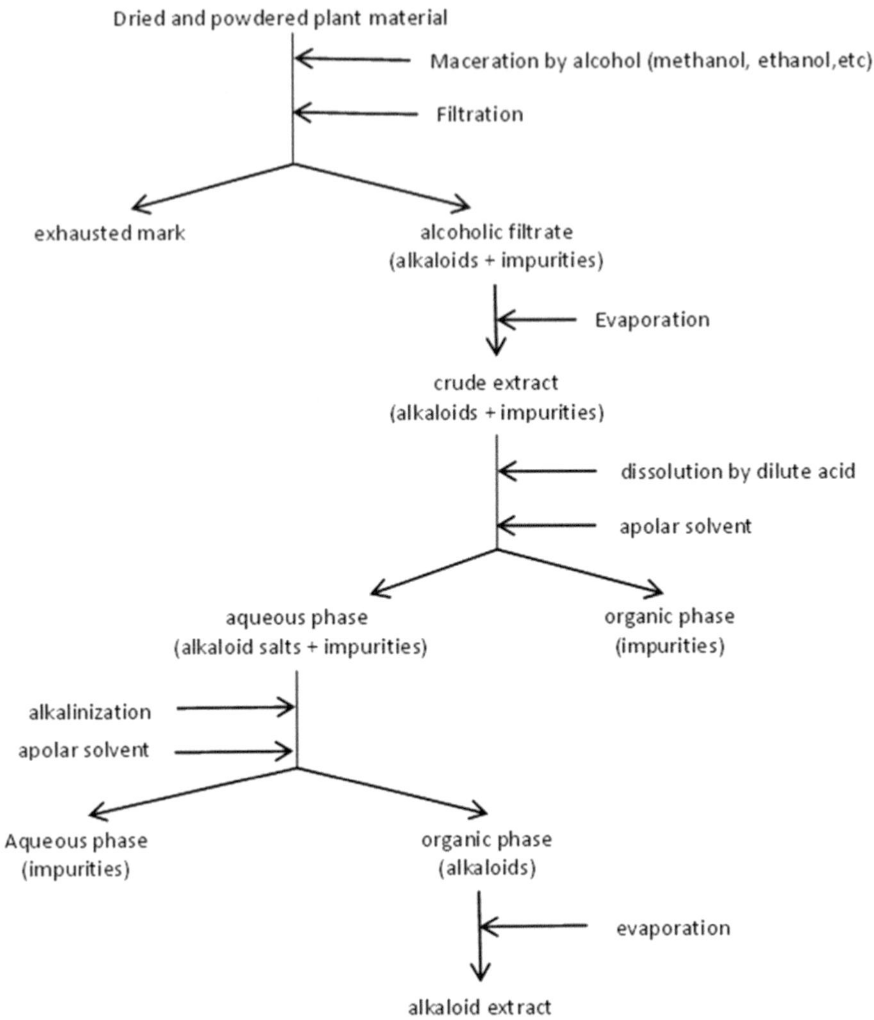

Fig. 13.5 General procedure to obtain alkaloidal extract from crude plant material by polar solvents

13.5.4 Extraction by Surfactant Substances

The extraction of alkaloïds by surfactants is oriented toward the ultrasound technique. The advantage of this method is a better yield in the alkaloïd extraction, the consumption of small quantities of solvents, and the extraction speed (Djilani et al., 2006). The use of ultrasound has the advantage of a kinetic gain, a better extraction yield (Wang & Weller, 2006). However, it should be noted that the yield and the kinetics are linked to the nature of the plant matrix (Amor, 2008b). Its principle is based on the phenomenon of cavitation. The propagation of an ultrasonic wave

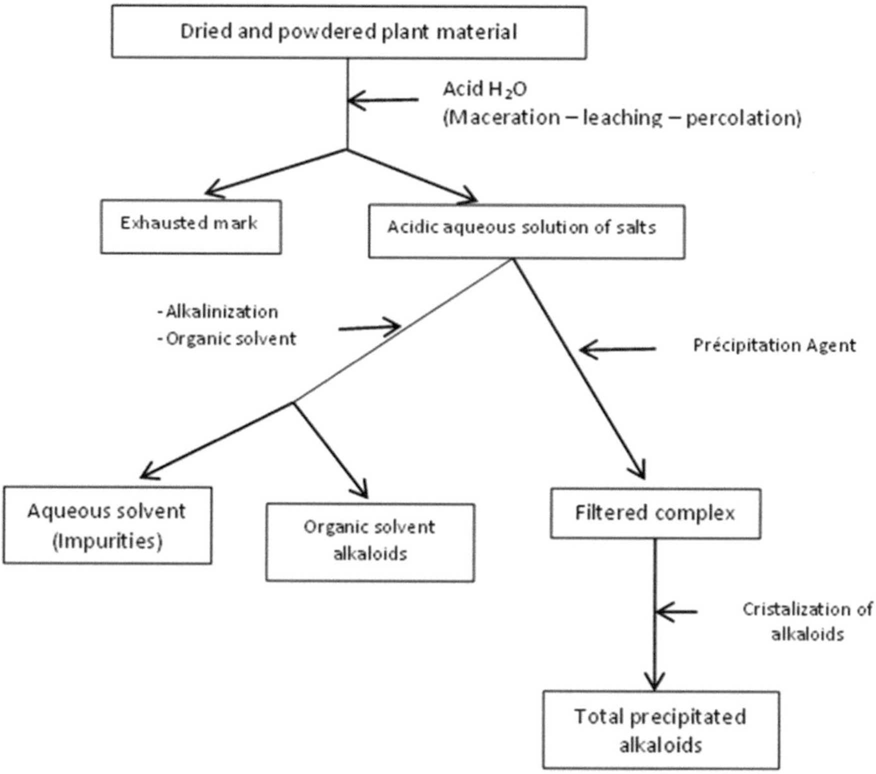

Fig. 13.6 General procedure to obtain alkaloïdal extract from water extraction

induces mechanical vibrations in the medium, which gives rise to regions of compression and decompression (which can also increase the contact surface between the solid phase and the liquid phase). When the pressure reaches very low values, air bubbles or liquid vapors form "cavitation phenomenon") (Chemat, 2005; Shirsath et al., 2012). Due to cavitation, cracks develop in the cell wall, which increases the permeability of the extraction solvent into plant tissues (Shirsath et al., 2012).

The powdered plant material is suspended in a surfactant solution [the surfactant can be anionic (SDS: sodium dodecyl sulfate), as it can be non-ionic (Emulgen: Polyethyleneglycoldodecyl ether)], treated in an ultrasonic bath.

The extract is separated by simple filtration, and the residual material is washed with pure water.

The solution of combined filtrates is acidified with a solution of sulfuric acid (2%), and the alkaloïds are precipitated by Mayer's reagent.

The precipitate is dissolved in an alkaline solution of sodium carbonate (5%) and extracted with CH_2Cl_2. The organic layer is washed with water until neutral pH,

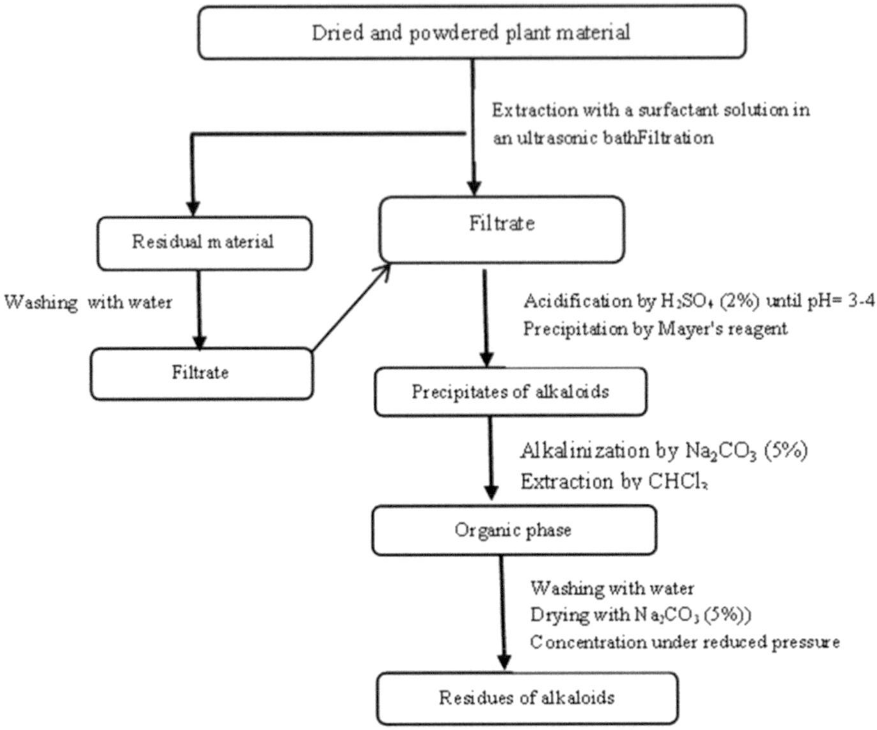

Fig. 13.7 Extraction protocol by active tension substance (Djilani et al., 2006)

dried with Na_2SO_4, and concentrated to dryness under reduced pressure to obtain the alkaloids (Fig. 13.7) (Djilani et al., 2006).

13.5.5 Preparation of Extracts Using Non-ionic Surfactant-Mediated Extraction (SME)

Among the methods of extraction of alkaloïds, we can use those developed by Takla et al. (2018), as summarized below.

Three samples of chopped bulbs of *C. powellii* (2.0 g each) were accurately weighed and placed in 15 ml centrifuge tubes; 5 ml non-ionic surfactant (Genapol X-080, Triton X-100, and Triton X-114) solution (10%, v/v) was separately added to the tubes. The tubes were capped and blended adequately and then placed in an ultrasonic cleaning bath for ultrasonic extraction. In ultrasonic-assisted micelle-mediated extraction (UAMME), the extraction time was 1 h, and the experiment was carried out at 40 °C. After ultrasonic-assisted extraction, the extracts of *C. powellii* bulbs were centrifuged at 4000 rpm for 20 min. The supernatant was transferred into a 10 mL centrifuge tube. The cloud-point extraction procedure was processed at an equilibration (60 °C) temperature of the surfactant. The sample solution was then kept in a thermostatic water bath at 60 °C for 30 min until the solution completely separated into two distinct phases—the upper phase was the small volume of surfactant-rich phase, and the lower phase was the large volume of aqueous phase. The aqueous phase was separated from the sticky surfactant-rich phase. No salting-out agent was used as complete phase separation occurred. The volume of the surfactant-rich phase was adjusted to 5 ml using water, which was added to lower the viscosity of the surfactant-rich phase. The pre-concentrated alkaloïds were then analyzed by applying the surfactant-rich phase on HPTLC plates and scanned densitometrically at 254 nm (Takla et al., 2018).

13.5.6 Preparation of Extracts Using NDES (Surfactants and Natural Deep Eutectic Solvent)

The two-component mixture NDES was obtained by mixing hydrogen acceptors (HBAs) and hydrogen-bond donors (HBDs) at a proper molar ratio (Table 13.1) with calculated amounts of water and placed in a bottle with a stirring bar and cap and heated in a water bath below 50 °C with agitation till a clear homogenous liquid was formed (about 30–90 min). Ten samples of *C. powellii* bulbs (2 g each) were accurately weighed and placed in 10 ml centrifuge tubes; 5 ml of the different prepared NDES solutions (Table 13.1) were separately added to these tubes. The tubes were capped and blended adequately and then placed in an ultrasonic cleaning bath (50 °C) for ultrasonic extraction (60 min). During the extraction, mixing was provided by an ultrasonic bath. After ultrasonic-assisted extraction, the different extracts of *C. powellii* bulbs were centrifuged at 3000 rpm for 10 min, and then the extracts were collected and diluted with 5 ml methanol. Amaryllidaceae alkaloids were then analyzed by applying the solutions on HPTLC plates (Takla et al., 2018).

13.6 Identification of Alkaloids

The detection methods currently used call upon general and specific reactions. The general reactions are preceded by an extraction and consist, in the most general case, of a precipitation of alkaloid salts in aqueous solution based on the ability of the

Table 13.1 Interpretation of the results of two tests: Mayer and Dragendorff (Castle)

Test	Reading	
Negative	0	Absence of precipitate
Positive	+	Doubtful precipitated
	++	Average rush
	+++	Abundant precipitate

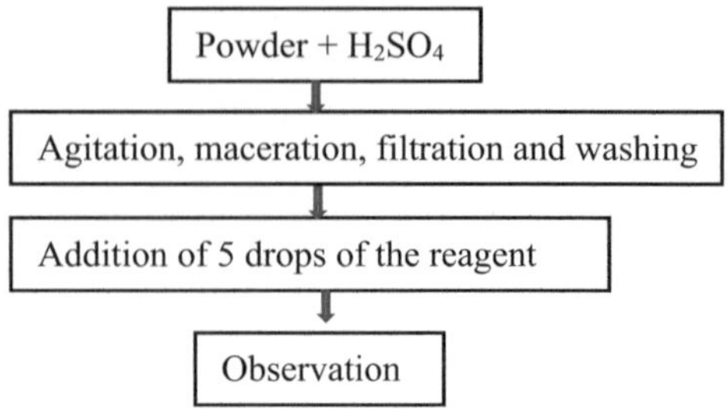

Fig. 13.8 Alkaloid detection steps

latter to combine with metals and metalloids: bismuth, mercury, tungsten, iodine, etc. (Bruneton, 2015).

The obtained precipitates provide an indication of the alkaloid content in the extract. Several general reagents are commonly used for this purpose:

General reagents for TLC:

- Valser-Mayer's reagent: it is a solution of potassium tetraiodomercurate (yellowish white precipitate).
- Dragendorff's reagent: potassium tetraiodobismuthate solution (orange-red precipitate).
- The silico-tungstic reagent or Bertrand's reagent: mixture of tungsten and silicon oxides (yellowish-white precipitate).
- Bouchardat's reagent: Potassium iodo-iodide solution (brown precipitate).
- Solutions of alkaline iodo-platinates.

Dragendorff's reagent, iodo-iodide (iodine vapor) solution, potassium iodo-platinate, or cerium ammonium sulfate are commonly used for developing TLC plates (Bruneton, 2015).

According to Daoudi et al. (2015) (Fig. 13.8), a sulfuric extract was prepared based on 10% sulfuric acid (H_2SO_4) and powder (material for which the alkaloid is being sought). On which applies agitation, maceration, filtration, and washing with distilled water. The filtrate is treated with five drops of Dragendorff's reagent in one tube, and five drops of Mayer's reagent in another tube. This test presents a preliminary analysis method.

The obtained results are presented as follows in Table 13.1:

The specific reactions are carried out on an alkaloid extract by colored reactions characteristic of a chemical type of alkaloid:

Specific reagents: Characteristic coloring of certain alkaloids (Vercauteren, 2011):

- The Vitali-Morin reagent is made up of concentrated nitric acid (HNO_3) and potassium hydroxide. This reagent is used to characterize the presence of tropane alkaloids in a sample and gives a purplish color.

- Urk's reagent is made up of p-dimethylaminobenzaldehyde (p-DMAB). It's used to highlight the presence of indoloisoprenic alkaloids and gives a purplish blue coloration.
- Another characteristic reagent is the mixture of cerium sulfate and ammoniacal. This reagent reacts with indolomonoterpene alkaloïds to give yellow or orange coloration.
- We also have alkaline iodoplatinate reagent used to characterize quinoline alkaloids, and muroxide reagent is used to highlight purine bases (caffeine).

13.7 Conclusion

In the search for bioactive molecules, nitrogen compounds have a place of choice, and the extraction process for this family of compounds is constantly evolving. With the development of science and technology, the development of extraction processes remains topical with the aim of having a better extraction yield, of course, based on the chemical properties of these nitrogen compounds. So to do this the protocols to be set up for extractions and particularly that of nitrogenous compounds, must seek a better yield, a low cost and efficiency in obtaining alkaloid extracts. The amelioration of alkaloids extraction methods could be useful for extracting alkaloïds for pharmaceutical interest by optimizing extraction yield. These methods are also used for the extraction of active principles from vegetable drugs or natural products. Methods of extracting alkaloïds present some inconvenience because other compounds can also be extracted during the process. Future research for extracting exclusively alkaloïds with new methods from a sample is a challenge to be overcome.

Conflict of Interest The authors declare no conflict of interest. The donors had no role in the design of the study; in the collection, analysis, or interpretation of data; in the writing of the manuscript or in the decision to publish the results.

References

Amor, B. B. (2008a). *Maîtrise de l'aptitude technologique de la matière végétale dans les opérations d'extraction de principes actifs: texturation par détente instantanée contrôlée (DIC). Génie des procédés.* Université de La Rochelle. Français. NNT: tel-00399131f. p. 3

Amor, B. B. (2008b). *Maîtrise de l'aptitude technologique de la matière végétale dans les opérations d'extraction de principes actifs: texturation par détente instantanée contrôlée (DIC). Génie des procédés.* Université de La Rochelle. Français. NNT: tel-00399131. p. 42.

Bérdy, J. (2005). Bioactive microbial metabolites. *The Journal of Antibiotics, 58*, 1–26.

Brielmann, H. R., Setzer, W. N., Kaufman, P. B., Kirakosyan, A., & Cseke, L. J. (2006). Phytochemicals: the chemical components of plants. In L. J. Cseke, A. Kirakosyan, P. B. Kaufman, S. L. Warber, J. A. Dule, & H. R. Brielmann (Eds.), *Natural Product from Plants* (2nd ed., pp. 1–50). CRC Press. ISBN 978-0-8493-2976-0.

Bruneton, J. (2009). *Pharmacognosie, phytochimie-plantes, médicinales. Lavoisier (4e éd.)* (128p). revue et augmentée, Tec & Dac-Editions médicinales internationales, Paris.

Bruneton, J. (2015). *Pharmacognosie, phytochimie plantes medicinales* (5th ed., pp. 1147–1148). Lavoisier Tec & DOC.

Chemat S. (2005). *Contribution à l'étude de l'extraction de la carvone et du limonène à partir des graines de carvi selon des procédés conventionnels, ultrasons et chauffage microondes: Application à l'extraction de polluants organiques de type PCBs et à l'oxydation des acides humiques*. Thèse de Doctorat, Université des Sciences et de la Technologie HOUARI BOUMEDIENE, Republique Algérienne Démocratique et Populaire. pp. 32–34.

Daoudi, A., Sabiri, M., Bammou, M., Zair, T., Ibijbijen, J., & Nassiri, L. (2015). Valorisation des extraits de trois espèces du genre Urtica: Urtica urens L., Urtica membranacea Poiret et Urtica pilulifera L. *Journal of Applied Biosciences, 87*, 8094–8104.

Djilani, A., Legseir, B., Soulimani, R., Dicko, A., & Younos, C. (2006). New extraction technique for alkaloïds. *Journal of the Brazilian Chemical Society, 17*(3), 518–520.

Guttiérrez-Grijalva, E. P., Lopez-Martinez, L. X., Contreras-Angulo, L. A., Elizalde-Romero, C. A., & Heredia, J. B. (2020). *Plant alkaloids: Structures and bioactive properties*. Plant-derived Bioactives. https://doi.org/10.1007/978-981-15-2361-8_5

Jones, W. P., & Kinghorn, A. D. (2012). Extraction of plant secondary metabolites. *Natural Products Isolation*, 341–366. https://doi.org/10.1007/978-1-61779-624-1_13

Mera, I. F. G., Falconi, D. E. G., & Córdova, V. M. (2019). Secondary metabolites in plants: Main classes, phytochemical analysis and pharmacological activities. *Revista Bionatura, 4*, 1000–1009. https://doi.org/10.21931/RB/2019.04.04.11

Pagare, S., Bhatia, M., Tripathi, N., Pagare, S., & Bansal, Y. K. (2015). Secondary metabolites of plants and their role: Overview. *Current Trends in Biotechnology and Pharmacy, 9*, 293–304.

Ranjitha, D., & Sudha, K. (2015). Alkaloïds in foods. *International Journal of Pharmaceutical and Chemical Sciences, 5*(4), 896–906.

Razafindrambao R. S. (1973). *Etude d'une plante médicinale malgache Buxus Madagascarica Baill. et ses variétés*. Edition de l'office de la Recherche Scientifique d'Outre-Mer (ORSTM), Th.3e cycle: Pharm.: Paris: 1971/11/22.

Sharad, V., & Swati, C. (2012). Isolation of natural products. *Current Pharma Research, 2*(3), 584–599. ISSN: 2230-7842.

Shirsath, S., Sonawane, S., & Gogate, P. (2012). Intensification of extraction of natural products using ultrasonic irradiations—A review of current status. *Chemical Engineering and Processing: Process Intensification, 53*, 10–23.

Suau, R., Cabezudo, B., Rico, R., Najera, F., & López-Romero, J. (2002). Direct determination of alkaloid contents in Fumaria species by GC-MS: Phytochemical analysis. *An International Journal of Plant Chemical and Biochemical Techniques, 13*(6), 363–336.

Takla, S. S., Shawky, E., Hammoda, H. M., & Darwish, F. A. (2018). Green techniques in comparison to conventional ones in the extraction of Amaryllidaceae alkaloids: Best solvents selection and parameters optimization. *Journal of Chromatography A, 1567*, 99–110.

Tidjani, S., & Rhouati, S. (2016). *Etude phytochimique et evaluation biologique de l'espèce senecio delphinifolius Vahl. Thèse de doctorat*. Université des frères MENTOURI CONSTANTINE. http://depot.umc.edu.dz/handle/123456789/809

Vercauteren, J. (2011). *Plan du cours de Pharmacognosie – Université Montpellier I, Laboratoire de Pharmacognosie* (p. 208). Formation Commune de Base en pharmacognosie générale.

Wang, L., & Weller, C. L. (2006). Recent advances in extraction of nutraceuticals from plants. *Trends in Food Science & Technology, 17*(6), 300–312.

Wangchuk, P. (2019). Plant alkaloïds: Classification, isolation and drug development. In M. Swamy, J. Patra, & G. Rudramurthy (Eds.), *Medicinal plants: Chemistry, pharmacology and therapeutic applications* (pp. 131–137). Taylor & Francis Ltd.

Yang L, Stöckigt J. (2010).Trends for diverse production strategies of plant medicinal alkaloids. *Natural Product Reports, 27*,1469–79.

Chapter 14
Some African Medicinal Plants for Syphilis and Gonorrhea: Ethnobotanical Knowledge and Ethno-diagnosis

Ester Innocent, Siri Abihudi, Alphonce Ignace Marealle, and Albert Shayo

Abstract Syphilis is a rising global health concern affecting approximately 5.6 million people annually. Primary syphilis lesions, which are typically painless, cannot be easily recognized, particularly in less visible areas such as the cervix or rectum. In addition to syphilis, gonorrhea presents a significant global health issue, with the WHO estimating about 87 million new cases each year, making it one of the most common STIs worldwide. The global prevalence of gonorrhea is 0.9% with more men affected than women.

In many rural African regions, medicinal plants are widely used to manage and treat syphilis and gonorrhea. However, in African traditional settings, treatment is challenging because it is based on the appearance of symptoms to identify the cause of the illness. Further, diagnosis of some symptoms takes a long time to manifest, often when the disease has progressed. Thus, the plants used to manage the two diseases are often the same or confused.

The book chapter explores current traditional methods for diagnosing syphilis and gonorrhea and identifies about 159 plants from 127 genera and 53 families used in Africa for their treatment.

E. Innocent (✉)
Department of Biological and Pre-clinical Studies, Institute of Traditional Medicine, Muhimbili University of Health and Allied Sciences, Dar es Salaam, Tanzania
e-mail: ester.innocent@muhas.ac.tz

S. Abihudi · A. Shayo
Department of Medical Botany, Agronomy and Plant Breeding, Institute of Traditional Medicine, Muhimbili University of Health and Allied Sciences, Dar es Salaam, Tanzania
e-mail: siri.abihudi@muhas.ac.tz

A. I. Marealle
Department of Clinical Pharmacy and Pharmacology, School of Pharmacy, Muhimbili University of Health and Allied Sciences, Dar es Salaam, Tanzania

A. Dekebo (ed.), *Bioactive Secondary Metabolites from Medicinal Plants of Africa*, Interdisciplinary Biotechnological Advances,
https://doi.org/10.1007/978-981-96-8590-5_14

Keywords Syphilis · Gonorrhea · Africa · Medicinal plants · Ethnobotanical knowledge · Ethno-diagnosis

Abbreviations

GIS	Geographical Information System
ITM	Institute of Traditional Medicine
MIC	Minimum inhibitory concentration
MUHAS	Muhimbili University of Health and Allied Sciences
STI	Sexually transmitted disease
UTI	Urinary tract infection
WHO	World Health Organization

14.1 Background

14.1.1 Prevalence of Syphilis and Gonorrhea

Gonorrhea and syphilis are among public health problems among the sexually transmitted infections (STIs) in both developing and developed world (Fig. 14.1)

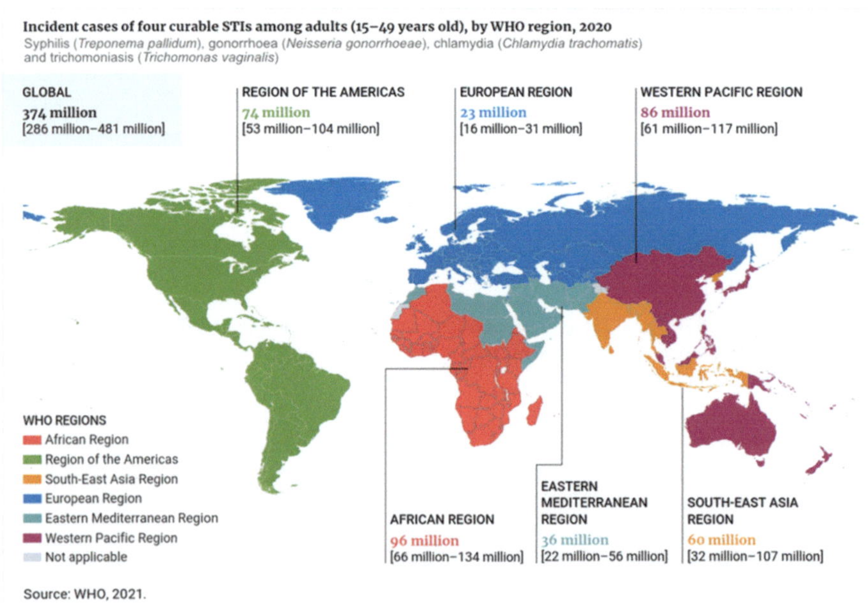

Fig. 14.1 Incident cases of some STIs, including syphilis and gonorrhea diseases (WHO, 2024)

(WHO, 2021; Rowley et al., 2019). Syphilis is a global public health threat, affecting about 5.6 million people annually (Newman et al., 2015). Timely and accurate diagnosis is key for appropriate clinical management and effective prevention of disease transmission. In Tanzania, the risk of syphilis infection has been reported to be higher among women in rural and semi-urban settings where access to healthcare might be limited (Manyahi et al., 2015). It has varied and often subtle symptoms that make it hard to diagnose and can lead to many infections going unnoticed, as the typical painless lesions of primary syphilis may be overlooked, particularly when they appear in hidden areas like the cervix or rectum (Peeling et al., 2017). The estimated number of new cases of syphilis globally in adults aged 15–49 years increased from 7.1 million (5.1–9.1 million) in 2020 to 8.0 million (5.6–10.4 million) in 2022, indicating a potential resurgence of the disease. This increase raises concerns about the effectiveness of current public health interventions and the need for more comprehensive strategies to curb the spread of STIs globally. The estimated number of congenital syphilis cases in 2022 was 700,000, and the congenital syphilis case rate was 523 per 100,000 live births. This is over 10 times the World Health Organization (WHO) elimination of vertical transmission of syphilis threshold of 50 per 100,000 live births (WHO, 2024).

The World Health Organization (WHO) estimated that around 87 million new cases of gonorrhea occur annually, making it one of the most widespread STIs globally. Its prevalence is especially high among sexually active young adults aged 15–24. Gonorrhea rates vary significantly by region, with higher prevalence seen in low- and middle-income countries (Fig. 14.2). Factors such as sexual behavior, population growth, and socioeconomic conditions contribute to its incidence. There is an increase in incidence rate in men (WHO, 2024).

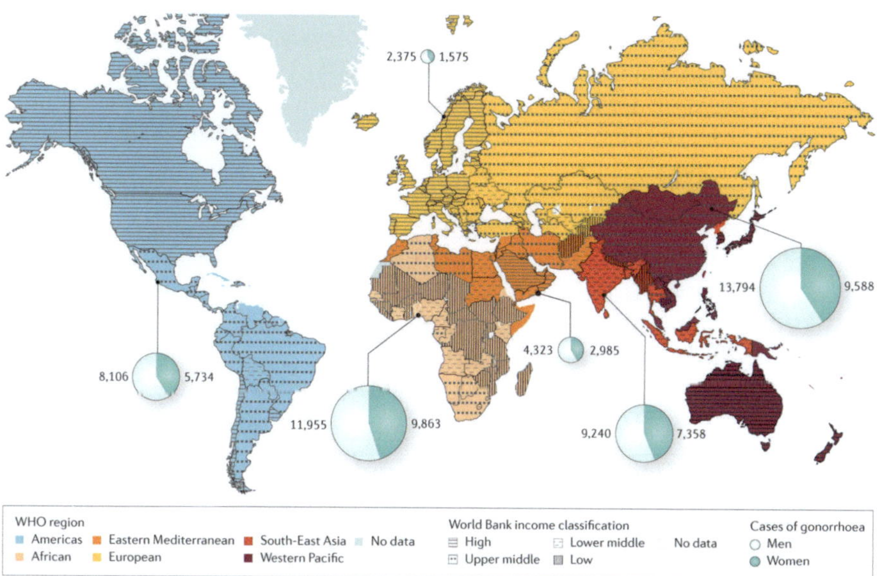

Fig. 14.2 Estimated new global cases of gonorrhea in 2016 (Rowley et al., 2019)

14.1.2 Ethno-diagnosis of Gonorrhea and Syphilis Among the STI Diseases

Gonorrhea and syphilis, like many of the STI diseases, are difficult to differentiate during diagnosis due to similarity in signs exhibited and of disease progression. For example, lower abdominal pain associated with genital itching and odor-free discharge are prominent signs of UTI, STI, and opportunistic infections (Hinkkanen, 2009; Marealle et al., 2020). Diseases like gonorrhea and chlamydia can have similarities in infections of the babies at birth. Usually, the infection gets into the baby's eyes and can cause blindness. However, in adults, the two infections can be differentiated because disease progression is usually fast and is more painful in gonorrhea. Nevertheless, treatment of gonorrhea or chlamydia may prevent the first symptoms of syphilis disease but may not cure it. This is because syphilis usually starts with a painless dirty sore, called a chancre. If the chancre occurs inside the vagina, a woman may not know that she has syphilis but will continue spreading to other people until the advanced signs are vivid.

In such times of infections, before turning to conventional medicine, some people seek treatment from the environment or else from the traditional healers. However, the diagnosis made by traditional healers is subjective because they mostly use symptomatic diagnosis (Ndubani & Höjer, 1999). In most of Bantu culture, a patient first seeks to recognize the *causal* agent and therapies that address the problems in social relations. In African traditional medicine, consulting the spiritual forces is an aspect of diagnosing the diseases. Traditional Healers seek guidance from the "spirit world" to determine the cause of the disease or to uncover any violations of customary laws by the patient (Ozioma & Chinwe, 2019). The "spirit world" not only helps to diagnose illnesses but also dictates the medicinal plants. Thus, in African traditional medicine there is an emphasis on visible symptoms, to get to the choice of treatment (Hinkkanen, 2009; Innocent, 2024). For example, the Nyamwezi and sukuma diagnosis for stomach problems in women is associated with ancestral presence in the human body and ancestral interference in the form of different *nzoka* disorders. For many young and middle-aged women *nzoka ja Buhale has* the connection of fertility problems, unless there is discharge then the illness may be associated with STI (Hinkkanen, 2009). In young men gonorrhea infection common signs are blood in the urine and pain in the lower belly, often associated with difficulty or pain in passing urine, symptoms like prostate in old men (Werner et al., 2011). Thus, in African traditional medicine treatment, it is often the symptoms that are used to judge the medicinal plant to use.

14.1.3 Plants Used for Gonorrhea and Syphilis Infections

In many rural areas of African countries, the treatment and management of syphilis and gonorrhea rely heavily on medicinal plants (Kacholi & Amir, 2022; Mongalo &

Raletsena, 2022). There are some plant species which are used for multiple disease alleviation, including *Olea europaea* L. (Oleaceae) and *Toddalia asiatica* (L) Lam (Rutaceae) which is reported to be used in the treatment of lower stomach ache (Hinkkanen, 2009; Marealle et al., 2020). *Kigelia africana* (Lam.) Benth. and *Securidaca longipedunculata* Fresen are being used for treatment of various bacterial infections such as tuberculosis, pneumonia, coughing, dysentery, and female gynecological problems, including gonorrhea (Innocent et al., 2022; Assanti et al., 2022; Neuwinger, 1998; Hedberg et al., 1983b; Hedimbi & Chinsembu, 2012; Hedberg & Hedberg, 1982).

14.2 Methodology Utilized in Collection of Information

The following methodologies were used.

- The ethnobotanical information was gathered from 1990 to 2024 using PubMed, SpringerLink, ScienceDirect, Google Scholar, Scopus, Web of Science, and ResearchGate.
- Key search terms including "syphilis" or "gonorrhea" in combination with terms like "Africa," "medicinal plants," "ethnobotanical," "ethnobotanical knowledge," "ethno-diagnosis," "savannah," "tropical," etc., were used.
- Coordinates were determined and then mapped for the medicinal plant families primarily used in treating syphilis and gonorrhea, distributed across different vegetation types in the African region using ArcGIS version 10.8. This mapping process used data from national herbarium databases information from University of Dar es Salaam (DSM), Muhimbili University of Health and Allied Sciences (ITMH), Tanzania Plant Health and Pesticides Authority (NHT), Sokoine University of Agricultur (SUA), and international herbarium databases, including those from the Addis Ababa University (ETH), Albany Museum (GRA), Arizona State University (ASU), Boise State University (SRP), Botanic Garden, Botanical Research Institute of Texas (BRIT), Harare (SRGH), California Botanic Garden (RSA), California Department of Food and Agriculture (CDA), Croatian Mycological Society (CNF) Direcção Geral do Ambiente, Cabinet of Environment, Carnegie Museum of Natural History (CM), CENAREST (LBV), Centre de Recherche en Sciences Naturelles (LWI), Conservatoire et Jardin botaniques de la Ville de Genève (G),Centre National de la Recherche Appliquée au Developement Rural (TEF),Centre National d' Application de Recherches Pharmaceutiques (CNARP), Centre Suisse de Recherches Scientifiqucs cn Côte d'Ivoire (CSRS), Ministry of Natural Resources and Environment (STPH), Eduardo Mondlane University (LMU), Eszterhazy Karoly University of Applied Sciences (EGR), Fairchild Tropical Botanic Garden (FTG), Fan Memorial Institute of Biolog (FM), Forestry Research Institute of Nigeria (FHI), French Institute, French Ministry of Foreign Affairs (HIFP), Herbarium of the University of Coimbra (COI), Herbarium of the University of Zürich (Z), Institut de Recherche

Agronomique de Guinée (SERG), Institute for Agricultural Research of Mozambique (LMA), Kunming Institute of Botany, Chinese Academy of Sciences (KUN), Longwood Gardens (KEN), Museum of Evolution (UPS), Ludwig-Maximilians-Universität (MSB), University of Lagos (LUH), Lund University (LD), Museo Friulano di Storia Naturale (UDM), Maharaja Sayajirao University of Baroda (BARO), Makerere University (MHU), Meise Botanic Garden (BR), Missouri Botanical Garden (MO), Museo Nacional de Costa Rica (CR), Museo Nacional de Historia Natural (MNHN), Museu Nacional de História Natural e da Ciência (LISU), Muséum National d'Histoire Naturelle (P), National Herbarium & Botanic Gardens of Malawi (MAL), Michigan State University (MSC), National Herbarium of Cameroon (YA), National Museums of Kenya (EA), National Taiwan University (TAI), Naturalis Biodiversity Cente (L), Naturalis Biodiversity Center (WAG), Naturhistorisches Museum Wien (W), Nicolaus Copernicus University (TRIN), Norwegian University of Science and Technology (TRH), Government (DNA Parc Botanique et Zoologique de Tsimbazaza (TAN), Pittsburg State University (KSP), Queensland Herbarium & Biodiversity Science (BRI), Real Jardín Botánico (MA), Royal Botanic Garden Edinburgh (E), Royal Botanic Gardens Victoria (MEL), San Diego State University (SDSU), Royal Botanic Gardens (K), Senckenberg Gesellschaft für Naturforschung: Senckenberg Forschungsinstitut und Naturmuseum (FR), Smithsonian Institution (US), South African National Biodiversity Institute (PRE) Staatliche Naturwissenschaftliche Sammlungen Bayerns (M), The Natural History Museum (BM), The New York Botanical Garden (NY), Staatliches Museum für Naturkunde Karlsruhe (KR), Swedish Museum of Natural History (S), Taiwan Forestry Research Institute (TAIF), Technische Universität Dresden (DR), Universidad Nacional de Colombia (COL), Universidad Nacional Mayor de San Marcos (SMF), University of Calgary (UAC), Università di Pisa (Italy) (PI), Universität Wien WU Université du Lomé (TOGO), Université Libre de Bruxelles (BRLU), University of California (UC), Université Gamal Abdel Nasser de Conakry (HNG), University of Bayreuth (UBT), University of Bergen (BG), University of Calgary (UAC), University of California, Davis (DAV), University of Copenhagen (C), University of Michigan (MICH), University of Illinois at Urbana-Champaign (ILL), University of Lagos (LUH), University of Lethbridge (LEA), University of Minnesota (MIN), University of Oslo (O), University of Oxford (FHO), Xiamen University (AU) and Botanischer Garten und Botanisches Museum, Freie Universität Berlin (B).

14.3 Results

About 53 plant families belonging to 127 genera were reported to have plant species used in the management of gonorrhea and syphilis. The highest ranked family with high number of plant species were Fabaceae (34), Rubiaceae (13), Euphorbiaceae (9), Asteraceae (9), Phyllanthaceae (8), Rutaceae (5), Anacardiaceae (4),

Combretaceae (4), Apocynaceae (4), Rhamnaceae (4), and Asparagaceae (4), and the rest of the families have less than 2 number of plant species (as shown in Fig. 14.3). The genus *Senna* (5) and *Asparagus* (4) have a high number of plant species, followed by *Combretum* (3), *Albizia* (3), *Euphorbia* (3), *Indigofera* (3), *Vachellia* (3), and *Zanthoxylum* (3) (Tables 14.1, 14.2, and 14.3).

The medicinal plants grow in a variety of habitats, including in savanna grassland, swamp forests and mangroves, dry forests, and the thickest and tropical lowland rain forests. These environments promote plants with distinct adaptations that produce medicinal compounds used in traditional medicine. The regions are primarily found in Eastern, Western, and Southern Africa. The Fabaceae family, which has high list of plant species, is widely used for treatment of venereal diseases; it is well adapted to a variety of African habitats, including savannas, tropical rainforests, coastal forests, and montane areas (Fig. 14.4). Furthermore, the distribution of families with the large number of species utilized in the management and treatment of gonorrhea and syphilis is shown on the various maps (Figs. 14.5, 14.6, 14.7, and 14.8).

About 43 plant species are used for treatment of only syphilis (Table 14.1); 93 plant species for only gonorrhea (Table 14.2), while 23 plant species are used for both syphilis and gonorrhea (Table 14.3) and are also used for associated other STIs or lower abdominal pains. Among plant species highly cited include *Flueggea virosa* (Roxb. ex Willd.) Royle (4 publications) which is reported for treatment of syphilis and gonorrhea and other diseases such as dysmenorrhea, oedema, epilepsy, convulsions, female infertility, stomach aches, rectal and uterine prolapse, skin conditions, diarrhea, wound, laxatives, aphrodisiac, constipation, and headaches. *Kigelia africana* (Lam.) Benth. (3) is reported for syphilis and gonorrhea, treatment of wounds, ulcers, abscesses, dysentery, rheumatism, pneumonia, bilharzia,

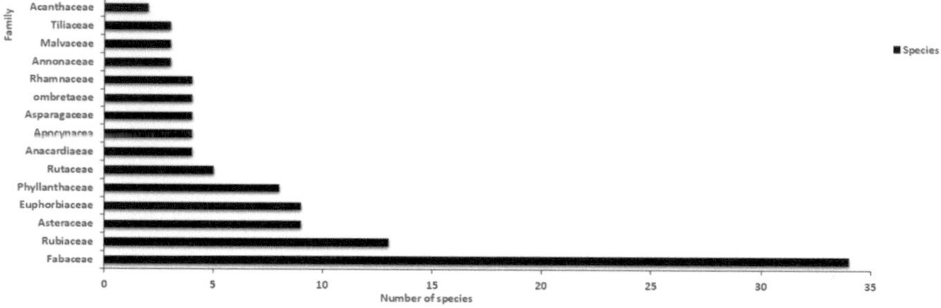

Fig. 14.3 Plant family with Plant species used in management of gonorrhea and syphilis

Table 14.1 African medicinal plants species used in management of plant species used in management of gonorrhea

*Family Name	Plant name	Habits	Part used	Preparation	Book or journal reference
1. Anacardiaceae	*Ozoroa mucronata* (Krauss) R. & A. Fern.	Tree	Roots	nd	(Kokwaro, 1976; Chhabra et al., 1987)
	Rhus natalensis Bernh. ex Krauss.	Shrub	Roots And Leaves	nd	(Watt & Breyer-Brandwijk, 1962; Hedberg & Hedberg, 1982)
	Sclerocarya caffra Sond.	Tree	Leaves	nd	(Chhabra et al., 1987)
2. Annonaceae	*Annona senegalensis* Pers.	Tree	Roots	Decoction mixed with coconut oil	(Chhabra et al., 1987)
	Uvaria leptocladon Oliv.	Tree or shrub	Roots	Decoction of roots	(Chhabra et al., 1987; Kokwaro, 1976; Hedberg & Hedberg, 1982)
3. Apiaceae	*Steganotaenia araliacea* Hochst.	Tree	Roots	Root is mixed with other drugs in the preparation of a decoction	(Hedberg et al., 1983b)
4. Apocynaceae	*Holarrhena febrifuga* Klotzsch.	Tree Or Shrub	Roots	nd	(Chhabra et al., 1987)
	Rauwolfia mombasiana Stapf.	Tree Or Shrub	Roots	nd	(Kokwaro, 1976; Chhabra et al., 1987)
5. Araliaceae	*Cussonia zimmermannii* Harms	Tree	Roots	nd	(Kokwaro, 1976; Chhabra et al., 1987)
6. Asparagaceae	*Asparagus buchananii* Bak.	Shrub	Roots	Infusion or decoction	(Kokwaro, 1976; Hedberg et al., 1983a)
	Asparagus racemosus Willd.	Herb	Roots	nd	(Kokwaro, 1976;

(continued)

Table 14.1 (continued)

*Family Name	Plant name	Habits	Part used	Preparation	Book or journal reference
					Chhabra et al., 1990a)
7. Asteraceae	*Aspilia mossambicensis* (Oliv.) Wild.	Herb	Roots	nd	(Kokwaro, 1976; Chhabra et al., 1989)
	Baccharoides adoensis (Sch.Bip. ex Walp.) H. Rob.	Herb	Roots	nd	(Burkill, 1985; Toyang & Verpoorte, 2013; Maregesi et al., 2007)
	Brachylaena hutchinsii Hutch.	Tree	Leaves	Leaf decoction	(Chhabra et al., 1989)
	Crassocephalum vitellinum (Benth.) S. Moore	Herb	nd	nd	(Watt & Breyer-Brandwijk, 1962; Hedberg & Hedberg, 1982)
	Launaea cornuta (Oliv. & Hiern) C. Jeffrey	Herb	Roots	Decoction	(Hedberg & Hedberg, 1982)
	Mikania cordata (Burmann f.) B. L. Robinson	Herb	Leaves	Leaf juice	(Neuwinger, 1998)
	Vernonia colorata (Willd.) Drake	Shrub	Roots	Decoction	(Hedberg & Hedberg, 1982)
	Vernonia glabra (Steetz) Vatke	Herb	Roots		(Chhabra et al., 1989)
8. Boraginaceae	*Ehretia amonea* Klotzsch	Tree	Roots	Decoction of roots	(Kokwaro, 1976; Hedberg & Hedberg, 1982)
	Ehretia bakeri Britten	Tree or shrub	Roots	nd	(Kokwaro, 1976; Chhabra et al., 1987)
9. Burseraceae	*Commiphora africana* (A. Rich.) Engl.	Tree	Roots	Decoction of roots	(Hedberg & Hedberg, 1982; Chhabra et al., 1987)

(continued)

Table 14.1 (continued)

*Family Name	Plant name	Habits	Part used	Preparation	Book or journal reference
10. Capparidaceae	*Capparis tomentosa* Lam.	Tree or shrub	Roots	Root decoction is drunk	(Chhabra et al., 1987)
	Ritchiea capparoides (Andr.) Britten var. capparoides	Tree or shrub	Leaves	nd	(Chhabra et al., 1989)
11. Combretaceae	*Combretum pentagonum* Laws	Tree	Roots	Grounded powders in Porridge	(Hedberg & Hedberg, 1982)
	Terminalia kaiserana F. Hoffm.	Tree or shrub	Roots	Decoction for roots	(Kokwaro, 1976; Chhabra et al., 1989; Hedberg & Hedberg, 1982)
12. Commelinaceae	*Commelina benghalensis* L.	Herb	Plant	Root decoction is warmed with camphor and drunk	(Chhabra et al., 1989)
13. Connaraceae	*Byrsocarpus orientalis* (Baill.) Baker	Shrub	Roots	nd	(Kokwaro, 1976; Chhabra et al., 1989)
14. Dilleniaceae	*Tetracera boiviana* Baill.	Tree or shrub	Roots	Root decoction	(Chhabra et al., 1989)
15. Ebenaceae	*Diospyros fischeri* Gürke	Tree	Roots	nd	(Kokwaro, 1976; Chhabra et al., 1989)
16. Euphorbiaceae	*Acalypha fruticosa* Forssk.	Shrub or small tree	Roots	nd	(Kokwaro, 1976; Watt & Breyer-Brandwijk, 1962; Chhabra et al., 1990a; Hedberg et al., 1983a)
	Bridelia cathartica Bertol.	Shrub	Root	Decoction of the entire root is drunk	(Chhabra et al., 1990a)

(continued)

Table 14.1 (continued)

*Family Name	Plant name	Habits	Part used	Preparation	Book or journal reference
	Croton pseudopulchellus Pax.	Tree	Leaves and twigs	Decoction of leaves and twigs	(Kokwaro, 1976; Hedberg et al., 1983a)
	Euphorbia hirta L.	Herb	Whole plant, root	Root decoction is drunk	(Watt & Breyer-Brandwijk, 1962; Chhabra et al., 1990a)
	Suregada zanzibariensis Baill.	Shrub	Roots	Root decoction	(Chhabra et al., 1990a)
17. Fabaceae	*Senna siamea* (Lam.) Irwin & Barneby	Tree	Roots	Roots are peeled, pounded, and boiled; the decoction is drunk	(Maregesi et al., 2007)
	Senna petersiana Bolle	Tree	Whole plant	nd	(Watt & Breyer-Brandwijk, 1962; Chhabra et al., 1987)
	Piliostigma thonningii (Schumach.) Milne-Redh.	Tree	Bark	nd	(Kokwaro, 1976; Watt & Breyer-Brandwijk, 1962; Chhabra et al., 1987)
	Senna singueana Del.	Shrub	Roots	nd	(Kokwaro, 1976; Hedberg & Hedberg, 1982)
	Senna didymobotrya (Fresen.) H.S. Irwin & Barneby	Shrub	Leaves	Decoction of leaves	(Kokwaro, 1976; Hedberg & Hedberg, 1982)

(continued)

Table 14.1 (continued)

*Family Name	Plant name	Habits	Part used	Preparation	Book or journal reference
	Abrus precatorius L.	Climber	Roots/ leaves	nd	(Kokwaro, 1976; Hedberg et al., 1983b)
	Indigofera arrecta A. Rich.	Herb	Roots	nd	(Chhabra et al., 1990b)
	Indigofera lupatana Baker f.	Shrub	Roots	nd	(Kokwaro, 1976; Chhabra et al., 1990b)
	Indigofera rhynchocarpa Baker	Tree	Roots	nd	(Augustino & Gillah, 2005)
	Lonchocarpus bussei Harms.	Tree	Roots	Root decoction	(Kokwaro, 1976; Hedberg et al., 1983b)
	Rhynchosia minima (L.) DC var. minima	Herb	Root	Root decoction	(Chhabra et al., 1990b)
	Tephrosia villosa (L.) Pers. subsp. *ehrenbergiana* (Schweinf.) Brummitt	Herb	Root	Root decoction	(Chhabra et al., 1990b)
	Cajanus cajan (L.) Millsp.	Herb	Roots	Decoction	(Kokwaro, 1976; Hedberg et al., 1983b)
	Dichrostachys cinerea (L.) Wight et Arn. *ssp. africana* Brenan & Brummitt	Tree	Leaves	nd	(Kokwaro, 1976; Hedberg et al., 1983b)
	Vahellia nilotica (L.) Willd. ex Del.	Tree	Roots	nd	(Kokwaro, 1976; Chhabra et al., 1990b)
	Senegalia polyacantha Willd. spp. *campylacantha* (Hochst. ex A. Rich.) Brenan.	Shrub	Undefined Plant Parts	nd	(Watt & Breyer-Brandwijk, 1962; Hedberg et al., 1983a)
	Vachellia robusta Burch.	Tree	Stembark Or Leaves	Decoction of the stembark or leaves	(Chhabra et al., 1990b)

(continued)

Table 14.1 (continued)

*Family Name	Plant name	Habits	Part used	Preparation	Book or journal reference
	Albizia lebbeck (L.) Benth.	Tree	Seeds	Seed oil against leprosy, seeds against gonorrhea	(Kokwaro, 1976; Chhabra et al., 1990b)
	Albizia versicolor Welw. ex Oliv.	Tree	Roots/ stem bark	Decoction	(Hedberg et al., 1983a)
	Crotalaria Caudata Welw. ex Baker.	Herb Or Shrub	Leaves	Leaf decoction	(Maregesi et al. 2007)
	Dichrostachys cinerea (L.) Wight & Arn.	Tree	Barks		(Augustino & Gillah, 2005)
18. Flacourtiaceae	*Trimeria grandiflora (Hochst.) Warb* ssp. *tropica (Burk.) Sleum.*	Shrub	Roots	Roots may be shredded, and the infusion made from them drunk	(Kokwaro, 1976; Hedberg et al., 1983a)
	Xylotheca tettensis (Klotzsch) Gilg var. kirkii (Oliv.) Wild	Tree Or Shrub	Roots	Root decoction	(Chhabra et al., 1990a)
19. Clusiaceae	*Harungana madagascariensis Poir.*	Tree or shrub	Leaves	nd	(Watt & Breyer-Brandwijk, 1962)
20. Lamiaceae	*Hoslundia opposite Vahl.*	Shrub Herb or shrub	Entire Plant Roots Roots And Leaves	Root decoction	(Watt & Breyer-Brandwijk, 1962; Hedberg et al., 1983a; Chhabra et al., 1990a; Gesssler et al., 1994)
21. Malvaceae	*Thespesia danis Oliv.*	Shrub	Roots	nd	(Kokwaro, 1976; Chhabra et al., 1990a)
	Hibiscus surrattensis L.	Herb	Leaves	nd	(Kokwaro, 1976; Chhabra et al., 1990a)
22. Meliaceae	*Turrea nilotica* Kotschy et Peyr.	Tree or shrub	Root	Root decoction	(Chhabra et al., 1990a)
23. Moraceae	*Ficus gnaphalocarpa*	Tree	Leaves	The leaves are pounded,	(Chhabra et al., 1990b)

(continued)

Table 14.1 (continued)

*Family Name	Plant name	Habits	Part used	Preparation	Book or journal reference
	(Miq.) STeud. ex A. Rich.			mixed with water, then filtered, and the juice is drunk	
24. Olacaceae	*Olax dissitiflora Oliv.*	Tree or Shrub	Roots	The root decoction is drunk	(Chhabra et al., 1990b)
25. Passifloraceae	*Adenia gummifera (Harv.) Harms.*	Climber	Roots	Decoction	(Kokwaro, 1976; Hedberg et al., 1983b)
26. Phyllanthaceae	*Antidesma venosum Tul.*	Tree or shrub	Roots	Root decoction, Cold-water extract of the roots mixed with roots of other plants	(Chhabra et al., 1993; Hedberg et al., 1983b)
	Cyathogyne bussei Pax.	Tree or shrub	Stem bark	Stembark decoction	(Chhabra et al., 1990a)
	Margaritaria discoidea (Baill.) Webster	Tree	Roots	Decoction	(Hedberg et al., 1983a)
	Phyllanthus muellerianus (O. Kuntze) Exell	Shrub	Roots	Root decoction	(Chhabra et al., 1990a)
	Phyllanthus reticulatus Poir.	Tree or shrub	Roots	Infusion	(Kokwaro, 1976; Hedberg et al., 1983a)
	Pseudolachnostylis maprouneifolia Pax	Tree	Bark	nd	(Chhabra et al., 1990a)
27. Rhamnaceae	*Rhamnus prinoides L'Herit.*	Tree or shrub	Roots	nd	(Kokwaro, 1976; Chhabra et al., 1991)
	Ziziphus mucronata Willd. ssp. mucronata	Tree	Roots	Decoction	(Watt & Breyer-Brandwijk, 1962; Hedberg et al., 1983b)

(continued)

Table 14.1 (continued)

*Family Name	Plant name	Habits	Part used	Preparation	Book or journal reference
28. Rubiaceae	*Chassalia umbraticola Vatke*	Shrub	Roots	nd	(Chhabra et al., 1991)
	Catunaregam spinosa (Thunb.) Tirvengadum	Tree or shrub	Roots	nd	(Kokwaro, 1976; Chhabra et al., 1991)
	Crossopteryx febrifuga (G. Don) Benth.	Tree	Bark	Cold-water extract	(Hedberg et al., 1983b)
	Pavetta crassipes K. Schum.	Tree or shrub	Leaves	nd	(Kokwaro, 1976; Chhabra et al., 1991)
	Oxyanthus pyriformis (Hochst.) Skeels subsp. *tanganyikensis* Bridson	nd	Leaves and roots	nd	(Gessler et al., 1994)
	Vangueria madagascariensis J. F. Gmel.	Tree or shrub	Stem bark	Stem bark decoction	(Chhabra et al., 1991)
29. Rutaceae	*Citrus aurantifolia (Christm.) Swingle*	Tree	Roots	nd	(Kokwaro, 1976; Chhabra et al., 1991; Augustino & Gillah, 2005)
	Zanthoxylum chalybeum Engl.	Shrub	Barks	nd	(Augustino & Gillah, 2005)
30. Salvadoraceae	*Salvadora persica L.*	Shrub	Roots	nd	(Kokwaro, 1976; Chhabra et al., 1991)
31. Sapindaceae	*Deinbollia borbonica* Scheff.	Tree or shrub	Roots	Decoction of roots	(Hedberg et al., 1983b; Chhabra et al., 1991)
	Paullinia pinnatta L.	Wooddy vine or scandent shrub	Leaves and roots	nd	(Kokwaro, 1976; Chhabra et al., 1991)

(continued)

Table 14.1 (continued)

*Family Name	Plant name	Habits	Part used	Preparation	Book or journal reference
32. Sapotaceae	*Mimusops fruticose* Bojer ex DC.	Tree	Root	Root decoction	(Chhabra et al., 1993)
33. Simaroubaceae	*Harrisonia abyssinica Oliv*	Shrub	Roots	Root decoction	(Chhabra et al., 1993)
34. Tiliaceae	*Grewia plagiophylla* K. Schum.	Shrub	Roots	Decoction of roots	(Kokwaro, 1976; Hedberg et al., 1983b)
	Grewia villosa Willd.	Shrub	Stembark	nd	(Chopra; Nadkarni in Chhabra et al., 1993)
	Triumfetta rhomboidea Jacq.	Herb	Roots	nd	(Kokwaro, 1976; Chhabra et al., 1993)
35. Zingiberaceae	*Costus afer* Ker-Gawl.	Herb	Roots	nd	(Chhabra et al., 1993)
36. Zygophyllaceae	*Tribulus terrestris* L.	Herb	Leaves	nd	(Augustino & Gillah, 2005)

*The plant family is arranged alphabetically
nd Not documented.

coughing, and female gynecological problems. *Securidaca longepedunculata* Fresenius (4), *Hoslundia opposita* Vahl (4) while other species have less than 2 reported (Tables 14.1, 14.2, and 14.3).

14.4 Discussion

The listed plants in Tables 14.1, 14.2, and 14.3 clearly show both representation of trees (32%), shrubs (23%), herbs (18%), shrubs or small trees (25%), and climbers (2%) being used. Trees and shrubs are preferred since they are available all year around, unlike herbs and climbers, which only grow briefly before drying and regrowing. Plant species are capable of producing varieties of compounds that enable them to adapt to unfavorable environments but also aid in the treatment of infections (Alvarez & Rodríguez, 2024; Kant et al., 2015). In environments such as savanna grasslands, arid woodlands, and tropical lowland rainforests, plant species have adapted to arid conditions by synthesizing chemical compounds, including essential oils and resins, to minimize water loss and deter organisms (Skarpe et al.,

Table 14.2 African medicinal plants species used in the management of syphilis

*Family	Plant name	Habit	Part used	Preparation	Reference
1. Acanthaceae	*Isoglossa lactea* Lindau	Herb	Leaves	The leaves are boiled and the decoction drunk or applied topically	(Moshi et al., 2009)
2. Anacardiaceae	*Lannea schweinfurthii* (Engl.) Engl.	Tree	Stem bark	Stem bark is boiled in water and drunk	(Maregesi et al., 2007)
3. Annonaceae	*Friesodielsia obovata* (Benth.) Verdc.	Tree	Roots	nd	(Augustino & Gillah, 2005)
4. Asparagaceae	*Asparagus cochinchinensis (Lour.) Merr.*	Herb	Roots	nd	(Watt & Breyer-Brandwijk, 1962; Chhabra et al., 1990a)
	Asparagus africanus Lam.	Scrambling shrub	Unripe fruit	External use	(Watt & Breyer-Brandwijk, 1962; Hedberg & Hedberg, 1982)
	Asparagus falcatus L.	Herb	Roots	nd	(Kokwaro, 1976; Chhabra et al., 1990a)
5. Asteraceae	*Senecio discifolius Oliv.*	Herb	Whole plant	Decoction of the whole plant is drunk; the preparation is said to be more effective when combined with *Harrisonia abyssinica* oliv.	(Maregesi et al., 2007)
6. Apocynaceae	*Diplorhynchus condylocarpon* (Müll. Arg.) Pichon.	Tree	Roots, unripe fruit	External use	(Watt & Breyer-Brandwijk, 1962; Hedberg & Hedberg, 1982)
7. Bignoniaceae	*Markhamia acuminata* (Klotzsch) K. Schum.	Tree or shrub	Bark	nd	(Kokwaro, 1976; Chhabra et al., 1987)
8. Burseraceae	*Canarium schweinfurthii* Engl.	Tree	Stem bark, roots	The bark is ground into powder and boiled and decoction drunk	(Moshi et al., 2009; Hedberg & Hedberg, 1982; Chhabra et al., 1987)

(continued)

Table 14.2 (continued)

*Family	Plant name	Habit	Part used	Preparation	Reference
9. Fabaceae	*Tamarindus indica* L.	Tree	Fruits	nd	(Hedberg & Hedberg, 1982)
	Chamaecrista absus (L.) H.S. Irwin & Barneby	Subshrub or shrub	Leaves	nd	(Kokwaro, 1976; Chhabra et al., 1987)
10. Celastraceae	*Elaeodendron buchananii* (Loes.) Loes.	Tree or shrub	Roots	nd	(Kokwaro, 1976; Chhabra et al., 1989)
	Gymnosporia heterophylla (Eckl. & Zeyh.) Loes.	Shrub	Roots	Decoction of roots	(Kokwaro, 1976; Hedberg & Hedberg, 1982)
11. Combretaceae	*Terminalia argentea Mart.*	Shrub or tree	Roots	Poultice from root pulp	(Neuwinger, 1998)
	Combretum adenogonium Steud. ex A. Rich.	Tree	Roots	nd	(Chhabra et al., 1989; Kokwaro, 1976; Chhabra et al., 1989)
12. Cucurbitaceae	*Zehneria scabra* (L.f.) Sond.	Herb	Roots	nd	(Kokwaro, 1976; Chhabra et al., 1989)
13. Euphorbiaceae	*Euphorbia tirucalli* L.	Tree or shrub	Latex	nd	(Kokwaro, 1976; Chhabra et al., 1990a)
	Jatropha curcas L.	Shrub or tree	Roots	Root decoction, decoction	(Neuwinger, 1998; Chhabra et al., 1990a; Hedberg et al., 1983a)
	Spirostachys africana Sond.	Tree	Roots	Decoction	(Kokwaro, 1976; Hedberg et al., 1983a)
14. Fabaceae	*Erythrina abyssinica* DC.	Tree	Roots	nd	(Kokwaro, 1976; Chhabra et al., 1990b)
	Neorautanenia mitis (A. Rich.) Verdc.	Herb	Roots	nd	(Kokwaro, 1976; Chhabra et al., 1990b)

	Senegalia mellifera (Vahl) Seigler & Ebinger	Tree	Bark	The bark is boiled in water, and the liquid is used as a remedy	(Kokwaro, 1976; Hedberg et al., 1983a)
	Afzelia quanzensis Welw.	Tree	Barks and root	nd	(Augustino & Gillah, 2005; Watt & Breyer-Brandwijk, 1962; Hedberg et al., 1983a)
	Sesbania macrantha Welw. ex E. Phillips & Hutch.	Shrub	Leaves/roots	The leaves are boiled, and one cup of the decoction is taken three times a day. The leaves can also be mixed with those of *Zehneria scabra*, and the roots of *Sapium ellipticum* and *Alchornea cordifolia* boiled, and the decoction taken orally	(Moshi et al., 2009)
15. Hypericaceae	*Psorospermum febrifugum* Spach	Tree or shrub	Roots	Nd	(Chhabra et al., 1990a)
16. Loganiaceae	*Strychnos spinosa* Lam.	Tree	Roots	Powdered roots in combination with other drugs	(Hedberg et al., 1983a)
17. Melastomataceae	*Dissotis brazzae* Congn.	Herb	Aerial part	The aerial parts are mixed with those of *solanum nigrum*, boiled, and the decoction administered orally	(Moshi et al., 2009)
18. Moraceae	*Ficus natalensis* Hochst.	Tree or shrub	Plant	nd	(Kokwaro, 1976; Chhabra et al., 1990b)
19. Olacaceae	*Ximenia caffra* Sond.	Tree	Root or leaves	Root decoction	(Kokwaro, 1976; Watt & Breyer-Brandwijk, 1962; Hedberg et al., 1983a)
20. Ranunculaceae	*Clematis brachiata* Thunb.	Woody climber	Roots	nd	(Kokwaro, 1976; Chhabra et al., 1991)
	Clematis hirsuta Guill. & Perr.	Herb	Leaves	nd	(Neuwinger, 1998)
21. Rhamnaceae	*Ziziphus abyssinica* Hochstetter ex A. Richard	Shrub	Stem bark	Stem bark decoction	(Neuwinger, 1998)

(continued)

Table 14.2 (continued)

*Family	Plant name	Habit	Part used	Preparation	Reference
22. Rosaceae	*Rubus rigidus* Sm.	Shrub	Aerial parts	nd	(Moshi et al., 2009)
23. Rubiaceae	*Agathisanthemum bojeri* Klotzsch	Herb	Roots	nd	(Kokwaro, 1976; Chhabra et al., 1991)
	Canthium oligocarpum Hiern.	Shrub	Roots	nd	(Augustino & Gillah, 2005)
	Chassalia violacea K.Schum.	Shrub	Roots and barks	nd	(Augustino & Gillah, 2005)
	Gardenia ternifolia Schumach. & Thonn. ssp. *jovistonantis* (Welw.) Verdc.	Tree or shrub	Leaves	nd	(Chhabra et al., 1991)
	Vangueria infausta Burchell ssp. *rutandata* (Robyns) Verdc.	Tree	Barks	nd	(Augustino & Gillah, 2005)
24. Rutaceae	*Clausena anisata* (Willd.) Benth.	Tree or shrub	Roots	nd	(Kokwaro, 1976; Chhabra et al., 1991)
	Toddalia asiatica (L.) Lam.		Leaves	nd	(Kokwaro, 1976; Chhabra et al., 1991; Augustino & Gillah, 2005)
25. Sterculiaceae	*Waltheria indica* L.	Shrub	Roots/ leaves	Decoction of the roots/juice of the leaves	(Hedberg et al., 1983b)
26. Vitaceae	*Cyphostemma adenocaule* (Steud. ex A. Rich.) Descoings.	Herb	Roots	Decoction of the tubers is drunk warm	(Kokwaro, 1976; Hedberg et al., 1983b)

*The plant family is arranged alphabetically
nd not documented

Table 14.3 African medicinal plants species used in the management of both syphilis and gonorrhea

*Family	Plant name	Habits	Part used	Other claims for disease condition treatment in	Preparation	Reference
1. Acanthaceae	*Acanthus polystachyus* Delile	Shrub	Roots, leaves	nd	The roots are cut into small pieces and boiled with water or broth/meat. Two bowls are taken per day.	(Moshi et al., 2009)
2. Asteraceae	*Emilia discifolia (Oliv.) C.Jeffrey*	Tree or shrub	Root	nd	Root decoction is drunk; sometimes the preparation is done by a combination of *Senna siamea* (Lam.) Irwin & Barneby	(Maregesi et al., 2007)
3. Bignoniaceae	*Kigelia africana* (Lam.) Benth.	Tree	Roots, stem bark	Wounds, ulcers, abscesses, dysentery	Stem bark is boiled in water and drunk	(Watt & Breyer-Brandwijk, 1962; Hedberg & Hedberg, 1982; Maregesi et al., 2007)
4. Capparaceae	*Maerua edulis* (Gilg & Bened.) De Wolf	Tree or shrub	Roots	Venereal diseases	The peeled roots infusion is taken orally	(Maregesi et al., 2007)
		Tree or shrub	Leaves	Skin diseases, cleansing uterus Before a child is delivered, Malaria	Leaves are boiled, and decoction is drunk	(Moshi et al., 2009)
5. Combretaceae	*Combretum molle* R.Br. ex G.Don	Tree	Roots/ leaves	Influenza, edema, skin diseases, wounds	A decoction also containing roots of *Securinega virosa* (roxb. Etx willd.) Pax et k. Hoffm., and *premna* cfr. *Senensis* klotsch	(Fyhrquist et al., 2002; Hedberg & Hedberg, 1982)
6. Fabaceae	*Cassia abbreviata* Oliv.	Tree	Roots	Pneumonia, chest complaints, uterine complaints, stomach troubles, and fever of malaria	Decoction of roots	(Hedberg & Hedberg, 1982)

(continued)

Table 14.3 (continued)

*Family	Plant name	Habits	Part used	Other claims for disease condition treatment in	Preparation	Reference
	Cassia afrofistula Brenan	Tree or shrub	Roots	Bilharzia, hernia, abdominal pain	nd	(Chhabra et al., 1987)
	Albizia anthelmintica A. Brogn.	Shrub	Bark	Rheumatism, nervous complaints, swelling of the body, during childbirth	nd	(Watt & Breyer-Brandwijk, 1962; Hedberg et al., 1983a)
	Vachellia tortilis (Forssk.) Galasso & Banfi	Tree	Roots/ barks	Roots: Late labor pains, stroke, convulsion, infertility, women's stomachache, Barks: Toothache, high fever	nd	(Augustino & Gillah, 2005)
	Dalbergia melanoxylon Guill. & Perr.	Tree	Leaves	Dysentery, throat inflammations, cardiac trouble, amoebic dysentery	Juice	(Watt & Breyer-Brandwijk, 1962; Hedberg et al., 1983b)
7. Lauraceae	*Cassytha filiformis* L.	Parasitic herb	Plant	Dysentery, ulcers, vermin, snake bites	Nd	(Kokwaro, 1976; Chhabra et al., 1990a)
8. Meliaceae	*Trichilia emetica* Vahl	Tree	Stembark, leaves, roots	Tuberculosis, epilepsy, malaria	Stem bark/leaves are boiled in water, and the decoction is administered. In the treatment of gonorrhea and syphilis, the roots are boiled together with the roots of *Alchornea cordifolia*, *Sapium ellipticum*, *Zehneria scabra*, *Acanthus pubescens*, *Emilia javanica*, and *Oxygonum sinuatum*, and the decoction is drunk	(Moshi et al., 2009; Augustino & Gillah, 2005)
9. Phyllanthaceae	*Flueggea virosa* (Willd.) Voigt	Tree or shrub	Roots	nd	The roots are cooked together with *Vigna unguiculata* (l.) Walp	(Chhabra et al., 1990a; Moshi et al., 2009; Watt & Breyer-Brandwijk, 1962; Hedberg et al., 1983a)

10. Phytolaccaceae	*Phytolacca dodecandra* L'Hérit.	Shrub	Roots	Bilharzia	Infusion of the roots	(Chhabra et al., 1991)
11. Ranunculaceae	*Securidaca longepedunculata* Fresenius	Shrub or tree	Root bark, stem bark	nd	Decoction of root bark, stem bark infusion, root decoction Hot water taken orally	(Neuwinger, 1998; Watt & Breyer-Brandwijk, 1962; Hedberg et al., 1983b; Hedimbi & Chinsembu, 2012)
12. Rhamnaceae	Scutia myrtina (Burm.f.) Kurz	Shrub	Roots/ stem bark	Lepra, mental diseases, dysentery, anthelmintic, diuretic, bilharzia	nd	(Hedberg et al., 1983b)
13. Rubiaceae	*Dolichopentas longiflora (Oliv.) Kårehed & B. Bremer* Oliv.	Herb	Root	nd	Root decoction	(Chhabra et al., 1991)
	Pentas zanzibarica (Klotzsch) Vatke	Herb	Roots	nd	nd	(Kokwaro, 1976)
14. Rutaceae	*Zanthoxylum capense* (Thunb.) Harv.	Tree	Root	Burning stomach aches, infertility, impotence, epilepsy Jaundice, hernia, stomach ache, hemorrhoids	Root decoction	(Chhabra et al., 1990b; Augustino & Gillah, 2005)
15. Sapotaceae	*Manilkara sansibarensis* (Engl.) Dubard	Tree	Roots	Constipation	nd	(Kokwaro, 1976; Chhabra et al., 1993)
16. Solanaceae	*Solanum incanum* L.	Shrub	Roots/ leaves	Cough, colic, sore throat	Nnd	(Watt & Breyer-Brandwijk, 1962; Hedberg et al., 1983b)

(continued)

Table 14.3 (continued)

*Family	Plant name	Habits	Part used	Other claims for disease condition treatment in	Preparation	Reference
17. Verbenaceae	*Lantana camara* L.	Shrub	Roots, leaves	Swollen legs, cough, to dilate vagina during labor	The roots/leaves are boiled together with the stem bark of *Mangifera indica* and *Ocimum basillicum*, and the decoction is taken	(Moshi et al., 2009)

*The plant family is arranged alphabetically
nd Not documented.

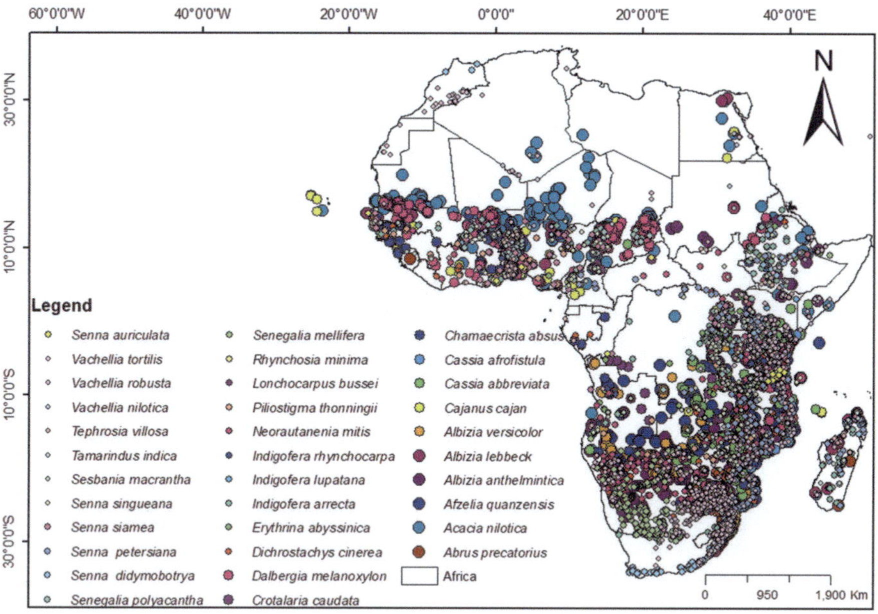

Fig. 14.4 Distribution of plant species within the Fabaceae family

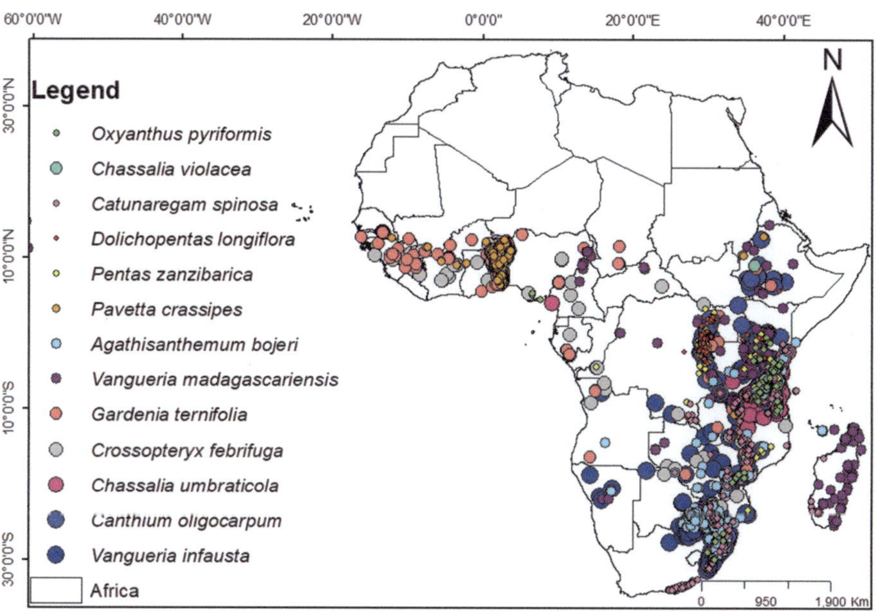

Fig. 14.5 Distribution of plant species within the Rubiaceae family

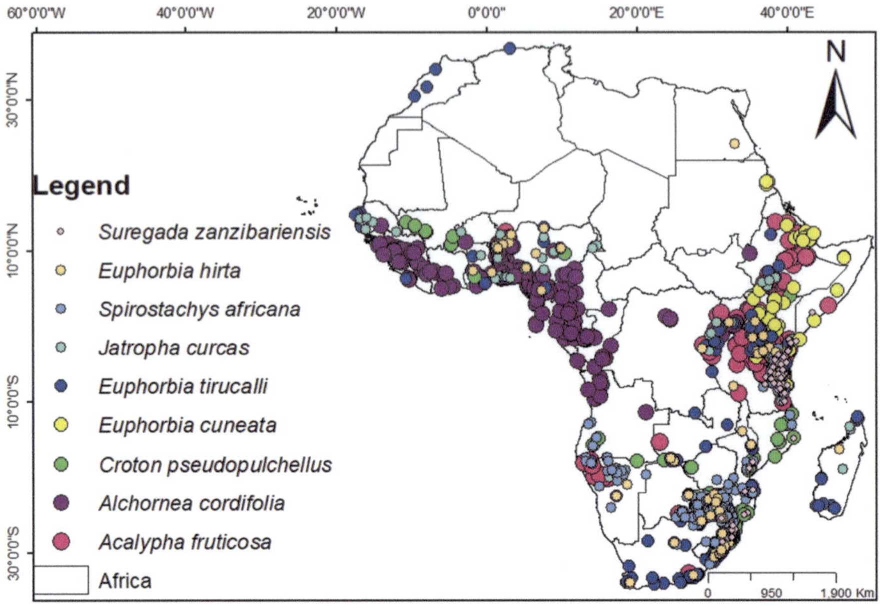

Fig. 14.6 Distribution of plant species within the Euphorbiaceae family

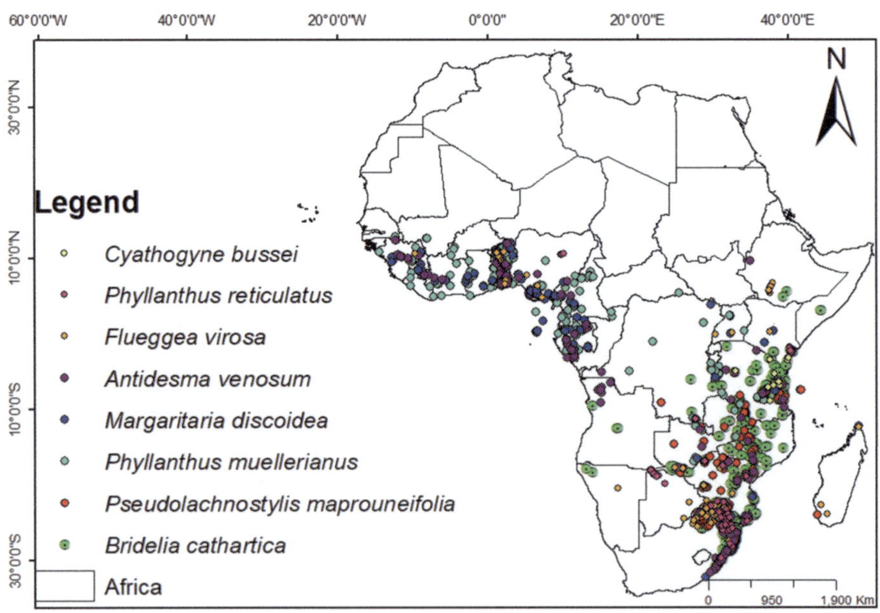

Fig. 14.7 Distribution of plant species within the Phyllanthaceae family

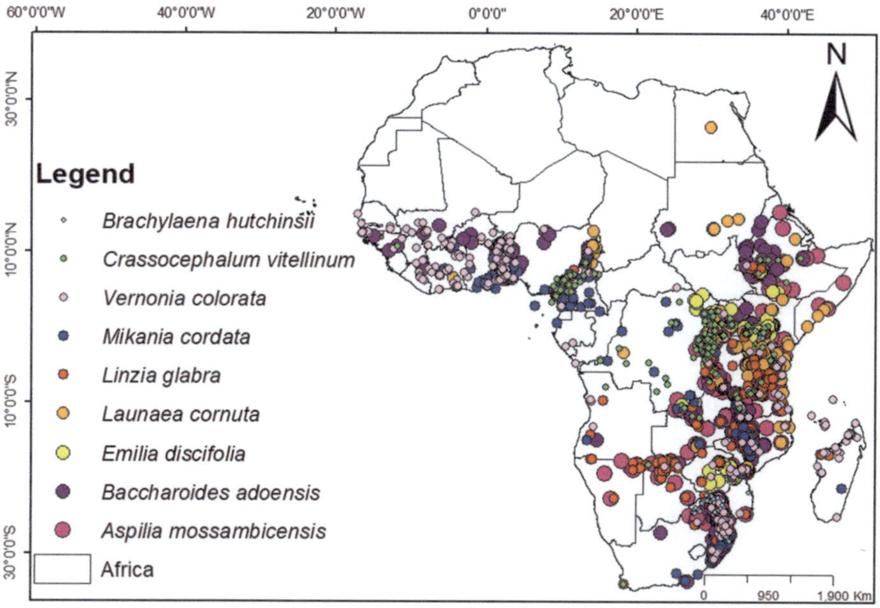

Fig. 14.8 Distribution of plant species within the Asteraceae family

2000). For example, plant species of the genus *Commiphora*, *Acacia*, *Dipterocarpus*, and *Croton* are found in such areas and are also known for their repellent, wound healing, and treating infections (Salatino et al., 2007; Chatuphonprasert et al., 2019; Moglad et al., 2020; Kankara et al., 2017). Conversely, plant species from swamp forests and mangroves adapted to high moisture and salinity conditions (Naidoo, 1985; Srikanth et al., 2016), leading to the synthesis of secondary metabolites with potential as antifungal and antibacterial, for example, *Kigelia africana* (Lam.) Benth (Idris et al., 2013). Additionally, mapping the plant families and potential medicinal species for treating syphilis and gonorrhea shows their geographical distribution and the vegetation types that support their conservation. On the other hand, plant parts used are roots or root barks (103) 64%, leaves (21) 12%, stem barks or barks (22) 13%, and whole plant (7) 4% (Tables 14.1, 14.2, and 14.3). The rarely used plant parts are barks and roots (4), aerial parts (2), fruits (3), latex (1), leaves and roots (3), leaves and twigs (1), and seeds (1). The roots contain a higher concentration of active ingredients compared to other parts of the plant, as reported by other researchers (Nankaya et al., 2020; Ozioma & Chinwe, 2019; Mbuni et al., 2020; Erasto & Majinda, 2011).

Flueggea virosa (Roxb. ex Willd.) Royle is one of the most important African medicinal plants, with all of its parts used to treat a variety of ailments including testicular inflammation, sterility, urinary disease, pneumonia, frigidity, liver, bile, kidney, contraceptive, and snake antidote. (Kamatenesi-Mugisha & Oryem-Origa, 2005; Maroyi, 2013; Nazer et al., 2019). It contains phytochemicals such as securinine, triterpenes, and alkaloids that have antibacterial, antioxidant, and antifungal properties (Maroyi, 2013).

Cassia abbreviata Oliv. (Fabaceae) contains proanthocyanidin compound cassinidin A, which showed activity against *E. coli* with MIC of 1 μg/mL, and against *B. subtilis* and *S. aureus* with MIC values of 0.5 μg/mL each. Another proanthocyanidin from the same plant, cassinidin B, demonstrated activity against *B. subtilis* and *S. aureus*, with MIC values of 5 mg/mL each (Erasto & Majinda, 2011). In another study, the roots of *Cassia abbreviata* Oliv were tested against *Neisseria gonorrhoeae*, revealing MIC of 125 μg/ml for the ethanol extract and 62.5 μg/ml for the aqueous extracts (Mudenda et al., 2024).

A decoction made from the stem bark of *Canarium schweinfurthii* Engl. is used as a remedy for roundworms, colic, stomach pains, pain after childbirth, gale, dysentery, and gonorrhea (Tcheghebe et al., 2016). The isolated compound, 3-β-hydroxylolean-12,18-diene, exhibited inhibitory activity at 6.25 μg/mL against *S. aureus*, *E. coli*, *S. dysenteria*, *B. subtlis*, *K. pneumonia*, and *C. stellatoidea.* Additionally, *S. typhi*, *C. albicans*, and *C. tropicalis* were inhibited at 12.5 μg/mL (Tcheghebe et al., 2016; Okoli et al., 2015).

Jatropha curcas L. is a plant species commonly utilized in traditional medicine, showing effectiveness against a variety of both gram-positive and gram-negative bacteria (Abdelgadir & Van Staden, 2013). Extracts obtained from hexane, ethyl acetate, and methanol demonstrated activity at a concentration of 200 mg/ml against *G. vaginalis*, *N. gonorrhoeae*, *E. coli*, *S. aureus*, *K. aerogenes*, *P. mirabills*, *P. aeruginosa* while MIC of MIC 0.75 μg ml was noted against *C. albicans* (Abdelgadir & Van Staden, 2013).

Highly used plant species included *Kigelia africana* (Lam.) Benth. and *Securidaca longipedunculata* Fresen, also being used for treatment of tuberculosis, pneumonia, coughing, dysentery, and female gynecological problems including gonorrhea (Innocent et al., 2022; Assanti et al., 2022; Neuwinger, 1998; Hedberg et al., 1983b; Hedimbi & Chinsembu, 2012; Hedberg & Hedberg, 1982). Previously, the same trend of ranking for the two species against respiratory bacterial infections was observed. However, none of the two species produced remarkable bioactive compounds (Innocent et al., 2022). This observation shows the plant species are being used for multiple symptoms that are caused by bacteria. Thus, it is recommended that the exploration of antimicrobial active extracts and compounds should divert from a too reductionist approach (Innocent et al., 2022). The exploration should also consider the concepts and description of symptoms underlying diseases in traditional African medicine.

14.5 Conclusion

A total of 159 plant species belonging to 127 genera and about 53 plant families were reported to be plant species used for the management of gonorrhea and syphilis. The highest ranked plant families were Fabaceae (34), Rubiaceae (13), Euphorbiaceae

(9), Asteraceae (9), Phyllanthaceae (8), Rutaceae (5), Anacardiaceae (4), Combretaceae (4), Apocynaceae (4) Rhamnaceae (4), and Asparagaceae (4), and the rest of the families have less than 2 number of plant species as shown in Fig. 14.3. This chapter highlights the importance of medicinal plants, including *Kigelia africana* (Lam.) Benth and *Securidaca longipedunculata* Fresen, in the treatment of syphilis and gonorrhea. Furthermore, investigations into their mechanisms of action, along with animal and human clinical trials, are crucial for validating their safety and efficacy for the management of gonorrhea and syphilis. Most of these plants are well distributed in Africa, especially in the savanna grassland, swamp forests and mangroves, dry forests, and the thickest and tropical lowland rainforests.

The list of mostly used plant species is often documented during ethnopharmacological surveys. The researchers use semi-structured questionnaires asking about conditions the traditional healer claims to treat and the plants used. However, in African traditional medicines, because of its holistic nature in treatment, consideration of the pattern of disease diagnosis is important. This is because, in most cases, the sick person seeks to recognize the *causal* agent and therapies that address the problems in social relations. Some of such conditions may be characterized as not symptoms rather can be used as signs of a certain disease. Thus, combining traditional diagnosis methods with modern medicine is crucial, given the challenging similarities between the two diseases and the traditional emphasis on the appearance of symptoms, in the choice of treatment. Nevertheless, some symptoms take a long time to manifest, often when the disease has progressed. Proper integration of traditional methods can support early identification and referral of such cases to conventional medicine facilities to avoid disease progression.

Acknowledgments Authors are thankful to the Institute of Traditional Medicine, Muhimbili University of Health and Allied Sciences, for availing time and resources for this work.

Competing Interests The authors confirm contribution to the paper as follows: study conception and design: EI, SA, AM, and AS wrote the main manuscript text, and SA and AS prepared figures. All authors reviewed and approved the manuscript. The authors have no conflicts of interest to declare that are relevant to the content of this chapter.

References

Abdelgadir, H. A., & Van Staden, J. (2013). Ethnobotany, ethnopharmacology and toxicity of Jatropha curcas L. (Euphorbiaceae): A review. *South African Journal of Botany, 88*, 204–218.

Alvarez, T. C. O., & Rodríguez, A. G. (2024). *How do plants protect themselves against animal attacks?* https://doi.org/10.3389/frym.2023.991723

Assanti, G., Kaur, R., Nizard, S., Pollack-Blackwood, E., Rafferty, B., Priano, C., Fernández Romero, J. A., & Koroch, A. R. (2022). Biology, chemistry, and pharmacological activity of Kigelia africana (Bignoniaceae) and Garcinia kola (Clusiaceae)-a review. *Journal of Medicinally Active Plants, 11*(1), 1.

Augustino, S., & Gillah, P. R. (2005). Medicinal plants in urban districts of Tanzania: Plants, gender roles and sustainable use. *International Forestry Review, 7*(1), 44–58.

Burkill, H. M. (Ed.). (1985). *The Useful Plants of West Tropical Africa, Vol 1 (second ed.)* (p. 510). Royal Botanic Gardens Kew, UK.

Chatuphonprasert, W., Sriset, Y., & Jarukamjorn, K. (2019). Continuous consumption of reused palm oil induced hepatic injury, depletion of glutathione stores, and modulation of cytochrome P450 profiles in mice. *Polish Journal of Food and Nutrition Sciences, 69*(1).

Chhabra, S. C., Mahunnah, B. L. A., & Mshiu, E. N. (1987). Plants used in traditional medicine in Eastern Tanzania. I. Pteridophytes and Angiosperms (Acanthaceae to Canellaceae). *Journal of Ethnopharmacology, 21*(3), 253–277.

Chhabra, S. C., Mahunnah, R. L. A., & Mshiu, E. N. (1989). Plants used in traditional medicine in Eastern Tanzania. II. Angiosperms (Capparidaceae to Ebenaceae). *Journal of Ethnopharmacology, 25*(3), 339–359.

Chhabra, S. C., Mahunnah, R. L. A., & Mshiu, E. N. (1990a). Plants used in traditional medicine in Eastern Tanzania. III. Angiosperms (Euphorbiaceae to Menispermaceae). *Journal of Ethnopharmacology, 28*(3), 255–283.

Chhabra, S. C., Mahunnah, R. L. A., & Mshiu, E. N. (1990b). Plants used in traditional medicine in Eastern Tanzania. IV. Angiosperms (Mimosaceae to Papilionaceae). *Journal of Ethnopharmacology, 29*(3), 295–323.

Chhabra, S. C., Mahunnah, R. L. A., & Mshiu, E. N. (1991). Plants used in traditional medicine in Eastern Tanzania. V. Angiosperms (Passifloraceae to Sapindaceae). *Journal of Ethnopharmacology, 33*(1–2), 143–157.

Chhabra, S. C., Mahunnah, R. L. A., & Mshiu, E. N. (1993). Plants used in traditional medicine in Eastern Tanzania. VI. Angiosperms (Sapotaceae to Zingiberaceae). *Journal of Ethnopharmacology, 39*(2), 83–103.

Erasto, P., & Majinda, R. T. (2011). Bioactive proanthocyanidins from the root bark of Cassia abbreviata. *International Journal of Biological and Chemical Sciences, 5*(5), 2170–2179.

Fyhrquist, P., Mwasumbi, L., Hæggström, C.-A., Vuorela, H. I. L. T. U. N. E. N., Hiltunen, R., & Vuorela, P. (2002). Ethnobotanical and antimicrobial investigation on some species of Terminalia and Combretum (Combretaceae) growing in Tanzania. *Journal of Ethnopharmacology, 79*(2), 169–177.

Gessler, M. C., Nkunya, M. H. H., Mwasumbi, L. B., Heinrich, M., & Tanner, M. (1994). Screening Tanzanian medicinal plants for antimalarial activity. *Acta Tropica, 56*(1), 65–77.

Hedberg, I., & Hedberg, O. (1982). Inventory of plants used in traditional medicine in Tanzania. I. Plants of the families Acanthaceae—Cucurbitaceae. *Journal of Ethnopharmacology, 6*, 29–60.

Hedberg, I., Hedberg, O., Madati, P. J., Mshigeni, K. E., Mshiu, E. N., & Samuelsson, G. (1983a). Inventory of plants used in traditional medicine in Tanzania. Part III. Plants of the families Papilionaceaevitaceae. *Journal of Ethnopharmacology, 9*(2–3), 237–260.

Hedberg, I., Hedberg, O., Madat, P. J., Mshigeni, K. E., Mshiu, E. N., & Samuelsson, G. (1983b). Inventory of plants used in traditional medicine in Tanzania. II. Plants of the families Dilleniaceae—Opiliaceae. *Journal of Ethnopharmacology, 9*(1), 105–127.

Hedimbi, M., & Chinsembu, K. C. (2012). Ethnomedicinal study of plants used to manage HIV/AIDS-related disease conditions in the Ohangwena region, Namibia. *International Journal of Medicinal Plants Research, 1*(1), 004–011.

Hinkkanen, R. (2009). *'Someone to welcome you home': Infertility, medicines and the Sukuma-Nyamwezi*. University of Helsinki.

Gessler, M. C., Nkunya, M. H. H., Mwasumbi, L. B., Heinrich, M., & Tanner, M. (1994). Screening Tanzanian medicinal plants for antimalarial activity. Acta Tropica 56(1), 65–77.

Idris, A.-m., Al-tahir, I., & Idris, E. (2013). Antibacterial activity of endophytic fungi extracts from the medicinal plant Kigelia africana. *Egyptian Academic Journal of Biological Sciences, G. Microbiology, 5*(1), 1–9.

Innocent, E. Bridging indigenous knowledge and the science: The role of indigenous peoples in health and pharmaceutical products development. Presented on 14-15th March, 2024 at The Pontifical Academy of Science, Rome, Vatican City.

Innocent, E., Marealle, A. I., Imming, P., & Moeller, L. (2022). An annotated inventory of Tanzanian medicinal plants traditionally used for the treatment of respiratory bacterial infections. *Plants, 11*(7), 931.

Kacholi, D. S., & Amir, H. M. (2022). Ethnobotanical survey of medicinal plants used by traditional healers in managing Gonorrhoea and Syphilis in Urambo District, Tabora Region, Tanzania. *Journal of Herbs, Spices & Medicinal Plants, 28*(2), 179–192.

Kamatenesi-Mugisha, M., & Oryem-Origa, H. (2005). Traditional herbal remedies used in the management of sexual impotence and erectile dysfunction in western Uganda. *African Health Sciences, 5*(1), 40–49.

Kankara, S. S., Sani, D., Ibrahim, M. H., Mustafa, M., & Go, R. (2017). Acacia nilotica pods' water extract enhances wound healing in Sprague-Dawley rats by alleviating oxidative stress and suppressing pro-inflammatory cytokines. *Nigerian Journal of Scientific Research, 16*(2), 202–210.

Kant, M. R., Wim Jonckheere, B., Knegt, F., Lemos, J., Liu, B. C. J., Schimmel, C. A., Villarroel, et al. (2015). Mechanisms and ecological consequences of plant defence induction and suppression in herbivore communities. *Annals of Botany, 115*(7), 1015–1051.

Kokwaro, J. O. (1976). *Medicinal plants of East Africa*. East African Literature Bureau.

Manyahi, J., Jullu, B. S., Abuya, M. I., Juma, J., Ndayongeje, J., Kilama, B., Sambu, V., et al. (2015). Prevalence of HIV and syphilis infections among pregnant women attending antenatal clinics in Tanzania, 2011. *BMC Public Health, 15*, 1–9.

Marealle, A. I., Innocent, E., Marobela, K. A., Qwarse, M., & Moshi, M. (2020). Ethnopharmacological survey of medicinal plants used in the management of HIV and AIDS-related conditions in Makete District, southern highlands of Tanzania. *European Journal of Medicinal Plants, 31*(14), 56–83.

Maregesi, S. M., et al. (2007) Ethnopharmacological survey of the Bunda district, Tanzania: Plants used to treat infectious diseases. *Journal of Ethnopharmacology, 113*(3), 457–470.

Maroyi, A. (2013). Traditional use of medicinal plants in south-central Zimbabwe: Review and perspectives. *Journal of Ethnobiology and Ethnomedicine, 9*, 1–18.

Mbuni, Y. M., Wang, S., Mwangi, B. N., Mbari, N. J., Musili, P. M., Walter, N. O., Guangwan, H., Zhou, Y., & Wang, Q. (2020). Medicinal plants and their traditional uses in local communities around Cherangani Hills, Western Kenya. *Plants, 9*(3), 331.

Moglad, E. H., Boon, S. K. A., & Ali, H. T. (2020). Various medicinal plants: A promising treatment for multidrug-resistant bacteria isolated from wound infection. *International Journal of Pharmaceutical Sciences and Research, 11*(2), 839–843.

Mongalo, N. I., & Raletsena, M. V. (2022). An inventory of South African medicinal plants used in the management of sexually transmitted and related opportunistic infections: An appraisal and some scientific evidence (1990–2020). *Plants, 11*(23), 3241.

Moshi, M. J., Otieno, D. F., Mbabazi, P. K., & Weisheit, A. (2009). The Ethnomedicine of the Haya people of Bugabo ward, Kagera Region, North Western Tanzania. *Journal of Ethnobiology and Ethnomedicine, 5*, 1–5.

Mudenda, S., Chalwe, R., Kabuka, R., & Chabalenge, B. (2024). Antibacterial activity of Cassia abbreviata oliv roots against Neisseria gonorrhoeae: A potential traditional medicine for the treatment of sexually transmitted infections. *Journal of Medicinal Plants Studies, 12*(1), 121–126.

Naidoo, G. (1985). Effects of waterlogging and salinity on plant-water relations and on the accumulation of solutes in three mangrove species. *Aquatic Botany, 22*(2), 133–143.

Nankaya, J., Nampushi, J., Petenya, S., & Balslev, H. (2020). Ethnomedicinal plants of the Loita Maasai of Kenya. *Environment, Development and Sustainability, 22*(3), 2569–2589.

Nazer, M., Abbaszadeh, S., Darvishi, M., Kheirollahi, A., Shahsavari, S., & Moghadasi, M. (2019). The most important herbs used in the treatment of sexually transmitted infections in traditional medicine. *Sudan Journal of Medical Sciences, 14*(2), 41–64.
Ndubani, P., & Höjer, B. (1999). Traditional healers and the treatment of sexually transmitted illnesses in rural Zambia. *Journal of Ethnopharmacology, 67*(1), 15–25.
Neuwinger, H. D. (1998). *Afrikanische Arzneipflanzen und Jagdgifte: Chemie, Pharmakologie, Toxikologie: eine afrikanische Ethnopharmakologie und Ethnobotanik.* Wissenschaftliche Verlagsgesellschaft.
Newman, L., Rowley, J., Hoorn, S. V., Wijesooriya, N. S., Unemo, M., Low, N., Stevens, G., Gottlieb, S., Kiarie, J., & Temmerman, M. (2015). Global estimates of the prevalence and incidence of four curable sexually transmitted infections in 2012 based on systematic review and global reporting. *PLoS One, 10*(12), e0143304.
Okoli, B. J., Ayo, R. G., Habila, J. D., Ndukwe, G. I., & Jummai, A. T. (2015). Inhibition of Methicillin ResistanceStaph. aureus (MRSA) and Fungi by *Canarium schweinfurthii* (Engl) Extracts. Scholars Acad. *Journal of Biosciences, 3*(5), 413–420.
Ozioma, E.-O. J., & Chinwe, O. A. N. (2019). Herbal medicines in African traditional medicine. *Herbal medicine, 10*, 191–214.
Peeling, R. W., Mabey, D., Kamb, M. L., Chen, X.-S., Radolf, J. D., & Benzaken, A. S. (2017). Syphilis (primer). *Nature Reviews: Disease Primers, 3*(1).
Rowley, J., Hoorn, S. V., Korenromp, E., Low, N., Unemo, M., Abu-Raddad, L. J., Matthew Chico, R., et al. (2019). Chlamydia, gonorrhoea, trichomoniasis and syphilis: Global prevalence and incidence estimates, 2016. *Bulletin of the World Health Organization, 97*(8), 548.
Salatino, A., Faria, M. L., & Salatino, and Giuseppina Negri. (2007). Traditional uses, chemistry and pharmacology of Croton species (Euphorbiaceae). *Journal of the Brazilian Chemical Society, 18*, 11–33.
Skarpe, C., Bergström, R., Bråten, A. L., & Danell, K. (2000). Browsing in a heterogeneons savanna. *Ecography, 23*(5), 632–640.
Srikanth, S., Lum, S. K. Y., & Chen, Z. (2016). Mangrove root: Adaptations and ecological importance. *Trees, 30*, 451–465.
Tcheghebe, O. T., Seukep, A. J., & Tatong, F. N. (2016). A review on traditional uses, phytochemical composition and pharmacological profile of Canarium schweinfurthii Eng. *Natural Science, 14*(11), 17–22.
Toyang, N. J., & Verpoorte, R. (2013). A review of the medicinal potentials of plants of the genus *Vernonia* (Asteraceae). *Journal of Ethnopharmacology, 146*, 681–723.
Watt, J. M., & Breyer-Brandwijk, M. G. (1962). The medicinal and poisonous plants of southern and Eastern Africa being an account of their medicinal and other uses, chemical composition. *Pharmacological Effects and Toxicology in Man and Animal*, xii+-1457.
Werner, D., Thuman, C., & Maxwell, J. (2011). *Where there is no doctor: A village health care handbook-rev*. Hesperian Health Guides.
WHO. (2021). https://www.who.int/teams/global-hiv-hepatitis-and-stis-programmes/stis/overview. Accessed on 11th November, 2024.
WHO. (2024). *Global HIV, Hepatitis and STIs Programmes (HHS).*

Ester Innocent specializes in natural product chemistry, having pursued her studies at the Department of Chemistry, University of Dar es Salaam. Since 2002, she is employed at the Institute of Traditional Medicine (ITM) under Muhimbili University of Health and Allied Sciences (MUHAS), where she currently holds the position of Associate Research Professor. She is active as supervisor for 5 PhD and 8 MSc students and an editor of *Journal of Natural Pesticide Research*. She has 2 patents and over 60 peer-reviewed international journal publications. Her research is centered on bio-prospecting, natural products chemistry, and biological testing with the goal of discovery of bioactive compounds for the development of herbal medicines and botanical insecticides.

Siri Abihudi is a lecturer at the Department of Medical Botany, Plant Breeding and Agronomy at MUHAS. She holds a PhD in Life Sciences with a focus on Biodiversity and Ecosystem Management, a Master's in Traditional Medicine Development, and a Bachelor's in Environmental Science. Abihudi mentors undergraduate and postgraduate students while conducting research in biodiversity conservation, plant ecology, ethnobotany, and climate-related studies. She has published eight papers, with another eight under review in reputable peer-reviewed journals. Her research interests center on plant species diversity, their use, and conservation amidst climate change.

Alphonce Ignace Marealle is a lecturer in the Department of Clinical Pharmacy and Pharmacology at MUHAS. He holds a PhD and a Master's degree in pharmacology, and a Bachelor's degree in pharmacy. Alphonce has participated in conducting numerous pre-clinical and clinical studies and mentoring both undergraduate and postgraduate students. His contributions to the field are well-documented, with over 50 articles published in reputable peer-reviewed journals. Additionally, he is skilled in isolating and elucidating the structures of compounds found in natural products, utilizing spectroscopic techniques.

Albert Shayo is a botanist with a Bachelor's degree in botanical sciences from the University of Dar es Salaam's Department of Botany. He is currently working as a laboratory scientist at MUHAS-ITM. He participated in several ethnobotanical, plant ecology, and conservation studies; he is adept at identifying plants and gathering ethnobotanical information for the development of herbal medicine, as well as educating society about medicinal plant conservation.

Chapter 15
Diversity and Uses of Medicinal Flora in the Vegetation of the Ankaratra Massif, Madagascar

Stéphan Richard Rakotonandrasana, Christian Marius Rakotonirina, Falitiana Marrino Rakotoarisoa, Andriamalala Rakotondrafara, Vonjison Rakotoarimanana, Miora Henintsoa Andrianarivelo, Vincent Emile Rasamison, and Rianasoambolanoro Rakotosaona

Abstract Madagascar, renowned for its unique biodiversity, harbors the Ankaratra Mountain, the third highest peak on the island. Despite its ecological significance, the ethnobotanical aspects of this region remain largely unexplored. This study aims to investigate the diversity of medicinal plant species, identify commonly used species in traditional medicine, and explore their applications in treating diseases. During the ethnobotanical survey, voucher specimens collected from the botanical inventory were presented to 26 traditional healers in the study area, selected based on their recognized expertise and extensive experience in traditional medicine. Data on plant usage, the specific parts harvested, and preparation methods for herbal remedies were gathered through semi-structured interviews. The study identified 203 medicinal plant species in the Ankaratra Mountain, with 139 (73.9%) endemic and 69 (33.9%) newly recorded in the ethnopharmacopeia. Savannah vegetation was the most cited habitat (35.5%), while endemic species dominated, comprising 77.8% of the most used and 71.6% of the most significant plants. These species were primarily used to treat locally prevalent diseases and those requiring specialized

S. R. Rakotonandrasana (✉) · C. M. Rakotonirina · F. M. Rakotoarisoa · A. Rakotondrafara
CNARP Centre National d'Application de Recherches Pharmaceutiques, Antananarivo, Madagascar
e-mail: stephan.rakotonandrasana@cnarp.mg

V. Rakotoarimanana
Faculty of Sciences, University of Antananarivo, Antananarivo, Madagascar

M. H. Andrianarivelo · V. E. Rasamison
Institut of Higher Education Antsirabe Vakinankaratra, Vakinankaratra, Madagascar

R. Rakotosaona
CNARP Centre National d'Application de Recherches Pharmaceutiques, Antananarivo, Madagascar

Laboratoire de Recherche en Sciences Alimentaires et Santé, Ecole Supérieure Polytechnique University of Antananarivo, Antananarivo, Madagascar

A. Dekebo (ed.), *Bioactive Secondary Metabolites from Medicinal Plants of Africa*, Interdisciplinary Biotechnological Advances,
https://doi.org/10.1007/978-981-96-8590-5_15

knowledge, underscoring their importance in traditional healthcare. This study provides the first comprehensive ethnobotanical report on the plant species of the Ankaratra Mountain. The findings bear implications for local healthcare practices and contribute to the potential conservation of this unique flora. Further avenues of research include investigating the ecological preferences of endemic species and conducting chemical and pharmacological analyses to uncover novel therapeutic compounds.

Keywords Madagascar · Ankaratra Mountain · Diversity · Medicinal uses of plants

15.1 Introduction

Medicinal plants represent an invaluable resource for human well-being, harboring vast untapped potential (Farnsworth & Soejarto, 1991). These botanical resources are integral to healthcare systems worldwide, particularly in developing countries, where they often serve as primary sources of remedies (Petrovska, 2012). In developing regions worldwide, more than 3.5 billion people depend on medicinal plants as their primary source of healthcare (WHO, 2008). This dependence is especially pronounced in Africa, where a significant majority (70–80%) of people seek out traditional healers to address health concerns (Mahomoodally, 2013).

The African continent houses these medicinal species within its lush forests, encompassing 216 million hectares. However, with an annual deforestation rate of 1%, this precious resource faces an imminent threat (Mahomoodally & Gurib-Fakim, 2013). These pressures are particularly pronounced in Madagascar, a recognized global biodiversity hotspot (Mittermeier et al., 2004), where over 80% of the population depends on natural resources for their livelihoods and healthcare. Tragically, between 2005 and 2013, human activities contributed to the degradation of 120,731 hectares of forested land per year (MEEMF, 2015). In this delicate ecological balance, numerous plant and animal species — including medicinal plants — are on the brink of extinction. These issues are further exacerbated by the declining use of local agricultural practices (Rakotonandrasana, 2013; Rakotoarisoa, 1986) and the aging population of traditional healers (De Beer &Van Wyk 2011). This dual threat—environmental degradation and loss of traditional knowledge—poses a significant risk to the future of medicinal plant use in Madagascar.

Madagascar's impressive vascular flora comprises 11,220 species (Callmander et al., 2011), among which 3245 species serve medicinal purposes (Rafidison et al., 2019). However, this roster is far from exhaustive, a point underscored by several authors (Ivan Norscia & Borgognini-Tarli, 2006b; Rakotonandrasana, 2013; Rakotoarivelo et al., 2015; Allorge, 2005). The unexplored realms of numerous protected areas and natural habitats add to the enigma of Madagascar's botanical treasures.

Among these unexplored areas, the Ankaratra Mountain stands out as Madagascar's third-highest peak, harboring a unique high-altitude ecosystem. The local communities surrounding this mountain have long relied on its rich flora for traditional medicine. However, the medicinal plants of this region remain largely undocumented, presenting a critical gap in our understanding of Madagascar's ethnobotanical wealth.

Against this backdrop, the present study sets forth an overarching objective: to conduct the first comprehensive survey of medicinal plants in the Ankaratra Mountain ecosystem. This research aims to enrich our understanding of Madagascar's medicinal plants through ethnobotanical surveys in a region where a dearth of recorded data exists. These endeavors not only seek to valorize the significance of medicinal species but also to actively contribute to their preservation.

The specific aims encompass:

1. Unraveling the diversity and endemism of medicinal species in the Ankaratra Mountain
2. Pinpointing the pivotal species in traditional medicine
3. Illuminating the crucial plants employed to combat ailments, leveraging the area's rich medicinal flora

By documenting this valuable knowledge, we hope to inform conservation efforts, contribute to potential drug discovery, and preserve the traditional medicinal practices of the local communities. This study thus represents a crucial step in safeguarding Madagascar's unique biodiversity and the invaluable traditional knowledge associated with it.

15.2 Methodology

15.2.1 Study Area

15.2.1.1 Geographic Location and Topography

The Ankaratra Mountain, situated in Madagascar's central highlands, lies about 70 km south of the capital city, Antananarivo. It extends across the southeastern Itasy Region and the northern Vakinankaratra Region, located at coordinates 19°20′ S and 47°15′E. The massif covers an area of about 2000 km^2, with its highest peak, Tsiafajavona, reaching 2643 meters above sea level, making it the third-highest mountain in Madagascar (Fig. 15.1) [1].

15.2.1.2 Geology and Soil Composition

The Ankaratra Mountain is a volcanic massif formed during the Quaternary period. It is composed of diverse volcanic formations, including trachytic, ordanchitic, basanitic, and ultra-vulcanian eruptions. The soil is predominantly volcanic in nature, characterized by its depth, clayey to clayey-silty texture, and black

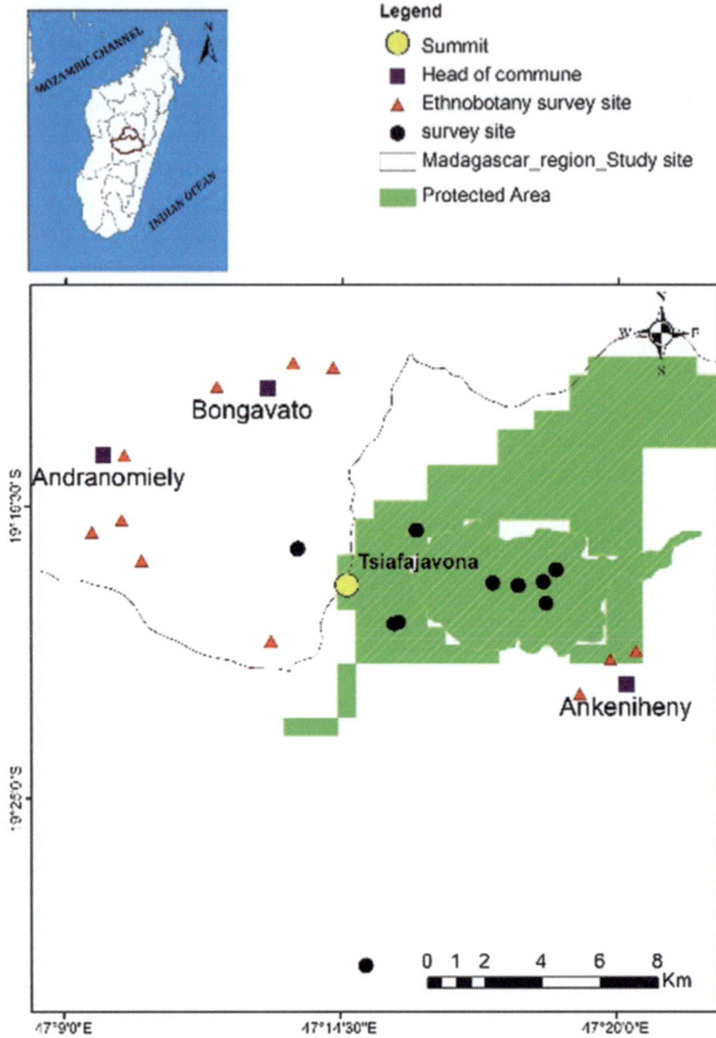

Fig. 15.1 Study area

coloration. These soils are known for their fertility but also for their tendency to become quite hard when dry (Rafidison et al., 2019; Zebrowski, 1971).

15.2.1.3 Climate

According to the Köppen climate classification, the area experiences a subtropical highland climate, characterized by two distinct seasons:

- Hot and Humid Season (November to April): This period is marked by high annual precipitation, ranging between 800 and 1000 mm, accompanied by warm and humid conditions.
- Cool and Dry Season (May to October): Cooler temperatures dominate this season, with significantly lower rainfall. The average monthly precipitation during this period is approximately 40 mm (Zebrowski, 1971).

The mountain creates a significant climatic dissymmetry between its eastern and western slopes, particularly in terms of rainfall distribution. The western slopes receive more rainfall from the northwest monsoon during the wet season, while the eastern slopes experience more frequent fine rains from the humid east trade winds during the dry season.

Average temperatures vary considerably with altitude. At lower elevations, the average temperature in January (warmest month) is about 26.7 °C, while in August (coolest month), it drops to around 7.1 °C. At higher elevations, temperatures can fall below freezing during winter nights (Raunet, 1981).

15.2.1.4 Vegetation and Biodiversity

The Ankaratra Mountain hosts a unique assemblage of vegetation types, largely influenced by its altitudinal gradient and climatic conditions:

- *Herbaceous Vegetation:* Species belonging to families such as Asteraceae (e.g., *Helichrysum* spp., *Senecio* spp.), Lamiaceae (e.g., *Tetradenia goudotii*), and Apiaceae (e.g., *Pimpinella* spp.) predominate the vegetation, exhibiting a high level of endemism. *Forest Formations:* These are relatively limited in extent and are primarily found at the base of the eastern slopes. Classified as evergreen humid forests, they host a rich diversity of endemic species. The upper canopy is rich in epiphytes, particularly orchids (Office Nationale de l'Environnement (ONE), 2007).
- *Reforested Areas:* Some parts of the mountain have been subject to reforestation efforts, primarily using species such as *Acacia dealbata*, *Eucalyptus robusta*, and *Cinnamomum camphora* (Rakotonandrasana et al., 2017).

15.2.1.5 Socio-Economic and Cultural Context

The area surrounding the Ankaratra Mountain is one of the most densely populated rural regions in Madagascar, with a population density reaching up to 125.6 people per square kilometer. The population of the Ankaratra Mountain, known as the Antakaratra, is part of the Merina ethnic group and maintains a strong connection to the Andriampenitra lineage. They primarily engage in agriculture and animal husbandry (Rakotonandrasana et al., 2017; Razafindralambo, 2016; Dez, 1967).

The mountain holds significant cultural and spiritual importance for the local communities. It houses two sacred sites where traditional healers and pilgrims conduct annual rituals and ceremonies. These sites, along with ancestral tombs and waterfalls within the forested areas, attract many traditional pilgrims during important dates in the Malagasy lunar calendar (Rakotonandrasana et al., 2017; Smith, 2007).

The combination of unique geological features, diverse microclimates, rich biodiversity, and cultural significance makes the Ankaratra Mountain an ideal location for ethnobotanical research, particularly in the realm of medicinal plants.

15.2.2 *Data Collection*

Our study followed a two-phase approach conducted during September–October 2018 and October–November 2019, corresponding to the dry and rainy seasons, respectively.

15.2.2.1 Botanical Inventory

We used the Braun-Blanquet method to inventory plant species in natural vegetation. Sixteen plots were established: five in forested areas (20 m x 50 m each) and eleven in savannah (20 m x 8 m each). In each plot, we recorded all plant species and created voucher specimens for later identification (Braun-Blanquet, 1965).

15.2.2.2 Ethnobotanical Surveys

Semi-structured interviews were conducted with 26 renowned traditional healers (24 men and 2 women) and selected in collaboration with local administrative officers and community elders. Each interview, lasting about 1–2 hours, was carried out in the local dialect, with translation assistance provided when required.

During the interviews, we showed voucher specimens of inventoried plants to the healers, asking about:

- Local names of the plants
- Medicinal uses
- Parts of the plant used
- Methods of preparation and administration
- Specific ailments treated

15.2.3 *Ethical Considerations*

This study received approval from the Advisory Scientific Committee of CNARP for the years 2018, 2019, and 2020. Research permits were secured from the Ministry of Environment and Sustainable Development, along with agreement from Madagascar's Access and Benefit Sharing focal point. Informed consent was obtained from all participants prior to interviews, ensuring confidentiality and their right to withdraw at any time.

Species Identification
Voucher specimens were identified using reference collections at the Department of Ethnobotany and Botany of CNARP, the Flora Department of the Botany and Zoology Parks in Tsimbazaza, and online resources from the Paris National Museum of Natural History. We consulted the Madagascar Vascular Plant Catalogue and the African Plants database for updated scientific names and phytogeographical distribution.

15.2.4 *Data Analysis*

15.2.4.1 Quantitative Analysis

We calculated two key metrics:

Frequency of Citation (FC)

$$\mathrm{FC}(\%) = (nu/Ni) \times 100$$

where *nu* is the number of informants citing a species and *Ni* is the total number of informants (Singh et al., 2012).

Fidelity Level (FL)

$$\mathrm{FL}(\%) = (Np/Nu) \times 100$$

where *Np* is the number of informants citing a species for a particular ailment and *Nu* is the total number of informants citing the species for any ailment (Friedman et al., 1986).

A high FL value indicates strong agreement among informants about a plant's use for a specific condition, suggesting it is the preferred species for treating that ailment in the local traditional medicine. This consensus can highlight promising candidates for further pharmacological investigation.

To ensure the reliability and significance of our findings, we implemented a methodological criterion in our analysis: only species that were independently cited by a minimum of three informants were included in the FL calculations and subsequent identification of important medicinal species.

15.2.4.2 Statistical Analysis

We performed analyses of variance (ANOVA) to assess the significance of differences in FC and FL based on species endemism and life forms. All statistical analyses were conducted using R software (version 4.0.3).

15.2.4.3 Database Creation

We created a comprehensive database in Microsoft Excel, compiling all collected data including species information, medicinal uses, FC, FL, and other relevant details. This database facilitated data sorting, proportion calculations, and the creation of visual representations of our findings.

15.3 Results

15.3.1 Diversity of Medicinal Plants

Our survey documented 203 plant species used for medicinal purposes in the Ankaratra Mountain region. These species are distributed across 172 genera and 82 families.

The medicinal plants identified during this study are detailed in the Annex section. This table includes scientific names, local vernacular names, and reported ethnomedicinal uses.

15.3.1.1 Family Diversity

Asteraceae was the most represented family with 52 species (25.6% of total).

Followed by Rubiaceae (14 species, 6.9%), Poaceae (12 species, 5.9%), Lamiaceae (11 species, 5.4%), and Apiaceae (10 species, 4.9%).

15.3.1.2 Genera Diversity

Helichrysum was the most diverse genus with 14 species (8.8% of total).

Followed by *Senecio* (7 species, 3.9%), *Salvia* (4 species, 2.3%), and *Pimpinella* and *Gerbera* (3 species each, 1.7%).

All species of Helichrysum, Salvia, and Pimpinella documented in this study are aromatic, a trait absent in the Senecio and Gerbera species.

15.3.1.3 Growth Forms

The medicinal plants documented in this study exhibited a diverse range of growth forms. Herbs were the most prevalent, comprising 108 species and representing 53.2% of the total medicinal flora. Trees formed the second largest group with 59 species, accounting for 29.1% of the documented plants. Shrubs were less common, with 22 species making up 10.8% of the total. Climbers were the least represented growth form, with 15 species constituting 7.4% of the medicinal plants inventoried (Fig. 15.2A).

15.3.1.4 Endemism

The study revealed a remarkably high level of endemism among the medicinal plants of the Ankaratra region. Endemic species, found nowhere else in the world outside of Madagascar, dominated the medicinal flora with 150 species, representing a

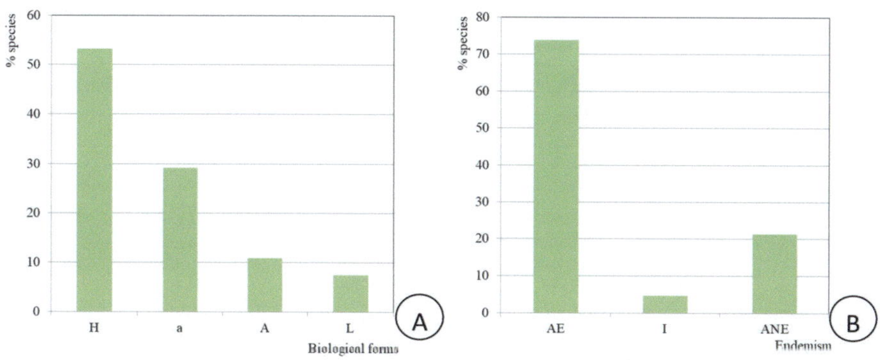

Fig. 15.2 Percentage of species according to the biological forms (**A**) and endemism (**B**); (H: grass, a: shrub, A: tree, L: liana, AE: indigenous endemic, I: introduced, NE: indigenous not endemic)

significant 73.9% of the total. Indigenous non-endemic species, which are native to Madagascar but also occur elsewhere, accounted for 38 species or 18.7% of the documented plants. Introduced species, those that have been brought to Madagascar from other regions, were the least common, with 15 species making up 7.4% of the medicinal plants recorded (Fig. 15.2B).

15.3.2 Most Cited Species in Traditional Medicine

The frequency of citation (FC) ranged from 4.2% to 100%. Table 15.1 presents the top 10 most cited species (Fig. 15.3).

Analysis of variance showed that endemic species were significantly more cited than non-endemic or introduced species ($p < 0.0001$) (Fig. 15.4a). Herbaceous and shrub species were more frequently cited than trees and lianas ($p < 0.0001$) (Fig. 15.4b).

Table 15.1 Top 10 most cited medicinal plant species in Ankaratra Mountain

Scientific names	Frequency of citation (%)
Tetradenia goudotii Briq.	100
Clematis pimpinellifolia Hook.	100
Billburttia capensoides Sales & Hedge	100
Micromeria flagellaris Baker	100
Agauria polyphylla Baker	100
Helichrysum benthamii R. Vig. & Humbert	100
Hubertia faujasioides (Baker) C. Jeffrey	100
Lycopodiella cernua (L.) Pic. Serm.	100
Pimpinella perrieri Sales & Hedge	95.8
Inulanthera brownii (Hochr.) Källersjö	95.8

Fig. 15.3 *Photos: Tetradenia goudotii* (**A**), *Billburttia capensoides* (**B**), *Helichrysum benthamii* (**C**), and Inulanthera brownii (**D**). (Crédit Photo: RAKOTONANDRASANA)

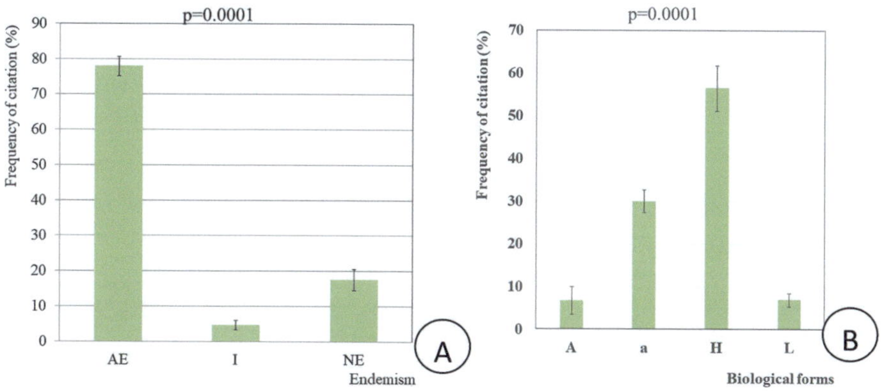

Fig. 15.4 Box plots showing frequency citation by endemism (**A**) and biological forms (**B**) (AE indigenous endemic, I: introduced, NE: indigenous not endemic, H: grass, a: shrub, A: tree, L: liana)

15.3.3 Fidelity Level Analysis

15.3.3.1 Important Species for Specific Diseases

The fidelity level (FL) of medicinal plant species in our dataset is ranging from 5.2% to 100%.

Eight endemic species stood out with a remarkable 100% FL for specific diseases:

- *Cantharanthus lanceus*, *Helichrysum bojerianum*, and *H. fulvescens* for infectious and parasitic diseases
- *Oncostemum bojerianum* and *Psiadia lucida* for symptoms and signs not elsewhere classified
- *Pteridium aquilinum* for skin diseases
- *Gerbera bojeri* for traumatic injuries and poisoning
- *Kniphofia ankaratrensis* and *Solanecio angulatus* for endocrine and metabolic diseases

Beyond these species with 100% FL, we identified numerous other plants with high fidelity levels for various disease categories. Table 15.2 presents the top five species with the highest FL for each of the 16 disease categories based on the International Classification of Diseases.

Table 15.2 Top five species with the highest fidelity level by MIP, GAP, INC, LTE, MPT, MAC and MAD diseases

Scientific name	Disease	Fidelity level (%)
Cantharanthus lanceus (Bojer ex A.DC.) Pichon	MIP	100
Helichrysum bojerianum DC.		100
Helichrysum fulvescens DC.		100
Helichrysum chermezonii Humbert		85.7
Helichrysum cf *fulvescens* DC.		80
Garcinia verrucosa Jum. & H. Perrier	GAP	40
Indigofera bojeri Baker		40
Hypericum japonicum Thunb.		33.3
Crassula ankaratrensis Desc.		25
Pyrostria italyensis (Cavaco) A.P. Davis & Govaerts		25
Stoebe pachyclada Humbert	INC	100
Aphloia theiformis (Vahl) Benn.		88.2
Croton mongue Baill.		87.5
Dioscorea hexagona Baker		83.3
Pityrogramma calomelanos (L.) Link		83.3
Gerbera bojeri (DC.) Sch. Bip.	LTE	100
Maesa lanceolata Forssk.		60
Plectaneia thouarsii Roem. & Schult.		50
Athrixia debilis DC		44.4
Micromeria sphaerophylla Baker		40
Pteridium aquilinum (L.) Kuhn	MPT	100
Kosteletzkya velutina Garcke		87.5
Asplenium cancellatum Alston		80
Indigofera thymoides Baker		66.7
Cyperus betafensis Cherm.		50
Hypericum japonicum Thunb.	MAC	66.7
Dombeyalucida Baill.		50
Melicopemadagascariensis (Baker) T.G. Hartley		50
Indigofera thymoides Baker		33.3
Pyrostria madagascariensis Lecomte		33.3
Euphorbia emirnensis Baker	MAD	66.7
Eragrostis cylindriflora Hochst.		66.7
Centella tussilaginifolia (Baker) Domin		57.1
Pimpinella ebracteata Baker		50
Aloe capitata Baker		50

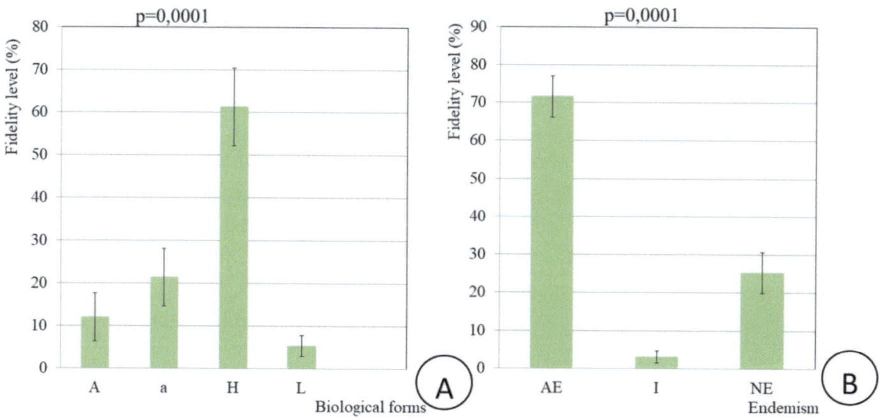

Fig. 15.5 Box plots showing fidelity level by endemism and biological forms (AE indigenous endemic, I: introduced, NE: indigenous not endemic, H: grass, a: shrub, A: tree, L: liana)

15.3.3.2 Relationship Between Fidelity Level, Endemism, and Biological Forms

To elucidate the relationships between fidelity level (FL), endemism, and biological forms of medicinal plants in the Ankaratra region, we conducted a comprehensive analysis of variance (ANOVA). The results, summarized in Fig. 15.5, reveal significant patterns that underscore the unique characteristics of the local ethnopharmacopoeia.

Endemism and Fidelity Level

Our analysis demonstrates a strong correlation between endemism and fidelity level (ANOVA, $p < 0.0001$, Fig. 15.5a). Endemic species exhibited significantly higher fidelity levels compared to both non-endemic indigenous species and introduced species. This finding highlights the paramount importance of endemic plants in the traditional medicinal practices of the Ankaratra region.

Biological Forms and Fidelity Level

The analysis also revealed a significant relationship between plant growth forms and fidelity levels (ANOVA, $p < 0.0001$, Fig. 15.5b). Herbaceous plants emerged as the most important group, showing substantially higher fidelity levels compared to other biological forms.

15.3.4 Disease Categories and Associated Medicinal Plants

We identified 16 disease categories based on the International Classification of Diseases. Figure 15.6 shows the distribution of plant species across these categories.

The category "Symptoms, signs and abnormal clinical and laboratory results, not classified elsewhere" (SYMP) was most prominent, with a 29.9% Frequency of Citation (FC) and 165 associated plant species (81.2% of total).

Infectious and parasitic diseases (FC 13.7%, 97 species, 47.7%) and digestive tract diseases (FC 10.7%, 93 species, 45.8%) followed.

In contrast, mental and behavioral disorders (2 species, 0.9%) and ear diseases (4 species, 1.9%) had minimal representation.

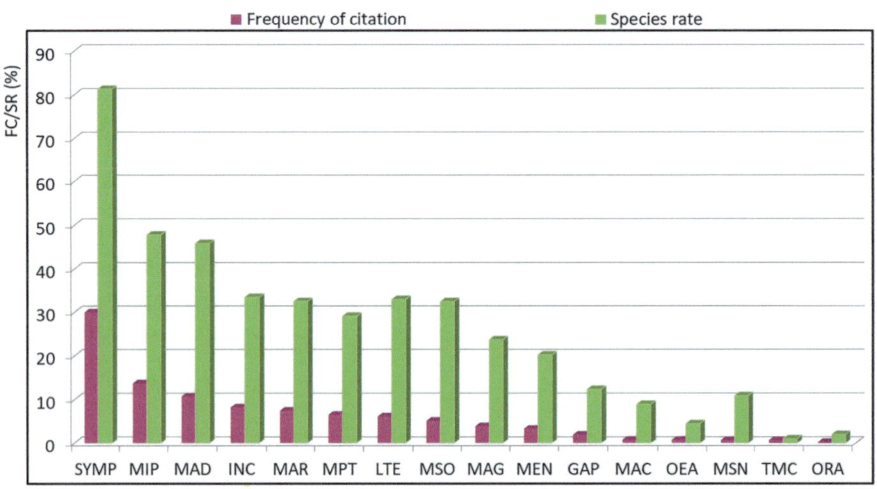

Fig. 15.6 Bar chart of disease categories and corresponding number of plant species. (SYMP: Symptoms, signs, and abnormal results of clinical and laboratory examinations, not elsewhere classified, MIP: Certain infectious and parasitic diseases, MAD: Disease of the digestive system, INC: Unclassifiable, MAR: Disease of the respiratory system, MPT: Diseases of the skin and subcutaneous cellular tissue, LTE: Traumatic injuries, poisonings and certain other consequences of external causes, MSO: Diseases of the osteoarticular system, muscles and connective tissue, MAG: Diseases of the genitourinary system, MEN: Endocrine, nutritional and metabolic diseases, GAP: Pregnancy, childbirth and the puerperium, MAC: Diseases of the circulatory system, OEA: Diseases of the eye and its appendices, MSN: Diseases of the nervous system, TMC: Mental and behavioral disorders, ORA: Diseases of the ear and mastoid process)

15.4 Discussion

This research constitutes the first medicinal plant survey on high summits and grasslands in Madagascar, and the third inventory conducted in key biodiversity areas after the *Baie de Rigny* and the *Zafimaniry*. Our study provides a comprehensive analysis of the medicinal plants used in the Ankaratra Mountain region, revealing a rich biocultural heritage characterized by high plant diversity, exceptional endemism, and sophisticated traditional knowledge.

15.4.1 Diversity and Abundance of Medicinal Plants

We recorded 203 plant species utilized for medicinal purposes, representing 172 genera and 82 families. This diversity is particularly noteworthy given the relatively limited geographic area of the study. Comparing our findings with previous studies in Madagascar offers valuable insights:

The number of species identified is similar to that found in the Zafimaniry region (Rakotondrafara et al., 2018), despite having only one-eighth the number of informants.

The species-to-informant ratio in Ankaratra (8.4 species per informant) is more than three times that of the Zafimaniry region (2.4 species per informant).

Our study focused solely on traditional healers, suggesting a concentration of medicinal plant knowledge among these specialists.

These comparisons underscore the remarkable depth of medicinal plant knowledge unique to the Ankaratra region. The high diversity may be attributed to several factors:

The high-altitude environment supports the findings of Byg et al. 2010, which suggest that villages located at higher elevations tend to possess greater knowledge of, and make more frequent use of, medicinal plants.

The Ankaratra Massif is home to sacred sites (Corrèges, 2016) that act as sanctuaries for preserving high-quality medicinal plants (Kunwar et al., 2019).

The unique ecological conditions of the area may have contributed to local endemism and the biosynthesis of specific secondary metabolites (Verpoorte et al., 2005).

Notably, 69 out of the 203 surveyed plants are new additions to the Malagasy ethnopharmacopoeia, underscoring the importance of the Ankaratra Mountain as a reservoir of both medicinal plants and traditional knowledge.

15.4.2 *Taxonomic Patterns and Growth Forms*

15.4.2.1 Family and Genera Diversity

The most diversified families in our study (Asteraceae, Rubiaceae, Poaceae, Lamiaceae, and Apiaceae) largely correspond with the most prolific families in the Madagascar flora (Gautier et al., 2012). This aligns with Moerman and Estabrook's (Moerman & Estabrook, 2003) observation that the proportion of medicinal species in a family is positively associated with the total species diversity within that family. .

The prominence of Apiaceae (11 species, approximately half of Madagascar's total Apiaceae species (Wyk et al., 2013)) is particularly noteworthy, reflecting the family's tendency to diversify in mountainous regions (Sales et al., 2004).

15.4.2.2 Biological Forms

Herbs dominated our findings, comprising 53.2% of the medicinal plant species. This prevalence of herbaceous plants in traditional medicine is consistent with findings from other ethnobotanical studies (Giday et al., 2009) and may be related to:

Ease of collection and processing (Tugume et al., 2016)
Rapid growth and regeneration, supporting sustainable use (Sher et al., 2015)
Often higher concentrations of bioactive compounds compared to woody plants (Ganjhu et al., 2015)

The distribution of biological forms (herbs: 53.2%, trees: 29.1%, shrubs: 10.8%, liana: 7.4%) reflects both the ecological characteristics of the region and the accessibility of different plant types for medicinal use. The predominance of herbs contrasts with studies in forested areas of Madagascar, where trees and shrubs are more prevalent (Rakotonandrasana et al., 2017).

15.4.3 *Endemism and Traditional Use*

A striking feature of our findings is the high level of endemism (73.9%) among the medicinal plants. This exceptional rate of endemism aligns with Madagascar's status as a biodiversity hotspot (Myers et al., 2000) and has several important implications:

Unique Pharmacological Potential: Endemic plants, having evolved in isolation, may possess unique chemical profiles that could lead to new drug discoveries (Kingston, 2011).

Traditional Knowledge and Endemic Species: Our study reveals that endemic species not only are more frequently used but also have higher fidelity levels in traditional medicine. This corroborates findings by Lyon and Hardesty (Lyon &

Hardesty, 2012) in the Anosy region and indicates that indigenous knowledge of medicinal plants is primarily held by traditional healers.

Ecological and Cultural Factors: The preference for endemic species may be influenced by the high-altitude environment of Ankaratra, supporting Byg et al.'s observation that higher altitude villages tend to use more autochthonous species (Byg et al., 2010). The presence of sacred sites in the massif further emphasizes the cultural importance of these endemic plants (Kunwar et al., 2019).

Conservation Implications: The reliance on endemic species for traditional medicine underscores the critical importance of conservation efforts in the region. Loss of these species would not only impact biodiversity but also threaten local healthcare practices and cultural heritage (Kingston, 2011).

15.4.4 Most Cited Species and Their Pharmacological Potential

The most frequently cited species (e.g., *Tetradenia goudotii*, *Clematis pimpinellifolia*, *Billburttia capensoides*) deserve special attention. These plants are often rich in secondary metabolites or biologically active compounds, which validate their traditional medicinal applications:

Tetradenia goudotii, one of the most frequently referenced species in our study, has demonstrated notable antimicrobial properties. Phytochemical analysis identified three key phenolic compounds: eriodictyol, rosmarinic acid, and cynaroside (Kingston, 2011). These compounds contribute to the plant's antimicrobial activity, which has been demonstrated through in vitro studies. Andrianjakaniaina et al. (2017) elucidated the antimicrobial properties of *T. goudotii*. Their study found that:

The dichloromethanic leaf extract exhibits bacteriostatic effects against *Escherichia coli* and *Yersinia enterocolitica*.

The butanolic leaf extract shows bactericidal effects on *Streptococcus pneumoniae*, *Yersinia enterocolitica*, and *Staphylococcus aureus*.

Both extracts demonstrate bacteriostatic effects on *Shigella flexneri* and *Bacillus cereus*.

In traditional medicine, *T. goudotii* is primarily used for treating angina, anthrax, and abscesses (Rakotonandrasana et al., 2017). Our study also revealed its use in treating abdominal colic, expanding our understanding of its traditional applications.

Billburttia capensoides is another species of particular interest due to its rich phytochemical profile and diverse traditional uses. Terpinolene has been identified as the predominant compound (40%) in the essential oil extracted from the aerial parts of *B. capensoides* (Stephan R. Rakotonandrasana et al., 2019), a compound known for its sedative properties (Ito & Ito, 2013). Further phytochemical investigations have identified two coumarins (xanthotoxin and imperatorin) and three prenylated chromones, two of which are novel compounds (Andrianjakaniaina et al., 2021). These findings are significant because:

Imperatorin and xanthotoxin have been reported to demonstrate antimicrobial and antiproliferative properties (Sumorek-Wiadro et al., 2024; Batiha et al., 2023).
Prenylated chromones have shown anti-HIV, antiproliferative, and antiinflammatory activities (Liu et al., 2020; Fu et al., 2020).
Traditionally, *B. capensoides* is used for fatigue relief and treating syphilis (Rakotonandrasana et al., 2017). Our study has expanded this list to include relief of abdominal pain and stomach aches, highlighting the plant's importance in gastrointestinal treatments.

Helichrysum benthamii is notable for its exceptionally high concentration of alpha-pinene (52%) in its essential oil. This concentration is rare within aromatic plants and has been identified for the first time at such levels in *H. benthamii* (Rabehaja et al., 2020). Several studies have focused on alpha-pinene to establish its biological profile, highlighting its properties such as antidepressant and anxiolytic effects, neuroprotective potential (especially in stroke and ischemia models), analgesic effects in inflammatory, migraine-associated, and neuropathic pain, and its potential for cognitive enhancement (Salehi et al., 2019; Alqudah et al., 2024; Weston-Green et al., 2021).

In traditional medicine, *H. benthamii* is used as a muscle relaxant and for treating anxiety (Rakotonandrasana et al., 2017). Our study has also identified its use in facilitating blood circulation, which aligns with its reported biological activities.

These findings highlight the potential for targeted pharmacological studies based on traditional knowledge.

15.4.5 Fidelity Level Analysis and Disease Categories

The fidelity level (FL) analysis provides crucial insights into the specificity of plant use for particular ailments. Notable findings include:

Species with 100% FL for specific diseases (e.g., *Cantharanthus lanceus* for infectious diseases, *Pteridium aquilinum* for skin diseases) represent prime candidates for targeted drug discovery efforts.
Some species show different uses in Ankaratra compared to other parts of Madagascar. For instance, *Drosera madagascariensis* is used for mental disorders in Ankaratra but for respiratory issues elsewhere in Madagascar (Novy, 1997).
Our study has made significant contributions to the Malagasy ethnopharmacopoeia through the discovery of new medicinal uses for known species and the identification of previously undocumented medicinal plants. This highlights the importance of conducting ethnobotanical research in biodiversity hotspots like Madagascar, as it can provide valuable insights and expand knowledge on medicinal plant resources.

We discovered novel medicinal applications for several known plant species in the Ankaratra region. For instance:

Cantharanthus lanceus was found to be used for treating abdominal colic, a use not previously documented in the scientific literature.

Drosera madagascariensis, known for its use in respiratory issues in other parts of Madagascar (Novy, 1997), is employed for mental disorders in Ankaratra.

These findings align with previous research demonstrating how local ethnobotanical studies can reveal new medicinal uses for known plants in Madagascar (Randrianarivelojosia et al., 2003).

15.4.6 Identification of New Medicinal Plants

Our study also identified previously undocumented medicinal plants. Sixty-nine (69) species are new in the Malagasy ethnopharmacopoeia. The identification of these species as medicinal plants adds to the growing body of knowledge about Madagascar's rich ethnobotanical heritage. This discovery is consistent with other studies that have highlighted the potential for identifying new medicinal species through focused ethnobotanical research in Madagascar's diverse ecosystems (Norscia & Borgognini-Tarli 2006a).

The distribution of plants across disease categories provides insights into local health priorities and the strengths of traditional medicine in addressing them. The prominence of plants used for symptomatic treatment and infectious diseases likely reflects common health issues in the region.

15.4.7 Traditional Knowledge and Conservation

The rich ethnomedicinal knowledge documented in this study underscores the importance of traditional ecological knowledge in biodiversity conservation and healthcare. However, it also poses challenges:

Overharvesting Risks: High-demand endemic species may face increased harvesting pressure, potentially leading to population declines.

Knowledge Erosion: As traditional lifestyles change, there is a risk of losing valuable medicinal plant knowledge if not properly documented and transmitted to younger generations.

Benefit Sharing: Ensuring that local communities benefit from any commercial developments arising from their traditional knowledge is crucial for ethical research and sustainable development.

15.4.8 Future Research Directions

This study lays the groundwork for several promising research avenues:

Phytochemical and Pharmacological Studies: Prioritize high-fidelity endemic species for detailed chemical analysis and bioactivity screening.

Ecological Research: Investigate the population dynamics and regeneration patterns of key medicinal plants to inform sustainable harvesting practices.

Clinical Validation: Design rigorous clinical trials to validate the efficacy and safety of traditional remedies, particularly those with high fidelity levels.

Comparative Studies: Extend the research to other regions of Madagascar to build a comprehensive picture of the island's ethnomedicinal practices and plant diversity.

Interdisciplinary Collaboration: Foster partnerships between traditional healers, botanists, ecologists, and pharmacologists to holistically study and preserve Madagascar's medicinal plant heritage.

15.4.9 Limitations

While comprehensive, this study has some limitations to consider:

Complex Factors in Plant Selection: The reasons behind the choice of specific plants for medicinal use are complex and dynamic, involving historical, cultural, and ecological factors (Akerreta et al., 2007; Savo et al., 2015; Heinrich et al., 2006). Our understanding of these processes remains rudimentary and warrants further investigation.

15.5 Conclusion

This investigation into the medicinal plants of the Ankaratra Mountain region highlights a remarkable biocultural heritage, marked by rich plant diversity, exceptional endemism, and advanced traditional knowledge. The close connection between endemic species and their medicinal applications emphasizes the unique significance of Madagascar's biodiversity and underscores the critical need for its conservation. .

The findings have significant implications for biodiversity conservation, drug discovery, and sustainable development. They highlight the potential for integrating traditional knowledge with modern scientific approaches to address health challenges and preserve valuable ecological and cultural resources.

As global biodiversity faces unprecedented threats, studies like this serve as crucial reminders of the inextricable links between human well-being, cultural knowledge, and the natural world. Conservation strategies in Madagascar and similar biodiversity hotspots must consider these complex relationships to be truly effective and sustainable.

Acknowledgments We sincerely thank the NGO Vondrona Ivon'ny Fampandrosoana (VIF) Manjakatompo, the promoter of the New Protected Area (NAP) Manjakatompo, for their open and collaborative support. We are also grateful to the local traditional healers and the mayors of the municipalities of Andranomiely, Ankeniheny, Mahatsinjo Est, and Manalalondo for their warm hospitality and cooperation.

Annex: Plant Used Medicinally in Ankaratra Massif (E: Indigenous endemic species, I: Introduced species, NE: Indigenous non endemic species, H: Herbaceous, A: Tree, a: Shrub)

Voucher	Local names	*Scientific names* (Family), endemism, growth forms	Therapeutic uses
RLL 1511	Fotsimavo, Kafotsimavo	*Abrahamia* sp. (ANACARDIACEAE), E, a	Abdominal colic, depurative, fatigue: drink the decoction of the leafy twigs
			Furunculosis, kwashiorkor, disease prevention, or fatigue relief: drink the leaf decoction
			Cough: macerate the leaves and drink
RLL 1506	Vatofosa, Tangongo, tangongolahy	*Achyranthes aspera* L. (AMARANTHACEAE), I, H	Gonorrhea, disease prevention, or fatigue relief; prostate: drink the leaf decoction
			Food supplement: cook the leaves and eat as a meal
			Wound and abscess: grind the leaves and make a poultice
			Persistent cough, urticaria: drink the decoction of the leafy twigs
RLL 1577	Kavodiana, Kavodiana antanety	*Agauria polyphylla* Baker (ERICACEAE), E, a	Angina, plague, abdominal colic: crush the dried leaves, swallow the powder
			Anthrax: Apply the powder of the crushed leaves after drying on the skin

(continued)

Voucher	Local names	*Scientific names* (Family), endemism, growth forms	Therapeutic uses
			Furunculosis, prevention of diseases or relief of fatigue, gastralgia; poisoning: drink the decoction of the leafy twigs
			Intestinal worm: drink root decoction
ST1322	Tongahoazy	*Ageratina riparia* (Regel) R.M. King & H. Rob. (ASTERACEAE), I, H	Abdominal colic: drink the decoction of the leafy twigs
RLL 1663	Anatsingita, Anangita, Tsingita, Karotitanety	*Agrocharis melanantha* Hochst (APIACEAE), NE, H	Make fertile: crush the leaves, drink the juice
			Food supplement, prevention of diseases or relief of fatigue; muscle relaxant: decoct the leaves and drink
RLL 1529	Vahotsahondra, Sahondra	*Aloe capitata* Baker (ASPHODELACEAE), E, H	Purgative, gastralgia, cough: grate the pulp of the leaves, drink
			Scabies: grate the pulp of the leaves, apply to the skin
			Dandruff: wash the hair with the pulp of the leaves
			Psychosis: burn leaves and inhale
RLL 1783	Tsevoka, Vendramanga, Antsoro	*Andropogon imerinensis* Bosser (POACEAE), E, H	Anthrax, chasing evil spirits: burning the leaves and inhaling
			Jaundice: macerate the leaves and drink
			Muscle relaxant: drink the leaves decoction
			Disease prevention or fatigue relief: decoct the roots and drink
			Cough, vertigo: drink the decoction of the whole plant
RLL 1720	Randemy, Ramasoandro	*Anthocleista madagascariensis* Baker (GENTIANACEAE), E, A	Loss of consciousness: crush the leaves, apply the poultice to the head
			Prevention of diseases or relief of fatigue: decoct leafy twigs and drink

(continued)

Voucher	Local names	*Scientific names* (Family), endemism, growth forms	Therapeutic uses
RLL 1761	Hazonorana	*Anthospermum emirnense* Baker (RUBIACEAE), E, a	Abdominal colic: drink a decoction of the roots
			Joint pain: drink the decoction of the leafy twigs
			Scabies: wash wounds with a decoction of leafy twigs
			Cough: chew or macerate the roots and drink
RLL 1510	Voafotsy, Ravimboafotsy	*Aphloia theiformis* (Vahl) Benn. (APHLOIACEAE), NE, A	Conjunctivitis: wash the eyes with a decoction of leafy reeds
			Gastralgia, disease prevention, or fatigue relief, rehydrating during illness: drink the decoction of leafy twigs
RLL 1737	Vazahakely	*Aristea kitchingii* Baker (IRIDACEAE), E, H	Anthrax: crush the fruit and apply the poultice to the skin
			Joint pain, Jaundice, disease prevention or fatigue relief, hives: drink whole plant decoction
			Prevention of disease or relief of fatigue
			Vitiligo: crush the leaves and apply to the skin
RLL 1780	Tsiota, Velokafo	*Asplenium apertum* C. Chr. (ASPLENIACEAE), E, H	Allergy: drink the decoction of the leaves
			Anthrax, urticaria: burn leaves and take a steam bath
			Drive out evil spirits: take a bath with the decoction of the whole plant
RLL 1711	Apangavy	*Asplenium cancellatum* Alston (ASPLENIACEAE), NE, H	Anthrax, hunt bad spirits, urticaria: burn leaves and take a steam bath
			Plague: drink a decoction of the leaves
RLL 1688	Apangavy	*Asplenium lividum* Mett. ex Kuhn (ASPLENIACEAE), NE, H	Anthrax, hunt bad spirits, urticaria: burn leaves and take a steam bath

(continued)

Voucher	Local names	*Scientific names* (Family), endemism, growth forms	Therapeutic uses
			Disinfectant after burial: decoct the leaves and take a bath
			Decreased visual acuity: burn the leaves and inhale
			Muscle relaxant: decoct the leaves and drink
RLL 1652	Voadimbana, Marorantsana	*Astropanax myrianthus* (Baker) Lowry, G.M. Plunkett, Gostel & Frodin (ARALIACEAE), E,a	Headache, poisoning, gastralgia, prevention of diseases or relief of fatigue: decoct leafy twigs and drink
			Gastralgia
			Insect bite: drink the decoction of the stems
RLL 1541	Kitotompanonto, Kelimandrà, Tsilakinahitra	*Athrixia debilis* DC. (ASTERACEAE), E, H	Asthenia, abdominal colic, gastralgia, low back pain, intestinal worm: decoct leafy twigs and drink
			Diarrhea: crush the leafy twigs and drink the juice
			Wounds and abscesses: chewing the leaves and making a poultice
			Cough: chew the leaves and swallow the juice
RLL 1545	Sarikaravola, Volotaratsintsina	*Billburttia capensoides* Sales & Hedge (APIACEAE), E, a	Headache, epistaxis: dry leaves and smoke
			Abdominal colic, convulsion, fever, furunculosis, gastralgia, jaundice, marasmus, liver pain, plague, disease prevention or relief of fatigue, soporific, cough: decoct leafy twigs or leaves and drink
RLL 1599	Mihorokorona	*Biophytum umbraculum* Welw. (OXALIDACEAE), I, H	Deodorant: take a bath with the decoction of the whole plant
			Prevention of diseases or relief of fatigue, cough: decoct the whole plant and drink
RLL 1514	Sevalahy, Longolongo	*Buddleja fusca* Baker (SCROPHULARIACEAE), E, a	Anthrax, chasing evil spirits: decoct the leaves and inhale

(continued)

Voucher	Local names	*Scientific names* (Family), endemism, growth forms	Therapeutic uses
			Abdominal colic: chew the leaves and swallow the juice
			Depurative: consume ripe fruits
			Furunculosis, disease prevention or fatigue relief, persistent cough: drink the leaf decoction
			Gastralgia: infuse the leaves and drink
RLL 1583	Sevavavy, seva	*Buddleja madagascariensis* Lam. (SCROPHULARIACEAE), E, a	Gonorrhea, dysentery, skin rash, disease prevention, or fatigue relief: decoct the leaves and drink
			Tooth decay: apply the fruits to damaged teeth
			Wound and abscess: make a poultice with the stem
RLL 1623	Vonenina	*Catharanthus lanceus* (Bojer ex A.DC.) Pichon (APOCYNACEAE), E, H	Abdominal colic, Diarrhea: grate or grind the root infuse and drink
RLL 1665	Mampahafaka, Hazomafaitra	*Cassinopsis madagascariensis* Baill. (ICACINACEAE), E, A	Abdominal colic, joint pain, poisoning, gastralgia, prevention of diseases or relief of fatigue: decoct the leaves or leafy twigs and drink
RLL 1746	Tsiandrikandriana, Fotsiavadika	*Catatia cordata* Humbert (ASTERACEAE), E, H	Gonorrhea, abdominal colic, constipation, diarrhea, gastralgia, hemorrhoid, liver pain, metrorrhagia, muscle relaxant, disease prevention or relief of fatigue, cough, vitiligo: decoct leaves or leafy twigs and drink
			Drive out evil spirits: inhale with the decoction of the leaves or the smoke of the leafy twigs
			Hair loss: wash the head with the infusion of leafy twigs

(continued)

Voucher	Local names	*Scientific names* (Family), endemism, growth forms	Therapeutic uses
			Anthrax, furunculosis, flu: decoct leaves or leafy twigs, drink and bathe
RLL 1754	Talapetraka	*Centella tussilaginifolia* (Baker) Domin (APIACEAE), NE, H	Gastralgia, liver pain, prevention of illnesses or relief of fatigue, anticoagulant: decoct the leaves and drink
			Eye trauma: crush the leaves and instill the juice
RLL 1784	Anapoza	*Cineraria anampoza* (Baker) Baker (ASTERACEAE), E, H	Aphrodisiac, convulsion, gastralgia, marasmus, prevention of diseases or relief of fatigue, dry cough: decoct the leaves and drink
			Food supplement: consume the leaves after cooking
RLL 1647	Voriravina	*Cissampelos pareira* L. (MENISPERMACEAE), NE, L	Gonorrhea, diarrhea, furunculosis, disease prevention, or fatigue relief: decoct the leafy twigs and drink
			Headache: decoct leafy twigs and inhale
RLL 1517	Sarobaratra	*Cissus microdonta* (Baker) Planch. (VITACEAE), NE, L	Tooth decay: crush the roots and apply to the teeth
			Drive out evil spirits: take a bath and inhale the decoction of the whole plant
			Psychosis: burn the stems and inhale
			Cough: decoct the leaves and drink
RLL 1664	Fotsiavadika	*Claoxylopsis perrieri* Leandri (EUPHORBIACEAE), E, a	Canker sore: crush the leaves and apply the juice to the wound
			Poisoning, gastralgia, plague, illnesses, or relief of fatigue: drink the leaf decoction
			Cough: crush the leaves, swallow the juice

(continued)

Voucher	Local names	*Scientific names* (Family), endemism, growth forms	Therapeutic uses
RLL 1610	Farimatilahy	*Clematis mauritiana* Lam. (RANUNCULACEAE), NE, L	Gonorrhea, fever, disease prevention, or fatigue relief: decoct leafy twigs and drink
RLL 1542	Tranokanohano, Fotsivolomanoka, Sarikaravola	*Clematis pimpinellifolia* Hook. (RANUNCULACEAE), E, H	Intestinal bloating, abdominal colic, diarrhea, prevention of disease or relief of fatigue, persistent cough: decoct the leaves and drink
			Headache, flu: burn or decoct leafy twigs and inhale
			Furunculosis, scabies, wounds and abscesses, edema, inflammation: crush the leaves and apply to the skin
			Plague: dry the leafy twigs under fire then crush, swallow the powders, or macerate the leafy twigs and drink
RLL 1527	Hafontsokina	*Commersonia madagascariensis* (Baker) C. F. Wilkins et Whitlock (MALVACEAE), E, H	Anthrax, edema, inflammation: scratching the root or stem, making a poultice
			Abdominal colic, disease prevention or fatigue relief, cough: decoct leafy twigs and drink
RLL 1625	Tanondrika	*Conyza garnieri* Klatt (ASTERACEAE) NE, H	Abdominal colic: dry the whole plant in a fire, crush, and swallow
RLL 1743	Kelimamenotany	*Crassula ankaratrensis* Desc. (), E, H	Avoid premature abortion: the infusion of the whole plant is used to massage the stomach
			Prevention of diseases or relief of fatigue: drink the decoction of the leafy twigs
RLL 1643	Hazomiavona	*Croton jennyanus* Gris ex Baill (EUPHORBIACEAE), E, a	Headache: decoct leafy twigs and inhale
			Plague: decoct the leafy twigs and inhale

(continued)

Voucher	Local names	*Scientific names* (Family), endemism, growth forms	Therapeutic uses
			Wounds and abscesses: decoct leafy twigs, drink and bathe
			Prevention of diseases or relief of fatigue: drink the decoction of the leafy twigs
RLL 1656	Mongo	*Croton mongue* Baill. (EUPHORBIACEAE), E, A	Headache: decoct leafy twigs and inhale
			Drive out evil spirits: decoct leafy twigs, drink and inhale, or burn leaves and inhale
RLL 1787	Dity	*Cyanthillium cinereum* (L.) H. Rob. (ASTERACEAE), NE, a	Inappetence, insect bite, disease prevention, or fatigue relief: decoct leafy twigs and drink
RLL 1714	Pelabody lahy	*Cyperus balfouri* C.B. Clarke (CYPERACEAE) NE, H	Anthrax, furunculosis, liver pain, muscle relaxant: decoct the leaves, drink and take a bath
			Liver pain, disease prevention or fatigue relief, cough: Ddecoct the leaves and drink
RLL 1770	Tsipoapoaka	*Cyperus betafensis* Cherm. (POACEAE), E, H	Anthrax, muscle relaxant: decoct the leaves, take a bath, and drink
RLL 1645	Laingomaimbo	*Danais microcarpa* Baker (RUBIACEAE), E, L	Gastralgia, liver aches, prevention of diseases, or relief of fatigue: decoct the leaves and drink
			Wound and abscess: grind the leaves and make a poultice
RLL 1565	Hazondreniomby	*Dialypetalum compactum* Zahlbr. (LOBELIACEAE), E, H	Intestinal bloating, gonorrhea, constipation, disease prevention or fatigue relief, cough: decoct the leaves and drink
RLL 1684	Tongotrombibe	*Dianella ensifolia* (L.) DC. (ASPHODELACEAE), NE, H	Gonorrhea, gastralgia, marasmus, prevention of diseases or relief of fatigue, postpartum care, mixing disorder: decoct the whole plant and drink

(continued)

Voucher	Local names	*Scientific names* (Family), endemism, growth forms	Therapeutic uses
			Furunculosis; decoct the whole plant, drink and take a bath
RLL 1651	Tsitsihana	*Dicoryphe buddleoides* Baker (HAMAMELIDACEAE), E, A	Gastralgia, disease prevention, or fatigue relief
			Wounds and abscesses: Ddecoct the leaves and drink
RLL 1546	Kitsahotsaho	*Dioscorea hexagona* Baker (DIOSCOREACEAE), E, H	Infant nutrition: eaten raw tubers
			Prevention of diseases or relief of fatigue, cough: decoct the tubers and drink
RLL 1699	Kijejalahy, Belohakely	*Distephanus garnierianus* (Klatt) H. Rob. & B. Kahn (ASTERACEAE), E, a	Decreased visual acuity: decoct leaves, wash eyes, and drink
			Gastralgia, liver pain, prevention of illnesses, or relief of fatigue: decoct the leaves and drink
RLL 1582	Kijejalahy, Belohakely	*Distephanus glutinosus* (DC.) H. Rob. & B. Kahn (ASTERACEAE), E, a	Gonorrhea, abdominal colic, diarrhea, furunculosis, indigestion, prevention of diseases or relief of fatigue: decoct the leafy twigs and drink
RLL 1723	Anadraisoa vavy	*Distephanus lastellei* (Drake) H. Rob. & B. Kahn (ASTERACEAE), E, a	Furunculosis, prevention of diseases or relief of fatigue, cough: drink the decoction of the leaves
RLL 1515	Hafomainty	*Dombeya ankaratrensis* Arènes (MALVACEAE), E, a	Gonorrhea, marasmus, disease prevention or fatigue relief, cough: decoct the leaves and drink
RLL1655	Hafobalo	*Dombeya lucida* Baill. (MALVACEAE), E, A	Conjunctivitis: macerate the leaves and wash the eyes
			Gastralgia, liver pain, muscle relaxant, disease prevention or fatigue relief: decoct leafy twigs and drink
			Dandruff: wash the hair with the leaf decoction

(continued)

Voucher	Local names	*Scientific names* (Family), endemism, growth forms	Therapeutic uses
RLL 1584	Mahatanando	*Drosera madagascariensis* DC. (DROSERACEAE), NE, H	Enuresis, lumbago, metrorrhagia: decoct the whole plant and drink
			Cough: infuse the whole plant and drink
RLL 1672	Sana, Tsitohaina	*Elaeocarpus hildebrandtii* Baill (ELAEOCARPACEAE), E, A	Joint pain, lower back pain, muscle relaxant, disease prevention or fatigue relief: Ddrink the leaf decoction
RLL 1570	Apalitany, Lelanomby	*Elaphoglossum* cf *poolii* Christ (ELAPHOGLOSSACEAE), E, H	Drive out evil spirits, urticaria: burn the whole plant and inhale
			Diarrhea, furunculosis, gastralgia, prevention of diseases or relief of fatigue: decoct leafy twigs and drink
RLL 1535	Antsoro	*Elionurus tristis* Hack. (POACEAE), E, H	Psychosis: infuse the leaves and take a bath
			Cough: decoct the whole plant and drink
RLL 1789	Takasina, Tanterakala	*Embelia concinna* Baker (PRIMULACEAE), E, L	Aphrodisiac, intestinal bloating, abnormal fetal development, furunculosis, gastralgia, indigestion, disease prevention or fatigue relief: decoct leafy twigs and drink
RLL 1607	Tsitoitoina	*Equisetum ramosissimum* Desf. (EQUISETACEAE), NE, H	Intestinal bloating, disease prevention or fatigue relief: decoct the whole plant and drink
			Convulsion: burn the aerial part and swallow the ashes
			Inappetence: dry and grind the aerial part and swallow the powders
RLL 1767	Tsirodroka	*Eragrostis cylindriflora* Hochst. (POACEAE), NE, H	Purgative, gastralgia, kwashiorkor: decoct the aerial parts and drink
			Fatigue: take a bath with a decoction of the aerial parts
RLL 1700	Anjavidy	*Erica baroniana* Dorr & E.G.H. Oliv. (ERICACEAE), E, a	Drive out evil spirits: burn the leafy twigs and take a steam bath

(continued)

Voucher	Local names	*Scientific names* (Family), endemism, growth forms	Therapeutic uses
RLL 1709	Anjavidy	*Erica madagascariensis* (H. Perrier) Dorr & E.G.H. Oliv. (ERICACEAE), E, a	Drive out evil spirits, urticaria: burn the whole plant and inhale
RLL 1698	Votsehatra	*Erythroxylum gerrardii* Baker (ERYTHROXYLACEAE), E, a	Jaundice, plague, cough: decoct the leaves and drink
RLL 1756	Tsimitrosa	*Euphorbia emirnensis* Baker (EUPHORBIACEAE), NE, H	Purgative: decoct the leaves and drink
RLL 1654	Fanalavorika	*Garcinia verrucosa* Jum. & H. Perrier(CLUSIACEAE), E, A	Poisoning: infuse the stems and drink
			Galactogen: mix powders from the stem with the meal
			Gastralgia, oxytocic, liver ailments, prevention of diseases or relief of fatigue: Ddecoct leafy twigs and drink
RLL 1666	Kabanja, Velon'arivo taona	*Gastrorchis francoisii* Schltr. (ORCHIDACEAE), E, H	Marasmus, disease prevention, or fatigue relief: decoct the aerial part and drink
RLL 1749	Talafoitra vavy, Talafoitra	*Geranium andringitrense* H. Perrier (GERANIACEAE), E, H	Abdominal colic, gastralgia, indigestion, kwashiorkor, plague, prevention of diseases, or relief of fatigue: decoct the leafy twigs and drink
			Food supplement: consume the leaves after cooking
			Cough: crush the leaves, swallow the juice
RLL 1620	Madiorano	*Gerbera bojeri* (DC.) Sch. Bip. (ASTERACEAE), E, H	Scorch: burn the leaves and apply the ashes to the leaves
RLL 1543	Fotsiavadika	*Gerbera hypochaeridioides* Baker (ASTERACEAE), E, H	Abdominal colic, indigestion, kwashiorkor, prevention of diseases or relief of fatigue, dry cough: decoct the whole plant and drink
			Furunculosis; decoct the whole plant, drink, and take a bath
			Wound and abscess: dry and grind the leaves, apply to the wound

(continued)

Voucher	Local names	*Scientific names* (Family), endemism, growth forms	Therapeutic uses
RLL 1620	Fotsiavadika	*Gerberapiloselloides* (L.) Cass., NE, H	Allergy, liver pain, gastralgia, intestinal worm: decoct the leaves and drink
			Canker sore, vitiligo: grind the leaves and apply to the skin
RLL 1585	Andriankazo	*Gnidia gnidioides* (Baker) Domke (THYMELIACEAE), E, a	Depurative, fortifying, prevention of diseases or relief of fatigue: decoct the leafy twigs and drink
			Plague: crush the leafy twigs, then decoct and drink
RLL 1608	Roarivotaona	*Gouania pannigera* Tul. (RHAMNACEAE), E, L	Prevention of diseases or relief of fatigue: drink the decoction of the leafy twigs
RLL 1678	Ambiaty	*Gymnanthemum appendiculatum* (Less.) H.Rob. (ASTERACEAE), E, a	Drive out evil spirits: burn leafy twigs and inhale
			Diarrhea, poisoning, disease prevention, or fatigue relief: decoct the leaves and drink
RLL 1676	Tsiota, ody fady	*Halleria ligustrifolia* Baker (STILBACEAE), E, a	Allergy: drink the decoction of the leaves
			Anthrax, wound: infuse leafy twigs and take a bath
			Depurative: drink the infusion of the stems
			Muscle relaxant, intestinal worm: decoct leafy twigs and drink
RLL 1739	Tsimonomonona	*Helichrysum benthamii* R. Vig. & Humbert (ASTERACEAE), E, H	Anthrax: infuse leafy twigs and take a bath
			Anxious, cast out evil spirits: burn leafy twigs and inhale
			Abdominal colic, diarrhea, fever, gastralgia, marasmus, muscle relaxant, disease prevention or fatigue relief: decoct leafy twigs and drink
			Poisoning: crush the leaves and swallow the juice

(continued)

Voucher	Local names	*Scientific names* (Family), endemism, growth forms	Therapeutic uses
			Urticaria: decoct leafy twigs, drink, and inhale
RLL 1616	Kelimandraha	*Helichrysum bojerianum* DC. (ASTERACEAE), E, H	Abdominal colic: chew the leaves and swallow the juice
			Wounds and abscesses: chewing the leaves and making a poultice
RLL 1718	Rambiazina	*Helichrysum bracteiferum* (DC.) Humbert (ASTERACEAE), E, a	Headache, drive out bad spirits: smoke the leaves
			Abdominal colic, jaundice, plague, prevention of diseases or relief of fatigue: decoct the leafy twigs and drink
			Scabies: wash wounds with a decoction of leafy twigs
RLL 1509	Bongobongo	*Helichrysum* cf *fulvescens* DC. (ASTERACEAE), E, H	Infant hydrops, abdominal colic, furunculosis, gastralgia, muscle relaxant,
			Headache, urticaria: decoct the leafy twigs and inhale
RLL 1597	Benandrongo	*Helichrysum chermezonii* Humbert (ASTERACEAE), E, H	Headache, urticaria: decoct the leafy twigs and inhale
			Abdominal colic, diarrhea: decoct the leafy twigs and drink
RLL 1551	Tsilakinondry	*Helichrysum cryptomerioides* Baker, (ASTERACEAE), E, H	Diarrhea, disease prevention or fatigue relief, dry cough: decoct leafy twigs and drink
			Fatigue: take a bath with a decoction of the aerial parts
			Urticaria: decoct leafy twigs, drink and inhale
RLL 1734	Benandrongo fotsy	*Helichrysum fulvescens* DC. (ASTERACEAE), E, H	Anthrax: burn leafy twigs and inhale
			Abdominal colic, diarrhea, dysentery, prevention of disease or relief of fatigue: decoct leafy twigs and drink

(continued)

Voucher	Local names	*Scientific names* (Family), endemism, growth forms	Therapeutic uses
			Wounds and abscesses: wash the wounds with a decoction of the leafy twigs, then drink
RLL 1710	Benandrongo fotsy	*Helichrysum hypnoides* (DC.) R. Vig. & Humbert (ASTERACEAE), E, a	Drive out evil spirits: burn leafy twigs and inhale
			Diarrhea, gastralgia, marasmus: decoct leafy twigs and drink
RLL 1708	Kely mavitrika	*Helichrysum mutisiaefolium* Less. (ASTERACEAE), E, L	Asthenia drink and massage with the leaf decoction
			Asthma, abdominal colic, diarrhea, disease prevention or relief of fatigue, cough: decoct leafy twigs and drink
			Headache: decoct leafy twigs and inhale
			Wound and abscess: grind the leaves and make a poultice
RLL 1539	Madiorano	*Helichrysum plantago* DC. (ASTERACEAE), E, H	Asthma, gastralgia, disease prevention or fatigue relief: decoct the whole plant and drink
			Burns, wounds, and abscesses: dry the leaves and apply the mash to the wound
			Conjunctivitis: wash the eyes with a decoction of the leaves
			Skin rash, furunculosis, scabies: decoct the whole plant, wash the wounds, then drink
RLL 1588	Kelimirandrana	*Helichrysum retrorsum* DC. (ASTERACEAE), E, H	Abdominal colic, diarrhea: chew the pimple and swallow the juice
			Furunculosis, disease prevention, or fatigue relief: decoct leafy twigs and drink
			Urticaria: decoct leafy twigs, drink and inhale

(continued)

Voucher	Local names	*Scientific names* (Family), endemism, growth forms	Therapeutic uses
RLL 1741	Benandrongo	*Helichrysum selaginifolium* (DC.) R. Vig. & Humbert (ASTERACEAE), E, H	Headache, flu: burn or decoct leafy twigs and inhale
			Marasmus, disease prevention, or fatigue relief: decoct the aerial part and drink
			Psychosis: burn the leafy twigs, pour the ashes into the water, drink and take a bath
RLL 1522	Benandrongo vavy	*Helichrysum stenoclinoides* (Baker) Humbert (ASTERACEAE), E, H	Infant hydrops, abdominal colic, furunculosis, diarrhea, disease prevention or fatigue relief: decoct leafy twigs and drink
			Conjunctivitis: wash the eyes with a decoction of the leaves
RLL 1644	Anadraisoa	*Hubertia alleizettei* (ASTERACEA), NE, L	Abdominal colic, disease prevention or fatigue relief, cough: decoct leafy twigs and drink
			Conjunctivitis: wash the eyes with a decoction of the leaves then drink
			Furunculosis: decoct the whole plant, drink and take a bath
RLL 1594	Anadraisoa	*Hubertia faujasioides* (Baker) C. Jeffrey (ASTERACEAE), E, a	Gonorrhea, fever, furunculosis, marasmus, prevention of disease or relief of fatigue, cough: decoct leafy twigs and drink
			Scabies, wounds, and abscesses: crush the leaves and apply to the wounds
			Otitis: crush the leaves, pour the juice into the ear
RLL 1719	Fanazava	*Hubertia myricifolia* (Bojer ex DC.) C. Jeffrey (ASTERACEAE), E, L	Furunculosis: take a bath with a decoction of the leaves
			Wound and abscess: dry and grind the leaves, apply to the wound
			Prevention of diseases or relief of fatigue: drink the decoction of the leafy twigs

(continued)

Voucher	Local names	*Scientific names* (Family), endemism, growth forms	Therapeutic uses
RLL 1548	Botra	*Hypericum bojerianum* H. Perrier (HYPERICACEAE), E, H	Asthenia drink and massage with a decoction of the whole plant
			Burn: dry and grind the leaves and apply the powder to the wounds
			Abdominal colic: chew the aerial parts and swallow the juice
			Enuresis, kwashiorkor, throat pruritus, asthma, headache, make fertile: decoct the whole plant and drink
			Leprosy: wash wounds with a decoction of the whole plant
RLL 1572	Anapetraka	*Hypochaeris radicata* L. (ASTERACEAE), I, H	Abdominal colic, avoid premature abortion, furunculosis, gastralgia, kwashiorkor, low back pain, prevention of diseases or relief of fatigue: decoct the leafy twigs and drink
			Food supplement: consume the leaves after cooking
RLL 1671	Lampivahatra	*Ilex mitis* (L.) Radlk. (AQUIFOLIACEAE), NE, A	Abdominal colic, expulsion of the placenta, furunculosis, disease prevention or relief of fatigue, dry cough: decoct the leaves and drink
RLL 1777	Benja, Besandry	*Impatiens baronii* Baker (ACANTHACEAE), E, H	Jaundice, marasmus: decoct the leaves drink
			Wound and abscess: crush the leaves and apply the juice to the wound
RLL 1716	Vendrana	*Imperata cylindrica* (L.) Raeusch. (POACEAE), I, H	Headache, joint pain: decoct the whole plant, take a bath, and drink
			Appetizing, dysentery, disease prevention or fatigue relief, cough: decoct the whole plant and drink

(continued)

Voucher	Local names	*Scientific names* (Family), endemism, growth forms	Therapeutic uses
RLL 1786	Kiramena	*Indigofera bojeri* Baker (FABACEAE), E, a	Allergy, diarrhea, oxytocic, disease prevention or relief of fatigue, inflammation of the vagina: decoct the leaves and drink
RLL 1596	Aikakely	*Indigofera pinifolia* Baker (FABACEAE), E, H	Blight: dry the leaves and apply the mash to the wound
			Tooth decay: rub the leaves on the teeth
			Abdominal colic, diarrhea, dysentery, hemorrhoid, jaundice, metrorrhagia, oxytocic, disease prevention or fatigue relief: decoct leafy twigs and drink
RLL 1729	Tsy hita fototra	*Indigofera thymoides* Baker (FABACEAE), E, H	Anthrax: burn the aerial parts and inhale
			Hair loss: wash the head with the decoction of the leaves
RLL 1592	Mahaibe	*Inulanthera brownii* (Hochr.) Källersjö (ASTERACEAE), E, a	Gonorrhea, abdominal colic, fever, disease prevention, or fatigue relief: decoct the leaves and drink
			Headache: decoct leafy twigs and smoke
			Chase away evil spirits, psychosis: burn leafy twigs and inhale
			Flu, cough: decoct leafy twigs, drink, and inhale
RLL 1524	Akondronjaza, Tsy maty ririnina	*Jumellea* sp. (ORCHIDACEAE), E, H	Gonorrhea, gastralgia, disease prevention, or fatigue relief: decoct the whole plant and drink
RLL 1526	Vatofosa lahy	*Justicia sp* (ACANTHACEAE), E, H	Wounds and abscesses, prevention of diseases or relief of fatigue: decoct leafy twigs and drink
RLL 1518	Kotrokotrobato	*Kalanchoe campanulata* (CRASSULACEAE), E, H	Depurative, gastroenteritis, inappetence, marasmus, prevention of diseases or relief of fatigue: decoct the leafy twigs and drink

(continued)

Voucher	Local names	*Scientific names* (Family), endemism, growth forms	Therapeutic uses
			High blood pressure: chew the leaves and swallow the juice
			Muscle relaxant: massage with a decoction of the leaves
			Otitis: crush the leaves, pour the juice into the ear
			Cough: decoct the crushed leaves and drink
RLL 1547	Kotrokotrobato fotsy	*Kalanchoe pumila* Baker (CRASSULACEAE), E, H	Gastroenteritis, edema, albuminuria, prevention of disease or relief of fatigue, Throat itching, cough: decoct leafy twigs and drink
			Otitis: crush the leaves, pour the juice into the ear
RLL 1571	Voandranobe	*Kniphofia ankaratrensis* Baker (ASPHODELACEAE), E, H	Marasmus: grate the roots, macerate, and drink
RLL 1534	Trongafotsy	*Kosteletzkya velutina* Garcke (MALVACEAE), E, a	Anthrax, wound, and abscess: grate the roots and make a poultice
RLL 1788	Bozadambo, haravolandambo	*Lasiorhachis perrieri* (A. Camus) Bosser (POACEAE), E, H	Anthrax, plague, chasing evil spirits: burn all plants and take a steam bath
			Scabies: burn ashes and apply to wounds
			Care after childbirth, depurative, urticaria: burn the leaves and inhale
RLL 1755	Ranofaritra, Madiorano lahy	*Launaea rarifolia* (Oliv. & Hiern) Boulos (ASTERACEAE), NE, H	Headache: decoct the leaves and take a steam bath
			Poisoning: spread the leaves on the fires, grind and pour into lukewarm water, drink
			Furunculosis, gastralgia, disease prevention, or fatigue relief: decoct the whole plant and drink
RLL 1554	Famakiasa	*Leptodesmia congesta* Benth. ex Baker f. (FABACEAE), I, H	Anthrax, burns, wounds, and abscesses: grind the leaves, apply to wounds
			Indigestion, prevention of diseases, or relief of fatigue: drink the decoction of the aerial part

(continued)

Voucher	Local names	*Scientific names* (Family), endemism, growth forms	Therapeutic uses
			Insect bite: grate the roots, infuse, and drink
RLL 1598	Tongotsokina	*Lycopodiella cernua* (L.) Pic. Serm. (LYCOPODIACEAE), NE, H	Anthrax, sores, and abscesses: wash the sores with a decoction of the leafy twigs and drink
			Gonorrhea, furunculosis, gastralgia, gout, kwashiorkor, marasmus, muscle relaxant, plague, disease prevention or fatigue relief, post-partum care: decoct leafy twigs and drink
			Drive out evil spirits: burn leafy twigs and inhale
			Drive out evil spirits: decoct leafy twigs, drink and inhale or burn leaves and inhale
			Malaria: fire the aerial parts, grind and suck the powders
			Cough, asthma cough, urticaria: decoct leafy twigs, drink and inhale
RLL 1658	Mokaranana	*Macaranga alnifolia* Baker (EUPHORBIACEAE), E, A	Diarrhea, disease prevention, or fatigue relief: decoct leafy twigs and drink
RLL 1705	Mokaranana	*Macaranga myriolepida* Baker (EUPHORBIACEAE), E, A	Abdominal colic: crush the leaves, infuse, and drink
			Dysentery, prevention of diseases or relief of fatigue, persistent cough: decoct leafy twigs and drink
RLL 1670	Rafy	*Maesa lanceolata* Forssk. (PRIMULACEAE), I, a	Drive out evil spirits, insect bite: burn leafy twigs and inhale
			Diarrhea: decoct the leaves and drink
RLL 1667	Fanazava	*Malleastrum perrieri* J.-F. Leroy (MELIACEAE), E, A	Joint pain: massage with a decoction of crushed leaves
			Muscle relaxant, disease prevention or fatigue

(continued)

Voucher	Local names	*Scientific names* (Family), endemism, growth forms	Therapeutic uses
			relief: decoct leafy twigs and drink
RLL 1690	Fatsik'ahitra	*Medinilla mandrakensis* H. Perrier (MELASTOMATACEAE), E, H	Decreased visual acuity, lower back pain, plague, disease prevention or fatigue relief: decoct leafy twigs, drink and massage
RLL 1646	Sandrinosy	*Melicope madagascariensis* (Baker) T.G. Hartley (RUTACEAE), E, A	Disease prevention or fatigue relief, hemorrhoid: decoct the leaves and drink
RLL 1549	Manitra andafitsaha	*Micromeria flagellaris* Baker (LAMIACEAE), E, H	Lymphadenopathy, abdominal colic, diarrhea, high blood pressure, prevention of diseases or relief of fatigue: crush leafy twigs and drink
			Blight: crush the leafy twigs, apply to the wound
			Headache, furunculosis, flu, cough: Ddecoct leafy twigs, drink, and inhale
RLL 1586	Manitra andafitsaha	*Micromeria sphaerophylla* Baker (LAMIACEAE), E, H	Furunculosis, wounds and abscesses, insect bite: crush the leafy twigs, macerate, and apply to wounds
			Abdominal colic, liver pain, headache, prevention of disease or relief of fatigue, make fertile, vertigo: decoct leafy twigs, drink
RLL 1726	Volotsangana	*Nastus aristatus* A. Camus (POACEAE), E, a	Anthrax, furunculosis, burn leafy twigs, apply ashes to wounds
			Prevention of diseases or relief of fatigue: drink the decoction of the leafy twigs
RLL 1618	Tsingila	*Neocussonia bojeri* (Seem.) Hutch. (ARALIACEAE), E, H	Fever, gastralgia: drink the decoction of the leafy twigs
RLL 1653	Voamatrakinify	*Oncostemum bojerianum* A. DC. (PRIMULACEAE), E, a	Headache, furunculosis, prevention of diseases, or relief of fatigue: drink the decoction of the leafy twigs

(continued)

Voucher	Local names	*Scientific names* (Family), endemism, growth forms	Therapeutic uses
			Wounds and abscesses: wash the wounds with a decoction of the leafy twigs, then drink
RLL 1601	Fenakoho	*Osmunda regalis* L. (OSMUNDACEAE), NE, H	Kwashiorkor, cough, muscle relaxant: drink the decoction of the leaves
RLL 1715	Ahi-boanjo	*Panicum andringitrense* A. Camus (POACEAE), E, H	Crévace of the palms of the feet, wounds and abscesses: burn the aerial parts, apply the ashes on the wounds
			Loss of consciousness: decoct air paries, take bath and drink
			Measles: drink and inhale the decoction from the aerial part
RLL 1713	Horokely	*Panicum dregeanum* Nees (POACEAE), NE, H	Chase away evil spirits: inhale with the decoction of air paries
			Low back pain, muscle relaxant: drink the decoction of the aerial part
RLL 1730	Horopotsy	*Panicum subalbidum* Kunth (POACEAE), NE, H	Anthrax: burn the aerial parts and inhale
			Loss of consciousness, vertigo: decoct the aerial part, drink, and inhale
			Prevention of diseases or relief of fatigue: drink the decoction of the aerial parts
RLL 1650	Tamarovy	*Pauridiantha paucinervis* (Hiern) Bremek. (RUBIACEAE), NE, a	Headache: crushing leaves and smoking
			Dysentery, gastralgia, disease prevention, or fatigue relief: decoct leafy twigs and drink
RLL 1668	Famaha hazo	*Peperomia imerinae* C. DC. (PIPERACEAE), E, H	Asthenia, joint pain, gastralgia, marasmus, muscle relaxant: decoct leafy twigs and drink
			Food supplement: consume the leaves after cooking
			Wounds and abscesses: crush the aerial parts, apply to wounds

(continued)

Voucher	Local names	*Scientific names* (Family), endemism, growth forms	Therapeutic uses
RLL 1724	Famaha hazo	*Peponidium buxifolium* (Baker) Razafim., Lantz & B. Bremer (RUBIACEAE), E, a	Headache, abdominal colic, gastralgia, disease prevention, or fatigue relief: scratch the stems, infuse then drink
			Cosmetics: scrape the stems, apply on the face
			Drive out evil spirits: decoct leafy twigs, drink, and inhale
RLL 1683	Rondro	*Peponidium horridum* (Baill.) Arènes (RUBIACEAE), E, a	Gonorrhea, muscle relaxant, plague, disease prevention, or fatigue relief: scrape the stems, infuse, then drink
RLL 1779	Tsileondrahoaho	*Phellolophium madagascariense* Baker (APIACEAE), E, H	Angina: crush the leaves, infuse, and gargle
			Furunculosis: take a bath with a decoction of the leaves and drink
			Gastralgia, disease prevention or fatigue relief, dry cough
			Hemostatic: infuse the roots and drink
			Muscle relaxant: take a bath with a decoction of the leaves
			Dandruff: wash the head with the decoction of the leaves
			Wounds and abscesses: crush the leaves, apply to wounds
			Otitis: crush the leaves, pour the juice into the ear
RLL 1611	Voatavondolo	*Pilogyne emirensis* (Baker) W.J. de Wilde & Duyfjes (CUCURBITACEAE), NE, L	Gonorrhea, abdominal colic: decoct the leafy twigs, drink, and inhale
			Drive out evil spirits: burn seeds and inhale
RLL 1622	Saonjorangahy	*Pimpinella ebracteata* Baker (APIACEAE), E, H	Appetizing: dry and grind the roots, consume the powder
			Intestinal bloating, abdominal colic: decoct the roots, drink, or chew the roots

(continued)

Voucher	Local names	*Scientific names* (Family), endemism, growth forms	Therapeutic uses
			Furunculosis: decoct the roots and apply to wounds
			Gastralgia, marasmus, poisoning: decoct the roots, drink and inhale
			Food supplement: eaten raw, the young leaves
RLL 1523	Farala	*Pimpinella perrieri* Sales & Hedge (APIACEAE), E, a	Appetizing: decoct the leaves and drink
			Knee arthralgia: massage with a decoction of the leaves and drink
			Gonorrhea, abdominal colic, diarrhea, furunculo-sis, gastralgia, marasmus, persistent cough, vitiligo: decoct the leaves or leafy twigs then drink
			Tooth decay: dry leaves and smoke
			Headache, drive out bad spirits, fever: inhale with a decoction of the leaves and drink
			Anthrax, scabies, leprosy: crush the leaves and apply to the wound
			Measles: take a bath with the decoction of the aerial part
RLL 1674	Ambovitsika	*Pittosporum viridiflorum* Sims (PITTOSPORACEAE), NE, A	Joint pain, gastralgia, plague, prevention of dis-eases or relief of fatigue: Ddrink the decoction of the leafy twigs
			Furunculosis: Ddrink the decoction of the leafy twigs and wash it on the wounds
RLL 1693	Apanga mangakely	*Pityrogramma calomelanos* (L.) Link (PTERIDACEAE), NE, H	Chase away evil spirits: burn leaves and inhale
			Prevention of diseases or relief of fatigue: decoct the aerial parts and drink
			Urticaria: decoct the leaves, drink, and inhale

(continued)

Voucher	Local names	*Scientific names* (Family), endemism, growth forms	Therapeutic uses
RLL 1682	Vahy	*Plectaneia thouarsii* Roem. & Schult. (APOCYNACEAE), E, L	Wound: wash wounds with a decoction of leafy twigs
			Poisoning, prevention of diseases or relief of fatigue: decoct the leafy twigs and drink
			Witchcraft
RLL 1661	Anan'ala, Borona	*Plectranthus membranaceus* (Scott Elliot) Hedge (LAMIACEAE), E, a	Furunculosis: crush the leaves, apply the juices to the wounds
			Gastralgia, kwashiorkor, low back pain, disease prevention, or fatigue relief: Iinfuse the leaves and drink
RLL 1669	Sangotraka	*Plectranthus secundiflorus* (Baker) Hedge (LAMIACEAE), E, a	Food supplement: consume the leaves after cooking
			Gastralgia: decoct the leaves and drink
RLL 1617	Antsoro	*Poa ankaratrensis* A. Canus & H. Perrier (POACEAE), E, H	Cough: chew the roots and swallow the juice
RLL 1600	Masonakora	*Polygala ankaratrensis* H. Perrier (POLYGALACEAE), E, H	Convulsion, marasmus, intestinal parasitosis: dry the whole plant on fire, infuse and drink
			Indigestion, prevention of diseases, or relief of fatigue: drink the decoction of the whole plant
RLL 1660	Anadraisoa	*Psiadia lucida* (Cass.) Drake (ASTERACEAE), E, L	Asthenia, furunculosis, abdominal colic, rash, fatigue, kwashiorkor, marasmus, cough, persistent cough, prevention of diseases or relief of fatigue: decoct the leafy twigs and drink
RLL 1681	Tambitsy	*Psorospermum androsaemifolium* Baker (HYPERICACEAE), E, a	Muscle pain, muscle relaxant, disease prevention or fatigue relief, lower urinary tract symptoms: decoct leafy twigs and drink
RLL 1706	Tambitsy	*Psorospermum ferrovestitum* Baker (HYPERICACEAE), E, a	Headache: infuse leafy twigs, drink, and inhale

(continued)

Voucher	Local names	*Scientific names* (Family), endemism, growth forms	Therapeutic uses
			Abdominal colic, joint pain, dysentery, gastralgia, kwashiorkor, disease prevention or fatigue relief, dry cough, mixing disorder: decoct leafy twigs and drink
RLL 1659	Marefolena	*Psychotria homolleae* Bremek. (RUBIACEAE), E, A	Headache, prevention of diseases or relief of fatigue, vertigo: decoct leafy twigs and drink
			Furunculosis: dry and crush the leaves, apply to the
			Plague: dry and crush the leaves, apply to the bubo
RLL 1528	Apanga be	*Pteridium aquilinum* (L.) Kuhn (PTERIDACEAE), NE, H	Anthrax, hunt bad spirits, urticaria: burn leaves and inhale
RLL 1692	Apanga antanety	*Pteridium esculentum* (G. Forst.) Cockayne (DENNSTAEDTIACEAE), E, H	Chase away evil spirits: burn the aerial parts and inhale
			Prevention of disease or relief of fatigue
RLL 1612	Voanjomanga lahy	*Pycnostachys coerulea* Hook (LAMIACEAE), NE, H	Prevention of diseases or relief of fatigue: decoct the aerial parts and drink
RLL 1702	Fantsikahatra, sandrinosy	*Pyrostria italyensis* (Cavaco) A.P. Davis & Govaerts (RUBIACEAE), E, A	Headache, diarrhea, placental expulsion, gastralgia, muscle relaxant, oxytocic, disease prevention or fatigue relief: decoct leafy twigs and drink
RLL 1677	Rovy	*Pyrostria madagascariensis* Lecomte (RUBIACEAE), E, A	Headache: infuse leafy twigs, drink and inhale
			Joint pain, disease prevention or fatigue relief: drink the decoction of the leafy twigs
RLL 1525	Tambarasaha, Voamasonomby	*Radamaea montana* Benth. (SCROPHULARIACEAE), E, a	Acne: scratch the roots, apply on the pimple
			Gonorrhea, diarrhea, jaundice, cough: decoct leafy twigs and drink
			Burn: chew stems and apply to wounds

(continued)

Voucher	Local names	*Scientific names* (Family), endemism, growth forms	Therapeutic uses
			Food supplement: edible fruits
			Conjunctivitis: decoct the leafy twigs, drink then wash the eyes
			Prevention of diseases or relief of fatigue: decoct the roots and drink
RLL 1562	Ahimafana vavy	*Ranunculus multifidus* Forssk. (RANUNCULACEAE), I, H	Angina: crush the leaves, infuse, and gargle
			Headache: infuse leafy twigs, drink, and inhale
			Abdominal colic, diarrhea, dyspnoea, gastralgia, gastroenteritis, marasmus, plague, disease prevention, or relief of fatigue: infuse the leafy twigs and drink
			Plague: dry and crush the leaves, drink the juice
RLL 1694	Malambovony	*Razafimandimbisonia minor* (Baill.) Kainul. & B. Bremer (RUBIACEAE), E, a	Decreased visual acuity: decoct leaves, wash eyes, and drink
			Fever, gastralgia, liver aches, prevention of disease or relief of fatigue, dry cough: drink the decoction of the leafy twigs
RLL 1531	Kelimavitrika	*Rochonia cinerarioides* DC. (ASTERACEAE), E, a	Headache: dry leaves and smoke
			Abdominal colic, strengthening, furunculosis, prevention of diseases or relief of fatigue: decoct the leafy twigs and drink
			Insect bite: chew the leaves and apply to the bite point
RLL 1530	Tsiparapanda fotsy	*Salvia cryptoclada* Baker (LAMIACEAE), E, a	Furunculosis, internal wound and abscess, prevention of diseases or relief of fatigue, cough: decoct leafy twigs and drink

(continued)

Voucher	Local names	*Scientific names* (Family), endemism, growth forms	Therapeutic uses
			Insect bite: chew the leaves and apply to the bite point
RLL 1508	Tsivaramonina lahy, Tsiparapanda	*Salvia parvifolia* Baker (LAMIACEAE), E, H	Infant hydrops, headache, abdominal colic, dyspnoea, strengthening, gastralgia, kwashiorkor, oxytocic, disease prevention or fatigue relief: decoct leafy twigs and drink
			Blight: chew the leaves and apply to wounds
			Furunculosis, jaundice, marasmus, muscle relaxant: decoct the leafy twigs, drink and inhale, then take a shower
			Insect bite, wound and abscess, wounds: chew the leaves and apply to the wound or to the point of sting
RLL 1728	Tsiparapandy	*Salvia porphyrocalyx* Baker (LAMIACEAE), E, a	Convulsion, intestinal worm: drink the decoction of the leaves
RLL 1512	Tsiparapandy	*Salvia sessilifolia* Baker (LAMIACEAE), E, a	Oxytocic, disease prevention or fatigue relief, cough: decoct leafy twigs and drink
			Insect bite, wound, and abscess: chew the leaves and apply them on the wound or on the point of sting
RLL 1687	Anamafana	*Sanicula elata* Buch.-Ham. ex D. Don (APIACEAE), NE, H	Abdominal colic, diarrhea: dry and crush the leaves, swallow the powder
			Kwashiorkor, marasmus: decoct the leaves and drink
			Disease prevention or fatigue relief: infuse the leaves and drink
RLL 1736	Madiorano	*Senecio adscendens* Bojer ex DC. (ASTERACEAE), E, H	Aphtha, frunculosis: crush the leafy twigs, pour the juice on the wounds

(continued)

Voucher	Local names	*Scientific names* (Family), endemism, growth forms	Therapeutic uses
			Avoid premature abortion, gastralgia, liver pain, prevention of diseases or relief of fatigue: decoct leafy twigs and drink
RLL 1550	Anadraisoa lahy	*Senecio* cf *hypargyraeus* DC. (ASTERACEAE), E, H	Gonorrhea, disease prevention or fatigue relief, vitiligo: decoct leafy twigs and drink
			Wounds and abscesses: crush the leaves, apply to wounds
RLL 1735	Benandrongo	*Senecio emirnensis* DC. (ASTERACEAE), E, H	Anthrax: burn the aerial parts and inhale
			Appetizing, gastralgia: infuse the leaves and drink
			Knee arthralgia, Kwashiorkor, disease prevention or fatigue relief, vertigo: decoct leafy twigs and drink
			Gastralgia: decoct the roots and drink
RLL 1775	Betefaka	*Senecio pleistophyllus* C. Jeffrey (ASTERACEAE), E, H	Anthrax: wash the wounds several times with a decoction of the leaves
			Post-diarrheal asthenia, post-infection asthenia, gonorrhea, gastralgia, marasmus: decoct, disease prevention or fatigue relief leaves and drink
			Insect bite, wound and abscess: dry the leaves, decoct, then wash the wounds or the bite point
			Measles: decoct the leaves, drink and then take a bath
RLL 1513	Tsivaramonina	*Senecio resectus* Bojer (ASTERACEAE), E, L	Anthrax: burn the aerial parts and inhale
			Asthenia, prevention of diseases or relief of fatigue, persistent cough: drink the decoction of the aerial parts
			Furunculosis: drink the decoction of the leafy

(continued)

Voucher	Local names	*Scientific names* (Family), endemism, growth forms	Therapeutic uses
			twigs and wash it from the wounds
			Jaundice, marasmus, muscle relaxant: Ttake a bath with a decoction of the leaves, then drink and inhale
RLL 1722	Avoatra	*Smilax anceps* Willd. (SMILACACEAE), NE, L	Decreased visual acuity: infuse leafy twigs, wash eyes
			Furunculosis, gastralgia, plague, prevention of diseases or relief of fatigue: decoct leafy twigs and drink
RLL 1629	Kelimavikitra	*Solanecio angulatus* (Vahl) C. Jeffrey (ASTERACEAE), NE, H	Kwashiorkor: decoct the leafy twigs and drink
RLL 1766	Mamoniavo	*Stachys trichophylla* Baker (LAMIACEAE), E, H	Gastralgia, low back pain, marasmus, prevention of disease or relief of fatigue, cough: drink decoction of aerial parts
			Wounds and abscesses: crush the leafy twigs, apply to the wounds
RLL 1555	Mparakindreniomby	*Stenocline inuloides* DC. (ASTERACEAE), E, H	Drive out evil spirits: burn leafy twigs, inhale
			Wound and abscess: dry and grind the leaves, apply to the wound
			Prevention of diseases or relief of fatigue: decoct the aerial parts and drink
RLL 1632	Tsimarary	*Stoebe pachyclada* Humbert (ASTERACEAE), E, a	Anthrax: wash the wounds several times with the decoction of the leafy twigs
			Drive out evil spirits: inhale with the decoction of the leaves or the smoke of the leafy twigs
RLL 1701	Votsehatra	*Syzygium emirnense* (Baker) Labat & G.E. Schatz (MYRTACEAE), E, A	Diarrhea: decoct the leaves and drink
			Poisoning: infuse the leafy twigs and drink
			Gastralgia, prevention of diseases or relief of

(continued)

Voucher	Local names	*Scientific names* (Family), endemism, growth forms	Therapeutic uses
			fatigue, vertigo: drink the decoction of the leafy twigs
			Plague: crush the leafy twigs swallow the powders
RLL 1649	Ambora	*Tambourissa purpurea* (Tul.) A. DC. (MONIMIACEAE), E, A	Fatigue: take a bath with a decoction of the leafy twigs
			Disease prevention or fatigue relief: infuse the leaves and drink
RLL 1778	Sandrinosy	*Tarenna alleizettei* (Dubard & Dop) De Block (RUBIACEAE), E, a	Headache, poisoning: decoct the leaves and drink
RLL 1776	Tadidy	*Tarenna angolensis* (RUBIACEAE), E, a	Poisoning, disease prevention, or fatigue relief: decoct the leaves, drink, and take a bath
RLL 1520	Borona	*Tetradenia goudotii* Briq. (LAMIACEAE), E, a	Angina: crush the leaves, infuse, and gargle
			Anthrax: burn leafy twigs and inhale
			Headache: dry leaves and smoke
			Abdominal colic, diarrhea: chew the leaves and swallow the juice or drink the leaf decoction
			Abdominal colic
			Fever, gastralgia, plague: drink the decoction of the leafy twigs
			Insect bite: chew the leaves and apply to the bite point
			Wound and abscess: dry and grind the leaves, apply to the wound
			Prevention of diseases or relief of fatigue, prostate: decoct the leaves and drink
RLL 1697	Voadimbana, Felaborona	*Tina striata* Radlk. (SAPINDACEAE), E, A	Joint pain, gastralgia, low back pain, prevention of diseases or relief of fatigue: drink the decoction of the leafy twigs

(continued)

Voucher	Local names	*Scientific names* (Family), endemism, growth forms	Therapeutic uses
RLL 1703	Voaramontsana	*Vaccinium madagascariense* (Thouars ex Poir.) Sleumer (ERICACEAE), E, a	Astringent, intestinal bloating, headache, abdominal colic, diarrhea, joint pain, fever, furunculosis, gastralgia, internal edema, care after childbirth, intestinal worm, disease prevention or fatigue relief: decoct leafy twigs and drink
			Plague: grind the leafy twigs, macerate and drink
RLL 1696	Rainisinda	*Vernonia delapsa* Baker (ASTERACEAE), E, a	Aphrodisiac, furunculosis, gastralgia, prevention of diseases or relief of fatigue: decoct the leafy twigs and drink
			Blight: crush the leaves and apply to the wound
RLL 1774	Avoko	*Vigna angivensis* Baker (FABACEAE), E, L	Anthrax, astringent, avoiding premature abortion, ovarian cyst: decoct the leafy twigs and drink
RLL 1680	Hazomiavona	*Viscum tsiafajavonense* Balle (SANTALACEAE), E, H	Joint pain: massage with crushed leaves
			Wounds and abscesses: decoct the leaves and drink and wash the wounds
			Prevention of diseases or relief of fatigue: drink the decoction of the leafy twigs
RLL 1782	Vendramanga	*Vittaria elongata* Sw. (VITTARIACEAE), NE, H	Headache, joint pain: Ddecoct the whole plant, take a bath and drink
RLL 1657	Lalona	*Weinmannia bojeriana* Tul. (CUNONIACEAE), E, A	Headache, abdominal colic, gastralgia, joint pain, furunculosis, prevention of diseases or relief of fatigue, dry cough: decoct the leafy twigs then drink
			Film: decoct the leafy twigs and wash the head

(continued)

Voucher	Local names	*Scientific names* (Family), endemism, growth forms	Therapeutic uses
RLL 1648	Tsy savihandahy	*Zanthoxylum madagascariense* Baker (RUTACEAE), E, A	Diarrhea: consume the ground bark
			Poisoning: decoct the bark and drink
			Gastralgia: infuse the leaves and drink
			Prevention of diseases or relief of fatigue, cough: decoct bark and drink

References

Akerreta, S., Cavero, R. Y., López, V., & Calvo, M. I. (2007). Analyzing factors that influence the folk use and phytonomy of 18 medicinal plants in Navarra. *Journal of Ethnobiology and Ethnomedicine, 3*(1), 16. https://doi.org/10.1186/1746-4269-3-16

Allorge, L. (2005). L'ethnopharmacologie à Madagascar de 1655 à nos jours. In *Ethnopharmacologia*. Société Française d'Ethnopharmacologie 36. http://www.ethnopharmacologia.org/boutique/ethnopharmacologia-36-2005/

Alqudah, A., Qnais, E., Gammoh, O., Bseiso, Y., & Wedyan, M. (2024). Protective effects of α-Pinene against carbon tetrachloride-induced cardiac injury in Wistar rats: Modulation of antioxidant and inflammatory responses. *Naunyn-Schmiedeberg's Archives of Pharmacology, 398*(3), 2845-2854. https://doi.org/10.1007/s00210-024-03434-8

Andrianjakaniaina, M., Grougnet, R., Ralambonirina, S., Michel, S., Rakotoarisoa, A. M., Rakotonandrasana, S. R., Rasoanaivo, L., Razanamehefa, B., Ratsimbason, M. A., & Rasamison, V. E. (2017). Etude chimique et pharmacologique de Tetradenia goudotii. (Lamiaceae). *Biodiversité et les Objectifs du Développement Durable*.

Andrianjakaniaina, M., Rakotondramanana, D., Benatrehina, P., Slebodnick, C., Rasoanaivo, L., Rakotosaona, R., Harinantenaina, L., Rakotonandrasana, S. R., & Rasamison, V. (2021). Prenylated chromones and coumarins from the leaves of Billburttia capensoides Sales & Hedge (Apiaceae). *GSC Biological and Pharmaceutical Sciences, 16*(août), 103–109. https://doi.org/10.30574/gscbps.2021.16.2.0233

Batiha, G. E.-S., Shaheen, H. M., Elhawary, E. A., Mostafa, N. M., Eldahshan, O. A., & Sabatier, J.-M. (2023). Phytochemical constituents, folk medicinal uses, and biological activities of genus Angelica: A review. *Molecules, 28*(1), 267. https://doi.org/10.3390/molecules28010267

Braun-Blanquet, J. (1965). *Plant sociology. The study of plant communities*. 439 p.

Byg, A., Salick, J., & Law, W. (2010). Medicinal plant knowledge among lay people in five eastern Tibet villages. *Human Ecology, 38*(2), 177–191. https://doi.org/10.1007/s10745-009-9300-z

Callmander, M. W., Phillipson, P. B., Schatz, G. E., Andriambololonera, S., Rabarimanarivo, M., Rakotonirina, N., Raharimampionona, J., Chatelain, C., Gautier, L., & Lowry, P. P. (2011). The endemic and non-endemic vascular flora of Madagascar updated. *Plant Ecology and Evolution, 144*(2), 121–125.

Corrèges, D. (2016). Les sites de culte ancestraux, entre histoires locales et mondialisation des ressources (Ankaratra, Madagascar). *Journal des Africanistes, 86-1*, 60–84. https://doi.org/10.4000/africanistes.4877

De Beer, J. J. J., & Van Wyk, B. E. (2011). An ethnobotanical survey of the Agter–Hantam, Northern Cape Province, South Africa. *South African Journal of Botany, 77*(3), 741–754. https://doi.org/10.1016/j.sajb.2011.03.013

Dez, J. (1967). Le Vakinankaratra: esquisse d'une histoire régionale. *Bulletin de Madagascar*.

Farnsworth, N. R., & Soejarto, D. D. (1991). Global importance of medicinal plants. *The Conservation of Medicinal Plants, 26*(26), 25–51.

Friedman, J., Yaniv, Z., Dafni, A., & Palewitch, D. (1986). A preliminary classification of the healing potential of medicinal plants, based on a rational analysis of an ethnopharmacological field survey among Bedouins in the Negev Desert, Israel. *Journal of Ethnopharmacology, 16*(2–3), 275–287.

Fu, Y.-H., Guo, J.-M., Xie, Y.-T., Xiao-Mei, Y., Qin-Ting, S., Qiang, L., Kong, L.-Y., & Liu, Y.-P. (2020). Prenylated Chromones from the fruits of *Artocarpus Heterophyllus* and their potential anti-HIV-1 activities. *Journal of Agricultural and Food Chemistry, 68*(7), 2024–2030. https://doi.org/10.1021/acs.jafc.9b06417

Ganjhu, R. K., Mudgal, P. P., Maity, H., Dowarha, D., Devadiga, S., Nag, S., & Arunkumar, G. (2015). Herbal plants and plant preparations as remedial approach for viral diseases. *VirusDisease, 26*(4), 225–236. https://doi.org/10.1007/s13337-015-0276-6

Gautier, L., Chatelain, C., Callmander, M. W., & Phillipson, P. B. (2012). Richness, similarity and specificity of Madagascar flora compared with sub-Saharan Africa. *Plant Ecology and Evolution, 145*(1), 55–64.

Giday, M., Asfaw, Z., & Woldu, Z. (2009). Medicinal plants of the Meinit ethnic group of Ethiopia: An ethnobotanical study. *Journal of Ethnopharmacology, 124*(3), 513–521. https://doi.org/10.1016/j.jep.2009.05.009

Heinrich, M., Kufer, J., Leonti, M., & Pardo-de-Santayana, M. (2006). Ethnobotany and Ethnopharmacology--Interdisciplinary Links with the Historical Sciences. *Journal of Ethnopharmacology, 107*(2), 157–160. https://doi.org/10.1016/j.jep.2006.05.035

Ito, K., & Ito, M. (2013). The sedative effect of inhaled Terpinolene in mice and its structure–activity relationships. *Journal of Natural Medicines, 67*(4), 833–837. https://doi.org/10.1007/s11418-012-0732-1

Kingston, D. G. I. (2011). Modern natural products drug discovery and its relevance to biodiversity conservation. *Journal of Natural Products, 74*(3), 496–511. https://doi.org/10.1021/np100550t

Kunwar, R. M., Shrestha, K., Malla, S., Acharya, T., Sementelli, A. J., Kutal, D., & Bussmann, R. W. (2019). Relation of medicinal plants, their use patterns and availability in the Lower Kailash Sacred Landscape, Nepal. *Ethnobotany Research and Applications, 18*, 1–14.

Liu, Y.-P., Xiao-Mei, Y., Zhang, W., Wang, T., Jiang, B., Tang, H.-X., Qin-Ting, S., & Yan-Hui, F. (2020). Prenylated chromones and flavonoids from Artocarpus Heterophyllus with their potential antiproliferative and anti-inflammatory activities. *Bioorganic Chemistry, 101*(août), 104030. https://doi.org/10.1016/j.bioorg.2020.104030

Lyon, L. M., & Hardesty, L. H. (2012). Quantifying medicinal plant knowledge among non–specialist Antanosy Villagers in Southern Madagascar. *Economic Botany, 66*(1), 1–11. https://doi.org/10.1007/s12231-011-9185-9

Mahomoodally, M. F. (2013). Traditional medicines in Africa: An appraisal of ten potent African medicinal plants. *Evidence-based Complementary and Alternative Medicine, 2013*(1), 617459. https://doi.org/10.1155/2013/617459

Mahomoodally, M. F., & Gurib-Fakim, A. (2013). Chapter 15 – Harnessing traditional knowledge to treat existing and emerging infectious diseases in Africa. In *Fighting multidrug resistance with herbal extracts, essential oils and their components, édité par Mahendra Kumar rai et Kateryna Volodymyrivna Kon* (pp. 223–235). Academic Press. https://doi.org/10.1016/B978-0-12-398539-2.00015-X

MEEMF, Ministère de l'Environnement, de l'Ecologie, de la Météorologie et des Forêts. (2015). *Changement de la couverture de forêts naturelles à Madagascar, 2005, 2010, 2013 Antananarivo*.

Mittermeier, R. A., Robles Gil, P., Hoffman, M., Pilgrim, J. T., Brooks, T., Mittermeier, C. G., Lamoureux, J., & da Fonseca, G. A. B. (2004). *Hotspots revisited: Earth's biologically richest and most threatened terrestrial ecosystems*. CEMEX. Anurans of APA Boqueirão da Mira.

Moerman, D. E., & Estabrook, G. F. (2003). Native Americans' choice of species for medicinal use is dependent on plant family: Confirmation with meta-significance analysis. *Journal of Ethnopharmacology, 87*(1), 51–59.

Myers, N., Mittermeier, R. A., Mittermeier, C. G., da Fonseca, G. A., & Kent, J. (2000). Biodiversity hotspots for conservation priorities. *Nature, 403*(6772), 853–858. https://doi.org/10.1038/35002501

Norscia, I., & Borgognini-Tarli, S. M. (2006a). Ethnobotanical reputation of plant species from two forests of Madagascar: A preliminary investigation. *South African Journal of Botany, 72*(4), 656–660. https://doi.org/10.1016/j.sajb.2006.04.004

Norscia, I., & Borgognini-Tarli, S. M. (2006b). Ethnobotanical reputation of plant species from two forests of Madagascar: A preliminary investigation. *South African Journal of Botany, 72*(4), 656–660.

Novy, J. W. (1997). Medicinal plants of the eastern region of Madagascar. *Journal of Ethnopharmacology, 55*(2), 119–126. https://doi.org/10.1016/S0378-8741(96)01489-4

Office Nationale de l'Environnement (ONE). (2007). *Tableau de bord environnemental Région Itasy*. https://www.pnae.mg/tbe/region-itasy.html

Petrovska, B. B. (2012). Historical review of medicinal plants' usage. *Pharmacognosy Reviews, 6*(11), 1–5. https://doi.org/10.4103/0973-7847.95849

Rabehaja, D. J. R., Bezert, G., Rakotonandrasana, S. R., Ramanoelina, P. A. R., Andrianjara, C., Bighelli, A., Tomi, F., & Paoli, M. (2020). Chemical composition of aerial parts essential oils from six endemic Malagasy Helichrysum species. *Plants, 9*(2), 265. https://doi.org/10.3390/plants9020265

Rafidison, V., Ratsimandresy, F., Rakotondrafara, A., Rakotondrajaona, R., Rasamison, V. E., Rakotoarisoa, F. M., & Rakotonandrasana, S. R. (2019). Synthèse et analyse de données sur les inventaires de plantes médicinales de Madagascar. *Ethnobotany Research and Applications, 18*, 1–19.

Rakotoarisoa, J. A. (1986). *Principaux aspects des formes d'adaptation de la société traditionnelle malgache* (pp. 89–106). Society and History.

Rakotoarivelo, N. H., Rakotoarivony, F., Ramarosandratana, A. V., Jeannoda, V. H., Kuhlman, A. R., Randrianasolo, A., & Bussmann, R. W. (2015). Medicinal plants used to treat the most frequent diseases encountered in Ambalabe rural community, Eastern Madagascar. *Journal of Ethnobiology and Ethnomedicine, 11*, 1–16.

Rakotonandrasana, S. R. (2013). Les plantes médicinales de l'aire protégée de Zahamena (Madagascar) et de ses environs: richesse floristique et endémicité. *Scripta Botanica Belgica, 50*, 356–362.

Rakotonandrasana, S. R., Rakotondrafara, A., & Rakotoarisoa, M. (2017). *Résultats des analyses préliminaires sur l'état de lieux des plantes médicinales de Madagascar* (pp. 197–200). Ministère de l'Enseignement Supérieur et de la Recherche Scientifique.

Rakotonandrasana, S. R., Rabehaja, D. J. R., Paoli, M., de Rocca, D., Serra, C. A., Bighelli, A., Panja, A. R., Ramanoelina, & Tomi, F. (2019). Chemical composition of the essential oils from the aerial parts of two Malagasy endemic species (Apiaceae): Billburttia capensoides Sales & Hedge and Billburttia vaginoides Sales & Hedge. *Natural Product Research, 33*(8), 1200–1203. https://doi.org/10.1080/14786419.2018.1457669

Rakotondrafara, A., Rakotondrajaona, R., Rakotoarisoa, M., Ratsimbason, M., Rasamison, V. E., & Rakotonandrasana, S. R. (2018). *Ethnobotany of medicinal plants used by the Zafimaniry clan in Madagascar* (p. 64). CONTRIBUTION A L'ETU MEDICINALES ET AROMA ETHNOBOTANIQUES, CHIMIQUES ET ACTIVIT.

Randrianarivelojosia, M., Rasidimanana, V. T., Rabarison, H., Cheplogoi, P. K., Ratsimbason, M., Mulholland, D. A., & Mauclère, P. (2003). Plants traditionally prescribed to treat Tazo (Malaria)

in the Eastern Region of Madagascar. *Malaria Journal, 2*(1), 25. https://doi.org/10.1186/1475-2875-2-25

Raunet, M. (1981). *Le milieu physique de la région volcanique Ankaratra-Vakinankaratra-Itasy (Madagascar).* Aptitude à la culture du blé pluvial. Ministère du Développement Rural et de la Réforme Agraire, Antananarivo.

Razafindralambo, L. (2016). L'Ankaratra au croisement de conceptions et d'intérêts divergents: conservation de la biodiversité et héritage ancestral. *Journal des africanistes, 86-1*, 86–104.

Salehi, B., Upadhyay, S., Orhan, I. E., Jugran, A. K., Jayaweera, S. L. D., Dias, D. A., Sharopov, F., et al. (2019). Therapeutic potential of α- and β-Pinene: A miracle gift of nature. *Biomolecules, 9*(11), 738. https://doi.org/10.3390/biom9110738

Sales, F., Hedge, I., Coutinho, A. X. P., Marques, A., van Wyk, B. E., & Tilney, P. M. (2004). Apiaceae subfamily Apioideae in Madagascar. *South African Journal of Botany, Special Issue — Apiales, 70*(3), 446–448. https://doi.org/10.1016/S0254-6299(15)30228-3

Savo, V., Joy, R., Caneva, G., & McClatchey, W. C. (2015). Plant selection for ethnobotanical uses on the Amalfi Coast (Southern Italy). *Journal of Ethnobiology and Ethnomedicine, 11*(1), 58. https://doi.org/10.1186/s13002-015-0038-y

Sher, H., Aldosari, A., Ali, A., & de Boer, H. J. (2015). Indigenous knowledge of folk medicines among tribal minorities in Khyber Pakhtunkhwa, northwestern Pakistan. *Journal of Ethnopharmacology, 166*(mai), 157–167. https://doi.org/10.1016/j.jep.2015.03.022

Singh, A. G., Kumar, A., & Tewari, D. D. (2012). An ethnobotanical survey of medicinal plants used in Terai Forest of Western Nepal. *Journal of Ethnobiology and Ethnomedicine, 8*(1), 19. https://doi.org/10.1186/1746-4269-8-19

Smith, Paul Philip (1964-) Directeur de publication. (2007). *Atlas of the vegetation of Madagascar.* Kew Publishing.

Sumorek-Wiadro, J., Zając, A., Skalicka-Woźniak, K., Rzeski, W., & Jakubowicz-Gil, J. (2024). Furanocoumarins as enhancers of antitumor potential of Sorafenib and LY294002 toward human glioma cells in vitro. *International Journal of Molecular Sciences, 25*(2), 759. https://doi.org/10.3390/ijms25020759

Tugume, P., Kakudidi, E. K., Buyinza, M., Namaalwa, J., Kamatenesi, M., Mucunguzi, P., & Kalema, J. (2016). Ethnobotanical survey of medicinal plant species used by communities around Mabira Central Forest Reserve, Uganda. *Journal of Ethnobiology and Ethnomedicine, 12*(1), 5. https://doi.org/10.1186/s13002-015-0077-4

Verpoorte, R., Choi, Y. H., & Kim, H. K. (2005). Ethnopharmacology and systems biology: A perfect holistic match. *Journal of Ethnopharmacology, 100th Volume Special Section: Perspectives of Ethnopharmacology, 100*(1), 53–56. https://doi.org/10.1016/j.jep.2005.05.033

Weston-Green, K., Clunas, H., & Naranjo, C. J. (2021). A review of the potential use of Pinene and Linalool as Terpene-based medicines for brain health: Discovering novel therapeutics in the flavours and fragrances of Cannabis. *Frontiers in Psychiatry, 12*(août). https://doi.org/10.3389/fpsyt.2021.583211

WHO, World Health Organization. (2008, décembre). *Traditional Medicine.*

Wyk, B. E. V., Tilney, P. M., & Magee, A. R. (2013). *African Apiaceae: A synopsis of the Apiaceae/Umbelliferae of sub-Saharan Africa and Madagascar.* Briza Academic Books.

Zebrowski, C. (1971). Propriétés des andosols de l'Itasy et de l'Ankaratra. *Cahier de l'ORSTOM, série Pédologie, 9*(1), 83–108.

Chapter 16
Contributions of African Medicinal Plants to Modern Drug Discovery

Abba Babandi

Abstract African cultures have long utilized medicinal plants for treating various illnesses, relying on traditional healers' extensive knowledge, passed down through generations. This ethnobotanical wisdom is a valuable resource for identifying medicinal plants with therapeutic potential for drug discovery. African medicinal plants are rich in a variety of bioactive agents such as alkaloids, flavonoids, terpenoids, and phenolic substances, which exhibit diverse pharmacological properties, making them excellent candidates for drug development. This chapter produced information through a systematic literature search across databases, such as PubMed, Scopus, and Google Scholar. Studies published between 2000 and 2023 were selected based on inclusion criteria emphasizing medicinal plant use, bioactive compound identification, and pharmacological properties. Data on traditional uses, phytochemical techniques, and therapeutic activities were critically analyzed. Advanced methods, including chromatography and spectroscopy, provided insights into isolating and characterizing bioactive compounds. Notable examples of medicinal plants include *Warburgia salutaris* and *Prunus africana* for their antimicrobial properties, and *Catharanthus roseus*, which has led to the development of anticancer drugs like vincristine and vinblastine. *Artemisia afra* has contributed significantly to malaria treatment, while *Sutherlandia frutescens* and *Harpagophytum procumbens* have shown anti-inflammatory and analgesic properties. Additionally, the development of Niprisan from anti-sickling herbs in Nigeria underscores the contributions of African plants to combating sickle cell anemia. By integrating traditional knowledge with modern phytochemical and pharmacological research, African medicinal plants offer significant therapeutic potential. Continued efforts in research, conservation, and equitable collaboration are critical to harnessing this potential and developing effective drugs for global health challenges.

A. Babandi (✉)
Department of Biochemistry, Bayero University, PMB 3011-Kano, Nigeria

Medical Biochemistry Unit, Faculty of Basic Medical Sciences, Federal University Dutse, PMB 7156-Dutse, Jigawa State, Nigeria
e-mail: ababandi.bch@buk.edu.edu.ng

A. Dekebo (ed.), *Bioactive Secondary Metabolites from Medicinal Plants of Africa*, Interdisciplinary Biotechnological Advances,
https://doi.org/10.1007/978-981-96-8590-5_16

Keywords Africa · Drug discovery · Medicinal plants · Phytochemicals · Traditional knowledge

16.1 Introduction: Overview of Medicinal Plants in Africa

16.1.1 African Medicinal Plants

Medicinal plants have been a fundamental portion of African traditional medication for eras. Africa, with its diverse ecosystems ranging from tropical rainforests to arid deserts, hosts projected 45,000 species of plants, with around 5000 are used medicinally across the continent (Okigbo & Mmeka, 2006; Mahomoodally, 2013; Van Wyk, 2020). These plants are employed in traditional healing practices to treat a wide range of ailments, from minor infections to chronic ailments like diabetes, cancers, and cardiovascular and other infectious sicknesses (Okigbo & Mmeka, 2006; Ibrahim et al., 2018; Babandi et al., 2019; Van Wyk, 2020).

A significant portion of biomedical knowledge, including drugs, vaccines, and bioresources, is derived from natural resources and is closely tied to traditional knowledge and medicine. This knowledge holds social, cultural, and scientific importance, particularly for indigenous peoples, local communities, genetic resources, and scientific research (Mbele et al., 2017). Historically, traditional knowledge has played a foundational role in shaping scientific thought and innovation. African medicinal plants are known for their diverse phytochemical profiles, including alkaloids, flavonoids, terpenoids, saponins, and glycosides. These compounds often exhibit a wide range of biological activities, such as antimicrobial, anti-inflammatory, antimalarial, and anticancer effects (Newman & Cragg, 2020). For instance, the alkaloid reserpine, isolated from *Rauvolfia vomitoria*, was identified as an effective antihypertensive agent and has been used in clinical practice (Kumar et al., 2010).

Many modern medications, such as aspirin, morphine, quinine, artemisinin, and Taxol, originate from medicinal plants, some of which have long been utilized in traditional medicine to address a wide range of ailments. Around 50% of present medications are plant-derived, and over 80% of the world population relies solely on traditional information for health self-care. In some nations of the world, up to 70% of the population uses traditional medicine as a primary healthcare option, in place of modern pharmaceuticals. However, the unsustainable harvesting of medicinal plants from the wild, due to over-exploitation by collectors and sellers, has led to the extinction or scarcity of many plant species (Van Wyk & Prinsloo, 2018). To address this, there is an urgent need for increased investment in growing medicinal plants for use in pharmaceutical research and development, especially post-COVID-19. Cultivation offers numerous benefits over wild harvesting and could contribute to the economies of many African nations through the export of medicinal plants (Street & Prinsloo, 2013).

Medicinal plants have historically played and remain essential in the pharmaceutical industry (Babandi et al., 2012). They have significantly contributed to the economic growth of numerous countries through the production and sale of pharmaceutical products. According to Paul-Marie (2008), the global annual sales of drugs, including those derived from natural products, semi-synthetic, and synthetic sources, amounted to approximately $300 billion. In 1976, the worldwide sales of medicines from natural origins were estimated at $5 billion, and in later years, this figure rose to around $20 billion, highlighting the substantial economic impact of medicinal plants and plant-based drugs on global markets.

16.1.2 Africa: Home of Bioresources Diversity and Economic Development

The African continent is rich in plant biodiversity, providing a huge reservoir of chemical agents that can be harnessed for drug development (Babandi et al., 2021a, b, c). It is estimated that Africa hosts an estimated 40,000 to 45,000 plant species, with about 5000 of them having potential therapeutic benefits (Mahomoodally, 2013). Nigeria alone has identified over 4600 plant species, placing it 11th in Africa for plant diversity (FAO, 1996). These medicinal plants, primarily found in the wild, represent a valuable resource for advancing pharmaceutical research and driving the growth of Africa's pharmaceutical industry. Given the widespread poverty in Africa, the commercialization of medicinal plants offers a significant opportunity to create jobs, particularly for rural and peri-urban populations (Okole & Odhav, 2004). This would provide many with a stable income, improving their standard of living. Additionally, commercial laboratories involved in this industry can help raise awareness about Africa's rich biodiversity. Cultivating medicinal plants not only reduces pressure on wild species, protecting the continent's flora, but also fosters technological collaboration across African countries. Despite the fact that 80% of South Africans rely on traditional medicine for primary healthcare, only a fraction of the country's medicinal plants has been fully utilized for commercialization (Street & Prinsloo, 2013). South Africa's untapped potential in developing pharmaceuticals from its medicinal plants mirrors broader trends across the continent. For instance, a report highlighted that 78 South African medicinal plants are currently sold in informal markets, providing income for vulnerable communities in rural and peri-urban areas (Marula et al., 2019).

In Nigeria, the economic benefits of biodiversity were estimated at over $6 billion in 1990, and by 2002, this figure had risen to over $8 billion annually, driven by bio-prospecting and biotechnology industries (Federal Republic of Nigeria, 2010). Additionally, the production of eucalyptus oil alone has been projected to generate substantial profits, showcasing the economic potential of medicinal plant cultivation (Ibrahim et al., 2019). In Ghana, about 951 tons of herbal medicine were sold in 2010, valued at $7.8 million (Tinde et al., 2012). Across Africa, 75% of people living with HIV/AIDS use complementary and alternative medicine, with the annual market for herbal drugs valued at nearly $43 billion, exceeding the total annual budget of several African nations (Enwonwu, 2003). The growing herbal medicine industry, particularly in South Africa, has created numerous jobs through the production of herbal mixtures by 50–100 private entrepreneurs (Ndhlala et al., 2011). The estimated value of traded medicinal plants across various African countries reflects the immense potential of this industry, with Johannesburg's market alone valued at $300,000, Ghana at $7.8 million, and Gabon and Kenya at $1.5 million and $25,900, respectively (Eziacka et al., 2021). In Tanzania, the Kariakoo Market in Dar es Salaam sees an estimated value of $200,000 (Posthouwer et al., 2018).

Aboelsoud (2010) reported that the ancient Egyptians are celebrated for their early use of medicinal plants. Written records from their civilization, dating back to around 1500 BCE, detail a wide array of herbal remedies. These documents offer valuable insights into the extensive variety of botanicals they employed for healing purposes. Their knowledge and techniques have influenced many modern herbal treatments and medical practices. The Ebers Papyrus, a notable ancient Egyptian medical text from approximately 1550 BCE, contains a wealth of information on herbal medicine and other therapeutic practices. With over 700 magical spells and traditional remedies, many of which are derived from plant sources and herbal extracts, this manuscript is considered one of the most detailed and well-preserved examples of ancient Egyptian medical knowledge (Forshaw, 2014; David & Forshaw, 2023).

16.1.3 Biodiversity and Distribution of Medicinal Plants in Africa

Africa's rich biodiversity is a result of its varied climates, soils, and ecological regions, which include the rainforests of Central Africa, the savannas of East and Southern Africa, and the deserts of the Sahara and Kalahari. Each region contributes uniquely to the continent's pharmacopoeia:

1. *Rainforests*: The rainforests of Central Africa are home to numerous medicinal plants, such as *Prunus africana*, which is used to treat prostate disorders, and *Garcinia kola*, known for its antimicrobial and anti-inflammatory properties (Ndung'u et al., 2024).
2. *Savannas*: In the savanna regions, plants such as *Sutherlandia frutescens* (cancer bush) and *Harpagophytum procumbens* (devil's claw) are widely used for their anti-cancer and anti-inflammatory effects, respectively (Mahomoodally, 2013).
3. *Deserts*: The arid regions of North and Southern Africa, though less diverse in flora, have unique medicinal plants like *Aloe ferox* and *Boswellia sacra* (Frankincense), which have been traditionally used for wound healing and anti-inflammatory purposes (Ofir, 2020).

16.1.4 Importance of Medicinal Plants in Drug Discovery and Development

The historical use of plants as therapeutic agents has significantly influenced the development of modern medicines derived from single chemical compounds. Plants, especially those with traditional medicinal applications, have played a pivotal role in early drug discovery processes. A recent study found that 80% of 122 plant-derived medications are used in ways consistent with their traditional ethnopharmacological applications. Today, drug discovery from terrestrial plants primarily employs bioactivity-guided isolation techniques, which have been instrumental in identifying critical anticancer compounds such as paclitaxel from *Taxus brevifolia* and camptothecin from *Camptotheca acuminata* (Kinghorn et al., 2021). With their thousands of chemical constituents, plants provide a vast reservoir of novel and biologically active molecules (Hostettmann, 1999; Babandi et al., 2012; Babandi et al., 2018; Babandi et al., 2019).

In the United States, around 25% of all prescriptions are directly derived from natural products, while another 25% are based on structural modifications of natural product components (Farnsworth, 1990). Farnsworth also reported that 119 drugs are still commercially produced from higher plants, with 74% of these based on ethnobotanical knowledge. Natural products play a crucial role in treating both human and veterinary diseases. Presently, at least 119 chemical compounds from 90 different plant species are recognized as essential medicines and are used in various countries worldwide (Newman & Cragg, 2020).

The use of natural products for medicinal purposes probably dates back before written history, with early humans using certain plants to heal ailments. In contrast, the practice of treating diseases with pure pharmaceutical compounds is a more modern innovation (Newman & Cragg, 2020). Nevertheless, traditional medicine has been crucial in discovering potent chemicals. For example, antimalarial compounds like quinine, quinidine, cinchonine, and cinchonidine were extracted from *Cinchona succirubra*. Quinine derivatives, such as hydroxychloroquine, are currently being investigated as potential treatments for COVID-19 (Samia et al., 2020). Pain relievers like morphine from *Papaver somniferum* and acetylsalicylic acid from *Salix alba* are also derived from plant sources (Viktorin & Sartorius, 1999; Dewick, 2002). In more recent times, the vinca alkaloids vinblastine and vincristine, which serve as antineoplastic agents, were derived from the Madagascan periwinkle (*Catharanthus roseus*). These compounds were later modified into vinorelbine and vindesine, which are now used for cancer treatment (Newman & Cragg, 2020). Similarly, artemisinin, a sesquiterpenoid endoperoxide, extracted from *Artemisia annua*, is employed to combat multidrug-resistant strains of *Plasmodium*. This plant has been traditionally used in Chinese medicine for its antimalarial properties for centuries. Derivatives of artemisinin, such as artemether and dihydroartemisinin, were synthesized to optimize their pharmacological properties and are now widely used globally (Newman & Cragg, 2020). Other notable drugs developed from ethnomedical knowledge, examples include reserpine, an antihypertensive derived from *Rauvolfia serpentina*; atropine, an anticholinergic extracted from *Atropa belladonna*; digoxin, a cardiotonic obtained from *Digitalis lanata*; physostigmine, a cholinesterase inhibitor sourced from *Physostigma venenosum*; and ephedrine, a sympathomimetic compound derived from *Ephedra sinica* (Fabricant & Farnsworth, 2001). These examples not only highlight the potential of plants as a valuable source of drugs and the strong connection between traditional medicine and modern drug development but also emphasize the importance of continued research into medicinal plants (Salim et al., 2008).

16.1.5 Some Indigenous African Drugs from Medicinal Plants

Approximately 52% of the drugs approved in the USA between 1981 and 2002 were derived from natural sources, primarily plants (Farnsworth, 1990). Among these, African medicinal plants contributed significantly, such as vincristine (I) and vinblastine (II), were derived from the Madagascar periwinkle, combretastatin (III) was extracted from the South African bush willow during the 1950s, and maytansine (IV) was obtained from *Maytenus serrata*, an Ethiopian plant, in the early 1970s. This unveil the immense potential of African medicinal plants for developing therapeutic agents. Many of these drugs, known for their potent anticancer properties, are still used today in various forms globally (Fig. 16.1 and Table 16.1) (Vicki, 2008).

Fig. 16.1 Some important medicinal plants from Africa: (**A**) *Boswellia sacra*; (**B**) *Prunus africana*; (**C**) *Harpagophytum procumbens*; and (**D**) *Artemisia afra* (https://en.wikipedia.org/wiki/)

Table 16.1 Early developed therapeutic agents from African medicinal plants

Drug	Medicinal plant	Current status	Reference
Vincristine and vinblastine	*Catharanthus roseus*	Both alkaloids are well-established chemotherapeutic agents. Vincristine is primarily used to treat leukemias and lymphomas, while vinblastine is utilized for Hodgkin's disease and testicular germ cell cancer. They are integral components of various chemotherapy regimens and continue to be widely used in clinical practice Vincristine and vinblastine have become mainstays in cancer treatment up to today.	Stolinsky et al. (1973)
Combretastatin	*Combretum caffrum*	Combretastatin and its analogs are under investigation as vascular-targeting agents in cancer therapy. Some derivatives have progressed to clinical trials, with ongoing research evaluating their efficacy and safety profiles. Combretastatin derivatives are progressing through clinical evaluations	Tozer et al. (2002) and Seddigi et al. (2017)
Maytansine	*Maytenus serrata*	Maytansine itself exhibited toxicity in early clinical trials, limiting its direct therapeutic use. However, its derivatives, particularly antibody-drug conjugates (ADCs) like trastuzumab emtansine (T-DM1), have been developed to target cancer cells more precisely, reducing systemic toxicity. T-DM1 is approved for treating HER2-positive metastatic breast cancer Maytansine-based ADCs are progressing through clinical evaluations	Chabner et al. (1978)
Niprisan	*Entandrophragma utile*, *Chenopodium ambrosioides*, and *Petiveria alliacea*	Niprisan, also known as Nicosan, was developed in Nigeria as a phytomedicine for managing sickle cell anemia. It showed promise in clinical studies; however, its production faced challenges, leading to its withdrawal from the market. Efforts are ongoing to address	Perampaladas et al. (2010)

(continued)

Table 16.1 (continued)

Drug	Medicinal plant	Current status	Reference
		these issues and potentially reintroduce the drug	
Ipomoeassin	*Ipomoea squamosa*	Ipomoeassin F is a natural product isolated from *Ipomoea squamosa*, exhibiting potent cytotoxicity against cancer cell lines. Research is in the preclinical stage, focusing on understanding its mechanism of action and therapeutic potential	Zong et al. (2016) and Zong et al. (2020)
Schweinfurthins	*Macaranga alnifolia*	Schweinfurthins are natural compounds with notable anti-cancer properties, particularly against central nervous system tumors and renal cell carcinoma. They are currently in the pre-clinical research phase, with studies aimed at elucidating their mechanisms and optimizing their efficacy	Koubek et al. (2018) and Zhang et al. (2022)

Niprisan, a drug used to treat sickle cell anemia, was formulated using herbs from Nigeria, drawing on Yoruba traditional medicine (Sunday et al., 2012). Nigeria is home to the largest population of individuals with sickle cell anemia, with approximately four million affected individuals and over 150,000 offspring born yearly with the disorder. Despite extensive research, only one FDA-approved drug, hydroxyurea, is available for treating the disease (Steinberg, 2003; Kumar et al., 2010). In the late 1980s, the National Institute for Pharmaceutical Research and Development (NIPRD) was established by Nigeria's Ministry of Science and Technology to advance research on traditional herbal medicines. In May 2011, NIPRD developed Niprisan, a formulation derived from three anti-sickling herbs (*Entandrophragma utile*, *Chenopodium ambrosioides*, and *Petiveria alliacea*), inspired by Yoruba traditional medicine. Currently, three additional herbal formulations are awaiting development (Sunday et al., 2012). Similarly, Madagascar's research institutions through collaboration with the National Cancer Institute (NCI), potent antiproliferative compounds, Ipomoeassin and Schweinfurthins, were discovered in the medicinal plants *Ipomoea squamosa* and *Macaranga alnifolia* in the 1990s (Kingston, 2012). More recently, the development of Covid-Organics by the Malagasy Institute of Applied Research (IMRA) and the National Pharmacology Research Centre, formulated from *Artemisia annua* and other plants, highlights the importance of utilizing African medicinal plants for drug research and development. These efforts showcase the potential of African medicinal plants for creating treatments for various diseases.

16.2 Review Methods

This chapter employed a systematic approach to produce and analyze information on contribution of African medicinal plants, their bioactive compounds, and their therapeutic potential to modern drug discovery and development. The methods included:

1. *Literature Review:* A comprehensive literature search was conducted using databases such as PubMed, Scopus, and Google Scholar. Keywords and phrases, including "African medicinal plants," "Extractions" "bioactive compounds," "traditional medicine and practices," "drug discovery and development," "pharmacological activities," "Mordern drugs," and "medicinal potentials and challenges" were used. The search focused on peer-reviewed articles, ethnobotanical surveys, phytochemical studies, and pharmacological evaluations published between 2000 and 2023.
2. *Inclusion and Exclusion Criteria:* Articles were included in the chapter if they documented the medicinal uses of African plants, identified their bioactive compounds, or reported therapeutic activities. Studies focusing on traditional knowledge and drug development processes were prioritized. Excluded were studies without peer review, articles unrelated to medicinal plants, those from non-African plant and continents, and those lacking detailed and clear methodologies.
3. *Data Collection and Analysis:* Information on traditional uses, bioactive compounds, and pharmacological activities of plants like *Warburgia salutaris*, *Prunus africana*, *Catharanthus roseus*, *Artemisia afra*, *Sutherlandia frutescens*, *Harpagophytum procumbens,* etc., was extracted. Data on drug discovery contributions, such as vincristine, vinblastine, and Niprisan, were also reviewed.
4. *Phytochemical Techniques:* Insights from studies employing advanced phytochemical techniques—such as chromatography, spectroscopy, and molecular characterization—were reviewed to highlight the isolation and structural analysis of bioactive compounds.
5. *Ethnobotanical Integration:* Traditional knowledge from African healers, as documented in ethnobotanical surveys, was analyzed to identify culturally significant plants with therapeutic potential.

16.3 Historical Background

The use of medicinal plants in Africa is deeply embedded in ancient traditions, forming a cornerstone of indigenous healthcare systems. Across the continent, the therapeutic application of plants predates recorded history, as evidenced by oral traditions and archaeological findings that highlight their centrality in addressing ailments and promoting well-being. This extensive heritage exemplifies the ingenuity of African societies in utilizing their natural environment to meet diverse healthcare needs (Gurib-Fakim, 2006; Hewson, 2021). The historical evolution of African medicinal plants emphasizes their enduring significance in both traditional and contemporary healthcare paradigms. As repositories of unparalleled biodiversity and cultural heritage, these plants continue to drive innovation in drug discovery. By fostering the integration of traditional knowledge with advanced research methodologies, Africa's medicinal flora remains poised to offer sustainable and effective therapeutic solutions to pressing global health challenges.

Traditional healers, including herbalists, spiritual healers, and diviners, play a central role in African communities, often serving as the primary healthcare providers. These healers use medicinal plants to treat a range of ailments, often guided by a holistic understanding of health that integrates physical, spiritual, and social well-being. The cultural importance of these plants is further highlighted by their use in ceremonies, rites of passage, and as protective charms against evil spirits (Gurib-Fakim, 2006; Van Wyk, 2020).

16.3.1 Traditional Knowledge and Cultural Significance

Medicinal plants are deeply embedded in the cultural practices of African societies. Traditional healers, known as herbalists, diviners, or shamans, play a crucial role in the community, relying on extensive knowledge passed down through generations. They are custodians of extensive botanical knowledge and play a vital role in community health care, using medicinal plants to treat a wide range of ailments from minor infections to chronic illnesses (Gurib-Fakim, 2006). These healers often undergo rigorous training that combines spiritual rituals and the study of medicinal plants and their applications. This training ensures that the knowledge is passed down accurately through generations, preserving both the medicinal and cultural significance of the practices (Cocks & Møller, 2002).

The knowledge encompasses not only the identification and use of medicinal plants but also the understanding of preparation methods and dosages tailored to individual patients (Gurib-Fakim, 2006). In many African cultures, traditional medicine is often the first line of healthcare due to its accessibility and affordability compared to modern pharmaceuticals. Moreover, traditional practices are often intertwined with spiritual beliefs, where illnesses are sometimes seen as a result of spiritual imbalance or ancestral displeasure, and medicinal plants are used in rituals to restore harmony (Hewson, 2021).

16.3.2 Holistic Healing Approaches

The holistic approach in traditional African medicine considers not just the symptoms of a disease, but the individual as a whole, including their environment, lifestyle, and spiritual state. Illnesses are often seen as a disturbance in the balance of physical and spiritual elements. Therefore, treatment may involve a combination of herbal remedies, spiritual rituals, and community involvement (Ozioma & Chinwe, 2019). For example, in West Africa, illnesses are sometimes believed to be caused by witchcraft or ancestral displeasure. Healers use plants like *Ficus exasperata* and *Ocimum gratissimum* in rituals to cleanse patients spiritually while treating physical symptoms (Githinji, 2020).

16.3.3 The Role of African Medicinal Plants in Primary Healthcare

In many African countries, traditional medicine is the first line of defense against illnesses due to its accessibility, affordability, and cultural acceptance. Medicinal plants are commonly used to treat both common ailments and more complex conditions, ranging from colds and digestive disorders to malaria and diabetes (Abdullahi, 2011). The knowledge of plant-based remedies is often localized, with specific plants and preparations used according to the needs and customs of individual communities.

The World Health Organization (WHO) reports that in certain African countries, as much as 80% of the population depends on traditional medicine as their main source of healthcare (WHO, 2013). This reliance is due in part to limited access to modern healthcare facilities and the high cost of pharmaceutical drugs. Medicinal plants offer a cost-effective and locally available resource for treating a wide range of conditions. For instance, plants like *Vernonia amygdalina* (bitter leaf) and *Azadirachta indica* (neem) are commonly used to manage diabetes and malaria, respectively (Kevin et al., 2023; Babandi et al., 2012).

16.4 Key Contributions of African Medicinal Plants to Modern Drug Discovery and General Health

African medicinal plants have played a crucial role in modern pharmacology, particularly in the treatment of malaria, cancer, infectious diseases, and inflammatory conditions. Their bioactive compounds have contributed to the development of life-saving therapies, demonstrating the continent's vast potential for drug discovery.

16.4.1 Antimalarial Drugs

The most notable contribution is artemisinin, derived from *Artemisia annua*, which remains the cornerstone of artemisinin-based combination therapies (ACTs) for malaria (Tu, 2011). Additionally, African medicinal plants such as *Artemisia afra*, *Azadirachta indica* (neem), *Cryptolepis sanguinolenta*, *Ficus sycomorus*, *Moringa oleifera*, and *Annickia chlorantha* have demonstrated potent antiplasmodial and antipyretic activities, providing promising leads for novel antimalarial treatments (Willcox & Bodeker, 2004; Asase et al., 2010;; Babandi et al., 2012, 2019). Some African plants such as *Ficus sycomorus*, *Calotropic procera*, etc., were reported to have insecticidal properties against the major malarial vector *Anopheles colluzzii* and hence capable of reducing malarial incidences in the region (Anosike et al., 2021; Babandi et al., 2021a, b, c, 2023).

16.4.2 Cancer Treatments

Cancer is a growing public health concern in Africa, with increasing incidence rates due to factors such as population aging, urbanization, and the prevalence of risk factors like smoking and infections (Bray et al., 2018). Many African countries face challenges in accessing modern cancer therapies, making traditional medicine a critical component of cancer care (Kuete & Efferth, 2015).

The Madagascar periwinkle (*Catharanthus roseus*) has yielded vincristine and vinblastine, essential chemotherapeutic agents used to treat leukemias, lymphomas, and solid tumors (Noble, 1990). Other African plants with anticancer potential include *Annona senegalensis*, *Hypoxis hemerocallidea*, and *Sutherlandia frutescens*, which exert effects through apoptosis induction, angiogenesis inhibition, and cell cycle modulation (Kuete, 2013). Also, *Indigofera hirsuta* and *Afrormosia laxiflora* (from Nigeria and other West African countries) extracts were reported to possess antioxidant, anti-inflammatory, and anticancer activities (Muhammad et al., 2019).

16.4.3 Antimicrobial Agents

The growing challenge of antibiotic resistance has intensified interest in African medicinal plants for novel antimicrobial agents. Plants such as *Combretum molle*, *Garcinia kola*, and *Zanthoxylum chalybeum* have shown significant antibacterial, antifungal, and antiviral properties, often through mechanisms such as cell wall disruption and inhibition of protein synthesis (Cowan, 1999). Their bioactive compounds present alternative strategies for addressing multidrug-resistant infections.

16.4.4 Anti-inflammatory and Analgesic Agents

African medicinal plants have also contributed to the development of anti-inflammatory and pain-relieving therapies. *Harpagophytum procumbens* (devil's claw) and *Boswellia sacra* (frankincense) are widely studied for their ability to inhibit cyclooxygenase (COX) and 5-lipoxygenase pathways, reducing inflammation and pain (Van Wyk, 2015). These plants offer alternatives for managing chronic inflammatory diseases such as arthritis.

16.4.5 Diabetes Management

With the rising incidence of diabetes in Africa, traditional medicine has increasingly focused on plants with hypoglycemic effects. *Vernonia amygdalina* (bitter leaf) and *Momordica charantia* (bitter melon) are popular choices for managing blood sugar levels (Afolayan & Sunmonu, 2010; Kasali et al., 2021). Also, *Clatropis procera*, *Azadritcha indica*, *Combretum micranthum*, and *Gymnema sylvestre* were reported for hypoglycemic and antidiabetic potentials (Babandi et al., 2012, 2016, 2018, Ibrahim et al., 2017).

16.4.6 Respiratory Conditions

Plants such as *Eucalyptus globulus* (blue gum tree) and *Zingiber officinale* (ginger) are used to treat respiratory conditions, including asthma and bronchitis. The leaves of *Eucalyptus* are inhaled as steam to relieve congestion, while ginger is used in teas to soothe coughs and sore throats (Van Wyk, 2020).

16.4.7 Role in Preventive Healthcare and Nutrition

In addition to their therapeutic uses, many African medicinal plants play a crucial role in preventive healthcare. They are often included in daily diets for their nutritional and health benefits. For example, *Moringa oleifera* leaves are rich in vitamins and minerals and are used to boost immunity and improve overall health (Bonomi et al., 2018). Similarly, the baobab fruit (*Adansonia digitata*) is known for its high vitamin C content and is consumed to prevent scurvy and enhance immune function (Abubakar et al., 2016). Medicinal plants are also used to prevent infections and maintain hygiene. For instance, *Allium sativum* (garlic) is widely used for its antimicrobial properties and is consumed regularly to ward off infections (Gurib-Fakim, 2006).

16.4.8 Spiritual and Ritualistic Uses

Medicinal plants in Africa are not only valued for their pharmacological properties but also for their spiritual significance. Plants are often employed in rituals to ward off evil spirits, cleanse the body and soul, or protect against witchcraft. For example, *Ocimum gratissimum* (African basil) is used in purification rituals and to ward off evil spirits, while *Ficus exasperata* leaves are used in spiritual cleansing (Githinji, 2020). In many African communities, the use of medicinal plants in rituals is believed to restore harmony and balance, which are seen as essential components of health. The integration of medicinal plants in spiritual practices underscores their holistic role in traditional medicine, where physical, emotional, and spiritual health are interconnected (Githinji, 2020; Hewson, 2021).

16.4.9 Challenges and Future Directions

Despite their promise, challenges such as bioactive variability, sustainability concerns, and limited research funding hinder the full integration of African medicinal plants into pharmaceutical development. Collaborative efforts among African researchers, traditional healers, and international stakeholders are essential to overcome these barriers. Advances in synthetic biology and metabolomics may further support the standardization and large-scale production of plant-derived pharmaceuticals (Isah, 2019).

16.5 Research Methodologies in African Medicinal Plant Drug Discovery

The systematic investigation of African medicinal plants for drug discovery requires an interdisciplinary approach that integrates traditional ethnobotanical knowledge with modern scientific methodologies (Sasidharan et al., 2011). This comprehensive framework encompasses ethnobotanical surveys, phytochemical analyses, bioassays, and pharmacological investigations, each contributing to the identification, isolation, and validation of bioactive compounds with therapeutic potential (Cowan, 1999).

16.5.1 Ethnobotanical Surveys

Ethnobotanical surveys are essential for documenting indigenous knowledge related to the medicinal applications of plants (Heinrich et al., 2020).

1. *Data Collection:* Structured interviews, participatory observations, and ethnopharmacological documentation with traditional healers and local communities facilitate the identification of medicinal plants used for treating inflammation, infections, and pain (Van Wyk, 2015).
2. *Cultural Context:* These surveys explore plant preparation methods, dosage formulations, and their cultural or spiritual significance, providing a holistic understanding of traditional medical systems.
3. *Prioritization:* Frequently utilized plants with consistent therapeutic claims across different communities are prioritized for scientific validation and pharmacological studies.

16.5.2 Phytochemical Analysis

Phytochemical analysis bridges traditional medicine and modern pharmacology by isolating and characterizing active constituents from medicinal plants (Kuete, 2013; Babandi et al., 2018, 2023).

1. *Extraction:* Various solvent-based techniques, including maceration, Soxhlet extraction, and supercritical fluid extraction, are employed to isolate bioactive compounds (Azwanida, 2015; Babandi et al., 2021a, b, c).
2. *Isolation:* Chromatographic techniques such as thin-layer chromatography (TLC) and high-performance liquid chromatography (HPLC) enable the separation of complex plant constituents (Ibrahim et al., 2017).
3. *Structural Elucidation:* Advanced spectroscopic techniques, including nuclear magnetic resonance (NMR), mass spectrometry (MS), and Fourier-transform infrared spectroscopy (FTIR), are utilized to determine the structural composition and functional groups of active compounds (Ibrahim et al., 2017; Babandi et al., 2018, 2023).

16.5.3 Bioassays

Bioassays are critical for evaluating the biological efficacy of crude extracts, fractions, and purified compounds (Cos et al., 2006).

1. In Vitro Assays:

    (a) Antimicrobial activity is assessed using disc diffusion and broth microdilution techniques to determine efficacy against bacterial and fungal pathogens (Cowan, 1999).
    (b) Cytotoxicity assays evaluate potential anticancer properties using established cancer cell lines (Kuete, 2013).

(c) Enzyme inhibition assays target specific pathways such as cyclooxygenase (COX) and acetylcholinesterase, which are relevant to inflammatory and neurodegenerative conditions.

2. In Vivo Assays:

   (a) Animal models assess pharmacokinetics, efficacy, and safety profiles of bioactive compounds.
   (b) Behavioral studies examine the analgesic, anxiolytic, and neuroprotective effects of plant-derived formulations (Van Wyk, 2015).

16.5.4 Pharmacological Studies

Pharmacological investigations elucidate the mechanisms of action, bioavailability, and therapeutic potential of plant-derived compounds (Sasidharan et al., 2011).

1. *Mechanistic Studies:* Research explores molecular pathways such as apoptosis induction, inflammatory mediator inhibition, immunomodulation, and enzyme inhibition (Ibrahim et al., 2017).
2. *Pharmacokinetics and Pharmacodynamics (PK/PD):* PK studies analyze the absorption, distribution, metabolism, and excretion (ADME) of bioactive compounds, while PD studies define dose-response relationships and therapeutic indices (Isah, 2019).
3. *Toxicological Evaluations:* Acute, sub-chronic, and chronic toxicity assessments provide critical insights into safety profiles and inform appropriate dosage guidelines.
4. *Computational Approaches:* Molecular docking and simulation techniques predict interactions between bioactive compounds and biological targets, accelerating the drug discovery process (Heinrich et al., 2020; Babandi et al., 2022).

16.5.5 Contributions to Contemporary Medicine

In recent decades, African medicinal plants have emerged as critical sources of novel therapeutic agents. Global health challenges, including malaria, cancer, and antibiotic resistance, have stressed the importance of plant-derived compounds. Artemisinin, extracted from *Artemisia annua*, revolutionized malaria treatment and stands as a testament to the successful integration of traditional knowledge with modern medical science (Tu, 2011). Collaborative initiatives involving African researchers, traditional healers, and international institutions have further demonstrated the value of equitable partnerships. A notable example is the development of Niprisan, a phytomedicine for sickle cell anemia, which highlights the potential of combining ethnobotanical insights with contemporary scientific methodologies to address unmet medical needs (Sunday et al., 2012).

16.6 Integration of Bioactive Compounds into Contemporary Drug Development

The integration of bioactive compounds from medicinal plants into modern drug development necessitates a multidisciplinary approach that merges traditional ethnobotanical knowledge with cutting-edge scientific methodologies. This process encompasses key stages, including compound identification, structural refinement, formulation, and large-scale production, while addressing economic and environmental considerations (Tu, 2011).

16.6.1 Incorporation into Drug Development Pipelines

16.6.1.1 Discovery and Preclinical Evaluation

1. *Target Identification and Validation:* High-throughput screening (HTS) and molecular docking techniques facilitate the identification of interactions between bioactive compounds and biological targets, such as enzymes and receptors (Tu, 2011; Babandi et al., 2022).
2. *Pharmacokinetic and Toxicological Profiling:* Studies on absorption, distribution, metabolism, and excretion (ADME), alongside toxicity evaluations, establish dosing regimens and safety parameters.

16.6.1.2 Structural Optimization and Clinical Development

1. Medicinal chemistry modifications enhance solubility, bioavailability, and therapeutic efficacy, as evidenced by the development of artemisinin derivatives (Méndez-Lucio & Medina-Franco, 2017; Kim et al., 2024).
2. Clinical trials (Phases I–III) systematically assess safety, efficacy, and comparative performance relative to existing treatments.
3. Regulatory agencies such as the FDA and EMA evaluate trial data before granting approval, ensuring adherence to stringent quality control measures (Isah, 2019).

16.6.2 Reproduction in the Pharmaceutical Industry

16.6.2.1 Plant-Based Extraction

Advantages Sustainable and cost-effective, particularly for widely cultivated species (Babandi et al., 2021a, b, c).

Challenges Overharvesting and habitat destruction threaten biodiversity, particularly in Africa's medicinal plant reserves (Okigbo et al., 2008).

16.6.2.2 Chemical and Semi-Synthetic Chemical Synthesis

Advantages Enables scalable, high-purity production with greater consistency in pharmacological properties.

Challenges Synthetic replication of structurally complex bioactive molecules is resource-intensive and requires specialized expertise (Méndez-Lucio & Medina-Franco, 2017; Kim et al., 2024).

Semi-synthetic approaches leverage plant-derived precursors, optimizing efficiency while reducing reliance on wild plant populations (e.g., paclitaxel production from *Taxus* species).

16.6.2.3 Economic Viability, Scalability, and Feasibility of African Medicinal Plants

Plant-Based Methods Cost-effective for abundant species but constrained by plant availability and environmental factors.

Synthetic Production High initial costs but economically advantageous for long-term, large-scale applications.

Scalability Chemical synthesis enables greater consistency and sustainability, whereas plant-based extraction faces ecological limitations.

Capacity Development Strengthening regional research infrastructure and fostering collaborations between African and global stakeholders can enhance the continent's role in pharmaceutical innovation (Isah, 2019).

16.6.2.4 Isolation of Bioactive Agents vs Synthesis: A Comparative Perspective

Isolation Retains the natural integrity of bioactive compounds, but yields are inconsistent due to environmental variability (Babandi et al., 2012, 2021a, b, c).

Synthesis Ensures reliable production but requires sophisticated technological capabilities and significant financial investment and thus not cost-effective.

Hybrid Strategies A combined approach utilizing plant-based isolation for compound discovery and chemical synthesis for large-scale production offers a balanced and sustainable solution to this challenge.

16.6.3 Recent Developments and Research in African Medicinal Plants

Recent scientific research has focused on validating the traditional uses of African medicinal plants and identifying bioactive compounds that could lead to new drug discoveries (Babandi et al., 2012). Advances in phytochemistry, genomics, and biotechnology are enabling researchers to better understand the mechanisms of action of these plants and their potential therapeutic applications. For example, compounds isolated from *Securidaca longipedunculata*, commonly used in traditional medicine for pain relief and inflammation, have shown promising anti-cancer properties in recent studies (Obasi et al., 2018). Similarly, research on *Annona senegalensis* has revealed its potential as a source of novel antimicrobial agents (Yakubu et al., 2020). Other plants, such as *Clatropis procera*, *Azadritcha indica*, and *Ficus sycomorus*, were tested for antidiabetics, antiplasmodium, and antimicrobials, respectively (Babandi et al., 2012, 2019).

Street and Prinsloo (2013) highlighted ten commercially significant medicinal plants in South Africa, including *Aspalathus linearis*, *Aloe ferox*, *Agathosma betulina*, *Harpagophytum procumbens*, *Sutherlandia frutescens*, *Merwilla natalensis*, *Pelargonium sidoides*, *Hypoxis hemerocallidea*, *Sclerocarya birrea*, and *Siphonochilus aethiopicus*. Similarly, Mahomoodally (2013) identified ten African medicinal plants with strong potential for development into phytopharmaceuticals for treating infectious diseases. These include *Cyclopia genistoides* (South Africa), Acacia (Senegal, Sub-Saharan, North and Southern part of Africa), *Pelargonium sidoides*, *Aloe ferox* (South Africa and Lesotho), *Artemisia herba-alba* (North Africa), *Centella asiatica* (Widely distributed in Africa), *Catharanthus roseus* (Madagascar), *Aspalathus linearis* (South Africa), *Harpagophytum procumbens* (South Africa, Bostwana, and Namibia), and *Momordica charantia* (Whole Africa).

Other economically valuable plants include *Fadogia agrestis*, *Griffonia simplicifolia*, *Monodora myristica*, *Xylopia aethiopica*, *Aframomum melegueta*, *Pteleopsis suberosa*, *Khaya senegalensis*, and *Voacanga africana* (Tinde et al., 2012). According to Kingston (2012) and Shilonga et al. (2019), the most traded plant species in Tanzania include *Albizia anthelmintica*, *Vachellia nilotica*, *Zanha africana*, *Zanthoxylum chalybaeum*, and *Warburgia stuhlmannii*. In Nigeria, according to the Federal Republic of Nigeria (2010), economically valuable plants that could generate substantial revenue if harnessed for their resources include *Adansonia digitata*, *Sesame indicum*, *Persia americana*, and *Vitellaria paradoxa*.

16.6.3.1 Major African Medicinal Plants and Their Uses

African medicinal plants have diverse applications in traditional medicine, ranging from treating common ailments to managing chronic diseases. This subsection will highlight some key medicinal plants, bioactive compounds, and discuss

Table 16.2 Key bioactive compounds from African medicinal plants and their sources

Compound/ drug	Plant source	Traditional uses	Therapeutic potential	References
Artemisinin	*Artemisia afra*	Malaria, fever	Antimalarial	Willcox and Bodeker (2004)
Reserpine	*Rauvolfia vomitoria*	Hypertension, mental disorders	Antihypertensive, antipsychotic	Kumar et al. (2010)
Cryptolepine	*Cryptolepis sanguinolenta*	Malaria, fever	Antimalarial, antimicrobial	Mensah et al. (2019) and Komlaga et al. (2015)
Vincristine, vinblastine	*Catharanthus roseus*	Cancer treatment in traditional medicine	Anticancer (chemotherapy)	Noble (2016)
Betulinic acid	*Ziziphus mucronata*	Wound healing, inflammation	Anticancer, anti-inflammatory	Fulda (2008)
Hypoxoside	*Hypoxis hemerocallidea*	Immune support, cancer	Immunomodulatory, anticancer	Ojewole (2006)
Ellipticine	*Ochrosia elliptica*	General therapeutic use in traditional medicine	Anticancer	Sathya et al. (2018) and Stiborova et al. (2006)
Ajugarin I	*Ajuga remota*	Inflammation, pain	Anti-inflammatory, analgesic	Njoroge and Bussmann (2006)
Michellamine B	*Ancistrocladus korupensis*	–	Anti-HIV	Kinghorn et al. (2021) and Thomford et al. (2018)
Sesquiterpene lactones	*Vernonia amygdalina*	Gastrointestinal ailments, diabetes, cancer	Anticancer, anti-inflammatory	Nerdy et al. (2021)

their traditional and therapeutic uses across different regions of Africa as suggested below (Table 16.2).

16.6.3.2 Examples of Key Plants

1. *Sutherlandia frutescens* (Cancer bush): Traditionally used in Southern Africa to treat cancer, diabetes, and viral infections. Recent studies have shown that it possesses anti-inflammatory and immune-boosting properties, supporting its traditional uses (Gouws et al., 2021).
2. *Artemisia afra* (African wormwood): Used widely for treating respiratory ailments, fever, and malaria. It is one of the few indigenous African species of the Artemisia genus and has shown antimalarial properties in preclinical studies (Willcox & Bodeker, 2004).

3. *Vernonia amygdalina* (Bitter leaf): Commonly used in West Africa to manage diabetes, gastrointestinal issues, and as a general tonic. Phytochemical studies have revealed that its leaves contain sesquiterpene lactones, which have hypoglycemic and anticancer effects (Nerdy et al., 2021).
4. *Prunus africana* (African cherry): Used traditionally to treat prostate issues in Central and East Africa. The bark of *Prunus africana* has been commercialized for its phytosterols, which are used in the treatment of benign prostatic hyperplasia (BPH) (Ndung'u et al., 2024).
5. *Borreria verticillate* (Shrubby falsebutton weed): The aqueous leaf extract of *B. verticillata* has demonstrated hepatocurative effects in rats with CCl4-induced hepatotoxicity, likely due to its phytochemical composition (Murtala et al., 2015). In Senegal, the plant is traditionally used to treat bacterial skin infections and leprosy, while in Nigeria, it serves various purposes, including analgesic, anti-inflammatory, antitumor, antimicrobial, larvicidal, antioxidant, gastrointestinal, anti-ulcer, and hepatoprotective applications, with alkaloids and iridoids identified as its primary active compounds (Murtala et al., 2015).
6. *Hypoxis hemerocallidea* (African potato): Known for its immunomodulatory and anticancer properties, it is widely used in Southern Africa for treating HIV/AIDS and cancer (Ojewole, 2006).

16.6.4 African Regional Variations in Medicinal Plant Use

The use of medicinal plants varies significantly across Africa due to differences in regional flora, cultural practices, and local health needs. This diversity is reflected in the different plants used and the various ailments they are believed to treat.

16.6.4.1 West Africa

In West Africa, the use of medicinal plants is often guided by a rich tradition of herbal medicine and spirituality. Plants like *Cryptolepis sanguinolenta* are used to treat malaria, while *Garcinia kola* (bitter kola) is used for its antimicrobial and anti-inflammatory properties (Komlaga et al., 2015; Mensah et al., 2019).

16.6.4.2 East Africa

East African traditional medicine often focuses on plants with antimicrobial and antiparasitic properties due to the region's high prevalence of infectious diseases. *Warburgia ugandensis* and *Azadirachta indica* (neem) are widely used to treat bacterial infections and malaria, respectively (Njoroge & Bussmann, 2006).

16.6.4.3 Southern Africa

Southern Africa has a diverse range of medicinal plants used in various ways. *Sceletium tortuosum* (kanna) is used for its mood-enhancing effects, while *Aloe ferox* is used for its laxative and skin-healing properties (Van Wyk, 2020).

16.6.4.4 North Africa

In North Africa, plants like *Nigella sativa* (black seed) and *Artemisia herba-alba* are commonly used for their digestive and antimicrobial properties. The use of these plants reflects the region's unique Mediterranean flora and the influence of Arabic and Mediterranean medicinal practices (Bouasla & Bouasla, 2017).

16.6.5 Contemporary Relevance and Integration with Modern Medicine

The significance of African medicinal plants extends beyond traditional medicine into modern healthcare. There is growing recognition of the potential of these plants to provide novel bioactive compounds for drug development (Babandi et al., 2018). Integrative medicine, which combines traditional knowledge with modern medical practices, is increasingly being promoted to enhance healthcare outcomes in Africa. For instance, *Vernonia amygdalina* (bitter leaf) is not only used in traditional medicine for diabetes management but has also been studied for its potential to reduce blood glucose levels in clinical settings (Kasali et al., 2021). Similarly, *Hypoxis hemerocallidea* (African potato) is used both traditionally and in modern contexts to boost the immune system and manage HIV/AIDS (Ojewole, 2006). The integration of traditional and modern medicine can improve healthcare delivery, especially in areas with limited access to conventional medical facilities. This approach also respects and preserves the cultural heritage of traditional medicine while leveraging scientific advancements to ensure safety and efficacy (Gurib-Fakim, 2006).

16.6.6 Transition from Traditional Use to Modern Drug Discovery

16.6.6.1 Transition

The transition from traditional use of African medicinal plants to modern drug discovery has been driven by a combination of cultural heritage, scientific curiosity, and the pressing need for new therapeutic agents. Traditional healers have long used plant-based remedies to treat various ailments, and many of these practices have caught the attention of researchers seeking novel compounds for drug development. This transition involves validating traditional knowledge through scientific methods, isolating bioactive compounds, and understanding their mechanisms of action (Newman & Cragg, 2020). Pioneering ethnobotanical studies from the nineteenth and early twentieth centuries facilitated the identification of bioactive compounds that would eventually influence pharmaceutical advancements. By the mid-twentieth century, the potential of traditional medicine to complement modern healthcare systems began to gain wider recognition. Advances in phytochemical analysis and pharmacological techniques enabled the isolation and characterization of active compounds from medicinal plants. This period saw transformative discoveries, such as the identification of vincristine and vinblastine from the Madagascar periwinkle (*Catharanthus roseus*), which have become indispensable in the treatment of various cancers (Noble, 1990).

16.6.6.2 The Drug Discovery Process: From Plant to Pill

Ethnopharmacology, the study of traditional knowledge and use of medicinal plants, has been instrumental in identifying potential lead compounds. By focusing on plants with documented therapeutic use in traditional medicine, researchers can prioritize candidates for bioactivity screening and further study (Mekonnen et al., 2022). For example, the study of *Cryptolepis sanguinolenta* in West Africa, traditionally used to treat malaria, led to the discovery of cryptolepine, a compound with significant antimalarial activity (Komlaga et al., 2015; Mensah et al., 2019).

The process of drug discovery from medicinal plants (Fig. 16.2) involves several stages, each requiring meticulous scientific methodologies to ensure safety, efficacy,

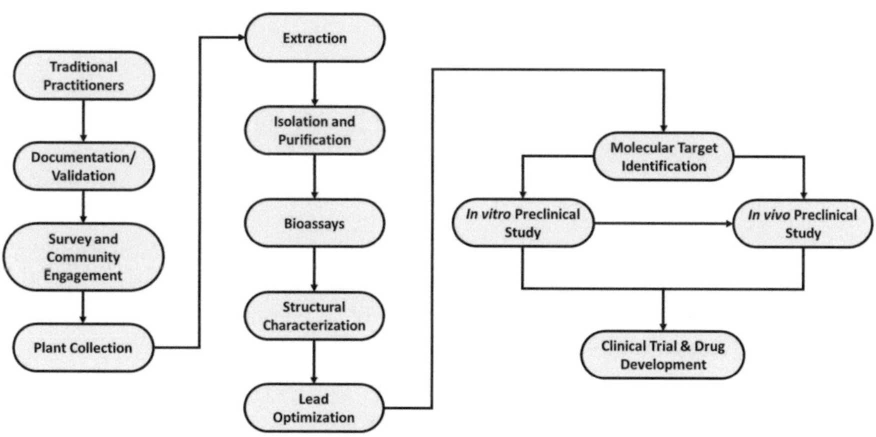

Fig. 16.2 Drug discovery and development process

and sustainability (Babandi et al., 2019). This section outlines the journey from plant selection to the development of a marketable drug, highlighting key steps and modern technological advancements.

16.6.6.3 Documentation and Validation of Traditional Knowledge

The first step in the transition process is often the documentation and validation of traditional medicinal knowledge. Ethnobotanical surveys play a critical role in capturing the wealth of knowledge held by traditional healers about the therapeutic uses of plants. This documentation is essential for identifying candidate plants for further study and ensuring the preservation of indigenous knowledge (Mekonnen et al., 2022).

16.6.6.4 Ethnobotanical Surveys and Their Importance

Ethnobotanical surveys involve systematic data collection on the local use of plants, including the parts used, preparation methods, dosages, and conditions treated. These surveys provide a foundation for identifying plants with potential medicinal properties that warrant scientific investigation. Researchers collaborate with traditional healers and local communities to gather information on the medicinal uses, preparation methods, and specific ailments treated by different plants (Mekonnen et al., 2022). For instance, research on the use of *Sutherlandia frutescens* (cancer bush) in South Africa for treating cancer led to the isolation of compounds that have been found to possess anti-cancer properties in preclinical studies (Gouws et al., 2021). Engaging communities ensures that the knowledge is preserved and that the benefits of drug discovery are shared. For example, the identification of *Cryptolepis sanguinolenta* as a potential antimalarial was based on ethnobotanical data from Ghana and Nigeria (Olorunnisola et al., 2013; Komlaga et al., 2015; Mensah et al., 2019).

16.6.6.5 Collaboration Between Traditional Healers and Scientists

Successful transition from traditional use to drug discovery often requires partnerships between traditional healers and scientists to foster collaboration, bridging the divide between traditional knowledge and modern scientific research, enabling the respectful and ethical use of indigenous knowledge. This collaboration is critical in guiding the selection of plants for scientific study and ensuring culturally sensitive approaches to research (Cox & Balick, 2020).

16.6.7 Phytochemical Analysis and Bioactive Compound Isolation

Once plants with potential therapeutic uses have been identified, the next step is phytochemical analysis to isolate and identify bioactive compounds. Advanced techniques in phytochemistry, such as chromatography and mass spectrometry, are used to isolate compounds, while bioassays help determine their biological activity (Ibrahim et al., 2018; Babandi et al., 2018).

16.6.7.1 Techniques for Compound Isolation

Modern drug discovery utilizes various techniques to isolate bioactive compounds from plants. Techniques like high-performance liquid chromatography (HPLC), gas chromatography-mass spectrometry (GC-MS) (Ibrahim et al., 2017, 2018), and nuclear magnetic resonance (NMR) spectroscopy are employed for the purification and characterization of these compounds.

For example, the isolation of alkaloids from *Rauvolfia vomitoria*, traditionally used in Nigeria for hypertension, has led to the development of reserpine, a drug used to treat high blood pressure and psychiatric disorders (Kumar et al., 2010).

16.6.7.2 Bioassay-Guided Fractionation

Bioassay-guided fractionation involves the use of biological assays to identify active fractions of plant extracts. This method allows researchers to pinpoint specific compounds responsible for observed biological effects. For instance, research on *Catharanthus roseus* (Madagascar periwinkle) identified vinca alkaloids, which led to the development of vinblastine and vincristine, potent anti-cancer drugs (Noble, 2016).

16.6.7.3 Mechanistic Studies and Preclinical Development

Following the identification of bioactive compounds, mechanistic studies are conducted to understand how these compounds exert their effects at the molecular and cellular levels. This step is crucial for elucidating the pathways and targets involved, which can guide the optimization of lead compounds for drug development.

16.6.7.4 Molecular Targets and Mechanisms of Action

Mechanistic studies often focus on identifying molecular targets such as enzymes, receptors, or ion channels that interact with bioactive compounds. Understanding these interactions helps in designing more potent derivatives and improving the efficacy and safety profile of potential drugs (Babandi et al., 2022). For example, studies on *Vernonia amygdalina* (bitter leaf) have shown that its sesquiterpene lactones inhibit NF-κB signaling, a pathway involved in inflammation and cancer (Nerdy et al., 2021). Ibrahim et al. (2017) revealed that *Gymenema sylevetre* inhibits alpha-glucosidase and amylase to control blood glucose. Also, the active fraction of *Ficus sycomorus* inhibits the activities of GSTs and P450s in malarial vectors (Anosike et al., 2021).

16.6.7.5 In Vitro and In Vivo Preclinical Studies

Preclinical studies, both in vitro (cell culture) and in vivo (animal models), are conducted to assess the efficacy, safety, and pharmacokinetics of the isolated compounds. These studies are essential to determine the therapeutic potential and possible toxicities of the compounds before advancing to clinical trials. The anti-diabetic potential of compounds isolated from *Momordica charantia* (bitter melon) has been validated through such studies, leading to further exploration in clinical settings (Krawinkel et al., 2018; Mohammed, 2023).

16.6.7.6 Clinical Trials, Toxicological Studies, and Drug Development

The progression from preclinical studies to clinical trials represents a pivotal stage in drug development. Clinical trials are carried out in phases to assess the safety, effectiveness, and appropriate dosage of new medications in humans. While relatively few African medicinal plants have reached this stage, there are notable examples where traditional knowledge has contributed to modern pharmacology. Toxicological studies are essential to determine the safety profile of potential drug candidates. These studies involve assessing acute, subacute, and chronic toxicity in multiple animal species, as well as evaluating potential effects on reproduction, development, and carcinogenicity (Balkrishna et al., 2024). For instance, toxicology studies on *Hypoxis hemerocallidea* extracts have shown that the plant is generally safe at therapeutic doses, but high doses can lead to hepatotoxicity and nephrotoxicity, necessitating careful dosing in clinical settings (Ojewole, 2006).

16.6.7.7 Case Study: Development of Antimalarial Drugs

The development of antimalarial drugs from traditional remedies is a prominent example of successful transition. *Artemisia afra*, a plant used traditionally in Africa for treating fever, led to the discovery of artemisinin derivatives, which are now widely used as first-line treatments for malaria. Although *Artemisia annua* (native to China) is more commonly associated with artemisinin, *Artemisia afra* has also been recognized for its potential in treating malaria in African contexts (Willcox & Bodeker, 2004).

16.6.7.8 Challenges and Opportunities in Clinical Development

The path from traditional use to modern drug development is fraught with challenges, including funding limitations, regulatory hurdles, and the need for ethical considerations in bioprospecting and benefit-sharing. However, opportunities abound in leveraging Africa's rich biodiversity and traditional knowledge to discover novel therapeutic agents. The integration of modern biotechnological tools, such as high-throughput screening and computational biology, can accelerate the discovery process and optimize lead compounds (Dehelean et al., 2021).

16.6.7.9 Ethical Considerations and Intellectual Property Rights

The transition from traditional medicine to modern drug discovery also raises important ethical issues, particularly concerning intellectual property rights (IPR) and benefit-sharing. The Nagoya Protocol emphasizes access and benefit-sharing (ABS) agreements, which require researchers to obtain prior informed consent from local communities and ensure fair compensation (Suneetha et al., 2021). There is a growing recognition of the need to protect the rights of indigenous communities and ensure fair compensation for their contributions to drug discovery (Suneetha et al., 2021).

16.6.7.10 Protecting Indigenous Knowledge

Efforts to protect indigenous knowledge involve legal frameworks and agreements that recognize the contributions of traditional healers and communities. This includes the implementation of Access and Benefit-Sharing (ABS) agreements under the Nagoya Protocol, which aims to ensure that communities receive equitable benefits from the commercialization of their traditional knowledge (Suneetha et al., 2021).

16.6.7.11 Benefit-Sharing Models and Community Involvement

Innovative benefit-sharing models are being developed to ensure that communities benefit directly from the commercialization of their traditional knowledge. This may include royalty payments, community development projects, and capacity-building initiatives. Involving communities in the research process helps build trust and ensures that the use of traditional knowledge is ethical and sustainable (Cunningham et al., 2018).

16.6.7.12 Clinical Trials and Regulatory Approval

After successful preclinical studies, compounds progress to clinical trials, where they are tested in humans to assess safety, efficacy, and optimal dosing.

Phases of Clinical Trials

Clinical trials are conducted in several phases:

1. *Phase I*: Tests the safety, tolerability, pharmacokinetics, and pharmacodynamics of a compound in a small group of healthy volunteers.
2. *Phase II*: Evaluates the efficacy and side effects in a larger group of patients with the targeted disease.
3. *Phase III*: Confirms efficacy and monitors adverse reactions in a much larger patient population, often comparing the new drug to standard treatments.
4. *Phase IV*: Conducted after a drug has been approved, these trials monitor long-term effects and effectiveness in the general population (FDA, 2021).

For example, clinical trials of etoposide, derived from *Maytenus buchananii*, followed this pathway, demonstrating its efficacy in treating lung and testicular cancers and leading to its regulatory approval (Newman & Cragg, 2020).

Regulatory Pathways and Approval

Regulatory approval is required before a new drug can be marketed. Regulatory agencies, such as the FDA, EMA, and African Medicines Agency (AMA), review clinical trial data to ensure the drug is safe and effective for its intended use (WHO, 2021). Regulatory approval involves a detailed evaluation of the drug's benefit-risk profile, quality control measures, and manufacturing standards to ensure that the drug is consistently produced and controlled according to rigorous standards (WHO, 2021).

16.6.7.13 The Role of Biotechnology in Enhancing Drug Discovery

Biotechnology has revolutionized the drug discovery process by introducing new tools and techniques that enhance the efficiency and precision of discovering and developing new drugs from natural sources.

Genomics, Metabolomics, and Bioinformatics

Genomics and metabolomics provide insights into the genetic and metabolic profiles of medicinal plants, allowing for the identification of biosynthetic pathways involved in the production of bioactive compounds (Tsugawa et al., 2021). Bioinformatics tools help analyze vast amounts of data, identifying gene clusters associated with compound biosynthesis and predicting the presence of novel bioactive compounds (Dehelean et al., 2021).

Genetic Engineering and Synthetic Biology

Genetic engineering and synthetic biology allow for the modification of plant genomes to boost the production of specific compounds or incorporate new biosynthetic pathways. These technologies can increase the yield of rare compounds and produce modified versions with improved pharmacological properties (Paddon & Keasling, 2014). For instance, the production of artemisinin in genetically engineered yeast has provided a sustainable and cost-effective source of this important antimalarial compound, reducing reliance on traditional plant cultivation (Paddon & Keasling, 2014).

High-Throughput Screening and Artificial Intelligence

High-throughput screening (HTS) and artificial intelligence (AI) have become indispensable tools in modern drug discovery. HTS allows for the quick screening of thousands of compounds, while AI algorithms can predict the activity of new compounds and optimize lead molecules based on vast chemical databases (Dehelean et al., 2021). For example, AI-driven platforms are being used to predict the pharmacokinetics and toxicity of novel agents, significantly reducing the period and cost related to traditional drug development processes (Ekins, 2017).

16.7 Compounds in Development

Numerous compounds derived from African medicinal plants are currently in various stages of development, from early-stage discovery to clinical trials.

16.7.1 Early-Stage Compounds

1. *Ajugarin* I from *Ajuga remota*: Traditionally used in East Africa for its anti-inflammatory properties, this compound has shown promise in preclinical models for its anti-inflammatory and analgesic effects (Njoroge & Bussmann, 2006).
2. *Michellamine B* from *Ancistrocladus korupensis*: Found in the rainforests of Cameroon, michellamine B has shown potent anti-HIV activity by inhibiting viral replication. This compound is in the early stages of development, with ongoing research focusing on its efficacy and safety profile (Kinghorn et al., 2021).

16.7.2 Clinical Trials and Advanced Development

1. *Etoposide* from *Maytenus buchananii*: Derived from the bark of this East African plant, etoposide is a semi-synthetic derivative used in chemotherapy to treat lung cancer, testicular cancer, and lymphomas. Clinical trials have demonstrated its effectiveness in various cancer treatment protocols (Newman & Cragg, 2020).
2. *Hypoxoside* from *Hypoxis hemerocallidea*: Used traditionally in Southern Africa for immune modulation, hypoxoside is currently being studied for its potential in cancer treatment and immune support in HIV/AIDS patients. Phase I and II clinical trials have shown promising results in enhancing immune function (Ojewole, 2006).

16.7.2.1 Case Study: The Example of *Voacanga africana,* Vinca Alkaloids, Betulinic Acid, etc.

(a) *Voacanga africana* is an example of a plant that has transitioned from traditional use to modern research, highlighting its potential in drug discovery.

Traditional Use

Traditionally, *Voacanga africana* has been used in West and Central Africa for its psychoactive properties, primarily in spiritual and ritualistic contexts. It is also used in traditional medicine to treat various ailments, including mental disorders, pain, and inflammation (Neuwinger, 2000).

Modern Research and Potential

Modern research has identified several indole alkaloids in *Voacanga africana* that display a series of pharmacological properties, such as antihypertensive, anti-inflammatory, and antimalarial effects. Recent studies have focused on the potential of these alkaloids as precursors for semi-synthetic derivatives in drug development, particularly for neurological and cardiovascular diseases (Bussmann & Sharon, 2006). Additionally, the plant's compounds have shown potential as lead compounds for novel antipsychotic drugs, targeting neurotransmitter systems involved in mental health disorders (Koroch et al., 2010).

(b) *Vinca Alkaloids* from *Catharanthus roseus*: Although not exclusive to Africa, the plant's origins on the continent are significant. Vinca alkaloids (vincristine and vinblastine) are essential in chemotherapy for various cancers, including leukemia and lymphoma (Noble, 2016).
(c) *Ellipticine* from *Ochrosia elliptica*: This compound, isolated from a Madagascan plant, has demonstrated potent cytotoxic effects against several cancer cell lines. Its mechanism involves DNA intercalation and topoisomerase inhibition, making it a promising candidate for anticancer drug development (Stiborova et al., 2006; Sathya et al., 2018).
(d) *Betulinic Acid* from *Ziziphus mucronata*: Found in several African medicinal plants, betulinic acid exhibits selective cytotoxicity against melanoma and neuroblastoma cells. It induces apoptosis through the mitochondrial pathway, highlighting its potential as a novel anticancer agent (Fulda, 2008).

Future Prospects and Challenges for Anti-cancer Drug Development.
The development of anticancer agents from African plants faces several challenges:

1. *Scientific Validation*: There is a need for more rigorous scientific studies to validate the traditional claims of anticancer properties. This includes understanding the mechanisms of action, optimizing extraction methods, and conducting clinical trials (Newman & Cragg, 2020).
2. *Sustainable Use and Conservation*: The increased interest in African medicinal plants for drug development must be balanced with sustainable use and conservation efforts to prevent overharvesting and ensure these resources remain available for future generations (Hamilton, 2020).
3. *Integration of Traditional and Modern Medicine*: Connecting the gap between traditional practices and modern medicine requires collaboration, capacity building, and a mutual respect for different knowledge systems. This integration can enhance the detection and development of new anti-cancer compounds (Gurib-Fakim, 2006).

16.8 Challenges in Drug Development from African Medicinal Plants

Despite these successes, developing drugs from African plants presents several challenges:

1. *Biopiracy and Intellectual Property Issues*: There is a risk of biopiracy, where the traditional knowledge of indigenous communities is used without fair compensation or acknowledgment. This has led to calls for stronger protections and benefit-sharing agreements (Suneetha et al., 2021).
2. *Sustainability and Conservation:* Overharvesting of medicinal plants, driven by increased demand for pharmaceutical use, poses a threat to biodiversity. Sustainable harvesting practices and conservation efforts are needed to protect these valuable resources (Hamilton, 2020).
3. *Regulatory and Standardization Challenges*: Developing a standardized extract that meets regulatory requirements can be difficult due to variations in plant chemistry based on growing conditions, harvesting time, and preparation methods (Dehelean et al., 2021).
4. *Complex Nature of Traditional Knowledge:* Traditional knowledge often exists as collectively held, orally transmitted wisdom deeply embedded in cultural practices. This complexity makes it challenging to fit within conventional intellectual property frameworks, which typically prioritize individual ownership and written documentation.
5. *Inadequate Research Facilities:* Advanced laboratories equipped for essential processes such as phytochemical analysis, bioassays, and pharmacological studies are lacking in many regions. This absence restricts the ability to identify, isolate, and characterize bioactive compounds derived from medicinal plants.
6. *Limited Pharmaceutical Manufacturing Capacity:* The pharmaceutical production infrastructure in Africa remains underdeveloped, with few facilities meeting Good Manufacturing Practice (GMP) standards required for the large-scale production of high-quality medicinal products.
7. *Chronic Underfunding of R&D:* Insufficient investment in research institutions and universities limits the capacity to conduct high-quality studies. This lack of funding also constrains access to cutting-edge technologies and the development of skilled personnel essential for innovative drug discovery.
8. *Restricted Access to Advanced Technologies:* Critical tools for modern drug discovery, such as high-throughput screening, metabolomics, and computational modeling, are often unavailable, significantly narrowing the scope of scientific inquiry.
9. *Weak Knowledge Transfer Mechanisms:* Ineffective infrastructure for collaboration among researchers, traditional knowledge holders, and industry stakeholders results in fragmented efforts and inefficiencies in translating discoveries into practical applications.

16.9 Future Perspectives

African medicinal plants hold immense potential for addressing global health challenges, yet their sustainable utilization requires targeted research and conservation efforts. Future research should prioritize mechanistic studies to elucidate the molecular pathways of bioactive compounds, facilitating the development of targeted therapies for diseases such as cancer, diabetes, and infectious conditions. Standardized herbal formulations must be developed to ensure consistent efficacy, safety, and dosage accuracy (Chinsembu, 2019). Additionally, integrating genomic and metabolomic approaches can accelerate the discovery of novel bioactive compounds by identifying biosynthetic pathways in underexplored plant species (Mekonnen et al., 2022). Rigorous clinical validation of traditional remedies is necessary to establish their therapeutic potential, requiring collaborative efforts between researchers and traditional healers (Newman & Cragg, 2020). Artificial intelligence (AI) and machine learning can further optimize lead compounds by predicting pharmacokinetics and toxicity, expediting drug development (Ekins, 2017).

Sustainable approaches must balance conservation and commercialization. Overharvesting poses a significant threat to biodiversity, making conservation strategies such as community-managed plantations and tissue culture techniques essential for maintaining medicinal plant populations (Hamilton, 2020). Fair and transparent benefit-sharing models, as outlined in the Nagoya Protocol, are crucial to ensuring equitable resource distribution and respecting indigenous knowledge (Suneetha et al., 2021). Strengthening local research infrastructure through investments in phytochemical analysis and bioassays can enhance Africa's capacity for drug discovery (Bachmann et al., 2021). Governments should also enact policies to protect biodiversity and integrate traditional medicine into healthcare systems (Cunningham et al., 2018). Biotechnology, including microbial fermentation and plant cell cultures, offers an innovative means of producing bioactive compounds sustainably while reducing reliance on wild plant populations (Paddon & Keasling, 2014). Ethical research practices that prioritize cultural respect and community involvement are essential for fostering long-term sustainability (Hewson, 2021). By combining traditional knowledge with modern scientific advancements, African medicinal plants can play a transformative role in global healthcare.

16.10 Conclusion

African medicinal plants represent a treasure trove of bioactive compounds with immense potential for addressing global health challenges. From their traditional use in treating common ailments to their role in the discovery of life-saving drugs like artemisinin and vincristine, these plants have demonstrated their therapeutic and economic value. Their unique phytochemical diversity, coupled with the wealth of ethnobotanical knowledge preserved in African cultures, positions them as critical resources for modern drug discovery and development.

However, harnessing this potential requires a delicate balance between conservation and commercialization. Unsustainable harvesting, habitat destruction, and climate change threaten the biodiversity that underpins Africa's medicinal plant heritage. At the same time, the commercialization of these resources presents opportunities for economic growth, job creation, and improved healthcare outcomes. Striking this balance necessitates a multifaceted approach:

1. *Conservation Efforts:* Sustainable cultivation, biodiversity protection, and the application of biotechnological solutions can mitigate the risks of overexploitation. Policies that promote environmental stewardship and community-driven conservation initiatives are essential to safeguarding these resources for future generations.
2. *Equitable Commercialization:* Transparent benefit-sharing models, as outlined in the Nagoya Protocol, ensure that local communities and traditional knowledge holders are fairly compensated for their contributions. Collaborative partnerships between researchers, industries, and indigenous groups can drive innovation while fostering trust and equity.
3. *Research and Innovation:* Continued investment in research and development is crucial for unlocking the full potential of African medicinal plants. Advances in genomics, metabolomics, and synthetic biology can accelerate the discovery of new drugs while reducing the environmental impact of production.

By integrating traditional knowledge with modern scientific approaches, Africa's medicinal plants can contribute significantly to global health and economic development. Achieving this vision requires a shared commitment to conservation, ethical research, and sustainable commercialization. Together, these efforts will ensure that African medicinal plants remain a source of healing and innovation for generations to come.

References

Abdullahi, A. A. (2011). Trends and challenges of traditional medicine in Africa. *African Journal of Traditional, Complementary and Alternative Medicines, 8*(5S), 115.

Aboelsoud, N. H. (2010). Herbal medicine in ancient Egypt. *Journal of Medicinal Plants Research, 2010*(4), 82–86.

Abubakar, S., Etim, V., Bwai, D., & Afolayan, M. (2016). Nutraceutical evaluation of baobab (*Adansonia digitate* L.) seeds and physicochemical properties of its oil. *Annals of Biological Sciences, 3*(2), 13–19.

Afolayan, A. J., & Sunmonu, T. O. (2010). *In vivo* studies on antidiabetic plants used in south African herbal medicine. *Journal of Clinical Biochemistry and Nutrition, 47*(2), 98–106.

Anosike, C. A., Babandi, A., & Ezeanyika, L. U. S. (2021). Potentiation effects of Ficus sycomorus active fraction against permethrin-resistant field-population of Anopheles coluzzii (Diptera: Culicidae). *Neotropical Entomology, 50*(3), 484–496.

Asase, A., Oteng-Yeboah, A. A., Odamtten, G. T., & Simmonds, M. S. J. (2010). Ethnobotanical study of some Ghanaian anti-malarial plants. *Journal of Ethnopharmacology, 99*(2), 273–279. https://doi.org/10.1016/j.jep.2005.02.020

Azwanida, N. N. (2015). A review on the extraction methods use in medicinal plants, principle, strength and limitation. *Medicinal and Aromatic Plants, 4*, 3. https://doi.org/10.4172/2167-0412.1000196

Babandi, A., Ibrahim, A., Murtala, Y., & Yakasai, M. H. (2012). Comparative assessment of phytochemical and trace element compositions of *Azadchiracta indica* and *Calotropis procera* roots. *BEST Journal, 9*(2), 123–126.

Babandi, A., Abubakar, S., Yakasai, H. M., Murtala, Y., Shehu, D., Muhammad, Y.Y., Kabara, H. T., Babagana, K., & Ibrahim, A. (2016). Co-administraton of *Azadirachta indica* and *Calotropis procera* Methanolic flower extracts ameliorate Hypozincemia and hyperglycemia on Alloxanized rats. *Bayero Journal of Biomedical Sciences.* 1 (1): 24–33. ISSN: 2635-3113 Available online: http://www.college.buk.edu.ng/bms/journal

Babandi, A., Musa, N. D., Ya'u, M., Yakasai, H. M., Shehu, D., Babagana, K., & Ibrahim, A. (2018). Lipid-lowering property of flavonoid-rich portion of *Combretum Micranthum* on High fat diet induced Hyperlipidemic rats. *International Journal of Bioscience, Biochemistry and Bioinformatics, 9*(1), 42–50. https://doi.org/10.17706/ijbbb.2019.9.1.42-50

Babandi, A. Muhammad, A. M. Anosike, C. A, & Ezeanyika L. U. S. (2019). *In vitro* anti-malarial potential of chloroform fruit crude extract of *Ficus sycomorus* L. against blood stage *Plasmodium falcifarum. Nigerian Research Journal of Chemical Sciences* 6:33–47. Available online: https://www.unn.edu.ng/nigerian-research-journal-of-chemical-sciences/.

Babandi, A., Anosike, C. A., Yakasai, H. M., & Ezeanyika, L. U. S. (2021a). Statistical optimization of alkaloids extraction from *Ficus sycomorus* leaves using response surface methodology. *Malaysian Journal of Biochemistry and Molecular Biology, 3*, 1–14.

Babandi, A., Anosike, C. A., Yakasai, H. M., Ibrahim, S., & Ezeanyika, L. U. S. (2021b). Statistical optimization of alkaloids extraction from *Ficus sycomorus* leaves using response surface methodology. *Malaysian Journal of Biochemistry and Molecular Biology, 2021*(3), 1–14.

Babandi, A., Anosike, C., & Ezeanyika, L. (2021c). Larvicidal and synergistic toxicity of *Ficus sycomorus* and *Calotropis procera* leaf extracts against malarial vector *Anopheles gambiae* complex from Kano-Nigeria: A green bio-control approach. *Asian Journal of Green Chemistry, 5*(3), 248–262. https://doi.org/10.22034/ajgc.2021.262853.1290

Babandi, A., Anosike, C. A., Ezeanyika, L. U. S., Yelekçi, K., & Uba, A. I. (2022). Molecular modeling studies of some phytoligands from *Ficus sycomorus* fraction as potential inhibitors of cytochrome *CYP6P3* enzyme of *Anopheles coluzzii. Jordan Journal of Pharmaceutical Sciences, 15*(2), 258–274.

Babandi, A., Anosike, C. A., Ezeanyika, L. U. S., Elijah, P. J., Yakasai, H. M., Muhammad, A., Muhammad, A., & Ibrahim, S. S. (2023). Chemical composition and Larvicidal efficacy of *Ficus sycomorus* leaf extract against major malaria vector *Anopheles coluzzii. Proceeding of Nigerian Academy of Science., 16*(2), 29–46.

Bachmann, M., Mbonyingabo, C., Mushagalusa, F. M., & Walker, R. (2021). Strengthening research capacity in Africa. *Global Health Action, 14*(1), 1886452. https://doi.org/10.1080/16549716.2021.1886452

Balkrishna, A., Sharma, N., Srivastava, D., Kukreti, A., Srivastava, S., & Arya, V. (2024). Exploring the safety, efficacy, and bioactivity of herbal medicines: Bridging traditional wisdom and modern science in healthcare. *Future Integrative Medicine, 3*(1), 35–49. https://doi.org/10.14218/FIM.2023.00086

Bonomi, Z. M., Al Amin, A. S., Gadanya, M. A., & Babandi, A. (2018). Characterization and proximate analysis of *Moringa oleifera* seed oil cultivated in Kano, Nigeria. *Nigerian Journal of Nutritional sciences, 39*(2), 175–178. https://www.ajol.info

Bouasla, A., & Bouasla, I. (2017). Ethnobotanical survey of medicinal plants in northeastern of Algeria. *Journal of Medicinal Plants Research, 11*(20), 317–324.

Bray, F., Ferlay, J., Soerjomataram, I., Siegel, R. L., Torre, L. A., & Jemal, A. (2018). Global cancer statistics 2018: GLOBOCAN estimates of incidence and mortality worldwide for 36 cancers in 185 countries. *CA: A Cancer Journal for Clinicians, 68*(6), 394–424. https://doi.org/10.3322/caac.21492

Bussmann, R. W., & Sharon, D. (2006). Traditional medicinal plant use in northern Peru: Tracking two thousand years of healing culture. *Journal of Ethnobiology and Ethnomedicine, 2*(1), 47.

Chabner, B. A., Levine, A. S., Johnson, B. L., & Young, R. C. (1978). Initial clinical trials of maytansine, an antitumor plant alkaloid. *Cancer Treatment Reports, 62*(3), 429–433.

Chinsembu, K. C. (2019). Research and development of medicinal plants in Africa. *Journal of Ethnobiology and Ethnomedicine, 15*(1), 68. https://doi.org/10.1186/s13002-019-0332-6

Cocks, M. L., & Møller, V. (2002). Use of indigenous and indigenised medicines to enhance personal Well-being: A south African case study. *Social Science and Medicine, 54*(3), 387–397.

Cos, P., Vlietinck, A. J., Berghe, D. V., & Maes, L. (2006). Anti-infective potential of natural products: How to develop a stronger in vitro 'proof-of-concept'. *Journal of Ethnopharmacology, 106*(3), 290–302.

Cowan, M. M. (1999). Plant products as antimicrobial agents. *Clinical Microbiology Reviews, 12*(4), 564–582.

Cox, P. A., & Balick, M. J. (2020). Ethnobotanical approaches to drug discovery: The collaborative intersection of ancient knowledge and modern science. *Phytochemistry Reviews, 19*(3), 685–693.

Cunningham, A. B., Garnett, S. T., Gorman, J. T., & Courtenay, K. R. (2018). African medicinal plants in the context of conservation and the law. *Frontiers in Plant Science, 9*, 577. https://doi.org/10.3389/fpls.2018.00577

David, R., & Forshaw, R. (2023). *Medicine and healing practices in ancient Egypt*. Liverpool University Press. Available online: https://books.google.com/books?hl=fr&lr=&id=gbjYEAAAQBAJ&oi=fnd&pg=PR1&dq=Ancient+Egyptian+Healing+Practices+&ots=TEZbQd8Fk6&sig=ZoI0NRKi_KCTGMEx3RuKtHF1Svw. Accessed on 8 Nov 2023.

Dehelean, C. A., Marcovici, I., Soica, C., Mioc, M., Coricovac, D., Iurciuc, S., Cretu, O. M., & Pinzaru, I. (2021). Plant-derived anticancer compounds as new perspectives in drug discovery and alternative therapy. *Molecules, 26*(4), 1109. https://doi.org/10.3390/molecules26041109

Dewick, P. M. (2002). *Medicinal natural products: A biosynthetic approach* (3rd ed., pp. 8–9). Wiley.

Ekins, S. (2017). The next era: Deep learning in pharmaceutical research. *Pharmaceutical Research, 34*(3), 212–216. https://doi.org/10.1007/s11095-016-2022-0

Enwonwu, C. O. (2003). Global trends in the use of complimentary medicine proceedings of the 2nd Dr. David Barmes' memorial public health symposium, 25th March 2003. The Regional Centre for Oral Health Research & Training for Africa Jos in collaboration with WHO Regional Office; 2003.

Eziacka, M. M., Jeremia, R. M., Olav, J. S., Kenneth, M., & Kitundu, B. (2021). The value chain of traded productsof medicinal plants in Tanzania: The emerging role of formulators. *African Journal of Economic and Management Studies, 2021*, 2040–0705.

Fabricant, S. D., & Farnsworth, R. N. (2001). (2001). The value of plants used in traditional medicine for drug discovery. *Environmental Health Perspectives., 109*(1), 65–75.

FAO. (1996). Nigeria-country report to the FAO international technical conference on plant genetic resources, Leipzig. 1996. p. 8.

Farnsworth, N. R. (1990). 1990. In D. J. Chadwick & J. Marsh (Eds.), *Bioactive compounds from plants* (pp. 2–21). Wiley.

FDA. (2021). *Clinical trials and drug development*. U.S. Food and Drug Administration. Available: https://www.fda.gov/drugs/development-approval-process-drugs/clinical-trials-and-drug-development

Federal Republic of Nigeria. (2010). *Fourth National Biodiversity Report Abuja 2010* (p. 12). Federal Ministry of Environment.

Forshaw, R. (2014). *Before Hippocrates. Healing practices in ancient Egypt*. In *Medicine, healing and performance* (pp. 25–41). Oxbow Books.

Fulda, S. (2008). Betulinic acid: A natural product with anticancer activity. *Molecular Nutrition and Food Research, 52*(1), 67–68.

Githinji, M. (2020). Ritual use of medicinal plants in Kenya. *Journal of Ethnobiology and Ethnomedicine, 16*(1), 45.

Gouws, C., Smit, T., Willers, C., Svitina, H., Calitz, C., & Wrzesinski, K. (2021). Anticancer potential of *Sutherlandia frutescens* and *Xysmalobium undulatum* in LS180 colorectal cancer mini-tumors. *Molecules, 26*(3), 605. https://doi.org/10.3390/molecules26030605

Gurib-Fakim, A. (2006). Medicinal plants: Traditions of yesterday and drugs of tomorrow. *Molecular Aspects of Medicine, 27*(1), 1–93.

Hamilton, A. C. (2020). Medicinal plants, conservation, and livelihoods. *Biodiversity and Conservation, 29*(5), 1323–1345. https://doi.org/10.1007/s10531-020-01953-2

Heinrich, M., Appendino, G., Efferth, T., Furst, R., Izzo, A. A., Kayser, O., Pezzuto, J. M., & Viljoen, A. (2020). Best practice in research: Consensus statement on ethnopharmacological field studies – ConSEFS. *Journal of Ethnopharmacology, 246*, 112230.

Hewson, M. G. (2021). Traditional medicine in contemporary contexts: Protecting and respecting indigenous knowledge and intellectual property. *Journal of Cultural Heritage Management and Sustainable Development, 11*(1), 17–30. https://doi.org/10.1108/JCHMSD-03-2020-0036

Hostettmann, K. (1999). (1999). Biodiversity and bioresources: Conservation and utilization. *Pure and Applied Chemistry., 70*(1), 50–56.

Ibrahim, A., Babandi, A., Sani, A. H., Wudil, A. M., Murtala, Y., & Umar, I. A. (2017). HPLC profile, *in-vitro* alpha- amylase, alpha-glucosidase inhibitory and antioxidant activities of *Gymnema sylvestre* ethyl acetate leaf extract. *Bayero Journal of Pure and Applied Sciences, 10*(1), 72–80. https://doi.org/10.4314/bajopas.v10i1.11

Ibrahim, A., Babaye, H., Ya'u, M., Babagana, K., Abubakar, S. M., et al. (2018). Nigerian *Citrullus Lanatus* fruit and seed juice reduces cardiovascular diseases modifiable risk biomarkers in Normal experimental rats. *Journal of Hypertension Management, 4*, 036. https://doi.org/10.23937/2474-3690/1510036

Ibrahim, H. M., Abdulrahman, A. Y., Warra, A. A., & Abdullahi, K. (2019). (2019). Entrepreneur in Pharmacognosy: An unexploited area in Nigeria; Gusau. *Journal of Entrepreneurship Development., 1*(1), 161–176.

Isah, T. (2019). Stress and defense responses in plant secondary metabolites production. *Biological Research, 52*(1), 39. https://doi.org/10.1186/s40659-019-0246-3

Kasali, F. M., Kadima, J. N., Peter, E. L., Mtewa, A. G., Ajayi, C. O., Tusiimire, J., Tolo, C. U., Ogwang, P. E., Weisheit, A., & Agaba, A. G. (2021). Antidiabetic medicinal plants used in the Democratic Republic of Congo: A critical review of ethnopharmacology and bioactivity data. *Frontiers in Pharmacology, 12*, 757090. https://doi.org/10.3389/fphar.2021.757090

Kevin, T. D. A., Cedric, Y., Nadia, N. A. C., Sidiki, N. N. A., Azizi, M. A., Guy-Armand, G. N., Sandra, T. N. J., Christian, M. N., Géraldine, E. S. E., Roméo, T. T., Payne, V. K., & Gustave, L. L. (2023). Antiplasmodial, antioxidant, and cytotoxic activity of *Bridelia micrantha*, a Cameroonian medicinal plant used for the treatment of malaria. *BioMed Research International, 2023*, 1219432. https://doi.org/10.1155/2023/1219432

Kim, M., Park, S., Kang, G., Jang, Y. S., Kim, K. H., Han, S., & Kim, C. S. (2024). Natural product anticipation via chemical synthesis: Discovery of two new securinega alkaloids. *iScience, 27*(8), 110495. https://doi.org/10.1016/j.isci.2024.110495

Kinghorn, A. D., et al. (2021). Discovery of bioactive natural products in the 21st century: Screening methods and trends. *Journal of Natural Products, 84*(5), 1304–1319.

Kingston, G. I. D. (2012). (2012). Modern natural products drug discovery and its relevance to biodiversity conservation. *Journal of Natural Product., 74*(3), 496–511.

Komlaga, G., Agyare, C., Dickson, R. A., Mensah, M. L. K., Annan, K., Loiseau, P. M., & Champy, P. (2015). Medicinal plants and finished marketed herbal products used in the treatment of malaria in the Ashanti region, Ghana. *Journal of Ethnopharmacology, 172*, 333–346. https://doi.org/10.1016/j.jep.2015.06.041

Koroch, A. R., Juliani, H. R., Kulakowski, D., Arthur, H., Asante-Dartey, J., & Simon, J. E. (2010). *Voacanga africana*: Chemistry, quality, and pharmacological activity. In H. Juliani et al. (Eds.), *African natural plant products: New discoveries and challenges in chemistry and quality* (pp. 363–380). American Chemical Society. https://doi.org/10.1021/bk-2010-1043.ch024

Koubek, E. J., Weissenrieder, J. S., Neighbors, J. D., & Hohl, R. J. (2018). Schweinfurthins: Lipid modulators with promising anticancer activity. *Lipids, 53*(8), 767–784. https://doi.org/10.1002/lipd.12088

Krawinkel, M. B., Ludwig, C., Swai, M. E., Yang, R. Y., Chun, K. P., & Habicht, S. D. (2018). Bitter gourd reduces elevated fasting plasma glucose levels in an intervention study among prediabetics in Tanzania. *Journal of Ethnopharmacology, 216*, 1–7. https://doi.org/10.1016/j.jep.2018.01.016

Kuete, V. (2013). Potential of African medicinal plants against microbial resistance. *Phytochemistry Reviews, 12*(3), 621–633.

Kuete, V., & Efferth, T. (2015). African flora has the potential to fight multidrug resistance of cancer. *BioMed Research International, 2015*(1), 914813.

Kumar, P., Hassan, M., Andrew, K., Ronak, S., Abdallah, S. D., & Peter, A. S. (2010). The road to commercialization in Africa: Lessons from developing the sickle-cell drug Niprisan. *BMC International Health and Human Rights., 10*(1), S11.

Mahomoodally, M. F. (2013). Traditional medicines in Africa: An appraisal of ten potent African medicinal plants. *Evidence-based Complementary and Alternative Medicine, 2013*, 617459.

Marula, T. R., Sebua, S. S., & Alfred, M. (2019). Medicinal plants traded in informal herbal medicine Markets of the Limpopo Province, South Africa. *Evidence-based Complementary and Alternative Medicine., 2019*, 11.

Mbele, M., Rodney, H., & Zodwa, D. (2017). African medicinal plants and their derivatives: Current efforts towards potential anti-cancer drugs. *Experimental and Molecular Pathology, 103*(2017), 121–134.

Mekonnen, A. B., Mohammed, A. S., & Tefera, A. K. (2022). Ethnobotanical study of traditional medicinal plants used to treat human and animal diseases in Sedie Muja District, South Gondar, Ethiopia. *Evidence-Based Complementary and Alternative Medicine, 2022*, 7328613. https://doi.org/10.1155/2022/7328613

Méndez-Lucio, O., & Medina-Franco, J. L. (2017). The many roles of molecular complexity in drug discovery. *Drug Discovery Today, 22*(1), 120–126. https://doi.org/10.1016/j.drudis.2016.08.009

Mensah, M. L. K., et al. (2019). Ethnobotanical survey of medicinal plants used for treating malaria in Ghana. *Journal of Ethnopharmacology, 238*, 111867.

Mohammed, A. (2023). Hypoglycemic potential of African medicinal plants in diabetic and non-diabetic human subjects: A review. *Clinical Complementary Medicine and Pharmacology, 3*(2), 100081. https://doi.org/10.1016/j.ccmp.2023.100081

Muhammad, B. Y., Adamu, R. M., Enemali, M. O., & Babandi, A. (2019). Antioxidant, Antiinflammatory, anti-proliferative and antimicrobial activities (*in vitro*) of *Indigofera hirsuta* and *Afrormosia laxiflora*. *Saudi Journal of Medical and Pharmaceutical sciences, 5*(11), 923–930. https://doi.org/10.36348/sjmps.2019.v05i11.002

Murtala, Y., Babandi, A., Sunusi, M. M., Shehu, D., Babagana, K., & Alhassan, A. J. (2015). Effect of aqueous leaf extract of *Borreria verticillata* species of Sudano-Sahelian savanna on CCl4 induced hepatotoxicity. *Journal of Natural Sciences Research, 5*(24), 73–78.

Ndhlala, A. R., Stafford, G. I., Finnie, J. F., & Staden, J. V. (2011). Commercial herbal preparations in KwaZulu-Natal, South Africa: The urban face of traditional medicine. *South African Journal of Botany., 77*, 830–843.

Ndung'u, J. K., Nguta, J. M., Mapenay, I. M., & Moriasi, G. A. (2024). A comprehensive review of ethnomedicinal uses, phytochemistry, pharmacology, and toxicity of *Prunus africana* (*Hook. f.*) Kalkman from Africa. *Scientifica, 2024*, 8862996. https://doi.org/10.1155/2024/8862996

Nerdy, N., Margata, L., Meliala, L., Purba, J. S., Sembiring, B. M., Ginting, S., & Bakri, T. K. (2021). *In Silico* evaluation of the physicochemical, pharmacokinetics, and toxicity profiles of Sesquiterpene lactones of south African leaf (*Vernonia amygdalina* Delile). *Tropical Journal of Natural Product Research (TJNPR), 5*(10), 1835–1840. https://doi.org/10.26538/tjnpr/v5i10.21. https://tjnpr.org/index.php/home/article/view/389

Neuwinger, H. D. (2000). *African traditional medicine: A dictionary of plant use and applications*. Medpharm Scientific Publishers.

Newman, D. J., & Cragg, G. M. (2020). Natural products as sources of new drugs over the nearly four decades from 1981 to 2019. *Journal of Natural Products, 83*(3), 770–803. https://doi.org/10.1021/acs.jnatprod.9b01285

Njoroge, G. N., & Bussmann, R. W. (2006). Traditional management of ear, nose and throat (ENT) diseases in Central Kenya. *Journal of Ethnobiology and Ethnomedicine, 2*(1), 54.

Noble, R. L. (1990). The discovery of the vinca alkaloids—Chemotherapeutic agents against cancer. *Biochemistry and Cell Biology, 68*(12), 1344–1351.

Noble, R. L. (2016). The discovery of the *Vinca* alkaloids—Chemotherapy for cancer: A story of serendipity. *The Oncologist, 21*(8), 1013–1017.

Obasi, T. C., Braicu, C., Iacob, B. C., Bodoki, E., Jurj, A., Raduly, L., Oniga, I., Berindan-Neagoe, I., & Oprean, R. (2018). *Securidaca*-saponins are natural inhibitors of AKT, MCL-1, and BCL2L1 in cervical cancer cells. *Cancer Management and Research, 10*, 5709–5724. https://doi.org/10.2147/CMAR.S163328

Ofir, R. (2020). "Desert chemotypes": The potential of desert plants-derived metabolome to become a sustainable resource for drug leads. *Medical Research Archives, 8*(7), 1–5.

Ojewole, J. A. O. (2006). Antinociceptive, anti-inflammatory, and antidiabetic properties of *Hypoxis hemerocallidea* Fisch. And C. A. Mey. (Hypoxidaceae) corm ('African potato') aqueous extract in mice and rats. *Journal of Ethnopharmacology, 103*(1), 126–134.

Okigbo, R. N., & Mmeka, E. C. (2006). An appraisal of phytomedicine in Africa. *Journal of Medicinal Plants Research, 2*(12), 335–342.

Okigbo, R. N., Anuagasi, C. L., & Amadi, J. E. (2008). Advances in selected medicinal and aromatic plants indigenous to Africa. *Journal of Medicinal Plants Research, 2*(2), 86–95. https://doi.org/10.5897/JMPR

Okole, B. N., & Odhav, B. (2004). (2004). Commercialisation of plants in Africa. *South African Journal of Botany., 70*(1), 109–115.

Olorunnisola, O. S., Adetutu, A., Balogun, E. A., & Afolayan, A. J. (2013). Ethnobotanical survey of medicinal plants used in the treatment of malaria in Ogbomoso, Southwest Nigeria. *Journal of Ethnopharmacology, 150*(1), 71–78. https://doi.org/10.1016/j.jep.2013.07.038

Ozioma, E. O., & Chinwe, O. A. (2019). *Herbal medicines in African traditional medicine*. Herbal Medicine, IntechOpen.

Paddon, C. J., & Keasling, J. D. (2014). Semi-synthetic artemisinin: A model for the use of synthetic biology in pharmaceutical development. *Nature Reviews Microbiology, 12*(5), 355–367. https://doi.org/10.1038/nrmicro3240

Paul-Marie, L. (2008). *Medicinal plant trade and opportunities for sustainable Management in Cape Peninsula, South Africa*, Master of science thesis (pp. 18–19). Department of Conservation Ecology and Forestry Sciences.

Perampaladas, K., Masum, H., Kapoor, A., Shah, R., Daar, A. S., & Singer, P. A. (2010). The road to commercialization in Africa: Lessons from developing the sickle-cell drug Niprisan. *BMC International Health and Human Rights, 10*(Suppl 1), S11. https://doi.org/10.1186/1472-698X-10-S1-S11

Posthouwer, C., Veldman, S., Abihudi, S., Otieno, J., Andel, T., & Boer, H. (2018). Quantitative market survey of non-woody plants sold at Kariakoo market in Dar Es Salaam, Tanzania. *Journal of Ethnopharmacology, 17*, 280–287.

Salim, A. A., Chin, Y. W., & Kinghorn, A. D. (2008). Drug discovery from plants. In K. G. Ramawat & J. M. Mérillon (Eds.), *Bioactive molecules and medicinal plants* (pp. 1–25). Springer. Available from: http://www.springer.com

Samia, A., Paul, K., Zohra, S. C., Gordon, J., Dee, D. W., Kylie, H., et al. (2020). Treatment of hydroxychloroquine, azithromycin, and combination in patients hospitalized with COVID-19. *International Journal of Infectious Diseases, 97*, 396–403.

Sasidharan, S., Chen, Y., Saravanan, D., Sundram, K. M., & Latha, L. Y. (2011). Extraction, isolation, and characterization of bioactive compounds from plant extracts. *African Journal of Traditional, Complementary and Alternative Medicines, 8*(1), 1–10.

Sathya, A., et al. (2018). Ellipticine: A potential antineoplastic agent for the treatment of human cancers. *Life Sciences, 199*, 31–39.

Seddigi, Z. S., Malik, M. S., Saraswati, A. P., Ahmed, S. A., Babalghith, A. O., Lamfon, H. A., & Kamal, A. (2017). Recent advances in combretastatin-based derivatives and prodrugs as antimitotic agents. *MedChemComm, 8*(8), 1592–1603. https://doi.org/10.1039/c7md00227k

Shilonga, S., Otieno, J. N., Ghorbani, A., Pereus, D., Kocyan, A., & Boer, H. (2019). Trade of wild-harvested medicinal plant species in local markets of Tanzania and its implications for conservation. *South African Journal of Botany., 2019*(122), 214–224.

Steinberg, M. H. (2003). (2003). Effect of hydroxyurea on mortality and morbidity in adult sickle cell anaemia. *Journal of the American Medical Association, 289*(13), 1645–1651.

Stiborova, M., Rupertova, M., Schmeiser, H. H., & Frei, E. (2006). Molecular mechanisms of antineoplastic action of an anticancer drug ellipticine. *Biomedicine, 150*(1), 13–23. https://doi.org/10.5507/bp.2006.002

Stolinsky, D. C., Bogdon, D. L., Pugh, R. P., Braunwald, J., Hestorff, R., & Bateman, J. R. (1973). Vinblastine (NSC-49842) plus vincristine (NSC-67574) given in intensive 3-day courses as therapy for lymphomas, sarcomas, and other neoplasms. *Cancer Chemotherapy Reports, 57*(4), 481–484.

Street, R. A., & Prinsloo, G. (2013). Commercially important medicinal plants of South Africa (a review). *Journal of Chemistry, 2013*(2013), 16.

Sunday, J. A., Florence, D. T., & Benjamin, U. E. (2012). Traditional herbal Management of Sickle Cell Anaemia: Lessons from Nigeria (review). *Anaemia, 2012*(2012), 9.

Suneetha, M. S., Pisupati, B., & Pauchard, N. (2021). Implementing the Nagoya protocol: Challenges and opportunities for access and benefit-sharing. *Biological Conservation, 261*, 109252. https://doi.org/10.1016/j.biocon.2021.109252

Thomford, N. E., Senthebane, D. A., Rowe, A., Munro, D., Seele, P., Maroyi, A., & Dzobo, K. (2018). Natural products for drug discovery in the 21st century: Innovations for novel drug discovery. *International Journal of Molecular Sciences, 19*(6), 1578. https://doi.org/10.3390/ijms19061578

Tinde, V. A., Britt, M., & Sabine, V. O. (2012). Ghana's herbal market. *Journal of Ethnopharmacology, 2012*(140), 368–378.

Tozer, G. M., Kanthou, C., Parkins, C. S., & Hill, S. A. (2002). The biology of the combretastatins as tumour vascular targeting agents. *International Journal of Experimental Pathology, 83*(1), 21–38. https://doi.org/10.1046/j.1365-2613.2002.00211.x

Tsugawa, H., Rai, A., Saito, K., & Nakabayashi, R. (2021). Metabolomics and complementary techniques to investigate the plant phytochemical cosmos. *Natural Product Reports, 38*, 1729. https://10.1039/d1np00014d

Tu, Y. (2011). The discovery of artemisinin (Qinghaosu) and gifts from Chinese medicine. *Nature Medicine, 17*(10), 1217–1220. https://doi.org/10.1038/nm.2471

Van Wyk, B.-E. (2015). A review of African medicinal plants with antimalarial activity. *Journal of Ethnopharmacology, 176*, 3–24. https://doi.org/10.1016/j.jep.2015.10.031

Van Wyk, B. E. (2020). A review of African medicinal plants with promising potential. *South African Journal of Botany, 130*, 1–20.

Van Wyk, A. S., & Prinsloo, G. (2018). Medicinal plant harvesting, sustainability and cultivation in South Africa. *Biological Conservation, 2018*(227), 335–342.

Vicki, B. (2008). Back to nature: Extinction of medicinal plants threatens drug discovery. *Journal of National Cancer Institute., 100*(12), 838–839.

Viktorin, M., & Sartorius, A. G. (1999). One-hundred-year anniversary of aspirin. The drug extracted from the willow bark, Göttingen, Germany. *Labor Praxis, 23*(7/8), 82–85.

Willcox, M. L., & Bodeker, G. (2004). Traditional herbal medicines for malaria. *BMJ, 329*(7475), 1156–1159. https://doi.org/10.1136/bmj.329.7475.1156

World Health Organization (WHO). (2013). Traditional medicine strategy 2014–2023.

World Health Organization (WHO). (2021). *Guidelines on good manufacturing practices (GMP) for herbal medicines*. World Health Organization. [Online]. Available: https://www.who.int/medicines/areas/traditional/gmp/en/

Yakubu, O. F., Metibemu, D. S., Adelani, I. B., Adesina, G. O., Edokwe, C. B., Oseha, O. E., & Adebayo, A. H. (2020). *Annona senegalensis* extract demonstrates anticancer properties in N-diethylnitrosamine-induced hepatocellular carcinoma in male Wistar rats. *Biomedicine & Pharmacotherapy, 131*, 110786. https://doi.org/10.1016/j.biopha.2020.110786

Zhang, R., Neighbors, J. D., Schell, T. D., & Hohl, R. J. (2022). Schweinfurthin induces ICD without ER stress and caspase activation. *Oncoimmunology, 11*(1), 2104551. https://doi.org/10.1080/2162402X.2022.2104551

Zong, G., Aljewari, H., Hu, Z., & Shi, W. Q. (2016). Revealing the pharmacophore of ipomoeassin F through molecular editing. *Organic Letters, 18*(7), 1674–1677. https://doi.org/10.1021/acs.orglett.6b00555

Zong, G., Hu, Z., Duah, K. B., Andrews, L. E., Zhou, J., O'Keefe, S., Whisenhunt, L., Shim, J. S., Du, Y., High, S., & Shi, W. Q. (2020). Ring expansion leads to a more potent analogue of ipomoeassin F. *Journal of Organic Chemistry, 85*(24), 16226–16235. https://doi.org/10.1021/acs.joc.0c01659

MIX
Papier aus verantwortungsvollen Quellen
Paper from responsible sources
FSC® C105338

If you have any concerns about our products,
you can contact us on
ProductSafety@springernature.com

In case Publisher is established outside the EU,
the EU authorized representative is:
Springer Nature Customer Service Center GmbH
Europaplatz 3, 69115 Heidelberg, Germany

Printed by Libri Plureos GmbH
in Hamburg, Germany